Fundamentals of Algebra

An Integrated Text-Workbook

Fundamentals of Algebra

An Integrated Text-Workbook

Robert Donaghey
Baruch College

JoAnna Ruddel
Baruch College

Harcourt Brace Jovanovich, Inc.
New York San Diego Chicago San Francisco Atlanta

Cover art by Joyce Kitchell

Library of Congress Catalog Card Number: 77-19337

ISBN: 0-15-529420-2

Printed in the United States of America

Preface

This book provides an extensive treatment of the fundamentals of algebra. The main body of the text contains all the basic material, while optional sections (''Observations and Reflections'') examine more difficult concepts in greater depth. Because many students who take this course also need some help with arithmetic, early chapters are paced so that the reader may acquire a firm grasp of arithmetic principles and see how algebra is developed from arithmetic. In most chapters the first section is written primarily as a review and may be skipped by better-prepared students.

ORGANIZATION

Fundamentals of Algebra is organized as a text-workbook, with each section addressing a small class of problems. The basic themes of algebra are developed gradually, beginning with simple computational drill-type problems and progressing to problems with more substance. Each section begins with an exploration of the basic concepts and definitions needed to approach the problems of the section. Clear, detailed procedural information follows, as well as numerous examples of worked problems. The introductory material, the illustrative examples, the frequent tables, and the discussions of special cases have all been designed to provide a text that students will find readable, instructive and valuable.

SPECIAL FEATURES

1. The textual material is accompanied by marginal exercises (with answers) that highlight and reinforce the concepts being developed by giving students immediate feedback in their understanding of what they read. These marginal exercises are suitable for classroom use.

2. Extensive exercise sets at the end of each section provide ample opportunity for drill. These exercise sets are graded in difficulty.
3. Each chapter concludes with a lengthy summary of problems (identified by section) to provide greater flexibility of use. There are 4,700 algebraic exercises in the text, one third of them in the chapter reviews. (Answers are provided for all problems at the back of the book.)
4. Cumulative tests (containing an additional 500 problems) appear at the end of most chapters. The problems in these self-tests are mixed (rather than organized and identified by sections) and often rely on concepts from previous chapters. (Answers are also provided for these tests.)
5. In-depth treatments of both evaluating algebraic expressions and ''Order of Operations'' are provided in eleven sections spread over ten chapters. These sections help students develop two important skills necessary for success in algebra — recognizing what expressions mean and deciding what to do with them.

WORD PROBLEMS AND PHRASES

Fundamentals of Algebra presents a variety of word-problem types and an extensive and careful treatment of the process of translating them into equations. Word problems are first integrated into the development of linear equations. The exercises of Chapter 10 include word problems that are solved using linear equations in one unknown; the exercises of Chapter 12 include word problems that require two equations in two unknowns for their solutions. Following these two chapters, an entire chapter (Chapter 13) is devoted to the process of translating word problems into equations. Chapter 13 provides a second set of word problems (the same number, 140, as in Chapters 10 and 12 combined), but in this chapter they are organized by type rather than by method of solution. This dual organization allows great flexibility for the instructor; word problems can be introduced concurrently with the development of equations, or the chapters on equations can be taught first (ignoring the word problems in these chapters), the word problems introduced afterward, one type at a time.

To prepare for translating word problems, word phrases are introduced in Chapter 1, and over 400 word phrases are included with the exercises of the first nine chapters. These word phrases are designed to serve two purposes: They provide students with ample opportunity to learn how to translate from English to algebraic notation, and they reinforce the algebraic skills the students are mastering.

ARRANGEMENT OF CHAPTERS

The twenty-one chapters in this text are organized into five parts:

PART I Chapters 1 to 5 — introduces signed number arithmetic, addition and multiplication of monomials, distribution and addition of polynomials.
PART II Chapters 6 to 9 — treats numerical fractions and algebraic fractions involving monomials.
PART III Chapters 10 to 13 — introduces linear equations in one and two unknowns, graphing straight lines, and word problems.
PART IV Chapters 14 to 18 — discusses operations with polynomials and fractional expressions involving polynomials, and factoring.
PART V Chapters 19 to 21 — treats nonlinear equations (both fractional and quadratic) and introduces square roots.

SEQUENCE OF TOPICS

The text allows the instructor great flexibility in selecting a sequence of topics. For example, PART III on linear equations may be started early or delayed until late. PART II on fractions may be postponed until after PART III or even until after the chapters on factoring polynomials. Well-prepared students can review Chapters 1, 6, and 7 quickly or skip them entirely. Any of Chapters 12, 13, 18, and 19, as well as sections of many other chapters, can be skipped without causing gaps in the remaining material. Possible pathways are shown in the diagram below:

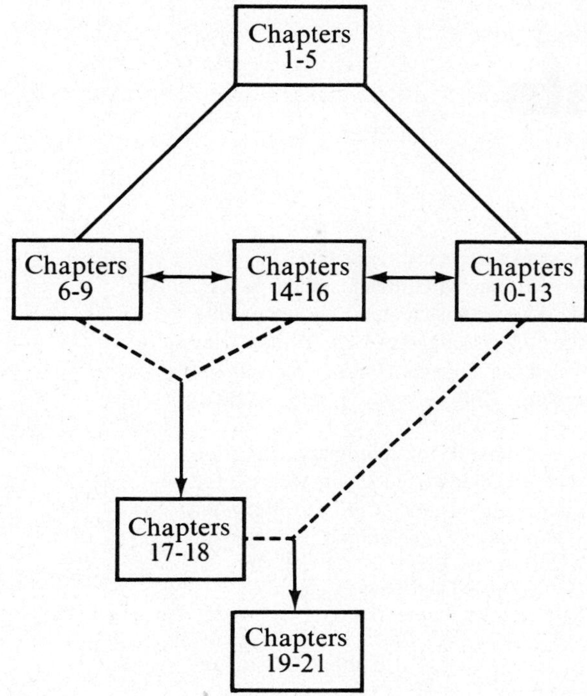

Chapter 6 requires only knowledge of the skills covered in Chapter 1. Prerequisite chapters for Chapter 7 are 1 and 6. For Chapter 8, prerequisite chapters are 1, 2 and 6. Chapter 9 requires previous study of all the preceeding chapters. Chapter 10* requires Chapters 1 and 3. Chapter 11* requires Chapters 1, 3 and 10. Chapter 12* requires Chapters 1, 3, 4, 10 and 11. Chapter 13* requires Chapters 1, 3, 4, and 10–12. Chapter 14 requires Chapters 1–4. Chapter 15 requires 1–4 and 14. Chapter 16 requires 1–4, 14 and 15. Chapter 17 requires 1–4, 6, 8 and 14–16. Chapter 18 requires all previous chapters except 10–13. Chapter 19 requires 1–4, 6, 10, 11, 14, 17 and 18. Chapter 20* requires 1–3, 10 and 14–16. Chapter 21 requires 1–4, 6–11, 14–11, 14–17 and 20.

*Indicates chapters for which reducing numerical fractions (section 6.1 or 6.2) is also needed.

Acknowledgments

We are very grateful for the many helpful suggestions and constructive criticisms given us by our colleagues. In particular, we wish to thank John Seely Brown, Gary M.C. Bean, and the faculty of the Mathematics Department at Baruch College. In addition, we wish to thank our reviewers, C.F. Blakemore, University of New Orleans; John Miller, City College of New York; Carla Oviatt, Montgomery College; Karen Butler, Metropolitan State College; Gregory McNeil, Foothill College; Bruce King, Schenectady County Community College; and Calvin A. Lathan, Monroe Community College.

We feel especially indebted to our typist, Marie Grossi; without her extraordinary dedication we would never have succeeded in producing this textbook and the two precursors we used for class testing. Thanks are also due to the Duplication Department at Baruch College for printing those earlier versions of this book. Finally, we wish to express our appreciation of everyone who has been involved with the production of this book at Harcourt Brace Jovanovich, Inc. We are indebted especially to our editor, Marilyn Davis, and our designer, Geri Davis.

Robert Donaghey
JoAnna Ruddel

Contents

List of Tables

Fundamentals of Algebra

An Integrated Text-Workbook

INTEGERS
TO
POLYNOMIALS

1 ALGEBRAIC ADDITION AND MULTIPLICATION OF INTEGERS

To study algebra we must begin with the study of numbers and the operations on them. The numerical foundation of algebra is the system of signed numbers, which includes both integers and fractions. We will begin our study by looking first at the integers.

Integers

The numbers

$$\ldots, -4, -3, -2, -1, 0, +1, +2, +3, +4, \ldots$$

are collectively called the *integers* (or, for emphasis, *signed* integers). The numbers +1, +2, +3, +4, +5, +6, . . . are called *positive integers*, whereas the numbers −1, −2, −3, −4, −5, −6, . . . are called *negative integers*. Note that 0 is neither positive nor negative. The positive integers are read "plus 1, plus 2," and so on or "positive 1, positive 2," and so on. The negative integers are read "minus 1, minus 2," and so on or "negative 1, negative 2," and so on.

The positive integers and 0 are *nonnegative integers*. You are used to seeing them written without signs, as

$$0, 1, 2, 3, 4, 5, 6, \ldots$$

When the positive integers are written without their plus signs, they are usually called *natural numbers*, and this way of writing them is called *natural number notation.*

Numerical Value

The *numerical value* (also called the *absolute value*) of a positive or negative integer equals the natural number formed by omitting the integer's sign. For example, the numerical value of +3 is 3, and the numerical value of −3 is also 3. Similarly, the numerical value of both +5 and −5 is 5. In general, the numerical value of a positive integer is the positive integer written in natural number notation, and the numerical value of a negative integer is the corresponding positive integer written in natural number notation. The numerical value of 0 is 0.

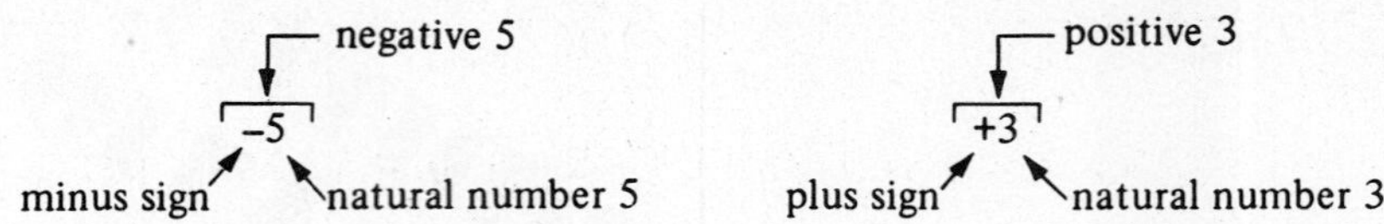

Order

Our experience with natural numbers tells us that they can be arranged in increasing order, as 1, 2, 3, 4, 5, ... These numbers are, of course, all larger than zero and are naturally ordered. We shall use the words *larger* and *smaller* when talking about the order of natural numbers and numerical values. For example,

5 is larger than 2 and 2 is smaller than 5

We would also expect the signed integers to be ordered. We can see how the positive and negative integers are ordered by representing them on the *number line* that extends infinitely in two directions with zero in the middle:

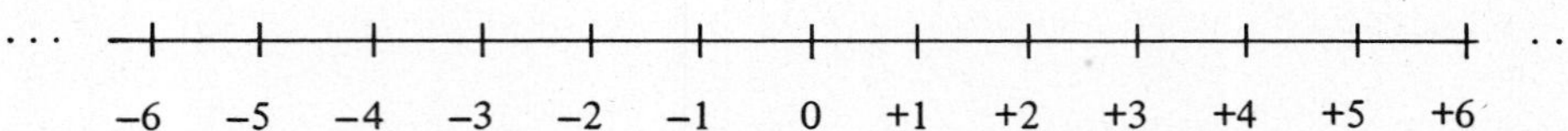

On this number line, the integers *increase* to the *right* and *decrease* to the *left*. Each integer is said to be *greater than* every number to its left. Similarly, each integer is said to be *less than* every number to its right. For example,

+5 is greater than −2 and −5 is less than +2

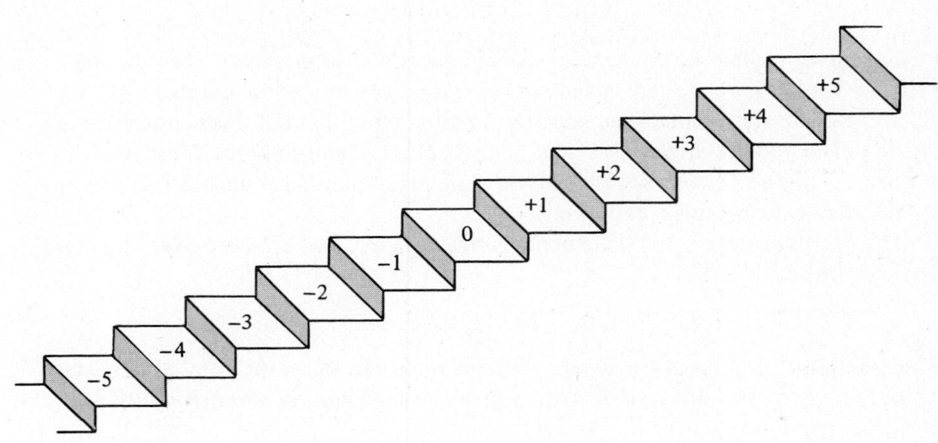

Simplifying Expressions

Any given expression equals many other expressions. For example, the expressions 2 + 3, 1 + 4, and 5 are all equal. Given an expression we frequently want to find the expression with the simplest form that has the same value. This is called *simplifying the expression.* For example,

2 + 3 simplifies to 5 and 4 + 5 + 6 simplifies to 15

1.0
ADDING TWO INTEGERS USING THE NUMBER LINE

Consider the following examples of *algebraic addition*:

+3 plus +2 = +5   +3 plus −2 = +1

−3 plus +2 = −1   −3 plus −2 = −5

Look at the pattern, and note that each integer can be positive or negative. To help us understand algebraic addition, we will look at this operation on the number line.

Look at the following example of *adding two integers using the number line*.

TYPICAL PROBLEM

−1 plus +4

We first find −1 (the first integer) on the number line, and mark the tail of an arrow above it:

−3 −2 −1 0 +1 +2 +3

From here, we count over four units, counting to the right since +4 (the second integer) is positive. As we count we draw the arrow representing +4:

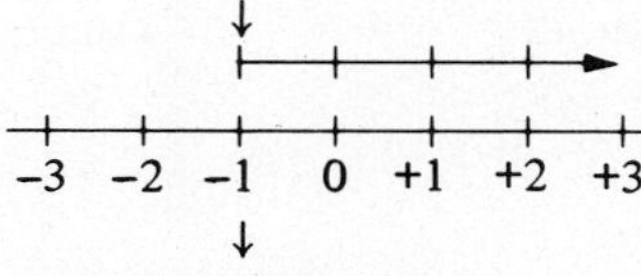

Then the integer below the arrowhead is the sum of the two integers:

−1 plus +4 = +3

Note that it is not necessary actually to draw an arrow. You need only count over from the first point the number of units given by the numerical value of the second integer, counting in the direction indicated by the sign of the second integer.

EXAMPLES

Add the following pairs of integers using the number line.

1. −2 plus −2
2. −5 plus +2
3. −4 plus +7
4. +3 plus +6

Solutions

1. From −2, move left 2 units (since −2 is negative)

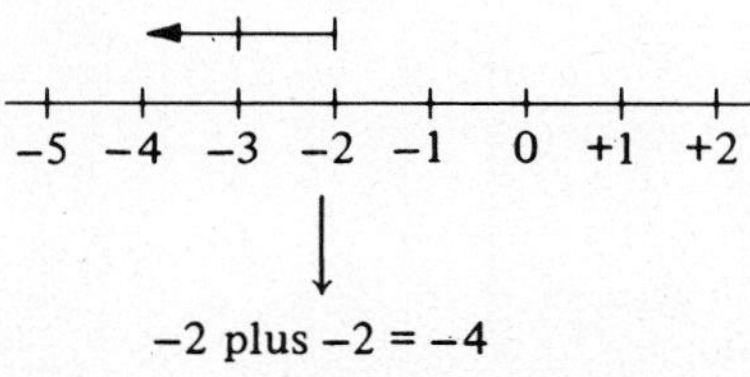

−2 plus −2 = −4

2. From −5, move right 2 units (since +2 is positive)

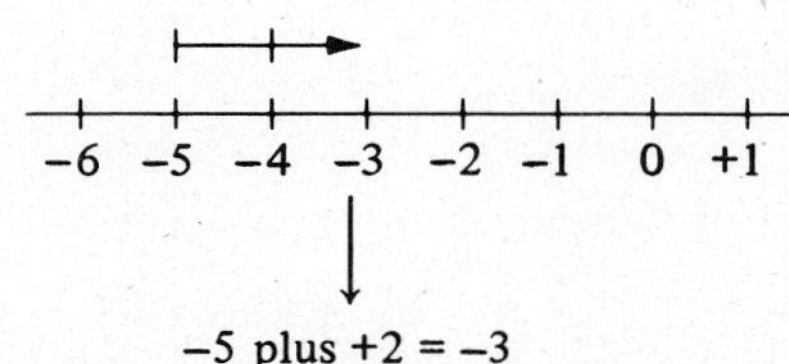

−5 plus +2 = −3

3. From −4, move right 7 units (since +7 is positive)

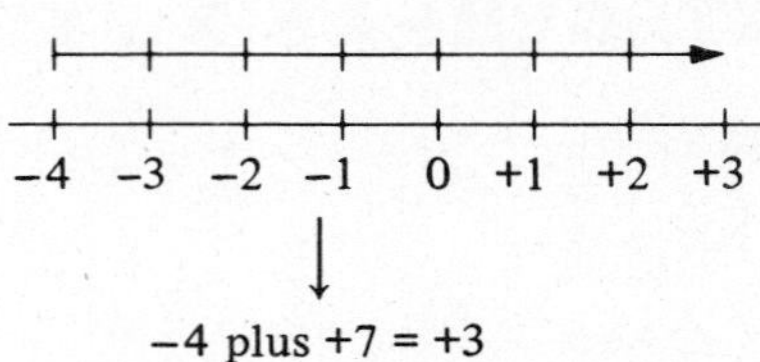

−4 plus +7 = +3

4. From +3, move right 6 units (since +6 is positive)

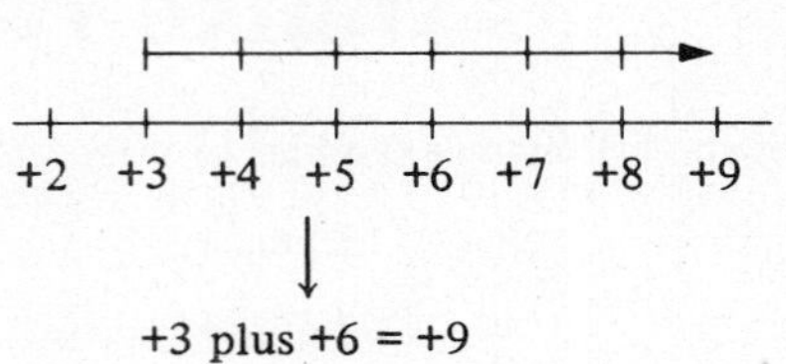

+3 plus +6 = +9

OBSERVATIONS AND REFLECTIONS

The examples we have seen demonstrate some important properties of algebraic addition. When we add *two positive integers* we start to the right of zero and move to the right, so the sum is *always positive* (Example 4). Similarly, when we add *two negative integers* we start to the left of zero and move to the left, so the sum is *always negative* (Example 1). On the other hand, the sum of two integers with *unlike* signs may be positive, negative, or zero, because we move from the first integer back toward (and possibly past) zero (Examples 2 and 3).

1. Compute
   a. +1 plus +2
   b. −1 plus −2
   c. +1 plus −2
   d. −1 plus +2

EXERCISES

Simplify each of the following expressions to an integer, using a number line.

1. −1 plus −2	2. −2 plus +3	3. −4 plus +1
4. +2 plus −4	5. +3 plus −1	6. +1 plus −2
7. +6 plus −4	8. −5 plus +3	9. +2 plus −6
10. −3 plus +7	11. +4 plus −8	12. −1 plus +6
13. +5 plus −5	14. −3 plus +9	15. +4 plus −9
16. −6 plus +6	17. +5 plus −10	18. −4 plus +7

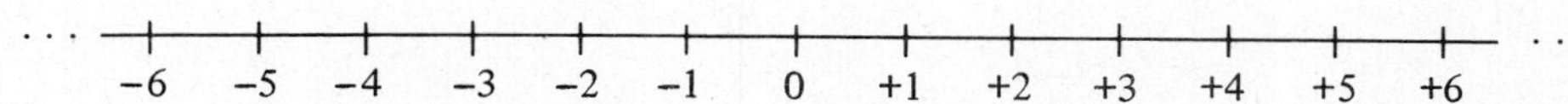

1.1 ADDING TWO INTEGERS

Look at the following examples of *algebraic addition.*

+5 plus +3 = +8   +5 plus −3 = +2

−5 plus +3 = −2   −5 plus −3 = −8

Here we have written the word *plus* to indicate algebraic addition, although the ordinary algebraic notation for these expressions is

$$+5 + 3 = +8 \qquad +5 - 3 = +2$$

$$-5 + 3 = -2 \qquad -5 - 3 = -8$$

Let us agree that when two signed integers are written side by side, algebraic addition is understood:

algebraic addition understood

↓

$-5\ +3$

negative 5 ↗ ↖ positive 3

For example,

$+5 + 3$	means	$+5$ plus $+3$	$+5 - 3$	means	$+5$ plus -3
$-5 + 3$	means	-5 plus $+3$	$-5 - 3$	means	-5 plus -3

To see how to use ordinary addition and subtraction to simplify algebraic sums, look at the pattern in the following columns.

$+2 + 2 = +4$	$-2 - 2 = -4$
$+2 + 1 = +3$	$-2 - 1 = -3$
$+2 + 0 = +2$	$-2 - 0 = -2$
$+2 - 1 = +1$	$-2 + 1 = -1$
$+2 - 2 = 0$	$-2 + 2 = 0$
$+2 - 3 = -1$	$-2 + 3 = +1$
$+2 - 4 = -2$	$-2 + 4 = +2$

Can you see how to add two negative integers together? Can you see how to add a positive and a negative integer? Look carefully at the signs and note that the sign of the sum is the sign of the integer with the larger numerical value.

Look at the following examples of *adding two integers.*

2. Write in algebraic notation
- a. -7 plus $-2 = -9$
- b. $+3$ plus $-6 = -3$

3. Insert the word *plus* into
- a. $-6 + 2 = -4$
- b. $+4 + 8 = +12$
- c. $-3 - 5 = -8$

4. Compute
- a. $+1 + 1$
  $+1 + 0$
  $+1 - 1$
  $+1 - 2$
- b. $-1 - 1$
  $-1 - 0$
  $-1 + 1$
  $-1 + 2$

TYPICAL PROBLEMS

1. $-5 - 3$ → add 5 and 3 → -8

We add the numerical values because the signs are the same:

We then attach the common sign:

Hence, $-5 - 3 = -8$.

(Since both integers are negative, the sum is negative.)

2.

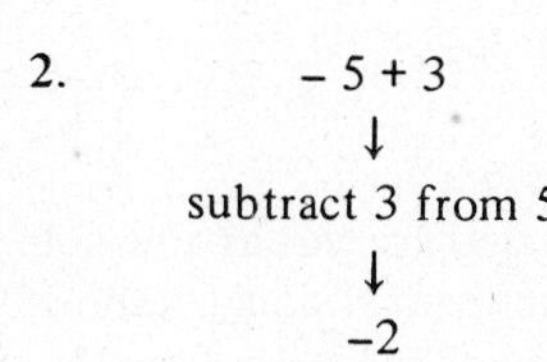

We take the difference of the numerical values because the signs are different:

We then attach the sign of the integer with the larger numerical value:

Hence, $-5 + 3 = -2$.

(Since 5 is the larger numerical value, the sum is negative.)

In both cases the sign of the sum is the sign of the number with the larger numerical value.

By making use of ordinary addition and subtraction, we can combine any two integers.

EXAMPLES

Combine the following pairs of integers.

1. $-12 - 7$
2. $+32 - 19$
3. $-27 + 16$
4. $+35 + 12$
5. $-12 + 38$
6. $+46 - 81$

Solutions

1. The signs agree, so add 12 and 7 → attach the sign → -19
2. The signs differ, so subtract 19 from 32 → attach the sign → $+13$
3. The signs differ, so subtract 16 from 27 → attach the sign → -11
4. The signs agree, so add 35 and 12 → attach the sign → $+47$
5. The signs differ, so subtract 12 from 38 → attach the sign → $+26$
6. The signs differ, so subtract 46 from 81 → attach the sign → -35

SPECIAL CASES

1. The number 0 is neither positive nor negative and has no sign preference. It is almost always written without a sign when it stands by itself.

-0 is written 0 and $+0$ is written 0

For example, when adding,

$$-5 + 5 = 0 \quad \text{and} \quad +8 - 8 = 0$$

2. When 0 appears as a term in a sum, it must be preceded by a sign if it is not the first term, although adding 0 (or +0 or −0) to an expression does not change the value of the expression. For example,

$$-5 + 0 = -5 \quad \text{and} \quad 0 - 5 = -5$$

$$+8 - 0 = +8 \quad \text{and} \quad 0 + 8 = +8$$

5. Compute
a. $-1 + 1$
b. $+1 - 1$
c. $-2 + 2$
d. $+2 - 2$
e. $-3 + 3$
f. $+3 - 3$

6. Compute
a. $+3 + 0$
b. $-3 + 0$
c. $+3 - 0$
d. $-3 - 0$
e. $0 + 3$
f. $0 - 3$

You are probably familiar with the addition table of ordinary arithmetic. Here is a modified addition table for signed integers.

plus	−6	−5	−4	−3	−2	−1	0	+1	+2	+3	+4	+5	+6
+6													
+5													
+4			0	1	2	3	4	5	6	7	8		
+3			−1	0	1	2	3	4	5	6	7		
+2			−2	−1	0	1	2	3	4	5	6		
+1			−3	−2	−1	0	1	2	3	4	5		
0			−4	−3	−2	−1	0	1	2	3	4		
−1			−5	−4	−3	−2	−1	0	1	2	3		
−2			−6	−5	−4	−3	−2	−1	0	1	2		
−3			−7	−6	−5	−4	−3	−2	−1	0	1		
−4			−8	−7	−6	−5	−4	−3	−2	−1	0		
−5													
−6													

7. Complete the table.

Verify the entries to make sure that you understand how to add two signed numbers.

OBSERVATIONS AND REFLECTIONS

Note that the sum of two *positive* integers is *positive* and the sum of two *negative* integers is *negative*, whereas the sum of two integers with unlike signs may be positive, negative, or zero. For example,

$$+2 + 3 = +5 \quad \text{and} \quad -2 - 3 = -5$$

whereas

$$-2 + 3 = +1, \quad -2 + 2 = 0, \quad \text{and} \quad -2 + 1 = -1$$

8. Compute
   a. $+1 + 2$
      $+1 + 1$
      $+1 - 1$
      $+1 - 2$
   b. $-1 + 2$
      $-1 + 1$
      $-1 - 1$
      $-1 - 2$

EXERCISES

Simplify each of the following expressions to an integer.

1. $-2 - 5$
2. $-6 - 4$
3. $-9 + 3$
4. $+5 - 8$
5. $+7 - 6$
6. $-3 - 11$
7. $0 + 10$
8. $-9 + 9$
9. $-14 - 0$
10. $+8 - 8$
11. $-5 + 0$
12. $0 - 6$

13. −4 − 9 14. −3 + 2 15. −8 + 6

16. +8 − 0 17. 0 − 16 18. −48 − 29

19. −19 + 53 20. −22 − 31 21. 64 + 13

22. −51 + 25 23. −64 + 95 24. 39 − 47

25. −92 + 50 26. −12 − 84 27. −78 + 25

28. 35 − 56 29. −43 − 59 30. 77 + 56

31. −51 − 51 32. 48 − 86 33. 21 − 194

34. −13 + 288 35. −72 + 463 36. −107 − 310

Express each of the following word phrases algebraically and then simplify it. (A review of translating word phrases is provided in Section 1.6.)

1. The sum of 8 and 7
2. 9 decreased by 13
3. 5 subtracted from 16
4. 17 added to 4
5. 11 reduced by 13
6. 6 less than 4
7. 5 plus 31
8. 15 is subtracted from 12
9. 11 more than 2
10. 1 less 5
11. 14 increased by 8
12. 16 is reduced by 19

1.2 ADDING SEVERAL INTEGERS

9. Compute
   a. −5 − 8
   b. −8 − 5
   c. +5 − 8
   d. −8 + 5
   e. −5 + 8
   f. +8 − 5

Commutative Law

From your experience with ordinary arithmetic you know that an indicated sum of two positive integers can be written in either order. For example, 3 + 4 = 4 + 3, 5 + 8 = 8 + 5, and so forth. This is true for signed numbers as well. For example,

$$-5 - 8 = -8 - 5 \quad \text{and} \quad +5 - 8 = -8 + 5 \quad \text{and} \quad -5 + 8 = +8 - 5$$

This property of integers is called the *Commutative Law of Addition.*

This law may seem somewhat obvious, but it needs to be stated because it will become very important for us later.

COMMUTATIVE LAW OF ADDITION

The sum of a and b is the sum of b and a

a plus b = b plus a

$$a + b = b + a$$

(where a and b represent signed numbers)

Associative Law

To simplify a sum such as 6 – 7 + 8, you probably expect that we can combine the integers two at a time. Does it matter, however, which two integers we add first? Compare the following computations:

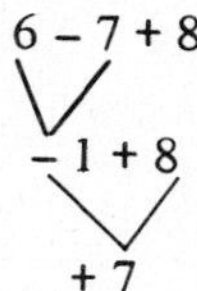

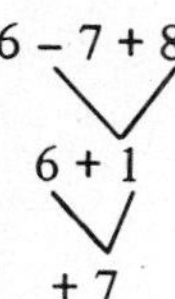

The intermediate sums –1 + 8 and 6 + 1 both equal 6 – 7 + 8, and both equal +7, and so have the same value. Hence, we can either add the first two numbers or the last two numbers first. This property of numbers is called the *Associative Law of Addition*.

10. Compute –5 + 2 + 4 both as
a. (–5 plus +2) plus +4
and as
b. –5 plus (+2 plus +4)

> **ASSOCIATIVE LAW OF ADDITION**
>
> (a plus b) plus c = a plus (b plus c)
>
> $(a + b) + c = a + (b + c)$
>
> (where a, b, and c represent signed numbers)

The Commutative and Associative Laws hold for all numbers, not just integers. In fact, as we shall see, they hold for all the algebraic expressions in this book.

Adding

To talk about sums, we need a new word. In a sum of integers, each integer is called a *term* of the sum. For example, the terms of 5 + 2 are 5 and 2, and the terms of 3 – 7 – 2 + 5 are 3, –7, –2, and 5.

11. What are the terms of
a. +5 – 2 + 9
b. –4 – 6 + 1

Combining the Commutative and Associative Laws of Addition, we see that we can write and combine the terms of a sum in any order. Hence, we can combine all the positive integers first regardless of their position in a sum and then combine the negative integers. As you might expect, this is often the fastest way to add signed integers.

Look at the following example of *adding several integers.*

TYPICAL PROBLEM

We can first rearrange the sum, writing the positive integers before the negative integers (this is optional):

Then we add the integers with like signs. We add the positive integers, and we add the negative integers:

Finally we combine the resulting two integers with unlike signs:

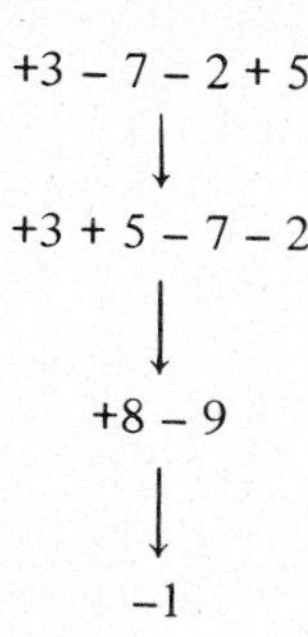

To combine integers with unlike signs we use the method in the previous section.

Hence, +3 – 7 – 2 + 5 = –1.

EXAMPLES

Every sum of integers can be simplified to a single integer.

Add the following integers.

1. $4 - 11 - 3 + 8$
2. $-9 - 6 - 7 - 3$
3. $-7 - 2 + 5 - 4 + 8$
4. $+3 - 14 + 21 + 7 - 16$

Solutions

1. $4 + 8 - 11 - 3$
   ↓
   $+12 - 14$
   ↓
   -2

2. -25

3. $+5 + 8 - 7 - 2 - 4$
   ↓
   $+13 - 13$
   ↓
   0

4. $+3 + 21 + 7 - 14 - 16$
   ↓
   $+31 - 30$
   ↓
   1

SPECIAL CASES

12. Algebraically express the sum of the terms
a. 7, −5, 3
b. −2, −8, 12

In a sum such as $5 - 9 + 2$ the first term, $+5$, may have its sign omitted. That is, the sum may be written as either $5 - 9 + 2$ or $+5 - 9 + 2$. Note, however, that the third term in this sum, $+2$, must have its sign expressed.

Remember that when a positive integer is written by itself, its sign may be expressed or omitted. Similarly, the number 0 by itself is always written without a sign. In an indicated algebraic sum, however, that contains more than one integer, only the *first* term (if it is not negative) may have its sign omitted.

EXERCISES

Simplify each of the following expressions to an integer.

1. $2 - 5 - 7$
2. $-11 - 13 + 15$
3. $-20 + 5 - 30$
4. $5 - 8 + 4$
5. $9 - 2 + 8$
6. $-5 + 7 - 8$
7. $-11 + 4 - 7$
8. $3 - 7 + 4$
9. $-5 + 11 - 6 + 4$
10. $+8 - 2 + 3 - 9$
11. $-13 + 8 + 4 - 11$
12. $16 - 11 + 19 - 10$
13. $15 - 8 - 3 + 4$
14. $5 - 11 - 6 + 8$
15. $-4 + 12 - 20 + 7$
16. $6 - 9 - 7 + 1$
17. $4 - 9 + 3 - 11$
18. $-4 + 8 + 7 - 12$
19. $-9 - 3 - 12 + 20 + 3$
20. $-8 + 12 - 9 + 16 - 11$
21. $7 + 15 - 21 + 4 - 8$
22. $16 - 9 - 4 + 7 - 12$
23. $12 - 19 - 7 + 13 + 9$
24. $-8 - 4 - 9 - 11 - 14$

1.3
ADDING TWO INTEGERS WITH PARENTHESES

Parentheses, ( ), are the symbols of grouping and any number of pairs of parentheses can occur in an expression. Expressions such as −(−3) and +(−7) are the simplest of many situations in which parentheses can and should be eliminated. These expressions have the structure:

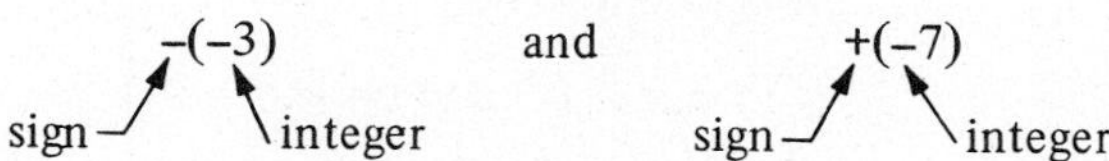

Removing Parentheses

To see how to simplify such expressions, look at the pattern in the following columns.

+(+2) = +2	−(−2) = +2
+(+1) = +1	−(−1) = +1
+(0) = 0	−(0) = 0
+(−1) = −1	−(+1) = −1
+(−2) = −2	−(+2) = −2

13. Compute
 a. +(+2) =
  +(−2) =
 b. −(+2) =
  −(−2) =

An integer does not change its value when it is placed inside parentheses or parentheses preceded by a plus sign. For example,

−2 = (−2) = +(−2) and +2 = (+2) = +(+2)

But what does a minus sign before parentheses mean? Placing an integer inside parentheses preceded by a minus sign is one way to form what is called the *negative* of the integer. The negative of an integer can also be formed by taking the corresponding integer with the same numerical value but the opposite sign. For example, the negative of +3 is −(+3) or −3, and the negative of −3 is −(−3) or +3. Keep in mind that the negative *of* an integer is not necessarily a negative integer.

Once we understand the meaning of parentheses, we can use integers inside parentheses to form more complex sums. Remember that in an algebraic sum each quantity is called a *term* of the sum. For example, in the sum −(−8) + (−7), the terms are −(−8) and (−7). Similarly, in the sum −(−9) − 5, the terms are −(−9) and −5.

Look at the following example of *simplifying sums containing parentheses.*

14. Remove parentheses:
 a. (−4)
 b. +(+5)
 c. +(−6)

15. What is the negative of
 a. +9
 b. −4
 c. 0

16. Remove parentheses:
 a. −(+5)
 b. −(−6)
 c. −(4)

17. What are the terms of
 a. −2 + (−7)
 b. −(+5) − (−6)

TYPICAL PROBLEM

−(+5) − (−9)

We first simplify each term containing parentheses by determining the sign and removing parentheses:

↓

−5 + 9

↓

Then we add the integers:

+4

Hence, −(+5) − (−9) = +4.

To add the integers we use the method in Section 1.1.

EXAMPLES

Simplify the following sums containing parentheses.

1. (+5) − (+8)
2. −(−16) + (−11)
3. −3 − (−7)
4. −(+10) − (−10)
5. −(+4) + (+9)
6. +(−8) − 15

Solutions

1. $+5 - 8 = -3$
2. $+16 - 11 = 5$
3. $-3 + 7 = 4$
4. $-10 + 10 = 0$
5. $-4 + 9 = 5$
6. $-8 - 15 = -23$

SPECIAL CASES

18. Simplify to an integer:
a. (−5)
b. −(3)
c. (+7)
d. (9)

Sometimes plus signs are omitted. For example,

$$(-5) = +(-5) \qquad (+7) = +(+7)$$

$$-(3) = -(+3) \qquad (9) = +(+9)$$

OBSERVATIONS AND REFLECTIONS

1. Remember that when signed integers are written side by side, algebraic addition is understood. Similarly, when signed integers and/or signed parenthetical expressions are written side by side, algebraic addition is understood.

19. Express algebraically the sum of the terms:
a. −2 and (−9)
b. −(+4) and 7
c. (−8) and −6
d. −(−2) and (−2)

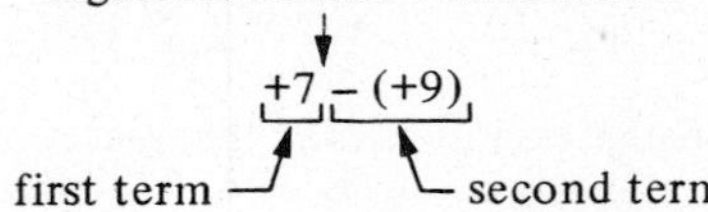

In an algebraic sum, every term (except possibly the first) begins with a sign, and every sign that is not inside parentheses begins a term of the sum.

2. The process of replacing an integer by its negative is called *negation.*

20. Negate
a. −5
b. +12
c. −16
d. 20

NEGATION

The negative of $+a$ is $-(+a)$ or $-a$.
The negative of $-a$ is $-(-a)$ or $+a$.

$$-(a) = -a$$

$$-(-a) = a$$

(where a represents a numerical value)

★ 3. Using negation, we can define *algebraic subtraction.* Just as we can add two signed integers to find their sum, we can subtract one signed integer from another to find their difference. To *subtract* one signed integer from a second, we add the negative of the first integer to the second integer. For example, to subtract −8 from −7, we add −(−8) to −7:

21. Subtract
a. +4 from +2
b. −4 from +2
c. +4 from −2
d. −4 from −2

$$-7 - (-8) = -7 + 8 = +1$$

ALGEBRAIC SUBTRACTION

To subtract a from b, add $-(a)$ to b

(where a and b represent (signed) integers)

Here is a table showing the difference of two signed integers. Check the entries in this table to make sure you understand the difference between algebraic addition and algebraic subtraction.

FROM THIS NUMBER: Subtract from ↓ / SUBTRACT THIS NUMBER: Subtract →	(−5)	(−4)	(−3)	(−2)	(−1)	(0)	(+1)	(+2)	(+3)	(+4)	(+5)
(+5)											
(+4)											
(+3)			6	5	4	3	2	1	0		
(+2)			5	4	3	2	1	0	−1		
(+1)			4	3	2	1	0	−1	−2		
(0)			3	2	1	0	−1	−2	−3		
(−1)			2	1	0	−1	−2	−3	−4		
(−2)			1	0	−1	−2	−3	−4	−5		
(−3)			0	−1	−2	−3	−4	−5	−6		
(−4)											
(−5)											

22. Complete the table.

EXERCISES

Simplify each of the following expressions to an integer.

1. +(−7)
2. −(−26)
3. −(+35)
4. −(−19)
5. +(+31)
6. +(+61)
7. +(−14)
8. −(+6)
9. −(−1)
10. −(+20)
11. −(−81)
12. +(−8)
13. (+3) − (+9)
14. (−24) + (−25)
15. −(−5) − (+7)
16. 3 − (−6)
17. 7 − (−2)
18. (+8) − (+12)
19. −(3) − (−11)
20. −(−6) − (+6)
21. −4 + (−8)
22. −3 + (−1)
23. (−24) − (−8)
24. −(7) − (−15)
25. (−7) + (−52)
26. −(37) + 37
27. −(+19) + 34
28. (−31) + (−29)
29. (+46) − (+61)
30. (+28) − (+28)

31. $-(63) - 58$ 32. $-(-75) - 75$ 33. $-(-98) - (+125)$

34. $-(-77) - (+113)$ 35. $-(69) - (-215)$ 36. $-56 - (+28)$

Express each of the following word phrases algebraically and then simplify it. (A review of translating word phrases is provided in Section 1.6.)

1. The sum of −9 and −8
2. −7 decreased by 5
3. −16 subtracted from −27
4. −3 added to −11
5. −14 reduced by −14
6. −2 less than −3
7. 5 is decreased by −1
8. −15 is added to −4
9. −7 is subtracted from −24
10. −9 less −6
11. −21 increased by −19
12. −13 is reduced by 22

1.4
MULTIPLYING INTEGERS

Products of Numerical Values

You are, of course, familiar with products of positive integers (numerical values). You also probably know that a product of two positive integers can be represented geometrically as the area of a rectangle. Since 2 times 3 is 6, for example, the area of a rectangle 2 units wide and 3 units long is 6 square units.

23. Represent geometrically
 a. $1 \cdot 5$
 b. $2 \cdot 4$

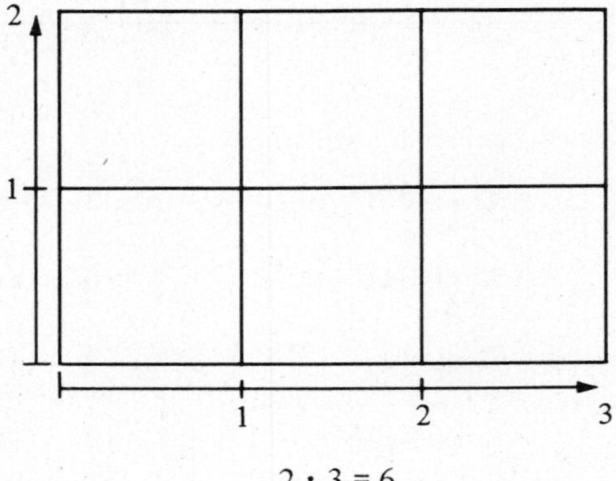

$2 \cdot 3 = 6$

When we study various properties of multiplication, we will use areas to illustrate these properties.

Algebraic Products

We need some new words to talk about products of signed numbers. In *algebraic products* such as $(-5)(+2)$ and $-2(+5)(-3)$ each quantity is called a *factor* of the product.

For example, in the product (−5) (+2), the factors are −5 and +2. Similarly, in the product −2(+5) (−3), the factors are −2, +5, and −3.

24. List the factors of
a. (−3) (+7)
b. 6(−6) (+2)

Algebraic multiplication is indicated in any of the following ways:

1. (+5) · (−2)	with a dot between sets of parentheses.
2. (+5) (+2)	with no symbol between sets of parentheses.
3. −5(−2)	with no symbol between factors and only the second number inside parentheses.
4. −5 · 2	with a dot between factors and no plus sign before the second number.

The first factor of a product may be inside parentheses, but need not be. (See 2 and 3 above.) Any positive factor *may* be written without parentheses if it is written in natural number notation with its plus sign omitted. (See 4 above.) All other factors of the product must be written inside parentheses.

Products of (Signed) Integers

Multiplication of signed integers closely parallels the multiplication of ordinary positive integers. The most obvious difference is in the handling of the signs of the integers.

To see the relationship between an algebraic product of two signed integers and an ordinary product of two natural numbers, look at the pattern in the following columns.

+3(+2) = +6	−3(−2) = +6
+2(+2) = +4	−2(−2) = +4
+1(+2) = +2	−1(−2) = +2
0(+2) = 0	0(−2) = 0
−1(+2) = −2	+1(−2) = −2
−2(+2) = −4	+2(−2) = −4
−3(+2) = −6	+3(−2) = −6

25. Compute
a. +2(+1)
+1(+1)
0(+1)
−1(+1)
−2(+1)
b. −2(−1)
−1(−1)
0(−1)
+1(−1)
+2(−1)

Can you see how these products relate to natural number products? Can you guess the rule for determining the sign?

Let us look at some more examples to see more generally how to determine the sign of a product. Look at the pattern in the following display:

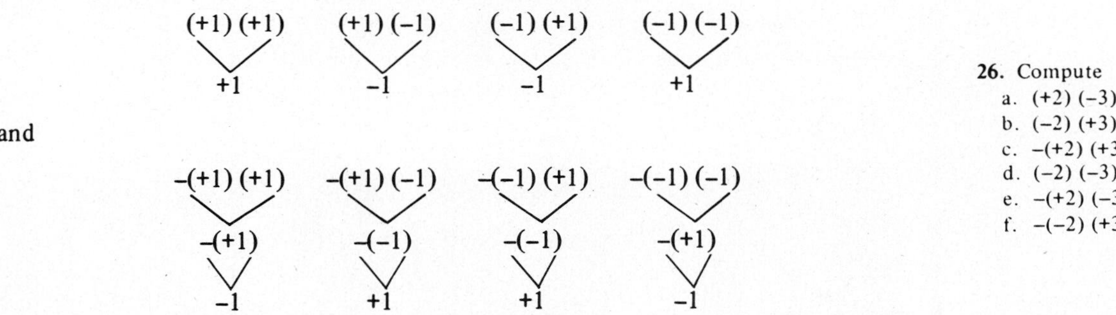

26. Compute
a. (+2) (−3)
b. (−2) (+3)
c. −(+2) (+3)
d. (−2) (−3)
e. −(+2) (−3)
f. −(−2) (+3)

27. Compute
a. (+1) (+1)
b. +1 + 1
c. (+1) (−1)
d. +1 − 1
e. (−1) (+1)
f. −1 + 1
g. (−1) (−1)
h. −1 − 1

As you can see, the product is positive when there is an even number of minus signs and negative when there is an odd number of minus signs.

As you probably have noticed, the rule for determining the sign of a product is different from the rule for determining the sign of a sum (which we saw in Section 1.1). For example,

$$(-2)(-2) = +4 \qquad \text{whereas} \qquad -2 - 2 = -4$$

Commutative and Associative Laws

28. Compute
a. (−2) (−3)
b. (−3) (−2)
c. (−2) (+3)
d. (+3) (−2)
e. (+2) (−3)
f. (−3) (+2)

We have seen that algebraic addition obeys two very important laws, the Commutative and Associative Laws of Addition. From your experience with arithmetic you know that ordinary multiplication obeys similar laws. Of course, algebraic products also obey these laws.

In an indicated product of two integers the integers may appear in either order without changing the value of the product. For example, (−2) (−3) equals (−3) (−2). This property of numbers is called the *Commutative Law of Multiplication.*

COMMUTATIVE LAW OF MULTIPLICATION

The product of a and b is the product of b and a

a times b = b times a

$a \cdot b = b \cdot a$

(where a and b represent signed numbers)

To simplify an indicated product of three integers, we can either multiply the first two numbers first or we can multiply the last two numbers first. For example, compare the following computations:

29. Compute (−1) (+2) (−3) both as
a. (−1 times +2) times −3
and as
b. −1 times (+2 times −3)

(−2) (+5) (−3)
(−10) (−3)
+30

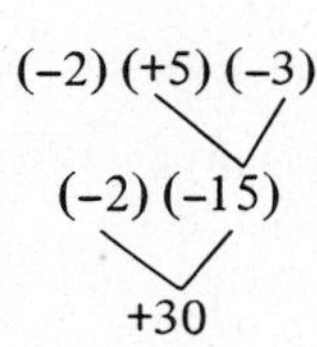

This property of numbers is called the *Associative Law of Multiplication.*

ASSOCIATIVE LAW OF MULTIPLICATION

(a times b) times c = a times (b times c)

$(a \cdot b) \cdot c = a \cdot (b \cdot c)$

(where a, b, and c represent signed numbers)

The Commutative and Associative Laws of Multiplication hold for all numbers, not just integers. In fact, as we shall see, they hold for all the algebraic expressions in this book. Together, they imply that the factors of an algebraic product may be written and multiplied in any order.

Look at the following example of *multiplying integers*.

TYPICAL PROBLEM

(−2) (+5) (−3)

We first determine the sign by counting the minus signs:

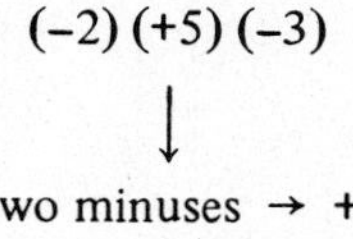

two minuses → +

Then we multiply the numerical values and attach the sign:

↓

+30

Hence, (−2) (+5) (−3) = +30.

If any factor is zero then the product is zero. Otherwise, the sign of the product is determined only by the signs of the factors.

EXAMPLES

Multiply the following integers.

Every product of integers can be simplified to a single integer.

1. (−16) (+4)
2. 6(−15)
3. (27) (6)
4. 5(−3) (−8)
5. −4(−2) (−5)
6. (−3) (−5) (0)

Solutions

1. one minus → −
↓
−64
2. one minus → −
↓
−90
3. no minuses → +
↓
+162
4. two minuses → +
↓
120
5. three minuses → −
↓
−40
6. one factor is 0
↓
0

SPECIAL CASES

1. Multiplying any integer by zero yields 0. For example,

$$0 \cdot 8 = 0 \quad \text{and} \quad 8 \cdot 0 = 0$$

$$0 \cdot (-8) = 0 \quad \text{and} \quad -8 \cdot 0 = 0$$

2. Multiplying any integer by 1 yields the integer. For example,

$$1 \cdot 8 = 8 \quad \text{and} \quad 8 \cdot 1 = 8$$

$$1 \cdot (-8) = -8 \quad \text{and} \quad -8 \cdot 1 = -8$$

30. Compute
a. 0(−2)
b. −2(0)
c. 0 · 4
d. 4 · 0

31. Compute
a. 1(−2)
b. −2(1)
c. 1 · 4
d. 4 · 1

32. Compute
a. −1(−2)
b. −2(−1)
c. (−1) (4)
d. (4) (−1)

3. Multiplying any integer by −1 yields the integer with the opposite sign; that is, the negative of the integer. For example,

$$-1 \cdot 8 = -8 \quad \text{and} \quad 8 \cdot (-1) = -8$$

$$-1 \cdot (-8) = 8 \quad \text{and} \quad -8 \cdot (-1) = 8$$

$0 \cdot a = 0$ $\quad 1 \cdot a = a$ $\quad -1 \cdot a = -a$
$0(-a) = 0$ $\quad 1(-a) = -a$ $\quad -1(-a) = a$
(where a represents a numerical value)

You are probably familiar with the multiplication table of arithmetic. Here is a modified multiplication table for signed integers. Verify the entries to make sure that you understand how to multiply two signed numbers.

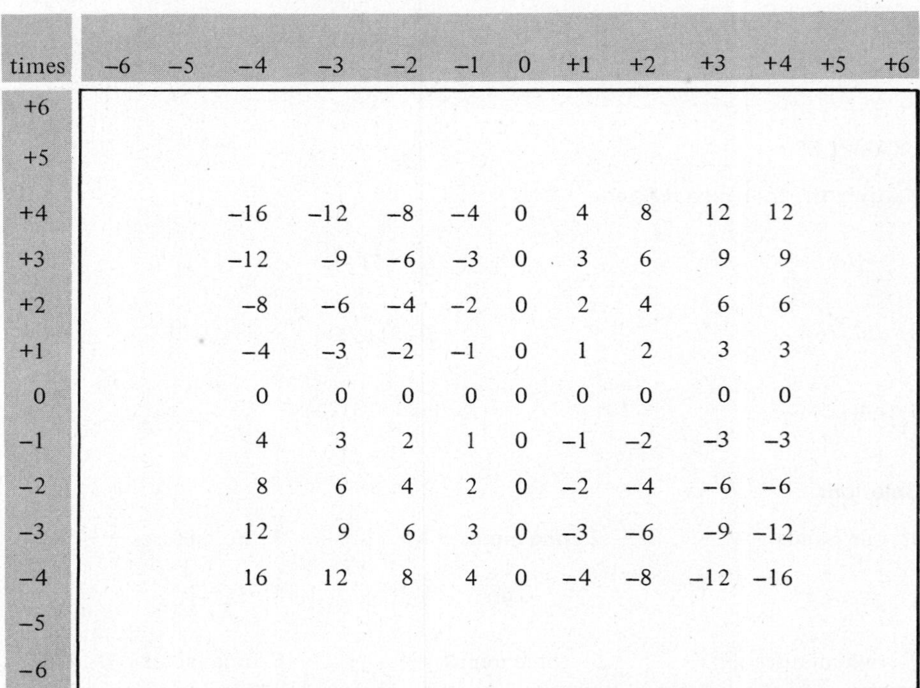

times	−6	−5	−4	−3	−2	−1	0	+1	+2	+3	+4	+5	+6
+6													
+5													
+4			−16	−12	−8	−4	0	4	8	12	12		
+3			−12	−9	−6	−3	0	3	6	9	9		
+2			−8	−6	−4	−2	0	2	4	6	6		
+1			−4	−3	−2	−1	0	1	2	3	3		
0			0	0	0	0	0	0	0	0	0		
−1			4	3	2	1	0	−1	−2	−3	−3		
−2			8	6	4	2	0	−2	−4	−6	−6		
−3			12	9	6	3	0	−3	−6	−9	−12		
−4			16	12	8	4	0	−4	−8	−12	−16		
−5													
−6													

33. Complete the table.

Contrast this table with the addition table in Section 1.1.

OBSERVATIONS AND REFLECTIONS

1. When working with integers (or any expressions), the operations of addition and multiplication are indicated by the placement of parentheses. Signs alone cannot be used to distinguish multiplication from addition since all integers have signs. The placement of signs inside or outside parentheses is used to distinguish sums from products. In a *sum*, every term (except possibly the first) begins with a sign outside parentheses and every sign outside of parentheses begins a term of the sum. In a *product*, no signs appear outside of parentheses, except possibly before the first factor.

Contrast −(−8) − (7) with −(−8) (−7). The expression −(−8) − (7) has a sign in the middle that is outside the parentheses and is therefore a sum of two terms. In contrast, the expression −(−8) (−7), whose only sign outside of parentheses is the sign beginning the expression, is a product.

34. Compute
a. −(−8) − (7)
b. −(−8) (−7)

2. Because the placement of signs inside or outside parentheses is used to distinguish sums from products, care must be used when applying the Commutative Law of Multiplication to signed numbers. For example, the product −4(+6) equals (+6) (−4) or +6(−4). It does not equal (+6) − 4, which is a sum. Only the first factor in a product may have a sign that is not inside parentheses.

35. Compute
a. (+6) (−4)
b. (+6) − 4

3. There is a straight-edge and compass method of multiplying numbers, which you may have seen before. For example, to multiply 2 by 3 we draw two parallel lines, as in the figure. We first draw a line diagonally from 1 to 2, then the parallel diagonal from 3 points to the product of 2 and 3.

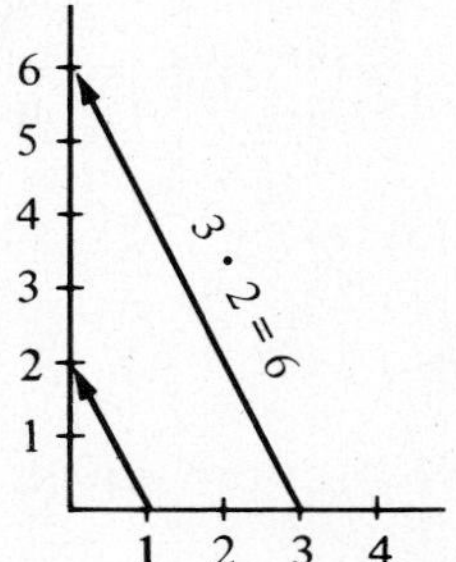

36. Compute 2 · 2 using this diagram.

This construction extends naturally to signed integers, as can be seen in the following diagrams:

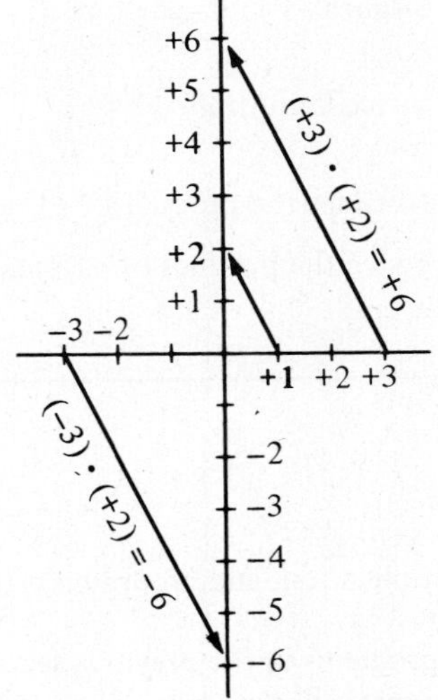

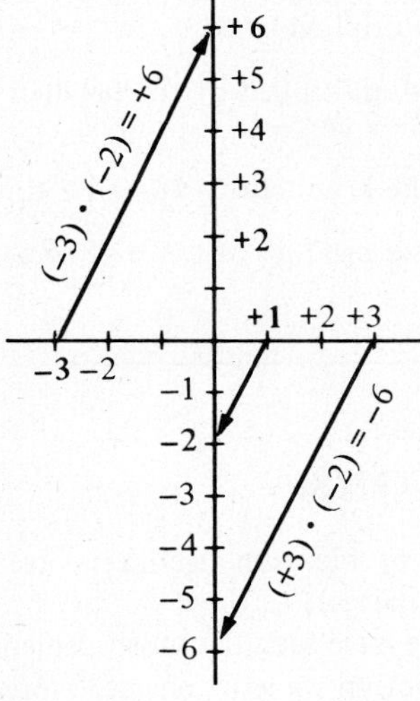

37. Using these diagrams, compute
a. (−2) (+2)
b. (−1) (−2)

For example, to multiply −2 by −3 we draw two parallel lines, as in the figure on the right above. We first draw a line diagonally from 1 to −2. Then the parallel diagonal from −3 points to the product of −2 and −3.

As you can see, this procedure again demonstrates that the product of two numbers with like signs is positive and that the product of two numbers with unlike signs is negative.

EXERCISES

Simplify each of the following expressions to an integer.

1. (−35) (2)	2. 37(−4)	3. −14(−9)	4. 18(4)
5. −3 · 8	6. −21 · 14	7. −(16) (4)	8. −(8) (−3)
9. −(−9) (−3)	10. −(−5) (−14)	11. (−7) (−8)	12. (0) (+7)
13. (0) (−9)	14. (−20) (+3)	15. 6(−8)	16. 0(−6)
17. +9(−7)	18. −4(−16)	19. 4(−7) (−6)	20. 3(2) (−1)
21. −2(−3) (−4)	22. −5(−3) (−8)	23. 14(−3) (−6)	24. −8(−6) (−6)
25. 3(−5) (−2)	26. −4(−6) (−3)	27. 4(−3) (+0)	28. 5(6) (−2)
29. 8(−9) (−1)	30. 3(0) (−2)	31. (−9) (−2) (−5)	32. (+4) (0) (−6)
33. −(7) (10) (−3)	34. −(0) (−3) (12)	35. (0) (−2) (+15)	36. −(6) (−7) (−3)

Express each of the following word phrases algebraically and then simplify it. (A review of translating word phrases is provided in Section 1.6.)

1. The product of 4 and −12
2. 3 times −7
3. Twice −16
4. −5 multiplied by −9
5. The product of −2 and −6 multiplied by 3
6. 8 times −1 multiplied by 5
7. −4 multiplied by the product of −2 and −3
8. Twice 9 multiplied by 6
9. The product of −10 and 7 times −4
10. 13 times the product of −1 and 2
11. The product of twice −8 and 10
12. Twice the product of −15 and −4

1.5 ORDER OF OPERATIONS

Many problems in algebra involve addition, multiplication, and grouping (using parentheses).

We have seen the most elementary of these problems in the previous sections. In this section we will consider more complicated expressions such as −8 − 2 (3), 5 (−3 + 12), (−12 + 5) (8 − 10), and +(6 − 11) + 3 (−9). Each expression involves addition, multiplication, and grouping of integers. But how do we determine the correct sequence of operations? For example, to simplify −8 − 2 (3) do we first combine −8 and −2? (No.) Or do we first multiply −2 and (3)? (Yes.)

To decide how to simplify an expression, we must learn when to add and when to multiply. The standard set of conventions for simplifying these expressions is called the *Order of Operations.*

ORDER OF OPERATIONS

(Multiplication and addition of integers)

A. *In an expression without parentheses:* We compute products before sums.

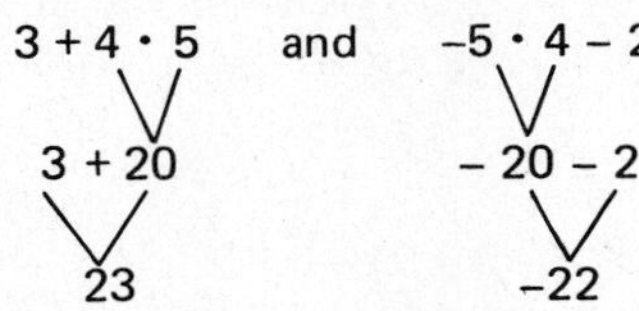

Multiplication is always performed before addition unless indicated otherwise by the presence of parentheses. For example, 3 + 4 · 5 means 3 plus 4 · 5 and not 3 + 4 multiplied by 5.

B. *In an expression containing parentheses:* We first simplify each expression *inside parentheses* to an integer *inside parentheses.*

(3 − 5) − (−3 · 5) and −(3 − 8) (−4) − (−8)

(−2) − (−15)   −(−5) (−4) − (−8)

Then we identify terms (each sign outside parentheses starts a new term). We simplify each term to an integer (with its sign expressed) by multiplying and/or removing parentheses.

(−2) − (−15) and −(−5) (−4) − (−8)

−2 + 15   −20 + 8

Finally, we combine terms to a single integer.

−2 + 15 and −20 + 8

13   −12

38. Compute
   a. 1 + 2 · 3
   b. (1 + 2) · 3
   c. 1 · 2 + 3
   d. 1 · (2 + 3)

WARNING 1. When simplifying inside parentheses, do not drop the parentheses. Removing parentheses from around a signed quantity may change an indicated multiplication to addition, because each sign outside parentheses starts a new term.

3(2 − 6) = 3(−4), **not** 3 − 4

2. When removing parentheses to simplify terms, do not omit plus signs. Signs are required before each integer (except the first) to indicate addition.

3 − 2(−6) = 3 + 12, **not** 312

39. Simplify
   a. 2(4 − 1)
   b. 2(1 − 4)

40. Simplify
   a. 1 −(−2)
   b. 3 −(+2) (−1)

EXAMPLES

Simplify the following expressions to integers, following the Order of Operations.

1. 8 − 3 · 7   2. 6 − 2(−8) (+2)   3. −(−3) (2 − 5)

4. $3-5(2-7)$
5. $8(-4+7)-8$
6. $-(-2)(-4)-6(7-10)$
7. $4-(-4-5)+(-5)$
8. $-5(7-12)+7(2-5)$
9. $-4\cdot 3+(6-3\cdot 7)$

Solutions

1. $8-3\cdot 7$
   $8-21$
   -13

2. $6-2(-8)(+2)$
   $6+32$
   38

3. $-(-3)(2-5)$
   $-(-3)(-3)$
   -9

4. $3-5(2-7)$
   $3-5(-5)$
   $3+25$
   28

5. $8(-4+7)-8$
   $8(+3)-8$
   $+24-8$
   16

6. $-(-2)(-4)-6(7-10)$
   $-(-2)(-4)-6(-3)$
   $-8+18$
   10

7. $4-(-4-5)+(-5)$
   $4-(-9)+(-5)$
   $4+9-5$
   8

8. $-5(7-12)+7(2-5)$
   $-5(-5)+7(-3)$
   $+25-21$
   4

9. $-4\cdot 3+(6-3\cdot 7)$
   $-4\cdot 3+(6-21)$
   $-4\cdot 3+(-15)$
   $-12-15$
   -27

EXERCISES

Simplify each of the following expressions to an integer.

1. $5+2(3)$
2. $-3-2(6)$
3. $-9+9(+2)$
4. $3-2(-6)$
5. $4+3(8+2)$
6. $4-5(4-5)$
7. $-7+2(-9+3)$
8. $-9+3(+4+9)$
9. $5-4\cdot 2$
10. $11-3\cdot 8$
11. $4\cdot 6-7$
12. $7\cdot 5-5$
13. $-2\cdot 6-4(-2)$
14. $-4\cdot 9-7(-1)$
15. $5-(-2)(-5)$
16. $6-(-7)(3)$
17. $4(-6)(-2)$
18. $3(+7)(-4)$
19. $6-3(8-1)$
20. $-9+5(-4+9)$
21. $-(-6)(8-5)$
22. $-(-6)(-4-2)$
23. $-5-(6-3\cdot 5)$
24. $-6-(-3+4\cdot 4)$
25. $(9-6)(8-9)$
26. $(-4+9)(8-6)$
27. $5-3(-2)(+6)$

28. $-14 - 7(-2)(+1)$
29. $-(-3 - 9) + (-4 + 11)$
30. $-(8 - 8) - (-4 - 4)$
31. $(-2)(-7) - (+2)(-7)$
32. $(+7)(-2) - (-5)(+3)$
33. $-(-3 + 8)(-4)$
34. $-(-4 + 11)(-5)$
35. $6(-9 + 2) - 4$
36. $11(-11 + 9) - 30$
37. $-(-9) - 3(-4 + 9)$
38. $-(-12) - 4(-12 + 4)$
39. $-3(7 - 15) + 5(7 - 3)$
40. $6(-4 + 1) - 2(-4 - 5)$
41. $-2(9 - 5) + 3(-6)$
42. $3(-4 - 5) - 8(-4)$
43. $4(-2 + 9) - 2(-4)(+2)$
44. $-5(-11 + 6) - 5(-2)(-5)$

Express each of the following word phrases algebraically and then simplify it. (A review of translating word phrases is provided in Section 1.6.)

1. The product of 5 and 12 is added to 18
2. 8 plus the product of 7 and 3
3. 15 added to the sum of 17 and 6
4. 50 decreased by 9 times 8
5. The product of 4 and 13 minus 19
6. The sum of 12 and 9 is subtracted from 19
7. 90 diminished by the product of 6 and 7
8. The sum of 20 and 15 is multiplied by −9
9. −8 subtracted from the sum of 15 and 3
10. −56 added to twice 27
11. The product of 3 and the sum of 9 and −18
12. −72 more than the product of 8 and 9
13. The product of 4 and 3 is reduced by 6
14. 43 more than −18 is multiplied by −3
15. −10 minus the product of 5 and 6
16. The product of 15 and 2 is subtracted from −80
17. The product of 4 and 16 is diminished by 32
18. Twice the sum of 16 and −24
19. 86 less the product of 16 and 9
20. The product of 25 and −3 plus −4

1.6
TRANSLATING WORD PHRASES

We have concerned ourselves so far with the mathematical notation used in arithmetic. However, the problems we considered can also be posed verbally. For example, the expression "19 − 30" can be expressed verbally as "19 decreased by 30," the expression "−6 + 15" can be expressed as "the sum of −6 and 15" or as "6 less than 15," and "$-7 \cdot 8$" can be expressed as "−7 multiplied by 8."

Several different word phrases can be used to describe any expression involving a single algebraic operation. Let a and b represent signed numbers and look at the following table to see how "$a + b$" can be represented:

41. Translate
a. 6 added to 19
b. 72 is increased by 8
c. −5 plus 3
d. The sum of −14 and 2

the sum of a and b a plus b a more than b a added to b a is added to b a increased by b a is increased by b	$a + b$

Similarly, "$a - b$" can be expressed verbally as

42. Translate
a. 18 minus 4
b. 95 decreased by 34
c. 3 less 11
d. 26 is reduced by 17

a minus b a less b a decreased by b a is decreased by b a reduced by b a is reduced by b a diminished by b a is diminished by b	$a - b$

On the other hand, "$b - a$" or "$-a + b$" can be represented verbally as

43. Translate
a. 12 subtracted from 16
b. 4 less than 20
c. 9 is subtracted from 6
d. subtract 8 from 11

subtract a from b a subtracted from b a is subtracted from b a less than b	$b - a$ or $-a + b$

The product "$a \cdot b$" can be represented verbally as

44. Translate
a. The product of 4 and 6
b. −9 times 3
c. 18 is multiplied by 2
d. −25 multiplied by −2

the product of a and b a times b a multiplied by b a is multiplied by b	$a \cdot b$ or $(a)(b)$

More Complex Phrases

More complicated problems involving two operations can also be posed verbally. For example, "$3 - 5 \cdot 8$" can be expressed as "3 minus the product of 5 and 8," and "$3 \cdot 5 - 8$" can be expressed as "8 less than 3 times 5." However, when a word phrase indicates more than one operation, it may not be clear how to translate the phrase into an algebraic expression. The word phrase "8 less than 3 times 5," for example, is not as precise as its algebraic counterpart. We may interpret this word phrase as "(8) *less than* (3 times 5)," but we can also interpret it as "(8 less than 3) *times* (5)," which corresponds to the algebraic expression "$(3 - 8) \cdot 5$."

To avoid ambiguity when translating word phrases containing two operations, we will designate one operation as the *major operation* and the other as the *minor operation* of the phrase. The minor operation involves only two of the three numbers and is computed before the major operation when the expression is simplified.

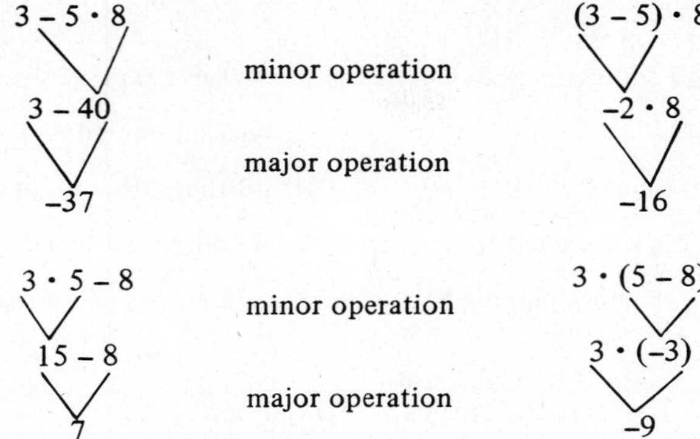

45. Identify the minor operation
a. $3 - 7 \cdot 8$
b. $-6 + 9(-2)$
c. $(7 - 4) \cdot 6$
d. $4 - (5 - 8)$

In some phrases the major operation is clearly indicated. For example, in the phrase "3 times the sum of 5 and 8" the major operation is "times," and the sentence is read "(3) *times* (the sum of 5 and 8)." Similarly, the phrase "4 times 7 is decreased by 9" clearly means "(4 times 7) *is decreased by* (9)," and "is decreased by" is the major operation. Frequently, however, the major operation is not clearly indicated by the word phrase, as we saw with the phrase "8 less than 5 times 3." With such ambiguous phrases we will always assume that the *first* operation is the major operation of the phrase. For example, we translate "8 less than 5 times 3" as "(8) *less than* (5 times 3)."

You should be aware, however, that our convention of taking the first operation as the major operation in ambiguous situations is an arbitrary one. Unfortunately, there is no universal agreement on how to translate such phrases.

Look at the following examples of *translating word phrases into algebraic expressions.*

TYPICAL PROBLEMS

1. 5 less than –9

We first read the expression and underline the operation:

↓ 5 less than –9

Then we translate the phrase using the tables given in the introduction. We place negative numbers in parentheses when the phrase involves subtraction or multiplication:

↓ $(-9) - 5$ or $-5 + (-9)$

Hence, "5 less than –9" is translated as $(-9) - 5$ or $-5 + (-9)$.

2. 4 diminished by the product of 2 and 6

First we read the expression and underline the operations. We identify the major operation and place the phrase containing the minor operation in parentheses:

↓ 4 diminished by (the product of 2 and 6)

We translate the minor operation using the tables given in the introduction, keeping parentheses around the expression:

↓ $(2 \cdot 6)$

Then we translate the major operation, treating the parenthetical expression as a unit:

↓ $4 - (2 \cdot 6)$

Hence, "4 diminished by the product of 2 and 6" is translated as $4 - (2 \cdot 6)$.

EXAMPLES

Express each of the following word phrases as an algebraic expression and simplify:

1. −12 multiplied by 9
2. −15 subtracted from 10
3. −11 less than 8 times 2
4. −3 times the sum of 5 and 4
5. 16 more than 7 is subtracted from −52
6. 89 decreased by 47 is added to −93
7. the product of −8 and 19 minus 12
8. −75 decreased by the sum of 23 and 7

Solutions

1. −12 multiplied by 9

↓

$-12 \cdot 9$ or $(-12)(9)$

↓

-108

2. −15 is subtracted from 10

↓

$10 - (-15)$ or $-(-15) + 10$

↓

25

3. −11 less than (8 times 2)

↓

$8 \cdot 2 - (-11)$ or $-(-11) + 8 \cdot 2$

↓

27

4. −3 times (the sum of 5 and 4)

↓

$-3(5 + 4)$ or $(-3)(5 + 4)$

↓

-27

5. (16 more than 7) is subtracted from −52

↓

$(-52) - (16 + 7)$ or $-(16 + 7) + (-52)$

↓

-75

6. (89 decreased by 47) is added to −93

↓

$(89 - 47) + (-93)$

↓

-51

7. the product of −8 and (19 minus 12)

↓

$(-8)(19 - 12)$

↓

-56

8. −75 decreased by (the sum of 23 and 7)

↓

$(-75) - (23 + 7)$

↓

-105

EXERCISES

Express each of the following word phrases algebraically and then simplify it.

1. The sum of 6 and 13
2. 5 decreased by 11
3. 16 reduced by 9
4. 8 less than 23
5. 26 increased by 8
6. 21 is reduced by 25
7. −13 subtracted from −31
8. −6 added to −17
9. −10 is decreased by 7
10. −3 is added to −19
11. −5 is subtracted from −25
12. 18 less −18
13. Twice −19
14. −8 multiplied by −7
15. 6 multiplied by the product of 3 and −5
16. Twice 7 multiplied by 5
17. The product of −2 and −3 times 9
18. −7 times twice −7

19. −8 multiplied by −4 times 5
20. The product of twice −15 and 4
21. The product of 4 and 5 is added to 6
22. 9 plus the product of 10 and 3
23. 18 added to the sum of 12 and 11
24. 63 decreased by 6 times 10
25. 27 diminished by 9 more than 12
26. The sum of 12 and 13 is subtracted from 14
27. 56 diminished by the product of 9 and 4
28. The sum of 8 and 22 is multiplied by −9
29. The product of 25 and 7 is reduced by 80
30. The sum of 42 and 11 is increased by 17
31. 12 times 9 is decreased by 99
32. 16 more than 31 is added to −62
33. −17 minus the product of 3 and 16
34. The product of 14 and 5 is subtracted from −18
35. −26 less the product of 8 and 7
36. The product of −11 and 2 plus 9

ANSWERS TO MARGIN EXERCISES

1. a. +3 b. −3 c. −1 d. +1

2. a. −7 − 2 = −9 b. +3 − 6 = −3

3. a. −6 plus +2 = −4 b. +4 plus +8 = +12 c. −3 plus −5 = −8

4. a. +2, +1, 0, −1 b. −2, −1, 0, +1

5. a. 0 b. 0 c. 0 d. 0 e. 0 f. 0

6. a. +3 b. −3 c. +3 d. −3 e. +3 f. −3

8. a. +3, +2, 0, −1 b. +1, 0, −2, −3

9. a. −13 b. −13 c. −3 d. −3 e. +3 f. +3

10. a. −3 + 4 = +1 b. −5 + 6 = +1

11. a. 5, −2, and 9 b. −4, −6, and 1

12. a. 7 − 5 + 3 b. −2 − 8 + 12

13. a. +2, −2 b. −2, +2

14. a. −4 b. +5 c. −6

15. a. −9 b. 4 c. 0

16. a. −5 b. +6 c. −4

17. a. −2 and (−7) b. −(+5) and −(−6)

18. a. −5 b. −3 c. +7 d. 9

19. a. −2 + (−9) b. −(+4) + 7 c. (−8) − 6 d. −(−2) + (−2)

20. a. 5 b. −12 c. 16 d. −20

21. a. −2 b. 6 c. −6 d. 2

23. a.

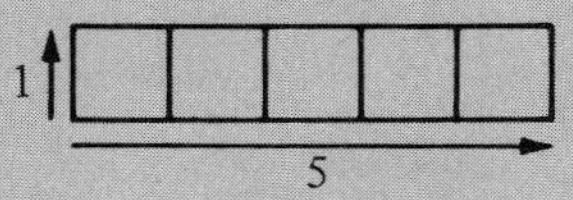

b.

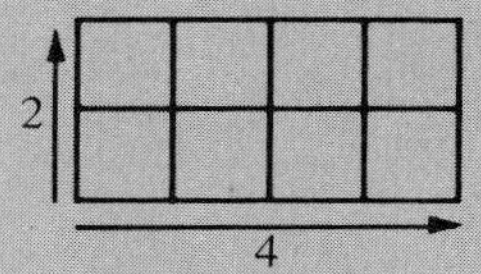

24. a. −3 and +7 b. 6, −6, and +2

25. a. 2, 1, 0, −1, −2 b. 2, 1, 0, −1, −2

26. a. −6 b. −6 c. −6 d. 6 e. 6 f. 6

27. a. 1 b. 2 c. −1 d. 0 e. −1 f. 0 g. 1 h. −2

28. a. 6 b. 6 c. −6 d. −6 e. −6 f. −6

29. a. (−2) (−3) = 6 b. (−1) (−6) = 6

30. a. 0 b. 0 c. 0 d. 0

31. a. −2 b. −2 c. 4 d. 4

32. a. 2 b. 2 c. −4 d. −4

34. a. 1 b. −56

35. a. −24 b. 2

36. Arrow from 2 to 4.

37. a. Arrow from −2 to −4 in diagram on left. b. Arrow from −1 to +2 in diagram on right.

38. a. 7 b. 9 c. 5 d. 5

39. a. 6 b. −6

40. a. 3 b. 5

41. a. 6 + 19 b. 72 + 8 c. −5 + 3 d. −14 + 2

42. a. 18 − 4 b. 95 − 34 c. 3 − 11 d. 26 − 17

43. a. 16 − 12 b. 20 − 4 c. 6 − 9 d. 11 − 8

44. a. (4) (6) b. (−9) (3) c. (18) (2) d. (−25) (−2)

45. a. −7 · 8 b. 9(−2) c. 7 − 4 d. 5 − 8

CHAPTER REVIEW

SECTION 1.0 Simplify to an integer, using the number line.

1. −2 plus +1
2. +1 plus −4
3. −3 plus −2
4. +6 plus −3
5. −5 plus +7
6. +4 plus −4
7. −3 plus +8
8. +6 plus −9
9. −4 plus +10

SECTION 1.1 Simplify to an integer.

10. −3 − 4
11. −8 + 2
12. +11 − 5
13. 0 + 7
14. −9 + 9
15. −5 + 0
16. −4 − 13
17. −18 + 13
18. −32 − 0
19. −25 + 43
20. 75 + 16
21. −62 + 96
22. −88 + 54
23. 43 − 72
24. −53 − 77
25. 65 − 124
26. −48 + 216
27. 436 − 89

SECTION 1.2 Simplify to an integer.

28. 4 − 7 + 9
29. 6 − 11 − 4
30. −5 + 12 − 4
31. 18 − 10 + 8
32. +5 − 7 − 2 + 6
33. −7 + 3 − 8 + 16
34. 4 − 9 − 11 + 22
35. −18 + 44 − 30 + 8
36. +7 − 4 + 15 − 6 − 3
37. 6 − 11 − 8 + 21 − 4
38. −7 + 3 + 9 − 16 + 11
39. −21 + 3 − 14 + 5 + 27

SECTION 1.3 Simplify to an integer.

40. −(−16)
41. +(+13)
42. −(42)
43. −(−40)
44. +(21)
45. −(+20)
46. (+4) − (+7)
47. −(−3) − (+9)
48. 9 − (−4)
49. −(4) − (−8)
50. −5 + (−2)
51. (−19) + (−11)
52. −(42) − (−16)
53. −(+26) + 26
54. (+32) − (+56)
55. −(−74) − 91
56. −(−73) − (+182)
57. −(133) − (−98)

SECTION 1.4 Simplify to an integer.

58. (−3) (−7)
59. −3(−20)
60. −6 · 11
61. −(5) (12)
62. +(−2) (0)
63. (−12) (+7)

64. $(+4)(-11)$ 65. $-2(+22)$ 66. $-5(+12)$

67. $-8(-12)(-3)$ 68. $-18(+3)(-2)$ 69. $16(3)(-3)$

70. $-3(-5)(-6)$ 71. $-6(+6)(2)$ 72. $+0(+2)(+4)$

73. $(6)(-7)(-1)$ 74. $-(4)(+9)(-2)$ 75. $(-1)(-2)(+15)$

SECTION 1.5

Simplify to an integer.

76. $9 - 4(7)$ 77. $1 + 8(-2)$ 78. $5 + 9(6 - 1)$

79. $(6 - 3)\,8 + 2$ 80. $+4 - 2 \cdot 7$ 81. $3 \cdot 4 - 12$

82. $+8 \cdot 3 - 2(-8)$ 83. $-9 -(8)(-5)$ 84. $-8(-4)(+5)$

85. $5 - 2(-4 + 7)$ 86. $-(+3)(5 + 2)$ 87. $-8 +(8 - 2 \cdot 4)$

88. $+(3 - 7)(5 - 11)$ 89. $18 - 4(-6)(-1)$ 90. $-(-7 + 3) - (-6 - 2)$

91. $(-3)(-9) - (-12)(-2)$ 92. $-(7 - 13)(-7)$ 93. $-3(5 - 7) + 3$

94. $-(+7) + 2(-7 + 2)$ 95. $8(-9 + 7) - 4(-10 + 6)$

96. $-7(-1 + 4) - 6(+4)$ 97. $-3(+11 - 7) + 3(+4)(-3)$

CUMULATIVE TEST

The following problems test your understanding of this chapter. Before taking this test, thoroughly review Sections 1.1, 1.2, 1.3, and 1.4.

Once you have finished the test compare your answers with the answers provided at the back of the book. Note the section number of each problem missed, and thoroughly review those sections again.

Simplify completely.

1. $48 - 21$
2. $5 - 7 + 8$
3. $(+8) - (+11)$
4. $(9)(-9)$
5. $-18 + 15$
6. $3 - 12 - 15 + 24$
7. $-6(-11)(-2)$
8. $14 - 4 - 17$
9. $-36 - 8$
10. $-(5) - (-9)$
11. $12 - 27 + 19 - 41$
12. $(-3)(0)$
13. $-5 + 31$
14. $-15(4)(-2)$
15. $8 - (-6)$
16. $4 - 9 - 3 + 5$
17. $-7(+7)(3)$
18. $-24 - 9$
19. $-6 + 9 - 11 + 3 + 8$
20. $(+7)(-12)$
21. $-(-6) - (+4)$
22. $-51 + 13$
23. $11 - 9 + 2 - 7$
24. $+0(+3)(+8)$
25. $-4 + 9 + 3 - 25 + 16$
26. $(-12) + (-13)$
27. $-9(+4)$
28. $-32 + 8 - 23 + 7 + 38$
29. $-(-15) - 17$
30. $(5)(-14)(-1)$

2 MULTIPLICATION OF MONOMIALS

To prepare for algebra we began with the study of signed integers. We are now ready to begin the study of algebra itself. (The word *algebra* comes from an Arabic word that means to reunite broken parts.) Algebra is a generalization of arithmetic. It is the study of numbers, including signed numbers and *literal numbers*, or *literals*, such as $a, b, C, e, i, K, L, m, n, S, T, x, y, \alpha, \beta, \pi$. The literals are usually represented by letters in the Latin and Greek alphabets.

The fundamental operations of *algebraic addition* and *algebraic multiplication* apply to literal numbers as well as to signed numbers. In the next four chapters, we will study how to apply these two operations to expressions containing literal numbers. We will first study products of expressions containing literals.

As we shall see, literal numbers in expressions behave like signed numbers *in parentheses*. Thus, for example, the product of x and y is indicated by $x \cdot y$ or xy, just as the product of -2 and -3 is indicated by $(-2) \cdot (-3)$ or $(-2)(-3)$.

Monomials

To talk about multiplication, we need to learn some new words. Products of integers and literals are called *monomials* or, more precisely, monomials with integer coefficients. (Integers or literals alone are also considered monomials.) For example, $-3x$, $+7y$, $-4xy$, $ABCD$, and -8 are monomials.

The numerical factor of a monomial is called the *coefficient* or the *numerical coefficient*. The product of the literal factors is frequently called the *literal component*.

$$\underline{-2}\ \underline{ab}$$

(numerical) coefficient ↗ ↖ literal component

When the numerical coefficient of a monomial is positive, we will call the monomial positive. When the numerical coefficient of a monomial is negative, we will say that the monomial is negative.

Conventions

The following conventions apply in writing monomials:

1. The plus sign of a positive monomial may be omitted.

   $5xy$ is the same as $+5xy$ just as 12 is the same as +12

2. When the numerical coefficient of a monomial is 1 or −1, the preferred way to write the monomial is without the numeral 1.

   $1b$ is the same as b and $-1b$ is the same as $-b$

   $+1xy$ is the same as xy and $-1xy$ is the same as $-xy$

3. The literals in a monomial are usually written in alphabetical order.

   xyz is preferred to zyx and $8ab$ is preferred to $8ba$

2.1 MULTIPLYING MONOMIALS (Without Exponential Notation)

When working with monomials we can use the same language we used to talk about integers. In *algebraic products* of monomials such as $(5x)(-8y)$ and $5(-3a)(-2x)$, each quantity is called a *factor* of the product. For example, in the product $(5x)(-8y)$ the factors are $5x$ and $-8y$. Similarly, in the product $5(-3a)(-2x)$ the factors are 5, $-3a$, and $-2x$.

1. List the factors of
a. $(-3x)(+7y)$
b. $6ad(-6c)(+2b)$

Multiplication is indicated in any of the following ways:

1. $(5x)\cdot(-8y)$	with a dot between sets of parentheses.
2. $(+5x)(+8y)$	with no symbol between sets of parentheses.
3. $-5x(-8y)$	with no symbol between factors and only the second factor inside parentheses.
4. $-5x\cdot 8y$	with a dot between factors and no plus sign before the second factor.

2. Compute
a.
$+2xyz(+3a)$
$+1xy(+3a)$
$0(+3a)$
$-1x(+3a)$
$-2(+3a)$
b.
$-2xyz(-3a)$
$-1xy(-3a)$
$0(-3a)$
$+1x(-3a)$
$+2(-3a)$

When monomials are factors of a product, they are frequently written inside parentheses. The first factor can be written inside or outside parentheses, but no other factor in a product can begin with an expressed sign outside of parentheses.

To see the relationship between an algebraic product of two monomials and an algebraic product of two integers, let us look at the pattern in the following columns.

$+2(+2x) = +4x$	$-2(-2x) = +4x$
$+1a(+2x) = +2ax$	$-1a(-2x) = +2ax$
$0(+2x) = 0$	$0(-2x) = 0$
$-1ab(+2x) = -2abx$	$+1ab(-2x) = -2abx$
$-2abc(+2x) = -4abcx$	$+2abc(-2x) = -4abcx$

Can you see how these products relate to products of integers? Can you guess the rule for multiplying monomials?

Commutative and Associative Laws

3. Compute
a. $x \cdot y$
b. $t \cdot vx \cdot w$
c. $ad \cdot bc$

In Chapter 1 we observed that Commutative and Associative Laws govern multiplication. These laws apply to signed numbers and literal numbers alike. In fact, they apply to all the algebraic expressions in this book.

Because multiplication is commutative, the factors in an algebraic product may be written in any order. For example, $(-3a)(-2x)$ equals $(-2x)(-3a)$ and $-6k(2m)(-3n)$ equals $-3n(-6k)(2m)$.

Furthermore, because multiplication is associative, the factors in an algebraic product may be multiplied in any order. For example, compare the following computations:

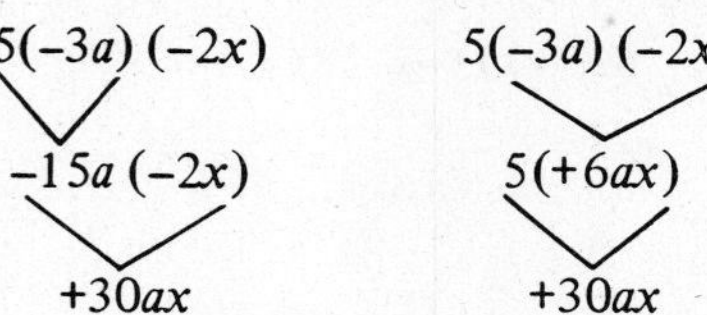

4. Compute
   a. $-2(3x)$
   b. $3x(-2)$
   c. $-w(-4xy)$
   d. $-4xy(-w)$
   e. $wz(2xy)$
   f. $2xy(wz)$

5. Compute $(-3x)(4w)(-z)$ both as
   a. ($-3x$ times $4w$) times $-z$
   and as
   b. $-3x$ times ($4w$ times $-z$)

In particular, when multiplying monomials, which are themselves products of integers and literals, these integers and literals may be multiplied in any order. Hence, the numerical coefficients may be multiplied first, regardless of which factors they are in. For example,

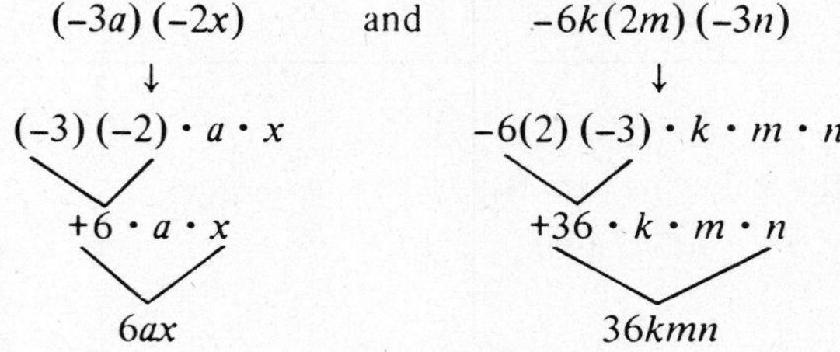

In this section we deliberately avoid products in which a literal appears more than once. When a literal appears more than once in a product of monomials, it is conventional to use *exponential notation*, which we will discuss in the next section.

Look at the following example of *multiplying monomials* (*without exponential notation*).

TYPICAL PROBLEM

$(-4ac)(9b)$

We first multiply the numerical coefficients. To do this we determine the sign of the product by counting the minus signs:

And then we multiply the numerical values and attach the sign:

We then attach the literal factors. (The literals are usually arranged in alphabetical order.)

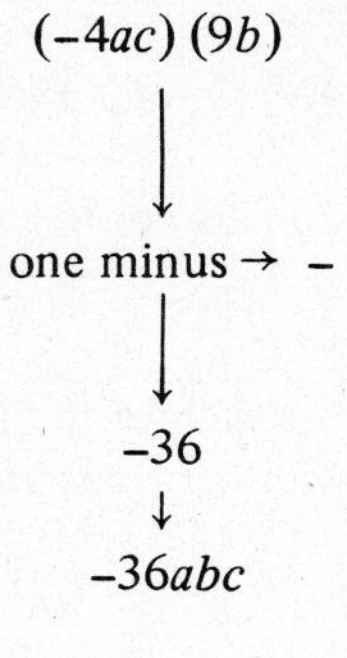

To multiply the numerical coefficients (integers), we use the method in Section 1.4. When multiplying the integers, the literal components are temporarily ignored.

Hence, $(-4ac)(9b) = -36abc$.

EXAMPLES

Multiply the following monomials.

Every product of monomials can be simplified to a single monomial.

1. $(5x)(6w)$
2. $(+7a)(-8b)$
3. $(-ab)(-9d)$
4. $-4x(6yz)(0)$
5. $(-x)(-2y)(8z)$
6. $-3(uv)(-2t)$

Solutions

1. $(5)(6) = 30$ ↓ $30wx$
2. $(+7)(-8) = -56$ ↓ $-56ab$
3. $(-1)(-9) = 9$ ↓ $9abd$
4. $-4(6)(0) = 0$ ↓ 0
5. $(-1)(-2)(8) = 16$ ↓ $16xyz$
6. $-3(1)(-2) = 6$ ↓ $6tuv$

6. Compute
a. $(3x)(0)$
b. $(-7w) \cdot (0)$

7. Simplify
a. $+0xy$
b. $-0tvz$

SPECIAL CASES

Zero times any literal is zero. Similarly, a monomial with a coefficient of zero is zero. For example,

$$0 \cdot x = 0 \quad \text{and} \quad (0)(ab) = 0$$
$$0x = 0 \quad \text{and} \quad 0ab = 0$$

EXERCISES

Simplify each of the following expressions to a monomial.

1. $5x \cdot 7y$	2. $9wx \cdot 3y$	3. $(2A)(3B)$
4. $(-mn)(-12p)$	5. $(4x)(+7yz)$	6. $(18x)(-2w)$
7. $(-8bd)(-6ac)$	8. $(-11y)(5xz)$	9. $(+6n)(-9m)$
10. $4A(-7B)$	11. $-6t(+6rs)$	12. $-4y(+8x)$
13. $-3r(-9s)$	14. $3xy(+2wz)$	15. $-8yz(4wx)$
16. $-3z(-7xy)$	17. $(-x)(-9y)(4z)$	18. $(-3x)(wy)(-8z)$
19. $(4N)(-7M)(+9K)$	20. $(5a)(-4bd)(-3c)$	21. $(-ad)(-7c)(-9b)$
22. $-6(+AB)(-3C)$	23. $+9(+8u)(-9t)$	24. $-5(-x)(-7)$

Express each of the following word phrases algebraically and then simplify it. (A review of translating word phrases is provided in Section 1.6.)

1. $5a$ times $-7b$
2. The product of $-12x$ and $3y$
3. Twice xy
4. $9c$ multiplied by $-6d$
5. $4x$ times the product of $-10z$ and $3y$
6. The product of twice $9x$ and $-y$
7. $-5w$ times $-9x$ multiplied by $-2u$
8. Twice the product of $5abc$ and $-13de$
9. abx multiplied by the product of $12w$ and z
10. The product of $8ad$ and $-2b$ times $-2c$
11. $-18a$ multiplied by $-5b$ times $2cd$
12. The product of $20xy$ and $-10z$ multiplied by $-9w$

2.2 MULTIPLYING POWERS WITH A COMMON BASE

Exponential Notation

Look at the monomials $xxyy$ and $-4abbb$. Each of them has a literal factor that appears more than once. Repeated literals are very common in monomials, and we can use *exponential notation* to write monomials with repeated literals more compactly. For example, the monomial $aaaaaa$ is written a^6. The expression a^6 is called a *power*. In exponential notation the repeated factor is called the *base*, and the numerical value indicating the number of repetitions of the base is called the *exponent*.

In b^4, the base is b and the exponent is 4.
In y^9, the base is y and the exponent is 9.
In 2^5, the base is 2 and the exponent is 5.

Hence, exponential notation is a shorthand for multiplication:

z^7 is shorthand for $z \cdot z \cdot z \cdot z \cdot z \cdot z \cdot z$

t^4 is shorthand for $t \cdot t \cdot t \cdot t$

5^3 is shorthand for $5 \cdot 5 \cdot 5$

When using exponential notation, we will use the following language to describe powers.

z^7 is read: "z to the 7th power" or "z is raised to the 7th power" (or informally: "z to the 7th").

t^4 is read: "t to the 4th power" or "t raised to the 4th power" (or informally: "t to the 4th").

In exponential notation it is conventional for *each exponent to have a single base*. For example, xy^2 means $x \cdot y^2$ and not $(xy)(xy)$, and $3x^2y^3$ means $(3)(x^2)(y^3)$. In order for a single exponent to have more than one base, parentheses must be used. For example, $(ab)^4$ means $(ab)(ab)(ab)(ab)$, and $(-2x)^3$ means $(-2x)(-2x)(-2x)$.

Exponential notation can be used for numerical values and (signed) integers inside parentheses. For example,

5^3	is shorthand for $5 \cdot 5 \cdot 5$	$= 125$
$(+5)^3$	is shorthand for $(+5)(+5)(+5)$	$= 125$
-2^4	is shorthand for $-(2 \cdot 2 \cdot 2 \cdot 2)$	$= -16$
$(-2)^4$	is shorthand for $(-2)(-2)(-2)(-2)$	$= +16$

Note that -2^4 is the negative of 2^4 (that is, the negative of 16), whereas $(-2)^4$ is a negative integer raised to the fourth power. Parentheses must be used to raise a signed integer to a power.

8. Identify the base and exponent of
a. x^5
b. y^8
c. 3^2

9. Identify the base(s) and exponent(s) of
a. x^4y^3
b. $7a^3bc^2$
c. $(3y)^8$
d. $(-4z)^7$

10. Identify the base and exponent of
a. 2^5
b. -3^2
c. $(-3)^2$
d. $-(-3)^3$

Conventions

When using exponential notation, the following special conventions apply:

1. When we have a power with the exponent 1, the exponent is not written.

   x^1 is x, k^1 is k, and 6^1 is 6.

2. If the base of a power is any nonzero expression, numerical or literal, then its zero power is defined to be 1.

   $a^0 = +1a^0 = +1$ (no a's) $\quad x^0 = 1 \quad 4^0 = 1$

Exception: 0^0 **is not defined.**

11. Rewrite without the exponent 1
a. $2k^1$
b. $x^1y^1z^1$
c. $4a^2b^1$

12. Simplify
a. $3x^0$
b. x^0y^2
c. $-2^0a^2b^0$

13. Determine the sign of
 a. $(-x)^9$
 b. $-(-1)^4$
 c. $-2(-2y)^3$

3. When a negative expression inside parentheses is raised to a power, the sign of the product is *minus* when the exponent is *odd*, and *plus* when the exponent is *even*.

$$(-x)^3 = (-x)(-x)(-x) \qquad = -x^3$$

$$(-x)^4 = (-x)(-x)(-x)(-x) = +x^4$$

In particular, the base (−1) to an *odd* power is −1, and (−1) to an *even* power is +1.

4. When we have a power with the exponent 2, the power is generally read as "the base squared" rather than "the base to the second power."

x^2 is read: "x squared" or "x to the second power"

5. When we have a power with the exponent 3, the power is frequently read as "the base cubed" rather than "the base to the third power."

x^3 is read: "x cubed" or "x to the third power"

14. Simplify
 a. $(1)^8$
 b. $-(1^6)$
 c. 1^5x^5

15. Simplify
 a. -0^3
 b. 0^4x^2y
 c. $(0)^8$

6. When we have a power whose base is 1, the power equals 1.

1^2 is 1 $\quad 1^5$ is 1 $\quad 1^0$ is 1

7. When we have a power whose base is zero and whose exponent is *larger* than zero, the power is equal to 0. (0^0 is **undefined.**)

0^2 is 0 $\quad$ and $\quad 0^5$ is 0

Multiplication

To understand how to multiply monomials containing powers, we must first consider the special case of products of powers with the same base, as for example $(a^5)(a^3)$. Two powers with the same base are called *like powers.* In the product $(a^5)(a^3)$, the first factor a^5 is a product of five a's, and the second factor a^3 is a product of three a's. So

16. Compute
 a. $(b^2)(b^3)$
 b. $(c^2)(c^4)$
 c. $(w^2)(w^5)$
 d. $(y^2)(y^6)$

$$(a^5)(a^3) = (aaaaa)(aaa) = a^8$$

and the sum of the exponents 5 and 3 counts the number of a's in the product. Note that to compute the product $(a^5)(a^3)$ we first wrote each power as a repeated product, before collecting the eight factors into a single power. As this example suggests, the exponents of the factors can be used to determine the exponent of the product. Do you see how? Check your ideas by looking at the pattern in the following table:

17. Complete the table.

times	1	x	x^2	x^3	x^4	x^5	x^6
1	1	x	x^2	x^3	x^4		
x	x	x^2	x^3	x^4	x^5		
x^2	x^2	x^3	x^4	x^5	x^6		
x^3	x^3	x^4	x^5	x^6	x^7		
x^4	x^4	x^5	x^6	x^7	x^8		
x^5							
x^6							

From this table it is clear that the exponent of the product of two like powers is the sum of the exponents of the factors. This is the *First Law of Exponents.*

FIRST LAW OF EXPONENTS

To multiply like powers, add their exponents.

$$x^n \cdot x^m = \underbrace{x \cdot x \cdot \ldots \cdot x \cdot x}_{n \text{ times}} \cdot \underbrace{x \cdot x \cdot \ldots \cdot x \cdot x}_{m \text{ times}} = x^{n+m}$$

(where n and m represent numerical values)

This law can easily be extended to the product of any number of like powers. For example, the product $(a^2)(a^3)(a^4)$ can be computed as follows:

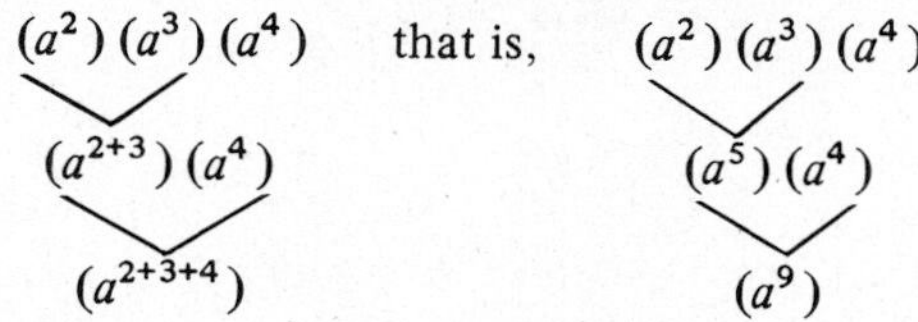

18. Compute
   a. $(t^2)(t^3)(t^4)$
   b. $(w^3)(w^3)(w^3)$
   c. $(k^1)(k^2)(k^4)(k^8)$
   d. $(x^1)(x^3)(x^5)(x^7)$

Hence, to multiply any number of like powers, we can add their exponents.

Bear in mind, however, that exponential notation only simplifies the appearance of a product. For example, x^3, $x^2 \cdot x$, $x^1 \cdot x^1 \cdot x^1$, and xxx are all products of three literals. The form x^3 is preferred because of its compactness.

Numerical Powers

The First Law of Exponents applies equally to numerical powers such as 2^3. Two numerical powers are *like powers* if their bases are the same integer.

Remember that when a numerical base has its sign expressed, it must be written in parentheses. For example, $(-2)^2$ is the square of -2 (i.e., $+4$), whereas -2^2 is the negative of "2 squared" (i.e., -4).

19. Compute
   a. $(-3)^2$
   b. -3^2
   c. $(-2)^3$
   d. -2^3

Usually exponential notation is more of a hindrance then a help when working with numbers, so we will ordinarily multiply numerical powers out and avoid exponential notation.

Look at the following examples of *multiplying like powers.*

TYPICAL PROBLEMS

1. $(a^4)(a^5)(a^8)$

We first add the exponents to get the exponent of the product: $a^{4+5+8(=17)}$

Then we attach this exponent to the common base: a^{17}

Hence, $(a^4)(a^5)(a^8) = a^{17}$.

2. $2^2 \cdot 2^3 \cdot 2$

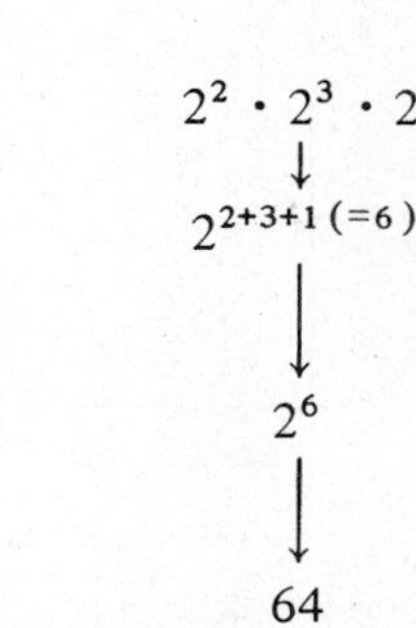

We first add the exponents to get the exponent of the product: $2^{2+3+1\,(=6)}$

Then we attach this exponent to the common base. *At this point we have a single power:* 2^6

Finally we multiply out the power, simplifying it to an integer: 64

Hence, $2^2 \cdot 2^3 \cdot 2 = 64$.

Remember $a = a^1$ and $2 = 2^1$

EXAMPLES

Every product of like powers can be simplified to a single power, and every numerical power can be simplified to an integer.

Multiply the following like powers.

1. $yy(yyy)(yy)$ 2. $(b^2)(b^8)(b^{12})$ 3. $x^2xx^3x^0x^6$

4. $4^2 \cdot 4 \cdot 4$ 5. $(2^2)(2)(2^4)$ 6. $(-1)^6(-1)^3(-1)^4(-1)$

Solutions

1. $y^{1+1+1+1+1+1+1\ (=7)} \rightarrow y^7$

2. $b^{2+8+12\ (=22)} \rightarrow b^{22}$

3. $x^{2+1+3+0+6\ (=12)} \rightarrow x^{12}$

4. $4^{2+1+1\ (=4)} \rightarrow 4^4 = 256$

5. $2^{2+1+4\ (=7)} \rightarrow 2^7 = 128$

6. $(-1)^{6+3+4+1\ (=14)} \rightarrow (-1)^{14} = +1$

SPECIAL CASES

20. Compute
a. $x \cdot x$
b. $x^1 \cdot x^1$
c. x^4xx^2
d. $x^4x^1x^2$
e. y^6yyy
f. $y^6y^1y^1y^1$

When an expression appears without an exponent, its exponent is understood to be 1. For example,

$$x^4x = x^4x^1 = x^{4+1} = x^5$$

You saw multiplication and addition tables with signed integers in Chapter 1. Now look at a power table for signed integers. Verify the entries to make sure that you understand how to determine powers of signed integers.

21. Complete the table.

BASE raised to the power	EXPONENT 0	1	2	3	4	5
(+5)						
(+4)	+1	+4	+16	+64	+256	
(+3)	+1	+3	+9	+27	+81	
(+2)	+1	+2	+4	+8	+16	
(+1)	+1	+1	+1	+1	+1	
(0)	*	0	0	0	0	
(−1)	+1	−1	+1	−1	+1	
(−2)	+1	−2	+4	−8	+16	
(−3)	+1	−3	+9	−27	+81	
(−4)	+1	−4	+16	−64	+256	
(−5)						

*undefined

OBSERVATIONS AND REFLECTIONS

★ Look at the first column in the power table, in which every entry is +1, except that for 0^0. Because every other number to the zero power is defined to be +1, you might think that 0^0 should be +1 as well. On the other hand, look at the middle row of the table, in which every entry is 0, except again that for 0^0. Because multiplying 0 by itself any number of times yields 0, you might think that 0^0 should be 0 as well. Both of these arguments, that 0^0 should equal +1 and that 0^0 should equal 0, seem equally plausible. In fact, each argument invalidates the other, so 0^0 cannot be defined.

EXERCISES

Simplify each of the following expressions to a monomial.

1. $xxxxx$
2. $nnnnnnn$
3. $r(r)(rr)$
4. $(cc)(ccc)$
5. xxx^2
6. ww^2ww
7. K^4K^2
8. $x^3x^5x^3$
9. $n^6n^3n^4$
10. bb^3b^6
11. $w^4(ww^6)$
12. $Y^4(Y^3Y^6)(YY)$
13. $(x^3x^0)(x^4x^1)$
14. $(s^6)(s^2s^1)(s^0)$
15. $(-z)(-z)^3(-z)$
16. $(-x)^2(-x)^3(-x)^2$
17. $(-t)^4(-t)^1(-t)^7$
18. $(-x)^6(-x)^0(-x)^2$

Simplify each of the following expressions to an integer.

19. $3 \cdot 3 \cdot 3 \cdot 3$
20. $2 \cdot 2 \cdot 2 \cdot 2 \cdot 2 \cdot 2 \cdot 2$
21. $4 \cdot 4 \cdot 4 \cdot 4$
22. $(3^2)(3^3)$
23. $2^2 2^3 2^5$
24. $6(6^2)$
25. $(4^2)(4^2)(4)$
26. $(3^2)(3^4)$
27. $5(5)(5^2)$
28. $(1^4)(1^6)(1^9)$
29. $(8)(8^3)$
30. $(2^5)(2^1)(2^2)$
31. $(-1)^3(-1)^2$
32. $(-1)(-1)^3(-1)^5$
33. $(-1)^4(-1)(-1)^5$
34. $(-2)(-2)(-2)^2$
35. $(-3)(-3)^3$
36. $(-2)^4(-2)^3(-2)^2$

Express each of the following word phrases algebraically and then simplify it. (A review of translating word phrases is provided in Section 1.6.)

1. a^3 times a^4
2. The product of m^9 and m^2.
3. The square of y
4. x^3 times the product of x^4 and x^5
5. b^9 times b^0 multiplied by b^6
6. x multiplied by the product of x^8 and x^9
7. The product of v^4 and v^7 times v^3
8. a^8 multiplied by a^{10} times a^{11}
9. The fifth power of c multiplied by c
10. 3^2 times 3^3
11. The product of 7^2 and 7
12. Twice 8 multiplied by 8^2

13. $(-1)^3$ times (-1) to the third power

14. 5 multiplied by the product of 5^2 and 5^0

15. Twice $(-3)^3$ times (-3) to the second power

16. The product of $(-2)^2$ and (-2) times $(-2)^4$

17. 4^2 multiplied by 4 times 4^0

18. The product of (-2) squared and $(-2)^4$

2.3 MULTIPLYING MONOMIALS

In the last two sections we have seen products of monomials without powers and products of powers with a single base. To multiply algebraic products of monomials, such as $(x^3y^2)(x^2y^4)(xy^3)$ and $(-6xy^4)(7x^4y^2)$, we must combine and extend the ideas and methods of these two sections. To see how to do this, let us look at the following tables.

times	0	x^2	$2x^2$	$3x^2$
0	0	0	0	0
x^2	0	x^4	$2x^4$	$3x^4$
$2x^2$	0	$2x^4$	$4x^4$	$6x^4$
$3x^2$	0	$3x^4$	$6x^4$	$9x^4$

and

times	x^3	x^2y	xy^2	y^3
x^3	x^6	x^5y	x^4y^2	x^3y^3
x^2y	x^5y	x^4y^2	x^3y^3	x^2y^4
xy^2	x^4y^2	x^3y^3	x^2y^4	xy^5
y^3	x^3y^3	x^2y^4	xy^5	y^6

22. Complete the tables.

a.

times	x	$2x$	$3x$
x			
$2x$			
$3x$			

b.

times	x^2	xy	y^2
x^2			
xy			
y^2			

As you can see from the table on the left, in computing a product of monomials the numerical factors are multiplied separately from the literal factors. Furthermore, the table on the right shows that the literal component of the product is computed by multiplying like powers together.

Look at the following two examples of *multiplying monomials.*

TYPICAL PROBLEMS

1. $(x^3y^2)(x^2y^4)(xy^3)$

We rearrange the product of literal powers, writing like powers together (this step is optional): $x^3x^2xy^2y^4y^3$

Then we multiply like powers together: x^6y^9

Hence, $(x^3y^2)(x^2y^4)(xy^3) = x^6y^9$.

2. $(-6xy^4)(7x^4y^2)$

First we multiply the integers together: $(-6)(7) = -42$

Then we rearrange the product of the literals, writing like powers together (this step is optional): $-42xx^4y^4y^2$

Then we multiply like powers together: $-42x^5y^6$

Hence, $(-6xy^4)(7x^4y^2) = -42x^5y^6$

To multiply integers together we use the method in Section 2.1. To simplify each group of like powers we use the method in the previous section.

EXAMPLES

Multiply the following monomials.

1. $(ax)(ax)(ax)$
2. $(-a^2x)(x^2y)(-ay^2)$
3. $-(-A^2B^3C^2)(-A^6B^2C)$
4. $(4x^3)(-9x^3)$
5. $-5y^4(+3x^2y^7)$
6. $(-8xy^5)(-x^4y^3)$
7. $(+6x^4y^9z^2)(-5x^3y^5)$
8. $-4y^2(x^3y^5)(-6x^4)$
9. $-t^3(-t)^4(t)^3$

Every product of monomials can be simplified to a single monomial whose numerical coefficient is the product of the coefficients of the factors and whose literal component is a product of unlike powers.

Solutions

1. $aaaxxx$
↓
a^3x^3
2. $+a^2axx^2yy^2$
↓
$a^3x^3y^3$
3. $-A^2A^6B^3B^2C^2C$
↓
$-A^8B^5C^3$
4. $-36x^3x^3$
↓
$-36x^6$
5. $-15x^2y^4y^7$
↓
$-15x^2y^{11}$
6. $+8xx^4y^5y^3$
↓
$8x^5y^8$
7. $-30x^4x^3y^9y^5z^2$
↓
$-30x^7y^{14}z^2$
8. $+24x^3x^4y^2y^5$
↓
$24x^7y^7$
9. $-t^3(-t)(-t)(-t)(-t)ttt$
↓
$-t^{10}$

EXERCISES

Simplify each of the following expressions to a monomial.

1. $-(ab)(-ac)(bc)$
2. $(-st)(-rt)(-rs)$
3. $(a^2b)(ab^4)$
4. $(R^2S)(R^2S^3)$
5. $-k^4(m)^3(-k^2m^4)$
6. $s^6(t)^6(-st)$
7. $(ar^2)(bs)(rs^2)$
8. $(ax^3)(by^2)(x^4y^8)$
9. $(-x^3yz)(xy^3z)$
10. $-(A^2B^3)(-A^3BC)$
11. $-(x^2y^3z^2)(-x^5y^2z^6)$
12. $-(-k^3r^2t)(k^2r^4t^2)$
13. $(5A^4)(6A)$
14. $(-10x)(12x^6)$
15. $(-3r^2)(2r^3)$
16. $(-6t^2)(-3t^4)$
17. $(-9A^3B)(-6B^5)$
18. $(3x^7y^2)(-8x^4)$
19. $-6B^4(5A^3B^2)$
20. $-8xy^3(-x^2y^2)$
21. $11r^2s(-4rs^2)$
22. $-9x^3y(-9x^2y^4)$
23. $(-4k^2m^9)(+15km^3)$
24. $(-18a^2b^6)(+12ab^3)$
25. $4x^9(-8x^3)(-2x)$
26. $-8a^4(-8a^5)(-a)$
27. $5y^2(-6y)(-2y^3)$
28. $-x(+12x^3)(3x)$
29. $(-m^3n^2)(-5n^7)(-7m^6)$
30. $(+5s^4)(+6t^7)(-8s^6t^4)$
31. $-(x^7)(-5x^4y)(+9y^9)$
32. $(+4u^2)(-7u^3v)(-uv^7)$
33. $(-3a^2b)(-b^3c)(ab^4c)$
34. $(-5w^4)(-x^3y)(-2wxy^3)$

35. $(9r^3s^2)(+7s^4t^9)(-2r^2t)$ 36. $(4w^3xy^2)(-3wx^2)(+6wy^4)$

37. $-t^4(-t)^2(t)^8$ 38. $-r^4(r)^6(-r)^7$ 39. $(-3k^3)(-k)^6$

40. $(-x)^6(-8x^3)$ 41. $-5t^4(+t)^6(-t)^3$ 42. $+8y^3(-y)^8(+y)^5$

Express each of the following word phrases algebraically and then simplify it. (A review of translating phrases is provided in Section 1.6.)

1. 5m times 7m^2
2. The product of 9cm and 12cm^2
3. Twice 30mm^3
4. 4km multiplied by 6km^2
5. $3x^2y$ times the product of $-4xyz$ and $7y^2z^4$
6. Twice $9x^3$ multiplied by $-9x^4z^5$
7. $-18a^3b^2$ times $a^6b^8c^4$ multiplied by $-2a^5bc^5$
8. $3abx$ multiplied by $7a^2b^5x^3$ times a^3bx
9. The product of $-8ad^5$ and $-3a^5d^3$ times $-5a^5d$
10. Twice the product of $4x^3y^2z^2$ and $25x^6y^8z^0$
11. $r^{10}s^8t^3$ times $r^6s^7t^4$ multiplied by r^5s^2t
12. $-6xyz$ times twice $-4x^3y^0z^0$

2.4 EVALUATING

Replacing a literal in an expression by some other quantity (and then simplifying) is called *evaluating the expression.* For example, we can replace the literal A in the expression $5A$ by the number 4 and then simplify the resulting numerical product 5(4) to 20. When we do this we say that we are "evaluating $5A$ when A equals 4." To say this in mathematical notation we simply write "$5A$, when $A = 4$." Note that we are using an equals sign between two expressions that do not have the same value. A statement such as $A = 4$ is called an *equation*, and we use it to state that the literal "represents" the number, or is another "name" for the number. Hence, when we replace the given literal by the given number in an expression we do not change the *intended value* of that expression.

In this section we will evaluate monomials for *integer values* of the literals. To evaluate a monomial, such as $5A$, given an equation, such as $A = 4$, we proceed as follows:

1. Replace the literal(s) by the given number(s) *in parentheses.*

   ($5A$ becomes 5(4).)

2. Simplify the resulting numerical expression.

   (5(4) becomes 20.)

For example, to evaluate $6x$, when $x = -4$, we first replace x by -4 in parentheses and then simplify:

$$6x, \text{ when } x = -4 \qquad \text{becomes} \qquad 6(-4) = -24$$

Similarly, to evaluate $3x^2$, when $x = -2$, we replace x by −2 in parentheses and then simplify:

$$3x^2, \text{ when } x = -2 \qquad \text{becomes} \qquad 3(-2)^2 = 3(+4) = 12$$

Note why parentheses were needed in these examples.

1. $6x$, when $x = -4$ becomes $6(-4)$, **not** $6 - 4$. Replacing the literal by a signed number without using parentheses can change an indicated product to a sum.

2. x^2, when $x = -2$ becomes $(-2)^2$, **not** (-2^2). Because $x^2 = x \cdot x$, it follows that x^2, when $x = -2$ is the same as $(-2) \cdot (-2)$. Remember that $(-2)^2 = (-2)(-2) = +4$, whereas $(-2^2) = -2 \cdot 2 = -4$.

3. $3(-2)^2$ equals $3(+4)$, **not** $3 + 4$ **or** $(-6)^2$. When simplifying a product containing powers, the powers must be computed first (because exponential notation is a shorthand for repeated multiplication). Further, we keep parentheses to avoid changing the indicated product to a sum.

To see some examples of evaluating a monomial, look at the following tables.

evaluated at	−3	−2	−1	0	+1	+2	+3
3A							
2A		2(−2) = −4	2(−1) = −2	2(0) = 0	2(+1) = 2	2(+2) = 4	
A		(−2) = −2	(−1) = −1	(0) = 0	(+1) = 1	(+2) = 2	
−A		−(−2) = 2	−(−1) = 1	−(0) = 0	−(+1) = −1	−(+2) = −2	
−2A		−2(−2) = 4	−2(−1) = 2	−2(0) = 0	−2(+1) = −2	−2(+2) = −4	
−3A							

and

evaluated at	−3	−2	−1	0	+1	+2	+3
$3A^2$							
$2A^2$		$2(-2)^2 = 8$	$2(-1)^2 = 2$	$2(0)^2 = 0$	$2(+1)^2 = 2$	$2(+2)^2 = 8$	
A^2		$(-2)^2 = 4$	$(-1)^2 = 1$	$(0)^2 = 0$	$(+1)^2 = 1$	$(+2)^2 = 4$	
$-A^2$		$(-2)^2 = -4$	$-(-1)^2 = -1$	$-(0)^2 = 0$	$-(+1)^2 = -1$	$-(+2)^2 = -4$	
$-2A^2$		$-2(-2)^2 = -8$	$-2(-1)^2 = -2$	$-2(0)^2 = 0$	$-2(+1)^2 = -2$	$-2(+2)^2 = -8$	
$-3A^2$							

23. Complete the tables.

Look at the following examples of *evaluating a monomial.*

TYPICAL PROBLEMS

Evaluate: 1. $-8xy$, when $x = -4, y = 3$

We first substitute the numbers given using parentheses: $-8(-4)(3)$

Then we simplify the resulting expression following the Order of Operations: 96

Hence, evaluating $-8xy$, when $x = -4$, $y = 3$ yields 96.

Evaluate: 2. $3s^2t^3$, when $s = 10, t = -2$

We first substitute the numbers given using parentheses: $3(10)^2(-2)^3$

Then we simplify the resulting expression following the Order of Operations: $3(100)(-8)$ → $-2{,}400$

Always place parentheses around each substituted number. The Order of Operations is given in Section 1.5.

Hence, evaluating $3s^2t^3$, when $s = 10$, $t = -2$ yields $-2{,}400$.

EXAMPLES

Evaluate the following monomials.

1. $-7w$, when $w = -4$
2. $3uv$, when $u = 9, v = -4$
3. $-6xy$, when $x = -12, y = 0$
4. $-8a^2$, when $a = -3$
5. $4k^3m$, when $k = -1, m = -12$
6. $6w^3x^2$, when $w = -2, x = 3$

Solutions

1. $-7(-4)$ → 28
2. $3(9)(-4)$ → -108
3. $-6(-12)(0)$ → 0
4. $-8(-3)^2$ → $-8(+9)$ → -72
5. $4(-1)^3(-12)$ → $4(-1)(-12)$ → 48
6. $6(-2)^3(3)^2$ → $6(-8)(9)$ → -432

SPECIAL CASES

Remember that multiplying any expression by zero yields zero. Hence, evaluating a monomial when any of its literal factors is zero yields zero. For example,

$-3xy$, when $x = 0$, $y = -2$ and $8x^3$, when $x = 0$

$-3(0)(-2)$ $8(0)^3$

↓ ↓

0 0

OBSERVATIONS AND REFLECTIONS

Remember that in a product of signed numbers each number (except possibly the first) must be inside parentheses. For example, the product of -3 and $+4$ can be written as $(-3)(+4)$ or as $-3(+4)$, but not as $-3 + 4$ or $(-3) + 4$. In a product of literals, the literals are written without parentheses, because they behave like signed numbers inside parentheses. For example, the product of x and y is written as xy.

Hence, in order to evaluate a monomial, the literals must be replaced by signed numbers inside parentheses. For example,

$-x$, when $x = -2$ becomes $-(-2)$, **not** -2 or $- -2$

$4x$, when $x = -2$ becomes $4(-2)$, **not** $4 - 2$

$-6x^2y$, when $x = -2$, $y = +3$ becomes $-6(-2)^2(+3)$, **not** $-6 - 2^2 + 3$

EXERCISES

Evaluate each of the following expressions, simplifying to an integer.

1. $4x$, when $x = 5$
2. $6x$, when $x = 3$
3. $-7a$, when $a = 4$
4. $-9b$, when $b = 8$
5. $3y$, when $y = -9$
6. $5z$, when $z = -6$
7. $-4w$, when $w = -4$
8. $-3x$, when $x = -3$
9. $4st$, when $s = 3$, $t = -9$
10. $6mn$, when $m = -8$, $n = 4$
11. $-6xy$, when $x = -3$, $y = -2$
12. $-7wx$, when $w = -1$, $x = -6$
13. $3uv$, when $u = 0$, $v = -4$
14. $5yz$, when $y = -8$, $z = 0$
15. $-8KL$, when $K = -5$, $L = 10$
16. $-6AB$, when $A = 3$, $B = -20$
17. $4x^2$, when $x = 3$
18. $8z^2$, when $z = 2$
19. $-6r^3$, when $r = -2$
20. $-12t^5$, when $t = -1$
21. $(-A)^3(-A)^5(-A)^3(-A)^0$
22. $-2k^2$, when $k = -12$
23. $10y^3$, when $y = -3$
24. $20x^3$, when $x = -2$
25. $3x^2y$, when $x = 2$, $y = 5$
26. $4uv^2$, when $u = 6$, $v = 2$
27. $5st^2$, when $s = 3$, $t = -4$
28. $2x^2z$, when $x = -3$, $z = 9$
29. $3x^2y^3$, when $x = -5$, $y = -2$
30. $4x^3y^2$, when $x = -2$, $y = -3$

ANSWERS TO MARGIN EXERCISES

1. a. $-3x$ and $+7y$
b. $6ad$, $-6c$, and $+2b$

2. a. $+6axyz$, $+3axy$, 0, $-3ax$, $-6a$
b. $+6axyz$, $+3axy$, 0, $-3ax$, $-6a$

3. a. xy
b. $tvwx$
c. $abcd$

4. a. $-6x$
b. $-6x$
c. $4wxy$
d. $4wxy$
e. $2wxyz$
f. $2wxyz$

5. a. $(-12wx)(-z) = 12wxz$
b. $-3x(-4wz) = 12wxz$

6. a. 0
b. 0

7. a. 0
b. 0

8.

	base	exponent
a.	x	5
b.	y	8
c.	3	2

9.

	base(s)	exponent(s)
a.	x and y	4 and 3
b.	a and c	3 and 2
c.	$(3y)$	8
d.	$(-4z)$	7

10.

	base	exponent
a.	2	5
b.	3	2
c.	(-3)	2
d.	(-3)	3

11. a. $2k$
b. xyz
c. $4a^2b$

12. a. 3
b. y^2
c. $-a^2$

13. a. minus
b. minus
c. plus

14. a. 1
b. -1
c. x^5

15. a. 0
b. 0
c. 0

16. a. b^5
b. c^6
c. w^7
d. y^8

18. a. t^9
b. w^9
c. k^{15}
d. x^{16}

19. a. 9
b. -9
c. -8
d. -8

20. a. x^2
b. x^2
c. x^7
d. x^7
e. y^9
f. y^9

CHAPTER REVIEW

Simplify to a monomial. SECTION 2.1

1. $3x \cdot 4y$
2. $(12m)(5n)$
3. $(8x)(+2yz)$
4. $(-6rt)(-4qs)$
5. $(-xy)(-3w)$
6. $-7b(+4ac)$
7. $-12k(-5m)$
8. $+5MN(-7KL)$
9. $(-A)(-3B)(2C)$
10. $-(-4x)(+9y)(-6z)$
11. $(2wy)(-az)(+4b)$
12. $+6(+12t)(-4r)$

Simplify to a monomial. SECTION 2.2

13. $bbbb$
14. $KKK(KKK)$
15. $w(ww)ww$
16. $t^2t^0t^3$
17. $x^7(x^4x)(x^6)$
18. $(p^4p^1)(p^0p^6p^0)$
19. $r^2(r^3r)r^6$
20. $(-b)^2(-b)(-b)^4(-b)$
21. $(-A)^3(-A)^5(-A)^3(-A)^0$

Simplify to an integer.

22. $5 \cdot 5 \cdot 5$
23. $(2^4)(2^3)$
24. $8^2(8)$
25. $3(3^3)$
26. $2(2^3)(2^2)(2^0)(2^1)$
27. $(1^7)(1^3)(1^5)$
28. $(-1)^4(-1)^3(-1)(-1)^8$
29. $(-2)^3(-2)(-2)$
30. $(-5)(-5)^2(-5)$

Simplify to a monomial. SECTION 2.3

31. $-(xy)(-wx)(yz)$
32. $(x^4y^2)(xy^3)$
33. $(x)^4(-x^2y^2)(xy^6)$
34. $-(AB^3)(+AB^5)(-A^5)$
35. $(x^3yz^4)(+x^3y^3z^2)$
36. $-(r^3s^2t^6)(-r^8s^7t^4)$
37. $(3x^5)(9x)$
38. $(-8t^4)(+2t^4)$
39. $(-5m^2n)(-15n^4)$
40. $-6x^3(4x^2y^3)$
41. $16a^2b(-8ab^2)$
42. $(-3x^3y^8)(+18xy^4)$
43. $3x^6(-4x^5)(-10x^7)$
44. $6t^3(-8t)(-2t^6)$
45. $(-x^2y^4)(-12x^6)(-8y^4)$
46. $-(y^6)(-15y^3z)(+15z^8)$
47. $(-9K^2L)(-L^5)(KL^4)$
48. $(2r^2s^4)(+20s^6t^8)(-4r^3t)$
49. $-A^6(-A)^4(A)^7$
50. $(-8k^4)(-k)^9$
51. $+14y^5(+y)^{12}(-y)^7$

Evaluate, simplifying to an integer. SECTION 2.4

52. $8x$, when $x = 7$
53. $-6A$, when $A = 9$
54. $4w$, when $w = -8$
55. $-5k$, when $k = -5$

56. $3xy$, when $x = 4, y = -7$

57. $-2st$, when $s = -6, t = -4$

58. $8rs$, when $r = 0, s = -6$

59. $-9AB$, when $A = -4, B = 10$

60. $3y^2$, when $y = 6$

61. $-3t^3$, when $t = -3$

62. $-12x^2$, when $x = -5$

63. $16w^5$, when $w = -1$

64. $4u^2v$, when $u = 2, v = 9$

65. $7st^2$, when $s = 5, t = -3$

66. $6m^2n^3$, when $m = -10, n = -2$

SUMMARY OF PHRASES

Express each of the following word phrases algebraically and then simplify it.

1. The product of $-5u$ and $-8x$
2. $-7a$ multiplied by $4c$
3. $11b$ times the product of $-2a$ and $4c$
4. Twice $-10xz$ multiplied by $5y$
5. $7uv$ multiplied by the product of $7w$ and z
6. The product of $3ab$ and $-10d$ times $-6c$
7. x^4 times x^2
8. a^9 multiplied by a to the twelfth power
9. u multiplied by the product of u^3 and u^8
10. a^7 multiplied by the cube of a
11. The product of x and x^7 multiplied by x^0
12. The product of 5 and 5^2
13. Twice 3 multiplied by the square of 3
14. The product of -2 and the fourth power of -2
15. $(-1)^9$ multiplied by (-1) times $(-1)^0$
16. The product of 7m and 11m
17. 13cm multiplied by 2cm^2
18. $2a^2y$ times the product of $-9ayz$ and $3yz^2$
19. $6axy$ multiplied by $9a^2x^3$ times x^4y^5
20. Twice the product of $14x^5y^3z^4$ and $3x^7y^0z$
21. $70x^3y$ times twice $-4x^3y^{10}z^0$

3 ADDITION OF MONOMIALS

When working with sums of monomials we can use the same language we used in Chapter 1 to talk about integers. Remember that in an algebraic sum, such as $-3x + 2x$ or $3x - 2y$, each quantity is called a *term* of the sum. For example, in the sum $-3x + 2x$ the terms are $-3x$ and $2x$. Similarly, in the sum $3x - 2y$ the terms are $3x$ and $-2y$. Note that the literal components of the two terms in the sum $-3x + 2x$ are the same, whereas the literal components of the two terms in the sum $3x - 2y$ are different.

To talk about sums of monomials we must introduce several new words. Monomials and sums of monomials with *different* literal components in each term are called *polynomials*, or, more precisely, polynomials with integer coefficients. For example,

$5xy$, a monomial, is a one-term polynomial.

$3x - 2y$ is a two-term polynomial and is called a *binomial.*

$2a + 3b - 4c$ is a three-term polynomial and is called a *trinomial.*

$x^3 + 5x^2 - 7x + 2$ is a four-term polynomial.

A polynomial with more than one term is also called a *multinomial.* Monomials are the only polynomials that are not multinomials. [*Mono* is Greek for one, *bi* is Latin for two, *tri* is French (derived from Latin or Greek) for three, *poly* is Greek for many, and *multi* is Latin for many.]

In the first two sections of this chapter we will consider the special case of sums of terms that simplify to monomials. In Section 3.3 we will consider the general case of sums that simplify to multinomials.

3.1 ADDING TWO LIKE TERMS

To add monomials we must be able to add signed integers, and we must know what to do with the literal components. Look at the following examples of algebraic addition of monomials:

$$3x + 2x = 5x \qquad -4x - 5x = -9x$$

$$-2x^3 + 6x^3 = 4x^3 \qquad xy^2 - 4xy^2 = -3xy^2$$

$$-x^4y - x^4y = -2x^4y \qquad -3xyz + 4xyz = xyz$$

1. Compute
a. $2x - 5x$
b. $-4x^2 + 7x^2$
c. $-8xy - xy$
d. $3yz^2 + 9yz^2$

Can you see any pattern in these examples? Did you notice that in each example the two terms of the indicated sum on the left have identical literal components? From these examples try to guess the rule for adding two monomials with the same literal component.

Because these sums are so important, we need a name for them. Two monomials are called *like terms* if they have identical literal components; that is, if they contain exactly the same powers (although the powers need not be written in the same order). Note that like terms need not have the same numerical coefficients. For example,

$-4a$ and $5a$ are like terms whereas -4 and $5a$ are not like terms

$-19w$ and w are like terms whereas $-19w$ and w^2 are not like terms

$12x^3$ and $37x^3$ are like terms whereas $12x^3$ and $37y^3$ are not like terms

$5x^2y$ and $9x^2y$ are like terms whereas $5x^2y$ and $9xy^2$ are not like terms

2. Which pairs are like terms?
a. $5z$ and $-z$
b. $3x^2y$ and $-3xy^2$
c. $-4t^4$ and $-2t^2$
d. $2a^2b$ and $5ba^2$

For two terms to be like terms, the literals and their exponents must be exactly the same.

In this section we will deal only with sums of two like terms.

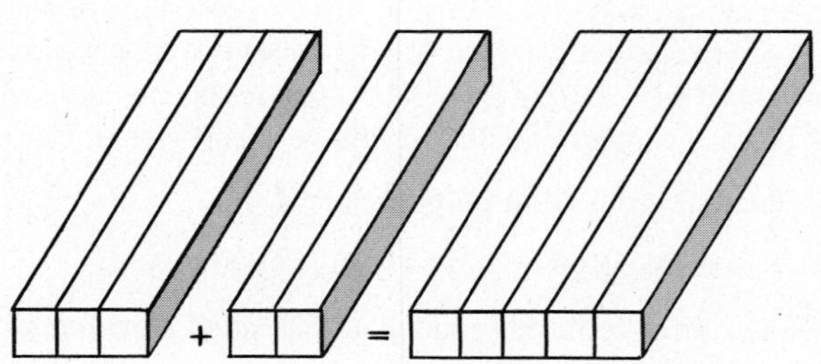

Look at the following example of *adding two like terms.*

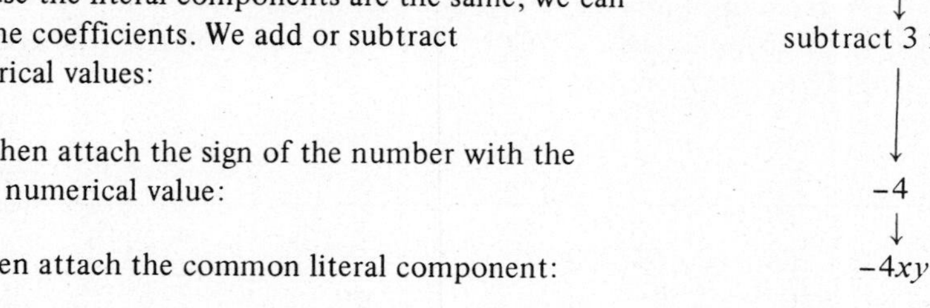

TYPICAL PROBLEM

$3xy - 7xy$

Because the literal components are the same, we can add the coefficients. We add or subtract numerical values: subtract 3 from 7

And then attach the sign of the number with the larger numerical value: -4

We then attach the common literal component: $-4xy$

To add the numerical coefficients (integers), we use the method in Section 1.1.

Hence, $3xy - 7xy = -4xy$.

WARNING This procedure can only be used to add like terms. If the literal components of the two terms are not the same, the terms cannot be combined into a single term.

EXAMPLES

Add the following pairs of like terms.

1. $-15y + 12y$ 2. $-2wx - 7wx$ 3. $+7xy^2 - 5xy^2$

4. $-8r^5 + 8r^5$ 5. $-18z^3 - 32z^3$ 6. $m^2 + m^2$

A sum of two like terms can be simplified to a monomial with the common literal component, whose coefficient is the sum of the coefficients of the two terms.

Solutions

1. $-3y$ 2. $-9wx$ 3. $2xy^2$

4. 0 5. $-50z^3$ 6. $2m^2$

SPECIAL CASES

1. Remember that a monomial such as m^2 is really $+1m^2$. Similarly, $-m^2$ is really $-1m^2$. For example,

$$7x^2 + x^2 \rightarrow \text{add 7 and 1} \rightarrow 8x^2$$

$$-x^2 + 3x^2 \rightarrow \text{subtract 1 from 3} \rightarrow 2x^2$$

3. Compute
a. $x^5 + x^5$
b. $-2ab^2 - ab^2$
c. $-6t^6 + t^6$

2. If the sum of the numerical coefficients is zero, then the sum of the monomials is zero. For example,

$$5ab^2 - 5ab^2 \rightarrow \text{subtract 5 from 5} \rightarrow 0ab^2 = 0$$

$$-12x + 12x \rightarrow \text{subtract 12 from 12} \rightarrow 0x = 0$$

4. Simplify
a. $0x^3$
b. $0wxyz$

Remember that zero times any literal component is zero. For example,

$0ab^2$ is 0 and $0x$ is 0

5. Compute
a. $4xy^2z - 4xy^2z$
b. $-x^3y^2 + x^3y^2$

We saw a signed number addition table for integers in Section 1.1. Now look at a modified addition table for like terms. Verify the entries to make sure that you understand addition of like terms.

plus	$-3x^2$	$-2x^2$	$-x^2$	0	$+x^2$	$+2x^2$	$+3x^2$
$+3x^2$	0	x^2	$2x^2$	$3x^2$	$4x^2$	$5x^2$	$6x^2$
$+2x^2$	$-x^2$	0	x^2	$2x^2$	$3x^2$	$4x^2$	$5x^2$
$+x^2$	$-2x^2$	$-x^2$	0	x^2	$2x^2$	$3x^2$	$4x^2$
0	$-3x^2$	$-2x^2$	$-x^2$	0	x^2	$2x^2$	$3x^2$
$-x^2$	$-4x^2$	$-3x^2$	$-2x^2$	$-x^2$	0	x^2	$2x^2$
$-2x^2$	$-5x^2$	$-4x^2$	$-3x^2$	$-2x^2$	$-x^2$	0	x^2
$-3x^2$	$-6x^2$	$-5x^2$	$-4x^2$	$-3x^2$	$-2x^2$	$-x^2$	0

6. Complete the tables.

a.

plus	$-x$	0	$+x$
$+x$			
0			
$-x$			

b.

plus	$+x^3$	$+2x^3$	$+3x^3$
$+x^3$			
$+2x^3$			
$+3x^3$			

In Section 1.2 we studied the Commutative Law of Addition, which applies to signed numbers and to literal numbers. Because addition is commutative, the terms in an algebraic sum may be written in any order. For example, $3xy - 7xy$ equals $-7xy + 3xy$ and $4x^3 + 9x^3 - 6x^3$ equals $-6x^3 + 9x^3 + 4x^3$.

7. Compute
a. $3xy - 7xy$
b. $-7xy + 3xy$
c. $-8t^4 - 3t^4$
d. $-3t^4 - 8t^4$

OBSERVATIONS AND REFLECTIONS

1. As we have seen, like terms are special because they can be combined to a monomial. In fact, *only* like terms can be combined.

   Consider the following sums, which are very similar in appearance:

   $$5x^2y + 9x^2y \quad \text{and} \quad 5x^2y + 9xy^2$$

   The first sum is of like terms and simplifies to the monomial $14x^2y$. The second sum is of two unlike terms (a binomial) and cannot be simplified to a monomial.

2. In a sum of like terms, the literal components in each term must be the same, but they need not be in the same order. For example,

   $$2cba + 3bac = 5abc \quad \text{and} \quad xy - yx = 0$$

8. Compute
a. $2ab^2 + 7b^2a$
b. $-4rs^2t - s^2rt$
c. $-xyz - zyx$

EXERCISES

Simplify each of the following expressions to a monomial.

1. $5y - 12y$
2. $4n - 9n$
3. $-3x - x$
4. $-6x - 6x$
5. $-15y + 7y$
6. $-16x + x$
7. $4yz + 12yz$
8. $9xy + 4xy$
9. $uv - 11uv$
10. $xz - 15xz$
11. $-3st - 11st$
12. $-6wx - 13wx$
13. $5x^2 - 3x^2$
14. $8x^2 - x^2$
15. $-4x^5 + x^5$
16. $-7y^2 + 4y^2$
17. $-3x^6 - 4x^6$
18. $-12x^3 - 5x^3$
19. $-16xy^2 + 8xy^2$
20. $-14v^2w + 9v^2w$
21. $-s^4t^8 - s^4t^8$
22. $-m^3n^9 - m^3n^9$
23. $-8x^4y^3 + 20x^4y^3$
24. $-6u^3v^2 + 19u^3v^2$

Express each of the following word phrases algebraically and then simplify it. (A review of translating word phrases is provided in Section 1.6.)

1. The sum of 9cm and 16cm
2. 11km decreased by 7km
3. 5g subtracted from 12g
4. 8L added to 13L
5. 19m reduced by 17m
6. 4mm less than 6mm
7. 19g plus 31g
8. 15km is subtracted from 18km
9. 2mL more than 27mL
10. 18g less 11g
11. 16cm increased by 6cm
12. 25L is reduced by 24L

3.2 ADDING SEVERAL LIKE TERMS

Once we know how to add two like terms, we can add any number of like terms. In the last section we saw that monomials obey the Commutative Law of Addition.

They obey the Associative Law as well. The Associative Law of Addition applies to signed numbers and to literal numbers.

Because addition is associative, the terms in an algebraic sum may be combined, two at a time, in any order. For example, compare the following computations:

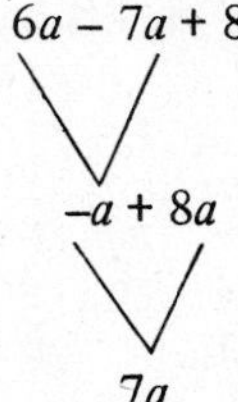

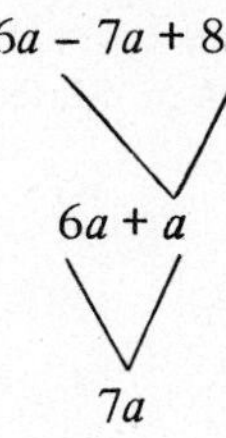

9. Compute $-4x^2 + x^2 - 7x^2$ both as
a. $(-4x^2 + x^2) - 7x^2$
and as
b. $-4x^2 + (x^2 - 7x^2)$

The Commutative and Associative Laws of addition imply that the terms in an algebraic sum can be written and combined in any order. Hence, we can combine all the positive terms first regardless of their position in the sum and then combine all the negative terms. (Remember that we call a monomial positive when its coefficient is positive and negative when its coefficient is negative.) This method is usually the fastest way to combine like terms to a monomial with the common literal component.

Look at the following example of *adding several like terms.*

TYPICAL PROBLEM

$-4B + 8B - 13B + 5B$

First we add the positive terms and then the negative terms: $\downarrow$ $+13B - 17B$

Then we combine terms with unlike signs: $\downarrow$ $-4B$

Hence, $-4B + 8B - 13B + 5B = -4B$.

To add like terms, we parallel the method in Section 1.2 for adding integers. If the sum has many terms, it can be rearranged with the positive terms written first, followed by the negative terms.

EXAMPLES

Add the following like terms.

Every sum of like terms can be simplified to a monomial with the common literal component.

1. $x^2 + x^2 + x^2$
2. $4z^3 - z^3 - 8z^3 - z^3$
3. $-8a^5 + 4a^5 - 10a^5$
4. $6xy + 7xy - 4xy + 8xy$
5. $-3x^2 - 7x^2 + 8x^2 - 6x^2 + 4x^2$
6. $7u^3v - 12u^3v + 6u^3v - u^3v$

Solutions

1. $3x^2$
2. $4z^3 - 10z^3 \downarrow -6z^3$
3. $4a^5 - 18a^5 \downarrow -14a^5$
4. $21xy - 4xy \downarrow 17xy$
5. $12x^2 - 16x^2 \downarrow -4x^2$
6. $13u^3v - 13u^3v \downarrow 0u^3v = 0$

10. Compute
a. $y^4 + y^4$
b. $y^4 \cdot y^4$
c. $xy + xy$
d. $xy \cdot xy$
e. $-b^5 - b^5$
f. $(-b^5)(-b^5)$
g. $2xy - 3xy$
h. $(2xy)(-3xy)$
i. $7st^3 + st^3$
j. $7st^3 \cdot st^3$
k. $-5w^2x^6 - 9w^2x^6$
l. $-5w^2x^6(-9w^2x^6)$

OBSERVATIONS AND REFLECTIONS

When working with expressions containing exponents, you may confuse the rules for addition and multiplication. Remember that to multiply like powers, add the exponents. To add like terms, add the numerical coefficients. Contrast the following columns to check your understanding of these operations.

$x + x = 2x$	$x \cdot x = x^2$
$x^2 + x^2 = 2x^2$	$x^2 \cdot x^2 = x^4$
$x^3 + x^3 = 2x^3$	$x^3 \cdot x^3 = x^6$

EXERCISES

Simplify each of the following expressions to a monomial.

1. $7A + A - 8A$
2. $-7x + 3x - x$
3. $-4y + 7y - 11y$
4. $-4x + 9x - 12x$
5. $n^2 - 9n^2 - 4n^2$
6. $-7A^3 + A^3 - 12A^3$
7. $3z^5 + 8z^5 - 10z^5$
8. $-6x^8 + 13x^8 - 15x^8$
9. $-w + 14w + 5w - 18w$
10. $12n + 7n - 8n + n$
11. $-11a - 4a + 13a - 6a$
12. $-3n + 7n - 8n + 6n$
13. $-5r^3 + 7r^3 + 11r^3 - r^3$
14. $-4x^6 + 10x^6 - 8x^6 + 6x^6$
15. $6t^7 - 5t^7 + 11t^7 - 4t^7 - t^7$
16. $-2y^2 + 7y^2 - y^2 - 8y^2 + 3y^2$
17. $6t^2 + 17t^2 - 13t^2 + 4t^2 - t^2$
18. $r^3 - 7r^3 - 3r^3 + 11r^3 - 6r^3$
19. $8ab + 4ab - 15ab - 3ab$
20. $-9xy - 12xy + 18xy - xy$
21. $4x^4y^3 - 7x^4y^3 + 15x^4y^3 - 12x^4y^3$
22. $-9a^2z^3 + 13a^2z^3 - 5a^2z^3 + a^2z^3$
23. $-11w^5x^2 - 25w^5x^2 + w^5x^2 - 14w^5x^2 + 18w^5x^2$
24. $s^7t^8 - 5s^7t^8 + 11s^7t^8 - 8s^7t^8 + 5s^7t^8$

3.3 SIMPLIFYING SUMS WITH UNLIKE TERMS

In the previous two sections we have considered the algebraic addition of like terms; that is, terms with identical literal components. In this section we will consider sums with some like terms and some unlike terms, as, for example, $-5x + 8y + 13x - 7y$ and $-8a^2b + 4ab^2 + 5a^2b - 2ab^2 + 3a^2b$.

Because addition is commutative and associative, we can always rearrange the terms of the sum to combine like terms. (Remember that we can only combine like terms, so a sum containing unlike terms does not, in general, simplify to a monomial. Instead it simplifies to a multinomial, a polynomial with more than one term.)

Look at the following examples of *simplifying sums with unlike terms.*

TYPICAL PROBLEMS

1. $-5x + 8y + 13x - 7y$

We first rearrange the terms, writing like terms together (this step is optional):

$$\downarrow$$
$$-5x + 13x + 8y - 7y$$

We then simplify to a polynomial by combining like terms:

$$\downarrow$$
$$8x + y$$

Hence, $-5x + 8y + 13x - 7y = 8x + y$.

2. $-8a^2b + 4ab^2 + 5a^2b - 2ab^2 + 3a^2b$

We first rearrange the terms, writing like terms together (this step is optional):

$$\downarrow$$
$$-8a^2b + 5a^2b + 3a^2b + 4ab^2 - 2ab^2$$

We then simplify to a polynomial by combining like terms:

$$\downarrow$$
$$0a^2b + 2ab^2$$
$$\downarrow$$
$$2ab^2$$

To combine each group of like terms we use the method in the previous section.

Hence, $-8a^2b + 4ab^2 + 5a^2b - 2ab^2 + 3a^2b = 2ab^2$.

EXAMPLES

Add the following monomials with unlike terms.

Every sum of monomials can be simplified to a polynomial.

1. $25 - 4k - 8 + 3$
2. $-6x^2 + 9x^3 - 8x^2$
3. $-8m^2 + 3m^3 + 16m^2 - 9m^3$
4. $4x^2 - 3y^3 + 5x^2 - 7y^2$
5. $12a^2 - 13b^3 + 4b^3 - 12a^2 + 9b^3$
6. $15x^2y - 8x^2y^2 - 3x^2y + 6xy^2$

Solutions

1. $25 - 8 + 3 - 4k$
$$\downarrow$$
$$20 - 4k$$

2. $-6x^2 - 8x^2 + 9x^3$
$$\downarrow$$
$$-14x^2 + 9x^3$$

3. $-8m^2 + 16m^2 + 3m^3 - 9m^3$
$$\downarrow$$
$$8m^2 - 6m^3$$

4. $4x^2 + 5x^2 - 3y^3 - 7y^2$
$$\downarrow$$
$$9x^2 - 3y^3 - 7y^2$$

5. $12a^2 - 12a^2 - 13b^3 + 4b^3 + 9b^3$
$$\downarrow$$
$$0a^2 + 0b^3 = 0$$

6. $15x^2y - 3x^2y - 8x^2y^2 + 6xy^2$
$$\downarrow$$
$$12x^2y - 8x^2y^2 + 6xy^2$$

Study the following tables and make sure you understand the rules for adding and multiplying monomials.

ADDITION

plus	$+1$	$+2x$	$+3x^2$	$+4x^3$	$+5x^4$	$+6x^5$
$+1$	2	$2x+1$	$3x^2+1$	$4x^3+1$		
$+2x$	$2x+1$	$4x$	$3x^2+2x$	$4x^3+2x$		
$+3x^2$	$3x^2+1$	$3x^2+2x$	$6x^2$	$4x^3+3x^2$		
$+4x^3$	$4x^3+1$	$4x^3+2x$	$4x^3+3x^2$	$8x^3$		
$+5x^4$						
$+6x^5$						

11. Complete the tables.

MULTIPLICATION

times	$+1$	$+2x$	$+3x^2$	$+4x^3$	$+5x^4$	$+6x^5$
$+1$	1	$2x$	$3x^2$	$4x^3$		
$+2x$	$2x$	$4x^2$	$6x^3$	$8x^4$		
$+3x^2$	$3x^2$	$6x^3$	$9x^4$	$12x^5$		
$+4x^3$	$4x^3$	$8x^4$	$12x^5$	$16x^6$		
$+5x^4$						
$+6x^5$						

Can you see which sums involve like terms in the addition table? Note that a sum of like terms reduces to a monomial, whereas a sum of unlike terms does not. In contrast, note that the product of two monomials is always a monomial.

OBSERVATIONS AND REFLECTIONS

12. Which of these sums are called polynomials?
   a. $3x - 7x^3 - x$
   b. $2w^2x - 5w^2x^2 + 4wx^2$
   c. $-9t^4 + 7t^4 - 4t^4$
   d. $4w^2 - 4x^2 - 4y^2 + 4z^2$

1. Remember that sums of monomials with *different* literal components in *each* term are called *polynomials* (or multinomials). If any two terms of a sum are like terms (contain the same literal component), then the sum is not called a polynomial because it can be simplified to a sum with fewer terms. For example, $8x + y$ is a *polynomial*, whereas $-5x + 8y + 13x - 7y$ is a *sum* that can be simplified to the polynomial $8x + y$.

2. In simplifying sums with unlike terms, we suggest first rearranging the terms, writing like terms together. The different groups of like terms may be written in any order. For example, the sum $3x - 7 - 5x - x + 8 - 6$ can be rearranged as

$$3x - 5x - x - 7 + 8 - 6 \quad \text{or} \quad -7 + 8 - 6 + 3x - 5x - x$$

When like terms are combined, the order of the terms in the resulting polynomial is determined by how the sets of like terms were arranged. For example, the sum above simplifies to $-3x - 5$ or $-5 - 3x$.

However, the order of the terms does not matter. The terms of a polynomial may be written in any order. The polynomials $-3x - 5$ and $-5 - 3x$ are equal. Two polynomials are equal if they contain the same terms, written in any order.

13. Pair equal polynomials
a. $-5x + 7y$
b. $5x - 7y$
c. $-5x - 7y$
d. $-7y - 5x$
e. $7y - 5x$
f. $-7y + 5x$

EXERCISES

Simplify each of the following expressions to a polynomial.

1. $3x - 5 + 7x$
2. $7x + 3xy - 5x$
3. $4A - 9A^2 + 6A^2$
4. $2w - 5w + 7w^2$
5. $-5n^3 + 12 + 5n^3$
6. $-3x^2 + 2x + 3x$
7. $4w - 10w + 17w^2$
8. $9x^2 + 6xy^2 - 8x^2$
9. $32A^3 - 10A^2 - 18A^3 + 8A^2$
10. $-6x^2 + 4x^3 + 6x^2 - 8x^3$
11. $3x^4 - 2x + 10x^4 + 4x$
12. $-8y^2 + 12xy^2 + 14y^2 - 16xy^2$
13. $-8a^5 + 7b^4 + 5a^5 + 13b^5$
14. $16T^3 - 5ST^3 - 9S^3 + 4ST^3$
15. $6x^3 - 4y^2 + 8x^3 + 9y^3$
16. $7y^4 - 4z^2 + 8y^4 + 7y^2$
17. $-4A^2 - 8 + 3A^3 + 7$
18. $-5u - 7v^3 + 6u + u^3$
19. $9x^2 - 12 + 3x^2 - 6 + 10$
20. $4n - 8m + 5m - 8n + 5m$
21. $-8k^3 + 7 + 4k - 6 - 3k$
22. $-4y + 6xy + 7y - 8xy - 5x$
23. $4x^2y - 9xy^3 + 6xy^3 - 9x^2y^3 - xy^3$
24. $-8A^2B^3 + 4A^3B - 9A^2B^3 - 3A^3B + 11AB^3$

3.4 ORDER OF OPERATIONS

Many kinds of algebraic expressions involve addition, multiplication, and grouping of monomials. We have seen the most elementary of these expressions in the previous sections of Chapters 2 and 3.

In this section we will consider more complicated expressions, such as $-(3a) - (-14a)$, $(-12n^2) - 6n(-3n)$, $3t^2(-2t^3) - (-4t^4)(-5t^5)$, and $6x^2 - (8x - x)$. In all these expressions, however, each sum inside parentheses involves like terms, so that when we simplify them we will compute products containing monomial factors. To simplify these expressions we must follow the Order of Operations.

ORDER OF OPERATIONS
(Multiplication and Addition of Monomials)

1. We first simplify each sum of like terms *inside parentheses* to a monomial *inside parentheses.*

$$2(-x^3 - 3x^3) + 7x^2 - (-7x^3 + 6x^3) \qquad (5x^2 - 8x^2)(-4x^3) - (-2x^5 - 9x^5)$$

$$2(-4x^3) + 7x^2 - (-x^3) \qquad (-3x^2)(-4x^3) - (-11x^5)$$

2. Then we identify terms (each sign outside parentheses starts a new term). We simplify each term to a monomial with its sign expressed by multiplying and/or removing parentheses.

$$2(-4x^3) + 7x^2 - (-x^3) \quad \text{and} \quad (-3x^2)(-4x^3) - (-11x^5)$$

$$-8x^3 + 7x^2 + x^3 \qquad +12x^5 + 11x^5$$

3. Finally we simplify to a polynomial by combining like terms.

$$-8x^3 + 7x^2 + x^3 \quad \text{and} \quad +12x^5 + 11x^5$$

$$-7x^3 + 7x^2 \qquad 23x^5$$

14. Compute
a. $2x(-3x + x)$
b. $-6(2x^2 - 7x^2)$

15. Compute
a. $3x - (-y^2)$
b. $5 + (x)$

WARNING 1. When simplifying inside parentheses, do not drop the parentheses. Removing parentheses from around a signed quantity may change an indicated multiplication to addition because each sign outside parentheses starts a new term.

$$4x^2(-9x^2 + 2x^2) = 4x^2(-7x^2), \quad \textbf{not} \quad 4x^2 - 7x^2$$

2. When removing parentheses to simplify terms, do not omit plus signs. Signs are required before each monomial (except the first) to indicate addition.

$$5x^3 - (-y) = 5x^3 + y, \quad \textbf{not} \quad 5x^3y$$

EXAMPLES

Simplify the following expressions to polynomials, following the Order of Operations.

1. $-(-3a) + (-12a)$
2. $4x^2 - (-3y^4) + (-12x^2) + 9y^4$
3. $(+2x)(+3y) - 10(xy^3)$
4. $3t + 2t(-4t) - (-7t^2)$
5. $(-5x^3 + 2x^3)(3x^2 - 9x^2)$
6. $(3x^3)(4y) - x(5x^2y - 2x^2y)$

Solutions

1. $-(-3a) + (-12a)$

$$+3a - 12a$$

$$-9a$$

2. $4x^2 - (-3y^4) + (-12x^2) + 9y^4$

$$4x^2 + 3y^4 - 12x^2 + 9y^4$$

$$-8x^2 + 12y^4$$

3. $(+2x)(+3y) + 10(xy^3)$

$$6xy + 10xy^3$$

4. $3t + 2t(-4t) - (-7t^2)$

$$3t - 8t^2 + 7t^2$$

$$3t - t^2$$

5. $(-5x^3 + 2x^3)(3x^2 - 9x^2)$

$(-3x^3)(-6x^2)$

$18x^5$

6. $(3x^3)(4y) - x(5x^2y - 2x^2y)$

$(3x^3)(4y) - x(3x^2y)$

$12x^3y - 3x^3y$

$9x^3y$

OBSERVATIONS AND REFLECTIONS

With monomials, the operations of addition and multiplication are indicated by the placement of parentheses. The placement of signs inside or outside parentheses is used to distinguish sums from products. Signs alone cannot be used to distinguish multiplication from addition because all monomials have signs. For example,

$$3x^2 - (7xy) \quad \text{and} \quad 3x^2(-7xy)$$

look similar, although the first expression is a sum and the second is a product. Remember that in a *sum* every term (except possibly the first) begins with a sign outside parentheses and every sign outside parentheses begins a term of the sum. In a *product*, on the other hand, no signs appear outside parentheses, except possibly before the first factor.

16. Compute
a. $x^2 - (4x^2)$
b. $x^2(-4x^2)$
c. $(-xy^2) - 3xy^2$
d. $(-xy^2)(-3xy^2)$
e. $3 - 4(-x)$
f. $3 - (4 - x)$

EXERCISES

Simplify each of the following expressions to a polynomial.

1. $-(+15x) - (8x)$
2. $(-15A) + (+6A)$
3. $(5n) - (4n)$
4. $(-9B) - (-19B)$
5. $-(-9x) - (16x)$
6. $-(18E) - (33E)$
7. $3x^4 + (-4x^4) - 6x^4$
8. $(-6y^3) - 4y^3 + (-y^3)$
9. $(-12n^6) - 6n^6 + (-3n^6)$
10. $3x^2 + (5x^2) - (-2x^2) - 4x^2$
11. $(k) - 10k - (-7k) + (-2k)$
12. $(-4xy) + 6xy + (-7xy) - 9xy$
13. $(-3x^4) + 7x^4 + (-6x^4) - 8x^4$
14. $-(2x^2y) + 4x^2y - 6x^2y + (5x^2y)$
15. $(5x^2) - 6 + (-7x^2) - 4x^2 + (-8)$
16. $(4x) - 5y + (-3x) - 4x + (-6y)$
17. $8a - (7b) + (+3b) - 7a - (-6b)$
18. $5p + (-8p^3) + (+2p^3) - 6p - (-4p^3)$
19. $2x + (-3x^2) - 4x + (+5x^2) + (-6x^2)$
20. $-(-5p) + (-6q) - (+6p) - 9p - (-3q)$
21. $(-3a) - (7a^2) + (8a) - 4a^2 - 5a$
22. $6y - (8y^2) + 6y^2 + (-3y) + 7y$
23. $(4x^2)(-2y^3) - (-4x^2y^3)$
24. $5x^3(-4x^2y) - 10x^5y$
25. $(-3x^7)(-2x^2) - (12x^9)$
26. $-4w^6(-wx^2) + (-8w^7x^2)$
27. $(12x^6y^2) - 3xy^2(-2x^6)$
28. $(+7x^4y^2) - 5x^4(-2y^2)$

29. $-9x^4 - 3x(-4x^3) - 7x^3$

30. $8w^6 - 7w^2(-3w^4) + 6w^2$

31. $9x^4 + (-5x^2)(+x^2) + 5x^2$

32. $-(-9x^7) - (3x^2)(-4x^5) + (9x^7)$

33. $(9x^2)(4x) - (7x)(3x^5)$

34. $(6y^2)(2y^4) + (3y)(5y^8)$

35. $(5t^6)(3t^9) - (6t^7)(7t^8)$

36. $(-3w^6)(4w) + (8w^2)(2w^4)$

37. $(5x^4)(-2x) - (7x^6)(-4x)$

38. $(4A^6)(-3A^3) - (3A^5)(-6A^4)$

39. $8x - 3(5x - x)$

40. $5y^2 - 4(-6y^2 + 3y^2)$

41. $-(-6x^3)(8x^4 - 3x^4)$

42. $-(-3y^6)(5y^3 - 7y^3)$

43. $(9x^2y - 6x^2y)(7y^3 - 9y^3)$

44. $(3xy^2 - 7xy^2)(4y^2 - y^2)$

45. $-(-2x^6 - 9x^6) + (5y^6 - 8y^6)$

46. $-(-7z^3 + 2z^3) - (-6z^3 - 4z^3)$

47. $(4x^3 - 7x^3) - (5y^3 - 8y^3)$

48. $-(3x^5 - 11x^5) + (-5x^3 + x^3)$

49. $x^2(-8x^2 + 2x^2) - 4x^4$

50. $-4x^6(5x - 11x) + 3x^7$

51. $-(-8x^3) - 3x(-4x^3 + 6x^3)$

52. $-(+3y^4) + 2y(-10y^4 + 6y^4)$

53. $-y^6(5x^4 - 9x^4) + 3x^2y^4(-6x^2y^2)$

54. $-x^2z(-x^4z^3 - 4x^4z^3) - z^4(+4x^6)$

Express each of the following word phrases algebraically and then simplify it. (A review of translating phrases is provided in Section 1.6.)

1. The product of 5 and 8m is added to 15m

2. 50cm^3 plus the product of 4cm and 10cm^2

3. 18km added to the sum of 12m and 9m

4. 20mm decreased by 8 times 2mm

5. The product of 12 and 18g minus 6g

6. 3cm times 7cm is subtracted from 8m^2

7. 17m diminished by 3cm more than 7cm

8. The sum of 13mL and 9mL is subtracted from 19L

9. 90cm^3 diminished by the product of 6cm and 7cm^2

10. The sum of $-20a$ and $15a$ is multiplied by $-9b^2$

11. $10w^2$ subtracted from the sum of $15w$ and $-8w$

12. $56a^2$ added to twice $-27a^2$

13. The produce of $3x^2y$ and the sum of $9xy$ and $-18xy$

14. $16c^2$ more than the product of $8c$ and $-9c^2$

15. The product of $-7x$ and $-3y$ is reduced by $6xy$

16. The sum of $-55a^2$ and $25a^2$ is increased by $19a$

17. $84x^3$ less 0 times $27x^3$

18. The product of $15w^2$ and $-2w$ is subtracted from $80w$

19. $88x^2y^3$ reduced by twice $44x^2y^3$

20. $60k^3$ added to $-56k^3$ is decreased by $14k$

21. The product of $-4x^2$ and $-16x$ is diminished by $90x^2$

22. Twice the sum of $73x^2$ and $-18x^2$

23. $86a^2b^3$ less than the product of $6ab^2$ and $9ab$

24. The product of $25x^5$ and $-3x^2$ plus $7x^2$

3.5 EVALUATING

In this section we will see how to evaluate polynomials for integer values of the literals. Look at the following table to see some examples of this procedure.

evaluated at	-3	-2	-1	0	+1	+2	+3
$1 + 2A$							
$1 + A$		$1 + (-2) = -1$	$1 + (-1) = 0$	$1 + (0) = 1$	$1 + (+1) = 2$	$1 + (+2) = 3$	
$1 - A$		$1 - (-2) = 3$	$1 - (-1) = 2$	$1 - (0) = 1$	$1 - (+1) = 0$	$1 - (+2) = -1$	
$-1 + A$		$-1 + (-2) = -3$	$-1 + (-1) = -2$	$-1 + (0) = -1$	$-1 + (+1) = 0$	$-1 + (+2) = 1$	
$-1 - A$		$-1 - (-2) = 1$	$-1 - (-1) = 0$	$-1 - (0) = -1$	$-1 - (+1) = -2$	$-1 - (+2) = -3$	
$-1 - 2A$							

17. Complete the table.

Remember that a literal expression is evaluated by replacing each literal by a number *inside parentheses* and then simplifying the resulting numerical expression. When evaluating a polynomial, we start with a completely simplified expression. When we replace the literals by numbers, however, we get an expression that may be far from simplified. To complete the evaluation, we must simplify the numerical expression completely according to the Order of Operations (which we studied in the previous section).

Note that when a literal in an expression is preceded by a sign, the sign must not be confused with the sign of the integer replacing the literal. For example,

$3 - x$, when $x = -2$ becomes $3 - (-2)$, **not** $3 - 2$.

Look at the following examples of *evaluating a polynomial.*

TYPICAL PROBLEMS

Evaluate: 1. $4x^2 + 9$, when $x = -3$

We first substitute the given number using parentheses:

$4(-3)^2 + 9$

Then we simplify the resulting expression according to the Order of Operations:

$4(9) + 9$

↓

$36 + 9$

↓

45

Hence, evaluating $4x^2 + 9$, when $x = -3$ yields 45.

To evaluate a polynomial we extend the method in Section 2.4. Always place parentheses around each substituted number. The Order of Operations is given in Section 1.5 and in the previous section.

Evaluate: 2. $3xy - 5y^2$, when $x = 0, y = 4$

We first substitute the given numbers using parentheses:

$3(0)(4) - 5(4)^2$

Then we simplify the resulting expression according to the Order of Operations:

↓

$0 - 80$

↓

-80

Hence, evaluating $3xy - 5y^2$, when $x = 0, y = 4$ yields -80.

EXAMPLES

Evaluate the following polynomials.

1. $5 - 4w$, when $w = 6$
2. $3A - 7B - 16$, when $A = -2, B = -5$
3. $8k^3 - 6k^2$, when $k = -2$
4. $5r^2 + 7s^2$, when $r = 7, s = -5$
5. $9x - 4x^2y$, when $x = -3, y = -6$
6. $3x^2 - 5xy - 9y^2$, when $x = 8, y = 0$

Solutions

1. $5 - 4(6)$
   $\downarrow$
   $5 - 24$
   $\downarrow$
   -19

2. $3(-2) - 7(-5) - 16$
   $\downarrow$
   $-6 + 35 - 16$
   $\downarrow$
   13

3. $8(-2)^3 - 6(-2)^2$
   $\downarrow$
   $8(-8) - 6(4)$
   $\downarrow$
   $-64 - 24$
   $\downarrow$
   -88

4. $5(7)^2 + 7(-5)^2$
   $\downarrow$
   $5(49) + 7(25)$
   $\downarrow$
   $245 + 175$
   $\downarrow$
   420

5. $9(-3) - 4(-3)^2(-6)$
   $\downarrow$
   $9(-3) - 4(9)(-6)$
   $\downarrow$
   $-27 + 216$
   $\downarrow$
   189

6. $3(8)^2 - 5(8)(0) - 9(0)^2$
   $\downarrow$
   $3(64) - 0 - 0$
   $\downarrow$
   192

OBSERVATIONS AND REFLECTIONS

Remember that in a numerical expression each sign outside parentheses begins a new term. For example, $4x - 5$ is the sum of $4x$ and -5, and $-x^2 + 2y^2$ is the sum of $-x^2$ and $+2y^2$. When we evaluate an expression, we replace the literals by signed numbers inside parentheses so that no new signs are introduced outside parentheses. Introducing new signs outside parentheses would introduce new terms by changing indicated products to sums. For example,

$4x - 5$, when $x = -2$	becomes	$4(-2) - 5$,	**not** $4 - 2 - 5$
$-x^2 + 2y^2$, when $x = 4, y = -3$	becomes	$-(4)^2 + 2(-3)^2$,	**not** $-4^2 + 2 - 3^2$

EXERCISES

Evaluate each of the following expressions, simplifying to an integer.

1. $4 - x$, when $x = 6$
2. $5 - y$, when $y = 8$
3. $3x + 5$, when $x = -2$
4. $8t - 6$, when $t = -3$

5. $6 - 3w$, when $w = -4$
6. $4 - 7x$, when $x = -2$
7. $3x + y$, when $x = 8$, $y = -9$
8. $4s + t$, when $s = 7$, $t = -12$
9. $-2m + 8n$, when $m = -4$, $n = 0$
10. $-6x + 4y$, when $x = -5$, $y = 0$
11. $4A - 7B - 6$, when $A = -4$, $B = -3$
12. $8u - 5v - 15$, when $u = -8$, $v = -11$
13. $4t^2 + 9$, when $t = 3$
14. $3r^2 + 7$, when $r = 4$
15. $5 - 2y^2$, when $y = -1$
16. $4 - 5z^2$, when $z = -2$
17. $3x^2 - 8x$, when $x = -3$
18. $6A^2 - 11A$, when $A = -1$
19. $3a^3 - 7$, when $a = -3$
20. $4x^3 - 10$, when $x = -3$
21. $x^2 - 4x + 9$, when $x = 5$
22. $x^2 - 7x + 12$, when $x = 6$
23. $3r^3 - 5r - 11$, when $r = -3$
24. $2t^3 - 4t - 17$, when $t = -4$
25. $7a^2 + 11b$, when $a = 2$, $b = -3$
26. $4s^2 + 9t$, when $s = 5$, $t = -6$
27. $-6x^2 + 11y$, when $x = -3$, $y = 0$
28. $-5A^2 + 8B$, when $A = -4$, $B = 0$
29. $3u^2 - v^2$, when $u = -7$, $v = -9$
30. $4m^2 - n^2$, when $m = -6$, $n = -11$
31. $4y^2 + 6z^2$, when $y = 0$, $z = 10$
32. $7A^2 + 6B^2$, when $A = 0$, $B = 9$
33. $3m^3 + 8n$, when $m = 3$, $n = -9$
34. $2x^3 + 11y$, when $x = 4$, $y = -11$
35. $-x^3 + 2y^2$, when $x = -1$, $y = 4$
36. $-r^3 + 3s^2$, when $r = -2$, $s = 3$
37. $3m^2 + 8mn$, when $m = -3$, $n = 2$
38. $5x^2 + 7xy$, when $x = -5$, $y = 4$
39. $6u^2v - 10uv^2$, when $u = -3$, $v = -2$
40. $8x^2y - 9xy^2$, when $x = -2$, $y = -1$
41. $x^2 - 8xy + 18y^2$, when $x = 5$, $y = -1$
42. $x^2 - 6xy + 21y^2$, when $x = 6$, $y = -1$

ANSWERS TO MARGIN EXERCISES

1. a. $-3x$ b. $3x^2$ c. $-9xy$ d. $12yz^2$

2. a and d have like terms.

3. a. $2x^5$ b. $-3ab^2$ c. $-5t^6$

4. a. 0 b. 0

5. a. 0 b. 0

7. a. $-4xy$ b. $-4xy$ c. $-11t^4$ d. $-11t^4$

8. a. $9ab^2$ b. $-5rs^2t$ c. $-2xyz$

9. a. $(-3x^2) - 7x^2 = -10x^2$ b. $-4x^2 + (-6x^2) = -10x^2$

10. a. $2y^4$ b. y^8 c. $2xy$ d. x^2y^2 e. $-2b^5$ f. b^{10} g. $-xy$ h. $-6x^2y^2$ i. $8st^3$ j. $7s^2t^6$ k. $-14w^2x^6$ l. $45w^4x^{12}$

12. b and d are polynomials, whereas a and c contain like terms.

13. a and e; b and f; c and d

14. a. $-4x^2$ b. $30x^2$

15. a. $3x + y^2$ b. $5 + x$

16. a. $-3x^2$ b. $-4x^4$ c. $-4xy^2$ d. $3x^2y^4$ e. $3 + 4x$ f. $-1 + x$

CHAPTER REVIEW

SECTION 3.1 Simplify to a monomial.

1. $-12y - 4y$
2. $2x - 9x$
3. $-16z + z$
4. $32wx + 56wx$
5. $xz - 29xz$
6. $11st - 2st$
7. $-30r^2 - 3r^2$
8. $-72A^5 + 6A^5$
9. $-4x^7 + 100x^7$
10. $-54kt^2 + 16kt^2$
11. $-A^3B^5 - A^3B^5$
12. $-8x^3z^6 + 22x^3z^6$

SECTION 3.2 Simplify to a monomial.

13. $6A + A - 9A$
14. $-3x + 4x - 7x$
15. $-4s^4 + 13s^4 - s^4$
16. $-12y^9 + 10y^9 - 4y^9$
17. $4x + 5x - 9x - x$
18. $-3A + 8A - 8A + 4A$
19. $8x^3 + 10x^3 - 18x^3 - 9x^3$
20. $3t^4 + 4t^4 - 9t^4 + 8t^4 - t^4$
21. $r^2 + r^2 - 3r^2 - 11r^2 + 7r^2$
22. $5xy - 17xy + 11xy - xy$
23. $-6y^5z^2 + 15y^5z^2 + 8y^5z^2 - 18y^5z^2 + y^5z^2$
24. $-7m^3n^8 - 11m^3n^8 + 15m^3n^8 - m^3n^8 + 23m^3n^8$

SECTION 3.3 Simplify to a polynomial.

25. $6x - 8 + 14x$
26. $2A - 6A^2 + 3A^2$
27. $-8m^3 + 6n^2 + 3m^3$
28. $3x^2 - 2xy^2 + 4x^2 + 5y^2$
29. $-2A^2 - 14A^4 + A^2 + 3A^4$
30. $6x^2y - 6y + 2x^2y - 3y + 5y$
31. $-10A - 14B^2 + 3A + 2A^2$
32. $8n^3 + 10m^4 - 9n^3 + 2m^3$
33. $-6x^3 - 14 + 7x^2 + 8$
34. $4s^2 + 2 - 9s - 8 + 2s$
35. $-4mn + 8n + 16mn - 2n - 6mn$
36. $8x^2y - 18xy^3 + 3xy^3 - 6x^2y^3 - 2xy^3$

SECTION 3.4 Simplify to a polynomial.

37. $(-3x) + (7x)$
38. $-(+12A) - (6A)$
39. $-(4x) + (-4x)$
40. $2x^2 + (-3x^2) - 4x^2$
41. $2x + (3x) - (-12x) - 4x$
42. $(6w) - 10w - (-3w) + (-5w)$
43. $-(4mn^2) + 3mn^2 - 7mn^2 + (+mn^2)$
44. $+(-3r) - 7t + (7t) - 4r + (+11t)$
45. $+(2y^2) - 6y + (4y) - 3y^2 + (+8y^2)$
46. $(-8y^3) - (5y^2) + 3y^2 - (-4y^3) + 7y^2$
47. $2x - (-8x^2) + 5x + (-3x) - 6x^2$
48. $(-3q^4r^6)(2qr^2) + (-6q^8r^5)$
49. $-3x^9(-4x^2) - 12x^{11}$
50. $(+5uvw) - 4v(+uw)$

51. $-4x^5 + 5x^2(-3x^5) + 2x^7$

52. $-3r^3 - 5r^3(+2r^3) - 9r^3$

53. $(3t^6)(-2t^3) + (-4t^2)(-5t^5)$

54. $(5b^4)(-6b) + (2b^3)(-6b^2)$

55. $(-3x^8)(-5x) - (8y^6)(4y^3)$

56. $4x^2 - 7x(3x - 6x)$

57. $-(-9y^2)(4y^3 - 5y^3)$

58. $(9y^3z - 2y^3z)(3z^2 - 8z^2)$

59. $-(-4x^5 - 3x^5) + (-7x^5 + x^5)$

60. $-(2x^2y - 5x^2y) - (6xy^2 - 11xy^2)$

61. $3x^3(-5x^2 + 3x^2) - 5x^5$

62. $-(-4x^2y^3) - 2y^2(-9xy + 12xy)$

63. $-xz^2(3x^6z - 4x^6z) + x^5(-2x^2z^3)$

SECTION 3.5

Evaluate, simplifying to an integer.

64. $6 - x$, when $x = 9$

65. $9y - 2$, when $y = -5$

66. $5 - 8n$, when $n = -3$

67. $6u + v$, when $x = 10$, $y = -8$

68. $-11A + 3B$, when $A = -6$, $B = 0$

69. $5x - 9y - 1$, when $x = -7$, $y = -3$

70. $2k^2 + 8$, when $k = 4$

71. $9 - 3y^2$, when $y = -2$

72. $3A^2 - 16A$, when $A = -6$

73. $5u^3 - 15$, when $u = -2$

74. $x^2 - 3x + 14$, when $x = 7$

75. $2r^3 - 7r^2 - 13$, when $r = -3$

76. $5x^2 + 15y$, when $x = 5$, $y = -8$

77. $-7A^2 + 12B$, when $A = 3$, $B = 0$

78. $3u^2 - v^2$, when $u = -6$, $v = -11$

79. $6m^2 + 9n^2$, when $m = 0$, $n = 10$

80. $4x^3 + 15y$, when $x = 5$, $y = -4$

81. $-s^3 + 2t^2$, when $s = -1$, $t = 6$

82. $3u^2 + 7uv$, when $u = -7$, $v = 2$

83. $7m^2n - 8mn^2$, when $m = -2$, $n = -3$

84. $x^2 - 5xy + 14y^2$, when $x = 8$, $y = -1$

SUMMARY OF PHRASES

Express each of the following word phrases algebraically and then simplify it.

1. The sum of 12km and 18km
2. 2mm subtracted from 19mm
3. 7m reduced by 4m
4. 14g less than 25g
5. 23L is subtracted from 35L
6. 11mL more than 42mL
7. The product of 4g and 21 is added to 14g
8. 39mm decreased by 5 times 7mm
9. 6cm times 9cm is subtracted from 90cm^2
10. The sum of 24mg and 17g is subtracted from 31g
11. The sum of $-18a$ and $7a$ is multiplied by $-8c^3$
12. $79a^2b$ added to twice $-42a^2b$
13. The product of $6ab^3$ and the sum of $5ab$ and $-17ab$
14. $13a^4$ more than the product of $-9a^2$ and $2a$

15. The product of $-3w$ and $-6y$ is reduced by $14wy$

16. $9xy$ times $-7x^3y^2$ is decreased by $5x^3y^2$

17. The product of $32a^3$ and $-ab^2$ is subtracted from $45a^4b^2$

18. The product of $-9x^5$ and $-11x^3$ is diminished by $78x^{15}$

CUMULATIVE TEST

The following problems test your understanding of this chapter and of the previous chapter. Before taking this test, thoroughly review Sections 2.1, 2.3, 3.1, 3.2, and 3.3.

Once you have finished the test, compare your answers with the answers provided at the back of the book. Note the section number of each problem missed, and thoroughly review those sections again.

Simplify.

1. $5x(-4y)$
2. $-15xy - 19xy$
3. $(5x^2)(-6x^6)(3x^3)$
4. $5a + 11a - 7a$
5. $2k - 9k^2 + 5k^2$
6. $7(3x^4)(-2x^8y^3)y^3$
7. $-27r^3 + 6r^3$
8. $(-3xz)(-sy)(-8t)$
9. $-5t^3 + t^3 - 8t^3$
10. $15x^3 - 4y + 3x^3 + 9y$
11. $-3m(-2n^2)m^4(-12n^3)$
12. $n^3 + n^3 - 4n^3 - 13n^3 + 7n^3$
13. $-54st^4 + st^4$
14. $7y^2z(-3z^4)(-y^2)z^4$
15. $5AB + 3AB - 12AB - 8AB$
16. $(-st)(-6r)$
17. $-y^9 + 7y^9 - 6y^9$
18. $-8m^2 + 7m^2 + m^3$
19. $-4y^3(-4x^2)x(-3xy^3)$
20. $x^4 + x^4$
21. $(6r^2s)(-4r^2s)(-9r^3s^4)$
22. $7AB + 3AC - 11AB - 18AC + 4AB$
23. $-12b^8 - 13b^8 + 19b^8 - b^8 + 16b^8$
24. $(-6b)(-2ac)(12d)$
25. $-6x + 3x - 9x$
26. $-4A + 8B + 7A - 11B$
27. $-2x^8(-2y^4)x^8(-2y^6)$
28. $-8yz^6 + 16yz^6 + 7yz^6 - 25yz^6 + yz^6$
29. $6u^5v(-8uv^2)u^4v^3(-3)$
30. $-4mn + 9 + 18mn - 22 - 5mn$

In the last two chapters we studied monomials, the building blocks of all algebraic expressions. In Chapter 3 we saw that sums of monomials frequently simplify to multinomials rather than monomials. Thus, in order to work with algebraic expressions we must learn how to add and multiply polynomials. In this chapter we will begin by studying the product of a monomial and a multinomial. The operation of multiplying a monomial and a multinomial is called *distribution.* Distribution is the process through which multiplication interacts with addition. It is the fundamental bridge between these two operations.

Distribution is indicated in any of the following ways:

1. $-3x(4x - 5)$ — with the monomial first and the multinomial in parentheses.
2. $(-3x)\ (4x - 5)$ — with both factors in parentheses.
3. $(4x - 5)\ (-3x)$ — with both factors in parentheses and with the monomial second. This form is rare and not preferred.
4. $(4x - 5)3x$ — with the monomial written second and with only the multinomial in parentheses. The monomial must be positive, and its plus sign must be omitted. This form cannot be used if the monomial is negative. It is rare and not preferred. (Remember that a monomial is said to be positive or negative depending on the sign of its coefficient.)

Note that no sign appears *between* the two factors in these forms. The multinomial factor of a product *must* be written inside parentheses. The monomial is almost always written first and may or may not be in parentheses.

4.1 REMOVING PARENTHESES PRECEDED BY A SIGN

To understand how to remove parentheses from around polynomials, we should first examine what we learned in Chapter 1 about removing parentheses. For example, we saw in Section 1.3 that

$$+(+2) = +2 \qquad -(-2) = +2$$
$$+(-2) = -2 \qquad -(+2) = -2$$

We must now generalize to more than one term inside parentheses. To see how to do this, look at the pattern in the following columns.

$$+(+2+1) = +2+1 \qquad -(-2-1) = +2+1$$
$$+(+3) = +3 \qquad -(-3) = +3$$

$$+(+2-1) = +2-1 \qquad -(-2+1) = +2-1$$
$$+(+1) = +1 \qquad -(-1) = +1$$

1. Form a similar diagram for
   a. $+(+1-3)$
   b. $-(+1+3)$

$$+(-2+1) = -2+1 \qquad -(+2-1) = -2+1$$
$$+(-1) = -1 \qquad -(+1) = -1$$

$$+(-2-1) = -2-1 \qquad -(+2+1) = -2-1$$
$$+(-3) = -3 \qquad -(+3) = -3$$

In each example we have combined terms in going from the first row to the second, and in each row we have removed parentheses in going from left to right.

From these examples can you see the rule for removing parentheses preceded by a plus sign? Can you guess the rule for removing parentheses preceded by a minus sign?

Look at the following examples of *removing parentheses preceded by a sign.*

TYPICAL PROBLEMS

1. $+(-2+x)$ ↓ $-2+x$

Look at the sign before the parentheses. Because the parentheses are preceded by a plus sign, we simply drop the parentheses and the preceding plus sign:

Hence, $+(-2+x) = -2+x$.

2. $-(-3w^3 + w^2)$ ↓ $+3w^3 - w^2$

Every polynomial inside parentheses preceded by a sign can be simplified to a polynomial without parentheses.

Look at the sign before the parentheses. Because the parentheses are preceded by a minus sign, we change the sign of each term inside the parentheses as we drop them and the preceding minus sign:

Hence, $-(-3w^3 + w^2) = 3w^3 - w^2$.

EXAMPLES

Simplify the following expressions and remove parentheses.

1. $+(-5+3x)$ 2. $-(7x-2)$ 3. $-(3x^2 - 4x - 1)$ 4. $-(-9x^2 - 5x + 4)$

Solutions

1. $-5+3x$ 2. $-7x+2$ 3. $-3x^2 + 4x + 1$ 4. $9x^2 + 5x - 4$

SPECIAL CASES

The rule for negating a polynomial is sometimes incorrectly paraphrased as "change the signs," rather than "change the sign of every term." In an expression such as

$(5x - 4y + 2z)$, the first term $5x$ does not have its plus sign expressed, so following the rule "change the signs" might lead to our forgetting to change the sign of the first term when negating.

$$-(5x - 4y + 2z) = -5x + 4y - 2z, \quad \textbf{not} \quad 5x + 4y - 2z$$

2. Simplify
   a. $-(3x + 2)$
   b. $-(7 - 9x)$
   c. $-(4x^2 - 3x + 1)$

OBSERVATIONS AND REFLECTIONS

1. Placing an expression inside parentheses preceded by a minus sign forms what is called the *negative* of the expression.

   Remember that the negative of a monomial can be formed either by changing the sign of the monomial, or by placing the monomial within parentheses preceded by a minus sign. In general, the *negative* of a *polynomial* can be formed either by changing the sign of *each* term of the polynomial, or by placing the polynomial within parentheses preceded by a minus sign. A polynomial within parentheses preceded by negative one is also the negative of the polynomial. For example,

   $$-(3x - 4), \quad -1(3x - 4), \quad \text{and} \quad -3x + 4$$

   are all considered to be the negative of $3x - 4$ and are all equal. The forms that include parentheses can be changed to the third form by the process of distribution.

3. Form the negative of each of the following expressions in three ways
   a. $5x + 9$
   b. $7t^2 - 8t$
   c. $-x^3 + 6y^3$

2. Because each term of a polynomial has its own sign, a polynomial cannot be considered to be either positive or negative. However, two polynomials are *negatives* of each other if their sum is zero. For example, $3x - 2y$ and $-3x + 2y$ are negatives of each other because $3x - 2y - 3x + 2y = 0$.

4. Which of the following polynomials are negatives of each other?
   a. $3x + y$ c. $-3x + y$
   b. $3x - y$ d. $-3x - y$

3. Two polynomials are *negatives* of each other if they contain the same terms with opposite signs, regardless of the order in which the terms are written. Rearranging the terms of a polynomial yields a polynomial with the same value, and any polynomial equal to the *negative* of a polynomial is also considered to be the negative of that polynomial. Hence, $2y - 3x$ is the negative of $3x - 2y$. Note that the order of the terms in these polynomials is reversed.

5. Pair each polynomial with its negative.
   a. $x + y - z$
   b. $x - y - z$
   c. $y - x - z$
   d. $y + z - x$
   e. $z - x - y$
   f. $x + z - y$

4. The process of changing a polynomial to its negative is called *negation.*

NEGATION OF A BINOMIAL

The negative of $+a + b$ is $-(+a + b)$ or $-a - b$

The negative of $+a - b$ is $-(+a - b)$ or $-a + b$

The negative of $-a + b$ is $-(-a + b)$ or $+a - b$

The negative of $-a - b$ is $-(-a - b)$ or $+a + b$

$-(a + b) = -a - b \qquad -(a - b) = -a + b$

$-(-a + b) = a - b \qquad -(-a - b) = a + b$

(where a and b represent signless monomials)

EXERCISES

Simplify each of the following expressions to a polynomial without parentheses.

1. $+(2i - 5)$	2. $+(-7x + 2)$	3. $+(-5r + s)$
4. $+(+27 - 8A^2)$	5. $(-18 - 9x)$	6. $(-3w + w^2)$
7. $+(-3x^2 + 5x - 7)$	8. $+(8a + 7b - 3c)$	9. $+(7 - 8s - 9t)$
10. $-(-x - 13y)$	11. $-(6A - 2B)$	12. $-(-2 + 4w^3)$
13. $-(-7 - 7j^2)$	14. $-(+6 - 14n)$	15. $-(-x^2 + 4x - 7)$
16. $-(-7y^2 + 4y + 9)$	17. $-(3 - 7w - 2w^2)$	18. $-(9 - x - 5x^2)$

4.2 DISTRIBUTING AN INTEGER

To see how to multiply a sum by an integer, look at the pattern in the following computations. In the first row of each computation, the expression on the left is an indicated product, and the expression on the right is its distributed form:

$$3(4 + 1) = 3(4) + 3(1) = 12 + 3 \qquad 3(4 - 1) = 3(4) + 3(-1) = 12 - 3$$
$$3(5) \qquad = \qquad 15 \qquad\qquad 3(3) \qquad = \qquad 9$$

6. Form a similar diagram for
   a. $2(3 + 1)$
   b. $2(3 - 2)$

$$3(4 + 2) = 3(4) + 3(2) = 12 + 6 \qquad 3(4 - 2) = 3(4) + 3(-2) = 12 - 6$$
$$3(6) \qquad = \qquad 18 \qquad\qquad 3(2) \qquad = \qquad 6$$

$$3(4 + 3) = 3(4) + 3(3) = 12 + 9 \qquad 3(4 - 3) = 3(4) + 3(-3) = 12 - 9$$
$$3(7) \qquad = \qquad 21 \qquad\qquad 3(1) \qquad = \qquad 3$$

In each example we have combined terms in going from the first row to the second, and in each row we have multiplied in going from left to right.

Of course, in these examples we can either distribute or simplify the sums first. In products containing a multinomial factor (a sum of unlike terms) we must distribute to compute the product. Can you guess how to multiply a multinomial by an integer?

In general, we will call the product of an integer and any polynomial an *integer multiple* of the polynomial. (This includes one times the polynomial.)

Look at the following example of *distributing an integer.*

TYPICAL PROBLEM

$$-3(2x - 3y)$$
$$\downarrow$$
$$-6x + 9y$$

To multiply each term of the multinomial by the integer we use the method in Section 2.3.

We multiply each term of the multinomial by the integer and drop parentheses:

Hence, $-3(2x - 3y) = -6x + 9y$.

EXAMPLES

Every integer multiple of a polynomial can be converted to a polynomial.

Distribute the integer over the sum in parentheses of the following expressions.

1. $7(4 - 9A)$
2. $-4(3x + 7)$
3. $(+11)(-3y + 8y^2)$
4. $(-9)(-2 + 10k^3)$
5. $+2(8A - 16B + 1)$
6. $-6(-x^2 + 4xy - 9y^2)$
7. $(5i - 4)(+8)$
8. $(3 + n^2)(-15)$
9. $(8x^2 + 11x - 16)(2)$

Solutions

1. $28 - 63A$
2. $-12x - 28$
3. $-33y + 88y^2$
4. $18 - 90k^3$
5. $16A - 32B + 2$
6. $6x^2 - 24xy + 54y^2$
7. $40i - 32$
8. $-45 - 15n^2$
9. $16x^2 + 22x - 32$

We have seen several multiplication tables in the previous chapters. Here is a multiplication table showing some integer multiples of binomials. Verify the entries to check your understanding of distribution.

times	$(x + 1)$	$(x - 1)$	$(-x + 1)$	$(-x - 1)$
+4				
+3	$3x + 3$	$3x - 3$	$-3x + 3$	$-3x - 3$
+2	$2x + 2$	$2x - 2$	$-2x + 2$	$-2x - 2$
+1	$x + 1$	$x - 1$	$-x + 1$	$-x - 1$
0	0	0	0	0
−1	$-x - 1$	$-x + 1$	$x - 1$	$x + 1$
−2	$-2x - 2$	$-2x + 2$	$2x - 2$	$2x + 2$
−3	$-3x - 3$	$-3x + 3$	$3x - 3$	$3x + 3$
−4				

7. Complete the table.

Distributing an integer over a multinomial yields another multinomial. For example, distributing −3 over the sum $2x - 3y$ yields $-6x + 9y$. Instead of computing the product of −3 and $2x - 3y$ in a single step, we can first *indicate* this distribution as $-3(2x) - 3(-3y)$. This second form is called an *indicated distribution.* The *Distributive Law of Multiplication over Addition* states that an indicated product of a monomial over a sum and its indicated distribution have the same value. For example, the indicated product $-3(2x - 3y)$ equals the indicated distribution $-3(2x) - 3(-3y)$. (Of course, both equal $-6x + 9y$.)

8. Form the indicated distribution of
   a. $5(x + 4)$
   b. $-6(5y + 8)$
   c. $7(3x - 5)$
   d. $-8(7y - 2)$

DISTRIBUTIVE LAW OF MULTIPLICATION OVER ADDITION

a times (*b* plus *c*) = (*a* times *b*) plus (*a* times *c*)

$$a(b + c) = a \cdot b + a \cdot c$$

(where *a*, *b*, and *c* represent monomials)

OBSERVATIONS AND REFLECTIONS

9. Compute using distribution
 a. 7(49)
 b. 19(6)

The Distributive Law is used in ordinary arithmetic multiplication. The product of 8 and 35, for example, is actually computed as the sum of 8(30) and 8(5). Distribution can also be used to speed up the computation of products such as 8(29), by writing 29 as 30 − 1 and then distributing:

$$8(29) = 8(30 - 1) = 8(30) - 8(1) = 240 - 8 = 232$$

EXERCISES

Multiply to convert each of the following expressions to a polynomial.

1. $2(5 + A)$
2. $4(3k + 7)$
3. $9(-1 + 6x)$
4. $3(-8 + 4n)$
5. $-6(3x - 5)$
6. $-2(12a - 5b)$
7. $(+11)(-A - 2B)$
8. $(+20)(6 - 8x)$
9. $-8(-4 + 2A)$
10. $-9(-7G + H)$
11. $4(5x^2 - 6x)$
12. $12(-3x^4 - y^2)$
13. $(-4)(15x^3 - 1)$
14. $(-7)(-7x + 6x^2)$
15. $+9(-3x^5 + x^4)$
16. $+3(5m^3 - 2m^2)$
17. $(-2)(-40 - 4k^4)$
18. $(-5)(-15x^2 - 25y^2)$
19. $5(-4A - 11B + C)$
20. $10(-24 - 17x + 15x^2)$
21. $-4(-7 + 2a - 3b)$
22. $-7(6x^2 + 2x - 5)$
23. $(+8)(-4y^2 - 7y - 5)$
24. $(+16)(1 - 2w + 4x)$
25. $-3(-8 - 11y - 4y^2)$
26. $-10(-4 + 5k - 3k^2)$
27. $+3(2a - 7b + 10)$
28. $+5(10x^2 - 4y^2 - z^2)$
29. $(-6)(1 - 5x - x^2)$
30. $(-9)(8x^2 - 7xy - 3y^2)$
31. $(2x + 9)(-6)$
32. $(-3m + 4n)(-12)$
33. $(-x^3 + 1)(+18)$
34. $(-11r^2 - 1)(+4)$
35. $(-x^2 + 11x - 18)(-2)$
36. $(3x^2 + 7xy - y^2)(-3)$

Express each of the following word phrases algebraically and then simplify it. (A review of translating phrases is provided in Section 1.6.)

1. The product of 8 and 3cm more than 2m
2. 5 times the sum of 7km and 12m
3. 6 times 4g minus 15mg
4. The product of 7 and 12L increased by 2mL.
5. −2 times $18a^2$ subtracted from $35ab^2$
6. 20 multiplied by the sum of $-8x^2$ and $12xy$
7. The product of −4 and $9a^2$ diminished by $11b$
8. The sum of $-90x^2$ and $20w^2$ is multiplied by −3
9. −9 multiplied by $7ac$ subtracted from $7ab$
10. 15 times $8x$ decreased by $5y$
11. −1 multiplied by $47x$ added to $89y$
12. The product of −11 and $8a$ less than 11

4.3 DISTRIBUTING A MONOMIAL

Distribution with a monomial factor is the same process as distribution with an integer. Remember, however, that the literal factors in the monomial can involve the multiplication of powers (which we saw in Chapter 2).

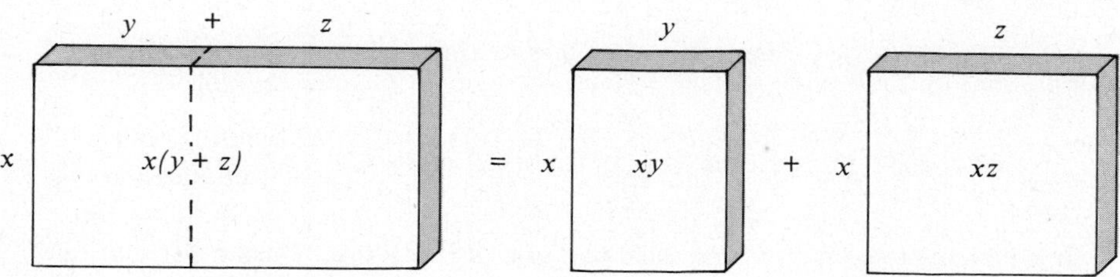

Look at the following example of *distributing a monomial.*

TYPICAL PROBLEM

We multiply each term of the multinomial by the monomial and drop parentheses:

$$7x^4(3x^2 - 9x^6)$$
$$\downarrow$$
$$21x^6 - 63x^{10}$$

Hence, $7x^4(3x^2 - 9x^6) = 21x^6 - 63x^{10}$

To distribute a monomial, we parallel the method in the previous section. To multiply each term of the multinomial by the monomial, we use the method in Section 2.3.

EXAMPLES

Distribute the monomial over the sum in parentheses in the following expressions.

Every product of a monomial and a polynomial can be converted to a polynomial.

1. $-x(8x - 3)$
2. $3xy(-6x + y)$
3. $u^3v^5(8uv^4 - 11u^2v^3)$
4. $4t(-2 + 5t + 6t^2)$
5. $(-5xy^2)(-8 - 7x + 3y)$
6. $-A^3B^7(-6A^5 + 3B^8 - A^4B^3)$

Solutions

1. $-8x^2 + 3x$
2. $-18x^2y + 3xy^2$
3. $8u^4v^9 - 11u^5v^8$
4. $-8t + 20t^2 + 24t^3$
5. $40xy^2 + 35x^2y^2 - 15xy^3$
6. $6A^8B^7 - 3A^3B^{15} + A^7B^{10}$

Here is a multiplication table showing some products of monomials and binomials. Verify the entries to check your understanding of distribution.

times	$(x^3 + 1)$	$(x^2 + y)$	$(x + y^2)$	$(1 + y^3)$
1	$x^3 + 1$	$x^2 + y$	$x + y^2$	$1 + y^3$
x	$x^4 + x$	$x^3 + xy$	$x^2 + xy^2$	$x + xy^3$
y	$x^3y + y$	$x^2y + y^2$	$xy + y^3$	$y + y^4$
xy	$x^4y + xy$	$x^3y + xy^2$	$x^2y + xy^3$	$xy + xy^4$

10. Complete the table.

OBSERVATIONS AND REFLECTIONS

1. The operations of distribution, multiplication, and addition are indicated by the placement of parentheses. Look at the following expressions to see how parentheses are used:

$-3x(4x - 5)$	means	$-3x$ times $(4x - 5)$	**Distribution**

whereas

$-3x(4x)(-5)$	means	$-3x$ times $4x$ times -5	**Multiplication**

and

$-3x(4x) - (5)$	means	$(-3x$ times $4x)$ minus (5)	**Addition and Multiplication**

11. Compute
   a. $-3x(4x - 5)$
   b. $-3x(4x)(-5)$
   c. $-3x(4x) - (5)$

★ 2. It is possible to convert a monomial to a sum of like terms by using distribution. Consider the monomial $8x^2y$. Because $3 + 5 = 8$, $8x^2y$ has the same value as $(3 + 5)x^2y$. Distributing the literal component yields the sum of like terms $3x^2y + 5x^2y$. Hence,

$$8x^2y = (3 + 5)x^2y = 3x^2y + 5x^2y$$

Similarly,

$$8x^2y = (2 - 7 + 13)x^2y = 2x^2y - 7x^2y + 13x^2y$$

This process can be viewed as the reverse of adding like terms (which we saw in Sections 3.1 and 3.2). Hence, when adding like terms, we could add an intermediate step:

$$3x^2y + 5x^2y = (3 + 5)x^2y = 8x^2y$$

and

$$2x^2y - 7x^2y + 13x^2y = (2 - 7 + 13)x^2y = 8x^2y$$

12. In a similar manner, compute in two steps
   a. $3x + 5x$
   b. $8x^2 - 2x^2$
   c. $-4xy + xy$

EXERCISES

Multiply to convert each of the following expressions to a polynomial.

1. $x^2(3 - 5x)$
2. $x^3(7x - 5)$
3. $7y^3(-5y^5 - 9y^9)$
4. $3k^6(-2k^2 - k^4)$
5. $-4a^4(9a^2 + 2)$
6. $-6t^3(8t^2 + 3)$
7. $-6x^3(x^2 - 7x - 14)$
8. $-8z^5(z^2 + 5z - 16)$
9. $x^3y^2(-3x^3 + xy)$
10. $a^4b^3(-6a^3 + ab^5)$
11. $-st^5(5s^6t - t^8)$
12. $-uv^4(9u^4v - v^7)$
13. $6x^3y^4(-3x^4y^3 - 8)$
14. $9A^3B^6(-5A^7B^2 - 10)$
15. $-12m^2n(-14m^6 + 6mn^4)$
16. $-20yz^6(-5z^3 + 3y^4z)$
17. $3ab^2c^5(5a^2b - 7b^3 + 9bc^3)$
18. $4x^3yz^2(3x^2z - 16xy^8 + 6y^3z)$

Express each of the following word phrases algebraically and then simplify it. (A review of translating phrases is provided in Section 1.6.)

1. The product of 4g and 2m increased by 5mm
2. 6m times the sum of 8mg and 9g

3. 9cm^2 multiplied by the sum of $3g$ and $4mg$

4. The product of $8mg$ and $6m^2$ reduced by $7cm^2$

5. $-3x^2y^3$ times $9x^2y$ minus $4x^3y^2$

6. $11abc$ multiplied by $3a$ subtracted from $7b$

7. $10x^3$ times $27x^4$ increased by $19x^5$

8. The product of $-xyz$ and the sum of x and y

9. $-2a^2b$ multiplied by $-7ab$ added to $-8ab^2$

10. The product of $13a^2b^3c$ and $2a^2$ less than a^3

11. $-45xy^2$ times $2x^3y^2$ less x^5y^3

12. $100w^3$ decreased by $50x^4$ is multiplied by $30w^3x^2$

4.4 EVALUATING EXPRESSIONS WITH LITERAL SUBSTITUTIONS

In Sections 2.4 and 3.5 we considered numerical evaluations of literal expressions. However, an expression can also be evaluated by replacing a literal by another literal quantity.

Replacing the literal x by $10 - y$ in the expression $3x - 5y$, to yield $3(10 - y) - 5y$, is called "evaluating $3x - 5y$ when x equals $10 - y$." In mathematical notation: $3x - 5y$, when $x = 10 - y$. The sentence $x = 10 - y$ contains a literal on each side of the equals sign and is called a *literal equation.* When we use a literal equation, such as $x = 10 - y$, to evaluate an expression containing x, we are asserting that x can be replaced by $10 - y$ without changing the *intended value* of the expression.

Using a literal equation, such as $x = 10 - y$, we can evaluate an expression containing two literals and thereby simplify it to an expression containing a single literal. For example, evaluating $3x - 5y$, when $x = 10 - y$ yields an expression in y. Similarly, evaluating the expression $5y - 7x + 12$, when $y = 10 - x$ yields an expression in x.

Look at the following example of *evaluating expressions in two literals with literal substitutions.*

TYPICAL PROBLEM Evaluate: $5y - 7x + 12$, when $y = 10 - x$

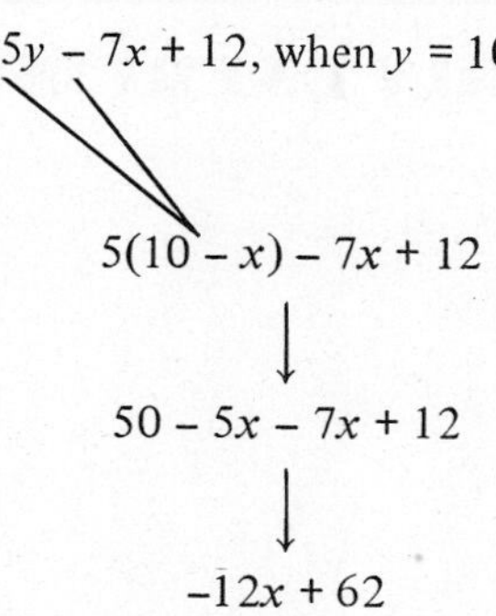

We first substitute the expression given by the equation for the literal, using parentheses: $5(10 - x) - 7x + 12$

Then we simplify the resulting expression in one literal: $50 - 5x - 7x + 12$

Hence, evaluating $5y - 7x + 12$, when $y = 10 - x$ yields $-12x + 62$. $-12x + 62$

This extends the method in Section 3.5 to literal substitutions. Always place parentheses around each substituted quantity.

EXAMPLES

Evaluate the following expressions in two literals with literal substitutions.

1. $4y + 2x - 8$, when $y = -x + 8$

2. $3x - 5y - 11$, when $x = 2y + 3$

3. $2x - y - 7$, when $y = 5 - x$

4. $3x - 2y - 13$, when $x = -5 - 4y$

Solutions

1. $4(-x + 8) + 2x - 8$
   $\downarrow$
   $-4x + 32 + 2x - 8$
   $\downarrow$
   $-2x + 24$

2. $3(2y + 3) - 5y - 11$
   $\downarrow$
   $6y + 9 - 5y - 11$
   $\downarrow$
   $y - 2$

3. $2x - (5 - x) - 7$
   $\downarrow$
   $2x - 5 + x - 7$
   $\downarrow$
   $3x - 12$

4. $3(-5 - 4y) - 2y - 13$
   $\downarrow$
   $-15 - 12y - 2y - 13$
   $\downarrow$
   $-14y - 28$

EXERCISES

Evaluate each of the following expressions, simplifying to a polynomial in x or y.

1. $-2x - 3y$, when $x = 2y - 5$
2. $8x + 10y$, when $y = 6x + 3$
3. $-10x - 5y$, when $y = 4x - 1$
4. $-4x - 6y$, when $x = 7y + 1$
5. $2x - 7y$, when $x = -3y - 8$
6. $-3x - 5y$, when $y = -4x + 3$
7. $2x + y - 4$, when $y = x + 3$
8. $6x + 7y - 9$, when $y = x - 5$
9. $4x + 5y + 2$, when $x = 6y - 1$
10. $-3x - 7y + 3$, when $x = -2y - 3$
11. $8x - y - 4$, when $y = 2x - 5$
12. $-3x - 11y + 14$, when $y = 4x + 6$
13. $3x - 7y - 9$, when $x = 2y - 7$
14. $-4x + 5y - 11$, when $x = 3y + 5$
15. $-3x - 7y + 2$, when $y = x - 7$
16. $8x - y + 2$, when $y = 3x + 1$
17. $-3x + 2y - 8$, when $x = -4y + 6$
18. $5x - 9y + 2$, when $x = -3y + 2$

ANSWERS TO MARGIN EXERCISES

1. a. $+(+1 - 3) = +1 - 3$
$+(-2) \quad = \quad -2$
b. $-(+1 + 3) = -1 - 3$
$-(+4) \quad = \quad -4$

2. a. $-3x - 2$
b. $-7 + 9x$
c. $-4x^2 + 3x - 1$

3. a. $-(5x + 9) = -1(5x + 9) = -5x - 9$
b. $-(7t^2 - 8t) = -1(7t^2 - 8t) = -7t^2 + 8t$
c. $-(x^3 + 6y^3) = -1(x^3 + 6y^3) = -x^3 - 6y^3$

4. a and d

5. a and e
b and d
c and f

6. a. $2(3 + 1) = 2(3) + 2(1) = 6 + 2$
$2(4) \quad = \quad 8$
b. $2(3 - 2) = 2(3) + 2(-2) = 6 - 4$
$2(1) \quad = \quad 2$

8. a. $5(x) + 5(4)$
b. $-6(5y) - 6(8)$
c. $7(3x) + 7(-5)$
d. $-8(7y) - 8(-2)$

9. a. $7(49) = 7(50 - 1)$
$= 350 - 7 = 343$
b. $19(6) = (20 - 1)6$
$= 120 - 6 = 114$

11. a. $-12x^2 + 15x$
b. $60x^2$
c. $-12x^2 - 5$

12. a. $3x + 5x = (3 + 5)x = 8x$
b. $8x^2 - 2x^2 = (8 - 2)x^2 = 6x^2$
c. $-4xy + xy = (-4 + 1)xy = -3xy$

CHAPTER REVIEW

SECTION 4.1

Simplify to a polynomial.

1. $+(-4x - 9y)$
2. $+(+7x + 11y)$
3. $(14A - 82)$
4. $+(25a + 11b - 8c)$
5. $-(4x^3 - 5x)$
6. $-(-3r + s^2)$
7. $-(-6A^2 - 7A)$
8. $-(2x^2 - 3xy + y^2)$
9. $-(-6 + 8t - 5t^2)$

SECTION 4.2

Multiply to convert to a polynomial.

10. $7(4x + 5)$
11. $15(-4n + 6)$
12. $-3(4x - 5y)$
13. $(+12)(3x + 12)$
14. $-6(-1 + 8A)$
15. $2(+18k^2 - 5k)$
16. $(-9)(2 - 5n^3)$
17. $+8(-5x + 3y)$
18. $(-6)(-15 - 7k^4)$
19. $8(-a - 5b + 4c)$
20. $-4(-2 + 5w - 10w^2)$
21. $(+3)(-21x^2 - x - 30)$
22. $-2(-24 - 5y + 12y^2)$
23. $+5(4p - 18q + 25)$
24. $(-11)(1 - 8x - 7x^2)$
25. $(3x + 8)(-8)$
26. $(-y^3 + 9)(+30)$
27. $(-x^2 + 12xy - 15y^2)(-2)$

SECTION 4.3

Multiply to convert to a polynomial.

28. $x^4(5 - 2x)$
29. $6t^3(-4t^6 - 3t^5)$
30. $-5k^5(15k^2 + 3)$
31. $-8x^3(x^2 - 12x - 20)$
32. $m^3n^5(-5m^6 + mn^3)$
33. $-xy^6(12x^8y - y^8)$
34. $4u^3v^2(-22u^6v^2 - 4)$
35. $-11x^3y(-11x^5 + 4xy^3)$
36. $2KL^4M^2(4K^6L - 18L^3 + L^5M^3)$

SECTION 4.4

Evaluate, simplifying to a polynomial in x or y.

37. $2x + 5y$, when $y = -5x + 4$
38. $-5x + 3y$, when $x = 5y - 6$
39. $-5x + 6y$, when $y = 3 - 7x$
40. $-3x + 9y - 4$, when $y = -x - 2$
41. $5x - 7y + 32$, when $y = -3x + 8$
42. $-2x - 5y - 4$, when $x = -3y + 8$
43. $-7x - 14y - 17$, when $x = -2y + 2$
44. $8x + 9y + 10$, when $y = -x - 9$
45. $+6x + 9y - 18$, when $x = -5y + 1$

SUMMARY OF PHRASES

Express each of the following word phrases algebraically and then simplify it.

1. 8 times the sum of 6cm and 9mm
2. The product of 12 and 5g increased by 8mg

3. 16 multiplied by the sum of $3x^3$ and $-4x^2$

4. The product of -6 and $7a^2$ diminished by $10a^5$

5. 25 times $4c$ decreased by $4c^2$

6. The product of -7 and 11 less than $7b$

7. The product of 4g and 6km increased by 21m

8. 3cm multiplied by the sum of 7g and 8mg

9. $7x^3$ times $4x^2$ minus $13x$

10. The product of $-abc$ and the sum of ab and ac

11. $-5x^2y$ multiplied by $-21x^2y$ added to $16xy^2$

12. $20a^3b$ times $15ab^2$ less than $6a^2b$

CUMULATIVE TEST

The following problems test your understanding of this chapter and of related subject matter in the previous chapter. Before taking this test, thoroughly review Sections 3.4, 4.1, 4.2, and 4.3.

Once you have finished the test, compare your answers with the answers provided at the back of the book. Note the section number of each problem missed, and thoroughly review those sections again.

Multiply or simplify.

1. $+(+9A + 16)$
2. $8(5k - 12)$
3. $-(+15w) - (27w) + (3w)$
4. $2x^2(x^4y - y^2)$
5. $-(-2x^4 - 15y^2)$
6. $(-5)(12x^3 - 11x^2)$
7. $-xy^3(-x^6y^2 - 17x^2y^5)$
8. $-(9x^2) + (-9x^2)$
9. $(-9)(-4 - 6w + 10w^2)$
10. $+(-4x^3 + 8x^2 - x)$
11. $-s^2t^3(st^6 - s^3t + s^8)$
12. $14(3 - x^2)$
13. $(6k) - 9m - (-3m) + (-4k)$
14. $-(5x^2 - 8xy + 18y^2)$
15. $4A(-9 + 3A + 2A^2)$
16. $(-11)(7s^2t - 8st^2 - 3t^3)$
17. $(7x^3y^2)(x^4 - y^2)$
18. $+(-7x) - 15y + (15y) - 9x - (-11y)$
19. $-3(-18 + 21a)$
20. $-8x^4(3x^5 - 7x^3 + 8x)$

Parentheses

Parentheses ( ) are the symbols of grouping in the language of algebra. *Brackets* [ ] and *braces* { } are used interchangeably with parentheses and have exactly the same meaning. Any number of pairs of parentheses may occur in an algebraic expression. They can occur side by side and/or one inside another, such as in $(3x)(4x-5)$, $(-3x^2)(-2(x-4))$, and $-\{-[3(-2x)+1]-x\}$.

In algebraic expressions, parentheses are essential to specify the intended operations and the order in which to perform them. For example,

$$3(-7x+2) \quad \text{and} \quad 3-(7x+2)$$

are very different expressions, although they differ only slightly in the placement of the opening parenthesis. The first expression $3(-7x+2)$ is a product that simplifies to $-21x+6$, whereas the second expression $3-(7x+2)$ is a sum that simplifies to $-7x+1$. Hence, the operation depends, in this case, on whether the opening parenthesis comes before or after the minus sign. Note that the expression $3(-7x+2)$ with the opening parenthesis before the minus sign is a single term, whereas $3-(7x+2)$ with the minus sign before the opening parenthesis has two terms.

In an algebraic sum, every term (except possibly the first) begins with a sign, and every sign that is not inside parentheses begins a term of the sum. When pairs of parentheses are one inside another, only signs outside *any* parentheses start new terms. To simplify the more complicated expressions in this chapter, we must learn to distinguish one term from another. Look at the following algebraic expressions and determine which of these are sums. If an expression is a sum, determine how many terms it has.

1. $(3x^2-4x+5)$ is one term. (The expression inside parentheses is a sum of three terms.)
2. $-4(5x-2)$ is one term.
3. $3(7x^2-9x+2)$ is one term.
4. $-(9x-4)-3x$ is the sum of the two terms $-(9x-4)$ and $-3x$.

5. $(-3x + 2) - (4x + 7)$	is	the sum of the two terms $(-3x + 2)$ and $-(4x + 7)$.
6. $6(2x - 5) + 8(-4x - 2)$	is	the sum of the two terms $6(2x - 5)$ and $8(-4x - 2)$.
7. $3 - (3x + 2) - 7x + (5)$	is	the sum of the four terms 3, $-(3x + 2)$, $-7x$, and (5).
8. $-(3x + 2y - 7)$	is	one term.
9. $(4x^2 - 9x) + (5)$	is	the sum of the two terms $(4x^2 - 9x)$ and (5).
10. $-(7x^2 - 9) + (8x) - 2$	is	the sum of the three terms $-(7x^2 - 9)$, $(8x)$, and -2.
11. $-6 - (7x - (4 - 3x) - 2)$	is	the sum of the two terms -6 and $-(7x - (4 - 3x) - 2)$.
12. $4(-3x - (7 - 5x))$	is	one term.
13. $-3(-2 + (4 - x) - 9x) - 3$	is	the sum of the two terms $-3(-2 + (4 - x) - 9x)$ and -3.
14. $3 - (7x - 4) - 5(3x - 2) + 9(-3x)$	is	the sum of the four terms 3, $-(7x - 4)$, $-5(3x - 2)$, and $9(-3x)$.
15. $8 - (x^2 - 9(x - 2)) - 7(3x - 4) - 4$	is	the sum of the four terms 8, $-(x^2 - 9(x -2))$, $-7(3x - 4)$, and -4.

Many of the terms with parentheses in the above examples contain sums within the parentheses. Example 3 is a single term, one of whose factors is the sum $7x^2 - 9x + 2$. In example 4, the first term is the negative of the sum $9x - 4$. The second term in example 11, $-(7x - (4 - 3x) - 2)$, which contains parentheses within parentheses, is the negative of the sum of the three terms $7x$, $-(4 - 3x)$, and -2. The middle term $-(4 - 3x)$ is the negative of a sum of two terms.

5.1 ADDING POLYNOMIALS

In Chapter 3 we studied sums of monomials. Monomials are a special type of polynomial, and the methods for adding monomials can be generalized to add polynomials.

In *algebraic sums of polynomials*, such as $-(12x + 6y) + 15x$ and $(-20m + 7n) - (15m - 8n)$, each polynomial is called a *term* of the sum. For example, in the sum $-(12x + 6y) + 15x$ the terms are $-(12x + 6y)$ and $15x$. Similarly, in the sum $(-20m + 7n) - (15m - 8n)$ the terms are $(-20m + 7n)$ and $-(15m - 8n)$.

1. What are the terms of
a. $3x - (-7 + 2x) + (-6)$
b. $-(x - 5y) + (-6x - 2y)$

In an algebraic sum, a plus sign before parentheses means add the polynomial in the parentheses, and a minus sign before parentheses means add the negative of the polynomial in parentheses. For example, $-(12x + 6y) + 15x$ means add $15x$ to the negative of $12x + 6y$, and $(-20m + 7n) - (15m - 8n)$ means add the negative of $15m - 8n$ to $-20m + 7n$.

2. Express algebraically
a. The sum of $3x - 7$ and $-4x - 9$
b. The sum of $5x$ and the negative of $2x - 3$

To simplify a sum containing parentheses, we must first identify the terms and then simplify each term to an expression that does not contain parentheses. For example, in the sum $-(7B + 3A) + (-7B + 8A)$, the first term is $-(7B + 3A)$, which simplifies to $-7B - 3A$. The second term is $(-7B + 8A)$, which simplifies to $-7B + 8A$.

In the problems in this section, the terms are polynomials inside parentheses. In Section 4.1 we saw how to remove parentheses from such expressions. Once the parentheses are removed we can simplify further using the methods given in Chapter 3.

Look at the following example of *adding polynomials.*

TYPICAL PROBLEM

$-(7B + 3A) + (-7B + 8A)$

We first identify the terms and remove parentheses within each term:

$-7B - 3A - 7B + 8A$

Then we rearrange the terms to bring like terms together (this step is optional):

$-3A + 8A - 7B - 7B$

We then combine like terms:

$5A - 14B$

Hence, $-(7B + 3A) + (-7B + 8A) = 5A - 14B$.

To remove parentheses, we use the methods in Sections 4.1 and 4.2. To combine like terms, we use the method in Section 3.3.

EXAMPLES

Add the following polynomials by removing parentheses and combining like terms.

Every sum of polynomials can be simplified to a single polynomial.

1. $-(12x + 6y) - 15x$
2. $5k - (11k + 6)$
3. $-(5x - 9y) + (4x + y)$
4. $(8s - 7t) - (3s - 9t)$
5. $(-20m + 7n) - (15m - 8n)$
6. $(6A - 12B) - 8B + (5B - 3A)$

Solutions

1. $-12x - 6y - 15x$
   $\downarrow$
   $-12x - 15x - 6y$
   $\downarrow$
   $-27x - 6y$

2. $5k - 11k - 6$
   $\downarrow$
   $-6k - 6$

3. $-5x + 9y + 4x + y$
   $\downarrow$
   $-5x + 4x + 9y + y$
   $\downarrow$
   $-x + 10y$

4. $8s - 7t - 3s + 9t$
   $\downarrow$
   $8s - 3s - 7t + 9t$
   $\downarrow$
   $5s + 2t$

5. $-20m + 7n - 15m + 8n$
   $\downarrow$
   $-20m - 15m + 7n + 8n$
   $\downarrow$
   $-35m + 15n$

6. $6A - 12B - 8B + 5B - 3A$
   $\downarrow$
   $6A - 3A - 12B - 8B + 5B$
   $\downarrow$
   $3A - 15B$

OBSERVATIONS AND REFLECTIONS

1. If the first term in a polynomial in parentheses is positive, its plus sign is almost always omitted, such as in $(x - y)$. However, when parentheses are dropped from

around polynomials in a sum, each term in each polynomial must have its sign expressed. For example,

$$+(x - y) = \quad +x - y \qquad \text{or} \qquad x - y$$

whereas

$$2 + (x - y) = 2 + x - y, \qquad \textbf{not} \qquad 2x - y$$

Similarly,

$$-(-x - y) = \quad +x + y \qquad \text{or} \qquad x + y$$

whereas

$$2 - (-x - y) = 2 + x + y, \qquad \textbf{not} \qquad 2x + y$$

2. We must remove parentheses before combining terms when adding polynomials because two terms cannot be combined if either is in a set of parentheses not containing the other. For example, in $3A - (-6A + 7B)$ the term $-6A$ inside parentheses cannot be combined with $3A$, which is not inside the same set of parentheses. We must remove parentheses and then combine $3A$ with $+6A$:

$$3A - (-6A + 7B) = 3A + 6A - 7B = 9A - 7B$$

★ 3. Using negation, we can define *algebraic subtraction* of polynomials. We can subtract one polynomial from a second polynomial, to find their difference, by adding the negative of the first polynomial to the second polynomial. For example, to subtract $3x - 4$ from $7x + 3$, we add $-(3x - 4)$ to $7x + 3$:

$$(7x + 3) - (3x - 4) = 7x + 3 - 3x + 4 = 4x + 7$$

3. Subtract
- a. $5x$ from $2x - 5$
- b. $x^2 - 9x$ from $-2x$
- c. $4a + 7b$ from $3a - 7b$
- d. $-3m - 5$ from $7 - 3m$

EXERCISES

Simplify each of the following expressions to a polynomial.

1. $-(9x + 2y) - 3x$
2. $4 - (-5 - 3x)$
3. $-(8A - 5B) - 11A$
4. $3n - (4n + 7)$
5. $-(-4x + 7) - 3x$
6. $3y - (9y + 4)$
7. $8s + (-4s + 11t)$
8. $-(12 - 5x^2) - x^2$
9. $-(y - 2x) + (-5x + 3y)$
10. $-(4i - 8j) + (2i + 9j)$
11. $(5a - 6b) - (4a - 2b)$
12. $-(7p - 4q) - (6p - 10q)$
13. $(-15m + 9n) - (13m - 7n)$
14. $(-4x - 5) - (-3 - 5x)$
15. $(-3 - 8r) - (+3 + 8r)$
16. $(x^2y - xy^2) - (4x^2y - xy^2)$
17. $-(-3a + 7b) - (+6b - 4a)$
18. $-(-5K + 7L) - (-7K + 5L)$
19. $(8u - 9v) - 6v + (4v - 2u)$
20. $(3m - 7n) - 9n + (4n - 6m)$
21. $7x - (4x - 3y) - (3y - 10x)$
22. $8A - (4A - 3B) - (9A - 3B)$
23. $-(6y^2 + 2z) - 3y^2 - (3z - 7y^2)$
24. $-(6x^2 + 6y) - 7y - (4y - 8x^2)$

5.2
ADDING INTEGER MULTIPLES OF POLYNOMIALS

In this section we will consider algebraic sums such as $12(4A - 5) - 7(6A - 9)$ and $5x - 2 - 4(5 + 7x)$. In these sums many of the terms are integer multiples of polynomials. In $12(4A - 5) - 7(6A - 9)$, for example, both $12(4A - 5)$ and $-7(6A - 9)$ are integer multiples. Similarly, in $5x - 2 - 4(5 + 7x)$, the third term $-4(5 + 7x)$ is an integer multiple of $5 + 7x$.

In such expressions each term can be simplified to a sum without parentheses by the process of distribution.

Look at the following example of *adding integer multiples of polynomials.*

TYPICAL PROBLEM

$12(4A - 5) - 7(6A - 9)$

We first identify terms and remove parentheses within each term, distributing where necessary. (Each sign outside parentheses begins a new term.):

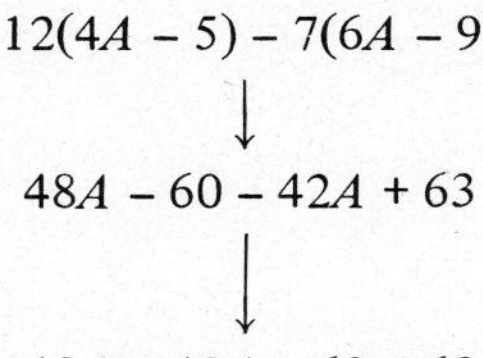

$48A - 60 - 42A + 63$

Then we rearrange terms to bring like terms together (this step is optional):

$48A - 42A - 60 + 63$

We then combine like terms:

$6A + 3$

Hence, $12(4A - 5) - 7(6A - 9) = 6A + 3$.

When distribution is required to remove parentheses within a term, we use the method in Section 4.2.

EXAMPLES

Add the following integer multiples of polynomials.

Every sum of polynomials and integer multiples of polynomials can be simplified to a polynomial.

1. $5x - 2 - 4(5 + 7x)$
2. $-5(6A + 2B) + 7(5A - 2B)$
3. $5(8 - 6r) - 3(4r - 9)$
4. $-3(7x - y) - (2x + 7y) + 8(3x + y)$

Solutions

1. $5x - 2 - 20 - 28x$ ↓ $-23x - 22$
2. $-30A - 10B + 35A - 14B$ ↓ $5A - 24B$
3. $40 - 30r - 12r + 27$ ↓ $67 - 42r$
4. $-21x + 3y - 2x - 7y + 24x + 8y$ ↓ $x + 4y$

EXERCISES

Simplify each of the following expressions to a polynomial.

1. $8 - 2(x + 9)$
2. $5 - 3(x + 10)$
3. $4A + 6(3A - 6)$
4. $7B + 2(8B - 2)$
5. $-3y + 5(-4y - 1)$
6. $-9x + 6(-3x - 2)$

7. $6 - 7k + 3(-5k + 4)$

8. $8 - 11n + 4(-6n + 3)$

9. $-4 + 8x - 2(-7 - 8x)$

10. $-9 + 3y - 6(-5 - 3y)$

11. $3b - 9 - 4(2b + 6) + 16$

12. $8a - 3 - 7(3a + 4) + 12$

13. $4(3x - 5y) + 2(7x + 3y)$

14. $2(4k + 5) + 6(5 - 2k)$

15. $6(3x + 7y) + 2(-4x - 6y)$

16. $9(u + 2v) + 8(v - 3u)$

17. $3(2a + 3b) - 4(5a + 2b)$

18. $-4(3m + 5n) + 7(3m - 8n)$

19. $8(-4s - 7t) - 3(7s + t)$

20. $-6(5x - 3y) - 2(-6y + 5x)$

21. $-4(9 - 7r) - 5(8 - 6r)$

22. $4(6 - 5P) - 8(2P + 3)$

23. $-2(5a - 7b) + 3(4a + 8c)$

24. $-9(3x - z) - 3(7x + 2y)$

25. $-6(4 - 5n) + 7(5 - 4m)$

26. $-8(1 - 2x) + 12(1 - 3x^2)$

27. $-8(7t^2 - 5) - 3(12t - 15)$

28. $-11(4w + 8) + 13(2v + 5)$

29. $3(2x - 5y) - 5(x - 3y) + 4(5x + y)$

30. $-5(5s - 7t) + 3(4s + 2t) - 3(9s + t)$

31. $4(5B + 3) - (7A - 4B) - 9(A - 2)$

32. $-6(3n - 7) + (24m - 3n) - (3m + 7)$

33. $-8(3x^2 - 5x - 11) + 3(-2x^2 - 4x + 12)$

34. $-10(7y^2 - 8y + 2) + 6(-4y^2 + 15y - 20)$

35. $4(3x^2 - 9xy - 8y^2) - 6(-2x^2 + 11xy - 7y^2)$

36. $3(-8x^2 + 4xy - 14y^2) - 2(7x^2 - 15xy - 24y^2)$

5.3 ORDER OF OPERATIONS

In Section 5.1 we saw expressions such as $-(5a - 9b) + (4b + 10a)$, which is a sum of polynomials. In Section 5.2 we dealt with expressions such as $7(3a - 2b) - 5(5a + b)$, which is a sum of integer multiples of polynomials.

Now we will work with more complicated expressions such as $-(-6 - (3 - 18n) + 3)$, $4[3(x - 2y) - 4x - 5(2x + 3y)]$, and $-(A - (3B + (7A - B)))$. Expressions such as these, which contain sets of parentheses within other sets of parentheses, occur frequently in algebra. Such expressions are said to have *nested parentheses.* We must be able to identify the pairs of parentheses (and brackets) in such expressions. In the above examples, the parentheses are paired as follows:

inner set: $(-3 - 18n)$ in $-(-6(-3 - 18n) + 3)$; outer set: $(-6(-3 - 18n) + 3)$

inner sets: $(x - 2y)$ and $(2x + 3y)$ in $4[3(x - 2y) - 4x - 5(2x + 3y)]$; outer set: $[3(x - 2y) - 4x - 5(2x + 3y)]$

innermost set: $(7A - B)$ in $-(A - (3B + (7A - B)))$; inner set: $(3B + (7A - B))$; outer set: $(A - (3B + (7A - B)))$

The third example would be easier to read if we used brackets and braces instead of only parentheses:

$$-(A - (3B + (7A - B))) \quad \text{is the same as} \quad -\{A - [3B + (7A - B)]\}$$

To simplify these expressions we follow the Order of Operations.

1. We identify (innermost) pairs of parentheses and simplify within parentheses first (working from the inside out).
2. We simplify the terms containing (innermost) parentheses to remove these parentheses.

We repeat these two steps alternately as needed until the expression has been simplified to a polynomial without parentheses.

4. Rewrite using brackets, braces, and parentheses:
a. $-(2a + (3b - 4a))$
b. $5(8(x - y) - 6(x + 7y))$
c. $-(7 - 3(5(n + 2) + 9))$

EXAMPLES

Simplify the following expressions to polynomials according to the Order of Operations.

1. $-(6m + (4m - 7n) - 8n)$
2. $-4A - 9B - (-(6B - 5A) - (3A - B))$
3. $2[-2(6B - 8A) + (9A + 8B)]$
4. $4[-3(x - 2y) - 4x + 5(2x - 3y)]$
5. $-4x - ((-8 + (6x - 7)) - 5x)$
6. $3A - (3A - ((4A - 5) - (6 - 4A))) + 11$

Solutions

1. $-(6m + (4m - 7n) - 8n)$

$-(6m + 4m - 7n - 8n)$

$-(10m - 15n)$

$-10m + 15n$

2. $-4A - 9B - (-(6B - 5A) - (3A - B))$

$-4A - 9B - (-6B + 5A - 3A + B)$

$-4A - 9B - (+2A - 5B)$

$-4A - 9B - 2A + 5B$

$-6A - 4B$

3. $2[-2(6B - 8A) + (9A + 8B)]$

$2[-12B + 16A + 9A + 8B]$

$2[25A - 4B]$

$50A - 8B$

4. $4[-3(x - 2y) - 4x + 5(2x - 3y)]$

$4[-3x + 6y - 4x + 10x - 15y]$

$4[+3x - 9y]$

$12x - 36y$

5. $-4x - ((-8 + (6x - 7)) - 5x)$

$-4x - ((-8 + 6x - 7) - 5x)$

$-4x - ((6x - 15) - 5x)$

$-4x - (+6x - 15 - 5x)$

$-4x - (x - 15)$

$-4x - x + 15$

$-5x + 15$

6. $3A - (3A - ((4A - 5) - (6 - 4A))) + 11$

$3A - (3A - (+4A - 5 - 6 + 4A)) + 11$

$3A - (3A - (8A - 11)) + 11$

$3A - (3A - 8A + 11) + 11$

$3A - (-5A + 11) + 11$

$3A + 5A - 11 + 11$

$8A$

ORDER OF OPERATIONS

(Multiplication and Addition of Polynomials)

1. Identify each pair of parentheses (or brackets). If there are nested parentheses, locate the innermost sets and *simplify* from the *inside out.* Apply step 2 to eliminate the inner sets of parentheses first before continuing in succession to outer sets of parentheses.
2. Simplify each sum (inside outer parentheses) to a polynomial (inside outer parentheses) as follows:

   2A. *Identify terms, and simplify* each term by multiplying and/or removing parentheses. At this point each term inside outer parentheses is a monomial *with its sign expressed.*

   2B. *Combine like terms.* If a sum is inside outer parentheses, do not remove these outer parentheses.

3. Repeat steps 1 and 2 until the expression is completely simplified.

EXERCISES

Simplify each of the following expressions to a polynomial.

1. $-[(5x - 7y) - 4x]$
2. $-[4n - (7m - 9n)]$
3. $-[-(4A - 9B) - 32B]$
4. $-[-7s + (3t - 4s)]$
5. $4x - (7x - (5 + 9x))$
6. $5n - (-(4m - 3n) - 5m)$
7. $-(3u - (4v + 9u) - 7v)$
8. $-(x^2 + (7x - 8x^2) - 5x)$
9. $-(-5A - (-4B + 2A) + 5B)$
10. $-(-6 - (3 - 18k) + 3)$
11. $3x - (4 - (2 - 5x) - 6)$
12. $-6 - (7i - (8 - 6i) - 1)$
13. $2y - [(4 - 6y) + (7y - 3)]$
14. $-5 - [-(A^2 - 2) + (-A^2 + 2)]$
15. $4z^2 - [(7z - 9z^2) - (-z^2 + z)]$
16. $2(6(2k - 5) + 3k)$

17. $4(6 + 7(2n + 1))$

18. $8(2(x + 2y) - 3x)$

19. $-3(2w - 5(w - 1))$

20. $-4(3(2t - 6) - 2)$

21. $6[2a - 3(4a - 7)] + 8$

22. $3[-2(4x + 7) - 5x] + 7$

23. $3A - [-4(3 - 5A) + 6A]$

24. $-2[6x - 3(4x - 5)] - 15$

25. $-5[-2(3i + 7) - (4i + 3)]$

26. $-3[-(n - 4) + 5(n - 7)]$

27. $2[3(2a + 3b) - 2(3a + 2b)]$

28. $4[3(7x + 2) - 3(4x + 3)]$

29. $-3[-2(4T + 5) + 3(2T + 4)]$

30. $-2[5(3x + 7y) - 2(4y - x)]$

31. $4 + \{6 - [8x - (4x - 3)]\}$

32. $3n + \{[(4n - 7) - 8n] - 5\}$

33. $-(3A - 1) + \{9 - [2 - (12A - 4)]\}$

34. $(2x - 7) - \{[(4x - 12) - 8x] - 4\}$

35. $-(4z + 7) - \{3 - [(2z - 9) + 18z]\}$

36. $-(n - 3) + \{-[15n - (7n + 12) - 11n]\}$

37. $(-2v - w) + \{-[(15v - 7w) - 9v] + 15w\}$

38. $(-2 + 8A) - \{-[-(12A - 25) + 10A] + 12\}$

39. $9t - (6t - ((10t - 4) - (12 + 5t)))$

40. $4 + (-(5i + 11) - (12i - (4i - 9)))$

41. $x^2 + 2 - (-((-2 - 9x^2) + 12) - (16 - 8x^2))$

42. $-(y^2 + 7) - (5y^2 - ((7y^2 - 9) + (11 + 3y^2)))$

ANSWERS TO MARGIN EXERCISES

1. a. $3x$, $-(-7 + 2x)$, and (-6)
b. $-(x - 5y)$ and $(-6x - 2y)$

2. a. $(3x - 7) + (-4x - 9)$
b. $5x - (2x - 3)$

3. a. $2x - 5 - (5x) = -3x - 5$
b. $-2x - (x^2 - 9x) = -x^2 + 7x$
c. $3a - 7b - (4a + 7b) = -a - 14b$
d. $7 - 3m - (-3m - 5) = 12$

4. a. $-[2a + (3b - 4a)]$
b. $5[8(x - y) - 6(x + 7y)]$
c. $-\{7 - 3[5(n + 2) + 9]\}$

CHAPTER REVIEW

SECTION 5.1

Simplify to a polynomial.

1. $-(22k + 14) - 8k$
2. $4x - (7x + 4)$
3. $3s + (-8s + t)$
4. $(4B - 11A) - (4A - 3B)$
5. $-(9K - 7) + (6K + 2)$
6. $(-40u + 20v) - (25u - 15v)$
7. $-(2x - 8y) + (4x + 3y)$
8. $-(20m + 4n) - (17m - 8n)$
9. $-(+4 + 9x^2) - (-3x^2 - 1)$
10. $3i - (2i - 4j) - (7j - 5i)$
11. $(4A - 9B) - 14B + (B - 8A)$
12. $-(7x + 2) - 3 - (3 - 5x)$

SECTION 5.2

Simplify to a polynomial.

13. $7 - 4(x + 7)$
14. $6A + 9(5A - 3)$
15. $-2y + 8(-3y - 1)$
16. $9 - 5k + 2(-9k + 8)$
17. $-7 + 6x - 5(-2 - 6x)$
18. $8a - 1 - 6(4a + 3) + 10$
19. $5(2i - 9j) - 3(7i + 4j)$
20. $-6(2A + 8B) + 8(4A - B)$
21. $3(9 - 5x) - 7(3x - 8)$
22. $6(9r - 10t) - 9(10r - 3t)$
23. $-2(12w - 14) - 4(9 - 3w)$
24. $9(6y - 5) - 3(7y - 9z)$
25. $12(4i - 6j) - 8(3j - 8k)$
26. $3(4A - 2) - 2(6 + 7B)$
27. $-5(3x - 4y) - (11x - 7y) + 9(2x - 3y)$
28. $-10(2a - 3b) - (14a + 6b) + 3(4a - 12b)$
29. $-6(4x^2 - 11x - 8) + 9(-3x^2 - 4x + 5)$
30. $3(5x^2 - 8xy - 13y^2) - 8(-4x^2 + 10xy - 6y^2)$

SECTION 5.3

Simplify to a polynomial.

31. $-[9A - (4B - 3A)]$
32. $-[(2x - 9x^2) - 7x]$
33. $3K - (2K - (-2K + L))$
34. $-2x - (2y - (5y - 2x) - 9x)$
35. $7u - [(3v - 4u) + (3u - 9v)]$
36. $-8t - 4p - [-(3t - p) + (2t - 3p)]$
37. $7A - [(5B - 7A) + (6A - 3B)]$
38. $4(-3(x - 5y) - 4y)$
39. $-5(-2(-6a + b) + 3a)$
40. $-3(-(-2r + 3) - 7)$
41. $4x + 2[-3(7x - 2) + 4]$
42. $-3 - 4[2y - 6(3 - 5y)]$

43. $2(-10p + 6(-2p + 5))$

44. $-7(-4K - 7(-2K - L))$

45. $-4[(8u - 4v) - 3(8u + 7v)]$

46. $-3k - \{3 - [(4k - 6) + 5k]\}$

47. $5 + \{-[3z - (14z + 5)] - 7z\}$

48. $2n - 5 + \{-[(3n - 12) - 5n] + 7\}$

49. $2 - t^2 + \{-[-(3t^2 - 11) + 5t^2] + 2\}$

50. $-(-(3 - 7n) - (-6n + 14) - (18n - (5n - 16)))$

51. $-(5x - ((-(6x - 7y) + 11x) - (4y - 9x)))$

CUMULATIVE TEST

The following problems test your understanding of this chapter and of related subject matter of previous chapters. Before taking this test, thoroughly review Sections 2.3, 3.2, 3.3, 3.4, 4.2, 5.1, and 5.2.

Once you have finished the test, compare your answers with the answers provided at the end of the book. Note the section number of each problem missed, and thoroughly review those sections again.

Simplify or multiply.

1. $-(25x + 13) - 14x$
2. $10 - 4(x + 9)$
3. $(4y)(-3y^4)(+3y^4)$
4. $-14s + 7s - 10s + 13s$
5. $(-5t^3) - (5t^2) + 7t^3 - (-11t^2) - t^3$
6. $-6(5x - 2y) - (13x - 8y) + 9(x - 4y)$
7. $-8x^2 - 19 + x^2 + 11$
8. $3w + (-6v + w)$
9. $-4(s^3 t^2)(-3s)$
10. $5x^4 + 9x^4 - 21x^4 - 8x^4$
11. $-7k + 6(-4k - 1)$
12. $6yz^2 - 8z^2 + 7yz^2 - 5z^2 + 2z^2$
13. $-3w^2(6wx^4) + (-9w^3x^4)$
14. $-(8A - 5) + (3A + 5)$
15. $(-2a^2)(+3b^3)a^4(-5b)$
16. $4(-15x + y - 26z)$
17. $-2(9n - 21) - 7(11 - 4n)$
18. $-28x - 16y^2 + 9x + 7y^2$
19. $(4x^2 - 12) - 15x^2 + (7 - 12x^2)$
20. $4(2rst) - 3s(+5rt)$
21. $(-45A + 21B) - (16A - 19B)$
22. $-7(3u^4 v)(-6v^4)$
23. $-6(-x - 6)$
24. $4(3x - 8y) - 5(6y - 3x)$
25. $4pq - 16pq + 9pq - pq + 3pq$
26. $3w - (4w - 11x) - (19x - 6w)$
27. $16s^3t^6 - 7st^3(-9s^2t^3)$
28. $6x^3(-2w^3)(-4wx^2)$
29. $-12 + 5x - 8(-3 - 4x)$
30. $6(6r^2 + 12s^2 - 9t^2)$
31. $3k^2 + 7k^2 - 8k^2 + 12k^2$
32. $-8x^5(-6x^3) + 4x^8(-6)$
33. $-(13S - 9T) + (14S - 22T)$
34. $8(x^3y^4)(-2y^2)x^4$
35. $(-10)(14x^2 - xy - 26y^2)$
36. $6m - 1 - 12(3m - 2) + 14$
37. $8x - 5yz + 11yz - 9x + 4yz$
38. $-6r^3 + 4r^2(-12r^2) - 12r^4$
39. $-(4u + 16v) - (18u - 16v)$
40. $4(i - 8) - 6(5i - 6) - (20i - 13)$

FRACTIONS

6 NUMERICAL FRACTIONS

In Chapter 1 we began the study of signed numbers with integers. In Chapters 6 and 7 we will return to signed numbers and study *fractions*. Look at

$$\frac{2}{3}, \frac{-3}{4}, \frac{7}{2}, \text{ and } -\frac{5}{8}$$

These expressions are called *fractions*, or *ratios*. The integer above the line is called the *numerator*, and the integer below the line is called the *denominator*. The line between them is called a *fraction bar*.

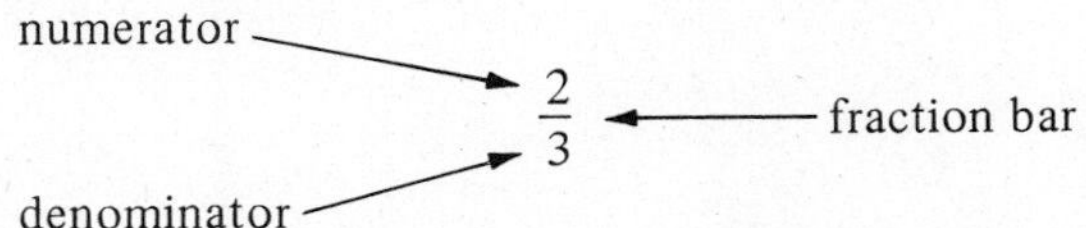

The ratio of any two algebraic expressions is called a fraction, even if the expressions themselves contain fractions. For example,

$$\frac{3-8}{2\cdot 5}, \quad \frac{2+\frac{3}{4}}{8}, \quad \frac{9-2^3}{7(-6)}, \quad \frac{(11-15)^3}{-(4)^2-(-3)^3}, \text{ and } \frac{\frac{5}{4}+\frac{1}{6}}{\frac{5}{6}-\frac{1}{4}}$$

are all fractions.

We will call fractions in which the numerator and denominator are both integers *simple fractions*. For example, $\frac{2}{3}$, $\frac{-3}{4}$, $\frac{7}{2}$, and $-\frac{5}{8}$ are simple fractions.

The Meaning of a Fraction

You are familiar from arithmetic with the idea that $\frac{1}{2}$ can be represented by splitting a unit into 2 equal parts, that $\frac{1}{3}$ can be represented by splitting a unit into 3 equal parts, that $\frac{1}{4}$ can be represented by splitting a unit into 4 equal parts, and so forth. Then $\frac{2}{3}$ is twice the size of $\frac{1}{3}$, $\frac{3}{4}$ is three times the size of $\frac{1}{4}$, and so forth.

Using this representation, we can locate the fractions on the number line:

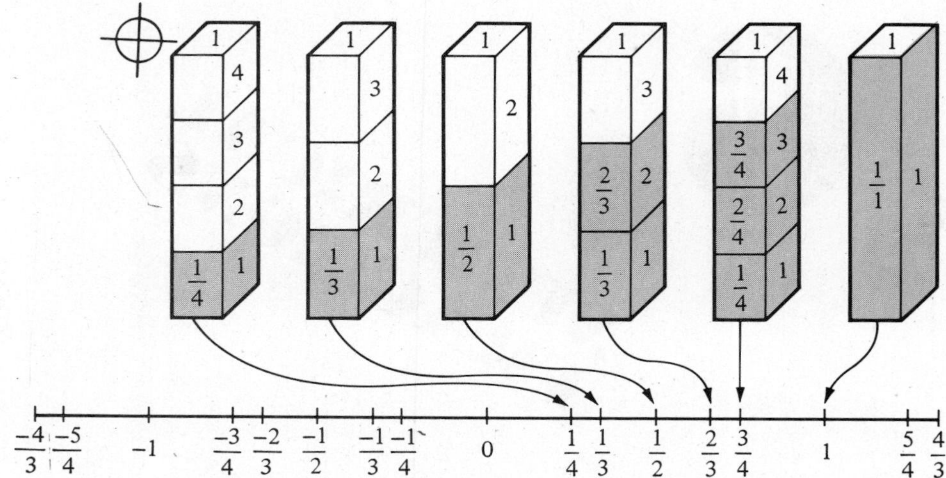

A fraction can also be represented by splitting the number of units represented by the numerator into the number of parts indicated by the denominator. For example, we can represent $\frac{2}{3}$ by splitting 2 units into 3 equal parts, and we can represent $\frac{3}{4}$ by splitting 3 units into 4 equal parts. Each part equals the fraction being represented. For example, when 2 units are split into 3 equal parts, each part is $\frac{2}{3}$ of a unit. Similarly, when 3 units are split into 4 equal parts, each part is $\frac{3}{4}$ of a unit.

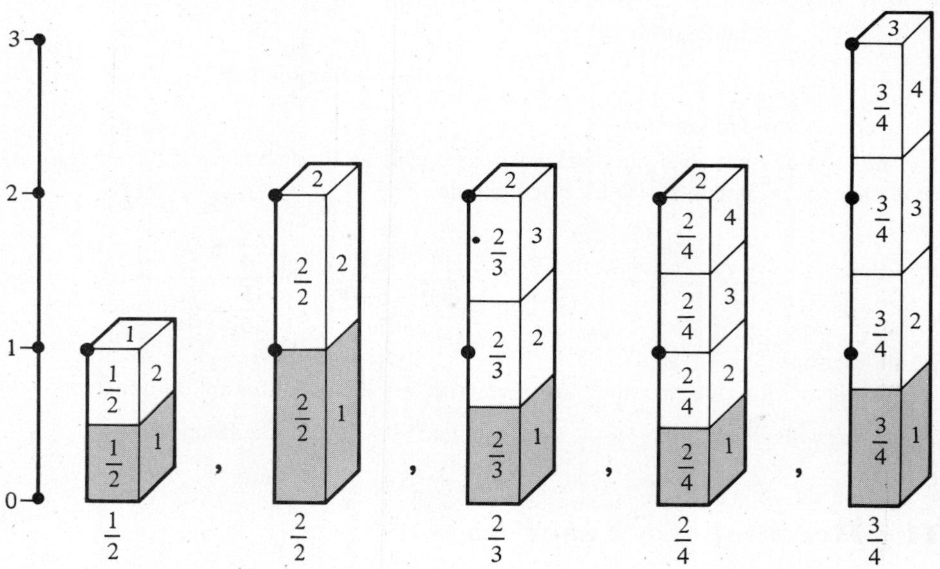

Integers

Integers can be considered fractions with a denominator of 1. For example,

$$2 \text{ is equivalent to } \frac{2}{1} \quad \text{and} \quad -7 \text{ is equivalent to } \frac{-7}{1}$$

Reciprocals of Integers

The fractions $\frac{1}{3}$, $\frac{1}{5}$, $\frac{1}{8}$, and so forth are *reciprocals* of integers (sometimes called *unit fractions*). The *reciprocal* of an integer is 1 over the integer. For example, the reciprocal of 3 is $\frac{1}{3}$, the reciprocal of 5 is $\frac{1}{5}$, and the reciprocal of 8 is $\frac{1}{8}$.

We can think of a fraction as the product of an integer (the numerator) and the reciprocal of a nonzero integer (the denominator). For example,

$$\frac{3}{8} = \frac{3 \cdot 1}{8} = 3 \cdot \frac{1}{8} \quad \text{and} \quad \frac{-2}{5} = \frac{-2 \cdot 1}{5} = -2 \cdot \frac{1}{5}$$

$$\frac{a}{b} = a \cdot \frac{1}{b}$$

(where a and b represent integers, with $b \neq 0$)

Zero

The number 0 does not ordinarily appear as part of a fraction. Zero can *never* appear as the denominator of a fraction. Any expression that appears to be a fraction with a denominator of 0 is **undefined**. (We say undefined because it makes no sense to talk about splitting a length into zero equal parts.) For example,

$$\frac{3}{0} \quad \text{and} \quad \frac{8}{0} \quad \text{are } \textbf{undefined}$$

In particular, $\frac{0}{0}$ is **undefined.**

If the numerator of a fraction (with a nonzero denominator) is zero, then the fraction equals zero. For example,

$$\frac{0}{3} = 0 \quad \text{and} \quad \frac{0}{8} = 0$$

After all, $\frac{0}{8} = \frac{0 \cdot 1}{8} = 0 \cdot \frac{1}{8}$, and multiplying any expression by 0 yields 0.

Multiples and Divisors

Multiplying an expression by any integer yields an *integer multiple* of that expression. For example, the integer multiples of the number 3 are

$$\cdots, -21, -18, -15, -12, -9, -6, -3, 0, 3, 6, 9, 12, 15, 18, 21, \cdots$$

Note the number 0 in the middle of this row. Multiplying any expression by 0 yields 0, so zero is a multiple of every expression. Similarly, the product of any expression and 1 equals the expression again, so every expression is an integer multiple of itself and every *integer* is an integer multiple of 1.

A number is said to be a *divisor* of each of its multiples. Hence one integer is a *divisor* of a second integer if the second integer is a *multiple* of the first. For example,

divisors of 12	1, 2, 3, 4, 6, 12 −1, −2, −3, −4, −6, −12
divisors of 30	1, 2, 3, 5, 6, 10, 15, 30 −1, −2, −3, −5, −6, −10, −15, −30
divisors of 32	1, 2, 4, 8, 16, 32 −1, −2, −4, −8, −16, −32

Every number is a divisor of zero and itself. On the other hand, zero is not a divisor of any other number, because zero has no multiples other than itself.

Common Divisors

Look now at the positive divisors of 12 and 30. As you can see, the numbers 1, 2, 3, and 6 are divisors of *both* 12 and 30. These numbers are therefore called *common divisors* of 12 and 30. Similarly, look at the positive divisors of 12 and 32. As you can see, the numbers 1, 2, and 4 are common divisors of 12 and 32.

In general, an integer is a *common divisor* of two given numbers if the given numbers are both multiples of the integer—that is, if the integer is a divisor of *both* given numbers. (Note that the number 1 is always a common divisor, whereas the number 0 is never a common divisor.) The negative of a common divisor of two integers is also a common divisor of the two numbers.

The largest *positive* common divisor of two integers is called their *greatest common divisor* (GCD). The GCD of 12 and 30, for example, is 6, and the GCD of 12 and 32 is 4. Look at the following table, which gives the GCD of pairs of numbers. Verify the entries to check your understanding.

GCD	1	2	3	4	5	6
1	1	1	1	1	1	1
2	1	2	1	2	1	2
3	1	1	3	1	1	3
4	1	2	1	4	1	2
5	1	1	1	1	5	1
6	1	2	3	2	1	6

Equivalent Fractions

From our experience with arithmetic, we know that different fractions can represent the same quantity. For example, we know that $\frac{1}{2}$, $\frac{2}{4}$, and $\frac{3}{6}$ are equal. Such fractions are called *equivalent fractions.*

6

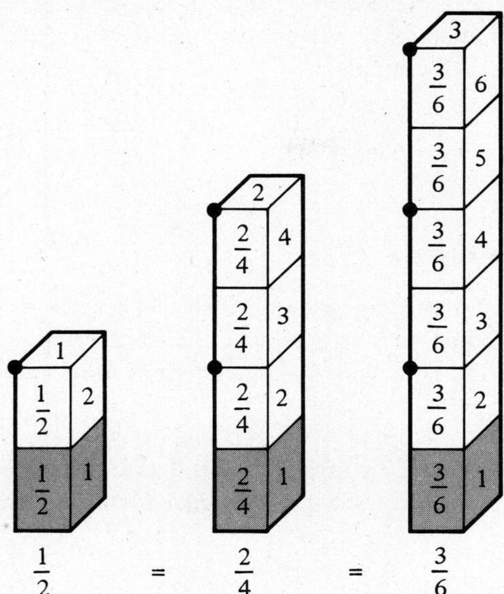

Look at the following sets of equivalent fractions:

$$\frac{1}{2} = \frac{2}{4} = \frac{3}{6} = \frac{4}{8} = \frac{5}{10} = \frac{6}{12} = \dots$$

$$\frac{1}{3} = \frac{2}{6} = \frac{3}{9} = \frac{4}{12} = \frac{5}{15} = \frac{6}{18} = \dots$$

$$\frac{2}{3} = \frac{4}{6} = \frac{6}{9} = \frac{8}{12} = \frac{10}{15} = \frac{12}{18} = \dots$$

$$\frac{1}{4} = \frac{2}{8} = \frac{3}{12} = \frac{4}{16} = \frac{5}{20} = \frac{6}{24} = \dots$$

$$\frac{3}{4} = \frac{6}{8} = \frac{9}{12} = \frac{12}{16} = \frac{15}{20} = \frac{18}{24} = \dots$$

$$\frac{1}{5} = \frac{2}{10} = \frac{3}{15} = \frac{4}{20} = \frac{5}{25} = \frac{6}{30} = \dots$$

$$\frac{2}{5} = \frac{4}{10} = \frac{6}{15} = \frac{8}{20} = \frac{10}{25} = \frac{12}{30} = \dots$$

$$\frac{3}{5} = \frac{6}{10} = \frac{9}{15} = \frac{12}{20} = \frac{15}{25} = \frac{18}{30} = \dots$$

$$\frac{4}{5} = \frac{8}{10} = \frac{12}{15} = \frac{16}{20} = \frac{20}{25} = \frac{24}{30} = \dots$$

Can you see the patterns in these sets? Note that multiplying both the numerator and denominator of a fraction by the *same* nonzero integer yields an equivalent fraction. For example, in the last row of the examples,

$$\frac{8}{10} = \frac{4 \cdot 2}{5 \cdot 2}, \quad \frac{12}{15} = \frac{4 \cdot 3}{5 \cdot 3}, \quad \frac{16}{20} = \frac{4 \cdot 4}{5 \cdot 4}, \quad \frac{20}{25} = \frac{4 \cdot 5}{5 \cdot 5}, \quad \frac{24}{30} = \frac{4 \cdot 6}{5 \cdot 6}$$

and so forth. If we read the rows from right to left, we can see that dividing the numerator and denominator of a fraction by any common divisor also results in an equivalent fraction. For example,

$$\frac{18}{24} = \frac{3 \cdot 6}{4 \cdot 6} = \frac{3}{4} \quad \text{and} \quad \frac{16}{20} = \frac{4 \cdot 4}{5 \cdot 4} = \frac{4}{5}$$

Reduced Fractions

The first fraction in each row in our example is special because its numerator and denominator have no common divisors except +1 and −1. A fraction is called *irreducible* if the numbers +1 and −1 are the only common divisors of its numerator and denominator. Thus, the fractions $\frac{1}{2}$, $\frac{1}{3}$, $\frac{2}{3}$, $\frac{1}{4}$, $\frac{3}{4}$, $\frac{1}{5}$, $\frac{2}{5}$, $\frac{3}{5}$, and $\frac{4}{5}$ are all irreducible. An irreducible fraction is also called *reduced* or is said to be in *lowest terms.*

The numerator and denominator of an *unreduced* fraction have a common divisor larger than 1. For example, the fractions $\frac{18}{24}$ and $\frac{16}{20}$ are unreduced.

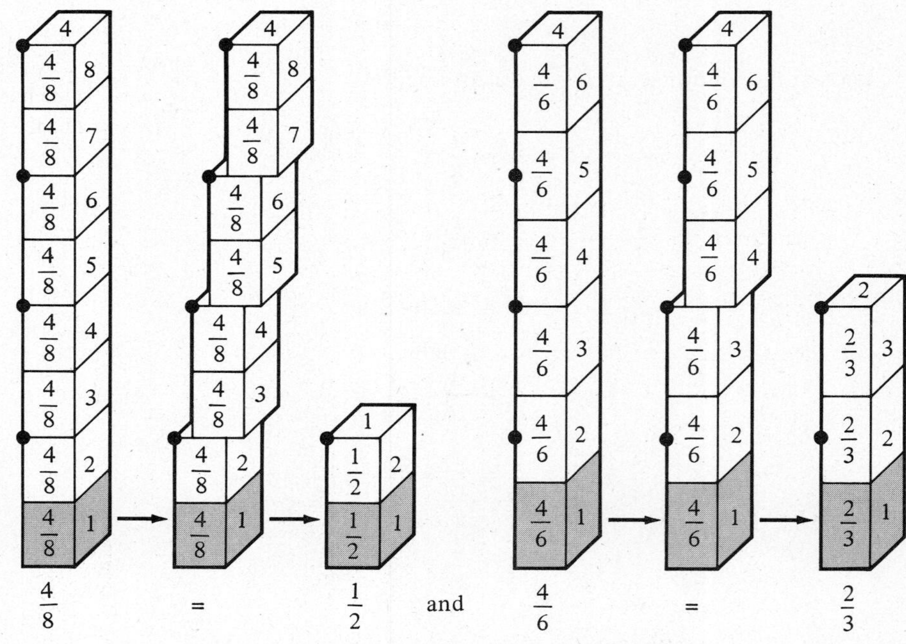

$$\frac{4}{8} = \frac{1}{2} \quad \text{and} \quad \frac{4}{6} = \frac{2}{3}$$

6.1 REDUCING FRACTIONS USING THE METHOD OF PRIMES

1. Find a prime divisor for each of the nonprime numbers between 4 and 12.

To reduce a simple fraction, we must first be able to find a common divisor of the numerator and denominator. In order to do this systematically, we should look at the divisors of the integers. In the study of the divisors of an integer, *prime numbers* play a key role. A *prime number* is a positive integer greater than one which is divisible only by itself and 1 (and their negatives). The prime numbers are also called *primes*, for short. The first ten prime numbers are 2, 3, 5, 7, 11, 13, 17, 19, 23, and 29.

By convention, the numbers +1 and −1 are called *units* and are not considered primes. Similarly, the negative of a prime is not considered a prime, but rather the product of a prime and the unit −1.

Every integer greater than one that is not itself a prime number can be expressed as a product of primes. Some numbers are products of two primes, some are products of three primes, whereas still others are products of four or more primes. (The primes need not be distinct.) For example, look at the following products of two primes:

$4 = 2 \cdot 2$	$9 = 3 \cdot 3$	$25 = 5 \cdot 5$	$49 = 7 \cdot 7$	$121 = 11 \cdot 11$
$6 = 2 \cdot 3$	$15 = 3 \cdot 5$	$35 = 5 \cdot 7$	$77 = 7 \cdot 11$	
$10 = 2 \cdot 5$	$21 = 3 \cdot 7$	$55 = 5 \cdot 11$		
$14 = 2 \cdot 7$	$33 = 3 \cdot 11$			
$22 = 2 \cdot 11$				

2. Which nonprimes less than 30 do not appear among these products?

Now look at the following products of three primes:

$8 = 2 \cdot 2 \cdot 2$	$18 = 2 \cdot 3 \cdot 3$	$27 = 3 \cdot 3 \cdot 3$	$125 = 5 \cdot 5 \cdot 5$
$12 = 2 \cdot 2 \cdot 3$	$30 = 2 \cdot 3 \cdot 5$	$45 = 3 \cdot 3 \cdot 5$	
$20 = 2 \cdot 2 \cdot 5$	$50 = 2 \cdot 5 \cdot 5$	$75 = 3 \cdot 5 \cdot 5$	

3. Compute the products
   a. $2 \cdot 2 \cdot 2 \cdot 2$
   b. $2 \cdot 2 \cdot 2 \cdot 3$
   c. $2 \cdot 2 \cdot 3 \cdot 3$
   d. $2 \cdot 3 \cdot 3 \cdot 3$
   e. $3 \cdot 3 \cdot 3 \cdot 3$

4. Avoiding exponential notation, list the prime factorizations of
   a. 36 b. 99
   c. 138 d. 168

When an integer is expressed as a product of primes, the indicated product is called the *prime factorization* of the integer. (Each prime may appear several times in such a factorization.) Look at the prime factorizations of the numbers 1 to 200 listed in the Table of Primes on the inside front cover.

Using this table, we can reduce fractions to lowest terms. Once the numerator and denominator of a fraction have been replaced by their prime factorizations, any common prime divisors of the two numbers can be seen by inspection. The fraction can then be reduced by dividing out the common divisors.

5. How many primes are there between
   a. 1 and 20?
   b. 20 and 100?
   c. 100 and 200?

Look at the following example of *reducing fractions using the Method of Primes.*

TYPICAL PROBLEM

We first factor the numerator and denominator into products of primes. Avoiding exponential notation, we write out repeated factors:

We then divide top and bottom by any common factors. *At this point, the fraction is completely reduced:*

Finally we simplify top and bottom by multiplying the remaining prime factors together. (If no factors remain, top or bottom, the product is 1.):

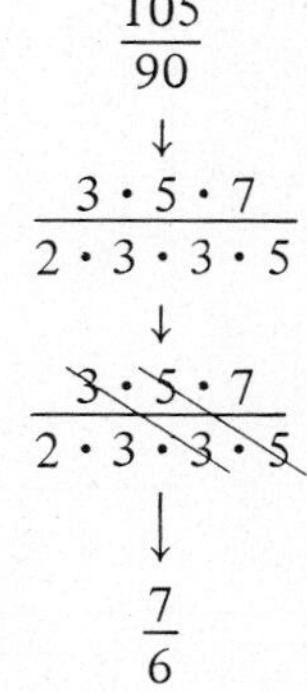

Hence, $\frac{105}{90} = \frac{7}{6}$.

The Table of Primes uses power notation, which was introduced in Chapter 2. You should always write out repeated factors yourself, however, and avoid exponents.

EXAMPLES

Reduce the following fractions using the Method of Primes.

1. $\frac{16}{72}$ 2. $\frac{42}{49}$ 3. $\frac{30}{60}$ 4. $\frac{48}{18}$

Every simple fraction can be reduced to a fraction in lowest terms.

Solutions

1. $\dfrac{\cancel{2}\cdot\cancel{2}\cdot\cancel{2}\cdot 2}{\cancel{2}\cdot\cancel{2}\cdot\cancel{2}\cdot 3\cdot 3} \rightarrow \dfrac{2}{9}$

2. $\dfrac{2\cdot 3\cdot\cancel{7}}{7\cdot\cancel{7}} \rightarrow \dfrac{6}{7}$

3. $\dfrac{\cancel{2}\cdot\cancel{3}\cdot\cancel{5}}{\cancel{2}\cdot 2\cdot\cancel{3}\cdot\cancel{5}} \rightarrow \dfrac{1}{2}$

4. $\dfrac{\cancel{2}\cdot 2\cdot 2\cdot 2\cdot\cancel{3}}{\cancel{2}\cdot 3\cdot\cancel{3}} \rightarrow \dfrac{8}{3}$

SPECIAL CASES

If all the prime factors cancel on top or bottom, then the reduced factor is 1. For example,

$$\frac{3}{3\cdot 5} = \frac{\cancel{3}}{\cancel{3}\cdot 5} = \frac{1}{5} \quad\text{and}\quad \frac{3\cdot 5}{3} = \frac{\cancel{3}\cdot 5}{\cancel{3}} = \frac{5}{1}$$

$$\frac{2\cdot 7}{2\cdot 3\cdot 7} = \frac{\cancel{2}\cdot\cancel{7}}{\cancel{2}\cdot 3\cdot\cancel{7}} = \frac{1}{3} \quad\text{and}\quad \frac{2\cdot 3\cdot 7}{2\cdot 7} = \frac{\cancel{2}\cdot 3\cdot\cancel{7}}{\cancel{2}\cdot\cancel{7}} = \frac{3}{1}$$

6. Reduce
a. $\frac{18}{54}$ b. $\frac{72}{36}$
c. $\frac{45}{15}$ d. $\frac{12}{60}$

Now look at a table of ratios of integers. The table entry is the reduced form of the ratio for each pair of numbers. Check the entries, using the Method of Primes.

7. Complete the table.

NUMERATOR over → DENOMINATOR	1	2	3	4	6	8	9	12
1	1	$\frac{1}{2}$	$\frac{1}{3}$	$\frac{1}{4}$	$\frac{1}{6}$	$\frac{1}{8}$	$\frac{1}{9}$	
2	2	1	$\frac{2}{3}$	$\frac{1}{2}$	$\frac{1}{3}$	$\frac{1}{4}$	$\frac{2}{9}$	
3	3	$\frac{3}{2}$	1	$\frac{3}{4}$	$\frac{1}{2}$	$\frac{3}{8}$	$\frac{1}{3}$	
4	4	2	$\frac{4}{3}$	1	$\frac{2}{3}$	$\frac{1}{2}$	$\frac{4}{9}$	
6	6	3	2	$\frac{3}{2}$	1	$\frac{3}{4}$	$\frac{2}{3}$	
8	8	4	$\frac{8}{3}$	2	$\frac{4}{3}$	1	$\frac{8}{9}$	
9	9	$\frac{9}{2}$	3	$\frac{9}{4}$	$\frac{3}{2}$	$\frac{9}{8}$	1	
12								

OBSERVATIONS AND REFLECTIONS

★ **1.** Remember that a fraction is unreduced if its numerator and denominator have *any* common divisor other than 1 (or –1). For example, in the fraction $\frac{18}{12}$, 18 and 12 are both divisible by 6 (which is not prime), and therefore $\frac{18}{12}$ is unreduced. However, if the numerator and denominator of a fraction have a common divisor other than 1 (or –1), they must have a common *prime* divisor. For example, the prime divisors of 6 are 2 and 3. So 2 and 3 are also common divisors of 12 and 18. Any prime divisor of a common divisor of two numbers is also a common divisor of these two numbers.

★ 2. In general, every divisor of a divisor of an integer is another divisor of the integer. (Equivalently, every divisor of a number is also a divisor of every multiple of that number.)

> If a is a divisor of b, and b is a divisor of c, then a is a divisor of c
>
> $$a|b \quad \text{and} \quad b|c \quad \Rightarrow \quad a|c$$
>
> (where a, b, and c are integers and $a|b$ is read "a is a divisor of b" and $\Rightarrow$ is read "implies").

Thus for example,

$$2|6 \quad \text{and} \quad 6|18 \quad \Rightarrow \quad 2|18$$

That is, 2 is a divisor of 6 and 6 is a divisor of 18, so 2 is a divisor of 18.

★ 3. If the numerator and denominator of a fraction do not have common primes in their prime factorizations, they do not have any common divisors, except +1 and −1. For example, $\frac{6}{35}$ is equivalent to $\frac{2 \cdot 3}{5 \cdot 7}$. So $\frac{6}{35}$ has no common divisors of its numerator and denominator except +1 and −1.

★ 4. An integer can be factored into primes in *only one way* (except for the order of the factors). It is impossible to have two different products of primes multiply to the same integer. This unique factorization property of integers is referred to as the *fundamental theorem of arithmetic.*

> FUNDAMENTAL THEOREM OF ARITHMETIC
>
> The prime factorization of an integer is unique (except for the order of the primes).

As a consequence, every prime that divides an integer is one of the primes in the unique prime factorization of that integer. For example, 2 divides 30 and 2 is a prime in the unique factorization of 30. Furthermore, *every positive nonprime* divisor of an integer (except 1) is itself a product of some (or all) of the primes in the integer's prime factorization. For example, 6 is a nonprime divisor of 30 and 6 is the product of 2 and 3, which are prime factors of 30. Hence, every positive divisor of an integer, except 1, is divisible by at least one of the prime factors of that integer. No prime can divide any integer whose prime factorization does not include that prime.

EXERCISES

Simplify each of the following fractions to lowest terms, using the Method of Primes.

1. $\frac{4}{10}$ 2. $\frac{9}{21}$ 3. $\frac{33}{15}$

4. $\frac{35}{21}$ 5. $\frac{12}{32}$ 6. $\frac{18}{81}$

7. $\frac{45}{75}$ 8. $\frac{28}{98}$ 9. $\frac{90}{48}$

10. $\frac{150}{40}$ 11. $\frac{130}{182}$ 12. $\frac{102}{114}$

13. $\frac{98}{154}$ 14. $\frac{75}{195}$ 15. $\frac{96}{104}$

16. $\frac{189}{162}$ 17. $\frac{72}{108}$ 18. $\frac{80}{200}$

19. $\frac{150}{120}$ 20. $\frac{90}{150}$ 21. $\frac{24}{198}$

22. $\frac{45}{150}$ 23. $\frac{138}{174}$ 24. $\frac{182}{154}$

6.2 REDUCING SIGNED FRACTIONS AND QUOTIENTS

The Signs of a Fraction

Signs can appear in simple fractions in three different places: before the fraction bar, in the numerator (top), and in the denominator (bottom). For example, look at the following fractions:

$$-\frac{+2}{+3}, \quad +\frac{-2}{+3}, \quad +\frac{+2}{-3}, \quad -\frac{-2}{+3}, \quad -\frac{-2}{-3}$$

The *plus* signs can, of course, be omitted. **Plus signs in the numerator and denominator are always omitted.**

Although a simple fraction can have as many as three minus signs, we rarely use fractions with more than one minus sign. To see how to work with minus signs in fractions, look at the pattern in the following table.

$+\frac{+2}{+3} = \frac{2}{3}$	$+\frac{+2}{-3} = -\frac{2}{3}$	$-\frac{+2}{+3} = -\frac{2}{3}$	$-\frac{+2}{-3} = \frac{2}{3}$
$+\frac{+1}{+3} = \frac{1}{3}$	$+\frac{+1}{-3} = -\frac{1}{3}$	$-\frac{+1}{+3} = -\frac{1}{3}$	$-\frac{+1}{-3} = \frac{1}{3}$
$+\frac{0}{+3} = 0$	$+\frac{0}{-3} = 0$	$-\frac{0}{+3} = 0$	$-\frac{0}{-3} = 0$
$+\frac{-1}{+3} = -\frac{1}{3}$	$+\frac{-1}{-3} = \frac{1}{3}$	$-\frac{-1}{+3} = \frac{1}{3}$	$-\frac{-1}{-3} = -\frac{1}{3}$
$+\frac{-2}{+3} = -\frac{2}{3}$	$+\frac{-2}{-3} = \frac{2}{3}$	$-\frac{-2}{+3} = \frac{2}{3}$	$-\frac{-2}{-3} = -\frac{2}{3}$

8. Which of these fractions are positive and which are negative?

a. $\frac{+4}{-9}$ b. $+\frac{-4}{-9}$

c. $\frac{-4}{+9}$ d. $-\frac{-4}{-9}$

e. $-\frac{4}{-9}$ f. $+\frac{+4}{+9}$

As you can see, a fraction with *two* minus signs can be simplified to a fraction with *no* minus signs, and a fraction with *three* minus signs can be simplified to a fraction with *one* minus sign.

For example, $\frac{-2}{-3}$, $-\frac{-2}{3}$, $-\frac{2}{-3}$, and $\frac{2}{3}$ are all positive, whereas $-\frac{-2}{-3}$, $\frac{-2}{3}$, $\frac{2}{-3}$, and $-\frac{2}{3}$ are all negative.

Looking at the table you can see that a minus sign means the same whether it is in the numerator, in the denominator, or before the fraction bar. For example,

$$\frac{2}{-3} = -\frac{2}{3} = \frac{-2}{3}$$

Note that changing any two signs in a simple fraction always yields an equivalent fraction. For example,

$$\frac{2}{-3} = +\frac{2}{-3} = -\frac{2}{+3} = -\frac{2}{3}$$

and

$$-\frac{2}{3} = -\frac{+2}{3} = +\frac{-2}{3} = \frac{-2}{3}$$

9. Simplify
a. $-\frac{-5}{+2}$ b. $-\frac{5}{-2}$
c. $-\frac{-5}{-2}$ d. $+\frac{-5}{-2}$

$$\frac{-a}{-b} = -\frac{-a}{b} = -\frac{a}{-b} = \frac{a}{b} \quad \text{and} \quad -\frac{-a}{-b} = \frac{a}{-b} = \frac{-a}{b} = -\frac{a}{b}$$

(where a and b represent numerical values)

Standard Form of a Simple Fraction

As we have seen, fractions have several equivalent forms. We say that a *positive* fraction is in *standard form* if it has no minus signs (and signs are omitted in the numerator and denominator). For example,

$\frac{2}{3}$ and $+\frac{2}{3}$ are in standard form

whereas $\frac{-2}{-3}$ and $-\frac{-2}{3}$ are **not** in standard form.

We say that a *negative fraction* is in *standard form* if it has only *one* minus sign, which is either in the *numerator* or before the fraction bar. For example,

$-\frac{2}{3}$, $\frac{-2}{3}$, and $+\frac{-2}{3}$ are in standard form

whereas $\frac{2}{-3}$ and $-\frac{-2}{-3}$ are **not** in standard form.

10. Place in standard form
a. $\frac{+4}{-9}$ b. $+\frac{-4}{-9}$
c. $\frac{-4}{+9}$ d. $-\frac{-4}{-9}$
e. $-\frac{4}{-9}$ f. $+\frac{+4}{+9}$

If a fraction is not in standard form, it will not be considered completely simplified.

Quotients

Look at the expressions $\frac{2}{3}$, 2/3, and 2 ÷ 3. Both $\frac{2}{3}$ and 2/3 are fractions, whereas 2 ÷ 3 is called a *quotient* (or an *indicated quotient*). These expressions all have the same meaning.

2 ÷ 3 is equivalent to 2/3 is equivalent to $\frac{2}{3}$

11. Rewrite quotients using fractional notation
a. 7 ÷ 3 b. −4 ÷ 9

12. Rewrite in standard form
a. $4 \div (-9)$
b. $-4 \div (+9)$

In a simple fraction using the "/" notation (e.g., $-2/3$) and in a quotient of integers (e.g., $-2 \div 3$), only two signs can appear (one before each integer). If the second integer has an expressed sign, it must be in parentheses; for example, $2 \div (-3)$. We say that the expression is in *standard form* if the *second* integer is positive and written in natural number notation with its plus sign omitted. For example,

$-2/3$ and $-2 \div 3$ are in standard form

whereas $2/(-3)$ and $2 \div (-3)$ are **not** in standard form.

Reducing Fractions

In the previous section we used the Method of Primes to reduce simple fractions. In this section we will study another method, which is perhaps faster to use (especially when a table of prime factorizations is not available for reference). This alternative procedure suggests looking for any common divisor, prime or otherwise. Often a divisor other than a prime (a product of primes) will be obvious and the need for dividing by one prime at a time can be avoided. For example, look at the following pairs of equivalent fractions:

13. a. Divide by 6 to reduce $\frac{18}{42}$
b. Divide by 8 to reduce $\frac{40}{24}$

$$\frac{8}{12} = \frac{2 \cdot \cancel{4}}{3 \cdot \cancel{4}} = \frac{2}{3} \qquad \frac{27}{36} = \frac{3 \cdot \cancel{9}}{4 \cdot \cancel{9}} = \frac{3}{4}$$

$$\frac{30}{75} = \frac{2 \cdot \cancel{15}}{5 \cdot \cancel{15}} = \frac{2}{5} \qquad \frac{42}{28} = \frac{3 \cdot \cancel{14}}{2 \cdot \cancel{14}} = \frac{3}{2}$$

Dividing the numerator and denominator by any common divisor larger than 1 always results in an equivalent fraction with a smaller numerator and denominator. If we divide by one common divisor after another, we will eventually arrive at the reduced form of the fraction. For example, the fraction $\frac{420}{252}$ can be reduced by the following sequence of computations:

$$\frac{420}{252} = \frac{105 \cdot \cancel{4}}{63 \cdot \cancel{4}} = \frac{35 \cdot \cancel{3}}{21 \cdot \cancel{3}} = \frac{5 \cdot \cancel{7}}{3 \cdot \cancel{7}} = \frac{5}{3}$$

It does not matter which common divisor of the numerator and denominator is found first, or in what order divisors are determined. The original fraction will always simplify to the same reduced fraction. For example, $\frac{80}{120}$ can be reduced to $\frac{2}{3}$ directly by the computation

$$\frac{80}{120} = \frac{2 \cdot \cancel{40}}{3 \cdot \cancel{40}} = \frac{2}{3}$$

or in two steps by the computation

$$\frac{80}{120} = \frac{8 \cdot \cancel{10}}{12 \cdot \cancel{10}} = \frac{2 \cdot \cancel{4}}{3 \cdot \cancel{4}} = \frac{2}{3}$$

Look at the following examples of *reducing signed fractions and quotients.*

TYPICAL PROBLEMS

1. $-\frac{16}{-24}$

We first determine the sign of the fraction: 2 minuses → +

Then we find common divisors and reduce the fraction: $+\frac{\overset{2}{\cancel{16}}}{\underset{3}{\cancel{24}}} \to \frac{2}{3}$

Hence, $-\frac{16}{-24} = \frac{2}{3}$.

2. $18 \div (-72)$

We must first form a fraction: $\frac{18}{-72}$

Then we determine the sign of the fraction: 1 minus → −

We then find common divisors and reduce the fraction. We repeat this step until the fraction is completely reduced: $-\frac{\overset{9}{\cancel{18}}}{\underset{36}{\cancel{72}}} \to -\frac{\overset{1}{\cancel{9}}}{\underset{4}{\cancel{36}}} \to -\frac{1}{4}$

Hence, $18 \div (-72) = -\frac{1}{4}$.

To determine the sign of the fraction, count all the minus signs (top, bottom, and before the fraction bar). If there is an odd number of minus signs, the quotient is negative; otherwise, the quotient is positive.

EXAMPLES

Reduce the following signed fractions and quotients.

Every simple fraction can be reduced to a fraction in lowest terms in standard form.

1. $\frac{-9}{54}$
2. $\frac{24}{-45}$
3. $\frac{-28}{-36}$
4. $-\frac{-55}{-22}$
5. $-80 \div (-52)$
6. $-48 \div 72$

Solutions

1. $-\frac{\overset{1}{\cancel{9}}}{\underset{6}{\cancel{54}}} \to -\frac{1}{6}$
2. $-\frac{\overset{8}{\cancel{24}}}{\underset{15}{\cancel{45}}} \to -\frac{8}{15}$
3. $+\frac{\overset{7}{\cancel{28}}}{\underset{9}{\cancel{36}}} \to \frac{7}{9}$
4. $-\frac{\overset{5}{\cancel{55}}}{\underset{2}{\cancel{22}}} \to -\frac{5}{2}$
5. $\frac{-80}{-52} \to +\frac{\overset{40}{\cancel{80}}}{\underset{26}{\cancel{52}}} \to +\frac{\overset{20}{\cancel{40}}}{\underset{13}{\cancel{26}}} \to \frac{20}{13}$
6. $\frac{-48}{72} \to -\frac{\overset{6}{\cancel{48}}}{\underset{9}{\cancel{72}}} \to -\frac{\overset{2}{\cancel{6}}}{\underset{3}{\cancel{9}}} \to -\frac{2}{3}$

SPECIAL CASES

When the denominator (bottom) of a fraction is 1, it can be omitted. For example,

$$\frac{2}{1} = 2, \quad \frac{-2}{1} = -2, \quad \text{and} \quad -\frac{2}{1} = -2$$

14. Simplify
  a. $\frac{18}{6}$
  b. $\frac{42}{14}$
  c. $\frac{105}{15}$

For example, look at the computation

$$\frac{10}{5} \rightarrow \frac{\overset{2}{\cancel{10}}}{\underset{1}{\cancel{5}}} \rightarrow \frac{2}{1} \rightarrow 2$$

In reducing simple fractions, express the reduced fraction as an integer if the denominator reduces to 1. Note that a simple fraction can be reduced to an integer whenever the numerator is a multiple of the denominator.

In the previous section we saw a table of reduced fractions with positive integers. Here is a similar table for signed fractions. Check the entries to make sure that you understand the rule for the sign of a fraction.

15. Complete the table.

NUMERATOR ↓ / DENOMINATOR over →	−6	−5	−4	−3	−2	−1	0	+1	+2	+3	+4	+5	+6
+6													
+5													
+4			−1	$-\frac{4}{3}$	−2	−4		+4	+2	$+\frac{4}{3}$	+1		
+3			$-\frac{3}{4}$	−1	$-\frac{3}{2}$	−3		+3	$+\frac{3}{2}$	+1	$+\frac{3}{4}$		
+2			$-\frac{1}{2}$	$-\frac{2}{3}$	−1	−2	U N	+2	+1	$+\frac{2}{3}$	$+\frac{1}{2}$		
+1			$-\frac{1}{4}$	$-\frac{1}{3}$	$-\frac{1}{2}$	−1	D E	+1	$+\frac{1}{2}$	$+\frac{1}{3}$	$+\frac{1}{4}$		
0			0	0	0	0	F I	0	0	0	0		
−1			$+\frac{1}{4}$	$+\frac{1}{3}$	$+\frac{1}{2}$	+1	N E	−1	$-\frac{1}{2}$	$-\frac{1}{3}$	$-\frac{1}{4}$		
−2			$+\frac{1}{2}$	$+\frac{2}{3}$	+1	+2	D	−2	−1	$-\frac{2}{3}$	$-\frac{1}{2}$		
−3			$+\frac{3}{4}$	+1	$+\frac{3}{2}$	+3		−3	$-\frac{3}{2}$	−1	$-\frac{3}{4}$		
−4			+1	$+\frac{4}{3}$	+2	+4		−4	−2	$-\frac{4}{3}$	−1		
−5													
−6													

OBSERVATIONS AND REFLECTIONS

1. When two fractions are equivalent, the product of the numerator of the first fraction and the denominator of the second fraction equals the product of the numerator of the second fraction and the denominator of the first fraction. For example,

$$\frac{3}{6} = \frac{5}{10} \quad \text{and} \quad 3 \cdot 10 = 6 \cdot 5$$

$$\frac{8}{24} = \frac{3}{9} \quad \text{and} \quad 8 \cdot 9 = 24 \cdot 3$$

$$\frac{6}{15} = \frac{10}{25} \quad \text{and} \quad 6 \cdot 25 = 15 \cdot 10$$

> $\frac{a}{b} = \frac{c}{d}$ if and only if $a \cdot d = b \cdot c$
>
> (where a, b, c, and d represent nonzero integers)

This can be justified as follows:

If $a \cdot d = b \cdot c$, then because $\frac{a}{b} = \frac{a \cdot d}{b \cdot d}$ and $\frac{c}{d} = \frac{b \cdot c}{b \cdot d}$,

because $a \cdot d = b \cdot c$

$$\frac{a}{b} = \frac{a \cdot d}{b \cdot d} = \frac{b \cdot c}{b \cdot d} = \frac{c}{d}$$

★ 2. As we have seen, dividing both the numerator and denominator of a fraction by any common divisor yields an equivalent fraction. Dividing by the *greatest common divisor* (GCD) of the numerator and denominator yields an equivalent *reduced* fraction. Hence, *every fraction is equivalent to a reduced fraction.* For example, the GCD of 80 and 120 is 40, and $\frac{80}{120} = \frac{2 \cdot \cancel{40}}{3 \cdot \cancel{40}} = \frac{2}{3}$, which is reduced. Note that the GCD of the numerator and denominator of a reduced fraction is 1.

★ 3. Dividing the numerator and denominator of a fraction by one common divisor after another brings us to the reduced form of the fraction. The product of these divisors is always the GCD of the original numerator and denominator of the fraction. However, is our result always the same as when we divide by the GCD? Specifically, does dividing a quantity first by one integer and then by another produce the same result as dividing all at once by the product of these two integers (divisors)? The answer is yes. To see that this is the case, look at the pattern in the following examples.

$6 \div 3 = 2$	and	$2 \div 2 = 1$	vs.	$6 \div 6 = 1$
$12 \div 3 = 4$	and	$4 \div 2 = 2$	vs.	$12 \div 6 = 2$
$18 \div 3 = 6$	and	$6 \div 2 = 3$	vs.	$18 \div 6 = 3$
$24 \div 3 = 8$	and	$8 \div 2 = 4$	vs.	$24 \div 6 = 4$

16. Determine which pairs of fractions are equivalent.

a. $\frac{4}{12}$ and $\frac{5}{15}$

b. $\frac{12}{16}$ and $\frac{8}{12}$

c. $\frac{20}{15}$ and $\frac{5}{4}$

d. $\frac{12}{18}$ and $\frac{20}{30}$

$256 \div 2 = 128$	and	$128 \div 2 = 64$	vs.	$256 \div 4 = 64$
$256 \div 4 = 64$	and	$64 \div 4 = 16$	vs.	$256 \div 16 = 16$
$256 \div 8 = 32$	and	$32 \div 8 = 4$	vs.	$256 \div 64 = 4$
$256 \div 16 = 16$	and	$16 \div 16 = 1$	vs.	$256 \div 256 = 1$

17. Compute
a. $36 \div 12$
b. $(36 \div 3) \div 4$
c. $96 \div 24$
d. $(96 \div 4) \div 6$

$$(a \div b) \div c = a \div (b \cdot c)$$
(where *a, b,* and *c* represent nonzero integers)

EXERCISES

Simplify each of the following expressions to a reduced fraction in standard form.

1. $\frac{4}{18}$
2. $\frac{120}{84}$
3. $\frac{18}{28}$
4. $\frac{16}{128}$
5. $\frac{120}{40}$
6. $\frac{24}{144}$
7. $15 \div 18$
8. $26 \div 169$
9. $50 \div 75$
10. $48 \div 64$
11. $\frac{-12}{56}$
12. $-\frac{72}{86}$
13. $\frac{-14}{-35}$
14. $\frac{15}{-48}$
15. $\frac{-8}{12}$
16. $-\frac{22}{-132}$
17. $\frac{144}{-56}$
18. $-\frac{-16}{72}$
19. $\frac{-27}{117}$
20. $-\frac{200}{-55}$
21. $-18 \div (-12)$
22. $24 \div (-32)$
23. $-(-32) \div (-36)$
24. $-(64) \div (-120)$

6.3 MULTIPLYING FRACTIONS

18. What are the factors of
a. $\left(\frac{-2}{7}\right)\left(\frac{1}{-5}\right)$
b. $\left(\frac{3}{8}\right)\left(\frac{-7}{2}\right)\left(\frac{4}{-9}\right)$

The language of algebraic multiplication extends naturally to fractions. In *algebraic products* of fractions, such as $\left(\frac{-3}{5}\right)\left(\frac{-4}{3}\right)$ and $\left(\frac{9}{-5}\right)\left(\frac{3}{8}\right)\left(\frac{-4}{15}\right)$, each quantity is called a *factor* of the product. For example, in the product $\left(\frac{-3}{5}\right)\left(\frac{-4}{3}\right)$ the factors are $\frac{-3}{5}$ and $\frac{-4}{3}$. Similarly, in the product $\left(\frac{9}{-5}\right)\left(\frac{3}{8}\right)\left(\frac{-4}{15}\right)$ the factors are $\frac{9}{-5}$, $\frac{3}{8}$, and $\frac{-4}{15}$.

Multiplication of fractions is indicated in any of the following ways:

1. $\frac{-3}{5} \cdot \frac{-4}{3}$ with a dot between fractions and no plus or minus sign between the fraction bars.
2. $\frac{-3}{5}\left(\frac{-4}{3}\right)$ with no symbol between the fractions and only the second fraction inside parentheses.
3. $\left(\frac{-3}{5}\right)\left(\frac{-4}{3}\right)$ with no symbol between sets of parentheses.
4. $\left(\frac{-3}{5}\right) \cdot \left(\frac{-4}{3}\right)$ with a dot between sets of parentheses.

When a fraction stands by itself or is the first factor in a product of fractions, it may be written with a sign before the fraction bar. However, no other fraction in a product may have a sign before the fraction bar unless the fraction is in parentheses. For example, the product of $-\frac{2}{3}$ and $+\frac{3}{4}$ can be written as $-\frac{2}{3} \cdot \frac{3}{4}$ or $\frac{-2}{3} \cdot \frac{3}{4}$ or $-\frac{2}{3}\left(+\frac{3}{4}\right)$, but **not** as $-\frac{2}{3} \cdot +\frac{3}{4}$.

Products

As we have seen, a simple fraction can be expressed as the product of an integer and the reciprocal of an integer. For example, $\frac{3}{4} = 3 \cdot \frac{1}{4}$ and $\frac{-9}{5} = -9 \cdot \frac{1}{5}$. Hence, a product of *fractions* can be expressed as a product of *integers* and *reciprocals of integers.* For example,

$$\frac{3}{-8} \cdot \frac{-4}{15} \quad \text{is equivalent to} \quad 3 \cdot \frac{1}{-8} \cdot (-4) \cdot \frac{1}{15}$$

Parentheses are implied in $\frac{-4}{15}$, but must be written in $(-4) \cdot \frac{1}{15}$.

Because multiplication is commutative and associative, the factors in a product of fractions may be written in any order and multiplied in any order. Hence, for example,

$$\frac{3}{-8} \cdot \frac{-4}{15} = 3 \cdot (-4) \cdot \frac{1}{-8} \cdot \frac{1}{15} = \frac{3 \cdot (-4)}{-8 \cdot 15}$$

$$\frac{a}{b} \cdot \frac{c}{d} = a \cdot c \cdot \frac{1}{b} \cdot \frac{1}{d} = \frac{a \cdot c}{b \cdot d}$$

(where a, b, c, and d represent nonzero integers)

19. Express as a product of integers and reciprocals of integers

a. $\frac{-8}{9} \cdot \frac{6}{28}$

b. $\frac{3}{5} \cdot \frac{-5}{18}$

c. $-\frac{14}{3} \cdot \frac{15}{-7}$

Reducing

Look at the pattern in the following examples:

$$\frac{1}{3} \cdot \frac{3}{4} = \frac{3}{3} \cdot \frac{1}{4} \qquad \frac{4}{3} \cdot \frac{9}{40} = \frac{9}{3} \cdot \frac{4}{40}$$

$$\frac{2}{5} \cdot \frac{3}{8} = \frac{3}{5} \cdot \frac{2}{8} \qquad \frac{6}{35} \cdot \frac{49}{27} = \frac{49}{35} \cdot \frac{6}{27}$$

$$\frac{12}{25} \cdot \frac{15}{16} = \frac{15}{25} \cdot \frac{12}{16} \qquad \frac{36}{65} \cdot \frac{55}{24} = \frac{55}{65} \cdot \frac{36}{24}$$

20. Reduce the fractions where possible.

Can you see how rewriting these products might help us to simplify them? In each example, at least one of the fractions on the right can be reduced. Reducing before multiplying simplifies the task of finding the product, because then we can work with smaller numbers.

It is not actually necessary, however, to rewrite a product of fractions in order to reduce the numerators and denominators. If one of the numerators and one of the denominators have a common divisor, we can divide them both by their common divisor without first rearranging the product.

Signs

Remember that a minus sign means the same whether it is in the numerator, in the denominator, or before the fraction bar. Hence, the sign of a product of fractions can be determined in the same way as the sign of a product of integers. If there is an odd number of minus signs (in the numerators, denominators, or before the fraction bars), the sign of the product is minus. If there is an even number of minus signs, the sign of the product is plus.

21. Similarly, indicate the product $\frac{1}{4} \cdot \frac{2}{3}$ in the diagram.

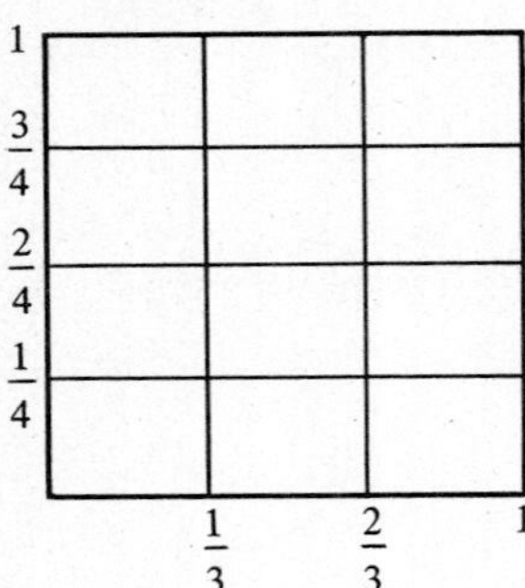

$$\frac{1}{4} \cdot \frac{1}{3} = \frac{1}{12}$$

and

$$\frac{3}{4} \cdot \frac{2}{3} = \frac{6}{12}$$

Look at the following example of *multiplying fractions.*

TYPICAL PROBLEM

$$\frac{25}{9} \cdot \frac{-3}{8} \cdot \frac{4}{35}$$

We first determine the sign of the product: 1 minus → −

Then we reduce the numerators and denominators:

$$-\frac{25}{\cancel{9}_{3}} \cdot \frac{\cancel{3}^{1}}{8} \cdot \frac{4}{35}$$

$$-\frac{25}{3} \cdot \frac{1}{\cancel{8}_{2}} \cdot \frac{\cancel{4}^{1}}{35}$$

$$-\frac{\cancel{25}^{5}}{3} \cdot \frac{1}{2} \cdot \frac{1}{\cancel{35}_{7}}$$

$$-\frac{5}{3} \cdot \frac{1}{2} \cdot \frac{1}{7}$$

To reduce the numerators and denominators we consider one numerator and denominator at a time. To reduce them, we use the method in the previous section.

Finally we multiply the numerators and then we multiply the denominators:

$$-\frac{5}{42}$$

Hence, $\frac{25}{9} \cdot \frac{-3}{8} \cdot \frac{4}{35} = -\frac{5}{42}$.

EXAMPLES

Every product of simple fractions can be simplified to a single reduced fraction in standard form.

Multiply the following numerical fractions.

1. $6\left(\frac{1}{3}\right)$

2. $\frac{1}{7} \cdot \frac{-1}{9}$

3. $\frac{-6}{7} \cdot \frac{5}{8} \cdot \frac{-4}{12}$

4. $15\left(-\frac{8}{45}\right)$

5. $\frac{-5}{3} \cdot \left(\frac{21}{-95}\right)$

6. $\frac{30}{-8} \cdot \frac{-18}{25} \cdot \frac{10}{21}$

Solutions

1. $\overset{2}{\cancel{6}} \cdot \frac{1}{\underset{1}{\cancel{3}}} \to 2$

2. $-\frac{1}{7} \cdot \frac{1}{9} \to -\frac{1}{63}$

3. $+\frac{\overset{1}{\cancel{6}}}{7} \cdot \frac{5}{\underset{2}{\cancel{8}}} \cdot \frac{\overset{1}{\cancel{4}}}{\underset{2}{\cancel{12}}} \to \frac{5}{28}$

4. $-\overset{1}{\cancel{15}} \cdot \frac{8}{\underset{3}{\cancel{45}}} \to -\frac{8}{3}$

5. $+\frac{\overset{1}{\cancel{5}}}{\underset{1}{\cancel{3}}} \cdot \frac{\overset{7}{\cancel{21}}}{\underset{19}{\cancel{95}}} \to \frac{7}{19}$

6. $-\frac{\overset{\overset{3}{\cancel{6}}}{\cancel{30}}}{\underset{\underset{\underset{1}{\cancel{2}}}{\cancel{4}}}{\cancel{8}}} \cdot \frac{\overset{\overset{3}{\cancel{6}}}{\cancel{18}}}{\underset{\underset{1}{\cancel{5}}}{\cancel{25}}} \cdot \frac{\overset{\overset{1}{\cancel{2}}}{\cancel{10}}}{\underset{7}{\cancel{21}}} \to \frac{-9}{7}$

SPECIAL CASES

1. Integer factors in a product of fractions are always treated as numerators. For example,

$$6\left(\frac{3}{4}\right) \quad \text{equals} \quad \frac{6}{1} \cdot \frac{3}{4}$$

When multiplying an integer by a fraction, we need not rewrite the integer as a fraction. The integer factor is treated as a numerator when the numerators and denominators are reduced. For example,

$$6\left(\frac{3}{4}\right) \xrightarrow{\text{Reduce}} \overset{3}{\cancel{6}}\left(\frac{3}{\underset{2}{\cancel{4}}}\right) \xrightarrow{\text{Multiply}} \frac{9}{2}$$

2. The product of any fraction and 1 equals the fraction. For example,

$$\left(\frac{3}{4}\right)(1) = \frac{3}{4} \quad \text{and} \quad (1)\left(\frac{3}{4}\right) = \frac{3}{4}$$

3. The product of any fraction and -1 equals the negative of the fraction. For example,

$$\left(\frac{3}{4}\right)(-1) = \frac{-3}{4} \quad \text{and} \quad (-1)\left(\frac{-3}{4}\right) = \frac{3}{4}$$

22. Compute

a. $3\left(\frac{2}{3}\right)$

b. $4\left(\frac{5}{8}\right)$

c. $6\left(\frac{14}{9}\right)$

23. Compute

a. $\frac{-3}{7} \cdot 1$

b. $1 \cdot \frac{4}{9}$

24. Compute

a. $\frac{-3}{7}(-1)$

b. $(-1)\left(\frac{4}{9}\right)$

Here is a simple multiplication table for the sixth's. Verify the entries to check your understanding of multiplication.

times	$\frac{1}{6}$	$\frac{2}{6}$	$\frac{3}{6}$	$\frac{4}{6}$	$\frac{5}{6}$	$\frac{6}{6}$
$\frac{1}{6}$	$\frac{1}{36}$	$\frac{1}{18}$	$\frac{1}{12}$	$\frac{1}{9}$	$\frac{5}{36}$	$\frac{1}{6}$
$\frac{2}{6}$	$\frac{1}{18}$	$\frac{1}{9}$	$\frac{1}{6}$	$\frac{2}{9}$	$\frac{5}{18}$	$\frac{1}{3}$
$\frac{3}{6}$	$\frac{1}{12}$	$\frac{1}{6}$	$\frac{1}{4}$	$\frac{1}{3}$	$\frac{5}{12}$	$\frac{1}{2}$
$\frac{4}{6}$	$\frac{1}{9}$	$\frac{2}{9}$	$\frac{1}{3}$	$\frac{4}{9}$	$\frac{5}{9}$	$\frac{2}{3}$
$\frac{5}{6}$	$\frac{5}{36}$	$\frac{5}{18}$	$\frac{5}{12}$	$\frac{5}{9}$	$\frac{25}{36}$	$\frac{5}{6}$
$\frac{6}{6}$	$\frac{1}{6}$	$\frac{1}{3}$	$\frac{1}{2}$	$\frac{2}{3}$	$\frac{5}{6}$	1

25. Complete the table.

times	$\frac{1}{4}$	$\frac{2}{4}$	$\frac{3}{4}$
$\frac{1}{4}$			
$\frac{2}{4}$			
$\frac{3}{4}$			

OBSERVATIONS AND REFLECTIONS

1. When multiplying fractions, we can form a single fraction before reducing. We can first multiply all the numerators to form a single numerator and then multiply all the denominators to form a single denominator. For example, the product

   $\frac{3}{6} \cdot \frac{2}{5} \cdot \frac{10}{12}$ can be converted to the single fraction $\frac{3 \cdot 2 \cdot 10}{6 \cdot 5 \cdot 12} = \frac{60}{360}$ before

   reducing. The resulting fraction can then be reduced by either of the methods in the previous two sections. The reduced fraction we obtain is, of course, the same as the one we get using the method of this section (see the following computations). The numerators and denominators in a product of fractions are all *factors of a product* and may be multiplied and reduced in any order.

2. To multiply fractions we do not specify any order in which to reduce the numerators and denominators. A product may be simplified in many different ways. For

   example, the product $\frac{3}{6} \cdot \frac{2}{5} \cdot \frac{10}{12}$ may be simplified in any of the following ways.

$$\frac{\overset{1}{\cancel{3}}}{6} \cdot \frac{2}{5} \cdot \frac{10}{\underset{4}{\cancel{12}}} \rightarrow \frac{1}{6} \cdot \frac{2}{\underset{1}{\cancel{5}}} \cdot \frac{\overset{2}{\cancel{10}}}{4} \rightarrow \frac{1}{\underset{3}{\cancel{6}}} \cdot \frac{\overset{1}{\cancel{2}}}{1} \cdot \frac{2}{4} \rightarrow \frac{1}{3} \cdot \frac{1}{1} \cdot \frac{\overset{1}{\cancel{2}}}{\underset{2}{\cancel{4}}} \rightarrow \frac{1}{3} \cdot \frac{1}{1} \cdot \frac{1}{2} = \frac{1}{6}$$

or

$$\frac{3}{\underset{3}{\cancel{6}}} \cdot \frac{\overset{1}{\cancel{2}}}{5} \cdot \frac{10}{12} \rightarrow \frac{\overset{1}{\cancel{3}}}{\underset{1}{\cancel{3}}} \cdot \frac{1}{5} \cdot \frac{10}{12} \rightarrow \frac{1}{1} \cdot \frac{1}{\underset{1}{\cancel{5}}} \cdot \frac{\overset{2}{\cancel{10}}}{12} \rightarrow \frac{1}{1} \cdot \frac{1}{1} \cdot \frac{\overset{1}{\cancel{2}}}{\underset{6}{\cancel{12}}} \rightarrow \frac{1}{1} \cdot \frac{1}{1} \cdot \frac{1}{6} = \frac{1}{6}$$

or

$$\frac{3}{6} \cdot \frac{\overset{1}{\cancel{2}}}{5} \cdot \frac{10}{\underset{6}{\cancel{12}}} \rightarrow \frac{\overset{1}{\cancel{3}}}{6} \cdot \frac{1}{5} \cdot \frac{10}{\underset{2}{\cancel{6}}} \rightarrow \frac{1}{\underset{3}{\cancel{6}}} \cdot \frac{1}{5} \cdot \frac{\overset{5}{\cancel{10}}}{2} \rightarrow \frac{1}{3} \cdot \frac{1}{\underset{1}{\cancel{5}}} \cdot \frac{\overset{1}{\cancel{5}}}{2} \rightarrow \frac{1}{3} \cdot \frac{1}{1} \cdot \frac{1}{2} = \frac{1}{6}$$

or

(multiplying first)

$$\frac{3 \cdot 2 \cdot 10}{6 \cdot 5 \cdot 12} \rightarrow \frac{60}{360} \rightarrow \frac{6}{36} \rightarrow \frac{1}{6}$$

There are other ways of reducing $\frac{3}{6} \cdot \frac{2}{5} \cdot \frac{10}{12}$, but the result is always $\frac{1}{6}$.

26. Find more ways to reduce this product.

★ **3.** Because integers can be considered fractions with a denominator of 1, multiplying integers is a special case of multiplying fractions. For example, $(-2)(3) = -6$ is equivalent to $\left(\frac{-2}{1}\right)\left(\frac{3}{1}\right) = \frac{-6}{1}$. In this special case, in which all fractions have a denominator of 1, our procedure is the same as the procedure in Section 1.4 for multiplying integers.

27. Express $-\frac{3}{1}\left(\frac{-4}{1}\right) = \frac{12}{1}$ as a product of integers.

EXERCISES

Simplify each of the following expressions to a reduced fraction.

1. $8 \cdot \frac{5}{6}$
2. $7 \cdot \frac{1}{21}$
3. $\frac{5}{2} \cdot 12$
4. $\frac{7}{36} \cdot 48$
5. $\frac{8}{9} \cdot \frac{1}{4}$
6. $\frac{1}{14} \cdot \frac{28}{9}$
7. $\frac{5}{3} \cdot \frac{-1}{4}$
8. $\frac{-1}{4} \cdot \frac{7}{2}$
9. $\frac{6}{5} \cdot \frac{15}{-8}$
10. $\frac{-5}{12} \cdot \frac{-18}{25}$
11. $\frac{-7}{24} \cdot \frac{16}{-42}$
12. $\frac{15}{-27} \cdot \frac{-72}{-35}$
13. $\frac{4}{9}\left(-\frac{27}{64}\right)$
14. $\frac{18}{56}\left(-\frac{42}{81}\right)$
15. $\frac{15}{16}\left(\frac{-9}{45}\right)$
16. $\frac{21}{-4}\left(\frac{-55}{35}\right)$
17. $-\frac{16}{25}\left(-\frac{90}{48}\right)$
18. $-\frac{7}{11}\left(-\frac{121}{77}\right)$

19. $\frac{3}{5} \cdot \frac{15}{16} \cdot \frac{4}{9}$ 20. $\frac{8}{5} \cdot \frac{35}{6} \cdot \frac{9}{14}$ 21. $\frac{7}{3} \cdot \frac{3}{8} \cdot \frac{16}{7}$

22. $\frac{-5}{12} \cdot \frac{-8}{7} \cdot \frac{21}{25}$ 23. $\frac{-6}{27} \cdot \frac{20}{21} \cdot \frac{18}{-8}$ 24. $\frac{-44}{16} \cdot \frac{18}{-39} \cdot \frac{-52}{66}$

Express each of the following word phrases algebraically and then simplify it.

1. $\frac{3}{7}$ times $\frac{5}{6}$

2. The product of $\frac{4}{5}$ and $-\frac{15}{16}$

3. $\frac{9}{10}$ of the product of $\frac{-5}{12}$ and -3

4. The product of half $\frac{13}{14}$ and $-\frac{8}{7}$

5. Twice $\frac{54}{25}$ multiplied by $-\frac{60}{63}$

6. The product of $\frac{-15}{35}$ and half -56

7. -5 multiplied by the product of $\frac{-8}{25}$ and $\frac{21}{12}$

8. The product of $\frac{-44}{30}$ and $\frac{-52}{16}$ times $\frac{9}{-66}$

9. 36 multiplied by $\frac{-13}{48}$ times $\frac{-2}{39}$

10. One-third the product of $\frac{60}{77}$ and $\frac{22}{35}$

11. Seven-eighths of $\frac{4}{21}$ times $\frac{-6}{13}$

12. $\frac{42}{15}$ multiplied by $\frac{25}{14}$ of $\frac{-27}{12}$

6.4 QUOTIENTS AND RATIOS OF FRACTIONS

Dividing By a Fraction

To divide by an integer we multiply by the reciprocal of that integer. For example,

$$8 \div 2 = \frac{8}{2} = 8 \cdot \frac{1}{2}$$

To extend this idea to fractions, compare the patterns in the following columns.

$$4 \div \frac{4}{1} = 4 \div 4 = 1 \quad \text{vs.} \quad 4 \cdot \frac{1}{4} = 1$$

$$4 \div \frac{2}{1} = 4 \div 2 = 2 \qquad 4 \cdot \frac{1}{2} = 2$$

$$4 \div \frac{1}{1} = 4 \div 1 = 4 \qquad 4 \cdot \frac{1}{1} = 4$$

$$4 \div \frac{1}{2} \stackrel{?}{=} 8 \qquad 4 \cdot \frac{2}{1} = 8$$

$$4 \div \frac{1}{4} \stackrel{?}{=} 16 \qquad 4 \cdot \frac{4}{1} = 16$$

28. Compute

a. $3 \div \frac{3}{1}$, $3 \div \frac{2}{1}$, $3 \div \frac{1}{2}$, $3 \div \frac{1}{3}$

b. $3 \cdot \frac{1}{3}$, $3 \cdot \frac{1}{2}$, $3 \cdot \frac{2}{1}$, $3 \cdot \frac{3}{1}$

These patterns suggest that we should use reciprocals when dividing by fractions as well: $4 \div \frac{1}{2} = 4 \cdot \frac{2}{1}$ and $4 \div \frac{1}{4} = 4 \cdot \frac{4}{1}$.

The Reciprocal of a Fraction

The *reciprocal of a fraction* is formed by inverting the fraction; that is, by interchanging the numerator and denominator. For example,

the reciprocal of $\frac{3}{8}$ is $\frac{8}{3}$ $\left[\frac{3}{8} \rightleftarrows \frac{8}{3}\right]$

and

the reciprocal of $\frac{-4}{5}$ is $\frac{5}{-4}$ $\left[\frac{-4}{5} \rightleftarrows \frac{5}{-4}\right]$

(However, a fraction such as $\frac{0}{4}$ has no reciprocal because $\frac{0}{4}$ is 0. Zero has no reciprocal.)

As a check on our reasoning, note that the product of a nonzero fraction and its reciprocal is 1. For example,

$$\frac{3}{8} \text{ times } \frac{8}{3} = 1 \quad \text{and} \quad \frac{-4}{5} \text{ times } \frac{5}{-4} = 1$$

29. Compute
a. $\frac{5}{4} \div \frac{5}{4}$
b. $\frac{5}{4} \cdot \frac{4}{5}$
c. $\frac{-1}{6} \div \frac{-1}{6}$
d. $\frac{-1}{6} \cdot \frac{-6}{1}$

Hence, dividing a fraction by itself gives the same result as multiplying the fraction by its reciprocal. (Dividing any nonzero expression by itself yields 1.)

$$\frac{3}{8} \div \frac{3}{8} \quad \text{is equivalent to} \quad \frac{3}{8} \cdot \frac{8}{3} = 1$$

Inverting and Multiplying

We can conclude that *dividing* one fraction by another is equivalent to *multiplying* the first fraction by the *reciprocal* of the second fraction. For example,

$$\frac{-4}{5} \div \frac{3}{8} \quad \text{is equivalent to} \quad \frac{-4}{5} \cdot \frac{8}{3} = \frac{-32}{15}$$

$$1 \div \frac{3}{8} \quad \text{is equivalent to} \quad 1 \cdot \frac{8}{3} = \frac{8}{3}$$

(A formal proof is provided at the end of the section.)

Complex Fractions

A fraction whose numerator and/or denominator is a fraction is a *ratio* of fractions and is called a *complex fraction.* Ratios of fractions are equivalent to quotients of fractions, but look very different.

Remember that the division symbol, ÷, and the fraction bar, —, are identical in meaning.

$$\frac{\frac{2}{7}}{14} \text{ is equivalent to } \frac{2}{7} \div 14$$

$$\frac{4}{\frac{7}{2}} \text{ is equivalent to } 4 \div \frac{7}{2}$$

$$\frac{\frac{1}{4}}{\frac{5}{2}} \text{ is equivalent to } \frac{1}{4} \div \frac{5}{2}$$

$$\frac{\frac{3}{5}}{\frac{2}{3}} \text{ is equivalent to } \frac{3}{5} \div \frac{2}{3}$$

(More complicated numerical complex fractions are discussed in Section 7.6.)

30. Express as quotients

a. $\dfrac{\frac{1}{3}}{6}$ b. $\dfrac{\frac{3}{8}}{\frac{9}{4}}$

A ratio of fractions can be simplified by changing the largest fraction bar to a *division* symbol and then simplifying the resulting quotient of fractions.

Look at the following example of *dividing fractions.*

TYPICAL PROBLEM

$$\frac{5}{-9} \div \frac{15}{2}$$

First we invert the fraction after the division sign and change the operation to multiplication:

$$\frac{5}{-9} \cdot \frac{2}{15}$$

Then we determine the sign of the product:

1 minus → −

We then multiply the fractions (first reducing the numerators and denominators):

$$-\frac{\overset{1}{\cancel{5}}}{9} \cdot \frac{2}{\underset{3}{\cancel{15}}}$$

$$-\frac{2}{27}$$

To multiply the fractions, we use the method in the previous section.

Hence, $\frac{5}{-9} \div \frac{15}{2} = -\frac{2}{27}$.

EXAMPLES

Every quotient or ratio of simple fractions can be simplified to a single reduced fraction in standard form.

Simplify the following quotients and ratios of fractions.

1. $10 \div \frac{-5}{8}$
2. $\frac{17}{-8} \div \left(-\frac{17}{27}\right)$
3. $\dfrac{\frac{2}{9}}{5}$
4. $\dfrac{\frac{-3}{-4}}{\frac{7}{-8}}$

Solutions

1. $10 \cdot \frac{8}{-5}$ ↓ $-\overset{2}{\cancel{10}} \cdot \frac{8}{\underset{1}{\cancel{5}}}$ ↓ -16

2. $\frac{17}{-8} \cdot \left(-\frac{27}{17}\right)$ ↓ $+\frac{\overset{1}{\cancel{17}}}{8} \cdot \frac{27}{\underset{1}{\cancel{17}}}$ ↓ $\frac{27}{8}$

3. $\frac{2}{9} \div 5$ ↓ $\frac{2}{9} \cdot \frac{1}{5}$ ↓ $\frac{2}{45}$

4. $\frac{-3}{-4} \div \frac{7}{-8}$ ↓ $\frac{-3}{-4} \cdot \frac{-8}{7}$ ↓ $-\frac{3}{\underset{1}{\cancel{4}}} \cdot \frac{\overset{2}{\cancel{8}}}{7}$ ↓ $-\frac{6}{7}$

SPECIAL CASES

1. Dividing any fraction by the number 1 yields the original fraction. For example,

$$\frac{\frac{3}{8}}{1} = \frac{3}{8} \div 1 = \frac{3}{8} \quad \text{and} \quad \frac{\frac{-4}{5}}{1} = \frac{-4}{5} \div 1 = \frac{-4}{5}$$

2. On the other hand, dividing 1 by any fraction yields the reciprocal of the fraction. For example,

$$\frac{1}{\frac{3}{8}} = 1 \div \frac{3}{8} = \frac{8}{3} \quad \text{and} \quad \frac{1}{\frac{-4}{5}} = 1 \div \frac{-4}{5} = \frac{5}{-4}$$

3. Dividing any fraction by itself yields the number 1. For example,

$$\frac{\frac{3}{8}}{\frac{3}{8}} = \frac{3}{8} \div \frac{3}{8} = 1 \quad \text{and} \quad \frac{\frac{-4}{5}}{\frac{-4}{5}} = \frac{-4}{5} \div \frac{-4}{5} = 1$$

Here is a simple division table for the sixth's. Contrast this table with the table provided at the end of the last section. Verify the entries to check your understanding of division.

NUMERATOR ↑ over → DENOMINATOR	$\frac{1}{6}$	$\frac{2}{6}$	$\frac{3}{6}$	$\frac{4}{6}$	$\frac{5}{6}$	$\frac{6}{6}$
$\frac{1}{6}$	1	$\frac{1}{2}$	$\frac{1}{3}$	$\frac{1}{4}$	$\frac{1}{5}$	$\frac{1}{6}$
$\frac{2}{6}$	2	1	$\frac{2}{3}$	$\frac{1}{2}$	$\frac{2}{5}$	$\frac{1}{3}$
$\frac{3}{6}$	3	$\frac{3}{2}$	1	$\frac{3}{4}$	$\frac{3}{5}$	$\frac{1}{2}$
$\frac{4}{6}$	4	2	$\frac{4}{3}$	1	$\frac{4}{5}$	$\frac{2}{3}$
$\frac{5}{6}$	5	$\frac{5}{2}$	$\frac{5}{3}$	$\frac{5}{4}$	1	$\frac{5}{6}$
$\frac{6}{6}$	6	3	2	$\frac{3}{2}$	$\frac{6}{5}$	1

31. Compute

a. $\frac{7}{5} \div 1$

b. $1 \div \frac{7}{5}$

c. $\frac{7}{5} \div \frac{7}{5}$

32. Complete the table.

over → / ↑	$\frac{1}{4}$	$\frac{2}{4}$	$\frac{3}{4}$
$\frac{1}{4}$			
$\frac{2}{4}$			
$\frac{3}{4}$			

OBSERVATIONS AND REFLECTIONS

1. We claimed in the introduction to this section that a quotient of two fractions is equivalent to the product of the first fraction times the reciprocal of the second fraction. For example, $\frac{2}{3} \div \frac{4}{3}$ is equivalent to $\frac{2}{3} \cdot \frac{3}{4}$. A quotient of two fractions is also equivalent to the ratio of the two fractions. For example, $\frac{2}{3} \div \frac{4}{3}$ is equivalent to $\dfrac{\frac{2}{3}}{\frac{4}{3}}$. Hence, we can show that the quotient $\frac{2}{3} \div \frac{4}{3}$ is equivalent to the product $\frac{2}{3} \cdot \frac{3}{4}$ by showing that the product is equivalent to the ratio $\dfrac{\frac{2}{3}}{\frac{4}{3}}$. To see that a ratio of fractions is equivalent to the product of the fraction on top and the reciprocal of the fraction on the bottom, look at the following argument:

   Consider the complex fraction $\dfrac{\frac{a}{b}}{\frac{c}{d}}$, where a, b, c, and d represent nonzero integers. Multiplying both the numerator and denominator of this fraction by $\frac{d}{c}$ will produce an equivalent fraction:

$$\frac{\frac{a}{b}}{\frac{c}{d}} = \frac{\frac{a}{b} \cdot \frac{d}{c}}{\frac{c}{d} \cdot \frac{d}{c}} = \frac{\frac{a}{b} \cdot \frac{d}{c}}{1} = \frac{a}{b} \cdot \frac{d}{c}$$

   The denominator of the middle complex fraction simplifies to 1, and dividing an expression by 1 does not change its value.

$$\frac{\frac{a}{b}}{\frac{c}{d}} = \frac{a}{b} \div \frac{c}{d} = \frac{a}{b} \cdot \frac{d}{c}$$

(where a, b, c, and d represent nonzero integers)

★ 2. Note that the fraction bars in complex fractions are different sizes. The meaning of a complex fraction is indicated by the relative lengths of the bars. For example,

$$\frac{\frac{7}{2}}{\frac{3}{8}} \text{ means } \frac{\left(\frac{7}{2}\right)}{\left(\frac{3}{8}\right)} \text{ and is equivalent to } \frac{7}{2} \div \frac{3}{8}$$

   whereas

$$\cfrac{7}{\cfrac{\frac{2}{3}}{8}} \text{ means } \frac{7}{\left(\frac{\left(\frac{2}{3}\right)}{8}\right)} \text{ and is equivalent to } 7 \div \left(\frac{2}{3} \div 8\right) \text{ which equals}$$

   $7 \cdot \frac{3}{2} \cdot 8$.

EXERCISES

Simplify each of the following expressions to a reduced fraction.

1. $\frac{2}{3} \div 4$
2. $\frac{6}{11} \div 3$
3. $6 \div \frac{2}{3}$
4. $4 \div \frac{2}{5}$
5. $\frac{4}{5} \div (-6)$
6. $\frac{8}{9} \div (-12)$
7. $18 \div \frac{-15}{24}$
8. $35 \div \frac{28}{-5}$
9. $\frac{5}{6} \div \frac{25}{8}$
10. $\frac{15}{7} \div \frac{10}{21}$
11. $\frac{12}{5} \div \frac{-8}{9}$
12. $\frac{16}{9} \div \frac{48}{13}$
13. $\frac{-3}{8} \div \frac{-3}{4}$
14. $\frac{4}{-9} \div \frac{-2}{27}$
15. $-\frac{5}{4} \div \frac{25}{-2}$
16. $-\frac{6}{35} \div \frac{-14}{15}$
17. $\frac{21}{40} \div \left(-\frac{35}{52}\right)$
18. $\frac{105}{-128} \div \left(-\frac{63}{64}\right)$
19. $\dfrac{12}{\frac{8}{9}}$
20. $\dfrac{6}{\frac{3}{4}}$
21. $\dfrac{18}{\frac{-9}{4}}$
22. $\dfrac{28}{\frac{-7}{8}}$
23. $\dfrac{\frac{14}{-15}}{\frac{-4}{25}}$
24. $\dfrac{\frac{14}{-36}}{\frac{28}{24}}$
25. $\dfrac{\frac{-15}{35}}{\frac{-24}{-56}}$
26. $\dfrac{\frac{-16}{42}}{\frac{-18}{49}}$
27. $\dfrac{\frac{18}{-49}}{\frac{-27}{56}}$
28. $-\dfrac{\frac{-60}{7}}{\frac{70}{-3}}$
29. $-\dfrac{\frac{-4}{72}}{\frac{40}{24}}$
30. $-\dfrac{\frac{-108}{15}}{\frac{-39}{75}}$

Express each of the following word phrases algebraically and then simplify it.

1. -18 over $\frac{6}{5}$
2. $\frac{48}{64}$ divided by -144
3. The ratio 125 over $\frac{250}{800}$
4. The quotient $\frac{42}{50}$ divided by -84
5. $\frac{5}{4}$ over $\frac{25}{-2}$
6. 45 divided by $\frac{-9}{10}$
7. The ratio $\frac{15}{-27}$ over $\frac{-35}{-72}$
8. The quotient -390 divided by $-\frac{260}{3}$

9. $\frac{35}{36}$ over -45

10. The ratio $-\frac{22}{35}$ over 121

11. $\frac{-60}{7}$ divided by $\frac{70}{-3}$

12. The quotient $\frac{25}{16}$ divided by $\frac{-90}{48}$

6.5 ORDER OF OPERATIONS

Many problems with fractions involve addition, multiplication, and exponentiation, and the grouping of integers. In this section we will treat expressions such as $2^5 \div 6^2$, $\frac{6-2^2}{3\,(5)}$, $\dfrac{-3+11}{\frac{-2-8}{-5+1}}$, and $\frac{-2(-9)}{-3-7} \div \frac{10-16}{-5}$. (In Section 7.6 we will discuss other expressions involving addition of fractions.)

Every such expression can be simplified to a simple reduced fraction (or possibly an integer). There is a standard set of rules for simplifying these expressions called the Order of Operations.

These rules can be summarized as follows:

1. First simplify any expressions inside parentheses.
2. Compute powers before products, because exponential notation is a shorthand for repeated multiplication.
3. Simplify each term to a simple fraction (or an integer) by multiplying and dividing before computing sums.

(Because this summary does not treat all the details of our notational system, we provide a more comprehensive set of rules at the end of the section.)

33. Compute
a. $5(-3)^2$
b. $-6(-1)^3$

34. Compute
a. $5-(-8)$
b. $6-(+6)\,(-1)$

WARNING: 1. Parentheses cannot be dropped when a parenthetical expression is raised to a power. Removing parentheses from around a signed quantity may change an indicated multiplication to addition.

$$-4(-2)^3 = -4(-8), \quad \textbf{not} \quad -4\ -8$$

2. Signs cannot be omitted when removing parentheses, because signs are required before each integer (except the first) to indicate addition.

$$3 + (4) - (2)(-3) = 3 + 4 + 6, \quad \textbf{not} \quad 346$$

EXAMPLES

To compute numerical powers we use the method in Section 2.2.

Simplify the following expressions to simple fractions, following the Order of Operations.

1. $\frac{4-6^2}{4\,(-8)}$

2. $\frac{2^3-2^2}{-(3)^2-(-3)^3}$

3. $\frac{-(-3+1)^4}{-2^2(-2^3)}$

4. $\dfrac{3\left(\frac{1}{9}\right)}{8\left(\frac{1}{6}\right)}$ 5. $\dfrac{(3-6)^2}{2^4} \cdot \dfrac{(-3+11)^2}{(7-1)^2}$ 6. $\dfrac{4(-6)}{-8-7} \div \dfrac{3-15}{5^2}$

Solutions

1. $\dfrac{4-6^2}{4(-8)} \to \dfrac{4-36}{4(-8)} \to \dfrac{-32}{-32} \to 1$

2. $\dfrac{2^3-2^2}{-(3)^2-(-3)^3} \to \dfrac{8-4}{-(9)-(-27)} \to \dfrac{8-4}{-9+27} \to \dfrac{4}{18} \to \dfrac{2}{9}$

3. $\dfrac{-(-3+1)^4}{-2^2(-2^3)} \to \dfrac{-(-2)^4}{-4(-8)} \to \dfrac{-(+16)}{-4(-8)} \to \dfrac{-16}{+32} \to \dfrac{-1}{2}$

4. $\dfrac{3\left(\frac{1}{9}\right)}{8\left(\frac{1}{6}\right)} \to \left(3\left(\frac{1}{9}\right)\right) \div \left(8\left(\frac{1}{6}\right)\right) \to \left(\frac{1}{3}\right) \div \left(\frac{4}{3}\right) \to \frac{1}{3} \cdot \frac{3}{4} \to \frac{1}{4}$

5. $\dfrac{(3-6)^2}{2^4} \cdot \dfrac{(-3+11)^2}{(7-1)^2} \to \dfrac{(-3)^2}{2^4} \cdot \dfrac{(8)^2}{(6)^2} \to \dfrac{+9}{+16} \cdot \dfrac{+64}{+36} \to 1$

6. $\dfrac{4(-6)}{-8-7} \div \dfrac{3-15}{5^2} \to \dfrac{-24}{-15} \div \dfrac{-12}{+25} \to \dfrac{-24}{-15} \cdot \dfrac{25}{-12} \to \dfrac{-10}{3}$

SPECIAL CASES

Any power of a fraction should be expressed as a product, avoiding exponential notation. For example,

$$\left(\frac{2}{3}\right)^2 = \left(\frac{2}{3}\right)\left(\frac{2}{3}\right) \quad \text{and} \quad \left(\frac{3}{4}\right)^3 = \left(\frac{3}{4}\right)\left(\frac{3}{4}\right)\left(\frac{3}{4}\right)$$

35. Express as a product and simplify

a. $\left(\frac{1}{2}\right)^4$

b. $\left(\frac{4}{3}\right)^2$

c. $\left(\frac{2}{5}\right)^3$

ORDER OF OPERATIONS
(Numerical Expressions)

1. Change each quotient of expressions not containing fractions to fractional notation. Change each complex fraction to the division format, placing parentheses around the numerator and denominator of the original fraction. (A fraction bar implies parentheses around its numerator and denominator.) If there is more than one term, each quotient must be placed in brackets.
2. Simplify each expression, including numerators and denominators, which involves addition, multiplication, and/or exponentiation of *integers.* Proceed as follows:

   2A. *Compute numerical powers.* Simplify each numerical power to an integer inside parentheses.
   2B. *Simplify products.* Multiply each product of integers to an integer with its sign expressed, and remove parentheses.
   2C. *Simplify sums.*

   At this point each term has been simplified to an integer, a simple fraction, or a product or quotient of fractions.
3. Simplify each term to a simple reduced fraction (or possibly an integer).
4. Add. Simplify to a single fraction (or possibly an integer).

EXERCISES

Simplify each of the following expressions to an integer or a reduced fraction.

1. $8 \div 4^2$
2. $12 \div 4^2$
3. $\frac{4+2}{2}$
4. $\frac{4}{4+9}$
5. $\frac{3+9}{6+2}$
6. $\frac{15+5}{5+10}$
7. $\frac{14-7}{9-2}$
8. $\frac{3-9}{9+6}$
9. $\frac{2-3^2}{4(-7)}$
10. $\frac{3(4)-8}{2-8(3)}$
11. $\frac{5-3^2}{2^2(3)}$
12. $\frac{3^3-3^2}{(2)^4-(2)^3}$
13. $\frac{(3+2)^2}{5(-4)}$
14. $\frac{-(-9+5)^2}{3(-2)^2}$
15. $3\left(\frac{1}{2}\right) \cdot 6\left(\frac{1}{2}\right)$
16. $9\left(\frac{1}{4}\right) \cdot 5\left(\frac{2}{3}\right)$
17. $8\left(\frac{3}{4}\right) \cdot 5\left(\frac{1}{3}\right)$
18. $5\left(\frac{2}{9}\right) \cdot 6\left(\frac{1}{5}\right)$
19. $2\left(\frac{3}{5}\right) \div 10\left(\frac{7}{9}\right)$
20. $9\left(\frac{4}{3}\right) \div 6\left(\frac{1}{7}\right)$
21. $7\left(\frac{3}{2}\right) \div 4\left(\frac{7}{3}\right)$
22. $8\left(\frac{3}{5}\right) \div 5\left(\frac{3}{8}\right)$
23. $\frac{6\left(\frac{3}{4}\right)}{5\left(\frac{3}{2}\right)}$
24. $\frac{9\left(\frac{1}{5}\right)}{5\left(\frac{1}{3}\right)}$

25. $\dfrac{3-7}{2+8} \cdot \dfrac{2-7}{3(-8)}$

26. $\dfrac{5-8}{7-11} \cdot \dfrac{-3-13}{-1-11}$

27. $\dfrac{3-2(-2)}{5(-2)(-1)} \cdot \dfrac{7-8(-1)}{4-5}$

28. $\dfrac{5-4(-1)}{-2(-6)} \cdot \dfrac{-3(-2-7)}{5(-9)(+9)}$

29. $\dfrac{(6-4)^3}{3^3} \cdot \dfrac{(-2+8)^2}{(6-10)^2}$

30. $\dfrac{(7-10)^3}{(-2-6)^2} \cdot \dfrac{(-4-16)^2}{(-2+11)^2}$

31. $\dfrac{\dfrac{2+3}{7-9}}{8-4}$

32. $\dfrac{\dfrac{2+13}{3-8}}{7-10}$

33. $\dfrac{3+9}{\dfrac{3-6}{1-6}}$

34. $\dfrac{5-11}{\dfrac{-9+17}{-3-5}}$

35. $\dfrac{\dfrac{3-8}{4+5}}{\dfrac{7+3}{-5-1}}$

36. $\dfrac{\dfrac{2+7}{3-7}}{\dfrac{5-8}{-7-5}}$

37. $\dfrac{3}{8-5} \div \dfrac{3}{8-3}$

38. $\dfrac{2}{5+9} \div \dfrac{5}{2-9}$

39. $\dfrac{2}{9(-2)} \div \dfrac{3(-4)}{-2-5}$

40. $\dfrac{3(-6)}{-4-8} \div \dfrac{27}{3-6}$

41. $\left(\dfrac{3}{4}\right)^2 \left(\dfrac{5}{18}\right)$

42. $\left(\dfrac{5}{6}\right)^2 \left(\dfrac{9}{10}\right)$

43. $\left(\dfrac{8}{21}\right)\left(\dfrac{7}{2}\right)^2$

44. $\left(\dfrac{20}{21}\right)\left(\dfrac{9}{4}\right)^2$

45. $\left(\dfrac{3}{8}\right)^2 \left(\dfrac{4}{3}\right)^2$

46. $\left(\dfrac{7}{6}\right)^2 \left(\dfrac{9}{7}\right)^2$

47. $\left(\dfrac{9}{10}\right)^2 \left(\dfrac{2}{3}\right)^3$

48. $\left(\dfrac{14}{15}\right)^2 \left(\dfrac{3}{2}\right)^3$

49. $\left(\dfrac{3}{8}\right)^2 \div \dfrac{15}{16}$

50. $\left(\dfrac{4}{9}\right)^2 \div \dfrac{20}{27}$

51. $\dfrac{21}{25} \div \left(\dfrac{9}{10}\right)^2$

52. $\dfrac{35}{36} \div \left(\dfrac{7}{12}\right)^2$

53. $\left(\dfrac{5}{12}\right)^2 \div \left(\dfrac{15}{8}\right)^2$

54. $\left(\dfrac{4}{15}\right)^2 \div \left(\dfrac{8}{9}\right)^2$

55. $\dfrac{\left(\dfrac{11}{6}\right)^2}{\dfrac{11}{12}}$

56. $\dfrac{\left(\dfrac{13}{8}\right)^2}{\dfrac{13}{16}}$

57. $\dfrac{\dfrac{15}{16}}{\left(\dfrac{9}{2}\right)^2}$

58. $\dfrac{\dfrac{20}{27}}{\left(\dfrac{8}{3}\right)^2}$

59. $\dfrac{\left(\dfrac{14}{15}\right)^2}{\left(\dfrac{7}{30}\right)^2}$

60. $\dfrac{\left(\dfrac{15}{8}\right)^2}{\left(\dfrac{5}{16}\right)^2}$

ANSWERS TO MARGIN EXERCISES

1. The prime αι.
4 : 2
6 : 2 and 3
8 : 2
9 : 3
10: 2 and 5
12: 2 and 3

2. 16, 24, 26 and 28.

3. a. 16
b. 24
c. 36
d. 54
e. 81

4. a. $2 \cdot 2 \cdot 3 \cdot 3$
b. $3 \cdot 3 \cdot 11$
c. $2 \cdot 3 \cdot 23$
d. $2 \cdot 2 \cdot 2 \cdot 3 \cdot 7$

5. a. 8
b. 17
c. 21

6. a. $\frac{1}{3}$
b. $\frac{2}{1} = 2$
c. $\frac{3}{1} = 3$
d. $\frac{1}{5}$

8. a, c, and d are negative.

9. a. $\frac{5}{2}$
b. $\frac{5}{2}$
c. $\frac{-5}{2}$
d. $\frac{5}{2}$

10. a. $\frac{-4}{9}$
b. $\frac{4}{9}$
c. $\frac{-4}{9}$
d. $\frac{-4}{9}$
e. $\frac{4}{9}$
f. $\frac{4}{9}$

11. a. $\frac{7}{3}$
b. $\frac{-4}{9}$

12. a. $-4 \div 9$
b. $-4 \div 9$

13. a. $\frac{3 \cdot \cancel{6}}{7 \cdot \cancel{6}} = \frac{3}{7}$
b. $\frac{5 \cdot \cancel{8}}{3 \cdot \cancel{8}} = \frac{5}{3}$

14. a. 3
b. 3
c. 7

16. a and d contain equivalent fractions.

17. a. 3
b. $12 \div 4 = 3$
c. 4
d. $24 \div 6 = 4$

18. a. $\frac{-2}{7}$ and $\frac{1}{-5}$
b. $\frac{3}{8}$, $\frac{-7}{2}$, and $\frac{4}{-9}$

19. a. $-8 \cdot \frac{1}{9} \cdot 6 \cdot \frac{1}{28}$
b. $3 \cdot \frac{1}{5} \cdot (-5) \cdot \frac{1}{18}$
c. $-14 \cdot \frac{1}{3} \cdot 15 \cdot \frac{1}{-7}$

22. a. 2
b. $\frac{5}{2}$
c. $\frac{28}{3}$

23. a. $\frac{-3}{7}$
b. $\frac{4}{9}$

24. a. $\frac{3}{7}$
b. $\frac{-4}{9}$

27. $-3(-4) = 12$

28.

a.	b.
1	1
$\frac{3}{2}$	$\frac{3}{2}$
6	6
9	9

29. a. 1
b. 1
c. 1
d. 1

30. a. $\frac{1}{3} \div 6$
b. $\frac{3}{8} \div \frac{9}{4}$

31. a. $\frac{7}{5}$
b. $\frac{5}{7}$
c. 1

33. a. $5(+9) = 45$
b. $-6(-1) = 6$

34. a. 13
b. 0

35. a. $\left(\frac{1}{2}\right)\left(\frac{1}{2}\right)\left(\frac{1}{2}\right)\left(\frac{1}{2}\right) = \frac{1}{16}$
b. $\left(\frac{4}{3}\right)\left(\frac{4}{3}\right) = \frac{16}{9}$
c. $\left(\frac{2}{5}\right)\left(\frac{2}{5}\right)\left(\frac{2}{5}\right) = \frac{8}{125}$

CHAPTER REVIEW

SECTION 6.1

Simplify to lowest terms using the Method of Primes.

1. $\frac{4}{14}$ 2. $\frac{65}{10}$ 3. $\frac{20}{64}$

4. $\frac{63}{147}$ 5. $\frac{84}{18}$ 6. $\frac{154}{70}$

7. $\frac{63}{105}$ 8. $\frac{192}{80}$ 9. $\frac{160}{200}$

10. $\frac{60}{90}$ 11. $\frac{24}{126}$ 12. $\frac{102}{186}$

SECTION 6.2

Simplify to a reduced fraction in standard form.

13. $\frac{28}{56}$ 14. $\frac{12}{30}$ 15. $\frac{112}{21}$

16. $70 \div 98$ 17. $18 \div 72$ 18. $\frac{-27}{48}$

19. $\frac{14}{-32}$ 20. $-\frac{8}{56}$ 21. $-\frac{-15}{80}$

22. $\frac{-75}{-45}$ 23. $72 \div (-96)$ 24. $-12 \div (-56)$

SECTION 6.3

Simplify to a reduced fraction.

25. $4 \cdot \frac{3}{2}$ 26. $\frac{7}{5} \cdot 15$ 27. $\frac{1}{4} \cdot \frac{8}{7}$

28. $\frac{5}{6} \cdot \frac{-1}{4}$ 29. $\frac{-7}{6} \cdot \frac{18}{35}$ 30. $\frac{-8}{15} \cdot \frac{-25}{40}$

31. $\frac{25}{12}\left(-\frac{84}{65}\right)$ 32. $\frac{14}{-9}\left(\frac{-32}{56}\right)$ 33. $-\frac{7}{36}\left(-\frac{9}{70}\right)$

34. $\frac{-7}{8} \cdot \frac{4}{9} \cdot \frac{3}{-5}$ 35. $\frac{8}{5} \cdot \frac{-9}{8} \cdot \frac{5}{6}$ 36. $\frac{42}{9} \cdot \frac{-12}{21} \cdot \frac{2}{-3}$

SECTION 6.4

Simplify to a reduced fraction.

37. $\frac{4}{6} \div 8$ 38. $9 \div \frac{5}{12}$ 39. $\frac{8}{15} \div (-24)$

40. $28 \div \frac{-8}{21}$ 41. $\frac{18}{5} \div \frac{27}{20}$ 42. $\frac{15}{16} \div \frac{-21}{20}$

43. $\frac{-5}{9} \div \frac{-5}{18}$ 44. $-\frac{15}{42} \div \frac{35}{-6}$ 45. $\frac{21}{40} \div \left(-\frac{70}{104}\right)$

46. $\dfrac{36}{\dfrac{3}{10}}$

47. $\dfrac{24}{\dfrac{20}{-3}}$

48. $\dfrac{\dfrac{28}{-30}}{\dfrac{-7}{50}}$

49. $\dfrac{\dfrac{-36}{84}}{\dfrac{-32}{-28}}$

50. $-\dfrac{\dfrac{14}{-75}}{\dfrac{-77}{45}}$

51. $-\dfrac{\dfrac{-16}{54}}{\dfrac{-64}{18}}$

SECTION 6.5

Simplify to an integer or a reduced fraction.

52. $6^2 \div 3^2$

53. $\dfrac{8+4}{8}$

54. $\dfrac{8+4}{4+12}$

55. $\dfrac{8-2}{7+4}$

56. $\dfrac{5^2-3^2}{4+2^2}$

57. $\dfrac{35-4(3^2)}{3^2-2^2(3)}$

58. $\dfrac{-(3+1)^2}{-2^2(2^3)}$

59. $6\left(\dfrac{2}{5}\right)\cdot 10\left(\dfrac{7}{2}\right)$

60. $5\left(\dfrac{1}{8}\right)\cdot 6\left(\dfrac{3}{5}\right)$

61. $7\left(\dfrac{2}{5}\right)\div 21\left(\dfrac{2}{3}\right)$

62. $12\left(\dfrac{3}{8}\right)\div 8\left(\dfrac{12}{5}\right)$

63. $\dfrac{6\left(\dfrac{3}{4}\right)}{15\left(\dfrac{3}{5}\right)}$

64. $\dfrac{-3-6}{-3+11}\cdot\dfrac{-10(-2)}{-18+3}$

65. $\dfrac{(2-7)(-4)}{(-3)-4(-2)}\cdot\dfrac{-7-3}{-5(-6)+5}$

66. $\dfrac{(3-13)^3}{(-7+3)^2}\cdot\dfrac{(-5+11)^3}{15^2}$

67. $\dfrac{\dfrac{4-11}{-7-1}}{-5+12}$

68. $\dfrac{-6+27}{\dfrac{-11-3}{-3-7}}$

69. $\dfrac{\dfrac{7-13}{-2-5}}{\dfrac{-4+13}{4+10}}$

70. $\dfrac{6-9}{8}\div\dfrac{-7-2}{-6}$

71. $\dfrac{-4(-2)}{8}\div\dfrac{-5(+3)}{-13-7}$

72. $\left(\dfrac{6}{5}\right)^2\left(\dfrac{7}{18}\right)$

73. $\left(\dfrac{32}{33}\right)\left(\dfrac{11}{4}\right)^2$

74. $\left(\dfrac{3}{2}\right)^2\left(\dfrac{8}{9}\right)^2$

75. $\left(\dfrac{7}{12}\right)^2\left(\dfrac{4}{7}\right)^3$

76. $\left(\dfrac{5}{9}\right)^2\div\dfrac{35}{36}$

77. $\dfrac{4}{75}\div\left(\dfrac{3}{10}\right)^2$

78. $\left(\dfrac{8}{15}\right)^2\div\left(\dfrac{4}{9}\right)^2$

79. $\dfrac{\left(\dfrac{12}{7}\right)^2}{\dfrac{12}{35}}$

80. $\dfrac{\dfrac{25}{16}}{\left(\dfrac{15}{2}\right)^2}$

81. $\dfrac{\left(\dfrac{16}{3}\right)^2}{\left(\dfrac{8}{9}\right)^2}$

SUMMARY OF PHRASES

Express each of the following word phrases algebraically and then simplify it.

1. The product of $-\frac{7}{8}$ and $\frac{16}{21}$
2. $\frac{2}{3}$ of the product of $\frac{-6}{21}$ and 7
3. The product of $\frac{-18}{48}$ and half -64
4. $\frac{-12}{51}$ times $\frac{17}{18}$ of $\frac{-27}{32}$
5. Half the product of $\frac{11}{13}$ and $-\frac{26}{55}$
6. 52 multiplied by $\frac{8}{26}$ of $\frac{10}{12}$
7. $-\frac{18}{48}$ divided by 72
8. The quotient, $\frac{25}{36}$ divided by -75
9. $\frac{14}{33}$ over $\frac{42}{-55}$
10. The ratio $\frac{-72}{-25}$ over $\frac{-54}{35}$
11. $\frac{-24}{25}$ over -48
12. $\frac{70}{-24}$ divided by $\frac{40}{16}$

CUMULATIVE TEST

The following problems test your understanding of this chapter. Before taking this test, thoroughly review Sections 6.2, 6.3, and 6.4.

Once you have finished the test, compare your answers with the answers provided at the back of the book. Note the section number of each problem missed, and thoroughly review those sections again.

Simplify.

1. $\frac{24}{42}$

2. $5 \cdot \frac{1}{3}$

3. $\frac{9}{4} \div 6$

4. $\frac{-4}{9} \cdot \frac{15}{28} \cdot \frac{7}{-6}$

5. $\frac{-15}{55}$

6. $\frac{\frac{8}{14}}{18}$

7. $\frac{3}{4} \cdot \frac{28}{-21}$

8. $\frac{-18}{56} \div \frac{-12}{49}$

9. $-18 \div (-54)$

10. $\frac{\frac{15}{48}}{\frac{10}{-9}}$

11. $21\left(\frac{-4}{56}\right)$

12. $16 \div \frac{-12}{5}$

13. $\frac{3}{4} \cdot 6$

14. $\frac{36}{-16}$

15. $-\frac{\frac{32}{-81}}{\frac{48}{18}}$

16. $\frac{15}{8} \cdot \frac{44}{-7} \cdot \frac{7}{20}$

17. $\frac{-52}{27} \div \frac{39}{-63}$

18. $\frac{39}{52} \cdot \frac{8}{-3}$

19. $\frac{\frac{-22}{-45}}{\frac{132}{-40}}$

20. $\frac{45}{-16} \cdot \frac{-12}{35} \cdot \frac{4}{-3}$

7 ADDITION OF NUMERICAL FRACTIONS

In this chapter we will study the addition of numerical fractions. The language of algebraic addition extends naturally to fractions. In *algebraic sums* of fractions such as

$$\frac{2}{3} + \frac{3}{4}, \quad -\frac{2}{3} - \frac{3}{4}, \quad \text{and} \quad \frac{-2}{3} + \frac{-3}{4},$$

each fraction is called a *term* of the sum. For example, in the sum $\frac{2}{3} + \frac{3}{4}$, the terms are $\frac{2}{3}$ and $\frac{3}{4}$. Similarly, in the sum $-\frac{2}{3} - \frac{3}{4}$, the terms are $-\frac{2}{3}$ and $-\frac{3}{4}$.

In a sum of fractions, each sign between fractions begins a term, and a plus or minus sign must appear before every fraction except the first one. Because each fraction in a sum may have as many as three signs, when adding fractions it is wise to rewrite the sum first, placing each fraction in *standard form.* Remember that a *positive* fraction is in *standard form* if it has no minus signs. For example, $\frac{2}{3}$ is the standard form of $-\frac{-2}{3}$. Similarly, a *negative* fraction is in *standard form* if it has only one minus sign, either in the numerator or before the fraction. For example, $\frac{-2}{3}$ or $-\frac{2}{3}$ is the standard form of $-\frac{-2}{-3}$.

7.1 ADDING FRACTIONS WITH COMMON DENOMINATORS

Fractions with identical denominators are said to have *common denominators* or *like denominators.* These fractions are the easiest ones to add, and we will discuss them first. We will consider the addition of fractions with *unlike* denominators in Sections 7.3 and 7.5.

Because integers can be considered fractions with a denominator of 1, adding integers can be considered a special case of adding fractions with common denominators.

For example, look at the following computations.

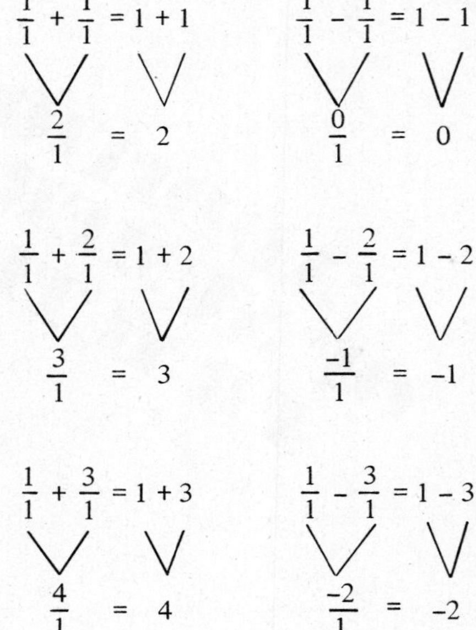

1. Form similar diagrams for

a. $\frac{2}{1} + \frac{4}{1}$

b. $\frac{2}{1} - \frac{4}{1}$

In each example we have combined terms in going from the first row to the second, and in each row we have converted from fractions to integers in going from left to right. These examples show that, at least in this special case, the rule for adding fractions with a common denominator is to add the numerators and keep the common denominator.

Let's look at some computations that include fractions with a denominator other than 1. We will use fractions that reduce to integers so that you can check the results.

$$\frac{3}{3} + \frac{3}{3} = 1 + 1 \qquad \frac{3}{3} - \frac{3}{3} = 1 - 1$$

$$\frac{6}{3} = 2 \qquad \frac{0}{3} = 0$$

$$\frac{3}{3} + \frac{6}{3} = 1 + 2 \qquad \frac{3}{3} - \frac{6}{3} = 1 - 2$$

$$\frac{9}{3} = 3 \qquad \frac{-3}{3} = -1$$

$$\frac{3}{3} + \frac{9}{3} = 1 + 3 \qquad \frac{3}{3} - \frac{9}{3} = 1 - 3$$

$$\frac{12}{3} = 4 \qquad \frac{-6}{3} = -2$$

2. Form similar diagrams for

a. $\frac{12}{6} - \frac{24}{6}$

b. $-\frac{12}{6} - \frac{24}{6}$

Again, we have added the numerators and kept the common denominator.

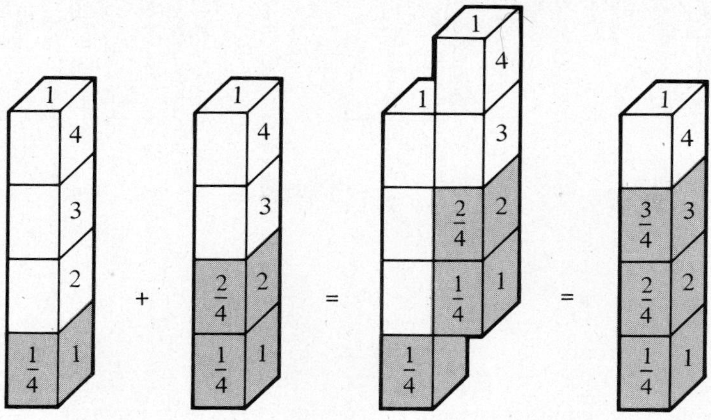

Look at the following example of *adding fractions with common denominators.*

TYPICAL PROBLEM

$$\frac{8}{15} - \frac{11}{15}$$

To avoid sign errors, we first move all minus signs to the numerators (so that no minus signs are in the denominators or before the fraction bars):

$$\downarrow$$
$$\frac{8}{15} + \frac{-11}{15}$$

We then write the numerators as an indicated sum over the common denominator:

$$\downarrow$$
$$\frac{8-11}{15}$$

Then we simplify the numerator:

$$\downarrow$$
$$\frac{-3}{15}$$

Finally we reduce to lowest terms:

$$\downarrow$$
$$\frac{-1}{5}$$

Hence, $\frac{8}{15} - \frac{11}{15} = \frac{-1}{5}$.

To reduce to lowest terms we use the method in Section 6.1 or 6.2.

EXAMPLES

Add the following numerical fractions with common denominators.

Every sum of simple fractions with common denominators can be simplified to a single reduced fraction.

1. $\frac{3}{5} + \frac{12}{5}$
2. $\frac{4}{7} - \frac{10}{7}$
3. $\frac{-21}{10} + \frac{14}{10}$
4. $\frac{33}{7} - \frac{22}{7}$
5. $\frac{7}{4} + \frac{-11}{4} - \frac{3}{4}$
6. $\frac{5}{9} + \frac{11}{9} - \frac{7}{9}$

Solutions

1. $\frac{3+12}{5} \downarrow \frac{15}{5} \downarrow 3$

2. $\frac{4}{7} + \frac{-10}{7} \downarrow \frac{4-10}{7} \downarrow \frac{-6}{7}$

3. $\frac{-21+14}{10} \downarrow \frac{-7}{10}$

4. $\frac{33}{7} + \frac{-22}{7} \to \frac{33-22}{7} \to \frac{11}{7}$

5. $\frac{7}{4} + \frac{-11}{4} + \frac{-3}{4} \to \frac{7-11-3}{4} \to \frac{-7}{4}$

6. $\frac{5}{9} + \frac{11}{9} + \frac{-7}{9} \to \frac{5+11-7}{9} \to \frac{9}{9} \to 1$

SPECIAL CASES

1. If the numerator simplifies to zero, the fraction is zero. For example,

$$\frac{3}{8} + \frac{-3}{8} = \frac{3-3}{8} = \frac{0}{8} = 0$$

and

$$\frac{-2}{3} + \frac{7}{3} + \frac{-5}{3} = \frac{-2+7-5}{3} = \frac{0}{3} = 0$$

Any fraction whose numerator is zero equals zero (except of course $\frac{0}{0}$, which is not defined).

3. Compute

a. $\frac{7}{12} + \frac{-7}{12}$

b. $\frac{-5}{9} + \frac{5}{9}$

c. $\frac{8}{5} + \frac{-2}{5} - \frac{6}{5}$

2. Adding 0 to an expression does not change the value of the expression. That is, the sum of any expression and zero equals the expression. For example,

$$\frac{3}{8} + 0 = \frac{3}{8} \quad \text{and} \quad \frac{3}{8} + \frac{0}{8} = \frac{3+0}{8} = \frac{3}{8}$$

4. Compute

a. $\frac{5}{9} - \frac{0}{9}$

b. $\frac{0}{9} - \frac{5}{9}$

Here is a simple addition table for the sixth's. Verify the entries to check your understanding of addition. (You saw a multiplication table for the sixth's in Section 6.3.)

plus	$\frac{1}{6}$	$\frac{2}{6}$	$\frac{3}{6}$	$\frac{4}{6}$	$\frac{5}{6}$	$\frac{6}{6}$
$\frac{1}{6}$	$\frac{1}{3}$	$\frac{1}{2}$	$\frac{2}{3}$	$\frac{5}{6}$	1	$\frac{7}{6}$
$\frac{2}{6}$	$\frac{1}{2}$	$\frac{2}{3}$	$\frac{5}{6}$	1	$\frac{7}{6}$	$\frac{4}{3}$
$\frac{3}{6}$	$\frac{2}{3}$	$\frac{5}{6}$	1	$\frac{7}{6}$	$\frac{4}{3}$	$\frac{3}{2}$
$\frac{4}{6}$	$\frac{5}{6}$	1	$\frac{7}{6}$	$\frac{4}{3}$	$\frac{3}{2}$	$\frac{5}{3}$
$\frac{5}{6}$	1	$\frac{7}{6}$	$\frac{4}{3}$	$\frac{3}{2}$	$\frac{5}{3}$	$\frac{11}{6}$
$\frac{6}{6}$	$\frac{7}{6}$	$\frac{4}{3}$	$\frac{3}{2}$	$\frac{5}{3}$	$\frac{11}{6}$	2

5. Complete the table.

plus	$\frac{1}{4}$	$\frac{2}{4}$	$\frac{3}{4}$
$\frac{1}{4}$			
$\frac{2}{4}$			
$\frac{3}{4}$			

OBSERVATIONS AND REFLECTIONS

1. Compare the sums $-\frac{2}{3} - \frac{4}{3}$ and $\frac{-2}{3} + \frac{-4}{3}$. The first sum is read "minus $\frac{2}{3}$ minus $\frac{4}{3}$." The second sum is read "$\frac{-2}{3}$ plus $\frac{-4}{3}$." Yet the sums are equal because $-\frac{2}{3} = \frac{-2}{3}$ and $-\frac{4}{3} = \frac{-4}{3}$. In any sum of fractions, the terms may be written so that minus signs appear only in the numerators. In an indicated sum of fractions, a negative fraction is usually written with its minus sign before the fraction bar, but it can be written with a plus sign before the fraction bar and its minus sign in the numerator.

★ 2. Consider the indicated sum $\frac{-8}{5} + \frac{7}{5}$, which equals the expression $\frac{-8+7}{5}$. This expression is also equal to the product $\frac{1}{5}(-8+7)$. If we were to start with the product $\frac{1}{5}(-8+7)$, we could return to the original sum $\frac{-8}{5} + \frac{7}{5}$ by distributing the fraction $\frac{1}{5}$ over the sum:

$$\frac{1}{5}(-8+7) = \frac{1}{5}(-8) + \frac{1}{5}(7) = \frac{-8}{5} + \frac{7}{5}$$

Hence, adding fractions with common denominators could be considered the reverse of distribution.

$$\frac{1}{c}(a+b) = \frac{a}{c} + \frac{b}{c} = \frac{a+b}{c}$$

(where a, b, and c represent numerical values with $c \neq 0$)

EXERCISES

Simplify each of the following expressions to a single reduced fraction.

1. $\frac{1}{3} + \frac{2}{3}$
2. $\frac{5}{8} + \frac{1}{8}$
3. $\frac{3}{11} + \frac{5}{11}$
4. $\frac{3}{5} + \frac{4}{5}$
5. $\frac{-3}{4} + \frac{3}{4}$
6. $\frac{5}{9} - \frac{5}{9}$
7. $\frac{2}{5} - \frac{4}{5}$
8. $-\frac{9}{11} + \frac{8}{11}$
9. $\frac{3}{8} - \frac{7}{8}$
10. $\frac{-10}{3} + \frac{8}{3}$
11. $-\frac{11}{12} - \frac{7}{12}$
12. $\frac{21}{4} + \frac{-9}{4}$
13. $\frac{-14}{15} + \frac{9}{15}$
14. $\frac{17}{18} + \frac{-8}{18}$
15. $-\frac{25}{21} + \frac{-3}{21}$
16. $\frac{-33}{36} + \frac{-19}{36}$
17. $\frac{-42}{25} - \frac{43}{25}$
18. $\frac{52}{15} - \frac{87}{15}$
19. $\frac{5}{8} + \frac{3}{8} + \frac{7}{8}$
20. $\frac{5}{9} + \frac{2}{9} + \frac{7}{9}$
21. $\frac{-3}{7} + \frac{5}{7} - \frac{6}{7}$

22. $\frac{-7}{6} - \frac{1}{6} + \frac{8}{6}$ 23. $\frac{4}{3} + \frac{-7}{3} - \frac{5}{3}$ 24. $\frac{7}{11} - \frac{3}{11} + \frac{-9}{11}$

Express each of the following word phrases algebraically and then simplify it.

1. The sum of $\frac{2}{7}$ and $\frac{4}{7}$
2. $\frac{7}{8}$ decreased by $\frac{3}{8}$
3. $\frac{13}{14}$ subtracted from $\frac{19}{14}$
4. $\frac{-8}{5}$ added to $\frac{-4}{5}$
5. $\frac{-12}{11}$ reduced by $\frac{10}{11}$
6. $\frac{8}{3}$ less than $\frac{10}{3}$
7. $\frac{-5}{12}$ plus $\frac{3}{12}$
8. $\frac{16}{25}$ is subtracted from $\frac{7}{25}$
9. $\frac{8}{9}$ more than $\frac{-5}{9}$
10. $\frac{27}{35}$ less $\frac{13}{35}$
11. $\frac{-33}{40}$ increased by $\frac{8}{40}$
12. $\frac{-3}{4}$ is reduced by $\frac{15}{4}$

7.2 FORMING EQIVALENT FRACTIONS

In the last section we saw how to add fractions with like denominators. Fractions with unlike denominators cannot be combined in the same direct manner. To add fractions with unlike denominators, we must first convert them to fractions with like denominators. Hence, we must learn how to take a fraction and form an equivalent fraction with a given new denominator.

Remember that multiplying the numerator and the denominator of a fraction by any nonzero integer results in an equivalent fraction. For example,

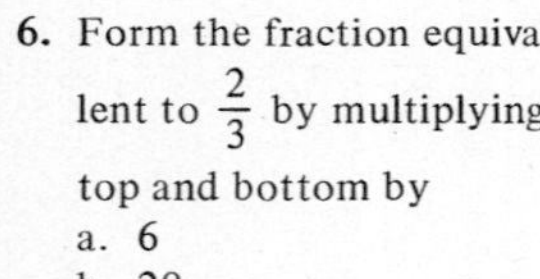

6. Form the fraction equivalent to $\frac{2}{3}$ by multiplying top and bottom by
a. 6
b. 20
c. 48

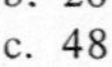

$$\frac{2}{3} = \frac{2 \cdot 2}{3 \cdot 2} = \frac{2 \cdot 3}{3 \cdot 3} = \frac{2 \cdot 4}{3 \cdot 4} = \cdots$$

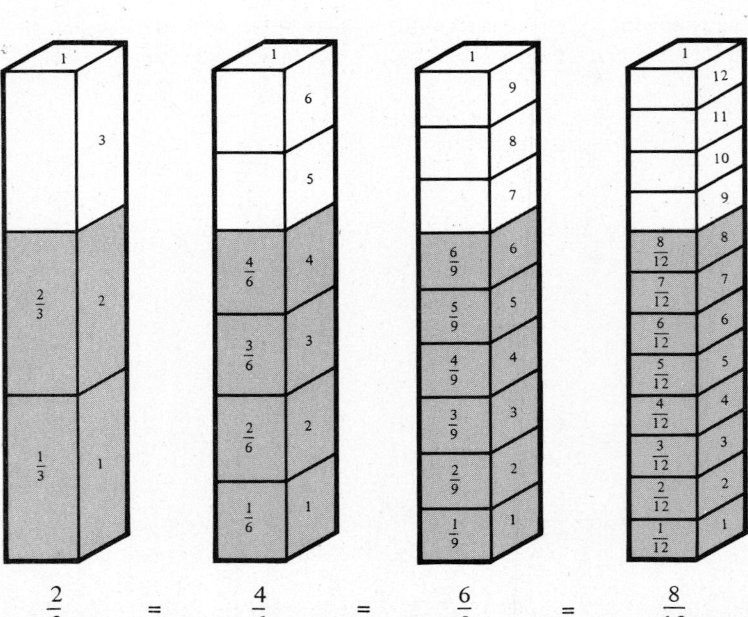

However, our problem is to form an equivalent fraction with a specified denominator. Consider the problem

$$\frac{2}{3} = \frac{?}{144}$$

How can we find the new numerator?

In the fraction $\frac{2}{3}$ the numerator is two-thirds the denominator, and this relation is true for any fraction equivalent to $\frac{2}{3}$. For example, in each of the equivalent fractions $\frac{4}{6}$, $\frac{6}{9}$, $\frac{8}{12}$, $\frac{10}{15}$, and $\frac{12}{18}$, the numerator is two-thirds the denominator. A fraction cannot be equivalent to $\frac{2}{3}$ unless its numerator is two-thirds its denominator. Thus, to find the numerator of the fraction equivalent to $\frac{2}{3}$ whose denominator is 144, we take two-thirds of 144 (i.e., $\frac{2}{3}$ times 144).

$$\text{new denominator} \longrightarrow 144\left(\frac{2}{3}\right) = 96 \longleftarrow \text{new numerator}$$

$\uparrow$ original fraction

Thus, $\frac{2}{3} = \frac{96}{144}$. Multiplying the original fraction by the new denominator always yields the new numerator.

Look at the following example of *forming an equivalent fraction with a given denominator.*

TYPICAL PROBLEM

$$\frac{3}{5} = \frac{?}{20}$$

We first multiply the fraction by the given new denominator to compute the new numerator:

$$20\left(\frac{3}{5}\right) \to \overset{4}{\cancel{20}}\left(\frac{3}{\cancel{5}}\right) \to 12$$

We then place our new numerator over the given new denominator:

$$\frac{12}{20}$$

Hence, $\frac{3}{5} = \frac{12}{20}$.

To multiply the fractions by the given new denominator we use the method in Section 6.3. The given new denominator must be a multiple of the original denominator, or the product will not be an integer.

EXAMPLES

Convert the following numerical fractions to equivalent fractions with the given denominators.

Every simple fraction can be converted to an equivalent fraction with a denominator that is any given multiple of the original denominator.

1. $\frac{1}{5} = \frac{?}{25}$ 2. $\frac{3}{8} = \frac{?}{16}$ 3. $\frac{2}{3} = \frac{?}{27}$

4. $\frac{7}{4} = \frac{?}{24}$ 5. $\frac{4}{1} = \frac{?}{4}$ 6. $1 = \frac{?}{30}$

Solutions

1. $\overset{5}{\cancel{25}}\left(\frac{1}{\cancel{5}}\right) = 5$ $\downarrow$ $\frac{5}{25}$

2. $\overset{2}{\cancel{16}}\left(\frac{3}{\cancel{8}}\right) = 6$ $\downarrow$ $\frac{6}{16}$

3. $\overset{9}{\cancel{27}}\left(\frac{2}{\cancel{3}}\right) = 18$ $\downarrow$ $\frac{18}{27}$

4. $\overset{6}{\cancel{24}}\left(\frac{7}{\cancel{4}}\right) = 42$ $\downarrow$ $\frac{42}{24}$

5. $4\left(\frac{4}{1}\right) = 16$ $\downarrow$ $\frac{16}{4}$

6. $30(1) = 30$ $\downarrow$ $\frac{30}{30}$

SPECIAL CASES

An integer can be thought of as a fraction with the denominator 1 and can be converted to an equivalent fraction with any other denominator. For example,

$$2, \frac{2 \cdot 1}{1}, \frac{2 \cdot 2}{2}, \frac{2 \cdot 3}{3}, \frac{2 \cdot 4}{4}, \frac{2 \cdot 5}{5}, \frac{2 \cdot 6}{6}, \frac{2 \cdot 7}{7}, \ldots$$

are equivalent.

In particular, the number 1 is equivalent to every fraction with an identical numerator (top) and denominator (bottom). For example, the fractions

$$\frac{1}{1}, \frac{2}{2}, \frac{3}{3}, \frac{4}{4}, \frac{5}{5}, \frac{6}{6}, \frac{7}{7}, \ldots$$

are equivalent to 1.

7. Form an equivalent fraction with the given denominator.
   a. $6 = \frac{?}{18}$
   b. $3 = \frac{?}{25}$
   c. $12 = \frac{?}{60}$
   d. $10 = \frac{?}{10}$

We need to convert fractions to new fractions with a given denominator in order to relate fractions with unlike denominators. Here is a table of fractions giving the corresponding fractions with the common denominator 60.

NUMERATOR over → / ↑	DENOMINATOR 1	2	3	4	5	6
1	$\frac{60}{60}$	$\frac{30}{60}$	$\frac{20}{60}$	$\frac{15}{60}$	$\frac{12}{60}$	$\frac{10}{60}$
2	$\frac{120}{60}$	$\frac{60}{60}$	$\frac{40}{60}$	$\frac{30}{60}$	$\frac{24}{60}$	$\frac{20}{60}$
3	$\frac{180}{60}$	$\frac{90}{60}$	$\frac{60}{60}$	$\frac{45}{60}$	$\frac{36}{60}$	$\frac{30}{60}$
4	$\frac{240}{60}$	$\frac{120}{60}$	$\frac{80}{60}$	$\frac{60}{60}$	$\frac{48}{60}$	$\frac{40}{60}$
5	$\frac{300}{60}$	$\frac{150}{60}$	$\frac{100}{60}$	$\frac{75}{60}$	$\frac{60}{60}$	$\frac{50}{60}$
6	$\frac{360}{60}$	$\frac{180}{60}$	$\frac{120}{60}$	$\frac{90}{60}$	$\frac{72}{60}$	$\frac{60}{60}$

Verify the entries to make sure that you understand how to convert to equivalent fractions.

8. Complete the table, converting each fraction to twelfths.

over → / ↑	2	3	4
2			
3			
4			

OBSERVATIONS AND REFLECTIONS

★ One reason for converting fractions to a common denominator is, of course, to add them. Another reason is to compare the size of fractions. Remember that you can tell which of two fractions with the same denominator is larger by comparing their numerators. For example, $\frac{3}{6}$ is larger than $\frac{2}{6}$, because 3 is larger than 2. We can compare fractions with unlike denominators by first converting them to equivalent fractions with like denominators. Thus, $\frac{1}{2}$ is larger than $\frac{1}{3}$, because $\frac{3}{6}$ $\left(\text{which equals } \frac{1}{2}\right)$ is larger than $\frac{2}{6}$ $\left(\text{which equals } \frac{1}{3}\right)$.

9. Place the following fractions in increasing order.

a. $\frac{3}{10}, \frac{9}{35}, \frac{5}{14}$

$\left(\text{Change each to } \frac{?}{70}\right)$

b. $\frac{11}{20}, \frac{8}{15}, \frac{13}{24}$

$\left(\text{Change each to } \frac{?}{120}\right)$

EXERCISES

Form an equivalent fraction with the given denominator.
In these exercises, the denominator on the left always divides the denominator on the right.

1. $\frac{3}{5} = \frac{?}{10}$
2. $\frac{5}{7} = \frac{?}{21}$
3. $\frac{3}{8} = \frac{?}{24}$
4. $\frac{4}{9} = \frac{?}{18}$
5. $\frac{7}{2} = \frac{?}{18}$
6. $\frac{10}{3} = \frac{?}{24}$
7. $\frac{9}{5} = \frac{?}{45}$
8. $\frac{12}{11} = \frac{?}{55}$
9. $\frac{5}{1} = \frac{?}{13}$
10. $\frac{8}{1} = \frac{?}{7}$
11. $\frac{3}{5} = \frac{?}{35}$
12. $\frac{4}{3} = \frac{?}{36}$
13. $\frac{3}{7} = \frac{?}{63}$
14. $\frac{8}{5} = \frac{?}{60}$
15. $2 = \frac{?}{4}$
16. $4 = \frac{?}{4}$
17. $5 = \frac{?}{15}$
18. $6 = \frac{?}{2}$

7.3 ADDING FRACTIONS GIVEN THE COMMON DENOMINATOR

Our ability to combine fractions with unlike denominators is derived from two facts:

1. Any fraction in a sum can be replaced by an equivalent fraction without changing the value of the sum. For example,

$$\frac{1}{2} = \frac{2}{4}, \quad \text{so} \quad \frac{1}{2} + \frac{3}{4} = \frac{2}{4} + \frac{3}{4}$$

2. Any two fractions are equivalent to fractions with a common denominator. For example, the fractions $\frac{1}{3}$ and $\frac{1}{2}$ are equivalent to the fractions $\frac{2}{6}$ and $\frac{3}{6}$, respectively.

10. Pair equal sums

a. $\frac{4}{8} + \frac{3}{4}$ b. $\frac{9}{12} + \frac{2}{3}$

c. $\frac{2}{3} + \frac{3}{6}$ d. $\frac{3}{4} + \frac{6}{9}$

e. $\frac{1}{2} + \frac{6}{8}$ f. $\frac{8}{12} + \frac{1}{2}$

Can you see how these two properties of numbers enable us to add fractions with unlike denominators? Remember that we have seen how to add fractions with like denominators in Section 7.1.

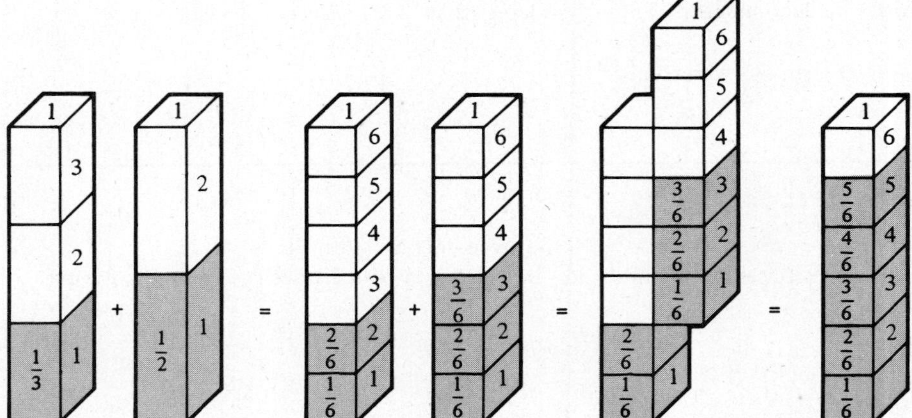

Look at the following example of *adding fractions given the common denominator.*

TYPICAL PROBLEM

$$3 + \frac{2}{5} - \frac{1}{2} = \frac{?}{10}$$

To avoid sign errors, we first move all minus signs to the numerators (so that no minus signs are in the denominators or before the fractions bars):

$$3 + \frac{2}{5} + \frac{-1}{2}$$

Then we convert the sum to a sum with like denominators. To do this we first multiply each term by the given new common denominator to determine the new numerators:

$$10(3) + 10\left(\frac{2}{5}\right) + 10\left(\frac{-1}{2}\right)$$

And then we form new fractions by placing each new numerator over the given new common denominator:

$$\frac{10(3)}{10} + \frac{\overset{2}{\cancel{10}}\left(\frac{2}{\cancel{5}}\right)}{10} + \frac{\overset{5}{\cancel{10}}\left(\frac{-1}{\cancel{2}}\right)}{10}$$

At this point we have a sum of fractions with common denominators:

$$\frac{30}{10} + \frac{4}{10} + \frac{-5}{10}$$

The given common denominator must always be a multiple of each original denominator.

To add fractions given the common denominator we combine the methods in the previous two sections.

Now we add the fractions. We first form a single fraction, writing the numerators as an indicated sum over the common denominator:

$$\frac{30 + 4 - 5}{10}$$

And then we simplify the numerator:

$$\frac{29}{10}$$

Hence, $3 + \frac{2}{5} - \frac{1}{2} = \frac{29}{10}$.

EXAMPLES

Add the following fractions, first converting them to the given common denominators.

Given a common denominator, every sum of simple fractions can be simplified to a single reduced fraction.

1. $\frac{7}{2} + \frac{3}{5} = \frac{?}{10}$

2. $\frac{-8}{27} - \frac{40}{9} = \frac{?}{27}$

3. $\frac{7}{4} - \frac{5}{6} = \frac{?}{12}$

4. $2 - \frac{5}{3} = \frac{?}{3}$

5. $\frac{3}{7} - \frac{5}{14} + \frac{1}{4} = \frac{?}{28}$

6. $\frac{3}{4} + \frac{1}{2} - \frac{5}{3} = \frac{?}{12}$

Solutions

1.
$$\frac{\overset{5}{\cancel{10}}\left(\frac{7}{\cancel{2}}\right)}{10} + \frac{\overset{2}{\cancel{10}}\left(\frac{3}{\cancel{5}}\right)}{10} \;\downarrow\; \frac{35}{10} + \frac{6}{10} \;\downarrow\; \frac{35+6}{10} \;\downarrow\; \frac{41}{10}$$

2.
$$\frac{\overset{1}{\cancel{27}}\left(\frac{-8}{\cancel{27}}\right)}{27} + \frac{\overset{3}{\cancel{27}}\left(\frac{-40}{\cancel{9}}\right)}{27} \;\downarrow\; \frac{-8}{27} + \frac{-120}{27} \;\downarrow\; \frac{-8-120}{27} \;\downarrow\; \frac{-128}{27}$$

3.
$$\frac{\overset{3}{\cancel{12}}\left(\frac{7}{\cancel{4}}\right)}{12} + \frac{\overset{2}{\cancel{12}}\left(\frac{-5}{\cancel{6}}\right)}{12} \;\downarrow\; \frac{21}{12} + \frac{-10}{12} \;\downarrow\; \frac{21-10}{12} \;\downarrow\; \frac{11}{12}$$

4.
$$\frac{3(2)}{3} + \frac{\overset{1}{\cancel{3}}\left(\frac{-5}{\cancel{3}}\right)}{3} \;\downarrow\; \frac{6}{3} + \frac{-5}{3} \;\downarrow\; \frac{6-5}{3} \;\downarrow\; \frac{1}{3}$$

5.
$$\frac{\overset{4}{\cancel{28}}\left(\frac{3}{\cancel{7}}\right)}{28} + \frac{\overset{2}{\cancel{28}}\left(\frac{-5}{\cancel{14}}\right)}{28} + \frac{\overset{7}{\cancel{28}}\left(\frac{1}{\cancel{4}}\right)}{28} \;\downarrow\; \frac{12}{28} + \frac{-10}{28} + \frac{7}{28} \;\downarrow\; \frac{12-10+7}{28} \;\downarrow\; \frac{9}{28}$$

6.
$$\frac{\overset{3}{\cancel{12}}\left(\frac{3}{\cancel{4}}\right)}{12} + \frac{\overset{6}{\cancel{12}}\left(\frac{1}{\cancel{2}}\right)}{12} + \frac{\overset{4}{\cancel{12}}\left(\frac{-5}{\cancel{3}}\right)}{12} \;\downarrow\; \frac{9}{12} + \frac{6}{12} + \frac{-20}{12} \;\downarrow\; \frac{9+6-20}{12} \;\downarrow\; \frac{-5}{12}$$

To add an integer and a fraction we take the denominator of the fraction as the common denominator. To see some examples of adding an integer and a fraction look at the following table. Verify the entries to check your understanding.

plus	$\frac{1}{2}$	$\frac{2}{3}$	$\frac{3}{4}$	$\frac{4}{5}$	$\frac{5}{6}$
1	$\frac{3}{2}$	$\frac{5}{3}$	$\frac{7}{4}$		
2	$\frac{5}{2}$	$\frac{8}{3}$	$\frac{11}{4}$		
3	$\frac{7}{2}$	$\frac{11}{3}$	$\frac{15}{4}$		
4					
5					

11. Complete the table.

OBSERVATIONS AND REFLECTIONS

In the problems of this section we provide a suitable denominator for each sum of fractions. However, many other denominators could be used. Any common multiple of the original denominators provides a suitable denominator. In particular, the product of the denominators is always a suitable new denominator. Look again at example 3, namely $\frac{7}{4} - \frac{5}{6} = \frac{?}{12}$. We could have used the *product* of the denominators 24. With this change the problem would read $\frac{7}{4} - \frac{5}{6} = \frac{?}{24}$. The solution to the original problem is $\frac{11}{12}$, whereas the solution to the second problem is $\frac{22}{24}$ (which is equivalent to $\frac{11}{12}$). Using any suitable denominator leads to an equivalent fraction before we reduce, and thus to the same fraction once we reduce.

12. Compute these sums using the product of the denominators as the common denominator.

a. $\frac{-8}{27} - \frac{40}{9}$

b. $\frac{3}{4} + \frac{1}{2} - \frac{5}{3}$

(Compare your solutions with examples 2 and 6.)

EXERCISES

Simplify each of the following expressions to a fraction with the given denominator.

In these exercises, all the original denominators are divisors of the given new denominator.

1. $\frac{7}{3} + \frac{5}{2} = \frac{?}{6}$
2. $\frac{3}{2} + \frac{9}{4} = \frac{?}{4}$
3. $\frac{13}{8} + \frac{7}{4} = \frac{?}{8}$
4. $-\frac{3}{4} + \frac{11}{16} = \frac{?}{16}$
5. $\frac{13}{9} - \frac{7}{6} = \frac{?}{18}$
6. $\frac{11}{10} - \frac{5}{4} = \frac{?}{20}$
7. $\frac{5}{9} - \frac{7}{6} = \frac{?}{18}$
8. $\frac{5}{6} - \frac{3}{4} = \frac{?}{12}$
9. $\frac{7}{3} + \frac{4}{5} = \frac{?}{15}$
10. $\frac{3}{2} - \frac{5}{6} = \frac{?}{6}$
11. $2 + \frac{3}{5} = \frac{?}{5}$
12. $5 + \frac{2}{7} = \frac{?}{7}$

13. $3 - \frac{9}{4} = \frac{?}{4}$ 14. $7 - \frac{2}{9} = \frac{?}{9}$ 15. $-\frac{3}{8} - 4 = \frac{?}{8}$

16. $-\frac{5}{6} - 2 = \frac{?}{6}$ 17. $\frac{8}{5} + \frac{1}{6} + \frac{7}{4} = \frac{?}{60}$ 18. $\frac{1}{2} + \frac{4}{9} + \frac{7}{12} = \frac{?}{36}$

19. $\frac{7}{8} + \frac{5}{6} - \frac{7}{9} = \frac{?}{72}$ 20. $\frac{8}{3} - \frac{5}{6} - \frac{7}{8} = \frac{?}{24}$ 21. $\frac{7}{5} - \frac{3}{4} - \frac{9}{10} = \frac{?}{20}$

22. $-\frac{4}{5} + \frac{4}{3} - \frac{5}{8} = \frac{?}{120}$ 23. $\frac{1}{2} - \frac{1}{3} + \frac{1}{5} = \frac{?}{30}$ 24. $\frac{3}{4} - \frac{5}{8} + \frac{7}{16} = \frac{?}{16}$

7.4 DETERMINING THE LEAST COMMON MULTIPLE OF TWO INTEGERS

Consider the positive multiples of the numbers 12, 30, and 32.

Multiples of 12	12, 24, 36, 48, 60, 72, 84, 96, 108, 120, 132, 144, 156, 168, 180, 192, 204, 216, 228, 240, 252, 264, 276, 288, 300, 312, 324, 336, 348, 360, 372, 384, . . .
Multiples of 30	30, 60, 90, 120, 150, 180, 210, 240, 270, 300, 330, 360, . . .
Multiples of 32	32, 64, 96, 128, 160, 192, 224, 256, 288, 320, 352, 384, . . .

If you look at the multiples of 12 and 30, you can see that 60, 120, 180, 240, 300, 360, . . . are multiples of both these numbers. Similarly, if you look at the multiples of 12 and 32, you can see that 96, 192, 288, 384, . . . are multiples of both these numbers. In general, a number is a *common multiple* of two integers if it is a multiple of each integer (that is, if both integers are divisors of the number). Hence, in particular, the product of two integers is always a common multiple of the two integers. For example, $12 \cdot 30 = 360$ is a common multiple of 12 and 30, and $12 \cdot 32 = 384$ is a common multiple of 12 and 32.

13. Determine the first three positive common multiples of 4 and 6.

In this section we will study two methods for determining the smallest positive common multiple of two positive integers.

The smallest *positive* common multiple of two integers is called their *least common multiple* (LCM). The LCM of 12 and 30, for example, is 60, and the LCM of 12 and 32 is 96.

Method A (Method of Primes)

To find the LCM of two integers, it is usually helpful to look at the divisors of the two integers. One way to do this systematically is to factor each integer into a product of primes, using the Method of Primes introduced in Section 6.1. (Refer to the table of prime factorizations of the numbers 1 to 200 on the inside front cover.)

Given the prime factorizations of two integers, we can quickly determine the prime factorization of their LCM. For example, look again at the numbers 12 and 30 with their LCM 60.

$$12 = 2 \cdot 2 \cdot 3$$
$$30 = 2 \cdot \quad 3 \cdot 5$$
$$\downarrow \quad \downarrow \quad \downarrow \quad \downarrow$$
$$60 = 2 \cdot 2 \cdot 3 \cdot 5$$

14. Form a similar diagram for the numbers 4 and 6 and their LCM 12.

Can you guess how to determine the prime factorization of the LCM from the prime factorizations of the two numbers?

Look at the following example of *determining the LCM of two integers using the Method of Primes* (Method A).

TYPICAL PROBLEM

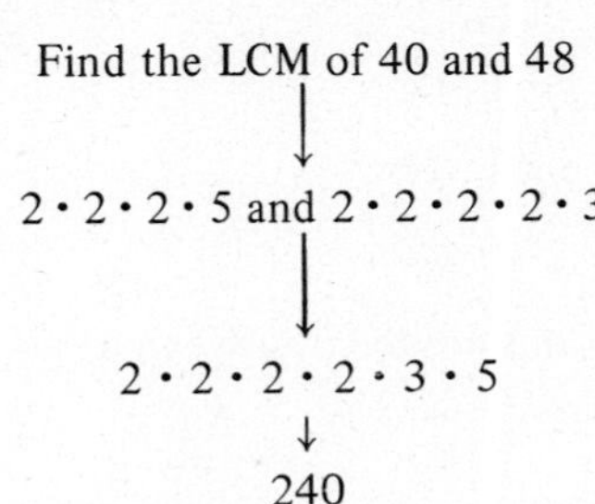

We first determine the prime factorizations of the two integers:

Then we write each prime factor the largest number of times it appears in either factorization:

We then multiply the factors to get the LCM:

Hence, the LCM of 40 and 48 = 240.

A table of prime factorizations of the numbers from 1 to 200 appears on the inside front cover.

EXAMPLES (Method A)

Every set of integers has a unique (positive) LCM.

Find the LCM of each pair of integers using the Method of Primes.

1. 54 and 84
2. 39 and 52
3. 24 and 90
4. 88 and 176

Solutions

1. $2 \cdot 3 \cdot 3 \cdot 3$ and $2 \cdot 2 \cdot 3 \cdot 7$
   $\downarrow$
   $2 \cdot 2 \cdot 3 \cdot 3 \cdot 3 \cdot 7 = 756$

2. $3 \cdot 13$ and $4 \cdot 13$
   $\downarrow$
   $3 \cdot 4 \cdot 13 = 156$

3. $2 \cdot 2 \cdot 2 \cdot 3$ and $2 \cdot 3 \cdot 3 \cdot 5$
   $\downarrow$
   $2 \cdot 2 \cdot 2 \cdot 3 \cdot 3 \cdot 5 = 360$

4. $2 \cdot 2 \cdot 2 \cdot 11$ and $2 \cdot 2 \cdot 2 \cdot 2 \cdot 11$
   $\downarrow$
   $2 \cdot 2 \cdot 2 \cdot 2 \cdot 11 = 176$

Method B

An alternative method for finding the LCM may be preferable when a table of primes is not available. This method parallels the method of reducing fractions in Section 6.2 (just as Method A parallels the method of reducing fractions in Section 6.1).

This method relies on a relationship between the LCM of two integers and their *greatest common divisor* (GCD). Remember that the GCD of two integers (defined in the introduction to Chapter 6) is the largest positive integer that is a common divisor of both numbers. For example, the GCD of 12 and 30 is 6.

Dividing each integer by the GCD yields another integer, which we shall call a *final quotient.* For example,

$$12 \div 6 = 2 \quad \text{and} \quad 30 \div 6 = 5$$

so 2 and 5 are the final quotients of 12 and 30.

Although it is not obvious, the product of the GCD and the two final quotients of two integers is always their LCM. For example, for 12 and 30 we have

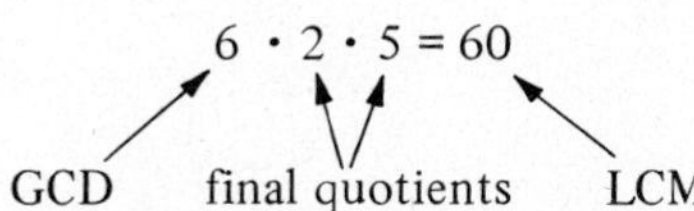

Our procedure is to find and divide the two numbers by one common divisor after another, eventually arriving at the final quotients. In this way we find the GCD at the same time, because the product of these divisors is always the GCD of the two integers.

15. Determine the final quotients of 12 and 32 by dividing by their GCD 4.

16. Determine the LCM of 12 and 32 by multiplying their GCD by their final quotients.

Look at the following example of *determining the LCM of two integers* (Method B).

TYPICAL PROBLEM

Find the LCM of 40 and 48

First we set up our format by drawing a line to the right of the two numbers:

40	48	

Then we find a common divisor of the two numbers. We write this divisor to the right of the line, and then, dividing by this number, we write the quotients below the two numbers:

40	48	
		4
10	12	

Then we find a common divisor of the two quotients, and repeat the above step until the GCD of the two quotients is 1:

40	48	
		4
10	12	
		2
5	6	

5 and 6: final quotients

We then multiply the numbers to the right of the line (the divisors). Their product is the GCD:

$$(4)(2) = 8$$

Finally we multiply the final quotients and the GCD. This product is the LCM:

$$(5)(6)(8) = 240$$

Hence, the LCM of 40 and 48 = 240.

EXAMPLES (Method B)

Find the LCM of each pair of integers.

Every set of integers has a unique (positive) LCM.

1. 45 and 21
2. 24 and 44
3. 81 and 54
4. 75 and 45

Solutions

1.

45	21	
		3
15	7	

GCD = 3
LCM = (15) (7) (3) = 315

2.

24	44	
		4
6	11	

GCD = 4
LCM = (6) (11) (4) = 264

3.

81	54	
		3
27	18	
		9
3	2	

GCD = (3) (9) = 27
LCM = (3) (2) (27) = 162

4.

75	45	
		5
15	9	
		3
5	3	

GCD = (5) (3) = 15
LCM = (5) (3) (15) = 225

SPECIAL CASES

17. Find the LCM of
a. 3 and 11
b. 5 and 13
c. 14 and 33
d. 12 and 35

1. If the two numbers have no common divisors greater than 1, their least common multiple is their product. For example, 10 and 21 have no common divisors greater than 1, and their LCM is $10 \cdot 21 = 210$.

18. Find the LCM of
a. 4 and 40
b. 12 and 84
c. 64 and 16
d. 22 and 2200

2. If one integer is a multiple of the other, then this integer is the LCM, because every integer is a multiple of itself. For example, 6 is a multiple of 2, so the LCM of 2 and 6 is 6.

3. Method A for finding the LCM of two integers, which uses the Method of Primes, can be extended without change to find the LCM of three (or more) integers. For example, we can find the LCM of 8, 12, and 20 by the computation

$$
\begin{array}{rcccccccccc}
8 = & 2 & \cdot & 2 & \cdot & 2 & & & & \\
12 = & 2 & \cdot & 2 & \cdot & & & 3 & & \\
20 = & 2 & \cdot & 2 & \cdot & & & & & 5 \\
& \downarrow & & \downarrow & & \downarrow & & \downarrow & & \downarrow \\
120 = & 2 & \cdot & 2 & \cdot & 2 & \cdot & 3 & \cdot & 5
\end{array}
$$

Method B, on the other hand, can only be applied to pairs of integers. Hence, to use Method B to find the LCM of 8, 12, and 20, for example, we first find the LCM of 8 and 12, which is 24. Then we find the LCM of 24 and 20, which is 120. (Note that we can also compute the LCM of 8, 12, and 20 by first computing the LCM of 12 and 20, getting 60, and then computing the LCM of 8 and 60, which, of course, is also 120.)

19. Compute the LCM of
a. 2, 3, and 4
b. 4, 6, and 15

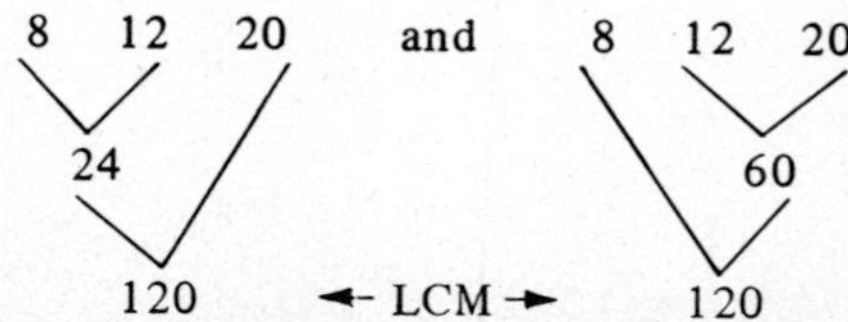

Here is a table giving the LCM of pairs of integers. Check the entries to verify your understanding of the LCM.

LCM	1	2	3	4	5	6	7	8	9
1	1	2	3	4	5	6			
2	2	2	6	4	10	6			
3	3	6	3	12	15	6			
4	4	4	12	4	20	12			
5	5	10	15	20	5	30			
6	6	6	6	12	30	6			
7									
8									
9									

20. Complete the table.

Compare this table with the table of GCD's in the introduction to Chapter 6.

OBSERVATIONS AND REFLECTIONS

★ 1. As we observed, the positive common multiples of 12 and 30 are 60, 120, 180, 240, 300, 360, ... and their LCM is 60. Similarly, we saw that the positive common multiples of 12 and 32 are 96, 192, 288, 384, ... and that their LCM is 96. Note that in both examples the least common multiple (LCM) is a divisor of every common multiple of the two numbers. In general, every common multiple of two integers is a multiple of their LCM. (This includes 0 and the LCM itself.)

21. Which of the following are multiples of
a. 4
b. 6
c. 12

{18, 20, 28, 32, 36, 38, 48, 54, 60}?

★ 2. In this section we have presented two different methods for finding the least common multiple (LCM). In Method A, using the Method of Primes, we determine the LCM directly from the prime factorizations of the two integers. Using Method B we first find the greatest common divisor (GCD) and the final quotients. Then we multiply the GCD and the final quotients to get the LCM.

Using the Method of Primes, we can also determine the GCD and the two final quotients. Using the prime factorizations of the two integers, we can write each prime factor the largest number of times it appears in *both* factorizations. The product of these common primes is the GCD. The product of the remaining primes in each factorization is a final quotient. For example, consider the numbers 40 and 48, which appear in the two Typical Problems:

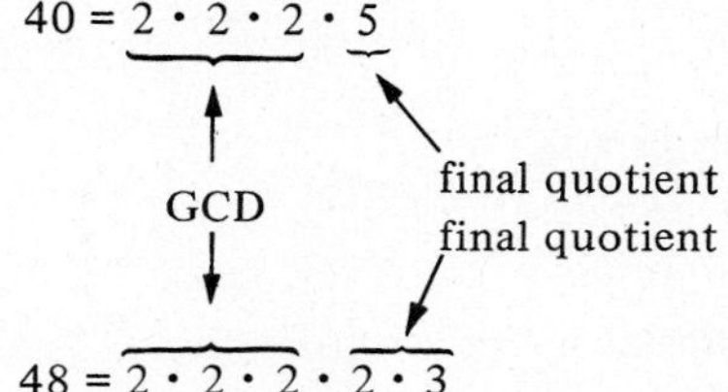

22. Using the Method of Primes determine the GCD of
a. 12 and 20
b. 45 and 75
c. 72 and 108

In this example the GCD is 2 · 2 · 2 = 8, and the final quotients are 5 and 2 · 3 = 6.

★ 3. The two positive integers, their GCD and LCM, and the two final quotients are related as follows:

1. The product of the two final quotients and the GCD is the LCM.
2. The product of the GCD and the LCM equals the product of the two integers.
3. The product of each integer and the final quotient of the other integer is the LCM.
4. Each integer is the product of its final quotient and the GCD.
5. Each integer equals the LCM divided by the final quotient of the other integer.

We can summarize these relationships as follows.

1. $\text{LCM} = \text{GCD} \cdot Q_1 \cdot Q_2$
2. $N_1 \cdot N_2 = \text{GCD} \cdot \text{LCM}$
3. $\text{LCM} = N_1 \cdot Q_2$ and $\text{LCM} = N_2 \cdot Q_1$
4. $N_1 = \text{GCD} \cdot Q_1$ and $N_2 = \text{GCD} \cdot Q_2$
5. $N_1 = \text{LCM} \div Q_2$ and $N_2 = \text{LCM} \div Q_1$

(where N_1 and N_2 represent the two integers, and Q_1 and Q_2 represent their respective final quotients)

EXERCISES

Find the LCM (Least Common Multiple).

1. 4, 6
2. 2, 12
3. 6, 9
4. 8, 14
5. 12, 5
6. 10, 6
7. 4, 18
8. 9, 12
9. 14, 35
10. 28, 18
11. 36, 16
12. 25, 45
13. 35, 28
14. 20, 18
15. 11, 22
16. 4, 15
17. 24, 36
18. 28, 48
19. 3, 4, 5
20. 4, 6, 10
21. 6, 10, 7
22. 6, 15, 21
23. 27, 15, 4
24. 40, 8, 7
25. 26, 54, 60
26. 54, 45, 72
27. 52, 20, 39
28. 14, 28, 35
29. 72, 84, 42
30. 18, 54, 72

7.5 ADDING NUMERICAL FRACTIONS

We first considered sums of fractions with unlike denominators in Section 7.3, where we simplified the problem by providing a common denominator. As we saw there, to add we must first convert the sum to an equivalent sum of fractions with common denominators.

The smallest common denominator, and hence usually the best one to use, is called the *lowest common denominator.* The lowest common denominator of a sum of fractions is the LCM (Least Common Multiple) of the denominators of the fractions.

In Section 7.3 the common denominator provided for each sum of fractions was always the LCM of the denominators.

Look at the following example of *adding fractions.*

TYPICAL PROBLEM

$$\frac{5}{6} - \frac{8}{15}$$

To avoid sign errors, we first move all minus signs to the numerators (so that no minus signs are in the denominators or before the fraction bars):

$$\frac{5}{6} + \frac{-8}{15}$$

We then determine the LCM of the denominators:

LCM of 6 and 15 is 30

Then we convert the sum to a sum with like denominators. To do this we first multiply each term by the LCM to determine the new numerators:

$$30\left(\frac{5}{6}\right) + 30\left(\frac{-8}{15}\right)$$

And then we form new fractions by placing each new numerator over the LCM:

$$\frac{\overset{5}{\cancel{30}}\left(\frac{5}{\cancel{6}}\right)}{30} + \frac{\overset{2}{\cancel{30}}\left(\frac{-8}{\cancel{15}}\right)}{30}$$

At this point we have a sum of fractions with common denominators:

$$\frac{25}{30} + \frac{-16}{30}$$

Now we add the fractions. We first form a single fraction, writing the numerators as an indicated sum over the common denominator:

$$\frac{25 - 16}{30}$$

We then simplify the numerator:

$$\frac{9}{30}$$

Finally, we reduce the fraction if possible:

$$\frac{3}{10}$$

Hence, $\frac{5}{6} - \frac{8}{15} = \frac{3}{10}$.

Except for the added step of determining the LCM of the denominators (the lowest common denominator), we use the method in Section 7.3. To determine the LCM we use the method in the previous section.

EXAMPLES

Add the following numerical fractions.

Every sum of simple fractions can be simplified to a single reduced fraction.

1. $\frac{2}{5} + \frac{2}{9}$
2. $-\frac{5}{9} + \frac{7}{6}$
3. $\frac{3}{7} + \frac{-1}{2}$
4. $\frac{3}{8} - \frac{7}{3}$
5. $\frac{-2}{7} - \frac{5}{2} + \frac{2}{3}$
6. $\frac{5}{6} + \frac{-1}{8} - 2$

Solutions

1. LCM = 45

$$\downarrow$$

$$\frac{\overset{9}{\cancel{45}}\left(\frac{2}{\cancel{5}}\right)}{45} + \frac{\overset{5}{\cancel{45}}\left(\frac{2}{\cancel{9}}\right)}{45}$$

$$\downarrow$$

$$\frac{18}{45} + \frac{10}{45}$$

$$\downarrow$$

$$\frac{18 + 10}{45}$$

$$\downarrow$$

$$\frac{28}{45}$$

2. LCM = 18

$$\downarrow$$

$$\frac{\overset{2}{\cancel{18}}\left(\frac{-5}{\cancel{9}}\right)}{18} + \frac{\overset{3}{\cancel{18}}\left(\frac{7}{\cancel{6}}\right)}{18}$$

$$\downarrow$$

$$\frac{-10}{18} + \frac{21}{18}$$

$$\downarrow$$

$$\frac{-10 + 21}{18}$$

$$\downarrow$$

$$\frac{11}{18}$$

3. LCM = 14

$$\downarrow$$

$$\frac{\overset{2}{\cancel{14}}\left(\frac{3}{\cancel{7}}\right)}{14} + \frac{\overset{7}{\cancel{14}}\left(\frac{-1}{\cancel{2}}\right)}{14}$$

$$\downarrow$$

$$\frac{6}{14} + \frac{-7}{14}$$

$$\downarrow$$

$$\frac{6 - 7}{14}$$

$$\downarrow$$

$$\frac{-1}{14}$$

4. LCM = 24

$$\downarrow$$

$$\frac{\overset{3}{\cancel{24}}\left(\frac{3}{\cancel{8}}\right)}{24} + \frac{\overset{8}{\cancel{24}}\left(\frac{-7}{\cancel{3}}\right)}{24}$$

$$\downarrow$$

$$\frac{9}{24} + \frac{-56}{24}$$

$$\downarrow$$

$$\frac{9 - 56}{24}$$

$$\downarrow$$

$$\frac{-47}{24}$$

5. LCM = 42

$$\downarrow$$

$$\frac{\overset{6}{\cancel{42}}\left(\frac{-2}{\cancel{7}}\right)}{42} + \frac{\overset{21}{\cancel{42}}\left(\frac{-5}{\cancel{2}}\right)}{42} + \frac{\overset{14}{\cancel{42}}\left(\frac{2}{\cancel{3}}\right)}{42}$$

$$\downarrow$$

$$\frac{-12}{42} + \frac{-105}{42} + \frac{28}{42}$$

$$\downarrow$$

$$\frac{-12 - 105 + 28}{42}$$

$$\downarrow$$

$$\frac{-89}{42}$$

6. LCM = 24

$$\downarrow$$

$$\frac{\overset{4}{\cancel{24}}\left(\frac{5}{\cancel{6}}\right)}{24} + \frac{\overset{3}{\cancel{24}}\left(\frac{-1}{\cancel{8}}\right)}{24} - \frac{24(2)}{24}$$

$$\downarrow$$

$$\frac{20}{24} + \frac{-3}{24} + \frac{-48}{24}$$

$$\downarrow$$

$$\frac{20 - 3 - 48}{24}$$

$$\downarrow$$

$$\frac{-31}{24}$$

Here is an addition table for signed fractions. Verify the entries to check your understanding of addition.

plus	-1	$-\frac{3}{4}$	$-\frac{1}{2}$	$-\frac{1}{4}$	0	$+\frac{1}{4}$	$+\frac{1}{2}$	$+\frac{3}{4}$	$+1$
$+1$									
$+\frac{3}{4}$		0	$\frac{1}{4}$	$\frac{1}{2}$	$\frac{3}{4}$	1	$\frac{5}{4}$	$\frac{3}{2}$	
$+\frac{1}{2}$		$\frac{-1}{4}$	0	$\frac{1}{4}$	$\frac{1}{2}$	$\frac{3}{4}$	1	$\frac{5}{4}$	
$+\frac{1}{4}$		$\frac{-1}{2}$	$\frac{-1}{4}$	0	$\frac{1}{4}$	$\frac{1}{2}$	$\frac{3}{4}$	1	
0		$\frac{-3}{4}$	$\frac{-1}{2}$	$\frac{-1}{4}$	0	$\frac{1}{4}$	$\frac{1}{2}$	$\frac{3}{4}$	
$-\frac{1}{4}$		-1	$\frac{-3}{4}$	$\frac{-1}{2}$	$\frac{-1}{4}$	0	$\frac{1}{4}$	$\frac{1}{2}$	
$-\frac{1}{2}$		$\frac{-5}{4}$	-1	$\frac{-3}{4}$	$\frac{-1}{2}$	$\frac{-1}{4}$	0	$\frac{1}{4}$	
$-\frac{3}{4}$		$\frac{-3}{2}$	$\frac{-5}{4}$	-1	$\frac{-3}{4}$	$\frac{-1}{2}$	$\frac{-1}{4}$	0	
-1									

23. Complete the table.

OBSERVATIONS AND REFLECTIONS

★ **1.** Two fractions are *negatives* of each other if their sum is zero. For example, $\frac{2}{3}$ and $\frac{-2}{3}$ $\left(\text{or } -\frac{2}{3}\right)$ are negatives of each other, because $\frac{2}{3} + \frac{-2}{3} = 0$ $\left(\text{or } \frac{2}{3} - \frac{2}{3} = 0\right)$. The *negative of a fraction* (which is not necessarily a negative fraction) is usually formed by changing the sign of the numerator or by changing the sign before the fraction bar. For example, the negative of $\frac{-2}{3}$ is $\frac{2}{3}$ $\left(\text{or } -\frac{-2}{3}\right)$, and the negative of $-\frac{5}{8}$ is $+\frac{5}{8}$ $\left(\text{or } -\frac{-5}{8}\right)$. Changing a fraction to its negative is called *negation*.

24. Form the negative of

a. $\frac{-5}{6}$

b. $\frac{2}{7}$

c. $-\frac{8}{3}$

★ **2.** Using negation, we can define *algebraic subtraction of fractions*. We can subtract one fraction from a second fraction by adding the negative of the first fraction to the second fraction. For example, to subtract $-\frac{2}{3}$ from $-\frac{5}{3}$, we add $+\frac{2}{3}$ to $-\frac{5}{3}$:

$$-\frac{5}{3} + \frac{2}{3} = \frac{-5+2}{3} = \frac{-3}{3} = -1$$

25. Subtract

a. $\frac{4}{5}$ from $\frac{1}{2}$

b. $\frac{2}{3}$ from $\frac{-3}{4}$

c. $\frac{-4}{9}$ from $\frac{5}{6}$

d. $\frac{-7}{8}$ from $\frac{-11}{12}$

EXERCISES

Simplify each of the following expressions to a single reduced fraction.

1. $\frac{1}{2} + \frac{1}{3}$
2. $\frac{2}{3} + \frac{1}{7}$
3. $\frac{2}{9} - \frac{3}{4}$
4. $-\frac{2}{3} + \frac{2}{5}$
5. $-\frac{3}{2} + \frac{5}{4}$
6. $-\frac{5}{6} + \frac{2}{3}$
7. $\frac{5}{6} + \frac{7}{12}$
8. $\frac{3}{7} + \frac{9}{14}$
9. $-\frac{7}{8} - \frac{3}{2}$
10. $-\frac{3}{5} - \frac{7}{15}$
11. $-\frac{1}{4} + \frac{1}{6}$
12. $\frac{1}{8} - \frac{3}{20}$
13. $\frac{-3}{10} + \frac{14}{15}$
14. $\frac{-8}{9} + \frac{7}{12}$
15. $\frac{-4}{21} - \frac{3}{14}$
16. $\frac{-1}{4} + \frac{-1}{18}$
17. $\frac{-25}{56} + \frac{-21}{40}$
18. $\frac{-13}{34} + \frac{-23}{51}$
19. $\frac{-27}{16} + \frac{35}{24}$
20. $\frac{-17}{12} + \frac{43}{30}$
21. $\frac{23}{15} + \frac{49}{80}$
22. $\frac{19}{48} + \frac{27}{80}$
23. $\frac{39}{56} - \frac{29}{63}$
24. $-\frac{15}{4} + \frac{49}{36}$
25. $-\frac{7}{30} + \frac{35}{24}$
26. $\frac{7}{12} - \frac{8}{15}$
27. $\frac{7}{2} + \frac{5}{3} + \frac{9}{10}$
28. $\frac{7}{4} + \frac{1}{6} + \frac{7}{8}$
29. $\frac{2}{9} - \frac{11}{6} + \frac{3}{4}$
30. $\frac{5}{18} - \frac{7}{12} + \frac{1}{8}$
31. $-\frac{6}{5} + \frac{11}{9} - \frac{7}{6}$
32. $-\frac{4}{11} + \frac{3}{4} - \frac{1}{6}$
33. $\frac{11}{15} + \frac{13}{21} + \frac{18}{35}$
34. $\frac{3}{10} + \frac{9}{22} + \frac{17}{55}$
35. $\frac{17}{30} - \frac{19}{35} + \frac{13}{42}$
36. $\frac{11}{14} - \frac{21}{22} + \frac{16}{77}$
37. $\frac{1}{2} + 8 + \frac{2}{3}$
38. $\frac{2}{3} + 12 + \frac{7}{8}$
39. $-\frac{4}{5} + 6 + \frac{1}{2}$
40. $\frac{5}{9} + 2 - \frac{3}{5}$
41. $7 + \frac{1}{2} - 4 - \frac{1}{4}$
42. $12 + \frac{3}{8} - 4 - \frac{3}{4}$

Express each of the following word phrases algebraically and then simplify it.

1. The sum of $\frac{-3}{5}$ and $\frac{7}{4}$
2. $\frac{2}{7}$ decreased by $\frac{2}{3}$
3. $\frac{-5}{9}$ plus $\frac{1}{2}$
4. $\frac{3}{8}$ reduced by $\frac{1}{6}$
5. $\frac{7}{15}$ subtracted from $\frac{-7}{12}$
6. $\frac{-9}{28}$ added to $\frac{4}{35}$

7. $\frac{1}{54}$ less than $\frac{1}{42}$

8. $\frac{-3}{11}$ less than $\frac{-2}{9}$

9. $\frac{-7}{18}$ increased by $\frac{15}{72}$

10. $\frac{-9}{56}$ is reduced by $\frac{5}{48}$

11. $\frac{-3}{20}$ is subtracted from $\frac{-7}{30}$

12. $\frac{3}{49}$ more than $\frac{-9}{77}$

7.6 ORDER OF OPERATIONS

In this section we will work with expressions requiring the operations of addition, multiplication, exponentiation, and grouping of integers and fractions. We will see expressions such as $\frac{3+11}{5-9} - \frac{8-14}{-3-9}$, $6\left(\frac{-5}{8}\right) + 5\left(\frac{6}{20}\right)$, $\frac{4^2-3^2}{3^2} + \frac{6^2}{2+4^2}$, and $22\left(\frac{7}{36}\right) - \left(9 - \frac{1}{9}\right)$.

The Order of Operations for simplifying these expressions is summarized as follows:

1. First simplify any expressions inside parentheses.
2. Compute powers before products, because exponential notation is a shorthand for repeated multiplication.
3. Simplify each term to a simple fraction (or an integer) by multiplying and dividing before computing sums.

(Remember that this summary does not handle all the details of our notational system, and that a more comprehensive set of rules is given at the end of Section 6.5.)

EXAMPLES

Simplify the following expressions to simple fractions, following the Order of Operations.

1. $\frac{3+11}{5-9} - \frac{8-14}{-3-9}$

2. $\left(\frac{3}{5}\right)^2 - \frac{3^2}{5}$

3. $\frac{4^2-3^2}{3^2} + \frac{6^2}{2+4^2}$

4. $\frac{-4(5-9)}{-4^2} - \frac{5(4^2)}{6^2}$

5. $\frac{5}{7}\left(\frac{8}{35}\right) - \left(\frac{15}{42} \cdot \frac{-12}{21}\right)$

6. $22\left(\frac{7}{36}\right) - \left(9 - \frac{1}{9}\right)$

7. $\dfrac{8 - \frac{4}{3}}{6 - \frac{2}{5}}$

8. $\dfrac{\frac{7}{3} - \frac{15}{8}}{\frac{42}{5} - \frac{82}{15}}$

Solutions

1. $\frac{3+11}{5-9} - \frac{8-14}{-3-9}$

$\frac{14}{-4} - \frac{-6}{-12}$

$\frac{-7}{2} - \frac{1}{2}$

-4

2. $\left(\frac{3}{5}\right)^2 - \frac{3^2}{5}$

$\left(\frac{3}{5}\right)\left(\frac{3}{5}\right) - \frac{3^2}{5}$

$\frac{9}{25} - \frac{9}{5}$

$\frac{9}{25} + \frac{-45}{25}$

$\frac{-36}{25}$

3. $\frac{4^2 - 3^2}{3^2} + \frac{6^2}{2+4^2}$

$\frac{16-9}{9} + \frac{36}{2+16}$

$\frac{7}{9} + \frac{36}{18}$

$\frac{7}{9} + 2$

$\frac{25}{9}$

4. $\frac{-4(5-9)}{-4^2} - \frac{5(4^2)}{6^2}$

$\frac{-4(-4)}{-16} - \frac{5(16)}{36}$

$\frac{+16}{-16} - \frac{80}{36}$

$-1 + \frac{-20}{9}$

$\frac{-29}{9}$

5. $\frac{5}{7}\left(\frac{8}{35}\right) - \left(\frac{15}{42} \cdot \frac{-12}{21}\right)$

$\frac{8}{49} - \left(\frac{-10}{49}\right)$

$\frac{8}{49} + \frac{10}{49}$

$\frac{18}{49}$

6. $22\left(\frac{7}{36}\right) - \left(9 - \frac{1}{9}\right)$

$\frac{77}{18} - \left(\frac{80}{9}\right)$

$\frac{77}{18} + \frac{-160}{18}$

$\frac{-83}{18}$

7. $\left(8 - \frac{4}{3}\right) \div \left(6 - \frac{2}{5}\right)$

$\left(\frac{20}{3}\right) \div \left(\frac{28}{5}\right)$

$\frac{20}{3} \cdot \frac{5}{28}$

$\frac{25}{21}$

8. $\left(\frac{7}{3} - \frac{15}{8}\right) \div \left(\frac{42}{5} - \frac{82}{15}\right)$

$\left(\frac{11}{24}\right) \div \left(\frac{44}{15}\right)$

$\frac{11}{24} \cdot \frac{15}{44}$

$\frac{5}{32}$

SPECIAL CASES

Any power of a fraction should be expressed as a product, avoiding exponential notation. For example,

$$\left(\frac{2}{3}\right)^2 = \left(\frac{2}{3}\right)\left(\frac{2}{3}\right) \quad \text{and} \quad \left(\frac{3}{4}\right)^3 = \left(\frac{3}{4}\right)\left(\frac{3}{4}\right)\left(\frac{3}{4}\right)$$

EXERCISES

Simplify each of the following expressions to a single reduced fraction.

1. $\frac{5-3}{5+3} + \frac{7-2}{6-2}$
2. $\frac{7-2}{7+2} + \frac{5+2}{5-2}$
3. $\frac{9-4}{7-4} - \frac{9-2}{7+2}$
4. $\frac{3-5}{3+5} - \frac{9-13}{11-9}$
5. $\frac{4}{5-11} + \frac{7}{7-9}$
6. $\frac{4}{9-12} + \frac{2}{2-11}$
7. $\frac{8-12}{5-8} - \frac{7-3}{7-13}$
8. $\frac{-4+9}{8-14} - \frac{12-20}{12-15}$
9. $\left(\frac{2}{3}\right)^2 - \frac{2}{3}$
10. $\left(\frac{5}{2}\right)^2 - \frac{5}{2}$
11. $\frac{2^3}{3} + \frac{2}{3^2}$
12. $\frac{5}{2^3} - \frac{5}{2^4}$
13. $\left(\frac{2}{5}\right)^2 - \frac{2^2}{5}$
14. $\left(\frac{6}{5}\right)^2 - \frac{6}{5^2}$
15. $5\left(\frac{2}{9}\right) + 3\left(\frac{5}{3}\right)$
16. $7\left(\frac{3}{5}\right) + 3\left(\frac{7}{5}\right)$
17. $-4\left(\frac{5}{6}\right) - 5\left(\frac{4}{9}\right)$
18. $4\left(\frac{5}{3}\right) - 5\left(\frac{4}{3}\right)$
19. $\frac{3-2^2}{3-2} + \frac{3}{2}$
20. $\frac{4^2-2^4}{3^2-1} - \frac{3^2}{2^3}$
21. $\frac{5^2-2^4}{3^2} + \frac{5^2}{6^2}$
22. $\frac{3}{4^3} - \frac{1^3}{2^6}$
23. $\frac{3(2^2)}{(2+3)^2} - \frac{3(5-2)}{1-4^2}$
24. $\frac{-6(3-6)}{3^2+3^2} + \frac{2^3-2^2}{3^2(2-3)}$
25. $\frac{-3(5-8)}{-3^2} - \frac{4(3^2)}{6^2}$
26. $\frac{3^2(6+2)}{3(6-2)^2} - \frac{5^2(5+3)}{5(5-3)^4}$
27. $30\left(\frac{2}{5} + \frac{5}{6}\right)$
28. $42\left(\frac{5}{6} - \frac{3}{14}\right)$
29. $\frac{8}{3}\left(5 + \frac{7}{4}\right)$
30. $\frac{7}{17}\left(\frac{6}{7} + 4\right)$
31. $\frac{12}{5}\left(\frac{3}{4} - \frac{5}{6}\right)$
32. $\frac{15}{8}\left(\frac{7}{3} + \frac{3}{5}\right)$
33. $\frac{-5}{3}\left(\frac{6}{7}\right) + \frac{2}{7}\left(\frac{9}{4}\right)$
34. $\frac{4}{2}\left(\frac{9}{2}\right) - \frac{3}{10}\left(\frac{7}{9}\right)$
35. $\left(5 + \frac{1}{3}\right)\left(8 + \frac{1}{4}\right)$
36. $\left(5 - \frac{4}{5}\right)\left(3 + \frac{4}{7}\right)$

37. $\left(4 + \frac{2}{3}\right) - \left(3 + \frac{1}{4}\right)$

38. $\left(3 - \frac{5}{7}\right) - \left(1 + \frac{4}{5}\right)$

39. $\left(2 + \frac{5}{6}\right) - 3\left(\frac{7}{12}\right)$

40. $-\left(3 - \frac{3}{4}\right) + 6\left(\frac{5}{18}\right)$

41. $8\left(\frac{11}{18}\right) - \left(3 - \frac{5}{12}\right)$

42. $14\left(\frac{3}{35}\right) - \left(5 - \frac{9}{10}\right)$

43. $\dfrac{4 + \frac{8}{3}}{7 + \frac{1}{5}}$

44. $\dfrac{8 - \frac{7}{4}}{9 + \frac{1}{6}}$

45. $\dfrac{-7 + \frac{7}{5}}{4 - \frac{3}{13}}$

46. $\dfrac{4 - \frac{4}{5}}{\frac{16}{3} + 8}$

47. $\dfrac{\frac{2}{3} - \frac{4}{5}}{\frac{8}{5} + \frac{5}{3}}$

48. $\dfrac{\frac{9}{2} - \frac{11}{6}}{\frac{35}{9} - \frac{7}{3}}$

49. $\dfrac{\frac{4}{9} + \frac{23}{6}}{\frac{15}{8} - \frac{13}{20}}$

50. $\dfrac{\frac{9}{4} - \frac{18}{25}}{\frac{18}{5} - \frac{87}{10}}$

51. $\dfrac{2}{4 + \frac{7}{2}} + \dfrac{7}{15}$

52. $\dfrac{8}{7 - \frac{11}{9}} - \dfrac{2}{13}$

53. $\dfrac{\frac{27}{10} - \frac{11}{6}}{\frac{13}{9}} - \dfrac{26}{35}$

54. $\dfrac{\frac{25}{3} + \frac{9}{4}}{\frac{15}{2}} - \dfrac{41}{18}$

Express each of the following word phrases algebraically and then simplify it.

1. The product of $\frac{1}{4}$ and $\frac{2}{7}$ is added to $\frac{1}{3}$
2. $\frac{9}{5}$ plus the product of $\frac{2}{3}$ and $\frac{-6}{7}$
3. $\frac{-2}{7}$ added to the sum of $\frac{3}{5}$ and $\frac{-8}{35}$
4. $\frac{-5}{12}$ decreased by $\frac{7}{8}$ times $\frac{-5}{3}$
5. The product of $\frac{-3}{7}$ and $\frac{1}{2}$ minus $\frac{5}{6}$
6. The sum of $\frac{4}{9}$ and $\frac{5}{6}$ is subtracted from $\frac{1}{12}$
7. $\frac{7}{16}$ diminished by the product of $\frac{-3}{4}$ and $\frac{-10}{27}$
8. The sum of $\frac{2}{15}$ and $\frac{3}{10}$ is multiplied by $\frac{-20}{39}$
9. $\frac{5}{8}$ subtracted from the sum of $\frac{12}{7}$ and $\frac{-9}{4}$
10. $\frac{-3}{11}$ added to twice $\frac{2}{7}$

11. The product of $\frac{-16}{7}$ and the sum of $\frac{-3}{8}$ and $\frac{-7}{12}$

12. $\frac{-13}{36}$ more than the product of $\frac{-35}{16}$ and $\frac{-14}{45}$

13. The product of $\frac{-35}{27}$ and $\frac{-18}{49}$ is reduced by $\frac{9}{14}$

14. The sum of $\frac{-8}{35}$ and $\frac{12}{25}$ times $\frac{10}{21}$

15. $\frac{12}{49}$ times $\frac{25}{16}$ is decreased by $\frac{-15}{28}$

16. The product of $\frac{-7}{51}$ and $\frac{-17}{21}$ is subtracted from $\frac{5}{18}$

17. $\frac{9}{25}$ reduced by twice $\frac{16}{15}$

18. Twice the sum of $\frac{13}{27}$ and $\frac{17}{18}$

ANSWERS TO MARGIN EXERCISES

1. a. $\frac{2}{1} + \frac{4}{1} = 2 + 4$; $\frac{6}{1} = 6$
b. $\frac{2}{1} - \frac{4}{1} = 2 - 4$; $\frac{-2}{1} = -2$

2. a. $\frac{12}{6} - \frac{24}{6} = 2 - 4$; $\frac{-12}{6} = -2$
b. $-\frac{12}{6} - \frac{24}{6} = -2 - 4$; $\frac{-36}{6} = -6$

3. a. 0
b. 0
c. 0

4. a. $\frac{5}{9}$
b. $\frac{-5}{9}$

6. a. $\frac{12}{18}$
b. $\frac{40}{60}$
c. $\frac{96}{144}$

7. a. $\frac{108}{18}$
b. $\frac{75}{25}$
c. $\frac{720}{60}$
d. $\frac{100}{10}$

9. a. $\frac{9}{35}, \frac{3}{10}, \frac{5}{14}$
b. $\frac{8}{15}, \frac{13}{24}, \frac{11}{20}$

10. a and e
b and d
c and f

12. a. $\frac{-1152}{243} = \frac{-128}{27}$
b. $\frac{-10}{24} = \frac{-5}{12}$

13. 12, 24, 36

14. $4 = 2 \cdot 2$
$6 = 2 \cdot \quad 3$
↓ ↓ ↓
$12 = 2 \cdot 2 \cdot 3$

15. 3 and 8

16. $4 \cdot 3 \cdot 8 = 96$

17. a. 33
b. 65
c. 462
d. 420

18. a. 40
b. 84
c. 64
d. 2200

19. a. 12
b. 60

21. a. 20, 28, 32, 36, 48, 60
b. 18, 36, 48, 54, 60
c. 36, 48, 60

22. a. 4
b. 15
c. 36

24. a. $\frac{5}{6}$
b. $\frac{-2}{7}$
c. $\frac{8}{3}$

25. a. $\frac{1}{2} - \frac{4}{5} = \frac{-3}{10}$
b. $\frac{-3}{4} - \frac{2}{3} = \frac{-17}{12}$
c. $\frac{5}{6} - \frac{-4}{9} = \frac{23}{18}$
d. $\frac{-11}{12} - \frac{-7}{8} = \frac{-1}{24}$

CHAPTER REVIEW

SECTION 7.1 Simplify to a single reduced fraction.

1. $\frac{15}{3} + \frac{20}{3}$
2. $\frac{8}{15} + \frac{2}{15}$
3. $\frac{30}{11} - \frac{15}{11}$
4. $-\frac{13}{14} + \frac{20}{14}$
5. $\frac{7}{2} + \frac{17}{2}$
6. $\frac{4}{15} - \frac{7}{15}$
7. $\frac{-3}{14} - \frac{7}{14}$
8. $\frac{-7}{15} + \frac{-3}{15}$
9. $-\frac{76}{35} + \frac{27}{35}$
10. $\frac{4}{5} + \frac{7}{5} + \frac{3}{5}$
11. $-\frac{4}{3} + \frac{8}{3} + \frac{-1}{3}$
12. $-\frac{3}{8} + \frac{9}{8} + \frac{-6}{8}$

SECTION 7.2 Form an equivalent fraction with the given denominator.

13. $\frac{1}{4} = \frac{?}{8}$
14. $\frac{5}{6} = \frac{?}{24}$
15. $\frac{7}{4} = \frac{?}{28}$
16. $\frac{15}{8} = \frac{?}{72}$
17. $\frac{6}{1} = \frac{?}{12}$
18. $\frac{5}{2} = \frac{?}{26}$
19. $\frac{2}{7} = \frac{?}{84}$
20. $3 = \frac{?}{6}$
21. $5 = \frac{?}{20}$

SECTION 7.3 Simplify to a fraction with the given denominator.

22. $\frac{2}{5} + \frac{7}{10} = \frac{?}{10}$
23. $\frac{5}{4} + \frac{8}{3} = \frac{?}{12}$
24. $\frac{18}{5} - \frac{1}{2} = \frac{?}{10}$
25. $\frac{7}{6} - \frac{3}{8} = \frac{?}{24}$
26. $\frac{7}{2} + \frac{4}{9} = \frac{?}{18}$
27. $\frac{5}{3} + 4 = \frac{?}{3}$
28. $\frac{6}{5} - 4 = \frac{?}{5}$
29. $-\frac{7}{3} - 3 = \frac{?}{3}$
30. $\frac{5}{6} + \frac{3}{10} + \frac{8}{15} = \frac{?}{30}$
31. $\frac{-2}{3} - \frac{1}{15} + \frac{1}{5} = \frac{?}{15}$
32. $\frac{5}{8} - \frac{3}{4} + \frac{7}{12} = \frac{?}{24}$
33. $\frac{15}{7} - \frac{9}{14} + \frac{1}{4} = \frac{?}{28}$

SECTION 7.4 Find the LCM (Least Common Multiple).

34. 24, 36
35. 60, 25
36. 14, 28
37. 21, 27
38. 84, 126
39. 32, 144
40. 18, 27
41. 60, 75
42. 288, 216
43. 2, 6, 9
44. 14, 4, 6
45. 32, 24, 40
46. 42, 36, 66
47. 33, 55, 66
48. 64, 96, 30

SECTION 7.5

Simplify to a single reduced fraction.

49. $\frac{1}{6} + \frac{1}{2}$ 50. $\frac{1}{2} - \frac{3}{5}$ 51. $-\frac{5}{6} + \frac{7}{18}$

52. $\frac{45}{32} - \frac{3}{8}$ 53. $-\frac{5}{4} + \frac{9}{28}$ 54. $\frac{3}{16} - \frac{1}{40}$

55. $\frac{-11}{42} + \frac{5}{18}$ 56. $\frac{-14}{45} - \frac{7}{20}$ 57. $\frac{-5}{18} + \frac{-7}{30}$

58. $-\frac{8}{45} + \frac{5}{72}$ 59. $\frac{33}{14} - \frac{53}{21}$ 60. $-\frac{7}{15} - \frac{17}{24}$

61. $-\frac{27}{56} + \frac{11}{70}$ 62. $\frac{4}{5} + \frac{9}{4} + \frac{7}{6}$ 63. $\frac{3}{7} - \frac{9}{2} + \frac{11}{3}$

64. $\frac{-3}{8} + \frac{1}{12} - \frac{5}{18}$ 65. $\frac{25}{36} + \frac{17}{20} + \frac{16}{45}$ 66. $\frac{7}{12} - \frac{18}{39} + \frac{27}{52}$

67. $\frac{1}{4} + 8 + \frac{5}{7}$ 68. $\frac{5}{9} + 2 - \frac{3}{5}$ 69. $-7 - \frac{3}{7} + 3 + \frac{4}{21}$

SECTION 7.6

Simplify to a single reduced fraction.

70. $\frac{6+8}{6+15} - \frac{8-3}{8-5}$ 71. $\frac{6+8}{2+8} + \frac{7-8}{7+8}$ 72. $\frac{8}{8-13} - \frac{2}{2-12}$

73. $\frac{8-11}{9-15} + \frac{6-11}{6-14}$ 74. $\frac{3}{4} + \left(\frac{3}{4}\right)^2$ 75. $\frac{3^2}{4} - \frac{3}{4^2}$

76. $\left(\frac{3}{2}\right)^2 + \frac{3}{2^2}$ 77. $6\left(\frac{5}{8}\right) - 4\left(\frac{5}{6}\right)$ 78. $8\left(\frac{-5}{9}\right) - 9\left(\frac{5}{18}\right)$

79. $\frac{3^2 - 2^3}{3 + 2^3} - \frac{2^2}{11}$ 80. $\frac{4^2 + 3^2}{5^2} - \frac{5^2 + 12^2}{13^2}$ 81. $\frac{2(3+1)}{8(3-2)} + \frac{5(3^3)}{3^3}$

82. $\frac{6(4^2)}{-4^3} - \frac{6 - 3^2}{2(6-3)^2}$ 83. $36\left(\frac{1}{6} - \frac{7}{9}\right)$ 84. $\frac{3}{7}\left(6 - \frac{5}{9}\right)$

85. $\frac{21}{10}\left(\frac{4}{3} - \frac{1}{7}\right)$ 86. $\frac{2}{9}\left(\frac{8}{12}\right) - \left(\frac{7}{3} \cdot \frac{4}{18}\right)$ 87. $\left(5 + \frac{5}{8}\right)\left(7 - \frac{7}{9}\right)$

88. $\left(4 + \frac{1}{8}\right) - \left(3 - \frac{2}{3}\right)$ 89. $\left(2 + \frac{7}{8}\right) + 15\left(\frac{-7}{20}\right)$ 90. $10\left(\frac{5}{36}\right) - \left(4 + \frac{1}{6}\right)$

91. $\dfrac{6 + \frac{3}{2}}{2 + \frac{14}{3}}$ 92. $\dfrac{\frac{15}{2} - 3}{11 + \frac{5}{8}}$ 93. $\dfrac{\frac{1}{8} + \frac{9}{16}}{\frac{26}{3} - \frac{9}{4}}$

94. $\dfrac{-\frac{7}{8} + \frac{11}{18}}{\frac{24}{15} + \frac{3}{10}}$ 95. $\dfrac{5}{7 - \frac{3}{5}} - \frac{9}{16}$ 96. $\dfrac{\frac{25}{3}}{\frac{13}{2} - \frac{19}{6}} - \frac{37}{10}$

SUMMARY OF PHRASES

Express each of the following word phrases algebraically and then simplify it.

1. The sum of $\frac{1}{8}$ and $\frac{5}{8}$
2. $\frac{7}{12}$ subtracted from $\frac{11}{12}$
3. $\frac{13}{5}$ less than $\frac{9}{5}$
4. $\frac{-3}{11}$ plus $\frac{18}{11}$
5. $\frac{25}{36}$ less $\frac{13}{36}$
6. $\frac{-29}{40}$ increased by $\frac{19}{40}$
7. $\frac{4}{5}$ decreased by $\frac{2}{3}$
8. $\frac{3}{8}$ reduced by $\frac{5}{6}$
9. $\frac{1}{15}$ subtracted from $\frac{3}{10}$
10. $\frac{9}{16}$ less $\frac{2}{9}$
11. $\frac{-25}{36}$ increased by $\frac{7}{24}$
12. $\frac{5}{42}$ more than $\frac{-6}{35}$
13. The product of $\frac{1}{3}$ and $\frac{2}{5}$ is added to $\frac{7}{15}$
14. $\frac{8}{9}$ times $\frac{3}{4}$ is subtracted from $\frac{7}{4}$
15. $\frac{-5}{12}$ diminished by the product of $\frac{-16}{21}$ and $\frac{35}{-24}$
16. $\frac{1}{10}$ subtracted from the sum of $\frac{4}{15}$ and $\frac{3}{10}$
17. The product of $\frac{45}{64}$ and $\frac{32}{54}$ is reduced by $\frac{8}{9}$
18. $\frac{18}{35}$ more than $\frac{-3}{14}$ is multiplied by $\frac{30}{49}$
19. $\frac{26}{21}$ times $\frac{14}{39}$ is decreased by $\frac{11}{6}$
20. The product of $\frac{-9}{20}$ and $\frac{15}{-4}$ is subtracted from $\frac{-11}{24}$
21. $\frac{3}{50}$ less than the product of $\frac{95}{-32}$ and $\frac{24}{75}$

CUMULATIVE TEST

The following problems test your understanding of this chapter and the previous chapter. Before taking this test, thoroughly review Sections 6.3, 6.4, 7.1, and 7.5.

Once you have finished the test, compare your answers with the answers provided at the back of the book. Note the section number of each problem missed, and thoroughly review those sections again.

Simplify.

1. $\frac{-7}{18} + \frac{4}{18}$
2. $\frac{17}{18} - \frac{2}{3}$
3. $\frac{3}{5}\left(-\frac{7}{8}\right)$
4. $\frac{16}{42} \div 12$
5. $\frac{-9}{14} - \frac{7}{14}$
6. $-\frac{4}{9} - \frac{11}{24}$
7. $\frac{1}{-9}\left(-\frac{132}{55}\right)$
8. $\frac{7}{4} - 6$
9. $\frac{-30}{72} \div \frac{-5}{6}$
10. $\frac{27}{16} - \left(\frac{33}{18}\right)$
11. $\frac{5}{14} \cdot \frac{-21}{-20}$
12. $-\frac{17}{28} + \frac{5}{28}$
13. $\frac{-54}{64} \div \frac{-27}{-32}$
14. $\frac{-25}{9} + \frac{3}{4}$
15. $\frac{-12}{-25} \cdot \frac{-45}{44}$
16. $5 - \frac{7}{10} - 7 + \frac{4}{15}$
17. $\frac{27}{40} \div \left(-\frac{63}{28}\right)$
18. $-\frac{7}{4} + \frac{4}{7}$
19. $-\frac{18}{35} \cdot \frac{-14}{-9}$
20. $\frac{5}{6} + \frac{7}{4} - \frac{2}{5}$
21. $-\frac{5}{6} + \frac{11}{6} + \frac{-17}{6}$
22. $-24 \div \frac{16}{9}$
23. $\frac{9}{14} - \frac{19}{21}$
24. $\frac{4}{7}\left(-\frac{91}{52}\right)$
25. $\frac{-7}{8} - \frac{5}{12} + \frac{1}{18}$
26. $-\frac{15}{32} \div \frac{-25}{24}$
27. $-\frac{1}{7} \cdot \frac{1}{8}$
28. $\frac{1}{45} + \frac{7}{10}$
29. $\frac{-16}{-21} \cdot \frac{49}{-80}$
30. $-\frac{19}{42} + \frac{1}{12} + \frac{7}{30}$

In Chapters 8 and 9 we begin the study of algebraic fractions. Any expression consisting of an algebraic quantity above a fraction bar and a nonzero algebraic quantity below the fraction bar is called an *algebraic fraction.* (It is also called a *ratio* of algebraic quantities.) In an algebraic fraction, the quantity on top is called the *numerator* and the quantity on the bottom is called the *denominator.*

Monomial Fractions

Ratios of monomials (with integer coefficients) are called *monomial fractions.* For example, the expressions $\frac{3a^2}{2b^3}$, $\frac{5}{7a}$, $\frac{13ab^2}{20xy}$, and $\frac{30xyz}{50a}$ are monomial fractions. Monomial fractions are the simplest algebraic fractions, and we will consider them here. A more general class of *polynomial* fractions will be discussed in Chapters 17 and 18.

Monomials

Monomials can be thought of as fractions with a denominator of 1. For example,

$$x = \frac{x}{1} \quad \text{and} \quad -7xy^2 = \frac{-7xy^2}{1}$$

Reciprocals of Monomials

The fractions $\frac{1}{x}$ and $\frac{1}{8xy^2}$ are *reciprocals* of monomials. The reciprocal of a monomial is one over the monomial. For example, the reciprocal of x is $\frac{1}{x}$ and the reciprocal of $8xy^2$ is $\frac{1}{8xy^2}$.

Every monomial fraction can be thought of as the product of a monomial (the numerator) and the reciprocal of a nonzero monomial (the denominator). For example,

$$\frac{3x}{4y} = \frac{3x \cdot 1}{4y} = 3x\left(\frac{1}{4y}\right) \quad \text{and} \quad \frac{4x^3}{y^2 z} = \frac{4x^3 \cdot 1}{y^2 z} = 4x^3\left(\frac{1}{y^2 z}\right)$$

Zero

The number 0 does not ordinarily appear as part of a monomial fraction, and zero can *never* appear as the denominator of a fraction. Any expression that appears to be a fraction with a denominator of 0 is **undefined**. For example,

$$\frac{3x}{0} \quad \text{and} \quad \frac{8y^2}{0} \quad \text{are } \textbf{undefined}$$

In particular, $\frac{0}{0}$ is **undefined**.

If the numerator of a monomial fraction (with a nonzero denominator) equals zero, the fraction equals zero. For example,

$$\frac{0}{3x} = 0 \quad \text{and} \quad \frac{0}{8y^2} = 0$$

Quotients

Look at the expressions $\frac{2x}{3y}$, $2x/3y$, and $2x \div 3y$. The first two are called fractions, whereas $2x \div 3y$ is called a *quotient.* All these expressions have the same meaning and can be used interchangeably.

Multiples and Divisors

Multiplying a monomial by any expression yields a *multiple* of the monomial. For example, $4x^2y$ and $6xy^2$ are multiples of $2xy$. In particular, multiplying a monomial by an integer yields an *integer multiple* of the monomial. For example, the integer multiples of $2xy$ are

$$\ldots, -8xy, -6xy, -4xy, -2xy, 0, 2xy, 4xy, 6xy, 8xy, \ldots$$

(Note that zero is an integer multiple of every expression, and every expression is an integer multiple of itself.)

A monomial is said to be a *divisor* of each of its multiples. One monomial is a *divisor* of a second if the second monomial is a *multiple* of the first. For example, $2xy$ is a divisor of $4x^2y$ and $6xy^2$.

Because $2xy$ divides both $4x^2y$ and $6xy^2$, it is called a *common divisor* of these two monomials. To see the common divisors of $4x^2y$ and $6xy^2$, look at their divisors:

Divisors of $4x^2y$	1	x	x^2	y	xy	x^2y
	2	$2x$	$2x^2$	$2y$	$2xy$	$2x^2y$
	4	$4x$	$4x^2$	$4y$	$4xy$	$4x^2y$

Divisors of $6xy^2$	1	x	y	y^2	xy	xy^2
	2	$2x$	$2y$	$2y^2$	$2xy$	$2xy^2$
	3	$3x$	$3y$	$3y^2$	$3xy$	$3xy^2$
	6	$6x$	$6y$	$6y^2$	$6xy$	$6xy^2$

As we can see, the common divisors of $4x^2y$ and $6xy^2$ are

Common divisors of $4x^2y$ and $6xy^2$	1	x	y	xy
	2	$2x$	$2y$	$2xy$

In general, a monomial is a *common divisor* of two expressions if both expressions are multiples of the monomial; that is, if the monomial is a divisor of both expressions.

Equivalent Fractions

The fractions $\frac{x}{3}, \frac{2x}{6}, \frac{3xy}{9y}, \frac{4ax}{12a}, \cdots$ are equivalent. Multiplying both the numerator (top) and the denominator (bottom) of a fraction by the *same* monomial (except 0) results in an equivalent fraction. For example,

$$\frac{2x}{3y} \quad \text{is equivalent to} \quad \frac{2x \cdot 4z}{3y \cdot 4z} = \frac{8xz}{12yz}$$

Similarly, dividing both the numerator and the denominator of a fraction by any common divisor results in an equivalent fraction. For example,

$$\frac{8xz}{12yz} \quad \text{is equivalent to} \quad \frac{2x \cdot \cancel{4z}}{3y \cdot \cancel{4z}} = \frac{2x}{3y}$$

Reduced Fractions

A monomial fraction is *irreducible* if +1 and −1 are the only common divisors of its numerator and denominator. For example, $\frac{2x}{3y}$ is irreducible, whereas $\frac{8xz}{12yz}$ is not. A fraction that is irreducible is also said to be *reduced* or in *lowest terms*.

The numerator and denominator of an *unreduced* fraction have a common divisor other than +1 and −1. For example, $\frac{8xz}{12yz}$ is unreduced, because $4z$ is a common divisor of its numerator and denominator.

8.1 REDUCING MONOMIAL FRACTIONS WITHOUT POWERS

To reduce a fraction, we must find common divisors of its numerator and denominator. An integer is a common divisor of the numerator and denominator of a monomial fraction if it is a common divisor of their numerical coefficients. For example, 4 is a common divisor of the numerator and denominator of $\frac{8xz}{12yz}$. A literal, on the other hand, is not a common divisor unless it appears in *both* the numerator and denominator. For example, z is a common divisor of the numerator and denominator of $\frac{8xz}{12yz}$ but no other literal is. Although an integer can have several different divisors, *a literal is divisible only by itself and 1* (and their negatives). Hence, a literal behaves very much like a prime number.

1. List the divisors of
   a. 8
   b. xy
   c. $8xy$

2. Express as the product of a simple fraction and a literal fraction

a. $\frac{9xy}{15y}$

b. $\frac{25uvw}{35wx}$

c. $\frac{abc}{14b}$

d. $\frac{16}{x}$

Because multiplication is commutative and associative, we can split a monomial fraction into two parts and think of it as the product of a simple fraction (a ratio of integers) and a literal fraction (a ratio of literal components). For example, we can think of $\frac{20x}{8xy}$ as $\left(\frac{20}{8}\right)\cdot\left(\frac{x}{xy}\right)$ and reduce the numerical and literal parts separately.

$$\frac{20x}{8xy} = \left(\frac{20}{8}\right)\left(\frac{x}{xy}\right) = \left(\frac{5}{2}\right)\left(\frac{1}{y}\right) = \frac{5}{2y}$$

The numerical coefficients of a monomial fraction can therefore be reduced by the methods in Chapter 6. The literal components are reduced by cancelling common literals in the numerator and denominator.

Look at the following example of *reducing monomial fractions (without exponential notation).*

TYPICAL PROBLEM

$$\frac{20x}{15xy}$$

We first reduce the numerical portion of the fraction:

$$\frac{\overset{4}{\cancel{20}}x}{\underset{3}{\cancel{15}}xy}$$

Then we cancel any literals common to the numerator and denominator:

$$\frac{4\cancel{x}}{3\cancel{x}y} \rightarrow \frac{4}{3y}$$

To reduce the numerical portion of the fraction we use the method in Section 6.2. The Method of Primes in Section 6.1 could also be used.

Hence, $\frac{20x}{15xy} = \frac{4}{3y}$.

EXAMPLES

Every monomial fraction can be reduced to a fraction in lowest terms.

Reduce the following monomial fractions without powers.

1. $\frac{xy}{wx}$ 2. $\frac{ace}{cde}$ 3. $\frac{3x}{5x}$

4. $\frac{6xy}{15xyz}$ 5. $\frac{4xy}{6y}$ 6. $\frac{48x}{12xy}$

Solutions

1. $\frac{\cancel{x}y}{w\cancel{x}} \rightarrow \frac{y}{w}$ 2. $\frac{a\cancel{c}\cancel{e}}{\cancel{c}d\cancel{e}} \rightarrow \frac{a}{d}$ 3. $\frac{3\cancel{x}}{5\cancel{x}} \rightarrow \frac{3}{5}$

4. $\frac{\overset{2}{\cancel{6}}\cancel{x}\cancel{y}}{\underset{5}{\cancel{15}}\cancel{x}\cancel{y}z} \rightarrow \frac{2}{5z}$ 5. $\frac{\overset{2}{\cancel{4}}x\cancel{y}}{\underset{3}{\cancel{6}}\cancel{y}} \rightarrow \frac{2x}{3}$ 6. $\frac{\overset{4}{\cancel{48}}\cancel{x}}{\underset{1}{\cancel{12}}\cancel{x}y} \rightarrow \frac{4}{y}$

SPECIAL CASES

If all the literals cancel on top or bottom and there is no numerical coefficient, the reduced coefficient is 1. For example,

$$\frac{a}{ab} \rightarrow \frac{\not{a}}{\not{a}b} \rightarrow \frac{1}{b} \quad \text{and} \quad \frac{ab}{a} \rightarrow \frac{\not{a}b}{\not{a}} \rightarrow \frac{b}{1} = b$$

$$\frac{xy}{4xyz} \rightarrow \frac{\not{x}\not{y}}{4\not{x}\not{y}z} \rightarrow \frac{1}{4z} \quad \text{and} \quad \frac{4xyz}{xy} \rightarrow \frac{4\not{x}\not{y}z}{\not{x}\not{y}} \rightarrow \frac{4z}{1} = 4z$$

3. Reduce
a. $\frac{st}{s}$ b. $\frac{uv}{uvw}$
c. $\frac{3mn}{mn}$ d. $\frac{w}{6twx}$

EXERCISES

Simplify each of the following expressions to a reduced fraction.

1. $\frac{x}{x}$ 2. $\frac{x}{wxy}$ 3. $\frac{mn}{km}$ 4. $\frac{mpt}{mt}$

5. $\frac{5x}{8x}$ 6. $\frac{7xy}{3yz}$ 7. $\frac{6x}{8x}$ 8. $\frac{3}{12k}$

9. $\frac{18t}{12st}$ 10. $\frac{24uv}{8uv}$ 11. $\frac{11y}{22}$ 12. $\frac{9rs}{15r}$

8.2 REDUCING FRACTIONS AND QUOTIENTS OF POWERS WITH ONE BASE

To understand how to reduce monomial fractions containing powers, we must first look at the special case of a ratio of powers with the same base, such as $\frac{a^5}{a^3}$. Remember that two powers with the same base are called *like powers.* To see how to reduce a ratio of like powers, look at the following examples:

$$\frac{a^5}{a^3} = \frac{aaaaa}{aaa} = \frac{\not{a}\not{a}\not{a}aa}{\not{a}\not{a}\not{a}} = \frac{aa}{1} = \frac{a^2}{1}$$

$$\frac{a^3}{a^5} = \frac{aaa}{aaaaa} = \frac{\not{a}\not{a}\not{a}}{\not{a}\not{a}\not{a}aa} = \frac{1}{aa} = \frac{1}{a^2}$$

4. Compute
a. $\frac{a^2}{a^3}$ c. $\frac{a^3}{a^4}$
b. $\frac{a^4}{a^2}$ d. $\frac{a^6}{a^3}$

In both examples the factor a^5 is a product of five a's and the factor a^3 is a product of three a's. We can see that subtracting the smaller exponent 3 from the larger exponent 5 gives the number of a's that do not cancel, and therefore the number of a's in the reduced fraction.

These examples demonstrate two facts. First, any quotient of like powers can be computed by first writing each power as a repeated product and then cancelling common literals. Second, the exponents of the powers can be used when reducing the fraction. Do you see how? Check your ideas by looking at the pattern in the following table.

NUMERATOR (over ↑) / DENOMINATOR (over →)	x	x^2	x^3	x^4	x^5	x^6	x^7	x^8
x	1	$\frac{1}{x}$	$\frac{1}{x^2}$	$\frac{1}{x^3}$	$\frac{1}{x^4}$	$\frac{1}{x^5}$		
x^2	x	1	$\frac{1}{x}$	$\frac{1}{x^2}$	$\frac{1}{x^3}$	$\frac{1}{x^4}$		
x^3	x^2	x	1	$\frac{1}{x}$	$\frac{1}{x^2}$	$\frac{1}{x^3}$		
x^4	x^3	x^2	x	1	$\frac{1}{x}$	$\frac{1}{x^2}$		
x^5	x^4	x^3	x^2	x	1	$\frac{1}{x}$		
x^6	x^5	x^4	x^3	x^2	x	1		
x^7								
x^8								

5. Complete the table.

The table shows that every ratio of like powers can be reduced to a fraction containing a single power in the numerator or denominator. If the exponent in the numerator of the original fraction is larger than that in the denominator, the power will be in the numerator of the reduced fraction. If the exponent in the denominator of the original fraction is larger, the power will be in the denominator of the reduced fraction.

SECOND LAW OF EXPONENTS

$$\frac{x^N}{x^n} = \frac{\overbrace{x \cdot x \cdot \cdots \cdot x \cdot x}^{N \text{ times}}}{\underbrace{x \cdot \cdots \cdot x}_{n \text{ times}}} = \frac{x^{N-n}}{1}$$

and

$$\frac{x^n}{x^N} = \frac{\overbrace{x \cdot \cdots \cdot x}^{n \text{ times}}}{\underbrace{x \cdot x \cdot \cdots \cdot x \cdot x}_{N \text{ times}}} = \frac{1}{x^{N-n}}$$

(where N and n represent numerical values with N larger than n)

Look at the following examples of *reducing fractions and quotients of powers with one base.*

TYPICAL PROBLEMS

1. $\dfrac{a^7}{a^9}$

We first subtract the smaller exponent from the larger exponent: ↓ Subtract 7 from 9 ↓

Then we replace the larger exponent by this difference and the smaller power by 1: $\dfrac{1}{a^2}$

Hence, $\dfrac{a^7}{a^9} = \dfrac{1}{a^2}$.

2. $x^8 \div x^5$ ↓

We first form a fraction: $\dfrac{x^8}{x^5}$ ↓

Then we subtract the smaller exponent from the larger exponent: Subtract 5 from 8 ↓

We then replace the larger exponent by this difference and the smaller power by 1: $\dfrac{x^3}{1}$

Hence, $x^8 \div x^5 = \dfrac{x^3}{1} = x^3$.

EXAMPLES

Reduce the following fractions and quotients of powers with one base.

Every ratio or quotient of like powers can be simplified to a single power over 1 or 1 over a single power.

1. $\dfrac{w^2}{w^4}$ 2. $\dfrac{z^{14}}{z^{10}}$ 3. $\dfrac{a}{a^3}$

4. $\dfrac{x^7}{x^8}$ 5. $a^{12} \div a^5$ 6. $w^{15} \div w^{23}$

Solutions

1. $\dfrac{1}{w^2}$ 2. $\dfrac{z^4}{1}$ or z^4 3. $\dfrac{1}{a^2}$

4. $\dfrac{1}{x}$ 5. $\dfrac{a^{12}}{a^5} = \dfrac{a^7}{1}$ or a^7 6. $\dfrac{w^{15}}{w^{23}} = \dfrac{1}{w^8}$

SPECIAL CASES

6. Compute

a. $\frac{t^2}{t^2}$

b. $\frac{a^8}{a^8}$

1. When reducing fractions with powers, if the numerator and denominator are exactly the same, the quotient is 1. For example,

$$\frac{x^4}{x^4} \quad \text{is a quantity divided by itself that reduces to 1}$$

7. Compute

a. $\frac{t^2}{t^3}$ b. $\frac{w^4}{w^3}$

c. $\frac{z^3}{z^2}$ d. $\frac{m^3}{m^4}$

2. When reducing fractions with powers, if one exponent is 1 larger than the other, the resulting power is written without the exponent 1. For example,

$$\frac{a^5}{a^4} = a \quad \text{and} \quad \frac{x^7}{x^8} = \frac{1}{x}$$

8. Compute

a. $\frac{s^6}{s}$

b. $\frac{z}{z^9}$

3. When reducing fractions with powers, if a power is written without an exponent, its exponent is understood to be 1. When reducing, subtract 1 from the other exponent. For example,

$$\frac{x^4}{x} \quad \rightarrow \quad \text{subtract 1 from 4} \quad \rightarrow \quad \frac{x^3}{1} \quad \rightarrow \quad x^3$$

EXERCISES

Simplify each of the following expressions to a reduced fraction or a monomial.

1. $\frac{x^7}{x^2}$
2. $\frac{A^3}{A^2}$
3. $\frac{w^9}{w^2}$
4. $\frac{k^8}{k^3}$
5. $\frac{r^6}{r^7}$
6. $\frac{T^8}{T^8}$
7. $x^2 \div x^{11}$
8. $R^5 \div R^6$
9. $n^7 \div n^4$
10. $y^7 \div y^{11}$
11. $P^6 \div P^6$
12. $w^4 \div w^3$

8.3
REDUCING MONOMIAL FRACTIONS AND QUOTIENTS WITH POWERS

We have prepared for reducing monomial fractions in two ways. In Section 8.1 we discussed ratios of monomials without exponents, and in the last section we studied ratios of like powers using exponential notation. To see how these two ideas can be combined to reduce monomial fractions containing powers, look at the following tables:

NUMERATOR ↓ over → DENOMINATOR	$2x^2$	$4x^2$	$6x^2$	$8x^2$
$2x^2$	1	$\frac{1}{2}$	$\frac{1}{3}$	$\frac{1}{4}$
$4x^2$	2	1	$\frac{2}{3}$	$\frac{1}{2}$
$6x^2$	3	$\frac{3}{2}$	1	$\frac{3}{4}$
$8x^2$	4	2	$\frac{4}{3}$	1

and

NUMERATOR ↓ over → DENOMINATOR	x^3	x^2y	xy^2	y^3
x^3	1	$\frac{x}{y}$	$\frac{x^2}{y^2}$	$\frac{x^3}{y^3}$
x^2y	$\frac{y}{x}$	1	$\frac{x}{y}$	$\frac{x^2}{y^2}$
xy^2	$\frac{y^2}{x^2}$	$\frac{y}{x}$	1	$\frac{x}{y}$
y^3	$\frac{y^3}{x^3}$	$\frac{y^2}{x^2}$	$\frac{y}{x}$	1

9. Complete the tables.

a.

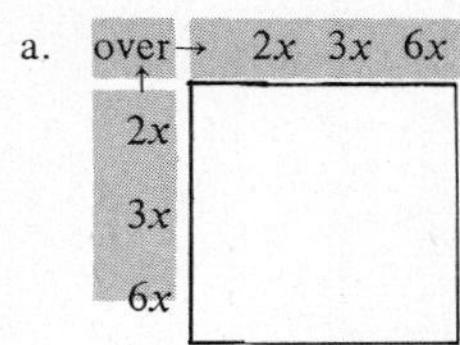

b.

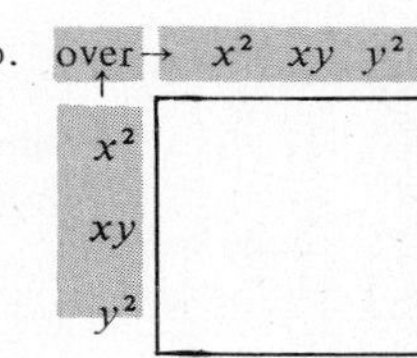

From the table on the left, we see that the numerical factors in a monomial fraction can be reduced separately from the literal factors. In addition, the table on the right shows that the literal components of the fraction are reduced by reducing like powers.

Look at the following example of *reducing monomial fractions and quotients.*

TYPICAL PROBLEM

$$6x^4yz^4 \div (-15x^8z^2)$$

$$\downarrow$$

We first form a fraction: $\dfrac{6x^4yz^4}{-15x^8z^2}$

$$\downarrow$$

Then we determine the sign of the fraction: 1 minus → −

$$\downarrow$$

We then rearrange the fraction, writing integers over integers and like powers over like powers: $-\dfrac{6x^4\ y\ z^4}{15x^8\ \ z^2}$

$$\downarrow$$

And then we reduce integers with integers and like powers with like powers. $-\dfrac{2yz^2}{5x^4}$

Hence, $6x^4yz^4 \div (-15x^8z^2) = -\dfrac{2yz^2}{5x^4}$.

To reduce monomial fractions and quotients we extend the method in Section 8.1. To reduce the numerical portion of the fraction we use the method in Section 6.1 or 6.2. Then to reduce like powers we use the method in the previous section.

EXAMPLES

Every ratio or quotient of monomials can be reduced to a monomial fraction in lowest terms.

Reduce each of the following monomial fractions and quotients with powers.

1. $\dfrac{-x^4y^7z^3}{x^5y^3z^9}$
2. $\dfrac{6xy}{-12x^7y^2}$
3. $36A^2B^4 \div (48AB^8)$
4. $(320x^7y^4) \div (-320x^7y^4)$

Solutions

1. $-\dfrac{x^4y^7z^3}{x^5y^3z^9} \downarrow -\dfrac{y^4}{xz^6}$
2. $-\dfrac{6x\ y}{12x^7y^2} \downarrow -\dfrac{1}{2x^6y}$
3. $+\dfrac{36A^2B^4}{48A\ \ B^8} \downarrow \dfrac{3A}{4B^4}$
4. $-\dfrac{320x^7y^4}{320x^7y^4} \downarrow -\dfrac{1}{1} = -1$

OBSERVATIONS AND REFLECTIONS

10. Which of the following pairs of fractions are equivalent?

a. $\dfrac{8x^3y}{12x^2y^2}$ and $\dfrac{12xy^3}{18y^4}$

b. $\dfrac{12x^2yz}{15xz^2}$ and $\dfrac{8xy^4z^2}{10y^3z^3}$

c. $\dfrac{56x^4y^2}{24x^3y^2}$ and $\dfrac{42x^2y^5}{18x^3y^5}$

d. $\dfrac{4wx^6y}{20w^6x^3}$ and $\dfrac{x^4y^2}{5w^5xy}$

1. When two fractions are equivalent, the product of the numerator of the first fraction and the denominator of the second fraction equals the product of the numerator of the second fraction and the denominator of the first fraction.

> $\dfrac{a}{b} = \dfrac{c}{d}$ if and only if $a \cdot d = b \cdot c$
>
> (where *a, b, c,* and *d* represent nonzero monomials)

We have already seen that this holds for numerical fractions (refer to the end of Section 6.2 where a proof is provided). We are merely restating it here to emphasize that it holds for all numerical and algebraic fractions.

★ **2.** The common divisor of two monomials with the largest positive numerical coefficient and the largest exponents is called their *greatest common divisor* (GCD). For example, the GCD of $4x^4y^2$ and $6x^3y^6$ is $2x^3y^2$. Every common divisor of two monomials is a divisor of their GCD. For example, the common divisors of $4x^4y^2$ and $6x^3y^6$ are divisors of $2x^3y^2$.

★ **3.** A monomial fraction is *not* reduced if its numerator and denominator have a common divisor other than 1 (and −1). Dividing both the numerator and denominator by any common divisor produces an equivalent fraction. Dividing the numerator and denominator by their greatest common divisor (GCD) produces a *reduced* fraction. For example,

$$\frac{4x^4y^2}{6x^3y^6} = \frac{2x \cdot 2x^3y^2}{3y^4 \cdot 2x^3y^2} = \frac{2x}{3y^4}$$

which is reduced.

Note that the GCD of the numerator and denominator of a reduced fraction is 1. For example, the GCD of $2x$ and $3y^4$ is 1.

11. Which of the following are divisors of

a. $4x^4y^2$

b. $6x^3y^6$

c. $2x^3y^2$

$\{2xy^3, 4x^3y^2, xy, 3x^4, 2x^2y^2, 6y^2, 2x^3y, xy^6, x^4, x^3y^2\}$?

EXERCISES

Simplify each of the following expressions to a reduced fraction.

1. $\dfrac{x^7y^3}{xy^8}$

2. $\dfrac{k^3m^2}{km^3}$

3. $\dfrac{-x^5y^7}{x^9y}$

4. $\dfrac{-a^5b^{10}}{-a^{10}b^{20}}$

5. $\dfrac{K^4L^2M^8}{-K^5LM^7}$

6. $\dfrac{p^3q^6r^9}{-p^9q^6r^9}$

7. $\dfrac{-15x^4}{12x^7}$

8. $\dfrac{-24y^7}{16y^8}$

9. $\dfrac{35A^4}{75A^2}$

10. $-\dfrac{25t^6}{175t^2}$

11. $\dfrac{-14x^6}{-7x^9}$

12. $\dfrac{-42k^4}{-56k^3}$

13. $-\dfrac{-12A^7B^4}{-28AB^2}$

14. $-\dfrac{-45x^7y^2}{-55x^2y^4}$

15. $\dfrac{-4x^7y^2}{18y^3z^2}$

16. $\dfrac{-7w^4x^7}{49w^4y^4}$

17. $\dfrac{28r^6s^2t}{63r^5s^2t^2}$

18. $\dfrac{-20a^7b^4c^2}{56a^5b^8c^3}$

19. $-12R^2 \div 16R^2$

20. $-9x^5 \div 27x^4$

21. $-8A^3B \div A^4B^6$

22. $-12m^4n^6 \div 60m^9n^3$

23. $16u^5v^4 \div (-2u^5v^3)$

24. $-15x^5y^6 \div (-90x^7y^6)$

8.4 MULTIPLYING MONOMIAL FRACTIONS

In *algebraic products* of fractions such as $\left(\dfrac{-3a}{5b}\right)\left(\dfrac{-4b^2}{3a^3}\right)$ and $\left(\dfrac{9x^4}{-5y}\right)\left(\dfrac{y^6}{8x}\right)\left(\dfrac{-4}{15xy}\right)$, each quantity is called a *factor* of the product. For example, in the product $\left(\dfrac{-3a}{5b}\right)\left(\dfrac{-4b^2}{3a^3}\right)$ the factors are $\dfrac{-3a}{5b}$ and $\dfrac{-4b^2}{3a^3}$. Similarly, in the product $\left(\dfrac{9x^4}{-5y}\right)\left(\dfrac{y^6}{8x}\right)\left(\dfrac{-4}{15xy}\right)$ the factors are $\dfrac{9x^4}{-5y}$, $\dfrac{y^6}{8x}$, and $\dfrac{-4}{15xy}$.

Multiplication of fractions is indicated in any of the following ways:

1. $\dfrac{-3x^2}{5y} \cdot \dfrac{-4x}{3y^5}$ with a dot between fractions and no plus or minus sign before the second fraction bar.
2. $\dfrac{-3x^2}{5y}\left(\dfrac{-4x}{3y^5}\right)$ with no symbol between the fractions and only the second fraction inside parentheses.
3. $\left(\dfrac{-3x^2}{5y}\right)\left(\dfrac{-4x}{3y^5}\right)$ with no symbol between sets of parentheses.
4. $\left(\dfrac{-3x^2}{5y}\right) \cdot \left(\dfrac{-4x}{3y^5}\right)$ with a dot between sets of parentheses.

When a fraction stands by itself or is the first factor in a product of fractions, it may be written with a sign before the fraction bar. However, no other fraction in a product may have a sign before the fraction bar (unless the fraction is in parentheses).

As we have seen, a monomial fraction can be expressed as the product of a monomial and the reciprocal of a monomial. For example, $\dfrac{3x^2y}{4w^6z}$ equals $3x^2y \cdot \left(\dfrac{1}{4w^6z}\right)$ and $\dfrac{8x^3z^4}{5wy^6}$ equals $8x^3z^4 \cdot \left(\dfrac{1}{5wy^6}\right)$. Similarly, a product of monomial fractions can be expressed as a product of monomials and reciprocals of monomials. For example,

$$\frac{3x^2y}{4w^6z} \cdot \frac{8x^3z^4}{5wy^6} \quad \text{equals} \quad 3x^2y \cdot \left(\frac{1}{4w^6z}\right) \cdot 8x^3z^4 \cdot \left(\frac{1}{5wy^6}\right)$$

12. Express as the product of monomials and reciprocals of monomials

a. $\dfrac{xz}{y^2} \cdot \dfrac{z^3}{5y}$

b. $\dfrac{4}{9w} \cdot \dfrac{3v}{28}$

Because multiplication is associative and commutative, the factors in a product of fractions may be written in any order and multiplied in any order. Hence, a product of monomial fractions can always be rewritten as a single fraction. For example,

$$\frac{3x^2y}{4w^6z} \cdot \frac{8x^3z^4}{5wy^6} \quad \text{is equivalent to} \quad \frac{3 \cdot 8 \cdot x^2 \cdot x^3 \cdot y \cdot z^4}{4 \cdot 5 \cdot w^6 \cdot w \cdot y^6 \cdot z}$$

Look at the following example of *multiplying monomial fractions.*

TYPICAL PROBLEM

$$\frac{-2a^2b^3}{5c^4d} \cdot \frac{10a^2c^3}{-3ab^2}$$

$\downarrow$

We first determine the sign of the product: 2 minuses $\rightarrow$ +

$\downarrow$

We then form a single fraction. We write integers over integers; then we multiply like powers in the numerator and the denominator and place like powers over like powers:

$$+\,\frac{2 \cdot \overset{2}{\cancel{10}}a^4b^3c^3}{\underset{1}{\cancel{5}} \cdot 3a\,b^2c^4d}$$

$\downarrow$

Then we reduce integers with integers and like powers with like powers.

$$+\,\frac{4a^3b}{3cd}$$

Hence, $\dfrac{-2a^2b^3}{5c^4d} \cdot \dfrac{10a^2c^3}{-3ab^2} = \dfrac{4a^3b}{3cd}$.

To determine the sign of the product and to reduce the numerical factors we use the method in Section 6.3. To multiply like powers we use the method in Section 8.2.

EXAMPLES

Multiply the following monomial fractions.

Every product of monomial fractions can be simplified to a single reduced fraction.

1. $\frac{6x^3}{11} \cdot \frac{-15x^2}{18}$
2. $\frac{w^4x^2}{x^3y^6} \cdot \frac{w^2x^7}{x^5y^9}$
3. $\left(\frac{8x^3y}{-18v^3}\right)\left(\frac{9vw^2}{12wx^3}\right)$
4. $\left(\frac{-8}{-6A^2}\right)\left(\frac{12B^2}{-15}\right)\left(\frac{-20B}{16A^2}\right)$

Solutions

1. $-\frac{6 \cdot 15x^5}{11 \cdot 18} \downarrow -\frac{5x^5}{11}$
2. $+\frac{w^6x^9}{x^8y^{15}} \downarrow \frac{w^6x}{y^{15}}$
3. $-\frac{8 \cdot 9v\ w^2x^3y}{18 \cdot 12v^3 w\ x^3} \downarrow -\frac{wy}{3v^2}$
4. $+\frac{8 \cdot 12 \cdot 20\ B^3}{6 \cdot 15 \cdot 16A^4} \downarrow \frac{4B^3}{3A^4}$

SPECIAL CASES

1. Monomial factors in a product of fractions are always treated as numerators. For example,

$$6x^3\left(\frac{3y^2}{4x^5}\right) \quad \text{is equivalent to} \quad \frac{6x^3}{1} \cdot \frac{3y^2}{4x^5}$$

When we multiply a fraction by a monomial, we need not rewrite the monomial as a fraction. We simply treat the monomial factor as a numerator when the numerators and denominators are reduced. For example, look at the computation

$$6x^3\left(\frac{3y^2}{4x^5}\right) \xrightarrow{\text{Reduce}} \overset{3}{\cancel{6x^3}}\left(\frac{3y^2}{\underset{2x^2}{\cancel{4x^5}}}\right) \xrightarrow{\text{Multiply}} \frac{9y^2}{2x^2}$$

2. The product of any fraction and 1 equals the fraction. For example,

$$\left(\frac{3y^2}{4x^5}\right)(1) = \frac{3y^2}{4x^5} \quad \text{and} \quad (1)\left(\frac{-3y^2}{4x^5}\right) = \frac{-3y^2}{4x^5}$$

3. The product of any fraction and -1 equals the negative of the fraction. For example,

$$\left(\frac{3y^2}{4x^5}\right)(-1) = \frac{-3y^2}{4x^5} \quad \text{and} \quad (-1)\left(\frac{-3y^2}{4x^5}\right) = \frac{3y^2}{4x^5}$$

4. The product of a monomial and its reciprocal is 1. For example,

$$3x\left(\frac{1}{3x}\right) = \frac{3x}{3x} = 1 \quad \text{and} \quad x^2y\left(\frac{1}{x^2y}\right) = \frac{x^2y}{x^2y} = 1$$

13. Compute
a. $3x\left(\frac{2}{3x}\right)$
b. $4x^2y\left(\frac{5y}{8x}\right)$
c. $6a^6b^2\left(\frac{14a^2}{9b^5}\right)$

14. Compute
a. $\frac{-3x}{7y} \cdot 1$
b. $1 \cdot \frac{4a^3}{9b^6}$

15. Compute
a. $\frac{-3x}{7y}(-1)$
b. $(-1)\left(\frac{4a^3}{9b^6}\right)$

16. Compute
a. $\frac{-3x}{7y} \cdot \frac{7y}{-3x}$
b. $\frac{4a^3}{9b^6} \cdot \frac{9b^6}{4a^3}$

17. Multiply to a single fraction before reducing

a. $\frac{3z}{y^2} \cdot \frac{5x^2}{12y^3z}$

b. $\frac{4u^2v}{9w} \cdot \frac{15vw}{28}$

18. Compute

a. $\left(\frac{3x^2}{1}\right)\left(\frac{4xy^3}{1}\right)$

b. $\left(\frac{-6u}{1}\right)\left(\frac{8v^3w}{1}\right)$

c. $\left(\frac{7ab}{1}\right)\left(-\frac{4b}{1}\right)$

OBSERVATIONS AND REFLECTIONS

1. When multiplying fractions, it is possible to form a single fraction before reducing. We can multiply all the numerators to form a single numerator and all the denominators to form a single denominator. We can then reduce the resulting fraction by the method in the previous section. For example, we can compute

$$\frac{3x^2y}{4w^6z} \cdot \frac{8x^3z^4}{5wy^6} = \frac{24x^5yz^4}{20w^7y^6z} = \frac{6x^5z^3}{5w^7y^5}$$

The reduced fraction obtained is, of course, the same as that obtained by the method of this section. They are identical because the numerators and denominators of the original fractions are *factors of a product* and may be multiplied and reduced in any order.

★ 2. Because monomials can be considered fractions with a denominator of 1, multiplying monomials is a special case of multiplying fractions. For example, $(-2x)(3x^2) = -6x^3$ is equivalent to $\left(\frac{-2x}{1}\right)\left(\frac{3x^2}{1}\right) = \frac{-6x^3}{1}$. In this special case, in which all fractions have a denominator of 1, the method of this section is the same as the method in Section 2.3 for multiplying monomials.

EXERCISES

Simplify each of the following expressions to a single reduced fraction.

1. $\frac{1}{x^5} \cdot \frac{5}{x^4}$

2. $\frac{7}{y^4} \cdot \frac{1}{y^7}$

3. $4t^3 \cdot \frac{3}{14t^6}$

4. $8x^4 \cdot \frac{5}{12x^9}$

5. $\frac{-3x^2}{5y^4} \cdot \frac{-7}{xy^3}$

6. $\frac{-4u^6}{9v^4} \cdot \frac{-5}{u^2v^3}$

7. $\frac{4x^5}{9y^6}\left(\frac{6y^4}{28x^3}\right)$

8. $\frac{8a^3}{14b^8}\left(\frac{35b^5}{24a^2}\right)$

9. $\frac{-15A^4}{12B^6} \cdot \frac{A^3B^4}{-60}$

10. $\frac{-4x^2}{15y^7} \cdot \frac{25xy^4}{-16}$

11. $\left(-\frac{42i^3}{64j^5}\right)\left(\frac{-8j^2}{63i^7}\right)$

12. $\left(-\frac{18x^4}{65w^8}\right)\left(\frac{55w^5}{-48x^7}\right)$

13. $\frac{v^2w^3}{x^7y^2}\left(\frac{x^5y^2}{v^5z^2}\right)$

14. $\frac{a^7b^4}{c^3d^3}\left(\frac{b^6c^4}{a^7d^3}\right)$

15. $\frac{55x^7y^2}{14z^9} \cdot \frac{21y^2z^8}{75x^5}$

16. $\frac{9k^{12}m^4}{49n^6} \cdot \frac{28k^8n^6}{63m^{12}}$

17. $\frac{(6A^4B^7)(9A^7B)}{(12A^5B)(8A^4B^2)}$

18. $\frac{(3x^3y)(-7xy^4)}{(18x^7y)(14x^3)}$

19. $\dfrac{(-9k^2)(15m^4n)}{(-6m^5)(-k^3n^2)}$

20. $\dfrac{(15u^7v^4)(-9w^7)}{(-45v^4)(-u^7w^6)}$

21. $\dfrac{-7a^4}{2b^2} \cdot \dfrac{-6a^7}{-5b^4} \cdot \dfrac{-20b^5}{14a^9}$

22. $\dfrac{5k^5}{-6} \cdot \dfrac{-4kt^4}{30} \cdot \dfrac{27k}{16t^2}$

23. $\left(\dfrac{-9}{14}\right)\left(\dfrac{r^3s^2}{-5t^4}\right)\left(\dfrac{-7s^2}{24rt^3}\right)$

24. $\left(\dfrac{8v^7}{14w^3}\right)\left(\dfrac{-9vx^3}{48w^4}\right)\left(\dfrac{7v^3w^4}{-72x^5}\right)$

Express each of the following word phrases algebraically and then simplify it.

1. $\dfrac{6\text{g}}{\text{m}}$ times $\dfrac{5\text{m}}{\text{sec}}$

2. The product of 9mg and $\dfrac{3\text{m}}{\text{sec}^2}$

3. Twice $\dfrac{8\text{g}}{\text{m}^2}$

4. $\dfrac{7\text{mg}}{5\text{m}^3}$ multiplied by $\dfrac{3\text{m}^3}{2\text{sec}^2}$

5. $\dfrac{5x^3}{-16}$ times the product of $\dfrac{-4xy^2}{-6}$ and $\dfrac{27y^5}{-15x^4}$

6. The product of half $\dfrac{10A^2B^3}{9C^7}$ and $\dfrac{-21}{25}A^2C^3$

7. $\dfrac{x^9y^7z^2}{w^4}$ times $\dfrac{-w^8x^2}{y^2z}$ multiplied by $\dfrac{z^8}{wxy}$

8. Half the product of $\dfrac{9a^3b^2}{-8c^4d}$ and $\dfrac{-16cd^8}{15ab}$

9. $\dfrac{7A^3B^8}{8C^4}$ multiplied by the product of $\dfrac{A^7C^2}{21B^2C}$ and $-4BC^5$

10. The product of $\dfrac{-6v^8z^4}{7w}$ and $\dfrac{10z^2}{-24w^3}$ times $\dfrac{-14w^9}{-5v^5}$

11. $\dfrac{A^8B^7C}{-D^3E^5}$ multiplied by $\dfrac{-B^2D^3E}{A^3C^4}$ times $\dfrac{C^3E^4}{A^5B^9}$

12. The product of $\dfrac{90x^3y^2}{18w^5}$ and $\dfrac{3w^2x^7}{-5y^2}$ multiplied by $\dfrac{-w^2}{21x^{10}y^2}$

8.5

QUOTIENTS AND RATIOS OF MONOMIAL FRACTIONS

Just as with numerical fractions, we form the *reciprocal* of a monomial fraction by inverting the fraction, interchanging the numerator and denominator. For example,

$$\text{the reciprocal of } \frac{3x}{8z} \text{ is } \frac{8z}{3x} \qquad \left[\frac{3x}{8z} \times \frac{8z}{3x}\right]$$

and

$$\text{the reciprocal of } \frac{-4}{5z^2} \text{ is } \frac{5z^2}{-4} \qquad \left[\frac{-4}{5z^2} \times \frac{5z^2}{-4}\right]$$

(However, a fraction such as $\frac{0}{4x}$ has no reciprocal because $\frac{0}{4x} = 0$. Zero has no reciprocal.)

As we saw with numerical fractions, *dividing* one fraction by another is equivalent to *multiplying* the first fraction by the *reciprocal* of the second fraction. For example,

$$\frac{-4}{5z^2} \div \frac{3x}{8z} \quad \text{is equivalent to} \quad \frac{-4}{5z^2} \cdot \frac{8z}{3x}$$

$$\frac{2x^2y}{3z^3} \div \frac{4x^3}{9z^2} \quad \text{is equivalent to} \quad \frac{2x^2y}{3z^3} \cdot \frac{9z^2}{4x^3}$$

$$\frac{x^3z^5}{y^8} \div \frac{xy^2}{z^2} \quad \text{is equivalent to} \quad \frac{x^3z^5}{y^8} \cdot \frac{z^2}{xy^2}$$

19. Form the reciprocal of

a. $\frac{2x}{9}$

b. $\frac{-3}{xz}$

c. $-\frac{4t^2}{7u}$

Complex Fractions

A fraction whose numerator and/or denominator is a fraction is a *ratio* of fractions and is called a *complex fraction.* Ratios of fractions are equivalent to quotients of fractions, but look very different. (Remember that the division symbol ÷ and the fraction bar — have identical meanings.)

$$\frac{\frac{-4}{5z^2}}{\frac{3x}{8z}} \quad \text{is equivalent to} \quad \frac{-4}{5z^2} \div \frac{3x}{8z}$$

$$\frac{\frac{2x^2y}{3z^3}}{\frac{4x^3}{9z^2}} \quad \text{is equivalent to} \quad \frac{2x^2y}{3z^2} \div \frac{4x^3}{9z^2}$$

$$\frac{\frac{x^3z^5}{y^8}}{\frac{xy^2}{z^2}} \quad \text{is equivalent to} \quad \frac{x^3z^5}{y^8} \div \frac{xy^2}{z^2}$$

A ratio of fractions can be simplified by changing the *largest* fraction bar to a *division* symbol and then simplifying the resulting quotient of fractions.

20. Convert to a quotient

a. $\frac{\frac{a}{2}}{b}$

b. $\frac{\frac{4}{x}}{3y}$

c. $\frac{\frac{x}{y^2}}{\frac{3x}{y}}$

Look at the following example of *simplifying quotients and ratios of monomial fractions.*

TYPICAL PROBLEM

$$\frac{\frac{12x^3}{25y^4}}{\frac{18x^5}{20y^2}}$$

$$\downarrow$$

We first form a quotient of fractions:

$$\frac{12x^3}{25y^4} \div \frac{18x^5}{20y^2}$$

$$\downarrow$$

Then we invert the fraction after the division sign and change the operation to multiplication:

$$\frac{12x^3}{25y^4} \cdot \frac{20y^2}{18x^5}$$

$$\downarrow$$

We then multiply the fractions. We first determine the sign of the product:

no minuses → +

$$\downarrow$$

Then we form a single fraction. We write integers over integers and like powers over like powers:

$$+\frac{12 \cdot 20x^3y^2}{25 \cdot 18x^5y^4}$$

$$\downarrow$$

We then reduce integers with integers and like powers with like powers:

$$+\frac{\overset{2}{\cancel{12}} \cdot \overset{4}{\cancel{20}}x^3y^2}{\underset{5}{\cancel{25}} \cdot \underset{3}{\cancel{18}}x^5y^4}$$

$$\downarrow$$

$$+\frac{8}{15x^2y^2}$$

Hence, $\dfrac{\frac{12x^3}{25y^4}}{\frac{18x^5}{20y^2}} = \dfrac{8}{15x^2y^2}$.

To simplify quotients and ratios of monomial fractions we parallel the methods in Section 6.4. To multiply fractions we use the method in the previous section.

EXAMPLES

Simplify the following quotients and ratios of monomial fractions.

Every ratio or quotient of monomial fractions can be simplified to a single reduced fraction.

1. $\dfrac{2x}{5y} \div \dfrac{7x^2}{20y}$

2. $\dfrac{7y}{-4x} \div \dfrac{12}{16xy^2}$

3. $\dfrac{5x^2y}{49w^2} \div \dfrac{9x^2y^3}{7w}$

4. $\dfrac{\frac{5x^6}{4y^5}}{\frac{3x^2}{8y^2}}$

5. $\dfrac{\frac{-8x^3}{12x^7}}{\frac{4x}{-24x^5}}$

6. $\dfrac{\frac{35x^2}{-18y^3}}{\frac{21x^5}{10y}}$

Solutions

1. $\dfrac{2x}{5y} \cdot \dfrac{20y}{7x^2}$
$\downarrow$
$+\dfrac{2 \cdot 20x\ y}{5 \cdot\ 7x^2y}$
$\downarrow$
$\dfrac{8}{7x}$

2. $\dfrac{7y}{-4x} \cdot \dfrac{16xy^2}{12}$
$\downarrow$
$-\dfrac{7 \cdot 16xy^3}{4 \cdot 12x}$
$\downarrow$
$-\dfrac{7y^3}{3}$

3. $\dfrac{5x^2y}{49w^2} \cdot \dfrac{7w}{9x^2y^3}$
$\downarrow$
$+\dfrac{5 \cdot 7w\ x^2y}{49 \cdot 9w^2x^2y^3}$
$\downarrow$
$\dfrac{5}{63wy^2}$

4. $\dfrac{5x^6}{4y^5} \cdot \dfrac{8y^2}{3x^2}$
$\downarrow$
$+\dfrac{5 \cdot 8x^6y^2}{4 \cdot 3x^2y^5}$
$\downarrow$
$\dfrac{10x^4}{3y^3}$

5. $\dfrac{-8x^3}{12x^7} \cdot \dfrac{-24x^5}{4x}$
$\downarrow$
$+\dfrac{8 \cdot 24x^8}{12 \cdot 4x^8}$
$\downarrow$
4

6. $\dfrac{35x^2}{-18y^3} \cdot \dfrac{10y}{21x^5}$
$\downarrow$
$-\dfrac{35 \cdot 10x^2\ y}{18 \cdot 21x^5y^3}$
$\downarrow$
$-\dfrac{25}{27x^3y^2}$

SPECIAL CASES

1. Dividing any fraction by 1 yields the original fraction. For example,

$$\frac{3x}{8y} \div 1 = \frac{3x}{8y} \quad \text{and} \quad \frac{-4}{5x^3} \div 1 = \frac{-4}{5x^3}$$

2. On the other hand, dividing 1 by any fraction yields the reciprocal of the fraction. For example,

$$1 \div \frac{3x}{8y} = \frac{8y}{3x} \quad \text{and} \quad 1 \div \frac{-4}{5x^3} = \frac{5x^3}{-4}$$

3. Dividing any fraction by itself yields the number 1. For example,

$$\frac{3x}{8y} \div \frac{3x}{8y} = 1 \quad \text{and} \quad \frac{-4}{5x^3} \div \frac{-4}{5x^3} = 1$$

21. Compute

a. $\dfrac{-4a^2}{9b^6} \div 1$

b. $\dfrac{8x^2y}{15z^4} \div 1$

22. Compute

a. $1 \div \dfrac{-4a^2}{9b^6}$

b. $1 \div \dfrac{8x^2y}{15z^4}$

23. Compute

a. $\dfrac{-4a^2}{9b^6} \div \dfrac{-4a^2}{9b^6}$

b. $\dfrac{8x^2y}{15z^4} \div \dfrac{8x^2y}{15z^4}$

OBSERVATIONS AND REFLECTIONS

A ratio of fractions is equivalent to a quotient of fractions, and a quotient of fractions is equivalent to the product of the first fraction and the reciprocal of the second fraction.

RULE OF QUOTIENTS

$$\frac{\frac{a}{b}}{\frac{c}{d}} = \frac{a}{b} \div \frac{c}{d} = \frac{a}{b} \cdot \frac{d}{c}$$

$$\frac{a}{b} = a \div b = a \cdot \frac{1}{b}$$

(where a, b, c, and d represent nonzero monomials)

24. Express as a product of fractions

a. $\dfrac{\frac{7x}{9y}}{\frac{5y^2}{6x}}$

b. $\dfrac{\frac{28ab^2}{51c}}{\frac{56a^3c^2}{17b^5}}$

We have already seen that this holds for numerical fractions (refer to the end of Section 6.4 where a proof is provided). We are merely restating it here to emphasize that it holds for all numerical and algebraic fractions.

EXERCISES

Simplify each of the following expressions to a single reduced fraction.

1. $\dfrac{x^2}{y} \div \dfrac{x}{y^2}$
2. $\dfrac{x^3}{y^4} \div \dfrac{x}{y^6}$
3. $\dfrac{-u^2}{vw} \div \dfrac{u^3}{v^2 w}$
4. $\dfrac{y^4 z}{x^2} \div \dfrac{y^3}{-x^3 z}$
5. $\dfrac{4x^3}{y^2} \div \dfrac{12x}{y^6}$
6. $\dfrac{15a^7}{12} \div \dfrac{35a^4}{9b}$
7. $\dfrac{-5x^2}{32y^6} \div \dfrac{15x^2}{-8y^3}$
8. $\dfrac{11x^2 y}{-25} \div \dfrac{-x^7}{55y}$
9. $\dfrac{6x^2 y}{-35} \div \dfrac{18y^2}{65x^2}$
10. $\dfrac{-6v^3}{-15w^2} \div \dfrac{-9v^3}{30w}$
11. $\dfrac{16xy^2}{27z^2} \div \dfrac{44x^3}{99z^3}$
12. $\dfrac{15x^2}{-48y^3} \div \dfrac{15y^2}{72z}$
13. $\dfrac{-36a^2}{42b^2 c^6} \div \dfrac{-9ab^2}{42c^6}$
14. $\dfrac{-38x^4}{81y^6 z^2} \div \dfrac{-38x^4}{-18y^4 z^6}$
15. $\dfrac{\frac{v^3}{u^2}}{\frac{u}{v^3}}$
16. $\dfrac{\frac{y^4 z}{-x^5}}{\frac{-yz^3}{x^4}}$
17. $\dfrac{\frac{t^5}{-s^3}}{\frac{t^3}{r^2 s}}$
18. $\dfrac{\frac{-x^7 y^2}{z^3}}{\frac{-y^4}{x^2 z^7}}$
19. $\dfrac{\frac{12x^4}{-5y^2}}{\frac{18x^6}{-25y^4}}$
20. $\dfrac{\frac{32m^6}{-9k^4}}{\frac{14m^3}{81k^2}}$
21. $\dfrac{\frac{12x^2}{63y^3}}{\frac{-30x^5}{-84y^7}}$
22. $\dfrac{\frac{14x^2 y}{-3z}}{\frac{21x}{15yz}}$
23. $\dfrac{\frac{25xy^3}{49z^3}}{\frac{-75y^4}{63x^7 z^2}}$
24. $\dfrac{\frac{-8a^3}{-33b^4 c^2}}{\frac{-40a^4}{-132b^8 c^2}}$

Express each of the following word phrases algebraically and then simplify it.

1. 8m^3 over $\dfrac{2\text{m}}{\text{sec}^2}$

2. The ratio $\dfrac{5\text{g}}{\text{sec}^2}$ over $\dfrac{9\text{m}}{5\text{sec}^2}$

3. $\dfrac{6\text{L}}{\text{sec}}$ divided by 3sec

4. The quotient 25cm^3 divided by $\dfrac{15\text{cm}^3}{\text{sec}}$

5. $\dfrac{-84y^8}{15x^6}$ over $\dfrac{63x^9}{12y^5}$

6. $\dfrac{72z^3}{-15x^4}$ divided by $\dfrac{-48z^5}{15x^7}$

7. The ratio $144x^{12}z^8$ over $\dfrac{-16x^5}{9z^8}$

8. The quotient $\dfrac{36b^3c^4}{-5a^9}$ divided by $-96a^3b^2c$

9. $\dfrac{-38x^9y}{7z^4}$ over $19y^3z^4$

10. The ratio $\dfrac{-A^{15}B^6}{C^8D^2}$ over $-A^{12}C^{10}E^3$

11. $\dfrac{-5a^3}{32b^7}$ divided by $\dfrac{-15a^4}{-8b^2}$

12. $\dfrac{-17A^8B^7}{29C^3D}$ divided by $\dfrac{17A^8B^7}{29C^3D}$

8.6 EVALUATING

In this section we extend the methods in Sections 2.4 and 3.5 to evaluate monomial fractions. We will consider monomial fractions with one or two literals and evaluate these fractions for both integer and fractional values of the literals.

To see some examples of evaluating a monomial fraction, look at the following table.

evaluated at → ↑	−3	−2	−1	0	+1	+2	+3
$\frac{4}{x}$		$\frac{4}{(-2)} = -2$	$\frac{4}{(-1)} = -4$	UNDEFINED	$\frac{4}{(+1)} = 4$	$\frac{4}{(+2)} = 2$	
$\frac{2}{x}$		$\frac{2}{(-2)} = -1$	$\frac{2}{(-1)} = -2$	UNDEFINED	$\frac{2}{(+1)} = 2$	$\frac{2}{(+2)} = 1$	
$\frac{x}{2}$		$\frac{(-2)}{2} = -1$	$\frac{(-1)}{2} = -\frac{1}{2}$	$\frac{(0)}{2} = 0$	$\frac{(+1)}{2} = \frac{1}{2}$	$\frac{(+2)}{2} = 1$	
$\frac{x}{4}$		$\frac{(-2)}{4} = -\frac{1}{2}$	$\frac{(-1)}{4} = -\frac{1}{4}$	$\frac{(0)}{4} = 0$	$\frac{(+1)}{4} = \frac{1}{4}$	$\frac{(+2)}{4} = 2$	

25. Complete the table.

Look at the following examples of *evaluating a monomial fraction.*

TYPICAL PROBLEMS

Evaluate: 1. $\frac{8A}{15B}$, when $A = -10, B = 12$

We first substitute the numbers given using parentheses:

$$\frac{8(-10)}{15(12)}$$

Then we simplify the resulting expression following the Order of Operations:

$$-\frac{\overset{2}{\cancel{8}} \cdot \overset{2}{\cancel{10}}}{\underset{3}{\cancel{15}} \cdot \underset{3}{\cancel{12}}}$$

$$-\frac{4}{9}$$

Hence, evaluating $\frac{8A}{15B}$, when $A = -10, B = 12$ yields $-\frac{4}{9}$.

Evaluate: 2. $\frac{-6}{5k}$, when $k = \frac{-9}{10}$

We first substitute the number given using parentheses:

$$\frac{-6}{5\left(\frac{-9}{10}\right)}$$

Then we simplify the resulting expression following the Order of Operations:

$$[-6] \div \left[5\left(\frac{-9}{10}\right)\right]$$

$$[-6] \div \left[\frac{-9}{2}\right]$$

$$+\overset{2}{\cancel{6}} \cdot \frac{2}{\underset{3}{\cancel{9}}}$$

$$\frac{4}{3}$$

Hence, evaluating $\frac{-6}{5k}$, when $k = \frac{-9}{10}$ yields $\frac{4}{3}$.

To evaluate a monomial fraction we extend the method in Section 2.4. Always remember to place parentheses around each number substituted. The Order of Operations is given in Section 6.5.

WARNING: No literal factor in the denominator of a monomial fraction can be evaluated for the value zero. A fraction whose denominator is zero is meaningless or undefined, and no expression can be evaluated for any values that make the resulting numerical expression meaningless or undefined.

EXAMPLES

Evaluate the following monomial fractions.

1. $\dfrac{15}{8t}$, when $t = 25$

2. $\dfrac{21x^2}{16}$, when $x = -6$

3. $-\dfrac{4uv}{9}$, when $u = -7$, $v = -6$

4. $\dfrac{18y^2}{35z^3}$, when $y = 0$, $z = -3$

5. $-\dfrac{3x}{16}$, when $x = \dfrac{-8}{9}$

6. $\dfrac{12A}{5B}$, when $A = \dfrac{10}{7}$, $B = -\dfrac{21}{4}$

Solutions

1. $\dfrac{15}{8(25)} \downarrow \dfrac{\overset{3}{\cancel{15}}}{8 \cdot \underset{5}{\cancel{25}}} \downarrow \dfrac{3}{40}$

2. $\dfrac{21(-6)^2}{16} \downarrow \dfrac{21 \cdot \overset{9}{\cancel{36}}}{\underset{4}{\cancel{16}}} \downarrow \dfrac{189}{4}$

3. $-\dfrac{4(-7)(-6)}{9} \downarrow -\dfrac{4 \cdot 7 \cdot \overset{2}{\cancel{6}}}{\underset{3}{\cancel{9}}} \downarrow -\dfrac{56}{3}$

4. $\dfrac{18(0)^2}{35(-3)^3} \downarrow 0$

5. $-\dfrac{3\left(\frac{-8}{9}\right)}{16} \downarrow -\left[3\left(\frac{-8}{9}\right)\right] \div [16] \downarrow +\dfrac{8}{3} \div 16 \downarrow \dfrac{\overset{1}{\cancel{8}}}{3} \cdot \dfrac{1}{\underset{2}{\cancel{16}}} \downarrow \dfrac{1}{6}$

6. $\dfrac{12\left(\frac{10}{7}\right)}{5\left(-\frac{21}{4}\right)} \downarrow \left[12\left(\frac{10}{7}\right)\right] \div \left[5\left(-\frac{21}{4}\right)\right] \downarrow \dfrac{120}{7} \div \dfrac{-105}{4} \downarrow -\dfrac{\overset{8}{\cancel{120}}}{7} \cdot \dfrac{4}{\underset{7}{\cancel{105}}} \downarrow -\dfrac{32}{49}$

SPECIAL CASES

1. Remember that multiplying or dividing zero by any other number yields zero. Hence, evaluating a (reduced) monomial fraction when any of the literal factors in its numerator is zero yields zero. For example,

$$\frac{3x}{5y}\text{, when } x = 0, y = -4 \rightarrow \frac{3(0)}{5(-4)} \rightarrow \frac{0}{-20} \rightarrow 0$$

and

$$\frac{-4xy^2}{9}\text{, when } x = -3, y = 0 \rightarrow \frac{-4(-3)(0)^2}{9} \rightarrow \frac{0}{9} \rightarrow 0$$

26. Evaluate

a. $\frac{-7x}{12}$, when $x = 0$

b. $\frac{2x^2}{15}$, when $x = 0$

2. On the other hand, remember that dividing any number by zero yields a meaningless or undefined answer. Hence, a (reduced) monomial fraction cannot be evaluated when any of the literal factors in its denominator is zero. For example,

$$\frac{3x}{5y}\text{, when } x = 4, y = 0 \rightarrow \frac{3(4)}{5(0)} \rightarrow \frac{12}{0} \rightarrow \text{undefined}$$

and

$$\frac{-4x}{9y^2}\text{, when } x = 0, y = 0 \rightarrow \frac{-4(0)}{9(0)^2} \rightarrow \frac{0}{0} \rightarrow \text{meaningless and undefined}$$

Here is a table of evaluations with fractional substitutions. Verify the entries to check your understanding of evaluations.

evaluated at →	$-\frac{3}{4}$	$-\frac{2}{3}$	$-\frac{1}{2}$	0	$+\frac{1}{2}$	$+\frac{2}{3}$	$+\frac{3}{4}$
$\frac{4}{x}$		$\frac{4}{\left(-\frac{2}{3}\right)} = -6$	$\frac{4}{\left(-\frac{1}{2}\right)} = -8$	UNDEFINED	$\frac{4}{\left(+\frac{1}{2}\right)} = 8$	$\frac{4}{\left(+\frac{2}{3}\right)} = 6$	
$\frac{2}{x}$		$\frac{2}{\left(-\frac{2}{3}\right)} = -3$	$\frac{2}{\left(-\frac{1}{2}\right)} = -4$	UNDEFINED	$\frac{2}{\left(+\frac{1}{2}\right)} = 4$	$\frac{2}{\left(+\frac{2}{3}\right)} = 3$	
$\frac{x}{2}$		$\frac{\left(-\frac{2}{3}\right)}{2} = -\frac{1}{3}$	$\frac{\left(-\frac{1}{2}\right)}{2} = -\frac{1}{4}$	$\frac{(0)}{2} = 0$	$\frac{\left(+\frac{1}{2}\right)}{2} = \frac{1}{4}$	$\frac{\left(+\frac{2}{3}\right)}{2} = \frac{1}{3}$	
$\frac{x}{4}$		$\frac{\left(-\frac{2}{3}\right)}{4} = -\frac{1}{6}$	$\frac{\left(-\frac{1}{2}\right)}{4} = -\frac{1}{8}$	$\frac{(0)}{4} = 0$	$\frac{\left(+\frac{1}{2}\right)}{4} = \frac{1}{8}$	$\frac{\left(+\frac{2}{3}\right)}{4} = \frac{1}{6}$	

27. Complete the table.

Compare this table with the table of evaluations given at the beginning of the section.

EXERCISES

Evaluate each of the following expressions, simplifying to a reduced fraction.

1. $\frac{8}{k}$, when $k = 14$
2. $\frac{14}{m}$, when $m = 35$
3. $\frac{12}{5x}$, when $x = -30$
4. $\frac{15}{4y}$, when $y = -20$
5. $\frac{16}{27m}$, when $m = 28$
6. $\frac{18}{25t}$, when $t = 42$
7. $-\frac{8r}{21}$, when $r = -9$
8. $-\frac{9v}{20}$, when $v = -16$
9. $\frac{-5w^2}{24}$, when $w = 6$
10. $\frac{-4n^2}{27}$, when $n = 9$
11. $\frac{-27}{8x^2}$, when $x = -6$
12. $\frac{-16}{9t^2}$, when $t = -12$
13. $\frac{7B^3}{24}$, when $B = -2$
14. $\frac{11m^3}{54}$, when $m = -3$
15. $-\frac{3x}{8y}$, when $x = 20, y = 6$
16. $-\frac{6s}{5t}$, when $s = 15, t = 27$
17. $\frac{-2xy}{39}$, when $x = 13, y = -6$
18. $\frac{-4mn}{55}$, when $m = 11, n = -10$
19. $\frac{48}{5AB}$, when $A = -20, B = -21$
20. $\frac{56}{3yz}$, when $y = -21, z = -40$
21. $\frac{14s^2}{3t^2}$, when $s = 6, t = -7$
22. $\frac{-9A^2}{10B^3}$, when $A = -4, B = -3$
23. $\frac{2u^2v^3}{45}$, when $u = 5, v = 3$
24. $\frac{-64}{3k^2m^2}$, when $k = -4, m = -6$
25. $\frac{15x}{16}$, when $x = -\frac{8}{9}$
26. $\frac{18r}{25}$, when $r = -\frac{15}{16}$
27. $-\frac{24k}{35}$, when $k = \frac{14}{27}$
28. $-\frac{24n}{55}$, when $n = \frac{22}{39}$
29. $\frac{-49}{6a}$, when $a = \frac{-7}{3}$
30. $\frac{-63}{4w}$, when $w = \frac{-9}{2}$
31. $\frac{21}{16n}$, when $n = -\frac{7}{12}$
32. $\frac{22}{15z}$, when $z = -\frac{11}{40}$
33. $\frac{8x}{15y}$, when $x = \frac{3}{4}, y = \frac{-6}{5}$
34. $\frac{-12k}{7m}$, when $k = -\frac{21}{16}, m = \frac{15}{28}$

35. $\frac{6st}{49}$, when $s = \frac{-7}{8}$, $t = \frac{-14}{15}$

36. $\frac{3AB}{80}$, when $A = \frac{24}{49}$, $B = \frac{-35}{6}$

37. $\frac{3k^2}{8m}$, when $k = \frac{4}{5}$, $m = \frac{6}{5}$

38. $\frac{5v^2}{9w}$, when $v = \frac{9}{4}$, $w = \frac{15}{8}$

39. $\frac{-42y}{25z^2}$, when $y = \frac{5}{6}$, $z = \frac{-7}{10}$

40. $\frac{-30r}{49s^2}$, when $r = \frac{7}{6}$, $s = \frac{-5}{14}$

41. $\frac{8u^2}{21v^2}$, when $u = -\frac{3}{4}$, $v = \frac{6}{7}$

42. $\frac{27A^2}{35B^2}$, when $A = -\frac{5}{9}$, $B = \frac{3}{7}$

ANSWERS TO MARGIN EXERCISES

1. a. 1, 2, 4, 8
b. 1, x, y, xy
c. 1, x, y, xy, 2, $2x$, $2y$, $2xy$, 4, $4x$, $4y$, $4xy$, 8, $8x$, $8y$, $8xy$ (and their negatives)

2. a. $\frac{9}{15} \cdot \frac{xy}{y}$
b. $\frac{25}{35} \cdot \frac{uvw}{wx}$
c. $\frac{1}{14} \cdot \frac{abc}{b}$
d. $\frac{16}{1} \cdot \frac{1}{x}$

3. a. t
b. $\frac{1}{w}$
c. 3
d. $\frac{1}{6tx}$

4. a. $\frac{1}{a}$
b. a^2
c. $\frac{1}{a}$
d. a^3

6. a. 1
b. 1

7. a. $\frac{1}{t}$
b. w
c. z
d. $\frac{1}{m}$

8. a. s^5
b. $\frac{1}{z^8}$

10. b and d contain equivalent pairs.

11. a. $4x^3y^2$, xy, $2x^2y^2$, $2x^3y$, x^4, x^3y^2
b. $2xy^3$, xy, $2x^2y^2$, $6y^2$, $2x^3y$, xy^6, x^3y^2
c. xy, $2x^2y^2$, $2x^3y$, x^3y^2

12. a. $xz \cdot \frac{1}{y^2} \cdot z^3 \cdot \frac{1}{5y}$
b. $4 \cdot \frac{1}{9w} \cdot 3v \cdot \frac{1}{28}$

13. a. 2
b. $\frac{5xy^2}{2}$
c. $\frac{28a^8}{3b^3}$

14. a. $\frac{-3x}{7y}$
b. $\frac{4a^3}{9b^6}$

15. a. $\frac{3x}{7y}$
b. $\frac{-4a^3}{9b^6}$

16. a. 1
b. 1

17. a. $\frac{15x^2z}{12y^5z} = \frac{5x^2}{4y^5}$
b. $\frac{60u^2v^2w}{252w} = \frac{5u^2v^2}{21}$

18. a. $\frac{12x^3y^3}{1}$
b. $\frac{-48uv^3w}{1}$
c. $\frac{-28ab^2}{1}$

19. a. $\frac{9}{2x}$
b. $\frac{xz}{-3}$
c. $-\frac{7u}{4t^2}$

20. a. $a \div \frac{2}{b}$
b. $\frac{4}{x} \div 3y$
c. $\frac{x}{y^2} \div \frac{3x}{y}$

21. a. $\frac{-4a^2}{9b^6}$
b. $\frac{8x^2y}{15z}$

22. a. $\frac{9b^6}{-4a^2}$
b. $\frac{15z^4}{8x^2y}$

23. a. 1
b. 1

24. a. $\frac{7x}{9y} \cdot \frac{6x}{5y^2}$
b. $\frac{28ab^2}{51c} \cdot \frac{17b^5}{56a^3c^2}$

26. a. 0
b. 0

CHAPTER REVIEW

Simplify to a reduced fraction.

SECTION 8.1

1. $\frac{st}{t}$ 2. $\frac{ace}{cde}$ 3. $\frac{15r}{4rs}$

4. $\frac{12t}{9}$ 5. $\frac{7x}{14ax}$ 6. $\frac{16xyz}{16xyz}$

SECTION 8.2

Simplify to a reduced fraction or a monomial.

7. $\frac{x}{x^8}$ 8. $\frac{r^5}{r^5}$ 9. $\frac{y^{12}}{y^3}$

10. $A^9 \div A^8$ 11. $x^6 \div x^7$ 12. $z^8 \div z^{10}$

SECTION 8.3

Simplify to a reduced fraction.

13. $\frac{x^{15}y^{18}}{x^3y^6}$ 14. $-\frac{s^4t^5}{s^2t^{10}}$ 15. $\frac{x^6y^4z^9}{x^2y^8z^9}$

16. $\frac{-5t^2}{15t^4}$ 17. $\frac{-9R^7}{-16R^7}$ 18. $\frac{36A^3B^2}{54AB^3}$

19. $\frac{16x^3y^4}{-24xy}$ 20. $\frac{20a^5b^6}{20a^7b^9}$ 21. $\frac{-18x^7y^4z^6}{-24x^7y^7z^3}$

22. $-33R^4 \div 27R^8$ 23. $36p^4q^5 \div 90p^7q^2$ 24. $24x^6y^5 \div (-60x^2y^6)$

SECTION 8.4

Simplify to a single reduced fraction.

25. $\frac{1}{w^6} \cdot \frac{6}{w^5}$ 26. $6x^5 \cdot \frac{7}{15x^8}$

27. $\frac{-8u^3}{3t^5} \cdot \frac{-7}{tu^4}$ 28. $\frac{6a^6}{25b^9}\left(\frac{15b^5}{27a^3}\right)$

29. $\frac{-18x^6}{28y^4} \cdot \frac{21xy^2}{-81}$ 30. $-\left(\frac{15i^3}{16j^9}\right)\left(\frac{-28j^5}{35i^8}\right)$

31. $\frac{A^3B^4}{C^7D^8}\left(\frac{A^9D^4}{B^5C^2}\right)$ 32. $\frac{6k^8m^7}{99n^5} \cdot \frac{27k^5n^6}{72m^{10}}$

33. $\frac{(4uv^2)(6vw^2)}{(-16w^3)(9u^3v)}$ 34. $\frac{(12a^3b)(-8c^4d^2)}{(-9a^7c^4)(-6b^8d)}$

35. $\frac{-k^5t^2}{k^7r^6} \cdot \frac{k^4r^3}{-r^8t^8} \cdot \frac{k^5r^2}{-t^3}$ 36. $\left(\frac{-7x^3}{-2y^4z}\right)\left(\frac{8y^6}{14z^5}\right)\left(\frac{-12z^8}{15x^9}\right)$

SECTION 8.5

Simplify to a single reduced fraction.

37. $\frac{x^3}{y} \div \frac{x}{y}$ 38. $\frac{x^5}{y^8} \div \frac{x^8}{y^4}$ 39. $\frac{s^3}{t^6} \div \frac{s^7}{t^3}$

40. $\frac{a^3}{16} \div \frac{a^5}{48}$ 41. $\frac{x^6}{32} \div \frac{x^3}{28}$ 42. $\frac{5x^4}{16y} \div \frac{10x}{12y^3}$

43. $\frac{9x^3}{32y^2} \cdot \frac{15x^5}{40y^3}$ 44. $\dfrac{\frac{m^5}{-n^3}}{\frac{m}{n^2}}$ 45. $\dfrac{\frac{x^2y^4}{-z^3}}{\frac{x^5y^3}{-z^2}}$

46. $\dfrac{\frac{5z^4}{-y^3}}{\frac{15z^2}{-4y^8}}$ 47. $\dfrac{\frac{18t^2}{56u^3}}{\frac{-60t^5}{-49u^7}}$ 48. $\dfrac{\frac{-24x^5}{35z^2}}{\frac{-32xy^2}{-15z^2}}$

SECTION 8.6

Evaluate, simplifying to a reduced fraction.

49. $\frac{12}{r}$, when $r = 28$ 50. $\frac{16}{3x}$, when $x = -36$

51. $\frac{14}{27s}$, when $s = 21$ 52. $-\frac{6t}{35}$, when $t = -25$

53. $-\frac{3v^2}{28}$, when $v = 4$ 54. $\frac{-36}{5m^2}$, when $m = -12$

55. $\frac{6A^3}{40}$, when $A = -2$ 56. $-\frac{9u}{10v}$, when $u = 35, v = 15$

57. $-\frac{3AB}{28}$, when $A = 7, B = -16$ 58. $\frac{60}{7xy}$, when $x = -35, y = -36$

59. $\frac{-9x^2}{8y^3}$, when $x = -4, y = -3$ 60. $\frac{-70}{9u^2v^2}$, when $u = -7, v = -10$

61. $\frac{16t}{21}$, when $t = -\frac{7}{12}$ 62. $-\frac{28y}{45}$, when $y = \frac{9}{49}$

63. $\frac{-42}{25y}$, when $y = \frac{-6}{5}$ 64. $\frac{20}{21k}$, when $k = -\frac{5}{14}$

65. $\frac{9r}{7s}$, when $r = \frac{7}{12}, s = \frac{-6}{11}$ 66. $\frac{3mn}{52}$, when $m = \frac{-13}{24}, n = \frac{-8}{9}$

67. $\frac{7u^2}{8v}$, when $u = \frac{4}{5}, v = \frac{7}{10}$ 68. $\frac{-70k}{9m^2}$, when $k = \frac{3}{7}, m = \frac{5}{12}$

69. $\frac{8x^2}{35y^2}$, when $x = -\frac{5}{3}, y = \frac{4}{7}$

SUMMARY OF PHRASES

Express each of the following word phrases algebraically and then simplify it.

1. The product of 18cm and $\frac{5\text{g}}{9\text{cm}^2}$

2. Twice $\frac{27\text{L}}{4\text{sec}}$

3. $\frac{72x^3}{35}$ times half $21x^4$

4. $\frac{22a^3b^2}{15c^4}$ times $\frac{-10ac^2}{33a^2}$ multiplied by $\frac{-a^3}{-2b^4}$

5. Half the product of $\frac{-12x^5y^2}{35y^3z}$ and $\frac{14xz}{-18xy^4}$

6. The product of $\frac{8ab^2c}{-9a^2bc^2}$ and $\frac{-15a^2b^3}{16c^5}$ multiplied by $-24a^2b^3c^8$

7. $\frac{25\text{km}^2}{\text{sec}}$ divided by $\frac{15\text{km}}{\text{sec}}$

8. The ratio $\frac{18\text{g}}{7\text{cm}^2}$ over $\frac{9\text{sec}}{4\text{cm}^2}$

9. The quotient $\frac{-54x^8}{55y^3}$ divided by $\frac{81x^6}{15y^5}$

10. $\frac{90a^8b^2}{-144c^8}$ over $\frac{-27b^5}{16ac^3}$

11. $\frac{49x^3y^2}{32z^9}$ divided by $\frac{42x^5y^2}{24z^9}$

12. The quotient $120x^3y^5z^9$ divided by $\frac{65x^2z^2}{26xy^8}$

CUMULATIVE TEST

The following problems test your understanding of this chapter. Before taking this test, thoroughly review Sections 8.1, 8.3, 8.4, and 8.5.

Once you have finished the test, compare your answers with the answers provided at the back of the book. Note the section number of each problem missed, and thoroughly review those sections again.

Simplify.

1. $\dfrac{18t}{28st}$

2. $\dfrac{36AB^4}{60A^3B^2}$

3. $x^5 \cdot \dfrac{y^4}{2x^3}$

4. $\dfrac{i^5}{j^9} \div \dfrac{i^7}{j^7}$

5. $-12t^6 \div 27t^{12}$

6. $\dfrac{\dfrac{18x^3}{-y^6}}{\dfrac{4x^6}{-9y^8}}$

7. $\dfrac{24xyz}{-8y}$

8. $\dfrac{(52r^3)(28s^2t^6)}{(35s^5)(16r^4t^7)}$

9. $\dfrac{16x^5y^6}{-44x^3y}$

10. $-\dfrac{36}{A^4} \div \dfrac{48}{A^7}$

11. $\dfrac{5a^4}{12b^9} \cdot 4b^3$

12. $\dfrac{21k^4m^6}{56k^5m^6}$

13. $-\dfrac{\dfrac{x^7z^4}{-y^5}}{\dfrac{y^8z^2}{-x^6}}$

14. $\left(\dfrac{-7z^4}{x^3y^8}\right)\left(\dfrac{x^4y}{-x^6z^2}\right)\left(\dfrac{-x^9}{y^4z^5}\right)$

15. $49u^6v^4 \div (-98u^7v^8)$

16. $\dfrac{6s^3}{77t^4} \div \dfrac{108s^{11}}{11t^9}$

17. $\dfrac{9s^4}{40t^9}\left(-\dfrac{16t^8}{15s^3}\right)$

18. $\dfrac{-35x^5y^4z^9}{-65x^3y^8z^3}$

19. $\dfrac{(-3x^4)(-21y^5z^7)}{(-7y^3z^2)(27x^3z)}$

20. $\dfrac{\dfrac{72A^3}{45B^2}}{\dfrac{175A^5}{56B^6}}$

9 ADDITION OF FRACTIONS WITH MONOMIAL DENOMINATORS

In this chapter we will study addition of monomial fractions. We will also discuss sums of fractions with binomial numerators and monomial denominators because they behave in a similar manner. In *algebraic sums* of fractions such as $\frac{-3}{4xy} + \frac{6}{7x^2}$, $\frac{7}{5} - \frac{3x-2}{4x}$, and $\frac{3x-4}{18} - \frac{3x+8}{3x}$ each fraction is called a *term* of the sum. For example, in the sum $\frac{-3}{4xy} + \frac{6}{7x^2}$, the terms are $\frac{-3}{4xy}$ and $\frac{6}{7x^2}$. Similarly, in the sum $\frac{3x-4}{18} - \frac{3x+8}{3x}$, the terms are $\frac{3x-4}{18}$ and $-\frac{3x+8}{3x}$.

9.1 ADDING FRACTIONS WITH COMMON DENOMINATORS

Fractions with identical denominators are said to have *common denominators* or *like denominators.* These fractions are the easiest ones to add, and we will consider them now. We will consider the addition of fractions with *unlike* denominators in Sections 9.2 and 9.4.

Any sum of fractions with common denominators can be simplified in the same way that we simplified numerical fractions with common denominators. That is, as we saw in Section 7.1, a sum of fractions with common denominators is equivalent to the sum of the numerators over the common denominator.

Look at the following example of *adding fractions with common monomial denominators.*

TYPICAL PROBLEM

$$\frac{8}{5x} - \frac{5x-3}{5x}$$

To avoid sign errors, we first move all minus signs to the numerators (so that no minus signs are in the denominators or before the fraction bars):

$$\downarrow$$

$$\frac{8}{5x} + \frac{-5x+3}{5x}$$

We then write the numerators as an indicated sum over the common denominator:

$$\downarrow$$

$$\frac{8-5x+3}{5x}$$

Then we simplify the numerator:

$$\downarrow$$

$$\frac{-5x+11}{5x}$$

(We do not try to reduce this fraction because it has a binomial numerator.)

To add fractions with common denominators we parallel the method in Section 7.1.

Hence, $\frac{8}{5x} - \frac{5x-3}{5x} = \frac{-5x+11}{5x}$.

WARNING: If the numerator simplifies to a binomial in the last step, do not try to reduce the fraction. Reducing fractions with multinomials is treated in Chapter 17.

EXAMPLES

Every sum of fractions with common monomial denominators can be simplified to a single fraction.

Add the following fractions with common monomial denominators.

1. $\frac{-17}{15w} - \frac{13}{15w}$
2. $-\frac{4y^2}{5x^2} + \frac{3y^2}{5x^2}$
3. $\frac{3xy}{2w} - \frac{5y^2}{2w}$
4. $-\frac{13x}{18y} + \frac{19x}{18y}$
5. $\frac{3x-7y}{4xy} + \frac{5x-2y}{4xy}$
6. $\frac{3x-8}{6x^2} - \frac{11x-8}{6x^2}$

Solutions

1. $\frac{-17-13}{15w} \downarrow \frac{-30}{15w} \downarrow \frac{-2}{w}$
2. $\frac{-4y^2+3y^2}{5x^2} \downarrow \frac{-y^2}{5x^2}$
3. $\frac{3xy-5y^2}{2w}$
4. $\frac{-13x+19x}{18y} \downarrow \frac{6x}{18y} \downarrow \frac{x}{3y}$
5. $\frac{3x-7y+5x-2y}{4xy} \downarrow \frac{8x-9y}{4xy}$
6. $\frac{3x-8-11x+8}{6x^2} \downarrow \frac{-8x}{6x^2} \downarrow \frac{-4}{3x}$

OBSERVATIONS AND REFLECTIONS

1. In a sum of fractions a plus or minus sign must appear before every fraction except the first. Remember that in a monomial fraction a sign can appear in three places: the numerator, the denominator, or before the fraction. Just as with numerical fractions, a monomial fraction is said to be in *standard form* if its only expressed sign is before the fraction, or if it has a minus sign in the numerator (and possibly a plus sign before the fraction).

   A fraction with a *binomial* numerator and a *monomial* denominator is in standard form if it has no expressed sign in the denominator. For example, the fractions

$$-\frac{3}{4xy}, \quad \frac{-3}{4xy}, \quad -\frac{3x-2}{4x}, \quad \text{and} \quad \frac{-3x+2}{4x}$$

are all in standard form.

2. In a fraction with a binomial numerator, each term of the numerator has its own sign. In a sum, such fractions frequently appear with a minus sign before the fraction bar. We can convert such a fraction to an equivalent fraction with a plus sign before the fraction bar by negating the numerator. For example,

$$-\frac{3x-2}{4x} = \frac{-(3x-2)}{4x} = \frac{-3x+2}{4x}$$

and

$$-\frac{3x+8}{3x} = \frac{-(3x+8)}{3x} = \frac{-3x-8}{3x}$$

Remember that the negative of a polynomial can be formed by either placing the polynomial within parentheses preceded by a minus sign or changing the sign of each term in the polynomial.

3. Look at the fraction $\frac{8x^2-20}{8x^2}$, which is reducible, but not by the methods of Chapter 8. In a fraction, only common *factors* in the numerator and denominator can be reduced. If the numerator of a fraction is a multinomial, the fraction **cannot** be reduced in this form.

Do not do this: $\frac{\cancel{8x^2}-20}{\cancel{8x^2}}$ **(illegal operation)**

(In fact, $\frac{8x^2-20}{8x^2} = \frac{2x^2-5}{2x^2}$, which is irreducible.) The *terms* in the multinomial are *not* themselves *factors* of the fraction, although the multinomial is a factor of the fraction. A method for reducing fractions of this type is given in Chapter 17.

$$\frac{ad}{bd} = \frac{a}{b}, \quad \text{but} \quad \frac{ad+c}{bd} \neq \frac{a+c}{b}$$

(where *a, b, c,* and *d* represent monomials)

1. Place in standard form
   a. $\frac{5}{-8x}$
   b. $\frac{3x-5}{-8x}$

2. Move the minus sign before the fraction to the numerator
   a. $-\frac{7x+2}{6y}$
   b. $-\frac{10x-5}{6y}$
   c. $-\frac{-8-9x}{6y}$

3. Reduce
   a. $\frac{8-20}{8}$
   b. $\frac{3+12}{6}$

★ **4.** Because polynomials can be considered fractions with a denominator of 1, adding polynomials is a special case of adding fractions with common monomial denominators. For example,

$$(3x - 8y) - (2x + 5y) = x - 13y \quad \text{is equivalent to} \quad \frac{3x - 8y}{1} - \frac{2x + 5y}{1} = \frac{x - 13y}{1}$$

(A fraction bar implies parentheses around the numerator of a fraction, so $-(2x + 5y) = \frac{-(2x + 5y)}{1} = -\frac{2x + 5y}{1}$.)

In this special case, in which all fractions have a denominator of 1, our method for adding fractions with common monomial denominators is the same as the method for adding polynomials in Section 5.1.

4. Compute

a. $\frac{3x}{1} + \frac{4x - 7}{1}$

b. $\frac{8x - y}{1} - \frac{2x - 9y}{1}$

EXERCISES

Simplify each of the following expressions to a single reduced fraction.

1. $\frac{3}{2x} + \frac{5}{2x}$
2. $\frac{11}{6x^2} - \frac{17}{6x^2}$
3. $-\frac{9}{14y} + \frac{27}{14y}$
4. $-\frac{18}{y^2z} - \frac{22}{y^2z}$
5. $\frac{4u}{7v} + \frac{5u}{7v}$
6. $-\frac{y^3}{6x^2} + \frac{5y^3}{6x^2}$
7. $-\frac{6z}{x^2y} - \frac{11z}{x^2y}$
8. $\frac{9a^2}{5b^3} + \frac{6a^2}{5b^3}$
9. $\frac{3s^2}{8} + \frac{5s^3}{8t}$
10. $\frac{5y^4}{14x^2} - \frac{5y^2}{14x^2}$
11. $-\frac{16y}{7x^3z} + \frac{2w}{7x^3z}$
12. $-\frac{5x^3}{8y^5} - \frac{11x^4}{8y^5}$
13. $\frac{7x + 5}{6} + \frac{1}{6}$
14. $-\frac{3x - 2}{7} + \frac{5}{7}$
15. $\frac{-5y - 2}{2y} - \frac{7}{2y}$
16. $\frac{9}{9z^2} + \frac{3z - 8}{9z^2}$
17. $\frac{3x}{5y} + \frac{4x - 5}{5y}$
18. $\frac{-2x - 9}{12x} + \frac{7x}{12x}$
19. $\frac{2x - 1}{3x} + \frac{3x + 7}{3x}$
20. $\frac{-2y^2 - 3}{8y} + \frac{3y^2 - 5}{8y}$
21. $\frac{6t - 9}{5} - \frac{6t - 4}{5}$
22. $\frac{5x^2 - 3x}{x^3} - \frac{5x^2 + 6x}{x^3}$
23. $\frac{4x - 7y}{3y^4} + \frac{2x + 9y}{3y^4}$
24. $\frac{18z - 7}{18z} + \frac{3z - 7}{18z}$

Express each of the following word phrases algebraically and then simplify it.

1. The sum of $\frac{2\text{g}}{3\text{sec}}$ and $\frac{8\text{g}}{3\text{sec}}$
2. $\frac{12\text{L}}{5\text{sec}}$ decreased by $\frac{3\text{L}}{5\text{sec}}$

3. $\frac{5\text{mg}}{3\text{cm}}$ subtracted from $\frac{23\text{mg}}{3\text{cm}}$

4. $\frac{5\text{m}^3}{8\text{sec}^2}$ added to $\frac{11\text{m}^3}{8\text{sec}^2}$

5. $\frac{-7a^3}{12b^2}$ reduced by $\frac{9a^3}{12b^2}$

6. $\frac{13x}{8w^2}$ less than $\frac{7x}{-8w^2}$

7. $\frac{2x-5}{18}$ plus $\frac{7x}{18}$

8. $\frac{7}{2a}$ is subtracted from $\frac{2-5a}{2a}$

9. $\frac{3x^2-2x}{5}$ more than $\frac{6x^2+4x}{5}$

10. $\frac{7a+3b}{12}$ less $\frac{5a-12b}{12}$

11. $\frac{15x-19y}{10x}$ increased by $\frac{6y-4x}{10x}$

12. $\frac{17a-16b}{4c}$ is reduced by $\frac{17a+16b}{4c}$

9.2 ADDING FRACTIONS GIVEN THE COMMON DENOMINATOR

Monomial fractions have many of the same properties as numerical fractions, and, as you might expect, we can parallel the methods of Chapter 7 when adding monomial fractions. To simplify a sum of fractions with unlike monomial denominators, we must first convert it to an equivalent sum with like denominators.

Remember that multiplying the numerator and denominator of a fraction by any nonzero monomial results in an equivalent fraction. For example, $\frac{2x+7}{4} = \frac{3(2x+7)}{3(4)} = \frac{5x(2x+7)}{5x(4)}$, and so forth. Furthermore, as we saw in Section 7.2, if we are given a fraction and the denominator of an equivalent fraction, we can find the numerator of the equivalent fraction by multiplying the given denominator and the original fraction. For example, to compute the numerator of the fraction equivalent to $\frac{2x+7}{4}$ with the denominator $12x$ we multiply $12x$ and $\frac{2x+7}{4}$:

$$12x\left(\frac{2x+7}{4}\right) = \overset{3x}{\cancel{12x}}\left(\frac{2x+7}{\cancel{4}}\right) = 3x(2x+7) = 6x^2 + 21x$$

That is,

$$\frac{2x+7}{4} = \frac{12x\left(\frac{2x+7}{4}\right)}{12x} = \frac{\overset{3x}{\cancel{12x}}\left(\frac{2x+7}{\cancel{4}}\right)}{12x} = \frac{6x^2+21x}{12x}$$

5. Form the equivalent fraction

a. $\frac{-3}{4xy} = \frac{?}{20xy}$

b. $\frac{6}{7x^2} = \frac{?}{14x^4}$

c. $\frac{3x-2}{4x} = \frac{?}{8x}$

d. $\frac{3x+8}{3x} = \frac{?}{12xy}$

Look at the following example of *adding fractions given the common denominator.*

TYPICAL PROBLEM

$$\frac{11}{6} - \frac{2x+7}{4} = \frac{?}{12}$$

To avoid sign errors, we first move all minus signs to the numerators (so that no minus signs are in the denominators or before the fraction bars):

$$\frac{11}{6} + \frac{-2x-7}{4}$$

Then we convert the sum to a sum with like denominators. To do this we first multiply each term by the given new common denominator to determine the new numerators:

$$12\left(\frac{11}{6}\right) + 12\left(\frac{-2x-7}{4}\right)$$

And then we form new fractions by placing each new numerator over the given new common denominator:

$$\frac{\overset{2}{\cancel{12}}\left(\frac{11}{\cancel{6}}\right)}{12} + \frac{\overset{3}{\cancel{12}}\left(\frac{-2x-7}{\cancel{4}}\right)}{12}$$

At this point we have a sum of fractions with common monomial denominators:

$$\frac{22}{12} + \frac{-6x-21}{12}$$

To add fractions given the common denominator we parallel the method in Section 7.3. To multiply the fractions by the given new denominator we use the method in Section 8.4. The given new denominator must always be a multiple of each original denominator.

Now we add the fractions. We first form a single fraction, writing the numerators as an indicated sum over the common denominator:

$$\frac{22-6x-21}{12}$$

And then we simplify the numerator.

$$\frac{-6x+1}{12}$$

Hence, $\frac{11}{6} - \frac{2x+7}{4} = \frac{-6x+1}{12}$.

WARNING: If in the last step the numerator simplifies to a binomial, do not try to reduce the fraction. Reducing fractions with multinomials is treated in Chapter 17.

EXAMPLES

Given a common denominator, every sum of fractions with monomial denominators can be simplified to a single fraction.

Add the following fractions, first converting them to the given common denominators.

1. $-\frac{13}{6y^6} + \frac{2}{9y^5} = \frac{?}{18y^6}$

2. $\frac{9}{8r} - \frac{15r^2}{16} = \frac{?}{16r}$

3. $\frac{4}{25a^2} - \frac{9a}{10b} = \frac{?}{50a^2b}$

4. $\frac{x+4}{6} + \frac{3}{8} = \frac{?}{24}$

5. $\frac{8x-5}{4} - \frac{3x-6}{7} = \frac{?}{28}$

6. $\frac{5x-9}{26} - \frac{-3x+2}{39} = \frac{?}{78}$

Solutions

1. $$\frac{\overset{3}{\cancel{18y^6}}\left(\frac{-13}{\cancel{6y^6}}\right)}{18y^6}+\frac{\overset{2y}{\cancel{18y^6}}\left(\frac{2}{\cancel{9y^5}}\right)}{18y^6}$$
$$\downarrow$$
$$\frac{-39}{18y^6}+\frac{4y}{18y^6}$$
$$\downarrow$$
$$\frac{-39+4y}{18y^6}$$

2. $$\frac{\overset{2}{\cancel{16r}}\left(\frac{9}{\cancel{8r}}\right)}{16r}+\frac{\overset{r}{\cancel{16r}}\left(\frac{-15r^2}{\cancel{16}}\right)}{16r}$$
$$\downarrow$$
$$\frac{18}{16r}+\frac{-15r^3}{16r}$$
$$\downarrow$$
$$\frac{18-15r^3}{16r}$$

3. $$\frac{\overset{2b}{\cancel{50a^2b}}\left(\frac{4}{\cancel{25a^2}}\right)}{50a^2b}+\frac{\overset{5a^2}{\cancel{50a^2b}}\left(\frac{-9a}{\cancel{10b}}\right)}{50a^2b}$$
$$\downarrow$$
$$\frac{8b}{50a^2b}+\frac{-45a^3}{50a^2b}$$
$$\downarrow$$
$$\frac{8b-45a^3}{50a^2b}$$

4. $$\frac{\overset{4}{\cancel{24}}\left(\frac{x+4}{\cancel{6}}\right)}{24}+\frac{\overset{3}{\cancel{24}}\left(\frac{3}{\cancel{8}}\right)}{24}$$
$$\downarrow$$
$$\frac{4x+16}{24}+\frac{9}{24}$$
$$\downarrow$$
$$\frac{4x+25}{24}$$

5. $$\frac{\overset{7}{\cancel{28}}\left(\frac{8x-5}{\cancel{4}}\right)}{28}+\frac{\overset{4}{\cancel{28}}\left(\frac{-3x+6}{\cancel{7}}\right)}{28}$$
$$\downarrow$$
$$\frac{56x-35}{28}+\frac{-12x+24}{28}$$
$$\downarrow$$
$$\frac{56x-35-12x+24}{28}$$
$$\downarrow$$
$$\frac{44x-11}{28}$$

6. $$\frac{\overset{3}{\cancel{78}}\left(\frac{5x-9}{\cancel{26}}\right)}{78}+\frac{\overset{2}{\cancel{78}}\left(\frac{3x-2}{\cancel{39}}\right)}{78}$$
$$\downarrow$$
$$\frac{15x-27}{78}+\frac{6x-4}{78}$$
$$\downarrow$$
$$\frac{15x-27+6x-4}{78}$$
$$\downarrow$$
$$\frac{21x-31}{78}$$

OBSERVATIONS AND REFLECTIONS

1. In the problems of this section we provide a suitable denominator for each sum of fractions. However, many other denominators could be used. Any common multiple of the original denominators is a suitable denonimator. In particular, the product of the denominators is always a suitable denominator.

Look again at the typical problem, namely $\frac{11}{6}-\frac{2x+7}{4}=\frac{?}{12}$. We could have used the *product* of the denominators, 24. The problem would then read $\frac{11}{6}-\frac{2x+7}{4}=\frac{?}{24}$. The solution to the original problem is $\frac{-6x+1}{12}$, whereas the solution to the second problem is $\frac{-12x+2}{24}$ (which is equivalent to $\frac{-6x+1}{12}$, although we will not be able to show this until we learn how to reduce such fractions in Chapter 17).

6. Combine using the product of the denominators

a. $\frac{5x-10}{6}+\frac{9x+6}{10}$

b. $\frac{4x-7}{8x}-\frac{6x-5}{12x}$

Using any suitable denominator leads to an equivalent fraction (before we reduce), and every suitable denominator leads to the same result when the fraction is completely reduced. (However, our procedure includes a warning against trying to reduce fractions containing multinomial numerators. The terms of the multinomial are not themselves factors of the fraction, so the methods of Chapter 8 do not apply.)

2. A fraction with a binomial numerator and a monomial denominator can be written as the sum of two monomial fractions with like denominators by reversing the method in the previous section. For example,

7. Write as a sum of reduced monomial fractions:

a. $\frac{x^2+4}{2x}$

b. $\frac{3x-12y}{6xy}$

c. $\frac{8y+9x}{3x}$

$$\frac{5x-3}{2x} = \frac{5x}{2x} + \frac{-3}{2x} \quad \left(\text{or } \frac{5x}{2x} - \frac{3}{2x}\right)$$

Hence, every sum of fractions with monomial denominators is equivalent to a sum of monomial fractions. For example,

$$\frac{x+6}{7x^2} + \frac{5x-3}{2x} = \frac{x}{7x^2} + \frac{6}{7x^2} + \frac{5x}{2x} + \frac{-3}{2x}$$

EXERCISES

Simplify each of the following expressions to a fraction with the given denominator. In these exercises, the denominators in the sum always divide the new denominator.

1. $\frac{4}{27} - \frac{7}{18t} = \frac{?}{54t}$

2. $\frac{11}{12} - \frac{5}{4x^2} = \frac{?}{12x^2}$

3. $\frac{4}{9k^2} - \frac{7k^4}{18} = \frac{?}{18k^2}$

4. $\frac{15}{28A^2} - \frac{17A}{7} = \frac{?}{28A^2}$

5. $\frac{1}{7x^4} - \frac{1}{2x^2} = \frac{?}{14x^4}$

6. $\frac{5}{12B^6} - \frac{1}{32B^3} = \frac{?}{96B^6}$

7. $\frac{r^4}{25s^2} - \frac{3r^2}{10s} = \frac{?}{50s^2}$

8. $\frac{5x^2}{12y^3} - \frac{7x}{20y^3} = \frac{?}{60y^3}$

9. $\frac{4s}{t} - \frac{t}{9s^2} = \frac{?}{9s^2t}$

10. $\frac{25}{14a^2} + \frac{15}{49b^8} = \frac{?}{98a^2b^8}$

11. $\frac{5}{18K^3} - \frac{7}{12R} = \frac{?}{36K^3R}$

12. $\frac{4y^6}{45x^6} + \frac{11x^2}{30y^2} = \frac{?}{90x^6y^2}$

13. $\frac{3x-4}{5} + \frac{3}{4} = \frac{?}{20}$

14. $\frac{5}{18} + \frac{4x+3}{12} = \frac{?}{36}$

15. $\frac{6x-5}{12} - \frac{7}{20} = \frac{?}{60}$

16. $\frac{-2z}{15} + \frac{8z-7}{5} = \frac{?}{15}$

17. $\frac{-3w-4}{28} - \frac{9}{14} = \frac{?}{28}$

18. $\frac{-7x}{36} + \frac{-5x-9}{12} = \frac{?}{36}$

19. $\frac{4z-1}{5} - \frac{-3z+2}{3} = \frac{?}{15}$

20. $\frac{7y+1}{4} + \frac{-4y+3}{6} = \frac{?}{12}$

21. $\frac{6x+1}{9} - \frac{-4x+3}{3} = \frac{?}{9}$

22. $\frac{3x+8}{6} + \frac{5x-9}{10} = \frac{?}{30}$

23. $\frac{7x-4}{18} - \frac{8x-3}{30} = \frac{?}{90}$

24. $\frac{-z+4}{12} + \frac{4z-11}{42} = \frac{?}{84}$

9.3 DETERMINING THE LEAST COMMON MULTIPLE OF MONOMIALS

Remember that any expression that is a multiple of each of two monomials is called a *common multiple* of the two monomials. For example, $12x^2y^3$, $24x^2y^4$, and $60x^5y^8$ are common multiples of $6x^2y$ and $4y^3$. The common multiple of two monomials with the smallest *positive* numerical coefficient and the smallest exponents is called their *least common multiple* (LCM). For example, the LCM of $6x^2y$ and $4y^3$ is $12x^2y^3$.

You learned how to find the LCM of two integers in Section 7.4. The LCM of two monomials is the product of the (numerical) LCM of their numerical coefficients and the (literal) LCM of their literal components. To see how to determine the LCM of the literal components, look at the pattern in the following tables.

LCM	x	x^2	x^3	x^4	x^5	x^6
x	x	x^2	x^3	x^4		
x^2	x^2	x^2	x^3	x^4		
x^3	x^3	x^3	x^3	x^4		
x^4	x^4	x^4	x^4	x^4		
x^5						
x^6						

and

LCM	x	x^2	x^3	x^4	x^5	x^6
y	xy	x^2y	x^3y	x^4y		
y^2	xy^2	x^2y^2	x^3y^2	x^4y^2		
y^3	xy^3	x^2y^3	x^3y^3	x^4y^3		
y^4	xy^4	x^2y^4	x^3y^4	x^4y^4		
y^5						
y^6						

8. Complete the tables.

The first table shows that the LCM of two like powers is the power with the larger exponent, and the second table shows that the LCM of two unlike powers is their product.

Look at the following example of *determining the least common multiple (LCM) of two monomials.*

TYPICAL PROBLEM

Find the LCM of $30x^2y^4$ and $40x^3y$

We first determine the LCM of the numerical coefficients: LCM of 30 and 40 = 120

We then determine the LCM of the literal components. We take the highest power of each literal: LCM of x^2y^4 and $x^3y = x^3y^4$

Then we attach the integer and the literal component of the LCM: LCM = $120x^3y^4$

Hence, the LCM of $30x^2y^4$ and $40x^3y$ is $120x^3y^4$.

To determine the LCM of the numerical coefficients we use either of the methods in Section 7.4.

EXAMPLES

Every set of monomials has a unique (positive) LCM.

Find the least common multiple (LCM) of each pair of monomials.

1. $28x^2$ and $35x^3$
2. $x^6y^2z^3$ and $x^4y^2z^5$
3. $72x^5y^4$ and $36x^3y^3$
4. $18u^3v$ and $81uv^5w$

Solutions

1. LCM of 28 and 35 = 140
   ↓
   LCM of x^2 and $x^3 = x^3$
   ↓
   LCM = $140x^3$

2. LCM of 1 and 1 = 1
   ↓
   LCM of $x^6y^2z^3$ and $x^4y^2z^5 = x^6y^2z^5$
   ↓
   LCM = $x^6y^2z^5$

3. LCM of 72 and 36 = 72
   ↓
   LCM of x^5y^4 and $x^3y^3 = x^5y^4$
   ↓
   LCM = $72x^5y^4$

4. LCM of 18 and 81 = 162
   ↓
   LCM of u^3v and $uv^5w = u^3v^5w$
   ↓
   LCM = $162u^3v^5w$

9. Which of the following are multiples of
   a. $6x^2y$
   b. $4y^3$
   c. $12x^2y^3$:
   $\{16x^3y^3, 24x^5y^4, 36x^2y^2, 48x^2y^8, 54x^5y^3, 60x^4y^3z^2\}$?

OBSERVATIONS AND REFLECTIONS

★ Just as with integers, every common multiple of two monomials is a multiple of their LCM. For example, the LCM of $6x^2y$ and $4y^3$ is $12x^2y^3$, and every common multiple of $6x^2y$ and $4y^3$ is also a multiple of $12x^2y^3$.

EXERCISES

Find the LCM (Least Common Multiple).

1. x^3, x^7
2. x^4, x^2
3. st^9, st^3
4. xy^3, x^3y^2
5. u^4v, u^3v
6. x^8y^9, x^6y
7. $m^6n^3p^5, m^4n^7p^5$
8. $a^8bc^3, a^6b^3c^2$
9. $15x^4, 18x^3$
10. $30y^2, 90y^3$
11. $14k^4, 21x^3$
12. $36p^5, 36q^7$
13. $32x^7, 28x^4y^2$
14. $72u^4, 60u^2v^3$
15. $96m^2n^2, 60mn^4$
16. $66x^4y^2, 132x^7y^5$
17. $84yz^3, 105y^5z^2$
18. $99x^7y^4, 44x^4y^8$
19. $18x^7y^2z^4, 108x^4y^6z$
20. $42ab^3c^2, 49a^8bc^3$
21. $56x^6y^2z^3, 50x^4yz^3$
22. $110r^2st^3, r^4s^3t^2$
23. $88x^4y^6z^2, 64y^6z$
24. $56x^4y^8z^9, 68x^7y^6z^7$

9.4 ADDING FRACTIONS WITH MONOMIAL DENOMINATORS

We first added fractions with unlike monomial denominators in Section 9.2, where we simplified the problem by providing a common denominator. As we saw there, to simplify a sum of fractions with unlike denominators we first convert the sum to an equivalent sum of fractions with common denominators.

The smallest common denominator, and usually the best one to use, is called the *lowest common denominator.* The lowest common denominator of a sum of fractions is the LCM of the denominators of the fractions. In Section 9.2 the common denominator provided for each sum of fractions was always the LCM of the denominators.

Look at the following example of *adding fractions with monomial denominators.*

TYPICAL PROBLEM

$$-\frac{5x-4}{7x^2y}-\frac{3x-2}{4xy}$$

To avoid sign errors, we first move all minus signs to the numerators (so that no minus signs are in the denominators or before the fractions bars):

$$\downarrow$$

$$\frac{-5x+4}{7x^2y}+\frac{-3x+2}{4xy}$$

We then determine the LCM of the denominators:

$$\downarrow$$

$$\text{LCM} = 28x^2y$$

Then we convert the sum to a sum with like denominators. To do this we first multiply each term by the LCM to determine the new numerators:

$$\downarrow$$

$$28x^2y\left(\frac{-5x+4}{7x^2y}\right)+28x^2y\left(\frac{-3x+2}{4xy}\right)$$

And then we form new fractions by placing each new numerator over the LCM:

$$\downarrow$$

$$\frac{\overset{4}{\cancel{28x^2y}}\left(\frac{-5x+4}{\cancel{7x^2y}}\right)}{28x^2y}+\frac{\overset{7x}{\cancel{28x^2y}}\left(\frac{-3x+2}{\cancel{4xy}}\right)}{28x^2y}$$

At this point we have a sum of fractions with common monomial denominators:

$$\downarrow$$

$$\frac{-20x+16}{28x^2y}+\frac{-21x^2+14x}{28x^2y}$$

Now we add the fractions by forming a single fraction and simplifying numerators.

$$\downarrow$$

$$\frac{-21x^2-6x+16}{28x^2y}$$

Hence, $-\dfrac{5x-4}{7x^2y}-\dfrac{3x-2}{4xy}=\dfrac{-21x^2-6x+16}{28x^2y}$.

Except for the added step of determining the LCM of the denominators (the lowest common denominator), this is the method in Section 9.2.

WARNING: If in the last step the numerator simplifies to a multinomial, do not try to reduce the fraction. Reducing fractions with multinomials is treated in Chapter 17.

EXAMPLES

Add the following fractions with monomial denominators.

Every sum of fractions with monomial denominators can be simplified to a single fraction with a monomial denominator.

1. $\dfrac{5}{3x}+\dfrac{4}{5x}$ 2. $\dfrac{y}{x^2}-\dfrac{1}{y}$ 3. $\dfrac{-3}{4xy}+\dfrac{6}{7x^2}$

4. $\frac{7}{5} - \frac{3x-2}{4x}$ 5. $\frac{3x-4}{18} - \frac{3x+8}{3x}$ 6. $\frac{-9x+4}{36x} + \frac{5x-3}{48x}$

Solutions

1. LCM = $15x$

$$\downarrow$$

$$\frac{\overset{5}{\cancel{15x}}\left(\frac{5}{\cancel{3x}}\right)}{15x} + \frac{\overset{3}{\cancel{15x}}\left(\frac{4}{\cancel{5x}}\right)}{15x}$$

$$\downarrow$$

$$\frac{25}{15x} + \frac{12}{15x}$$

$$\downarrow$$

$$\frac{25+12}{15x}$$

$$\downarrow$$

$$\frac{37}{15x}$$

2. LCM = x^2y

$$\downarrow$$

$$\frac{\overset{y}{\cancel{x^2y}}\left(\frac{y}{\cancel{x^2}}\right)}{x^2y} + \frac{\overset{x^2}{\cancel{x^2y}}\left(\frac{-1}{\cancel{y}}\right)}{x^2y}$$

$$\downarrow$$

$$\frac{y^2}{x^2y} + \frac{-x^2}{x^2y}$$

$$\downarrow$$

$$\frac{y^2-x^2}{x^2y}$$

3. LCM = $28x^2y$

$$\downarrow$$

$$\frac{\overset{7x}{\cancel{28x^2y}}\left(\frac{-3}{\cancel{4xy}}\right)}{28x^2y} + \frac{\overset{4y}{\cancel{28x^2y}}\left(\frac{6}{\cancel{7x^2}}\right)}{28x^2y}$$

$$\downarrow$$

$$\frac{-21x}{28x^2y} + \frac{24y}{28x^2y}$$

$$\downarrow$$

$$\frac{-21x+24y}{28x^2y}$$

4. LCM = $20x$

$$\downarrow$$

$$\frac{\overset{4x}{\cancel{20x}}\left(\frac{7}{\cancel{5}}\right)}{20x} + \frac{\overset{5}{\cancel{20x}}\left(\frac{-3x+2}{\cancel{4x}}\right)}{20x}$$

$$\downarrow$$

$$\frac{28x}{20x} + \frac{-15x+10}{20x}$$

$$\downarrow$$

$$\frac{28x-15x+10}{20x}$$

$$\downarrow$$

$$\frac{13x+10}{20x}$$

5. LCM = $18x$

$$\downarrow$$

$$\frac{\overset{x}{\cancel{18x}}\left(\frac{3x-4}{\cancel{18}}\right)}{18x} + \frac{\overset{6}{\cancel{18x}}\left(\frac{-3x-8}{\cancel{3x}}\right)}{18x}$$

$$\downarrow$$

$$\frac{3x^2-4x}{18x} + \frac{-18x-48}{18x}$$

$$\downarrow$$

$$\frac{3x^2-4x-18x-48}{18x}$$

$$\downarrow$$

$$\frac{3x^2-22x-48}{18x}$$

6. LCM = $144x$

$$\downarrow$$

$$\frac{\overset{4}{\cancel{144x}}\left(\frac{-9x+4}{\cancel{36x}}\right)}{144x} + \frac{\overset{3}{\cancel{144x}}\left(\frac{5x-3}{\cancel{48x}}\right)}{144x}$$

$$\downarrow$$

$$\frac{-36x+16}{144x} + \frac{15x-9}{144x}$$

$$\downarrow$$

$$\frac{-36x+16+15x-9}{144x}$$

$$\downarrow$$

$$\frac{-21x+7}{144x}$$

Here is an addition table for simple monomial fractions. Verify the entries to check your understanding of addition.

plus	1	$\frac{1}{x}$	$\frac{1}{x^2}$	$\frac{1}{x^3}$	$\frac{1}{x^4}$
1	2	$\frac{x+1}{x}$	$\frac{x^2+1}{x^2}$	$\frac{x^3+1}{x^3}$	
$\frac{1}{x}$	$\frac{x+1}{x}$	$\frac{2}{x}$	$\frac{x+1}{x^2}$	$\frac{x^2+1}{x^3}$	
$\frac{1}{x^2}$	$\frac{x^2+1}{x^2}$	$\frac{x+1}{x^2}$	$\frac{2}{x^2}$	$\frac{x+1}{x^3}$	
$\frac{1}{x^3}$	$\frac{x^3+1}{x^3}$	$\frac{x^2+1}{x^3}$	$\frac{x+1}{x^3}$	$\frac{2}{x^3}$	
$\frac{1}{x^4}$					

10. Complete the table.

OBSERVATIONS AND REFLECTIONS

★ **1.** Two fractions are *negatives* of each other if their sum is zero. For example, $\frac{2x-5}{3x}$ and $\frac{-2x+5}{3x}$ $\left(\text{or } -\frac{2x-5}{3x}\right)$ are negatives of each other, because $\frac{2x-5}{3x} + \frac{-2x+5}{3x} = 0$ $\left(\text{or } \frac{2x-5}{3x} - \frac{2x-5}{3x} = 0\right)$. The *negative of a fraction* is usually formed by either negating the numerator or changing the sign before the fraction bar. For example, the negative of $\frac{-2x+5}{3x}$ is $\frac{2x-5}{3x}$ $\left(\text{or } -\frac{-2x+5}{3x}\right)$, and the negative of $-\frac{5}{3x}$ is $+\frac{5}{3x}$ $\left(\text{or } -\frac{-5}{3x}\right)$. Changing a fraction to its negative is called *negation.*

11. Negate each of these fractions

a. $\frac{4x+5}{6x}$

b. $\frac{3x-7}{6x}$

c. $\frac{-2x+9}{6x}$

d. $\frac{-x-8}{6x}$

★ **2.** Using negation, we can define *algebraic subtraction* of fractions. We can subtract one fraction from a second fraction by adding the negative of the first fraction to the second fraction. For example, to subtract $-\frac{2x-5}{3x}$ from $-\frac{5}{3x}$, we add $+\frac{2x-5}{3x}$ to $-\frac{5}{3x}$:

$$-\frac{5}{3x} + \frac{2x-5}{3x} = \frac{-5+2x-5}{3x} = \frac{2x-10}{3x}.$$

12. Subtract

a. $\frac{4x+5}{6x}$ from $\frac{3x-7}{6x}$

b. $\frac{3x-7}{6x}$ from $\frac{-2x+9}{6x}$

c. $\frac{-2x+9}{6x}$ from $\frac{-x-8}{6x}$

d. $\frac{-x-8}{6x}$ from $\frac{4x+5}{6x}$

★ **3.** Here is a subtraction table for simple monomial fractions. Verify the entries to check your understanding of subtraction.

subtract → from ↓	SUBTRACT THIS $\frac{-1}{x^3}$	$\frac{-1}{x^2}$	$\frac{-1}{x}$	0	$\frac{1}{x}$	$\frac{1}{x^2}$	$\frac{1}{x^3}$
FROM THIS $\frac{1}{x^3}$							
$\frac{1}{x^2}$		$\frac{2}{x^2}$	$\frac{x+1}{x^2}$	$\frac{1}{x^2}$	$\frac{1-x}{x^2}$	0	
$\frac{1}{x}$		$\frac{x+1}{x^2}$	$\frac{2}{x}$	$\frac{1}{x}$	0	$\frac{x-1}{x^2}$	
0		$\frac{1}{x^2}$	$\frac{1}{x}$	0	$\frac{-1}{x}$	$\frac{-1}{x^2}$	
$\frac{-1}{x}$		$\frac{1-x}{x^2}$	0	$\frac{-1}{x}$	$\frac{-2}{x}$	$\frac{-x-1}{x^2}$	
$\frac{-1}{x^2}$		0	$\frac{x-1}{x^2}$	$\frac{-1}{x^2}$	$\frac{-x-1}{x^2}$	$\frac{-2}{x^2}$	
$\frac{-1}{x^3}$							

13. Complete the table.

EXERCISES

Simplify each of the following expressions to a single fraction.

1. $\frac{1}{2x^2} + \frac{2}{3x^2}$
2. $\frac{3}{4y^4} + \frac{7}{10y^4}$
3. $-\frac{5}{12x^2} + \frac{10}{21x^2}$
4. $\frac{-2}{9z} + \frac{5}{21z}$
5. $\frac{3}{5} + \frac{5}{7x}$
6. $-\frac{8}{9z^3} - \frac{3}{5}$
7. $-\frac{5}{3xy} + \frac{10}{11}$
8. $-\frac{7}{8w} + \frac{2}{15}$
9. $\frac{1}{a} + \frac{1}{b}$
10. $\frac{y^2}{x^3} - \frac{y}{x}$
11. $-\frac{z^2}{xy^2} + \frac{z^5}{x^3y}$
12. $\frac{-1}{A^8B^2} + \frac{1}{A^2B^5}$
13. $-\frac{8}{3x^3y} - \frac{9}{4y^2}$
14. $-\frac{15}{4s^2t^3} - \frac{3}{3s^5t}$
15. $\frac{5}{8x^2} + \frac{1}{6y}$
16. $-\frac{4}{15y^4} + \frac{3}{20yz}$
17. $\frac{9y^2}{16x^3} + \frac{3}{8xy^2}$
18. $\frac{n^3}{4m^2} + \frac{5n^2}{6m^8}$
19. $\frac{3}{5} + \frac{x+4}{2}$
20. $\frac{-7y-9}{9} - \frac{3}{2}$
21. $\frac{9}{7w} - \frac{2w+1}{4w}$

22. $-\dfrac{21x-16}{11xy}+\dfrac{11}{3xy}$ 23. $\dfrac{-2w+1}{4w}+\dfrac{3}{4}$ 24. $-\dfrac{9}{4}-\dfrac{4x+3}{3x}$

25. $-\dfrac{2x-1}{3x}+\dfrac{13}{1}$ 26. $-\dfrac{2}{15}+\dfrac{5k-2}{25k}$ 27. $\dfrac{9y+7}{4y}-\dfrac{-11y+4}{5y}$

28. $\dfrac{-4x+5}{6x^2}-\dfrac{6x+5}{5x^2}$ 29. $\dfrac{n+8}{8}+\dfrac{n+10}{10}$ 30. $\dfrac{15z-11}{20}-\dfrac{18z+7}{30}$

31. $\dfrac{-5x+8}{12x}+\dfrac{-3x-5}{15x}$ 32. $-\dfrac{10x-9}{36y^4}+\dfrac{25x-18}{60y^4}$

33. $\dfrac{w+1}{2w}+\dfrac{3w+4}{2}$ 34. $-\dfrac{3y-2}{4y}+\dfrac{8y-1}{7}$

35. $\dfrac{-x-9}{8}-\dfrac{3x-11}{9x}$ 36. $\dfrac{7t+1}{2t}-\dfrac{8t+5}{4}$

Express each of the following word phrases algebraically and then simplify it.

1. The sum of $\dfrac{4\text{L}}{9\text{sec}}$ and $\dfrac{7\text{L}}{15\text{sec}}$
2. $\dfrac{3\text{mg}}{8\text{cm}^2}$ decreased by $\dfrac{5\text{mg}}{21\text{cm}^2}$
3. $\dfrac{8\text{m}}{3\text{sec}^2}$ plus $\dfrac{10\text{km}}{4\text{sec}^2}$
4. $\dfrac{7\text{cm}}{5\text{sec}}$ reduced by $\dfrac{9\text{mm}}{8\text{sec}}$
5. $\dfrac{3x}{5}$ subtracted from $\dfrac{x+7}{2}$
6. $-\dfrac{6a-7}{9}$ added to $\dfrac{8a}{21}$
7. $\dfrac{13}{7y^2}$ less than $\dfrac{4y+2}{7y}$
8. $\dfrac{12a-7}{10}$ less $\dfrac{16a+15}{15}$
9. $\dfrac{25x-60y}{14}$ increased by $\dfrac{8y-2x}{35}$
10. $\dfrac{8x-11}{6x}$ is reduced by $\dfrac{12-15x}{33x^2}$
11. $\dfrac{3x-7y}{5x}$ more than $\dfrac{7y-3x}{5y}$
12. $\dfrac{10a+3b}{21a}$ is subtracted from $\dfrac{-8a-5b}{12b}$

9.5 EVALUATING

In this section we extend Section 8.6 to evaluate expressions requiring the addition of fractions.

Look at the following examples of *evaluating fractions.*

TYPICAL PROBLEMS Evaluate: 1. $\frac{8w-5x}{12}+\frac{7x}{2w}$ when $w=6, x=-4$

We first substitute the numbers given using parentheses:

$$\frac{8(6)-5(-4)}{12}+\frac{7(-4)}{2(6)}$$

Then we simplify the resulting expression following the Order of Operations:

$$\frac{48+20}{12}-\frac{7\cdot 4}{2\cdot 6}$$

$$\downarrow$$

$$\frac{\overset{17}{\cancel{68}}}{\underset{3}{\cancel{12}}}-\frac{7\cdot\overset{1}{\cancel{4}}}{\underset{1}{\cancel{2}}\cdot\underset{3}{\cancel{6}}}$$

$$\downarrow$$

$$\frac{17}{3}-\frac{7}{3}$$

$$\downarrow$$

$$\frac{10}{3}$$

Hence, evaluating $\frac{8w-5x}{12}+\frac{7x}{2w}$, when $w=6, x=-4$ yields $\frac{10}{3}$.

Evaluate: 2. $\frac{5k}{6}+\frac{8m}{11}$, when $k=\frac{-14}{9}$, $m=\frac{22}{27}$

We first substitute the numbers given using parentheses:

$$\frac{5\left(\frac{-14}{9}\right)}{6}+\frac{8\left(\frac{22}{27}\right)}{11}$$

Then we simplify the resulting expression following the Order of Operations:

$$\left(\left[5\left(\frac{-14}{9}\right)\right]\div[6]\right)+\left(\left[8\left(\frac{22}{27}\right)\right]\div[11]\right)$$

$$\downarrow$$

$$\left(\frac{-70}{9}\div 6\right)+\left(\frac{176}{27}\div 11\right)$$

$$\downarrow$$

$$\left(-\frac{\overset{35}{\cancel{70}}}{9}\cdot\frac{1}{\underset{3}{\cancel{6}}}\right)+\left(\frac{\overset{16}{\cancel{176}}}{27}\cdot\frac{1}{\underset{1}{\cancel{11}}}\right)$$

$$\downarrow$$

$$-\frac{35}{27}+\frac{16}{27}$$

$$\downarrow$$

$$\frac{-19}{27}$$

To evaluate fractions we extend the method in Section 8.6. Always remember to place parentheses around each number substituted. The Order of Operations is given in Section 6.5.

Hence, evaluating $\frac{5k}{6}+\frac{8m}{11}$, when $k=\frac{-14}{9}$, $m=\frac{22}{27}$ yields $\frac{-19}{27}$.

EXAMPLES

Evaluate the following fractions.

1. $\dfrac{3r^2 - 11}{8}$, when $r = 5$

2. $\dfrac{7s + 3t}{5st}$, when $s = -7$, $t = -2$

3. $-\dfrac{11A + 6}{4} + \dfrac{7A}{6}$, when $A = 8$

4. $\dfrac{2x - 5}{9} - \dfrac{7x - 2}{12}$, when $x = -4$

5. $\dfrac{3k - 10}{16}$, when $k = \dfrac{-2}{9}$

6. $\dfrac{4y}{15z} - \dfrac{11}{12}$, when $y = \dfrac{5}{6}$, $z = -\dfrac{4}{3}$

Solutions

1.
$$\frac{3(5)^2 - 11}{8}$$
$$\downarrow$$
$$\frac{3(25) - 11}{8}$$
$$\downarrow$$
$$\frac{75 - 11}{8}$$
$$\downarrow$$
$$\frac{64}{8}$$
$$\downarrow$$
$$8$$

2.
$$\frac{7(-7) + 3(-2)}{5(-7)(-2)}$$
$$\downarrow$$
$$\frac{-49 - 6}{70}$$
$$\downarrow$$
$$\frac{-55}{70}$$
$$\downarrow$$
$$\frac{-11}{14}$$

3.
$$-\frac{11(8) + 6}{4} + \frac{7(8)}{6}$$
$$\downarrow$$
$$-\frac{88 + 6}{4} + \frac{7 \cdot 8}{6}$$
$$\downarrow$$
$$-\frac{\overset{47}{\cancel{94}}}{\underset{2}{\cancel{4}}} + \frac{7 \cdot \overset{4}{\cancel{8}}}{\underset{3}{\cancel{6}}}$$
$$\downarrow$$
$$-\frac{47}{2} + \frac{28}{3}$$
$$\downarrow$$
$$-\frac{141}{6} + \frac{56}{6}$$
$$\downarrow$$
$$-\frac{85}{6}$$

4.
$$\frac{2(-4) - 5}{9} - \frac{7(-4) - 2}{12}$$
$$\downarrow$$
$$\frac{-8 - 5}{9} - \frac{-28 - 2}{12}$$
$$\downarrow$$
$$\frac{-13}{9} + \frac{\overset{5}{\cancel{30}}}{\underset{2}{\cancel{12}}}$$
$$\downarrow$$
$$\frac{-13}{9} + \frac{5}{2}$$
$$\downarrow$$
$$\frac{-26}{18} + \frac{45}{18}$$
$$\downarrow$$
$$\frac{19}{18}$$

5. $$\frac{3\left(\frac{-2}{9}\right) - 10}{16}$$

$$\downarrow$$

$$\left[\overset{1}{\cancel{3}}\left(\frac{-2}{\underset{3}{\cancel{9}}}\right) - 10\right] \div [16]$$

$$\downarrow$$

$$\left[\frac{-2}{3} - 10\right] \div [16]$$

$$\downarrow$$

$$\left[\frac{-2}{3} + \frac{-30}{3}\right] \div [16]$$

$$\downarrow$$

$$\frac{-32}{3} \div 16$$

$$\downarrow$$

$$-\frac{\overset{2}{\cancel{32}}}{3} \cdot \frac{1}{\underset{1}{\cancel{16}}}$$

$$\downarrow$$

$$-\frac{2}{3}$$

6. $$\frac{4\left(\frac{5}{6}\right)}{15\left(-\frac{4}{3}\right)} - \frac{11}{12}$$

$$\downarrow$$

$$\left(\left[\overset{2}{\cancel{4}}\left(\frac{5}{\underset{3}{\cancel{6}}}\right)\right] \div \left[\overset{5}{\cancel{15}}\left(-\frac{4}{\underset{1}{\cancel{3}}}\right)\right]\right) - \frac{11}{12}$$

$$\downarrow$$

$$\left(\frac{10}{3} \div [-20]\right) - \frac{11}{12}$$

$$\downarrow$$

$$-\frac{\overset{1}{\cancel{10}}}{3} \cdot \frac{1}{\underset{2}{\cancel{20}}} - \frac{11}{12}$$

$$\downarrow$$

$$-\frac{1}{6} - \frac{11}{12}$$

$$\downarrow$$

$$\frac{-2}{12} + \frac{-11}{12}$$

$$\downarrow$$

$$\frac{-13}{12}$$

EXERCISES

Evaluate each of the following expressions, simplifying to a reduced fraction.

1. $\frac{4t + 7}{9}$, when $t = -4$

2. $\frac{5k + 9}{8}$, when $k = -5$

3. $-\frac{2A - 9}{A}$, when $A = -3$

4. $-\frac{5n - 12}{n}$, when $n = -6$

5. $\frac{5 - 8a^2}{3}$, when $a = 2$

6. $\frac{7 - 11k^2}{4}$, when $k = 3$

7. $-\frac{2B - 9}{7B^2}$, when $B = -6$

8. $-\frac{3B - 4}{7B^2}$, when $B = -8$

9. $-\frac{7A - 10}{4B}$, when $A = 14, B = 11$

10. $-\frac{9m - 14}{5n}$, when $m = 11, n = 17$

11. $\frac{12 - 7A}{9B}$, when $A = 6, B = -5$

12. $\frac{15 - 8k}{6m}$, when $k = 12, m = -3$

13. $\frac{4s + 5t}{11s}$, when $s = 9, t = 6$

14. $\frac{6u + 7v}{11u}$, when $u = 8, v = 12$

15. $\frac{5}{8r} + \frac{11}{6}$, when $r = 15$

16. $\frac{4}{9a} + \frac{8}{15}$, when $a = 20$

17. $-\dfrac{3-4x}{5} - \dfrac{7x}{8}$, when $x = -4$

18. $-\dfrac{3-2A}{3} - \dfrac{7A}{10}$, when $A = -5$

19. $\dfrac{4m}{9} - \dfrac{7n}{18}$, when $m = -5, n = -6$

20. $\dfrac{5u}{8} - \dfrac{11v}{16}$, when $u = -7, v = -12$

21. $\dfrac{5A-7}{3B} + \dfrac{6A}{5B}$, when $A = 11, B = -4$

22. $\dfrac{9x-13}{4y} + \dfrac{6x}{7y}$, when $x = 5, y = -6$

23. $\dfrac{4s+7}{6} - \dfrac{3t-10}{8}$, when $s = 5, t = 9$

24. $\dfrac{6y+11}{10} - \dfrac{5z-2}{15}$, when $y = 4, z = 7$

25. $-\dfrac{9-3k}{14}$, when $k = \dfrac{2}{3}$

26. $-\dfrac{8-5w}{9}$, when $w = \dfrac{2}{5}$

27. $\dfrac{3A+11}{4A}$, when $A = \dfrac{-5}{9}$

28. $\dfrac{4x+15}{10x}$, when $x = \dfrac{-5}{12}$

29. $-\dfrac{m-5n}{12}$, when $m = \dfrac{-8}{3}, n = \dfrac{2}{3}$

30. $-\dfrac{A-7B}{16}$, when $A = \dfrac{-9}{5}, B = \dfrac{3}{5}$

31. $\dfrac{1-4k}{14m}$, when $k = \dfrac{5}{6}, m = -\dfrac{2}{9}$

32. $\dfrac{5-6u}{11v}$, when $u = \dfrac{2}{9}, v = -\dfrac{5}{6}$

33. $4w^2 + 3$, when $w = \dfrac{5}{6}$

34. $15x^2 + 2$, when $x = \dfrac{4}{5}$

35. $4 - 5m^2$, when $m = \dfrac{-4}{5}$

36. $5 - 7A^2$, when $A = \dfrac{-5}{7}$

37. $4n^2 - \dfrac{8}{25}$, when $n = \dfrac{-2}{5}$

38. $6k^2 - \dfrac{9}{8}$, when $k = \dfrac{-3}{2}$

39. $-2x^2 - \dfrac{7}{2}$, when $x = \dfrac{9}{4}$

40. $-3a^2 - \dfrac{11}{3}$, when $a = \dfrac{8}{9}$

41. $\dfrac{5m}{14} + 3n$, when $m = -\dfrac{7}{8}, n = \dfrac{5}{4}$

42. $\dfrac{4A}{15} + 2B$, when $A = -\dfrac{5}{9}, B = \dfrac{4}{3}$

43. $\dfrac{3x}{16} - \dfrac{7y}{20}$, when $x = \dfrac{8}{5}, y = \dfrac{-3}{7}$

44. $\dfrac{5m}{18} - \dfrac{11n}{24}$, when $m = \dfrac{3}{8}, n = \dfrac{-5}{2}$

45. $\dfrac{5A}{8} - \dfrac{7B}{12}$, when $A = \dfrac{7}{10}, B = \dfrac{5}{12}$

46. $\dfrac{7s}{6} - \dfrac{11t}{9}$, when $s = \dfrac{11}{14}, t = \dfrac{7}{8}$

47. $\dfrac{3x-8}{8} + \dfrac{7y}{12}$, when $x = -\dfrac{2}{5}, y = \dfrac{9}{5}$

48. $-\dfrac{5y-2}{12} + \dfrac{9z}{16}$, when $y = -\dfrac{8}{7}, z = \dfrac{5}{7}$

ANSWERS TO MARGIN EXERCISES

1. a. $-\frac{5}{8x}$
   b. $-\frac{3x-5}{8x}$

2. a. $\frac{-7x-2}{6y}$
   b. $\frac{-10x+5}{6y}$
   c. $\frac{8+9x}{6y}$

3. a. $\frac{-3}{2}$
   b. $\frac{5}{2}$

4. a. $\frac{7x-7}{1}$
   b. $\frac{6x+8y}{1}$

5. a. $\frac{-15}{20xy}$
   b. $\frac{12x^2}{14x^4}$
   c. $\frac{6x-4}{8x}$
   d. $\frac{12xy+32y}{12xy}$

6. a. $\frac{104x-64}{60}$
   b. $\frac{-44x}{96x^2} = \frac{-11}{24x}$

7. a. $\frac{x}{2}+\frac{2}{x}$
   b. $\frac{1}{2y}+\frac{-2}{x}$
   c. $\frac{8y}{3x}+3$

9. a. $24x^5y^4$, $36x^2y^2$, $48x^2y^8$, $54x^5y^3$, $60x^4y^3z^2$
   b. $16x^3y^3$, $24x^5y^4$, $48x^2y^8$, $60x^4y^3z^2$
   c. $24x^5y^4$, $48x^2y^8$, $60x^4y^3z^2$

11. a. $\frac{-4x-5}{6x}$
    b. $\frac{-3x+7}{6x}$
    c. $\frac{2x-9}{6x}$
    d. $\frac{x+8}{6x}$

12. a. $\frac{-x-12}{6x}$
    b. $\frac{-5x+16}{6x}$
    c. $\frac{x-17}{6x}$
    d. $\frac{5x+13}{6x}$

CHAPTER REVIEW

Simplify to a single reduced fraction. SECTION 9.1

1. $\frac{7x^2}{8y} - \frac{45x^2}{8y}$
2. $\frac{9y^2}{4z} - \frac{23y^2}{4z}$
3. $\frac{-8}{5t} + \frac{19}{5t}$
4. $\frac{3}{12x^2} - \frac{8}{12x^2}$
5. $\frac{5a}{12a^2b^2} - \frac{21b}{12a^2b^2}$
6. $-\frac{3y^2}{14xy^4} - \frac{8x}{14xy^4}$
7. $\frac{5}{8} - \frac{3z+7}{8}$
8. $-\frac{3x+5}{5xy} + \frac{8}{5xy}$
9. $\frac{9y^2-11}{4z^2} - \frac{5y^2}{4z^2}$
10. $\frac{4r-3}{5r^2} + \frac{7r-2}{5r^2}$
11. $\frac{3u-4v}{9uv} + \frac{3u+4v}{9uv}$
12. $\frac{5x-9}{7} - \frac{-2x+2}{7}$

Simplify to a fraction with the given denominator. SECTION 9.2

13. $\frac{3}{4} - \frac{5}{12t} = \frac{?}{12t}$
14. $-\frac{5}{6k^2} + \frac{7k^4}{12} = \frac{?}{12k^2}$
15. $\frac{9}{4x^4} - \frac{3}{8x^2} = \frac{?}{8x^4}$
16. $\frac{r^4}{24s^2} - \frac{3r^2}{16s} = \frac{?}{48s^2}$
17. $-\frac{2r}{t} + \frac{t}{3s^2} = \frac{?}{3s^2t}$
18. $\frac{9}{16K^3} - \frac{5}{12R} = \frac{?}{48K^3R}$
19. $\frac{6x+8}{10} + \frac{5}{8} = \frac{?}{40}$
20. $\frac{3w-2}{45} - \frac{7}{30} = \frac{?}{90}$
21. $\frac{-2x-3}{14} - \frac{7}{8} = \frac{?}{56}$
22. $\frac{3z+1}{2} + \frac{-2z+3}{3} = \frac{?}{6}$
23. $\frac{2k+4}{25} - \frac{5k-4}{10} = \frac{?}{50}$
24. $\frac{-7y+2}{7} - \frac{2y+5}{8} = \frac{?}{56}$

Find the LCM (Least Common Multiple). SECTION 9.3

25. t^7, t^5
26. x^3y^5, xy^8
27. a^5b^7, a^2b^{10}
28. $r^6s^5t^3, r^2s^4t^4$
29. $16x^4, 32x^2$
30. $9u^4, 12v^3$
31. $15m^2, 45m^3n^2$
32. $160x^4y^3, 24x^2y^5$
33. $35xy^2, 105y^2z$
34. $80u^6v^3w^6, 50u^4v^8w^4$
35. $135x^2y^5z, 45y^5z^2$
36. $48a^4b^2c^6, 72a^2b^4c^6$

Simplify to a single fraction. SECTION 9.4

37. $\frac{3}{5xy} + \frac{4}{7xy}$
38. $-\frac{3}{16xy^3} - \frac{9}{40xy^3}$
39. $\frac{5}{6} - \frac{4}{7t^2}$
40. $\frac{6}{5y^2} + \frac{5}{6}$
41. $\frac{y}{x^2} + \frac{1}{xy}$
42. $-\frac{z^2}{x^5y} - \frac{y^3z}{x^7}$

43. $\frac{11}{9z^4} - \frac{3}{2z^7}$

44. $\frac{4}{9x^2} + \frac{5}{12x}$

45. $\frac{3x^2}{4y^7} - \frac{8x^2}{14y^6}$

46. $-\frac{3t-8}{3} + \frac{3}{8}$

47. $-\frac{11}{8y} - \frac{3x-5}{5y}$

48. $\frac{2}{5} - \frac{3w-2}{5w}$

49. $\frac{-2x+3}{6x} + \frac{7}{8}$

50. $\frac{6x-1}{3y} + \frac{9x-7}{7y}$

51. $-\frac{5t-1}{12} - \frac{4t-3}{16}$

52. $\frac{-5x+4}{9xy} - \frac{14x-15}{21xy}$

53. $\frac{2w-7}{7} - \frac{18w-7}{7w}$

54. $\frac{4x+5}{15} + \frac{8x-3}{10x}$

SECTION 9.5

Evaluate, simplifying to a reduced fraction.

55. $\frac{5x+8}{6}$, when $x = -7$

56. $-\frac{3r-10}{r}$, when $r = -5$

57. $\frac{6-11n^2}{4}$, when $n = 2$

58. $-\frac{2x-7}{5x^2}$, when $x = -9$

59. $-\frac{8u-15}{9v}$, when $u = 12, v = 9$

60. $\frac{16-3y}{6z}$, when $y = 10, z = -7$

61. $\frac{7A+5B}{13A}$, when $A = 5, B = 6$

62. $\frac{25}{6A} + \frac{13}{21}$, when $A = 10$

63. $-\frac{5-3t}{7} + \frac{2t}{9}$, when $t = -3$

64. $\frac{3y}{10} - \frac{9z}{20}$, when $y = -7, z = -18$

65. $\frac{9r-11}{4s} + \frac{7r}{5s}$, when $r = 7, s = -8$

66. $\frac{3a+4}{12} - \frac{7b-1}{27}$, when $a = 11, b = 4$

67. $-\frac{7-4a}{12}$, when $a = \frac{3}{4}$

68. $\frac{5k+14}{3k}$, when $k = \frac{-7}{10}$

69. $-\frac{x-6y}{14}$, when $x = \frac{-11}{5}, y = \frac{4}{5}$

70. $\frac{4-9s}{15t}$, when $s = \frac{11}{6}, t = -\frac{5}{12}$

71. $8r^2 + 2$, when $r = \frac{5}{4}$

72. $4 - 7y^2$, when $y = \frac{-4}{7}$

73. $6A^2 - \frac{8}{21}$, when $A = \frac{-4}{3}$

74. $-4t^2 - \frac{9}{4}$, when $t = \frac{5}{8}$

75. $\frac{6u}{25} + 4v$, when $u = -\frac{5}{9}, v = \frac{11}{12}$

76. $\frac{9s}{14} - \frac{5t}{24}$, when $s = \frac{7}{8}, t = \frac{-11}{5}$

77. $\frac{3a}{10} - \frac{4b}{15}$, when $a = \frac{4}{9}, b = \frac{7}{15}$

78. $-\frac{4x-7}{6} - \frac{8y}{9}$, when $x = -\frac{3}{5}, y = \frac{6}{5}$

SUMMARY OF PHRASES

Express each of the following word phrases algebraically, then simplify it.

1. The sum of $\frac{10\text{L}}{3\text{sec}}$ and $\frac{11\text{L}}{3\text{sec}}$

2. $\frac{81\text{g}}{5\text{cm}^2}$ subtracted from $\frac{96\text{g}}{5\text{cm}^2}$

3. $\frac{-18a^2}{25b}$ added to $\frac{-17a}{25b}$

4. $\frac{13z}{16x^3y}$ less than $\frac{17z}{16x^3y}$

5. $\frac{6a + 7b}{9b}$ is reduced by $\frac{4a - 3b}{9b}$

6. $\frac{27 - 14x}{13y}$ less $\frac{-19 - 70x}{13y}$

7. $\frac{8\text{L}}{9\text{cm}^2}$ decreased by $\frac{5\text{L}}{6\text{cm}^2}$

8. $\frac{4\text{km}}{3\text{sec}^2}$ plus $\frac{19\text{m}}{2\text{sec}^2}$

9. $\frac{10x}{9y}$ reduced by $\frac{-7x}{8z}$

10. $\frac{12 - a}{4b}$ increased by $\frac{3c}{7b^3}$

11. $\frac{3a - 8b}{14a^2b}$ more than $\frac{-5a + 2b}{12ab^2}$

12. $\frac{3x - 5y}{45x}$ is subtracted from $\frac{10x - 9y}{36y}$

CUMULATIVE TEST

The following problems test your understanding of this chapter and the previous chapter. Before taking this test, thoroughly review Sections 8.4, 8.5, 9.1, and 9.4.

Once you have finished the test, compare your answers with the answers provided at the back of the book. Note the section number of each problem missed, and thoroughly review those sections again.

Simplify.

1. $-\frac{25x^3}{12y} + \frac{16x^3}{12y}$

2. $\frac{5}{3} - \frac{14}{9x^2}$

3. $\frac{-1}{4x^5} \cdot \frac{12}{x^4}$

4. $\frac{13z^4}{16x^3y} + \frac{19y^2}{72xz^3}$

5. $\frac{x^4}{y^2} \div \frac{x}{y}$

6. $\frac{3x - 2}{7x} - \frac{4x - 9}{7x}$

7. $-\frac{x^3}{y^2z} - \left(\frac{x^2y^3}{z^4}\right)$

8. $\frac{K^3L^6}{M^4N^9}\left(+\frac{K^8N^2}{L^7M^3}\right)$

9. $-\frac{-9x - 1}{12} - \frac{8x - 7}{16}$

10. $\frac{4x^4}{15y} \div \frac{14x}{45y^6}$

11. $\frac{14x}{9xy^2} - \frac{21y^2}{9xy^2}$

12. $\frac{63x^3}{-48y^8} \cdot \frac{-16y^4}{27x^9}$

13. $\frac{5}{8t} + \frac{3}{12t^2}$

14. $\frac{m^3}{n^4} \div \left(-\frac{m^7}{n^6}\right)$

15. $\frac{8x^3}{9} - \frac{3x}{8}$

16. $\frac{-4}{9k} + \frac{4k - 7}{6k}$

17. $\frac{-9t}{-20u^4} \cdot \frac{25u^6}{-12t^7}$

18. $-\frac{5x - 11y}{18xy} - \frac{9x - 2y}{18xy}$

19. $\frac{4A^2}{15B^3} - \frac{7A^3}{6B^2}$

20. $\frac{18A^6}{25B^3} \div \frac{81B^2}{10A^4}$

21. $\frac{3}{14x^6}\left(+\frac{21x^3}{8}\right)$

22. $\frac{-2x + 11y}{48xy} - \frac{16x - 3y}{72xy}$

23. $-\frac{t^4}{26} \div \left(-\frac{t^7}{39}\right)$

24. $\frac{3}{8} - \frac{8A - 1}{20}$

25. $\frac{-6x^2z^3}{49y^4} \cdot \frac{-35x^3y^6}{30z^8}$

26. $\frac{-x - 4}{2} - \frac{9x - 3}{7}$

27. $-\frac{16uv^3}{4w^6} \div \frac{24u^3v}{63w^4}$

28. $\frac{5y - 2}{16} + \frac{5y - 3}{40y}$

29. $\frac{65x^3}{24}\left(-\frac{18}{13x^4}\right)$

30. $\frac{-5x^2 + 6}{15x^2} + \frac{9}{10}$

EQUATIONS

Equations

An *equation* is an algebraic sentence containing two expressions called *members* separated by an equals sign =. For example, $1 + 2 = 3$ and $x + 2 = 3$ are equations. An equation is called an *identity* if its two members have the same value. For example, $1 + 2 = 3$ and $x + x = 2x$ are identities. An equation containing literals is called a *conditional equation* if its two members do *not* have the same value.

In this chapter we will study conditional equations. The simplest conditional equations have the form literal = number, such as $x = 3$ and $y = \frac{-3}{4}$. In previous chapters we have used this type of equation when evaluating literal expressions. We use these equations to state that the literal *represents* the number, so that we can replace the literal in any expression by the number without changing the *intended* value of the expression. We can use the equation $x = 3$, for example, to say that x is another name for 3.

In this chapter we will study more complicated conditional equations in which the number (or numbers) the literal represents is not clear. In such cases we usually say that x represents an unknown number, and we call x an *unknown*.

Linear Equations

There are many kinds of conditional equations; for example,

a. $5x - 12 = 12 + 7x$

b. $\frac{8y}{3} - 1 = y - \frac{3y - 2}{5}$

c. $2x^2 + 5x - 8 = x^2 + 6x + 4$

d. $\frac{5}{2x - 6} - \frac{7x}{3x - 9} = -\frac{1}{4}$

e. $3y - 5x + 2z = y - 2x + 8z$

f. $\frac{3B^2}{7C^3} = 2AB$

We will look at all these types. We will first study *linear equations in one unknown* (only a and b above). An equation is said to be in *one unknown* if it involves a single literal. An equation is *linear* if its members are *linear expressions,* that is, if its members are expressions of the following types:

1. A polynomial is linear if none of its terms contains more than one literal factor.
   For example, $-4x - 7$, $1 - 3y$, and $x + 2y - 5z$ are linear.
2. A reduced fraction is linear if its denominator is a nonzero *integer* and its numerator is linear.
   For example, $\frac{5x}{8}$ and $\frac{x-1}{2}$ are linear.
3. A sum of linear expressions is linear.
   For example, $\frac{x}{2} - \frac{1}{3}$ and $\frac{2x - y}{4} - \frac{7x + 3y}{5}$ are linear.
4. Any expression that has the same value as a linear expression is linear.
   For example, $3(4x - 9)$ and $4(x - 9) - (7x + 8)$ are linear.

WARNING A *product* or *quotient* of linear expressions is not necessarily linear. For example, x^2, $3xy$, $x(x + 3)$, and $\frac{2}{x}$ are *not* linear expressions.

Satisfying an Equation in One Unknown

Look at the equation $x = 3$. Replacing the literal x by the number 3 in this equation yields the identity $3 = 3$. Hence, the number 3 is said to *satisfy the equation* $x = 3$, and 3 is called a *root* of this equation. Similarly, note that replacing x by 3 in the equation $3x - 8 = 13 - 4x$ yields $3(3) - 8 = 13 - 4(3)$, which simplifies to $1 = 1$, an identity. Hence, 3 satisfies $3x - 8 = 13 - 4x$ and is called a root of this equation.

When we replace the literal in an equation by a number and then simplify each member we are *evaluating the equation.* A number is then a root of a conditional equation if evaluating the equation for this number yields an identity.

Equivalent Equations

In general, a conditional equation may have any number of roots. However, a conditional *linear equation* in *one* unknown has *only one root* (an integer or a fraction). For example, $3x - 8 = 13 - 4x$ has one root, the number 3. If we are given a conditional linear equation in one unknown, we call finding its root *solving the equation.* We will give the *solution* of the equation by setting its literal equal to its root. For example, we solve $3x - 8 = 13 - 4x$ by determining that its root is 3, and we give its solution as $x = 3$.

Any two conditional linear equations in one unknown are *equivalent equations* if they both contain the same literal and have the same root. Hence, two equations are equivalent if they have the same solution. For example, the equations $3x - 8 = 13 - 4x$ and $x = 3$ are equivalent, because they both contain x and have the root 3.

Observe that $x = 3$ and $x = -6$ are *not* equivalent because they have different roots, whereas $x = 3$ and $y = 3$ are not equivalent because they do not contain the same literal.

Transformation Rules

We can solve a conditional linear equation in one unknown algebraically by using the following rules in some sequence. Given such an equation, we can apply any of these rules to get an equivalent equation.

RULE 1

Replace either member by any expression with the same value.

$3(x-4)=7$ is equivalent to $3x-12=7$

RULE 2

Reverse the order of the members.

$2=5x-6$ is equivalent to $5x-6=2$

RULE 3

Add any linear expression, not containing a new literal, to both members of the equation.

$3x=4$ is equivalent to $3x-8=4-8$

RULE 4

Multiply or divide both members of the equation by any nonzero number (which may be a fraction). **Do not multiply or divide by any literal expression.**

$\frac{3x}{2}=\frac{1}{3}$ is equivalent to $6\left(\frac{3x}{2}\right)=6\left(\frac{1}{3}\right)$

RULE 5

Negate both members of the equation.

$-x=7$ is equivalent to $x=-7$

and

$-x=-7$ is equivalent to $x=7$

Applying these transformation rules in any sequence will yield an equivalent equation (which may or may not be a solution). The transformation rules only state what *can* be done to an equation. We must learn how to apply these rules effectively to solve an equation.

10.1
SOLVING EQUATIONS BY SHIFTING TERMS

Shifting Terms Between Members

Look at the patterns in the following columns:

$x = 6$	$x = 6$
$x + 1 = 7$	$2x = x + 6$
$x + 2 = 8$	$3x = 2x + 6$
$x + 3 = 9$	$4x = 3x + 6$
$x + 4 = 10$	$5x = 4x + 6$

All these equations are equivalent. Starting with $x = 6$ at the top in each column we can use transformation rule 3 to obtain each of the other equations. The equations on the left are formed from $x = 6$ by adding an integer to both members, whereas the equations on the right are formed by adding a multiple of x to both members. To solve any of these equations we must show that it is equivalent to $x = 6$, and we do this by applying transformation rule 3 again.

Look once more at the columns above and then compare them with these columns:

$x = 6$	$x = 6$
$x + 1 - 1 = 7 - 1$	$2x - x = x + 6 - x$
$x + 2 - 2 = 8 - 2$	$3x - 2x = 2x + 6 - 2x$
$x + 3 - 3 = 9 - 3$	$4x - 3x = 3x + 6 - 3x$
$x + 4 - 4 = 10 - 4$	$5x - 4x = 4x + 6 - 4x$

1. Simplify both sides in each of these equations.

Can you see from these examples what quantity we should add to the members of an equation such as $x + 4 = 10$ or $5x = 4x + 6$ to solve it?

To solve such an equation we first isolate the unknown in the left member of the equation by using transformation rules 1 and 3. Consider, for example, the equation

$$-8 + 2x = x + 13$$

To isolate x we must eliminate -8 from the left side and shift it to the right side. Further, we must shift x from the right side to the left. To do this we add the *negatives* of -8 and x to both members of the equation using rule 3. That is, we add $+8$ and $-x$:

$$-8 + 2x = x + 13$$
$$\downarrow$$
$$-8 + 2x + 8 - x = x + 13 + 8 - x$$

Then, using rule 1, we combine each pair of negatives. Their sum is zero:

$$-8 + 2x + 8 - x = x + 13 + 8 - x$$
$$\downarrow$$
$$(-8 + 8) + 2x - x = (x - x) + 13 + 8$$
$$\downarrow$$
$$2x - x = 13 + 8$$

2. What should be added to both sides to solve these equations?
   a. $x - 5 = 4$
   b. $3x = 2x + 7$
   c. $-x - 6 = -2x - 9$

We call these two steps *shifting terms.*

Solving an Equation

To solve a linear equation in x we must determine the equivalent equation of the form x = root. To do this we first isolate x in the left member by shifting terms. Then we simplify the equation further by combining terms on each side (using rule 1 again). For example, $2x - x = 13 + 8$ simplifies to $x = 21$. If, as in this example, the left member should simplify to x (with coefficient +1), the equation has the form x = number and is the solution to the original equation. In this section we will only discuss equations in which the left member simplifies immediately to x once we have shifted terms to isolate x.

In most equations the left member will simplify to a multiple of x. We will discuss such equations in the next section.

Look at the following example of *solving equations by shifting terms.*

TYPICAL PROBLEM

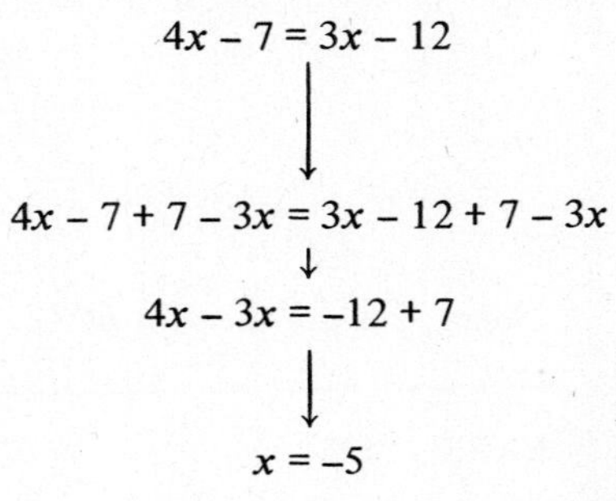

$4x - 7 = 3x - 12$

We first isolate the unknown by shifting terms. To do this we add the negatives of the terms to be shifted to both sides (rule 3):

$4x - 7 + 7 - 3x = 3x - 12 + 7 - 3x$

And then we add negatives to zero (rule 1):

$4x - 3x = -12 + 7$

Then we simplify each member of the equation (rule 1):

$x = -5$

To simplify each member of the equation we combine terms using the methods in Sections 1.1, 1.2, 3.1, and 3.2.

Hence, $4x - 7 = 3x - 12 \quad \Rightarrow \quad x = -5$.

EXAMPLES

Solve the following equations by shifting terms.

Every linear equation in one unknown can be simplified to the form literal = root.

1. $18 + x - 25 = -12$
2. $12x - 40 = 26 + 11x$
3. $13 - 2x = 20 - 3x$
4. $-6x - 8 = -7x + 16$
5. $-5 = 7 - x$
6. $16 - 18 = 8 - x$

Solutions

1. Add −18 and +25
   $\downarrow$
   $x = -12 - 18 + 25$
   $\downarrow$
   $x = -5$

2. Add +40 and $-11x$
   $\downarrow$
   $12x - 11x = 26 + 40$
   $\downarrow$
   $x = 66$

3. Add −13 and $+3x$
   $\downarrow$
   $-2x + 3x = 20 - 13$
   $\downarrow$
   $x = 7$

4. Add +8 and $+7x$
   $\downarrow$
   $-6x + 7x = 16 + 8$
   $\downarrow$
   $x = 24$

5. Add +5 and $+x$
   $\downarrow$
   $x = 7 + 5$
   $\downarrow$
   $x = 12$

6. Add −16, +18, and $+x$
   $\downarrow$
   $x = 8 - 16 + 18$
   $\downarrow$
   $x = 10$

Checking

When we solve an equation, we can make computational errors and arrive at the wrong answer. Because we are solving for the root of the equation, we can check our solution by evaluating the equation. If the equation reduces to an identity, our solution is correct. If the equation reduces to a falsehood, our solution is incorrect. For example, we can check the solution $x = 3$ of the equation $3x - 8 = 13 - 4x$ as follows:

$$3x - 8 = 13 - 4x, \text{ when } x = 3$$
$$\downarrow \qquad \downarrow$$
$$3(3) - 8 = 13 - 4(3)$$
$$\downarrow$$
$$1 = 1, \text{ an identity.}$$

3. Evaluate, and determine which yield identities:
A. $3x - 9 = 2x + 3$
  a. when $x = 6$
  b. when $x = 12$
B. $-2x + 4 = -3x - 6$
  a. when $x = -2$
  b. when $x = -10$
C. $-x + 5 = -2x + 5$
  a. when $x = 0$
  b. when $x = -8$
D. $7x + 12 = 6x + 7$
  a. when $x = -3$
  b. when $x = -5$

Hence, 3 is the root of $3x - 8 = 13 - 4x$ and $x = 3$ is the correct solution.

On the other hand, if we were to evaluate this equation using the incorrect solution $x = 2$, we would get

$$3x - 8 = 13 - 4x, \text{ when } x = 2$$
$$\downarrow \qquad \downarrow$$
$$3(2) - 8 = 13 - 4(2)$$
$$\downarrow$$
$$-2 = 5, \text{ a falsehood.}$$

Hence, 2 is not the root of $3x - 8 = 13 - 4x$.

EXERCISES

Solve each of the following equations.
In these exercises, only shifting terms and addition are needed.

1. $3 + x = 9$
2. $x - 4 = 5$
3. $8 + x = -2$
4. $14 + x = 15$
5. $11 + x = 7$
6. $-4 + x = 4$
7. $2 + x - 4 = 9$
8. $-5 + 9 + x = -3$
9. $-3 + x - 5 = 2$
10. $-4 + x = 8 - 3$
11. $-6 + x = -6 + 8$
12. $8 + x = -8 + 6$
13. $3 + 4x = 3x$
14. $8x = 7x + 4$
15. $-6x = 8 - 7x$
16. $6 - 7x = -8x$
17. $5 + 2x = 9 + x$
18. $16 + 12x = 11x + 6$
19. $-x - 4 = -2x - 4$
20. $5x + 7 = 4x + 7$
21. $-4 + 12 = 7 - x$
22. $-8 - 4 = -10 - x$
23. $-8 = -x$
24. $6 = -x$

Express each of the following word problems algebraically and then solve it.
These problems can be translated by the methods in Section 13.1.

1. Twice a certain number equals the sum of 42 and the number. Find the number.
2. A certain number multiplied by 2 is 17 more than the number. What is the number?
3. When 18 added to a certain number is reduced by 29, the result is 5. What is the number?
4. Fifty-one less twice a certain number is 121 less three times the number. Find the number.
5. Five times a number is 88 less than the product of the number and 4. What is the number?
6. When 2 times a certain number is decreased by 6, the result is the sum of the number and 21. What is the number?

7. When 7 times a certain number is decreased by 78, the result is the number multiplied by 6. Find the number.
8. The sum of 70 and 10 times a number is 42 increased by 9 times the number. Find the number.
9. When a product of a certain number and 17 is reduced by 5, the result is 19 increased by 16 times the number. What is the number?
10. If 18 decreased by 12 times a number equals 11 minus 13 times the number, what is the number?

10.2 SOLVING EQUATIONS BY ISOLATION AND DIVISION

Division

Using transformation rule 4, we can multiply or divide both members of an equation by any nonzero number. To see how to use this rule, look at the pattern in the following columns:

$x = 2$	$x = \frac{2}{3}$
$3x = 6$	$3x = 2$
$6x = 12$	$6x = 4$

The three equations in each column are equivalent. Starting with $x = 2$ on the left and $x = \frac{2}{3}$ on the right, we can use transformation rule 4 to obtain the other two equations in each column. The equations in the second row are obtained from the equations above them by multiplying by 3 and the equations in the third row are obtained by multiplying by 6.

4. Multiply both members of $x = \frac{3}{4}$ by
a. 4
b. 8
c. 12

Because multiplication and division are inverse operations, we can solve any of these equations by applying transformation rule 4 again. Look once more at the columns above and then compare them with these columns:

$x = 2$	$x = \frac{2}{3}$
$\frac{3x}{3} = \frac{6}{3}$	$\frac{3x}{3} = \frac{2}{3}$
$\frac{6x}{6} = \frac{12}{6}$	$\frac{6x}{6} = \frac{4}{6}$

5. Simplify each fraction in these equations.

Can you see from these examples what quantity we divide by in an equation such as $6x = 12$ or $6x = 4$ to solve it?

Solving by Isolation and Division

In this section we will discuss the general class of linear equations in x in which each term is an integer or an integer times x. (These equations contain no parentheses or fractions.) To solve such an equation we first isolate x by shifting terms. Then we combine terms on each side of the equation. If the left member simplifies to x, the equation is solved as we saw in the previous section. Otherwise we divide both members of the equation by the coefficient of x to obtain an equation of the form x = root.

Look at the following example of *solving equations by isolation and division.*

TYPICAL PROBLEM

$$5x - 12 = 12 + 7x$$

We first isolate the unknown by shifting terms. To do this we add the negatives of the terms to be shifted to both sides (rule 3):

$$5x - 12 + 12 - 7x = 12 + 7x + 12 - 7x$$

And then we add the negatives to zero (rule 1):

$$5x - 7x = 12 + 12$$

Then we simplify completely (rule 1):

$$-2x = 24$$

At this point each member of the equation is a single term. The unknown is on the left and the constant term on the right of the equals sign.

If the coefficient of the unknown is negative, we must negate both sides of the equation (rule 5):

$$2x = -24$$

Then we divide by the coefficient of the unknown (rule 4):

$$\frac{2x}{2} = \frac{-24}{2}$$

And reduce (rule 1):

$$x = -12$$

Hence, $5x - 12 = 12 + 7x \Rightarrow x = -12$.

Every linear equation in one unknown can be simplified to the form literal = root.

EXAMPLES

Solve the following equations by isolation and division.

1. $6x = -9$
2. $-3x = 12$
3. $-8x = -18$
4. $8x = 12x + 1$
5. $6 - 4x = 5x + 6$
6. $2 - 5x + 7 + 9x - 4 = 0$

Solutions

1. $\frac{6x}{6} = \frac{-9}{6} \downarrow x = \frac{-3}{2}$
2. $3x = -12 \downarrow \frac{3x}{3} = \frac{-12}{3} \downarrow x = -4$
3. $8x = 18 \downarrow \frac{8x}{8} = \frac{18}{8} \downarrow x = \frac{9}{4}$
4. $8x - 12x = 1 \downarrow -4x = 1 \downarrow 4x = -1 \downarrow x = \frac{-1}{4}$
5. $-4x - 5x = 6 - 6 \downarrow -9x = 0 \downarrow x = 0$
6. $-5x + 9x = -2 - 7 + 4 \downarrow 4x = -5 \downarrow x = \frac{-5}{4}$

SPECIAL CASES

If any nonzero multiple of x is zero, then x must be zero. For example,

If $4x = 0$, then $x = 0$
If $-3x = 0$, then $x = 0$
If $-x = 0$, then $x = 0$

If x is not zero, then no multiple of x can be zero, except of course $0 \cdot x$.

OBSERVATIONS AND REFLECTIONS

Transformation rule 5 states that we can negate both members of an equation. This rule is equivalent to a special case of rule 4, because negation is equivalent to multiplying by -1. For example,

$$-x = 4 \longrightarrow x = -4$$

is equivalent to

$$-x = 4 \rightarrow (-1)(-x) = (-1)(4) \rightarrow x = -4$$

6. Solve
a. $-x = 9$
b. $-x = -16$

EXERCISES

Solve each of the following equations.

1. $-x = 21$
2. $-x = -1$
3. $3 - x = -8$
4. $-x + 5 = -6$
5. $15 - x = 15$
6. $9 = 3 + x$
7. $11 = 2 + x$
8. $-4 = x - 7$
9. $4x = 3 + 5x$
10. $11 - 3x = 5 - 2x$
11. $6x - 7 = 7x - 8$
12. $16 + 9x = 6 + 10x$
13. $7x = 21$
14. $2x = -14$
15. $-6x = 3$
16. $-15x = -5$
17. $12 = 8x$
18. $27 = -6x$
19. $8 + 3x = 14$
20. $5 + 7x = 12$
21. $-4 + 8x = 52$
22. $3 + 8x = 5$
23. $7 + 9x = 10$
24. $-14 + 6x = -11$
25. $4x + 9 = 7$
26. $6 + 8x = 2$
27. $4 + 9x = -2$
28. $5 = 7x - 9$
29. $-13 = 3 - 4x$
30. $6 = -2 + 12x$
31. $4x - 7 = 2 - 2x$
32. $3 - 7x = 15 - 4x$
33. $5x + 9 = -3x - 3$
34. $-8 - 16 = 3x - 16$
35. $-3x - 4 = 7x + 8$
36. $-5 + 11x = -5x + 7$
37. $5 + 7x = -9 - 7x$
38. $3x - 7 = 8x - 7$
39. $2x - 11 = 8 - 4x$
40. $15x - 7 = 8 - 5x$
41. $-3x - 9 = 2x - 9$
42. $5 - 7x = 3x - 5$
43. $5x - 4 = 0$
44. $12x - 1 = 0$
45. $3x - 9 = 0$
46. $-4x - 6 = 0$
47. $2 - 5x + 7 - 4x = 0$
48. $7 - 5x - 2x + 7 = 0$

Express each of the following word problems algebraically and then solve it. These problems can be translated by the methods in Section 13.1.

1. Ten is 30 less 4 times a certain number. Find the number.
2. If 9 more than the product of a certain number and 7 is 37, what is the number?
3. When 6 times a certain number is subtracted from 19, the result is 73. What is the number?
4. When the product of 15 and a certain number is reduced by 12, the result is 30. Find the number.
5. If 31 reduced by a certain number equals the number diminished by 13, what is the number?
6. The sum of a certain number and 55 is 11 times the number. What is the number?
7. If 95 plus a certain number equals the product of 21 and the number, what is the number?
8. If 12 less than a certain number equals 9 times the number, what is the number?
9. If the product of a certain number and 7 is 24 reduced by the number, what is the number?
10. Four times a certain number is 85 less than the product of the number and 9. Find the number.
11. When 17 times a certain number is decreased by 24, the result is the sum of the number and 40. Find the number.
12. When 14 times a certain number is decreased by 6, the result is the number multiplied by 23. What is the number?

10.3 SOLVING EQUATIONS CONTAINING PARENTHESES

7. Simplify
   a. $-3(x - 7) + 1$
   b. $8 - 2(3x + 1)$

The members of the equations in this section are sums containing parentheses, such as those we have seen in Sections 5.1 and 5.2. We can use the methods of those sections to simplify the members, and then we can solve the simplified equations by the methods in the previous section.

Look at the following example of *solving equations containing parentheses.*

TYPICAL PROBLEM

$$3 - (x + 2) = 12 - (5 - 2x)$$

First we simplify each member of the equation. We eliminate parentheses from each expression (rule 1):

$$3 - x - 2 = 12 - 5 + 2x$$

And then we combine like terms within each member of the equation (rule 1):

$$1 - x = 7 + 2x$$

At this point each member of the equation contains at most two terms.

Then we solve the equation:

$$-3x = 6$$

$$x = -2$$

Hence, $3 - (x + 2) = 12 - (5 - 2x) \quad \Rightarrow \quad x = -2.$

To eliminate parentheses from each expression we use the methods in Chapter 5. The Order of Operations is given in Section 5.3. To solve the simplified equation we use the methods in the previous section.

Every linear equation in one unknown can be simplified to the form literal = root.

EXAMPLES

Solve the following equations containing parentheses.

1. $(6x - 9) - 12 = 3 - (9x - 3)$
2. $-(4x - 12) = -(7x + 5) - 2x + 3$
3. $4 - (2 - 3x) - 6 = (8x - 6) - (7x + 3)$
4. $6(3x - 1) = 4(3x - 2)$
5. $2(7x - 6) - 3 = 5(6 - 4x) - 2x$
6. $-3(x - 3) + 5 = -4(3x - 2) + 7 + 13x$

Solutions

1. $6x - 9 - 12 = 3 - 9x + 3$
$$\downarrow$$
$$15x = 27$$
$$\downarrow$$
$$x = \frac{9}{5}$$

2. $-4x + 12 = -7x - 5 - 2x + 3$
$$\downarrow$$
$$5x = -14$$
$$\downarrow$$
$$x = \frac{-14}{5}$$

3. $4 - 2 + 3x - 6 = 8x - 6 - 7x - 3$
$$\downarrow$$
$$2x = -5$$
$$\downarrow$$
$$x = \frac{-5}{2}$$

4. $18x - 6 = 12x - 8$
$$\downarrow$$
$$6x = -2$$
$$\downarrow$$
$$x = \frac{-1}{3}$$

5. $14x - 12 - 3 = 30 - 20x - 2x$
$$\downarrow$$
$$36x = 45$$
$$\downarrow$$
$$x = \frac{5}{4}$$

6. $-3x + 9 + 5 = -12x + 8 + 7 + 13x$
$$\downarrow$$
$$-4x = 1$$
$$\downarrow$$
$$x = \frac{-1}{4}$$

OBSERVATIONS AND REFLECTIONS

Transformation rule 1 is very different from the other transformation rules. It is the only transformation rule that preserves the value of each member of the equation. For example, consider the equation $-3(2x - 5) = 7$. Because $-3(2x - 5)$ equals $-6x + 15$, we can simplify the equation to $-6x + 15 = 7$. Transformation rule 1 can be paraphrased as: Equals can be replaced by equals.

TRANSITIVE RELATION OF EQUATIONS

If A equals B, and B equals C, then A equals C.

$$A = B \quad \text{and} \quad B = C \quad \Rightarrow \quad A = C$$

(where A, B, and C represent linear expressions)

8. Solve for y
   a. $x = 7$ and $y = x$
   b. $y = 3x - 8$ and $3x - 8 = 5$

EXERCISES

Solve each of the following equations.

1. $5 - (x - 3) = -4 - (6 + 7x)$
2. $5x - (3 + 9x) = (7x - 2) - 12$

3. $(6x - 7) - 13 = 10 - (7x + 4)$
4. $-(4x - 8) = (9x + 4) - 3x + 2$
5. $9 - (2x - 5) = 1 - (7 + 6x)$
6. $4x - (6 + 3x) = (5x - 7) - 7$
7. $(3x - 8) - 11 = 5 - (6x + 3)$
8. $-(6x - 9) = -(3x + 2) - 5x + 3$
9. $6 - (x - 7) = (6x + 7) - (3x + 5)$
10. $-5 - (3x - 1) = (7x + 4) - (4x + 5)$
11. $(4 - x) - (5 - 2x) = 1 - (11x - 6)$
12. $7 - (8 - 10x) - 4 = (7x - 9) - (5x + 4)$
13. $5 - (9 - 2x) = (3x - 9) - (-2x - 7) + 11$
14. $-4 - (8x + 3) = (7x + 5) - (6 - 3x)$
15. $(3x - 7) - (4 - 3x) = (-4 - x) - (5x - 1)$
16. $(3x - 1) - (x - 9) + (4x - 7) = (-2x - 3)$
17. $(4x - 9) = (2x + 7) - (4x - 11) + (3 + x)$
18. $(4 + x) - (3 + 5x) = (9 + x) - (14 - 3x)$
19. $(3x - 7) - (2x - 5) = 0$
20. $-(3x - 9) + (2x + 11) = 0$
21. $-(3 - 9x) + (5 - 8x) = 0$
22. $(7x - 8) + (2x + 8) = 0$
23. $4(3 - x) = 0$
24. $-4(5x - 1) = 0$
25. $8(3x - 4) = 9(x - 1)$
26. $5(2x - 7) = -2(8 - 3x)$
27. $3(3x - 3) + 2 = 4(5 - 4x) + 4x$
28. $-4(x - 9) + 12 = -6(3x - 2) + 6 + 2x$
29. $6(3x + 5) - 10(7x - 6) = 6 - 17x$
30. $6(5x - 7) - 3 = 5(3 - 12x) + 15x$
31. $-4(5x - 5) + 3 = -8(2x - 4) + 2x$
32. $7(1 - x) - 8(4x - 3) = 21x + 31$
33. $-2(7 - x) = 4(5x - 2) - (9x - 12)$
34. $7(3x - 1) + x = 8(2x + 1) + (5x - 2)$
35. $-4(8 - 2x) + 8 = -5(5 + x) - (2 - x)$
36. $4x - 3(8 - x) = 8(x - 2) - 3(5x - 9) + 1$

Express each of the following word problems algebraically and then solve it. These problems can be translated by the methods in Section 13.1.

1. Six times the sum of a certain number and 8 is 21. What is the number?
2. When the sum of 27 and a certain number is multiplied by 3, the result is 42. Find the number.
3. If the sum of a certain number and 8 is multiplied by 11, the result is 12 diminished by the number. Find the number.
4. The product of 9 and a certain number is 20 times the sum of the number and 3. What is the number?
5. Forty more than the product of a certain number and 7 is 4 less than the sum of 8 and the number. What is the number?

6. Ten times the sum of a certain number and 18 equals the product of 4 and the number. Find the number.

These problems can be translated by the methods in Section 13.2.

7. The sum of two consecutive integers is 39. What are the integers?
8. The sum of two consecutive integers is 107. Find the integers.
9. The sum of two consecutive odd integers is 84. Find the integers.
10. If the sum of two consecutive even integers is 114, what are the integers?
11. The sum of three consecutive integers is −81. What are the integers?
12. The sum of three consecutive odd integers is 201. Find the integers.
13. If the sum of three consecutive even integers is 10 more than twice the largest integer, what are the integers?
14. Find the four consecutive even integers such that 8 less than the sum of the first two integers is 10 more than the third integer.
15. Find the three consecutive integers such that 3 times the third integer is 6 more than the sum of the first and second integers.

These problems can be translated by the methods in Section 13.6.

16. If 450 kilograms of grass seed worth $1.00 a kilogram are mixed with 300 kilograms of seed worth 70¢ a kilogram, how much is a kilogram of the mixture worth?
17. If 5 pounds of ground chuck at $1.08 a pound are mixed with 7 pounds of ground sirloin at $1.56 a pound, how much should a butcher charge for a pound of the mixture?
18. If 10 pounds of candy worth 70¢ a pound are mixed with 30 pounds of candy worth 90¢ a pound, how much is a pound of the mixture worth?
19. A manufacturer wishes to mix 28 kilograms of tea worth $2.40 a kilogram with 42 kilograms of tea worth $3.80 a kilogram. How much will a kilogram of the mixture be worth?
20. If 200 pounds of brass worth 80¢ a pound are mixed with 100 pounds of copper worth $1.40 a pound, how much is a pound of the alloy (mixture) worth?
21. How much should be charged per kilogram for a tea produced from 80 kilograms of China tea at $5.95 a kilogram and 120 kilograms of Ceylon tea at $6.50 a kilogram?
22. How many pounds of peanuts worth 80¢ a pound should be mixed with 10 pounds of cashews worth $1.20 a pound to produce a mixture worth 96¢ a pound?
23. How many bags of sand costing 50¢ a bag must be mixed with 100 bags of cement costing $1.50 a bag to produce concrete worth $1.30 a bag?
24. A fabric manufacturer wants to produce fine cotton-dacron material to sell at a cost of $2.80 a meter. If the cotton costs $3.00 a meter and the dacron fiber costs $2.50 a meter, how many meters of cotton must be blended with 40 meters of dacron fiber?
25. How many kilograms of Colombian coffee at $4.50 a kilogram should be mixed with 60 kilograms of Brazilian coffee at $3.30 a kilogram to make a blend worth $4.02 a kilogram?

26. How many liters of wine vinegar worth \$2.00 a liter should be mixed with 20 liters of white vinegar worth \$1.25 a liter to produce a blend worth \$1.70 a liter?

27. How much wheat flour costing 33¢ a pound should be mixed with 3 pounds of soya flour costing \$1.10 a pound to produce a flour worth 54¢ a pound?

10.4
SOLVING EQUATIONS CONTAINING FRACTIONS

9. Which terms are linear?
   a. $\frac{4x}{5y}$
   b. $\frac{4x - 5y}{9}$
   c. $\frac{3}{6x - 1}$
   d. $\frac{7x}{5}$

Remember that a fraction is linear if its denominator is a nonzero integer and its numerator is linear. Hence, for example,

$$\frac{x}{2} - \frac{1}{5} = \frac{3}{10} \quad \text{and} \quad \frac{8y}{3} - 1 = y - \frac{3y - 2}{5}$$

are linear equations.

Clearing Denominators

To solve an equation containing fractions, we first simplify it to an equation without fractions. To see how to do this, recall the following facts:

1. We can multiply both members of an equation by any nonzero number by transformation rule 4.
2. Multiplying a (polynomial) fraction by any multiple of its denominator yields a polynomial without fractions. For example,

10. Compute
   a. 8 times $\frac{3x + 2}{4}$
   b. 12 times $\frac{5x - 9}{3}$

$$3\left(\frac{2x - 5}{3}\right) = 2x - 5 \quad \text{and} \quad 8\left(\frac{9x - 2}{4}\right) = 18x - 4$$

In order to simplify an equation containing fractions to an equivalent equation without fractions, we multiply both members of the equation by a *common multiple* of all the denominators. For example, the equation $\frac{x}{2} = x + \frac{1}{3}$ can be cleared of fractions by multiplying both members by 6:

$$6\left(\frac{x}{2}\right) = 6\left(x + \frac{1}{3}\right)$$
$$\downarrow$$
$$6\left(\frac{x}{2}\right) = 6(x) + 6\left(\frac{1}{3}\right)$$
$$\downarrow$$
$$3x = 6x + 2$$

This process is called *clearing the denominators* of the equation. That is, an equation is *cleared of fractions* by multiplying every term of both members of the equation by a common multiple of its denominators and then simplifying to an equation without fractions. The most economical multiplier is the *least common multiple* (LCM) of the denominators. (Refer back to Section 7.4 for the definition of the LCM.)

10.4

Look at the following example of *solving equations containing fractions with numerical denominators.*

TYPICAL PROBLEM

$$\frac{8y}{3} - 1 = y - \frac{3y-2}{5}$$

As we do when adding fractions, we first move all minus signs to the numerators (rule 1). To move a minus sign to a multinomial numerator we change the sign of each term of the numerator:

$$\frac{8y}{3} - 1 = y + \frac{-3y+2}{5}$$

Then we clear fractions. We first determine the LCM (least common multiple) of the denominators:

LCM of 3 and 5 is 15

And then we multiply each term by the LCM on both sides of the equation:

$$15\left(\frac{8y}{3}\right) + 15(-1) = 15(y) + 15\left(\frac{-3y+2}{5}\right)$$

$$\overset{5}{\cancel{15}}\left(\frac{8y}{\cancel{3}}\right) + 15(-1) = 15(y) + \overset{3}{\cancel{15}}\left(\frac{-3y+2}{\cancel{5}}\right)$$

At this point each member of the equation is a sum of terms containing no fractions and no parentheses.

$$40y - 15 = 15y - 9y + 6$$

Having cleared fractions, we can now solve the equation:

$$34y = 21$$

$$y = \frac{21}{34}$$

Once we clear fractions we solve the resulting equation by isolation and division, using the method in Section 10.2.

Hence, $\frac{8y}{3} - 1 = y - \frac{3y-2}{5} \Rightarrow y = \frac{21}{34}$.

EXAMPLES

Solve the following equations containing fractions.

Every linear equation in one unknown can be simplified to the form literal = root.

1. $\frac{3x}{8} = \frac{4}{3}$

2. $\frac{3x-4}{5} - \frac{3-5x}{8} = 0$

3. $2x - \frac{1}{4} = \frac{3}{2}$

4. $\frac{8-12x}{9} + \frac{4x}{7} = -\frac{20x-8}{21}$

5. $\frac{x}{3} - \frac{1}{5} = \frac{5x}{4}$

6. $\frac{6x-3}{4} - \frac{9x-2}{5} = 1 - \frac{9x}{2}$

Solutions

1. LCM = 24

$$\downarrow$$

$$\overset{3}{\cancel{24}}\left(\frac{3x}{\cancel{8}}\right) = \overset{8}{\cancel{24}}\left(\frac{4}{\cancel{3}}\right)$$

$$\downarrow$$

$$9x = 32$$

$$\downarrow$$

$$x = \frac{32}{9}$$

2. LCM = 40

$$\downarrow$$

$$\overset{8}{\cancel{40}}\left(\frac{3x-4}{\cancel{5}}\right) + \overset{5}{\cancel{40}}\left(\frac{-3+5x}{\cancel{8}}\right) = 40(0)$$

$$\downarrow$$

$$24x - 32 - 15 + 25x = 0$$

$$\downarrow$$

$$49x = 47$$

$$\downarrow$$

$$x = \frac{47}{49}$$

3. LCM = 4

$$\downarrow$$

$$4(2x) + \overset{1}{\cancel{4}}\left(\frac{-1}{\cancel{4}}\right) = \overset{2}{\cancel{4}}\left(\frac{3}{\cancel{2}}\right)$$

$$\downarrow$$

$$8x - 1 = 6$$

$$\downarrow$$

$$x = \frac{7}{8}$$

4. LCM = 63

$$\downarrow$$

$$\overset{7}{\cancel{63}}\left(\frac{8-12x}{\cancel{9}}\right) + \overset{9}{\cancel{63}}\left(\frac{4x}{\cancel{7}}\right) = \overset{3}{\cancel{63}}\left(\frac{-20x+8}{\cancel{21}}\right)$$

$$\downarrow$$

$$56 - 84x + 36x = -60x + 24$$

$$\downarrow$$

$$12x = -32$$

$$\downarrow$$

$$x = \frac{-8}{3}$$

5. LCM = 60

$$\downarrow$$

$$\overset{20}{\cancel{60}}\left(\frac{x}{\cancel{3}}\right) + \overset{12}{\cancel{60}}\left(\frac{-1}{\cancel{5}}\right) = \overset{15}{\cancel{60}}\left(\frac{5x}{\cancel{4}}\right)$$

$$\downarrow$$

$$20x - 12 = 75x$$

$$\downarrow$$

$$x = \frac{-12}{55}$$

6. LCM = 20

$$\downarrow$$

$$\overset{5}{\cancel{20}}\left(\frac{6x-3}{\cancel{4}}\right) + \overset{4}{\cancel{20}}\left(\frac{-9x+2}{\cancel{5}}\right) = 20(1) + \overset{10}{\cancel{20}}\left(\frac{-9x}{\cancel{2}}\right)$$

$$\downarrow$$

$$30x - 15 - 36x + 8 = 20 - 90x$$

$$\downarrow$$

$$84x = 27$$

$$\downarrow$$

$$x = \frac{9}{28}$$

OBSERVATIONS AND REFLECTIONS

11. Multiply by 6
   a. $2x + 3 = 6x$
   b. $\frac{x}{2} + \frac{1}{3} = \frac{x}{6}$

1. Using transformation rule 4, we can multiply both members of an equation by any nonzero number. When a member of the equation contains more than one term, we must use distribution. For example,

$$3 = 2x - 5 \quad \Rightarrow \quad 6(3) = 6(2x - 5) \quad \Rightarrow \quad 6(3) = 6(2x) + 6(-5)$$

★ 2. Clearing fractions in an equation and converting a sum of fractions to common denominators are very closely related. Consider the problems:

Solve $\frac{x}{2} = \frac{2}{3}$

$\downarrow$

$6\left(\frac{x}{2}\right) = 6\left(\frac{2}{3}\right)$

$\downarrow$

$3x = 4$

Add $\frac{x}{2} + \frac{2}{3}$

$\downarrow$

$\frac{6\left(\frac{x}{2}\right)}{6} + \frac{6\left(\frac{2}{3}\right)}{6}$

$\downarrow$

$\frac{3x}{6} + \frac{4}{6}$

12. Add

a. $\frac{x}{5} + \frac{3}{4}$

b. $\frac{5x}{6} - \frac{3}{2}$

13. Solve

a. $\frac{x}{5} = \frac{3}{4}$

b. $\frac{5x}{6} = -\frac{3}{2}$

The equation $\frac{x}{2} = \frac{2}{3}$ is equivalent to $3x = 4$, whereas the sum $\frac{x}{2} + \frac{2}{3}$ equals $\frac{3x}{6} + \frac{4}{6}$. In both problems, we multiply $\frac{x}{2}$ and $\frac{2}{3}$ by 6. When *solving*, we multiply to clear denominators; when *adding*, we multiply to get the *numerators* of the equivalent fractions with like denominators.

Although these computations are similar, our reasons for performing them are very different. When we clear fractions in an equation, we must find an equation with the same root as the original equation (that is, an equivalent equation). Hence, we must multiply both members of the equation by the same nonzero number. On the other hand, when we add fractions, we must form equivalent fractions with new common denominators. Hence, we must multiply and divide each fraction by the same nonzero number to preserve its value.

EXERCISES

Solve each of the following equations.

1. $\frac{4x}{3} = 5$
2. $\frac{7x}{4} = \frac{5}{2}$
3. $\frac{6x}{5} = -3$
4. $-\frac{2x}{9} = \frac{4}{3}$
5. $\frac{7x}{2} = \frac{-5}{8}$
6. $-\frac{11x}{3} = -\frac{1}{2}$
7. $\frac{8x}{3} - \frac{1}{2} = \frac{7}{6}$
8. $\frac{4x}{9} - \frac{8}{3} = \frac{x}{6}$
9. $\frac{7}{3} - \frac{5x}{6} = \frac{x}{2}$
10. $\frac{8x}{5} - \frac{x}{2} = \frac{21}{10}$
11. $\frac{8x}{3} - \frac{4}{5} = 2$
12. $\frac{6x}{5} + \frac{1}{5} = \frac{x}{2}$
13. $\frac{2x}{3} + \frac{5}{4} = \frac{7}{6} - x$
14. $\frac{7x}{8} - \frac{3}{4} = x - \frac{7}{8}$
15. $\frac{5}{8} - \frac{2}{3} = \frac{7}{6} + \frac{x}{12}$
16. $\frac{4}{5} - \frac{7x}{6} = \frac{2x}{15} + \frac{11}{10}$
17. $\frac{4x}{5} = \frac{7}{3} - \frac{8x}{15} + 1$
18. $2 - \frac{5x}{2} = 2x + \frac{7}{2}$
19. $\frac{2x}{3} - \frac{1}{4} = \frac{3x}{5}$
20. $\frac{3}{8} + \frac{5x}{6} = \frac{2}{7}$
21. $\frac{3x}{4} = \frac{-3x}{10} - \frac{7}{6}$
22. $\frac{9}{14} + \frac{x}{4} = \frac{5x}{6}$
23. $\frac{9x}{14} + \frac{2}{5} = \frac{5x}{6}$
24. $\frac{3}{10} - \frac{4x}{15} = \frac{7}{25}$
25. $\frac{5x}{7} = \frac{8 - 9x}{2}$
26. $\frac{7x + 4}{4} = \frac{3x + 1}{2}$
27. $\frac{2x + 9}{6} = -\frac{4x - 7}{4}$

28. $\dfrac{8x+4}{9} = \dfrac{3x+7}{3}$

29. $\dfrac{3x-5}{8} = \dfrac{4}{3} - \dfrac{x}{12}$

30. $\dfrac{5x-1}{4} = 3 + \dfrac{x}{6}$

31. $\dfrac{2x-13}{7} - x = \dfrac{5x}{2} + 2$

32. $\dfrac{12x-8}{18} - \dfrac{4}{3} = \dfrac{x}{6} - 1$

33. $\dfrac{x-7}{4} + \dfrac{9-2x}{3} = \dfrac{7x}{6}$

34. $\dfrac{5-2x}{7} + \dfrac{x}{3} = -\dfrac{14x-9}{21}$

35. $\dfrac{5x-9}{3} + \dfrac{5x-2}{5} = -1 - \dfrac{2x}{15}$

36. $\dfrac{x-4}{8} + \dfrac{3x-14}{12} = \dfrac{25x}{24}$

Express each of the following word problems algebraically and then solve it. These problems can be translated by the methods in Section 13.1.

1. If 8 less a certain number is divided by 13, the result is 2. What is the number?
2. Three more than a certain number over 7 equals $\frac{2}{3}$. Find the number.
3. Seven multiplied by $\frac{4}{5}$ a certain number is $\frac{10}{3}$. What is the number?
4. One-ninth of the sum of a certain number and 16 is 22. What is the number?
5. If $\frac{3}{4}$ a certain number equals the number decreased by 17, what is the number?
6. The sum of 11 and a certain number is $\frac{4}{7}$ the number. Find the number.
7. When a certain number is divided by 6, the result is 14 diminished by the number. What is the number?
8. A certain number increased by 21 equals $\frac{8}{5}$ of the number. What is the number?
9. The product of $\frac{7}{12}$ and a certain number is $\frac{9}{7}$ reduced by $\frac{1}{14}$ the number. Find the number.
10. Three-tenths of a certain number is $\frac{2}{15}$ less $\frac{5}{6}$ times the number. What is the number?
11. One-third diminished by a certain number is $\frac{8}{33}$ increased by the number over 11. What is the number?
12. A certain number divided by 15 equals $\frac{2}{5}$ times the sum of the number and 20. Find the number.

These problems can be translated by the methods in Section 13.2.

13. One-third the sum of two consecutive integers is 11. Find the integers.
14. If the sum of three consecutive integers is divided by 9, the result is 9. What are the integers?

15. If $\frac{1}{4}$ the second consecutive integer is added to the third integer, the result is $\frac{47}{2}$. Find the three consecutive integers.

16. One-third times the sum of the second and third integers is $\frac{7}{12}$ the first integer. What are the three consecutive integers?

17. Find the three consecutive integers such that if the second integer divided by 5 is added to the third integer divided by 3, the result is $\frac{1}{2}$ the first integer.

18. If $\frac{1}{3}$ the first of four consecutive integers is subtracted from the third integer, the result is 18 more than the fourth integer. Find the integers.

10.5 SOLVING LINEAR INEQUALITIES

A *linear inequality* is an algebraic sentence containing two linear expressions separated by the inequality sign $<$, which means *less than*, or $>$, which means *more than*. The direction of the inequality sign is referred to as the *sense* of the inequality. Replacing one inequality sign by the other is called *reversing the sense* of the inequality. For example, $x < 3$ is read x is less than 3, whereas $x > 3$ is read x is greater than 3, and these two inequalities have opposite senses.

14. Reverse the sense of
a. $x > 5$
b. $x < -8$
c. $x > \frac{-3}{4}$

The Corresponding Equation

We can also change an inequality sign to an equals sign, of course, and form what is called the *corresponding equation.* The inequality is said to be a *linear inequality in one unknown* if its corresponding equation is a linear equation in one unknown.

The language of linear equations extends naturally to linear inequalities. The two expressions separated by an inequality sign ($<$ or $>$) are called *members of the inequality.* Further, a number is said to *satisfy the inequality* if evaluating the inequality for this number yields a true statement. For example, -1 and 2 satisfy $x < 3$, whereas 5 does not.

15. Form the equation corresponding to
a. $x > 5$
b. $x < -8$
c. $x > \frac{-3}{4}$

16. Which of the numbers $-8, -4, 0, +4, +8$ satisfy
a. $x > 3$ b. $x < 5$
c. $x > -1$ d. $x < -5$

Conditional Inequalities

An inequality is called a *conditional inequality* if the corresponding equation is a conditional equation. Just as with linear equations, two linear inequalities in one unknown are *equivalent* if they contain the same unknown and are satisfied by the same set of numbers. For example, the inequalities $x + 4 > 6$ and $x > 2$ are *equivalent.*

Further, just as every linear equation in x is equivalent to an equation of the form x = number, every linear inequality in x is equivalent to an inequality of the form $x >$ number or $x <$ number. For example, just as $x + 4 = 6$ is equivalent to $x = 2$, so $x + 4 > 6$ is equivalent to $x > 2$ and $x + 4 < 6$ is equivalent to $x < 2$.

The solution of a linear inequality in x and the solution of its corresponding equation are closely related. If the solution of the corresponding equation is x = number, then every *larger* number or else every *smaller* number satisfies the inequality. For example, if the solution of the corresponding equation is $x = 2$, then the solution of the inequality is either $x > 2$ or $x < 2$.

17. Indicate on the number line
a. $x < 1$ b. $x = 1$ c. $x > 1$

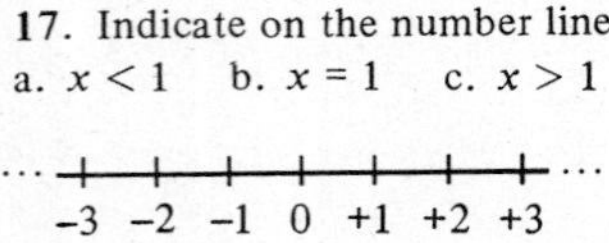

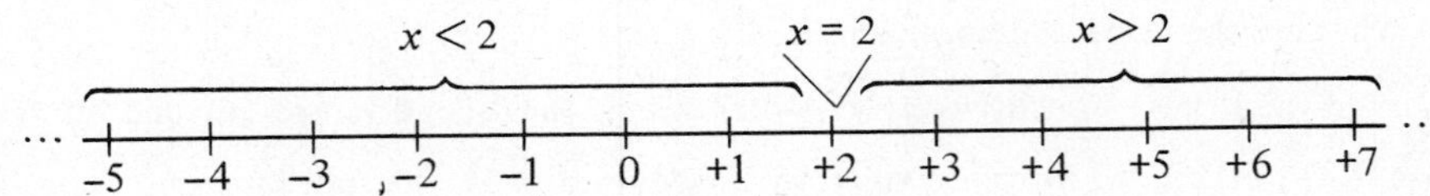

Transformation Rules

We can solve a conditional linear inequality in one unknown algebraically by using the following rules in some sequence. Given such an inequality, we can apply any of these rules to get an equivalent inequality.

RULE 1′

Replace either member by another expression with the same value.

$3(x - 4) < 7$ is equivalent to $3x - 12 < 7$

RULE 2′

Reverse the order of the members *and reverse the sense of the inequality.*

$3 > x$ is equivalent to $x < 3$

RULE 3′

Add any linear expression, not containing a new literal, to both members of the inequality. This step includes shifting any term from one side of the inequality to the other.

$3x > 4$ is equivalent to $3x - 8 > 4 - 8$

and

$3x - 7 < 4x + 4$ is equivalent to $3x - 4x < 4 + 7$

RULE 4′

Multiply or divide both members by any *strictly positive* number (which may be a fraction). **Do not multiply or divide by any literal expression.**

$\frac{x}{2} < \frac{1}{3}$ is equivalent to $6\left(\frac{x}{2}\right) < 6\left(\frac{1}{3}\right)$

RULE 5′

Negate both members of the inequality *and reverse the sense of the inequality.*

$-x > 7$ is equivalent to $x < -7$

and

$-x < -4$ is equivalent to $x > 4$

Compare these transformation rules with the rules for solving a linear equation given in the chapter introduction. Rules 1′ and 3′ are the same as rules 1 and 3, whereas rules 2′, 4′, and 5′ differ in small but significant details from rules 2, 4, and 5.

Keeping these differences in mind, we can solve inequalities by essentially the same methods we used to solve equations. Remember that once we isolate the unknown, if its coefficient is negative, we must reverse the sense of the inequality when we negate the members. Hence, the sense of the original inequality and its solution need not be the same.

Look at the following example of *solving linear inequalities in one unknown.*

TYPICAL PROBLEM

$3x - 7 < 11 + 7x$

We first isolate the unknown by shifting terms. To do this we add the negatives of the terms to be shifted to both sides (rule 3′):

$3x - 7 + 7 - 7x < 11 + 7x + 7 - 7x$

And then we add negatives to zero (rule 1′):

$3x - 7x < 11 + 7$

Then we simplify each member of the inequality (rule 1′).

$-4x < 18$

At this point each member of the inequality is a single term. The unknown is on the left and the constant term is on the right.

If the coefficient of the unknown is negative, we must negate both members and reverse the sense of the inequality (rule 5′):

$4x > -18$

Then we divide both sides by the coefficient of the unknown (rule 4′):

$$\frac{4x}{4} > \frac{-18}{4}$$

$$x > \frac{-9}{2}$$

Hence, $3x - 7 < 11 + 7x \quad \Rightarrow \quad x > \frac{-9}{2}$.

EXAMPLES

Solve the following linear inequalities in one unknown.

Every linear inequality in one unknown can be simplified to the form literal < number or literal > number.

1. $6x - 8 < 4x$
2. $7x > -4 + 3x$
3. $8 - 9x < -3 - 2x$
4. $-7x + 11 > 7x + 11$

Solutions

1. $6x - 4x < 8$
   ↓
   $2x < 8$
   ↓
   $x < 4$

2. $7x - 3x > -4$
   ↓
   $4x > -4$
   ↓
   $x > -1$

3. $-9x + 2x < -3 - 8$
$\downarrow$
$-7x < -11$
$\downarrow$
$7x > 11$
$\downarrow$
$x > \frac{11}{7}$

4. $-7x - 7x > 11 - 11$
$\downarrow$
$-14x > 0$
$\downarrow$
$14x < 0$
$\downarrow$
$x < 0$

Checking

When we solve an inequality, we can make computational errors and arrive at the wrong answer. As with equations, we can check our solution, but with inequalities we do it in two steps. Our first step is to determine whether the number is correct. Then we check the sense of the inequality. For example, to check the solution $x < 4$ of the inequality $6x - 8 < 4x$, we proceed as follows:

We form the corresponding equations to the original inequality and the presumed solution. Then we check whether these two equations are equivalent by evaluating. The two equations must be equivalent or we have the wrong number in our presumed solution:

$$6x - 8 = 4x, \text{ when } x = 4$$
$$6(4) - 8 = 4(4)$$
$$\downarrow$$
$$24 - 8 = 16$$

This is an identity. Hence, the solution must be either $x > 4$ or $x < 4$.

We next evaluate both the original inequality and the presumed solution for another value of x. (For example, we can use 0 or 1.) The two inequalities must reduce to two true statements or two falsehoods—it doesn't matter which. Otherwise the sense of the inequality in the presumed solution is wrong and must be reversed:

$$6x - 8 < 4x \text{ and } x < 4, \text{ when } x = 0$$
$$6(0) - 8 < 4(0) \quad (0) < 4$$
$$\downarrow \qquad\qquad \downarrow$$
$$-8 < 0 \qquad 0 < 4$$

These are both true statements. Hence, the solution is indeed $x < 4$. (The two statements must both be true or both false.)

18. Check these solutions
a. $7x > -4 + 3x \Rightarrow x > -1$
b. $-7x + 11 > 7x + 11 \Rightarrow x > 0$

OBSERVATIONS AND REFLECTIONS

★ 1. An inequality such as $x < 8$ can be thought of as an infinite set of equations of the form $x = a$, where a represents any number less than 8. Every number satisfying $x < 8$ is a root of exactly one of these equations.

★ 2. Remember that a conditional linear equation in one unknown has exactly one root Every other number is a root of one or the other of the two inequalities formed by replacing the = sign by either < or >. This follows from the *trichotomy property* of signed numbers.

> **TRICHOTOMY PROPERTY OF SIGNED NUMBERS**
>
> a is less than b, *or* a equals b, *or* a is greater than b
>
> $a < b$ or $a = b$ or $a > b$
>
> (where a and b represent signed numbers)

On the number line, every other number is either to the left or to the right of any given number b.

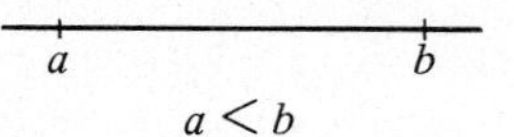

$a < b$

or

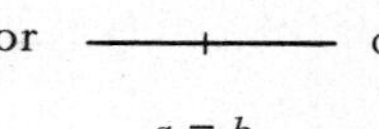
$a = b$

or

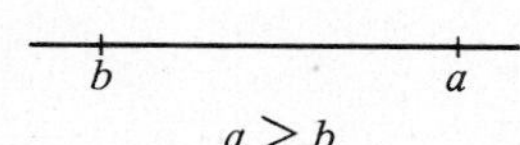

$a > b$

EXERCISES

Solve each of the following inequalities.

1. $x - 5 < 9$
2. $x + 7 > 4$
3. $5 < 4 - x$
4. $-2 > 8 - x$
5. $8x < -1$
6. $6x > -14$
7. $-12x > 20$
8. $-15x < -35$
9. $2x - 7 < 5x$
10. $3x > 3 - 5x$
11. $3 - 5x > -7$
12. $6x - 4 < -4$
13. $15 - 8x < -2x$
14. $3x > 12 - 6x$
15. $6 - 11x > -5x$
16. $5x - 14 < 12x$
17. $5x - 8 < 3x$
18. $4x > -13 - 9x$
19. $2x - 5 > 7x + 5$
20. $9x - 4 < -3x + 2$
21. $7x - 9 < -2x - 9$
22. $-4 + 5x > 3 - 2x$
23. $-5x + 2 > -2x + 14$
24. $-3x + 8 < 8 - 5x$

ANSWERS TO MARGIN EXERCISES

2. a. 5
b. $-2x$
c. $2x + 6$

3. A. a. falsehood
b. identity
B. a. falsehood
b. identity
C. a. identity
b. falsehood
D. a. falsehood
b. identity

4. a. $4x = 3$
b. $8x = 6$
c. $12x = 9$

6. a. $x = -9$
b. $x = 16$

7. a. $-3x + 22$
b. $6 - 6x$

8. a. $y = 7$
b. $y = 5$

9. b and d are linear.

10. a. $6x + 4$
b. $20x - 36$

11. a. $12x + 18 = 36x$
b. $3x + 2 = x$

12. a. $\frac{4x + 15}{20}$
b. $\frac{5x - 9}{6}$

13. a. $x = \frac{15}{4}$
b. $x = \frac{-9}{5}$

14. a. $x < 5$
b. $x > -8$
c. $x < \frac{-3}{4}$

15. a. $x = 5$
b. $x = -8$
c. $x = \frac{-3}{4}$

16. a. +4 and +8
b. −8, −4, 0, and +4
c. −8

17. $x < 1$, $x = 1$, $x > 1$

… −3 −2 −1 0 +1 +2 +3 …

18. Compare with examples 2 and 4.

CHAPTER REVIEW

SECTION 10.1

Solve the equations.

1. $5x = 4x - 9$
2. $-4x = -5 - 5x$
3. $14 - x = 7 - 2x$
4. $-x - 1 = -2x - 1$
5. $-4 = 7 - x$
6. $-2 = -x$
7. $12 + x - 14 = -16$
8. $21 - 3x = 31 - 4x$
9. $-8x - 12 = -9x + 20$
10. $15 - 24 = 6 - x$
11. $5 + x + 3 = 9$
12. $14x - 32 = 16 + 13x$

SECTION 10.2

Solve the equations.

13. $-x = -6$
14. $5 - x = -3$
15. $-8 = -3 + x$
16. $-4 = 4 + x$
17. $12 - 2x = 7 - x$
18. $12 - 9x = 9 - 8x$
19. $8x = 32$
20. $12x = -4$
21. $-25 = 10x$
22. $6x = 9x - 1$
23. $6 - 8x = 4$
24. $7 - 9x = 4$
25. $5 - 4x = 10 - 6x$
26. $5 - 3x = 9x - 3$
27. $8 + 6x = 15 - 4x$
28. $5 - 7x = 3x - 4$
29. $5 - 2x - 9 = 7x$
30. $6x + 5 = 8x - 5$
31. $-7 = 3x - 9 + 7x$
32. $-8x - 3 = -4x + 3$
33. $-8 - 7x = 5x - 8$
34. $-3x - 12 = 0$
35. $12x + 15 = 0$
36. $6 - 4x + 9 - x = 0$

SECTION 10.3

Solve the equations.

37. $6 - (x - 8) = -10 - (4 + 5x)$
38. $4x - (4 + 2x) = (8x - 5) - 8$
39. $(5x - 8) - 14 = 2 - (7x + 9)$
40. $5 - (2x + 6) = -8 - (3 + 4x)$
41. $-5 - (x - 8) = (4x + 5) - (6x - 8)$
42. $(8 - 2x) - (6 - 8x) = -13 - (4x - 3)$
43. $(5x - 7) - (3 - 8x) = (-2x - 2) - (6x - 19)$
44. $(4x - 1) - (9x - 4) + (8x + 6) = -(6x + 3)$
45. $-(2x + 10) = (12 + 4x) + (2x - 6) - (3x + 9)$
46. $(7x - 9) - (11x + 3) = 0$
47. $(1 - 11x) - 14x - 1 = 0$
48. $11(2x - 7) = 0$
49. $6(x - 7) = -4(2 - 3x)$
50. $4(x + 5) = -6(4 - 3x)$
51. $4(6x - 5) - 11 = 3(8 - 2x) - 3x$
52. $8(2x + 5) - 12(3x - 2) = 16 - 12x$
53. $-3(8 - 3x) + 16 = 4(5 + 4x) - 12(x - 1)$
54. $3(6x - 10) + 8 = 5(6x + 2) - (5 - 4x) - x$

SECTION 10.4

Solve the equations.

55. $\frac{5x}{6} = \frac{3}{4}$

56. $\frac{7x}{4} = \frac{-13}{8}$

57. $\frac{3x}{8} = \frac{-11}{16}$

58. $3 - \frac{x}{4} = \frac{2}{3}$

59. $3x - \frac{5}{6} = 5 - \frac{x}{3}$

60. $2x - \frac{5}{12} = 10 - \frac{4x}{3}$

61. $4x - \frac{1}{3} = \frac{2}{5}$

62. $\frac{9x}{4} - \frac{5}{3} = 6x$

63. $\frac{5}{8} + \frac{x}{5} = \frac{9x}{20} + 1$

64. $\frac{3x}{2} - \frac{4}{5} = \frac{-2}{3}$

65. $\frac{x}{6} = \frac{-3x}{10} - \frac{7}{22}$

66. $\frac{7x}{12} = \frac{11x}{30} + \frac{13}{18}$

67. $\frac{x-5}{2} = \frac{9-4x}{3}$

68. $\frac{x-4}{5} - \frac{3}{2} = \frac{x}{10}$

69. $\frac{3x-7}{4} = 2 + \frac{x}{3}$

70. $4 - \frac{5x-2}{6} = \frac{x}{3} - \frac{3}{2}$

71. $\frac{9x-10}{2} + \frac{5x}{6} = \frac{5}{12} + \frac{x}{3}$

72. $\frac{-2x-3}{5} = 2x - \frac{7x}{4} + 2$

SECTION 10.5

Solve the inequalities.

73. $x + 3 < 8$

74. $10 < -5 - x$

75. $9x < 12$

76. $-18x > 42$

77. $4 - 7x < 4x$

78. $9 + 4x > 7x$

79. $10 + 7x < -8x$

80. $-8x - 3 > -3$

81. $12x - 7 < -11$

82. $-7x + 4 > 3x - 2$

83. $9 - 7x < 11x - 9$

84. $-8x + 3 > 12x + 3$

CUMULATIVE TEST

The following problems test your understanding of this chapter and of related subject matter in Chapter 9. Before taking this test, thoroughly review Sections 9.4, 10.1, 10.2, 10.3, 10.4, and 10.5.

Once you have finished the test, compare your answers with the answers provided at the back of the book. Note the section number of each problem missed, and thoroughly review those sections again.

Solve or simplify.

1. $8 - x = 9 - 2x$
2. $8 - 10x = 3$
3. $8 - (x - 7) = 6 - (4 + 9x)$
4. $\frac{5x}{6} = \frac{-11}{2}$
5. $x - 8 > -6$
6. $-4(9 + 3x) + 10 = 6(7 + 4x) - 10(x - 2)$
7. $12x - 1 = -6x + 1$
8. $\frac{3x - 4}{6} = \frac{6 - x}{5}$
9. $\frac{-3}{5} + \frac{7x}{4}$
10. $-4x + 5 > 13x + 5$
11. $12(x + 3) = -8(5 - 2x)$
12. $\frac{4x - 5}{21} - \frac{5x - 1}{14} = -2 - \frac{11x}{6}$
13. $\frac{8x}{3} - \frac{16x}{7} + \frac{1 - 3x}{2}$
14. $-8x - 9 = -9x + 18$
15. $18 + 7x < -5x$
16. $6x = 30$
17. $4 - 7x = 8x - 11$
18. $-13 - (x - 6) = (7x + 3) - (11x + 1)$
19. $5x - \frac{4}{3} = \frac{3}{4}$
20. $8(3x + 2) - 11(4x - 1) = 9 - 14x$
21. $8 - 5x < 11x - 8$
22. $14 - 3x = 21 - 4x$
23. $\frac{12 - 5x}{18} + \frac{4x}{9} = -\frac{11x + 7}{12}$
24. $\frac{7x}{4} + \frac{11}{6} + \frac{3x - 1}{5}$
25. $8x = 11x - 1$
26. $5x - (2 + 3x) = (8x - 7) - 6$
27. $7x + 4 = 9x - 4$
28. $-11x + 8 > 7x - 6$
29. $6(9x - 2) - 2 = 5(8 - 3x) - 3x$
30. $16 + x - 18 = -15$
31. $\frac{3}{16} - \frac{13x}{40}$
32. $\frac{x - 9}{3} - \frac{5}{2} = \frac{11x}{6}$
33. $27 - 7x < 7x + 6$
34. $-8x = 36$

35. $(8x - 1) - (8x - 5) + (7x + 5) = -(6x + 5)$

36. $\frac{5x}{2} + \frac{1}{6} = -\frac{2}{3}$

37. $-4x > 28$

38. $-3 = 8x - 8 + 7x$

39. $5x - \frac{1}{8} = 6 - \frac{x}{4}$

40. $5 - 9x < x$

Literal Equations

An equation containing two or more unknowns is called a *literal equation.* For example,

$$3x + 5 = y, \qquad 7x - 9y + z = y - 4x, \qquad AB^2 = C, \qquad Ax - x^3 = 6y^2$$

are literal equations. A literal equation is called *linear* if its members are linear expressions. For example, $3x + 5 = y$ and $7x - 9y + z = y - 4z$ are linear. On the other hand, $AB^2 = C$ and $Ax - x^3 = 6y^2$ are nonlinear (and are frequently called *formula-type equations),* and we will discuss them in Section 19.3.

Satisfying a Literal Equation

Replacing each unknown in a literal equation by a number may yield either a numerical identity or a falsehood. For example, replacing x by 2 and y by 1 in $x = y + 1$ yields the identity $2 = 1 + 1$, whereas replacing x by 1 and y by 2 in $x = y + 1$ yields the falsehood $1 = 2 + 1$. If we get an identity, then the numbers that replaced the unknowns are said to *satisfy the equation* and are called collectively a *root* of the equation. (Keep track of which number replaces which unknown.) For example, the value 2 for x and 1 for y is a root of $x = y + 1$. In general, a literal equation has an unlimited number of roots.

Equivalent Equations

Two literal equations are *equivalent* if they contain the same unknowns and have the same roots. Any two equivalent *linear equations* can be transformed one to the other by the transformation rules introduced in Chapter 10.

Solving a Literal Equation

As we saw in Chapter 10, a linear equation in *one* unknown is solved by finding an equivalent equation of the form literal = root. We will say that a linear equation in two or more unknowns is *solved* by finding an equivalent equation of the form literal = algebraic expression, where the literal on the left does not appear in the expression on the right. For example, we solve $x - y = 1$ by finding either $x = y + 1$ or $y = x - 1$. Because a literal equation contains more than one unknown, the unknown to be solved for must be specified. For example, $x = y + 1$ is solved for x, whereas $y = x - 1$ is solved for y.

11.1 SOLVING LINEAR EQUATIONS BY SHIFTING TERMS

To solve a linear equation in several unknowns for x, we first isolate the terms containing x in the left member of the equation by shifting terms. For example, to solve

$$8x - 5y + z = 7x + 9z$$

we shift $7x$ to the left and $-5y + z$ to the right:

$$8x - 5y + z = 7x + 9z$$
$$\downarrow$$
$$8x - 5y + z - 7x + 5y - z = 7x + 9z - 7x + 5y - z$$
$$\downarrow$$
$$8x - 7x = 9z + 5y - z$$

Then we can simplify the equation further by combining like terms on each side. For example,

$$8x - 7x = 9z + 5y - z$$
$$\downarrow$$
$$x = 8z + 5y$$

1. Solve by shifting terms
   a. $8x - 5 = 7x + 9$
   b. $-3x - 4 = 9 - 4x$

If the left member should simplify to x (with coefficient +1), we have an equation of the form x = polynomial and the solution to the original equation. In this section, we will only discuss equations in which the left member simplifies immediately to x once we have shifted terms to isolate x.

In most equations the left member will simplify to a multiple of x. We will discuss such equations in the next section.

Look at the following example of *solving linear equations by shifting terms.*

TYPICAL PROBLEM Solve for x:

$$-4z + 9x = -10y + 8x$$

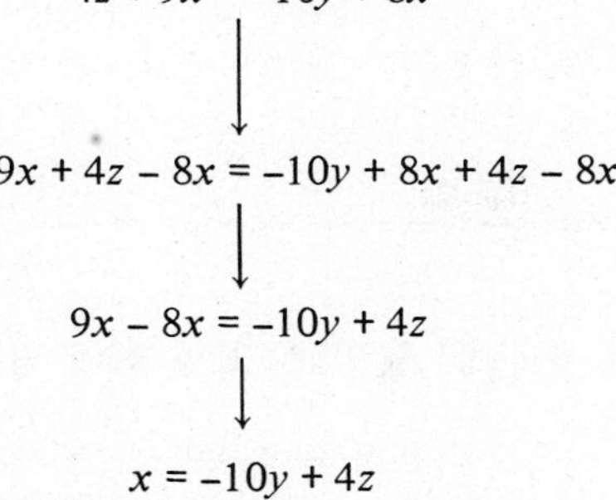

We first isolate x by shifting terms. To do this we add the negatives of the terms to be shifted to both sides (rule 3):

$$-4z + 9x + 4z - 8x = -10y + 8x + 4z - 8x$$

And then we add negatives to zero (rule 1):

$$9x - 8x = -10y + 4z$$

We then simplify each member of the equation (rule 1).

$$x = -10y + 4z$$

At this point the equation is solved if the coefficient of x *is +1.*

Hence, $-4z + 9x = -10y + 8x \Rightarrow x = -10y + 4z$.

To solve linear equations by shifting terms we parallel the method in Section 10.1 for solving equations in one unknown.

EXAMPLES

Solve the following equations for x by shifting terms.

1. $3y + x + 1 = 4y$
2. $3y - 7z = 2 - x$
3. $8y + x - 2 = -12y + 1$
4. $5x - 4y = 6y + 4x$
5. $-6x - 9y + 12 = 6 - 7x + 3y$
6. $4z + 7y = 7y - x - z$

Every linear equation in several unknowns can be solved for any of the unknowns. the equations in this section can be simplified to the form x = *polynomial.*

Solutions

1. $x = 4y - 3y - 1$
   $\downarrow$
   $x = y - 1$
2. $x = 2 - 3y + 7z$
3. $x = -12y + 1 - 8y + 2$
   $\downarrow$
   $x = -20y + 3$
4. $5x - 4x = 6y + 4y$
   $\downarrow$
   $x = 10y$
5. $-6x + 7x = 6 + 3y + 9y - 12$
   $\downarrow$
   $x = 12y - 6$
6. $x = 7y - z - 4z - 7y$
   $\downarrow$
   $x = -5z$

EXERCISES

Solve each of the following equations for x. In these exercises, only shifting terms and addition are needed.

1. $4y + x = 1$
2. $5 + x = 4y$
3. $x + y = 6y$
4. $x - 9y = 4y$
5. $6y + x = -11z$
6. $4z + x = 3y$
7. $5y - 2x = y - 3x$
8. $5x + 2y = 7y + 4x$
9. $x - 2y = 7y - 3z$

10. $5y - 7 + 2x = x$ 11. $5y - 7z = y - x$ 12. $3y - 9x = 8y - 10x$

13. $2y - 5z - 1 = -x$ 14. $3y - 9z + 4x = y + 3x$ 15. $7x - 9y + 4 = 6x$

16. $3y + x = -9z - 1$ 17. $z = 4y - 9 - x$ 18. $y = -3z + 7 - x$

11.2 SOLVING LINEAR EQUATIONS BY ISOLATION AND DIVISION

2. Solve by isolation and division.
   a. $7x - 8 = 3x - 5$
   b. $6 - 5x = 7 - 2x$

In this section we will discuss the general class of linear equations in several unknowns in which each term is an integer or an integer times an unknown. These equations parallel the equations in Section 10.2, and we will use the same method for solving them. In these equations, isolating x will lead to an equation of the form integer times x = polynomial, such as $2x = y - 3$. As in Section 10.2, we must then divide both members by the coefficient of x, using transformation rule 4, to get an equation of the form $x = \dfrac{\text{polynomial}}{\text{integer}}$.

Look at the following example of *solving linear equations by isolation and division.*

TYPICAL PROBLEM Solve for x: $5x - 6z = -2y + 8x$

We first isolate x by shifting terms. To do this we add the negatives of the terms to be shifted to both sides (rule 3): $5x - 6z + 6z - 8x = 2y + 8x + 6z - 8x$

And then we add negatives to zero (rule 1): $5x - 8x = -2y + 6z$

We then simplify each member of the equation (rule 1): $-3x = -2y + 6z$

At this point the left member is a single term containing x, *and the right member does not contain* x.

Then if the x term is negative, we negate both sides of the equation (rule 5): $3x = 2y - 6z$

We then place parentheses around the right side of the equation and divide each side by the coefficient of x (rule 4): $\dfrac{3x}{3} = \dfrac{(2y - 6z)}{3}$

To solve linear equations by isolation and division we parallel the method in Section 10.2 for solving equations in one unknown.

And we reduce the left side (rule 1): (We do not reduce the fraction containing a binomial on the right.) $x = \dfrac{2y - 6z}{3}$

Hence, $5x - 6z = -2y + 8x \Rightarrow x = \dfrac{2y - 6z}{3}$.

WARNING Do not try to reduce the fraction in the right member of the equation in the last step unless its numerator is a monomial. Reducing fractions containing polynomials is treated in Chapter 17.

EXAMPLES

Solve the following equations for x by isolation and division.

1. $-8x = 4y - 11z$
2. $3y - 2x = 7y - 11$
3. $7y + 4z = 7y - 5x$
4. $6y - 9x = 7y + 3z - 6x$

Every linear equation in several unknowns can be solved for any of the unknowns. When solved for x *such equations simplify to the form* $x = \frac{\textit{polynomial}}{\textit{integer}}$

Solutions

1. $8x = -4y + 11z$

$\downarrow$

$$x = \frac{(-4y + 11z)}{8}$$

2. $-2x = 7y - 11 - 3y$

$\downarrow$

$-2x = 4y - 11$

$\downarrow$

$2x = -4y + 11$

$\downarrow$

$$x = \frac{(-4y + 11)}{2}$$

3. $5x = 7y - 7y - 4z$

$\downarrow$

$5x = -4z$

$\downarrow$

$$x = \frac{-4z}{5}$$

4. $-9x + 6x = 7y + 3z - 6y$

$\downarrow$

$-3x = y + 3z$

$\downarrow$

$3x = -y - 3z$

$\downarrow$

$$x = \frac{(-y - 3z)}{3}$$

OBSERVATIONS AND REFLECTIONS

★ Look at the following equations and their solutions:

$4x - 3 = 3x + z$	$4x - 3 = 3x + 2$
$\downarrow$	$\downarrow$
$x = z + 3$	$x = 5$

3. Solve for x.
   a. $4x - 3 = 3x + z$
   b. $4x - 3 = 3x + 2$

Note that the equations on the left are in the two unknowns x and z. The equations on the right can be obtained by evaluating the equations on the left for $z = 2$. This relationship can be summarized as follows:

$$\begin{array}{ccc} 4x - 3 = 3x + z & \xrightarrow{\text{evaluate for } z = 2} & 4x - 3 = 3x + 2 \\ \text{solve} \downarrow \text{for } x & & \text{solve} \downarrow \text{for } x \\ x = z + 3 & \xrightarrow{\text{evaluate for } z = 2} & x = 5 \end{array}$$

4. Form a similar diagram, evaluating $4x - 3 = 3x + z$ for
   a. $z = 5$
   b. $z = -6$

Hence, evaluating a literal equation and then solving yields the same result as first solving the literal equation for x and then evaluating the solution.

EXERCISES

Solve each of the following equations for x.

1. $4x = 7y - 3$
2. $11x = 6y + 2z$
3. $-40x = y - 7z$
4. $5x - 9y = 2$
5. $3y - 4x = 7z - 20$
6. $5y - 7x = 4y - 9$
7. $3y + 4 - 7x = z$
8. $3x - 9y = 7z - 8$
9. $4y - 3x = 7y - 1$
10. $3x - 7y = 4y + z$
11. $7y + 9z = 2x - z$
12. $7y + 3z = 2x + 7z$
13. $-9x - 8 = 7 + 2y$
14. $2x + 4y = 5x - z$
15. $8x + 7z - 2y = 4x$
16. $-7y + 3x = -4x - z$
17. $-3x = 7y - x + 2z$
18. $11x - 9z = 2y - 8x$
19. $4x - 7y + 5z = 2x - 4z$
20. $4y - 9x - 2z = 7y - 6z$
21. $8y - 8x + 2z = 5x - 1$
22. $5x - 4y + 2z = 4 - 2x + y$
23. $8y - 4z = 7x + 3y - 7z$
24. $-7x - 5y + 2z = 7x + 9y - 3z$

11.3 DETERMINING ROOTS OF A LINEAR EQUATION IN TWO UNKNOWNS

In this section we will take another look at linear equations in two unknowns, and we will learn how to find the roots of these equations. Remember that a linear equation in x and y has an unlimited number of roots. Each root is a pair of values: a value for x and a corresponding value for y. For example, the equation $2x + y = 6$ is satisfied when x is 0 and y is 6, when x is 1 and y is 4, when x is 2 and y is 2, when x is 3 and y is 0, and so forth. Hence, each pair of numbers is a root of $2x + y = 6$.

5. Which of these pairs of values satisfy $3x - 4y = 12$?
a. x is 12 and y is 6
b. x is 3 and y is -4
c. x is 0 and y is -3
d. x is 6 and y is 0

To help us list the roots of an equation in x and y we will use a *table of values*. For example, we can list roots of $2x + y = 6$ in a table of values as follows:

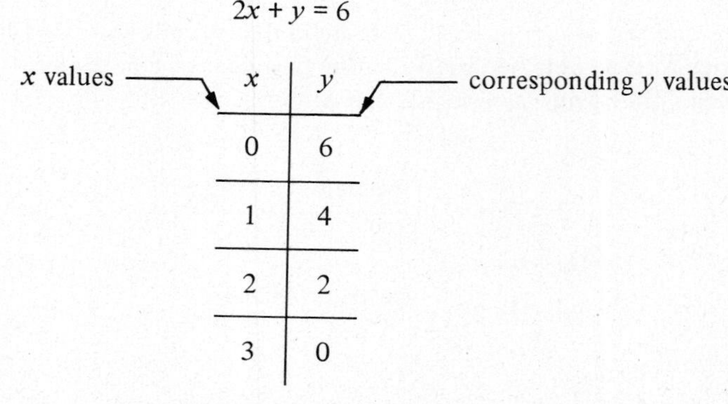

x	y
0	6
1	4
2	2
3	0

Table of Values

6. List these roots of $x + 3y = -9$ in a table of values
a. x is 0 and y is -3
b. x is 6 and y is -5
c. x is -3 and y is -2
d. x is -15 and y is 2

In the table of values we list a root by writing the value for x in the left column and the corresponding value for y opposite it in the right column.

Determining Roots

A conditional linear equation of the form

$$ax + by = c$$

where a, b, and c represent integers with a and b both nonzero has a root for every value of x. That is, we can find a root of the equation whose x value is any number we choose. For example, $2x + y = 6$ has a root for each value of x. To find the root for a particular x value, such as 6, we evaluate

$$2x + y = 6, \quad \text{when } x = 6$$

$$\downarrow$$

$$2(6) + y = 6$$

$$\downarrow$$

$$y = -6$$

$$\downarrow$$

$$2x + y = 6$$

x	y
6	-6

Similarly, we can find the root of such an equation for each value of y. We can find the root of $2x + y = 6$ for a particular y value, such as 3, by evaluating

$$2x + y = 6, \quad \text{when } y = 3$$

$$\downarrow$$

$$2x + (3) = 6$$

$$\downarrow$$

$$x = \frac{3}{2}$$

$$\downarrow$$

$$2x + y = 6$$

x	y
6	-6
$\frac{3}{2}$	3

Look at the following example of *determining corresponding values to find roots of a linear equation.*

TYPICAL PROBLEM Complete the table of values: $-4x + 9y = 12$

x	y
0	
	0
6	
	2
3	
	−4

First, we evaluate the equation when $x = 0$:

$$-4x + 9y = 12, \text{ when } x = 0$$
$$-4(0) + 9y = 12$$
$$y = \frac{4}{3}$$

Then we evaluate the equation when $y = 0$:

$$-4x + 9y = 12, \text{ when } y = 0$$
$$-4x + 9(0) = 12$$
$$x = -3$$

We then evaluate the equation for each of the other values given for x or y:

$$x = 6: \quad -4(6) + 9y = 12$$
$$y = 4$$

$$y = 2: \quad -4x + 9(2) = 12$$
$$x = \frac{3}{2}$$

$$x = 3: \quad -4(3) + 9y = 12$$
$$y = \frac{8}{3}$$

$$y = -4: \quad -4x + 9(-4) = 12$$
$$x = -12$$

As we do this we write the corresponding values in the table:

$-4x + 96 = 12$

x	y
0	$\frac{4}{3}$
–3	0
6	4
$\frac{3}{2}$	2
3	$\frac{8}{3}$
–12	–4

To evaluate the equation for each value of x *or* y *we extend the methods in Sections 3.5 and 9.5 for numerical evaluations.*

EXAMPLES

Complete the following tables of values, evaluating for each value of x or y and solving for the corresponding unknown.

1. $-2y = 2 - x$

x	y
0	
	0
4	
	–2
–4	
	2

2. $3x = 2y - 6$

x	y
0	
	0
2	
	–3
–1	
	2

3. $-2x + 3y = -9$

x	y
0	
	0
3	
	1
–1	
	–4

4. $4 - 12y = 4x$

x	y
0	
	0
3	
	1
–4	
	–1

Solutions

1.

$-2y = 2 - x$

$x = 0$: $-2y = 2 - (0) \Rightarrow y = -1$

$y = 0$: $-2(0) = 2 - x \Rightarrow x = 2$

$x = 4$: $-2y = 2 - (4) \Rightarrow y = 1$

$y = -2$: $-2(-2) = 2 - x \Rightarrow x = -2$

$x = -4$: $-2y = 2 - (-4) \Rightarrow y = -3$

$y = 2$: $-2(2) = 2 - x \Rightarrow x = 6$

x	y
0	−1
2	0
4	1
−2	−2
−4	−3
6	2

2.

$3x = 2y - 6$

$x = 0$: $3(0) = 2y - 6 \Rightarrow y = 3$

$y = 0$: $3x = 2(0) - 6 \Rightarrow x = -2$

$x = 2$: $3(2) = 2y - 6 \Rightarrow y = 6$

$y = -3$: $3x = 2(-3) - 6 \Rightarrow x = -4$

$x = -1$: $3(-1) = 2y - 6 \Rightarrow y = \frac{3}{2}$

$y = 2$: $3x = 2(2) - 6 \Rightarrow x = \frac{-2}{3}$

x	y
0	3
−2	0
2	6
−4	−3
−1	$\frac{3}{2}$
$\frac{-2}{3}$	2

3.

$-2x + 3y = -9$

$x = 0$: $-2(0) + 3y = -9 \Rightarrow y = -3$

$y = 0$: $-2x + 3(0) = -9 \Rightarrow x = \frac{9}{2}$

$x = 3$: $-2(3) + 3y = -9 \Rightarrow y = -1$

$y = 1$: $-2x + 3(1) = -9 \Rightarrow x = 6$

$x = -1$: $-2(-1) + 3y = 9 \Rightarrow y = \frac{-11}{3}$

$x = -4$: $-2x + 3(-4) = 9 \Rightarrow x = \frac{-3}{2}$

x	y
0	−3
$\frac{9}{2}$	0
3	−1
6	1
−1	$\frac{-11}{3}$
$\frac{-3}{2}$	−4

4.

$x = 0$: $4 - 12y = 4(0) \Rightarrow y = \frac{1}{3}$

$y = 0$: $4 - 12(0) = 4x \Rightarrow x = 1$

$x = 3$: $4 - 12y = 4(3) \Rightarrow y = \frac{-2}{3}$

$y = 1$: $4 - 12(1) = 4x \Rightarrow x = -2$

$x = -4$: $4 - 12y = 4(-4) \Rightarrow y = \frac{5}{3}$

$y = -1$: $4 - 12(-1) = 4x \Rightarrow x = 4$

$4 - 12y = 4x$

x	y
0	$\frac{1}{3}$
1	0
3	$\frac{-2}{3}$
-2	1
-4	$\frac{5}{3}$
4	-1

OBSERVATIONS AND REFLECTIONS

Remember that a pair of corresponding values for x and y is not a root of an equation such as $2x + y = 6$ unless evaluating the equation for these values yields a numerical identity. For example,

$$2x + y = 6, \text{ when } x = 1 \text{ and } y = 4$$
$$2(1) + (4) = 6$$
$$\downarrow$$
$$2 + 4 = 6 \quad \text{an identity}$$

so $2x + y = 6$ is satisfied when x is 1 and y is 4.

Rather than using $x = 1$ and $y = 4$ together, we can first evaluate $2x + y = 6$ when $x = 1$ and then evaluate the resulting equation when $y = 4$. That is,

$$2x + y = 6, \text{ when } x = 1$$
$$\downarrow$$
$$2 + y = 6$$

and

$$2 + y = 6, \text{ when } y = 4$$
$$\downarrow$$
$$2 + 4 = 6$$

Alternatively, we can first evaluate $2x + y = 6$ when $y = 4$ and then evaluate the resulting equation $2x + 4 = 6$ when $x = 1$ to get $2 + 4 = 6$. Either way we obtain the same numerical identity as when we evaluate in a single step.

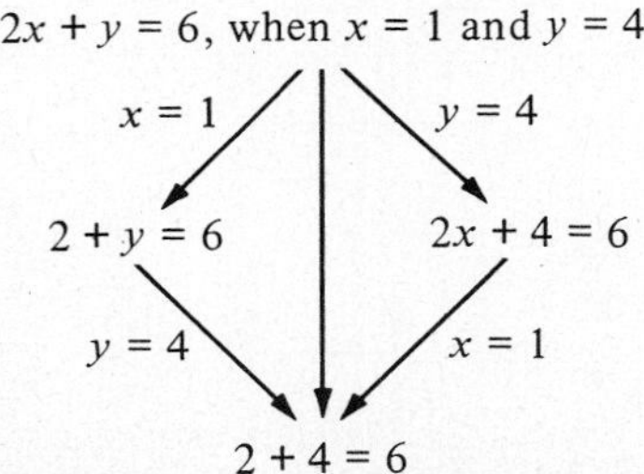

It follows that $2 + y = 6$ is satisfied when y is 4, and $2x + 4 = 6$ is satisfied when x is 1. In fact, solving $2 + y = 6$ yields $y = 4$ and solving $2x + 4 = 6$ yields $x = 1$. Thus, had we not known the x value corresponding to $y = 4$ (or the y value corresponding to $x = 1$) we could have found it by evaluating $2x + y = 6$ for the value known and then solving. That is,

$$2x + y = 6, \text{ when } x = 1 \quad \text{and} \quad 2x + y = 6, \text{ when } y = 4$$
$$\downarrow \qquad\qquad\qquad \downarrow$$
$$2 + y = 6 \qquad\qquad 2x + 4 = 6$$
$$\downarrow \qquad\qquad\qquad \downarrow$$
$$y = 4 \qquad\qquad x = 1$$

7. Pair corresponding values of x and y for $2x - 3y = 12$
   a. $x = 3$
   b. $x = 9$
   c. $x = 12$
   d. $y = 2$
   e. $y = -2$
   f. $y = 4$

In the same way we can find the x or y value corresponding to any given value of the other unknown. Hence, an equation such as $2x + y = 6$ has a root for every value of x or y.

EXERCISES

Complete the table of values for each of the following equations.

1. $3x + 7y = 21$

x	y
0	
	0
14	
	6
−28	
	−9

2. $5x + 4y = 20$

x	y
0	
	0
12	
	20
−8	
	−5

3. $4x - 12 = 6y$

x	y
0	
	0
12	
	8
−9	
	−12

4. $9x - 45 = 15y$

x	y
0	
	0
15	
	12
−10	
	−6

5. $-4x = 56 - 7y$

x	y
0	
	0
7	
	4
−4	
	−6

6. $-5x = 60 - 6y$

x	y
0	
	0
18	
	15
−9	
	12

7. $-12x - 15y = 60$

x	y
0	
	0
10	
	10
–8	
	–8

8. $-18x - 12y = 36$

x	y
0	
	0
9	
	12
–8	
	–10

9. $12x = 24 + 5y$

x	y
0	
	0
6	
	9
–8	
	–12

10. $-9y = 18 - 7x$

x	y
0	
	0
3	
	12
–12	
	–9

11. $6y = 18 - 15x$

x	y
0	
	0
5	
	–7
–8	
	–9

12. $8x = 40 - 12y$

x	y
0	
	0
7	
	6
–10	
	–5

11.4 GRAPHING A LINEAR EQUATION IN TWO UNKNOWNS

In the previous section we learned how to display roots of a linear equation in two unknowns using a table of values. For example, we saw that we can display roots of $2x + y = 6$ as

$2x + y = 6$

x	y
0	6
1	4
2	2
3	0

In this section we will learn how to display the roots of such an equation graphically.

The Cartesian Plane

Because a root of an equation in x and y is a pair of numbers (an x value and a corresponding y value), we need two number lines to display a root graphically. We place one number line horizontally in the plane and the other across it vertically to form the *rectangular (or Cartesian) coordinate system.*

8. Plot the points
   a. (−2, 1)
   b. (3, 3)
   c. (−1, −2)
   d. (2, −3)

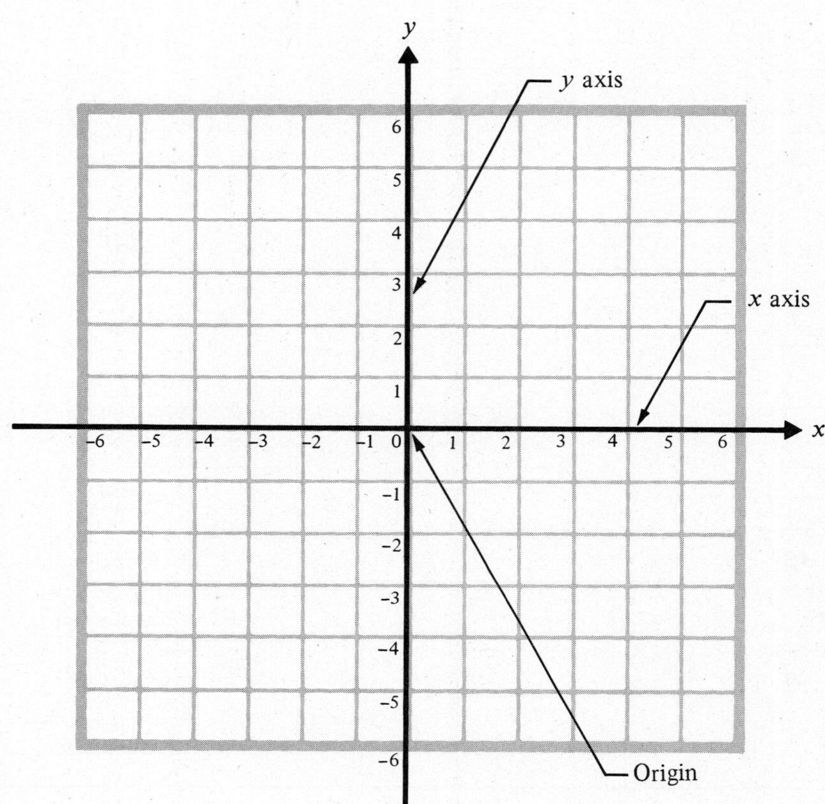

Rectangular Coordinate System

The horizontal number line is called by tradition the *x axis*, and the vertical number line is called the *y axis*. The two axes cross at their zero points, and this point in the plane is called the *origin.* Within the framework of these two axes, every point in the plane is designated by an *ordered pair of numbers* called *coordinates*, such as (3,4), (−6,1), (−2,−5), or (4,−2). The first number of the pair is called by tradition the *x coordinate*, and the second number is called the *y coordinate*.

In (3,4), 3 is the x coordinate and 4 is the y coordinate
In (−6,1), −6 is the x coordinate and 1 is the y coordinate
In (−2,−5), −2 is the x coordinate and −5 is the y coordinate
In (4,−2), 4 is the x coordinate and −2 is the y coordinate

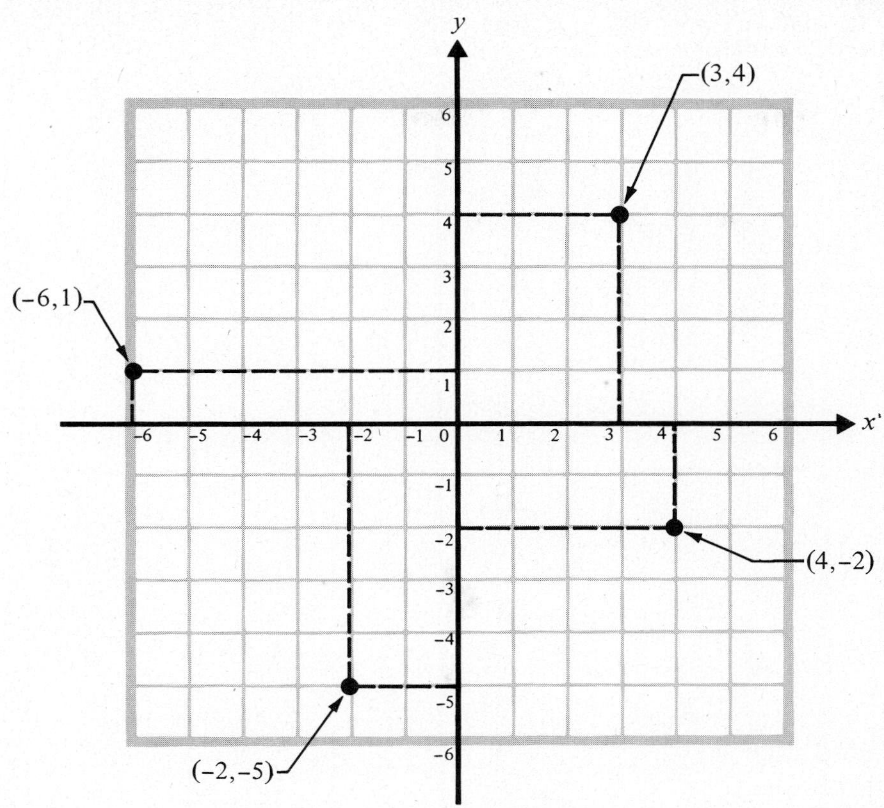

9. Plot the points
 a. (2, 0)
 b. (0, 3)
 c. (−1, 0)
 d. (0, 1)

Points with a positive x coordinate lie to the right of the y axis, and points with a negative x coordinate lie to the left of the y axis. Similarly, points with a positive y coordinate lie above the x axis and points with a negative y coordinate lie below the x axis. Thus, for example, the point (3,4) lies above 3 on the x axis and to the right of 4 on the y axis. Similarly, the point (−6,1) lies above −6 on the x axis and to the left of 1 on the y axis. In general, the point (a,b) lies above or below a on the x axis and to the left or right of b on the y axis.

Points with a zero x coordinate lie under the y axis, and points with a zero y coordinate lie under the x axis. For example, the point (3, 0) lies under 3 on the x axis, whereas (0,−2) lies under −2 on the y axis. The origin, the point (0,0) in the plane, lies at the intersection of the two axes.

Graphing an Equation

To graph an equation such as $2x + y = 6$, or more precisely to graph the roots of this equation, we assign a point on the plane for each root of the equation. To do this we simply form an ordered pair of numbers corresponding to each row in the equation's table of values. For example, to graph $2x + y = 6$ we form the correspondence

$$2x + y = 6$$

x	y		
0	6	↔	(0,6)
1	4	↔	(1,4)
2	2	↔	(2,2)
3	0	↔	(3,0)

10. Form corresponding ordered pairs for $2x + y = 6$

	x	y
a.	4	−2
b.	5	−4
c.	6	−6

11. Plot the ordered pairs formed in problem 10.

Then we plot the ordered pairs:

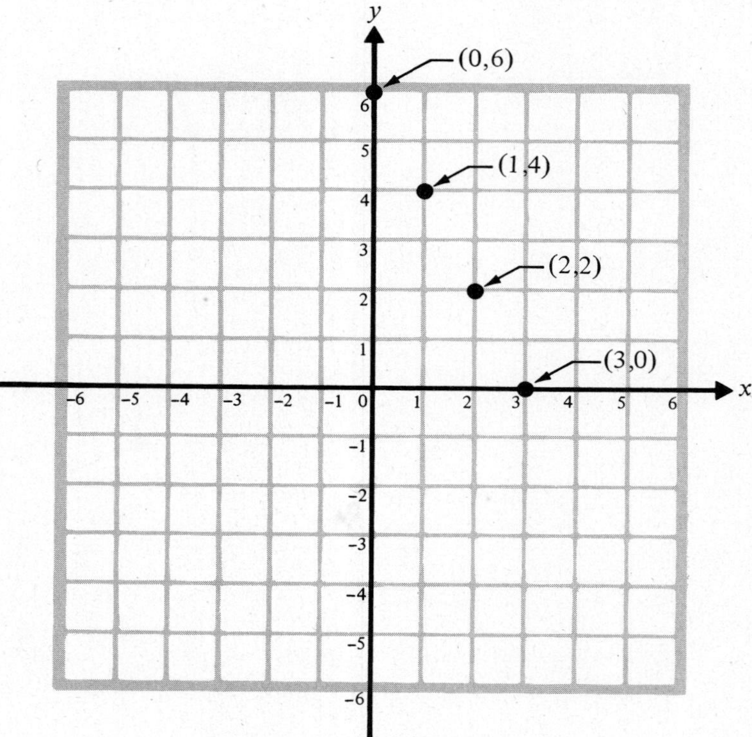

Note that these points all lie on a straight line. In fact, all the roots of $2x + y = 6$ correspond to points on this line and every point on this line corresponds to a root of the equation. This line in the plane is called the *graph of the equation.*

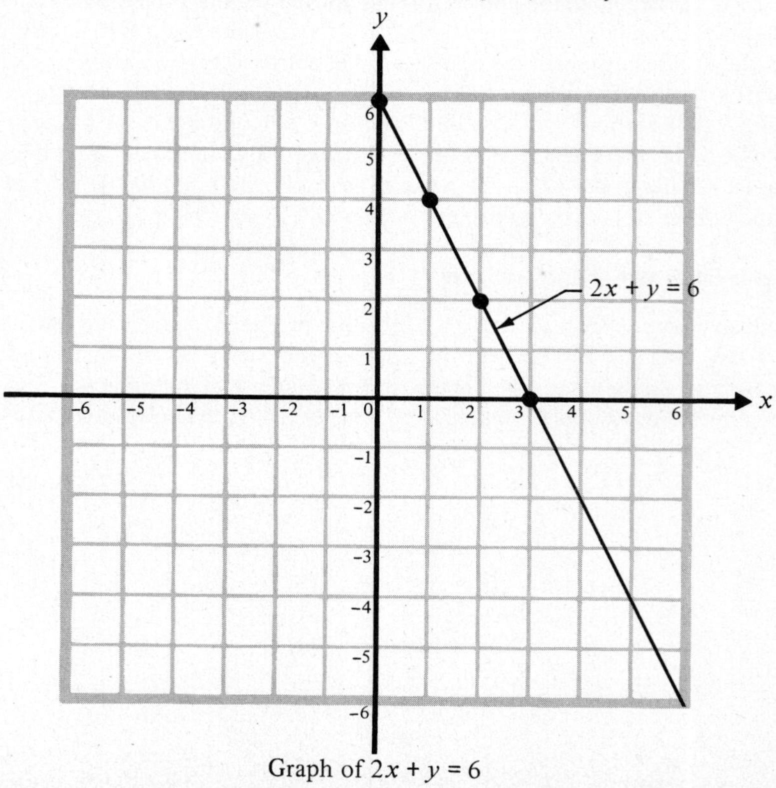

Graph of $2x + y = 6$

11.4

Linear equations in x and y are distinguished from all other equations in x and y because the graph of every linear equation is a straight line. Furthermore, every straight line is the graph of a linear equation. Hence, to graph a linear equation in x and y we need only determine a few roots and then draw a straight line through the corresponding points on the plane. In principle, two points are all that are needed to determine the straight line, but in practice it is best to plot at least three points to guard against errors.

Look at the following example of *graphing a linear equation in two unknowns.*

TYPICAL PROBLEM Graph: $-4x + 9y = 12$

We first form a table of values for the equation. We list the pairs of values with $x = 0$ and $y = 0$. Then we determine a third pair of values:

x	y
0	$\frac{4}{3}$
−3	0
6	4

Then we plot the corresponding ordered pairs on the graph:

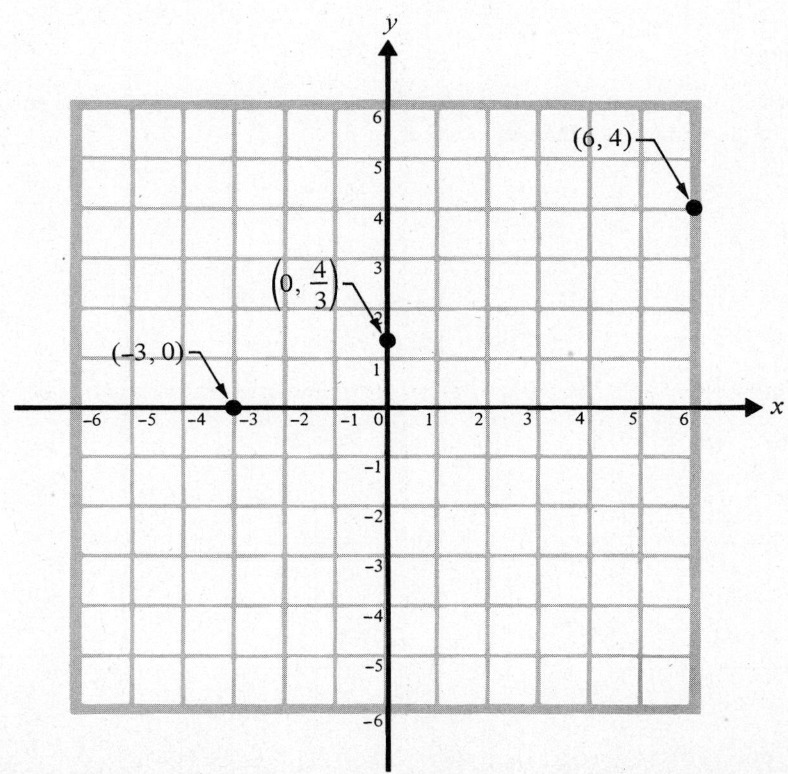

Finally we draw a straight line through the three plotted points:

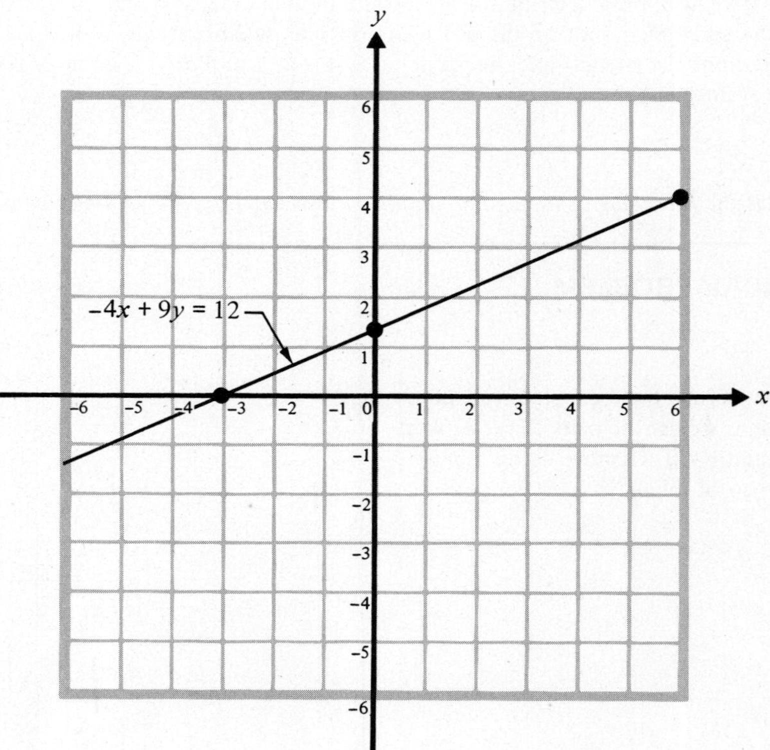

To form the table of values in the first step we use the method in the previous section. (The equation here also appears in the typical problem there.)

EXAMPLES

Graph the following equations, first determining tables of values. (The same equations appear in the examples of the previous section.)

1. $-2y = 2 - x$

2. $3x = 2y - 6$

3. $-2x + 3y = -9$

4. $4 - 12y = 4x$

Solutions

1. $-2y = 2 - x$

x	y
0	−1
2	0
4	1

↓

2. $3x = 2y - 6$

x	y
0	3
−2	0
2	6

↓

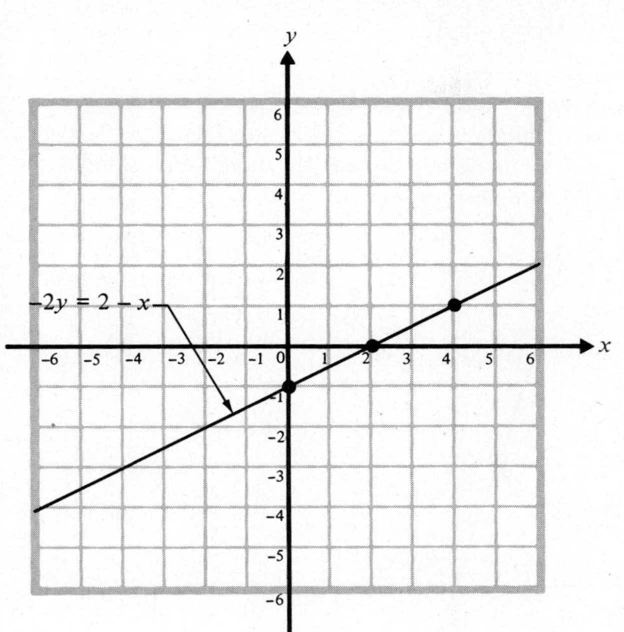

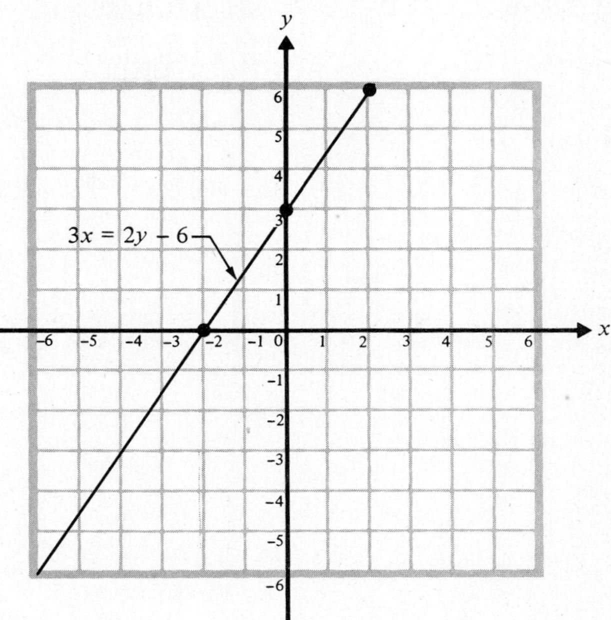

3. $-2x + 3y = -9$

x	y
0	-3
$\frac{9}{2}$	0
3	-1

↓

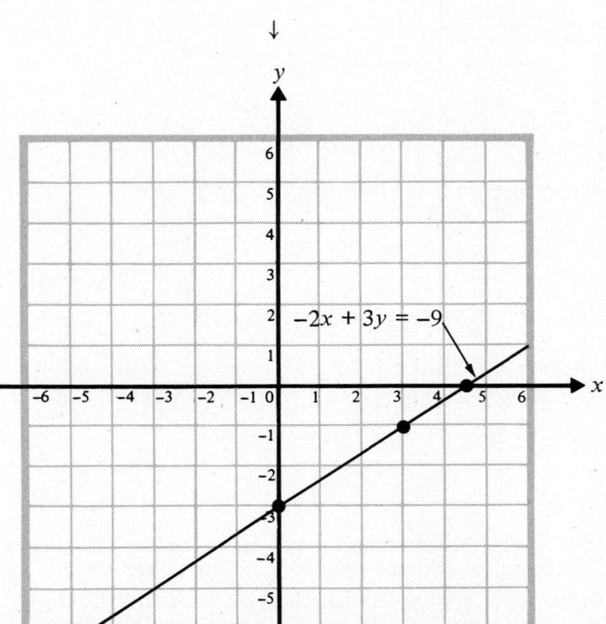

4. $4 - 12y = 4x$

x	y
0	$\frac{1}{3}$
1	0
3	$\frac{-2}{3}$

↓

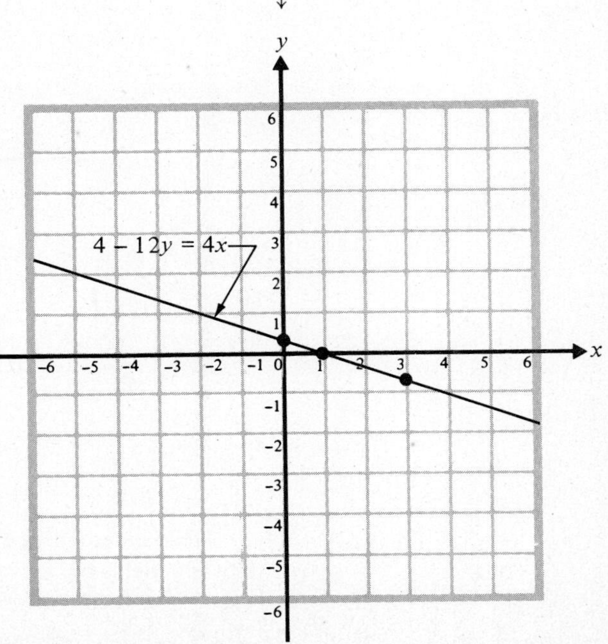

SPECIAL CASES

1. An equation of the special form

$$ax + by = 0 \quad \text{or} \quad ax = -by$$

where a and b represent nonzero numbers is satisfied when x is 0 and y is 0. Hence, the graph of such an equation is a line passing through the origin. For example, the graphs of $3x + y = 0$ and $2x = 5y$ are, respectively,

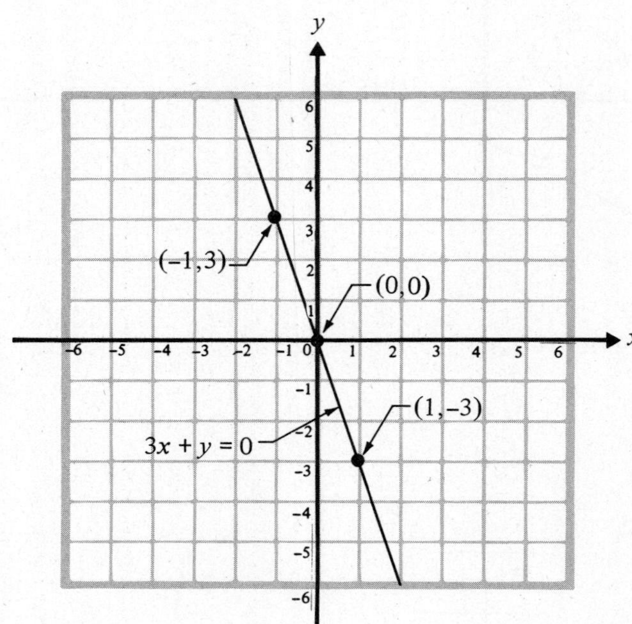

12. Graph
 a. $x + y = 0$
 b. $x = y$
 c. $2x - y = 0$
 d. $4x = 3y$
 e. $-5x + 6y = 0$
 f. $x = -4y$

and

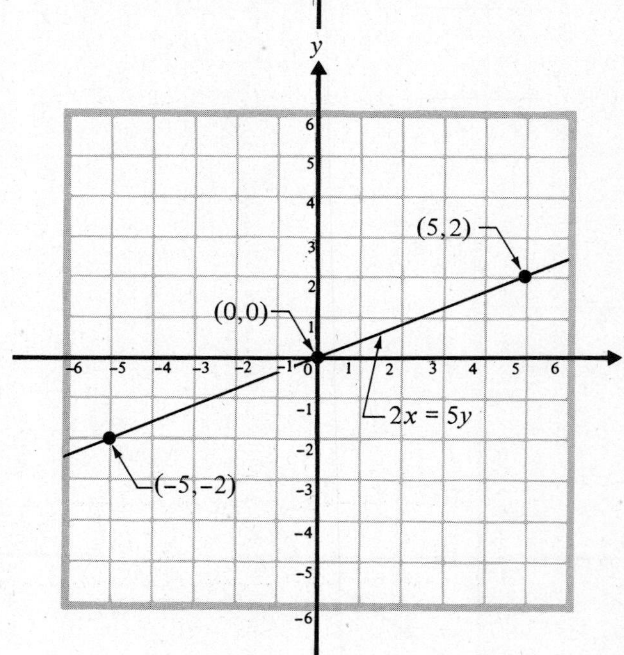

When determining a table of values for such an equation, setting either $x = 0$ or $y = 0$ yields the same root (corresponding to the origin). If we want to determine three roots before plotting such an equation, we must determine two roots with nonzero x and y values.

2. An equation of the form

$$ax = c \qquad \text{or} \qquad by = c$$

where a, b, and c represent numbers with a and b nonzero can be thought of as an equation in x and y for the purposes of graphing, even though it contains, in fact, only one unknown. For example, treating $x = 3$ as the equation $x + 0y = 3$, we can form a table of values and then graph as follows:

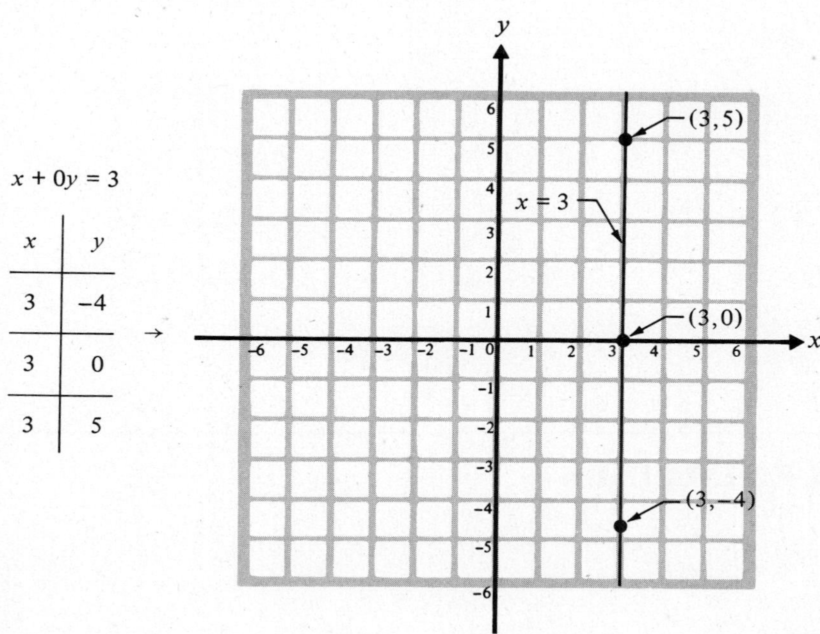

$x + 0y = 3$

x	y
3	−4
3	0
3	5

Note that the graph of x = 3 *is the* vertical *line of points above and below 3 on the* x *axis.*

13. Graph
    a. $x = -4$
    b. $x = 0$
    c. $x = 5$
    d. $y = -3$
    e. $y = 0$
    f. $y = 6$

Similarly, to graph $y = -2$, we can treat it as the equation $0x + y = -2$:

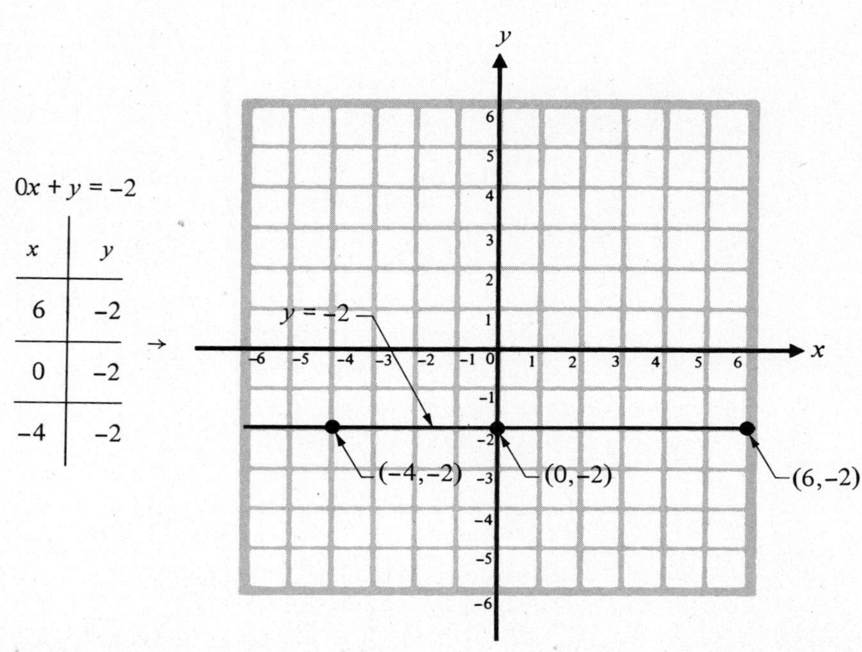

$0x + y = -2$

x	y
6	−2
0	−2
−4	−2

Note that the graph of y = −2 *is the* horizontal *line of points to the left and right of* −2 *on the* y *axis.*

EXERCISES

Graph each of the following equations.

1. $5x - y = 5$	2. $2x + y = 4$	3. $3y = -x - 6$
4. $4y = 4 - x$	5. $5x = 4y - 20$	6. $6x = 5y + 30$
7. $x + y = -4$	8. $-x - y = -3$	9. $2y - 6 = x$
10. $3y + 6 = x$	11. $-6x = 5y - 30$	12. $-5x = 3y - 15$
13. $3x - y = 4$	14. $4x - y = 6$	15. $4x = 3y + 8$
16. $3x = 2y + 3$	17. $6x + 5y = -20$	18. $5x + 2y = -12$
19. $3x - 9 = -2y$	20. $6x - 12 = -5y$	21. $4y = x - 3$
22. $3y = x - 5$	23. $-5x - 4y = -12$	24. $-3x - 5y = -5$

ANSWERS TO MARGIN EXERCISES

1. a. $x = 14$
b. $x = 13$

2. a. $x = \frac{3}{4}$
b. $x = \frac{-1}{3}$

3. a. $x = z + 3$
b. $x = 5$

4. a. $4x - 3 = 3x + z \rightarrow 4x - 3 = 3x + 5$
$\downarrow \qquad \downarrow$
$x = z + 3 \rightarrow x = 8$
b. $4x - 3 = 3x + z \rightarrow 4x - 3 = 3x - 6$
$\downarrow \qquad \downarrow$
$x = z + 3 \rightarrow x = -3$

5. a and c

6. $x + 3y = -9$

x	y
0	−3
6	−5
−3	−2
−15	1

7. a and e
b and d
c and f

CHAPTER REVIEW

Solve for x.

SECTION 11.1

1. $6y + x = 2$
2. $3y + x = -9y$
3. $x - 4y = 14y - 9z$
4. $10y - 14y = y - x$
5. $2z = 8y - 18 - x$
6. $10y - 12 + 2x = x$
7. $6y - 9z + 7x = 2y + 6x$
8. $13x - 18y + 8 = 12x$
9. $4y - 10z - 2 = -x$

Solve for x.

SECTION 11.2

10. $3x = 8y - 2$
11. $6y - 8x = 14z - 10$
12. $12y - 19x = 21y - 3$
13. $9x - 14y = 8y + 4z$
14. $3x - 16 = 14 + 16y$
15. $2y - 18x - 4z = 3y - 10z$
16. $14y + 6z = 7x + 14z$
17. $-10y - 5x = -10x - 2z$
18. $8x - 14y + 11z = 4x - 8z$
19. $16y - 16x + 4z = 10x - 3$
20. $18y - 7z = 14x + 6y - 14z$
21. $9x = 12y - 21z + 6 - x$

Complete the table of values.

SECTION 11.3

22. $6x + 5y = 30$

x	y
0	
	0
15	
	18
−5	
	−6

23. $8x - 40 = 10y$

x	y
0	
	0
20	
	8
−10	
	−20

24. $-7x = 63 - 3y$

x	y
0	
	0
6	
	9
−8	
	−14

25. $-20x - 30y = 60$

x	y
0	
	0
12	
	9
−4	
	−6

26. $8x = 7y + 16$

x	y
0	
	0
9	
	12
−5	
	−10

27. $14y = 28 - 18x$

x	y
0	
	0
6	
	9
−7	
	−7

SECTION 11.4

Graph the equations.

28. $4x - y = 4$

29. $5y = 5 - x$

30. $3x = 4y + 12$

31. $x + y = -5$

32. $3y - 6 = x$

33. $-5x = 4y - 20$

34. $5x - y = 3$

35. $5x = 3y + 10$

36. $4x + 3y = -15$

37. $5x - 5 = -2y$

38. $3y = x - 4$

39. $-3x - 4y = -16$

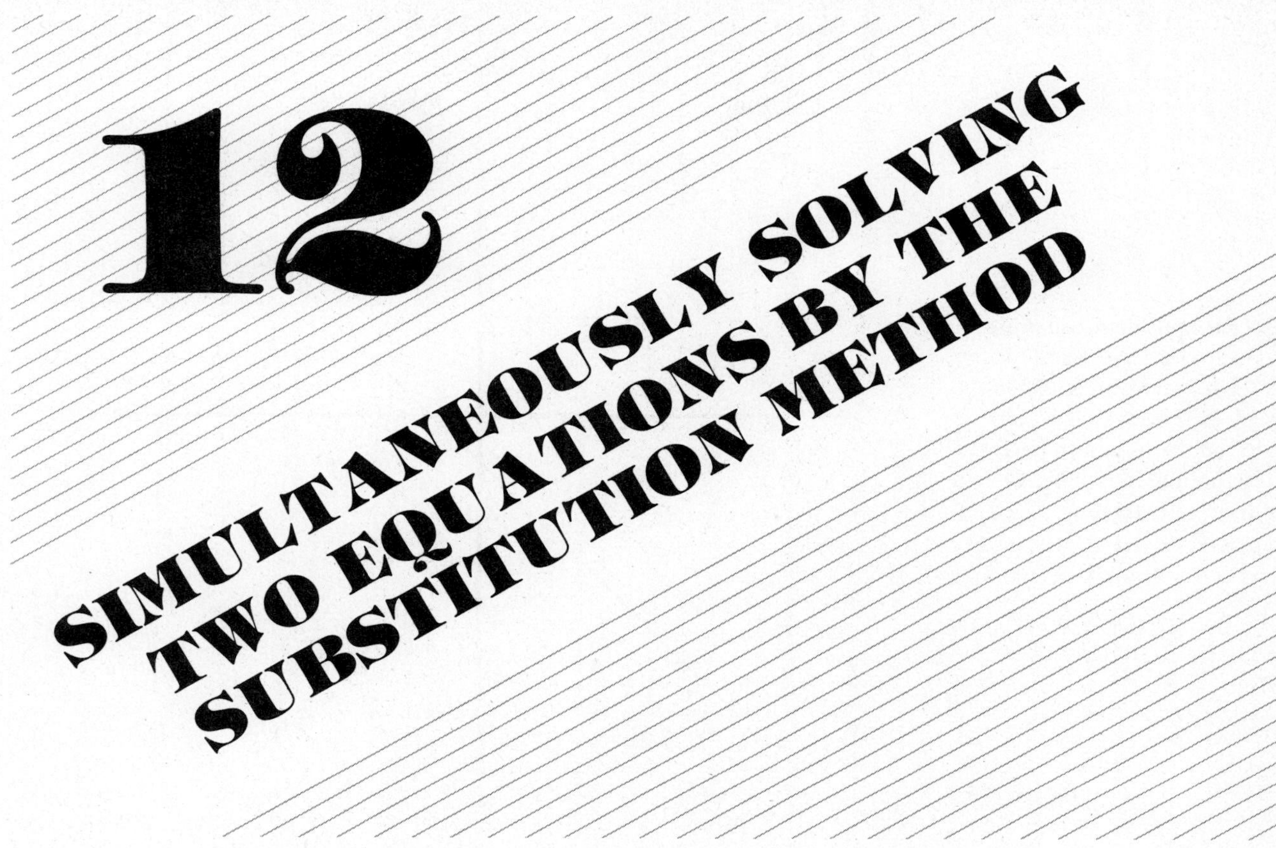

Common Roots of Two Equations

In this chapter we will consider pairs of linear equations in two unknowns. Consider, for example, the linear equations

$$x = y + 1 \qquad \text{and} \qquad 2x - y = 3$$

Each equation is a conditional equation in the two unknowns x and y. Remember that if an equation reduces to a numerical identity for particular values of x and y, then this pair of values is called a root of the equation. Although a single linear equation in two unknowns has an unlimited number of roots, a pair of linear equations in two unknowns *usually* has only *one* common root. For example, the equations $x = y + 1$ and $2x - y = 3$ are both satisfied when x is 2 and y is 1, whereas no other root of either equation satisfies the other.

There are two exceptions to the general rule that two linear equations in two unknowns have one root in common:

1. Two equivalent equations, such as $x = y + 1$ and $2x = 2y + 2$, are satisfied by exactly the same pairs of numbers, and so have an unlimited number of roots in common.

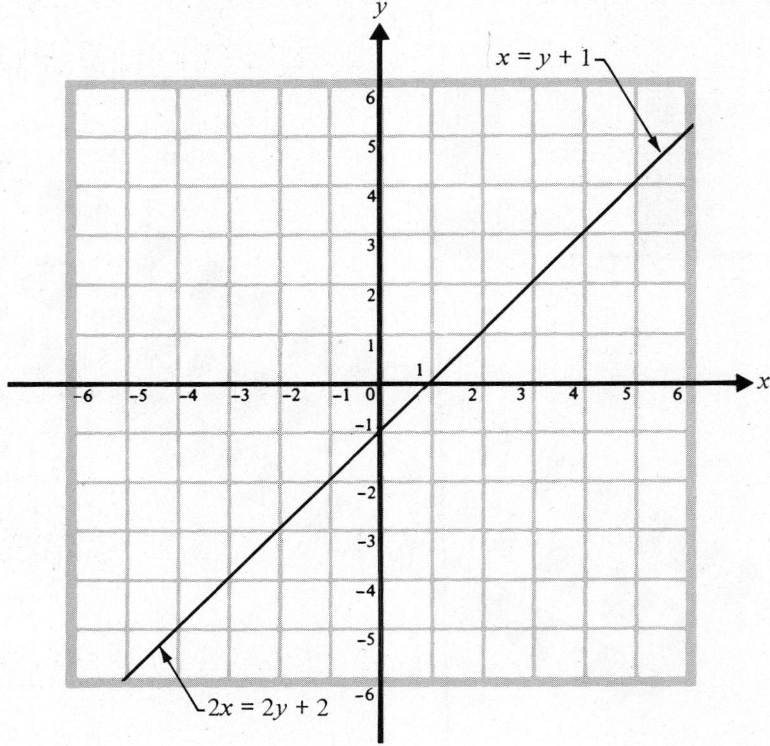

2. Two nonequivalent equations may have no roots in common. For example, $x = y - 1$ and $x = y + 1$ have no roots in common.

Two equations in two unknowns are called *consistent* if they are satisfied by at least one common pair of numbers. In this chapter we will only consider pairs of nonequivalent equations that are consistent and thus have a single root in common.

Solving a Pair of Equations

A pair of nonequivalent consistent linear equations in x and y is (simultaneously) solved by determining the unique common root. We will give the *solution of the pair of equations* by setting each literal equal to its value. For example, to solve the pair of equations $x = y + 1$ and $2x - y = 3$ we determine that their root is the value 2 for x and 1 for y, and we give its solution as the pair of equations $x = 2$ and $y = 1$.

Transformation Rules

Two consistent linear equations can be solved by the *Substitution Method* using the following transformation rules:

RULE 1

Replace either member with any other expression with the same value.

RULE 2

Reverse the order of the members.

RULE 3

Add any linear expression, not including a new literal, to both members of the equation.

RULE 4

Multiply or divide both members by any number except zero.

RULE 5

Negate both members of the equation.

(These are Rules 1–5 from Chapter 10.)

RULE 6

If one of the equations is solved for an unknown, say x, then evaluate the other equation using the first equation to eliminate the unknown x.

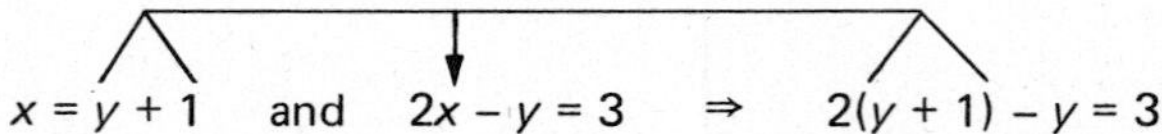

$$x = y + 1 \quad \text{and} \quad 2x - y = 3 \quad \Rightarrow \quad 2(y + 1) - y = 3$$

These transformation rules state what *can* be done to a pair of equations. We must learn how to apply these rules effectively in order to solve a pair of equations.

12.1 SOLVING TWO EQUATIONS BY SUBSTITUTION

In this section we will discuss pairs of equations in which one of the equations is solved for an unknown. For example, we will look at pairs of equations such as

$$3y - 2x = 1 \quad \text{and} \quad x = 2y - 8$$

in which the second equation is solved for x. We consider pairs of equations of this special form first in order to demonstrate the Substitution Method, which must start with an equation solved for an unknown. To solve a pair of linear equations in x and y, we must find some way to use the equations together.

Evaluating an Equation

Remember that in Section 4.4 we saw how to evaluate an expression in two unknowns using a literal equation solved for one of the unknowns. For example, we can evaluate $2x - y$, when $x = y + 1$ to yield $2(y + 1) - y \rightarrow y + 2$, an expression in y. In the same way we can evaluate an equation containing two unknowns and produce an equation containing only one literal. For example, evaluating $2x - y = 3$, when $x = y + 1$ yields $2(y + 1) - y = 3 \Rightarrow y + 2 = 3$.

If we solve the resulting equation in y, we get $y = 1$. But the solution of the pair of equations $x = y + 1$ and $2x - y = 3$ is $x = 2$ and $y = 1$. Evaluating $2x - y = 3$, when $x = y + 1$ and then solving has given us half the solution of this pair of equations.

1. Evaluate when $x = 2y - 1$
   a. $3x + y$
   b. $5 - x$
   c. $3x + y = 5 - x$

2. Evaluate and then solve when $y = 1$
   a. $2x - y = 3$
   b. $x = y + 1$

The Substitution Method relies on the (not obvious) fact that evaluating one of two equations using the other and then solving always yields half the (simultaneous) solution of these two equations.

Look at the following example of *solving two equations by substitution.*

TYPICAL PROBLEM

$3y - 2x = 1$ and $x = 2y - 8$

↓

We evaluate the first equation using the second equation:

$3y - 2x = 1$, when $x = 2y - 8$

$3y - 2(2y - 8) = 1$

↓

And then we solve this equation:

$3y - 4y + 16 = 1$

↓

$3y - 4y = 1 - 16$

↓

$-y = -15$

↓

At this point we have one of the two equations of the solution.

$\boxed{y = 15}$

↓

Then we evaluate the second equation using the solution found:

$x = 2y - 8$, when $y = 15$

$x = 2(15) - 8$

↓

And then we solve this equation:

$x = 30 - 8$

↓

At this point we have the other equation of the solution.

$\boxed{x = 22}$

↓

Finally we write the two equations of the solution together:

$x = 22$ and $y = 15$

Hence, $3y - 2x = 1$ and $x = 2y - 8 \Rightarrow x = 22$ and $y = 15$.

To evaluate in the first step we extend the method in Section 4.4 for literal evaluations. Then to evaluate the second equation, once we have one of the equations of the solution, we use the method in Section 11.3.

EXAMPLES

Solve the following pairs of equations for x and y by the Substitution Method.

1. $2x = 3y + 1$ and $x = 2y + 1$
2. $-3x + 6 = -y - 7$ and $y = 7 - x$
3. $3y = 4x + 5$ and $x = 3y + 1$
4. $y = 9 - x$ and $x = y + 1$
5. $-2x + 3y = 0$ and $x = 3y - 5$
6. $6x - 7 = 6y$ and $y = -7 - 4x$

Solutions

1. $2x = 3y + 1$, when $x = 2y + 1$

$2(2y + 1) = 3y + 1$

↓

$4y + 2 = 3y + 1$

↓

$\boxed{y = -1}$

↓

$x = 2y + 1$, when $y = -1$

$x = 2(-1) + 1$

↓

$x = -2 + 1$

↓

$\boxed{x = -1}$

↓

$x = -1$ and $y = -1$

2. $-3x + 6 = -y - 7$, when $y = 7 - x$

$-3x + 6 = -(7 - x) - 7$

↓

$-3x + 6 = -7 + x - 7$

↓

$\boxed{x = 5}$

↓

$y = 7 - x$, when $x = 5$

$y = 7 - (5)$

↓

$y = 7 - 5$

↓

$\boxed{y = 2}$

↓

$x = 5$ and $y = 2$

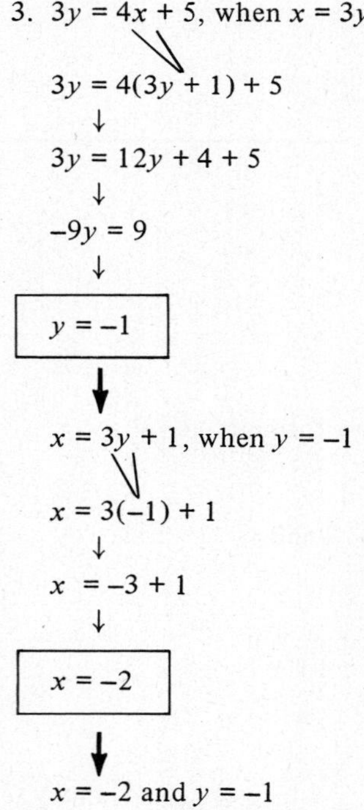

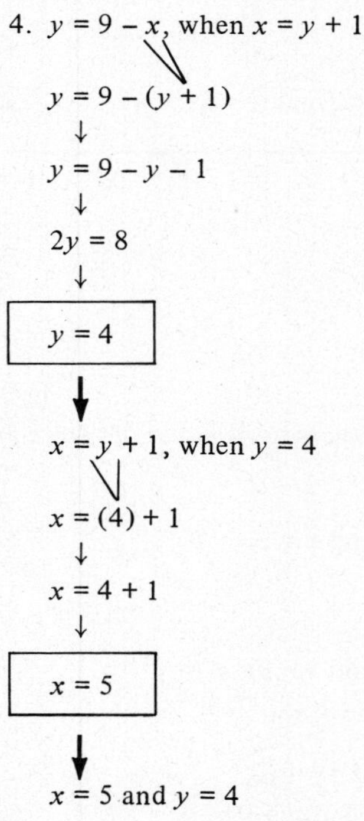

5. $-2x + 3y = 0$, when $x = 3y - 5$

$-2(3y - 5) + 3y = 0$

↓

$-6y + 10 + 3y = 0$

↓

$\boxed{y = \frac{10}{3}}$

↓

$x = 3y - 5$, when $y = \frac{10}{3}$

$x = 3\left(\frac{10}{3}\right) - 5$

↓

$x = 10 - 5$

↓

$\boxed{x = 5}$

↓

$x = 5$ and $y = \frac{10}{3}$

6. $6x - 7 = 6y$, when $y = -7 - 4x$

$6x - 7 = 6(-7 - 4x)$

↓

$6x - 7 = -42 - 24x$

↓

$\boxed{x = \frac{-7}{6}}$

↓

$y = -7 - 4x$, when $x = \frac{-7}{6}$

$y = -7 - 4\left(\frac{-7}{6}\right)$

↓

$y = -7 + \frac{14}{3}$

↓

$\boxed{y = \frac{-7}{3}}$

↓

$x = \frac{-7}{6}$ and $y = \frac{-7}{3}$

SPECIAL CASES

If the second equation is of the form unknown = number, we already have half of our solution, and we only need to evaluate once to obtain all of our solution. For example, to solve

$3x + 5y = -11$ and $x = 3$

↓

We evaluate: $3x + 5y = -11$, when $x = 3$

$3(3) + 5y = -11$

↓

Then we solve for y: $9 + 5y = -11$

↓

$5y = -20$

↓

$\boxed{y = -4}$

↓

And then we write the two equations of the solution together: $x = 3$ and $y = -4$

EXERCISES

Solve each of the following pairs of equations for x and y.
In these exercises, all pairs of equations have integer solutions.

1. $3x + 4y = 18$ and $x = 2$
2. $7x + 4y = 1$ and $y = 2$
3. $7y = 14 - 6x$ and $y = -4$
4. $5x = 24 - 4y$ and $x = 4$
5. $9x = -5y + 7$ and $y = 5$
6. $5y = 8x + 22$ and $x = -4$
7. $-7x + 8 = -9y + 5$ and $x = -6$
8. $-7x + 21 = -4y - 42$ and $y = 0$
9. $x + 2y = 5$ and $y = x + 1$
10. $3x - y = 10$ and $y = 2x - 6$
11. $3x = y + 10$ and $x = 2y + 5$
12. $-4x = -3y + 20$ and $x = y - 6$
13. $5y - 18 = 4x$ and $y = x + 5$
14. $-y + 19 = -6x$ and $y = 2x + 3$
15. $-4x + 7 = -y + 8$ and $y = 3x + 2$
16. $-7x - 11 = -3y + 10$ and $y = 2x + 5$
17. $-3x + 2y = 0$ and $x = 2y - 4$
18. $7x + 3y = 24$ and $x = 5y - 2$
19. $9 - 2x = 3y$ and $y = 11 - 2x$
20. $-8x + 64 = -5y$ and $y = 7 - 5x$
21. $9y = -7x - 88$ and $x = y + 8$
22. $7y = -4x - 51$ and $x = -14 - 2y$
23. $x + 14 = 2y + 26$ and $y = 2x + 3$
24. $7x + 4 = 4y - 11$ and $y = 4x - 3$

In these exercises, the solutions contain fractions.

25. $12x + 5y = 1$ and $y = 9$
26. $7x + 10y = 12$ and $x = 4$
27. $3x - 10 = 7y$ and $x = \frac{-3}{4}$
28. $2y - 13 = -7x$ and $y = \frac{-5}{3}$
29. $7y = 3x + 6$ and $y = \frac{3}{8}$
30. $6x = -3y + 1$ and $x = \frac{4}{15}$
31. $4x + 7 = -3y + 3$ and $x = \frac{-7}{10}$
32. $4y + 9 = -5x + 2$ and $y = \frac{-11}{12}$
33. $-2x + 5y = -1$ and $x = 10y + 1$
34. $3x - 4y = -2$ and $x = 4 - 2y$
35. $5x = -8y + 6$ and $y = 5x - 3$
36. $5x = 3y - 4$ and $y = -3x - 2$
37. $3 + 7x = 4y$ and $x = 4y - 1$
38. $-3 - 4x = 2y$ and $x = 4y + 3$
39. $8x - 7 = -3y - 4$ and $y = 4x - 7$
40. $3 - 5x = 7 - 7y$ and $y = 5x - 3$
41. $-3x + 2y = -6$ and $y = 5x - 8$
42. $5x + 4y = 20$ and $y = 4x - 9$
43. $9x - 3 = 5y$ and $x = -7y + 6$
44. $5x + 15 = 8y$ and $x = -8y + 5$
45. $-7x = 2y + 1$ and $y = -6x - 1$
46. $2x = 10y + 3$ and $y = -7x - 3$
47. $-9x - 7 = -2y - 4$ and $x = -6y + 5$
48. $5x - 8 = 4y - 9$ and $x = -9y + 11$

Express each of the following word problems algebraically and then solve it. These problems can be translated by the methods in Section 13.3.

1. The larger of two numbers equals the sum of 6 and the smaller number. If the smaller number equals 10 reduced by the larger number, what are the numbers?
2. The first number is 5 times the second. If the first number equals 12 decreased by the second number, what are the numbers?
3. The smaller number is equal to the larger number reduced by 8, and the larger number is equal to the smaller number multiplied by 3. Find the numbers.
4. The sum of two numbers is 3. If one number equals 7 added to the other number, what are the numbers?
5. The product of the first number and 2 equals the second number. If the second number is 3 less than the first number, what are the numbers?
6. One number is 8 less than the second number. Find the numbers if the product of the second number and 5 equals the first number.
7. The first number is 11 increased by the second number. If the sum of the two numbers is –3, what are the numbers?
8. The first number decreased by the second number is –12. If the first number is 7 times as great as the second number, what are the numbers?
9. Subtracting 15 from one number gives the other number. Find the numbers if adding one number to the other number yields –9.
10. The first number is 11 more than the second number. If the sum of the second number and the product of 4 and the first number is 39, what are the numbers?
11. If the first number subtracted from –16 equals the second number, and if –40 increased by the product of the second number and 7 equals the first number, what are the numbers?
12. The first number is the sum of 21 and 6 times the second number. If the second number added to 5 times the first number is –50, what are the numbers?

These problems can be translated by the methods in Section 13.4.

13. The length of a rectangle is 7 meters more than its width. What are the dimensions of the rectangle if its perimeter is 62 meters?
14. What are the dimensions of a rectangle whose width is 3 centimeters less than its length, and whose perimeter is 50 centimeters?
15. The length of a rectangle is 3 times its width, and its perimeter is 120 millimeters. What are its dimensions?
16. The width of a rectangle added to 12 meters equals its length. What are its dimensions if its perimeter is 44 meters?
17. The length of a rectangle decreased by 10 meters equals its width, and its perimeter is 80 kilometers. Find its dimensions.
18. The length of a rectangle is 24 millimeters longer than its width, and its perimeter is 76 millimeters. What are its dimensions?
19. Find the dimensions of a rectangle whose width is 4 meters less than its length, and whose perimeter is 28 meters.
20. What are the dimensions of a rectangle with a length 4 times its width and a perimeter of 60 centimeters?

21. Seven times the width of a rectangle equals its length. What are its dimensions if its perimeter is 64 kilometers?

These problems can be translated by the methods in Section 13.5.

22. If Fred has 2 more dimes than quarters for a total of \$1.95, how many coins does he have?

23. Ethel has 7 fewer nickels than dimes. If the coins are worth \$2.35, how many coins does she have?

24. If Bruce has 6 times as many dimes as pennies, and together they are worth \$9.15, how many coins does he have?

25. Kim has \$7.40 in change. If she has only nickels and pennies, and if she has 140 fewer nickels than pennies, how many coins does she have?

26. Sam has 9 times as many quarters as half-dollars. If he has \$8.25, how many coins does he have?

27. If Lisa has 3 more half dollars than pennies and the coins are worth \$16.80, how many coins does she have?

28. Luis has 12 times as many pennies as quarters, and together they are worth \$2.22. How many coins does he have?

29. Gwen has \$3.65. If she has 5 more dimes than 3 times her nickels, how many coins does she have?

30. If Chet has 19 fewer nickels than quarters and together the coins are worth \$7.75, how much are the nickels worth?

31. Elena has 13 fewer quarters than half-dollars. If the coins are worth \$11.00, how many coins does Elena have?

These problems can be translated by the methods in Section 13.6.

32. Both the customer elevator and the freight elevator bring people to the twentieth floor. The customer elevator can carry 20 fewer people than the freight elevator. If the customer elevator makes 12 trips and the freight elevator makes 8 trips, 460 people can get to the twentieth floor. How many people can each elevator carry per trip?

33. Jerry has 55¢ special delivery stamps and 15¢ air mail stamps. He has 17 more air mail stamps than special delivery stamps. How many of each does he have if together the stamps are worth \$13.05?

34. The larger of two pick-up trucks can carry 3,000 pounds more of sand than the smaller truck. If the larger truck makes 6 full-capacity deliveries and the smaller truck makes 8 full-capacity deliveries, they will deliver 228,000 pounds of sand. What is the capacity of each truck?

35. Glenn has 26 more 20¢ stamps than 15¢ stamps. The stamps are worth \$20.95. How many of each stamp does he have?

36. Anita has foreign coins worth \$8.44. The French coins are worth 9¢ each and the German coins are worth 22¢ each. If she has 18 more French coins than German coins, how many of each type of coin does she have?

37. A bricklayer can lay 12 more bricks than his helper can in an hour. If the bricklayer works for 5 hours and the helper works for 6 hours, they can lay 258 bricks. How many bricks can each man lay per hour?

12.2 SOLVING TWO EQUATIONS BY ISOLATION AND SUBSTITUTION

In this section we will consider pairs of equations in which neither of the equations is solved for an unknown. However, it is always possible to solve either equation for either unknown, using the methods in Chapter 11.

Look at the following example of *solving two equations by isolation and substitution.*

TYPICAL PROBLEM $-3x + 8y = -25$ and $x - 4y = 11$

In the second equation, solve for one of the unknowns. If one of the unknowns has the coefficient +1 or −1, then solve for this unknown:

$$x - 4y = 11$$
$$\downarrow$$
$$\boxed{x = 11 + 4y}$$

Then we evaluate the first equation using the equation just found:

$$-3x + 8y = -25, \text{ when } x = 11 + 4y$$
$$-3(11 + 4y) + 8y = -25$$

And then we solve this equation:

$$-33 - 12y + 8y = -25$$
$$\downarrow$$
$$-12y + 8y = -25 + 33$$
$$\downarrow$$
$$-4y = 8$$
$$\downarrow$$

At this point we have one of the two equations of the solution.

$$\boxed{y = -2}$$

Using the solution just found, we evaluate the equation in the first box above (which was derived from the second equation):

$$x = 11 + 4y, \text{ when } y = -2$$
$$x = 11 + 4(-2)$$
$$\downarrow$$

And then we solve this equation:

$$x = 11 - 8$$
$$\downarrow$$

At this point we have the other equation of the solution.

$$\boxed{x = 3}$$

Finally we write the two equations of the solution together:

$$x = 3 \text{ and } y = -2$$

Hence, $-3x + 8y = -25$ and $x - 4y = 11 \Rightarrow x = 3$ and $y = -2$.

To solve two equations by isolation and substitution we extend the method in the previous section. We first isolate an unknown by the methods in Chapter 11.

EXAMPLES

Solve the following equations in two unknowns for x and y by the Substitution Method.

1. $2x = -3y - 5$ and $3x = y + 9$
2. $3x + 10 = 13 - 7y$ and $4x = y - 27$
3. $2x + 7y = -4$ and $5y = 5 - 3x$
4. $-7x + 2y = -8$ and $2x = y + 3$

Solutions

1. $3x = y + 9$

↓

$\boxed{y = -9 + 3x}$

↓

$2x = -3y - 5$, when $y = -9 + 3x$

$2x = -3(-9 + 3x) - 5$

↓

$2x = 27 - 9x - 5$

↓

$\boxed{x = 2}$

↓

$y = -9 + 3x$, when $x = 2$

$y = -9 + 3(2)$

↓

$y = -9 + 6$

↓

$\boxed{y = -3}$

↓

$x = 2$ and $y = -3$

2. $4x = y - 27$

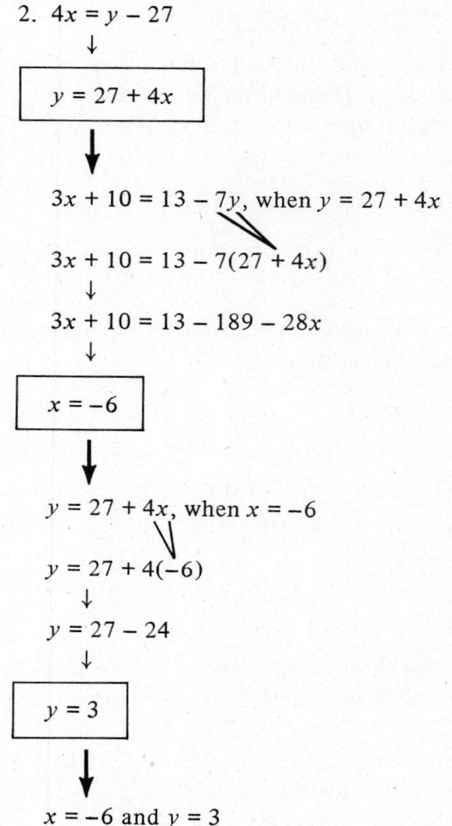

3. $5y = 5 - 3x$

$\downarrow$

$y = \dfrac{5 - 3x}{5}$

$\downarrow$

$2x + 7y = -4$, when $y = \dfrac{5 - 3x}{5}$

$2x + 7\left(\dfrac{5 - 3x}{5}\right) = -4$

$\downarrow$

$10x + 35 - 21x = -20$

$\downarrow$

$x = 5$

$\downarrow$

$y = \dfrac{5 - 3x}{5}$, when $x = 5$

$y = \dfrac{5 - 3(5)}{5}$

$\downarrow$

$y = -2$

$\downarrow$

$x = 5$ and $y = -2$

4. $2x = y + 3$

$\downarrow$

$y = 2x - 3$

$\downarrow$

$-7x + 2y = -8$, when $y = 2x - 3$

$-7x + 2(2x - 3) = -8$

$\downarrow$

$-7x + 4x - 6 = -8$

$\downarrow$

$x = \dfrac{2}{3}$

$\downarrow$

$y = 2x - 3$, when $x = \dfrac{2}{3}$

$y = 2\left(\dfrac{2}{3}\right) - 3$

$\downarrow$

$y = \dfrac{-5}{3}$

$\downarrow$

$x = \dfrac{2}{3}$ and $y = \dfrac{-5}{3}$

OBSERVATIONS AND REFLECTIONS

★ 1. Remember that two nonequivalent consistent linear equations have a single common root. This root is also common to many other equations, of course. Every equation that shares this common root is said to be *dependent* on the original pair. For example, $x = y$ and $x = 2y$ are both satisifed when x is 0 and y is 0. Hence, the equations $x = 5y$, $8x = y$, and $9x = 4y$ are dependent on this pair, because these equations are also satisfied when x is 0 and y is 0.

In particular, the solution of a pair of equations shares their common root, so each equation in the solution is dependent on the original pair. For example, the solution to $x = y$ and $x = 2y$ is $x = 0$ and $y = 0$, so $x = 0$ is dependent on $x = y$ and $x = 2y$, and $y = 0$ is dependent on $x = y$ and $x = 2y$.

★ 2. There are three rules for forming an equation that is dependent on a given pair of equations and therefore shares the common root. One of these is transformation rule 6. Evaluating one of two equations using the other always yields an equation that is dependent on these two equations. For example, evaluating $2x - y = 3$, when $x = y + 1$ yields

$$2(y + 1) - y = 3 \quad \Rightarrow \quad y + 2 = 3$$

3. Which of these equations are dependent on the pair of equations $x = 1$ and $y = 2$?
a. $x + y = 3$
b. $y = 2x$
c. $x - y = 1$
d. $y - x = 1$

and $y + 2 = 3$ is dependent on $x = y + 1$ and $2x - y = 3$. (All three equations are satisfied when x is 2 and y is 1.)

The other two transformation rules for working with pairs of equations are:

RULE 7

Equate the sum of the left members with the sum of the right members.

$$A = B \text{ and } C = D \quad \Rightarrow \quad A + C = B + D$$

(where A, B, C, and D represent linear expressions)

4. Using rule 7 form an equation dependent on
   a. $2x + 5y = -14$ and $3x - 5y = 29$
   b. $-4x - 7y = 29$ and $4x + 3y = -17$

For example, $2x = 3y + 2$ is dependent on $x = 2y$ and $x = y + 2$, because

$$\begin{array}{rl} & x = 2y \\ \text{“plus”} & x = \ \ y + 2 \\ \hline \text{yields} & 2x = 3y + 2 \end{array}$$

RULE 8

If the same expression appears as one member of two equations, then equate the other two members.

$$A = B \text{ and } A = C \quad \Rightarrow \quad B = C$$

(where A, B, and C represent linear expressions)

5. Using rule 8 form an equation dependent on
   a. $x = y + 7$ and $x = 2y - 9$
   b. $3y + 8 = x - 5$ and $3y + 8 = 12 + 2x$

For example, $2y = y + 2$ is dependent on $x = 2y$ and $x = y + 2$, because

$$x = 2y \text{ and } x = y + 2$$
$$\Rightarrow \quad 2y = \quad y + 2$$

EXERCISES

Solve each of the following pairs of equations for x and y.
In these exercises, all pairs of equations have integer solutions.

1. $2x + 3y = 13$ and $x + 2y = 8$
2. $3x - 2y = 8$ and $x - 3y = -2$
3. $2x - 9y = 42$ and $4x = 8 - y$
4. $4x + 7y = -23$ and $-2y = x + 6$
5. $-2y = -5x - 19$ and $-x - 4y = 17$
6. $-7x = -4y - 17$ and $3x - y = 8$
7. $-8x = 3y - 19$ and $6x = y + 11$
8. $-2x = -3y - 12$ and $3x = y + 4$
9. $8x - 4 = -7y$ and $3x + y = -5$
10. $4x + 9 = 11y$ and $-2x + y = 9$
11. $4y = -9x + 3$ and $-3x = y - 3$
12. $9y = 10 - 4x$ and $2x = 12 - y$
13. $3x + 15 = 7y - 16$ and $x + 5y = -3$
14. $7x - 15 = -8 - 4y$ and $4x + y = -5$
15. $7x + 19 = 5y - 9$ and $3y = x + 4$
16. $9x + 6 = 8y - 3$ and $2x = y + 5$

17. $5x + 4y = -5$ and $x - 11 = 3y + 7$
18. $3x - 8y = 21$ and $-x + 4 = 7 - 6y$
19. $11y = 3 - 2x$ and $2x + 5 = y - 4$
20. $5y = 7x - 15$ and $x - 8 = 3y + 1$
21. $7x = 5y + 19$ and $5x - 4 = y - 11$
22. $4x = 5y + 15$ and $x - 6 = 3y - 4$
23. $8x - 3 = -5 - 2y$ and $9 - 7x = y + 4$
24. $-8x + 5 = 3y - 6$ and $4x + 3 = 12 - y$
25. $4x + 9y = -7$ and $8x + 5y = 25$
26. $6x + 7y = 18$ and $3x + 10y = 48$
27. $3x - 14y = -7$ and $2x = 4 + 5y$
28. $5x - 2y = 2$ and $4x = -11 + 3y$
29. $5x + 3 = 2y$ and $-8x + 3y = 6$
30. $4x - 5 = 3y$ and $-6x + 5y = -11$
31. $11x - 18 = 2y$ and $-5x = 27 + 3y$
32. $9x - 24 = 3y$ and $-7x = 40 + 5y$
33. $5x - 9 = -4y - 10$ and $8x + 7y = -4$
34. $9x - 4 = -5y - 11$ and $7x + 4y = -6$
35. $-7x + 11 = 10 + 2y$ and $-5y = -17 + 3x$
36. $-8x + 7 = 6 + 3y$ and $-4y = -17 - 5x$

In these exercises, the solutions contain fractions.

37. $7x + 3y = 1$ and $x - 6y = 4$
38. $4x + 5y = -4$ and $x - 4y = 13$
39. $3x + 2y = -2$ and $4x = y - 1$
40. $4x + 6y = -2$ and $6x = y + 5$
41. $5x + 10 = 6y$ and $x + 3y = 1$
42. $4x + 8 = -7y$ and $4x = 1 - y$
43. $5y - 7 = -7x$ and $7y = x + 8$
44. $7y + 2 = 3x$ and $3y = -x - 6$
45. $4x + 3 = 3 - 5y$ and $8x + y = 3$
46. $2x - 5 = 3y - 8$ and $x + 6y = 2$
47. $-3y - 7 = -3x + 4$ and $-3y = x - 7$
48. $-3x + 5 = 5 - 4y$ and $-6x = 3 + y$

Express each of the following word problems algebraically and then solve it. These problems can be translated by the methods in Section 13.3.

1. The sum of two numbers is −37. Find the numbers if the smaller number decreased by the larger number is −11.
2. The larger number diminished by the smaller number is 23, and the larger number subtracted from 4 times the smaller number is −38. What are the numbers?
3. When the product of 3 and the first number is increased by the second number, the result is −26. If the sum of the first number and 5 times the second number equals 38, what are the numbers?
4. The smaller number increased by the larger number equals 0. If the product of 9 and the smaller number is subtracted from the larger number, the result is 10. Find the numbers.
5. Reducing 6 times the second number by the first number yields 27. If the product of 4 and the first number is reduced by 9 times the second number, the result is −3. What are the numbers?

6. When –1 times the first number is increased by the product of 4 and the second number, the result is 25. What are the numbers if 2 times the second number is 35 more than 3 times the first number?

7. When the product of 4 and a certain number is subtracted from 14 times another number, the result is 48. What are the numbers if the sum of the numbers is –3?

8. If the product of 3 and the first number is added to the product of 2 and the second number, the result is 18. Subtracting the first number from 5 times the second number yields –74. Find the numbers.

These problems can be translated by the methods in Section 13.4.

9. The length of a rectangle multiplied by 4 kilometers equals the width subtracted from 49 kilometers. What are the dimensions of the rectangle if its perimeter is 38 kilometers?

10. The perimeter of a rectangle is 56 meters, and its length increased by 11 meters equals its width added to 13 meters. Find its dimensions.

11. If the perimeter of a rectangle is 130 centimeters, and its length plus 35 centimeters equals 3 times its width, what are its dimensions?

12. If the length of a rectangle is subtracted from 11 meters, the result is its width subtracted from 6 meters. What are its dimensions if its perimeter is 58 meters?

13. Find the dimensions of a rectangle whose perimeter is 234 millimeters and whose length minus 16 millimeters is 37 millimeters more than its width.

14. The perimeter of a rectangle is 98 centimeters. Eighty-one centimeters more than its width is 4 times its length. What are its dimensions?

These problems can be translated by the methods in Section 13.5.

15. Mike has nickels and quarters worth $4.35. If he has 35 coins, how many quarters does he have?

16. Arthur saved dimes and half-dollars, and his collection was worth $7.70. If he had 29 coins, how many half-dollars did he have?

17. Juan saves only pennies and Marilyn saves dimes. Together they have 370 coins worth $5.50. How many coins has each of them saved?

18. Keith counted his money and found that five times the number of dimes was 12 more than the number of quarters. If the coins were worth $7.80, how many coins did he have?

19. If 3 times the number of Kristen's nickels is added to 12 times her pennies, the result is 189. How many coins does she have if they are worth 87 cents?

20. Twenty-one more than the number of Agatha's dimes is 7 times the number of her nickels. If the dimes and nickels are worth $2.40, how many coins does she have?

21. Estelle has nickels and quarters worth $3.90. If 5 times the number of her quarters is 8 less than the number of her nickels, how many quarters does Estelle have?

22. If Victor adds 13 times the number of dimes he has to 2 times his pennies, he gets 128. How many coins does he have if they are worth 71 cents?

23. If twice the number of Jason's quarters is subtracted from 15 times his pennies, the result is 10. How many coins does Jason have if they are worth $10.06?

24. Jenny has quarters and dimes worth $9.75. If she has 51 coins, how many quarters does Jenny have?

These problems can be translated by the methods in Section 13.6.

25. A furniture company sells desks and chairs. The desks sell for $210.00 each and the chairs for $55.00 each. If a total of 240 chairs and desks were sold for $27,150.00, how many desks were sold?

26. A carpenter ordered 195 pounds of nails and screws. The nails cost 50 cents a pound and the screws cost 85 cents a pound. How many pounds of screws did he order if the order cost him $116.75?

27. A total of 950 tickets were sold for a hockey game. Adults paid $7.50 each and senior citizens paid $6.00. If the total receipts for the game were $6,900.00, how many adult tickets were sold?

28. Sixty-five plumbers earn a total of $19,400.00 a week. If apprentices earn $200.00 a week and master plumbers earn $360.00 a week, how many apprentices are there?

29. Students paid $4.00 each for a school play and adults paid $6.00. If 1,260 tickets were sold for a total of $5,660.00, how many student tickets were sold and how many adult tickets were sold?

30. An order of 480 books costing $6,120.00 was delivered to a school bookstore. If the books were mathematics books costing $12.00 each and chemistry books costing $14.00, how many chemistry books were delivered?

31. One customer purchased 3 chairs and 1 table for $450.00. Another customer bought 2 tables and 10 chairs for $1,140.00. What is the price of one table and one chair?

32. Emily bought 3 pounds of apples and 5 pounds of pears for $3.55. At the same store her friend Beth paid $2.95 for 7 pounds of apples and 1 pound of pears. What is the cost of a pound of apples and what is the cost of a pound of pears?

33. Nine pounds of chocolate are mixed with 1 pound of nuts to make a candy selling for $1.62 a pound. If 16 pounds of chocolate were mixed with 4 pounds of nuts, the candy would sell for $1.64 a pound. What is the cost of a pound of chocolate?

34. Two kilograms of Colombian coffee are mixed with 1 kilogram of Brazilian coffee to produce a mixture selling for $5.40 a kilogram. If 3 kilograms of Colombian coffee were mixed with 2 kilograms of Brazilian coffee, the mixture would sell for $5.46 a kilogram. What is the value of a kilogram of each type of coffee?

35. Seven bags of foam rubber are mixed with 3 bags of polyfoam to produce a mixture costing $3.24 a bag. If 2 bags of foam rubber were mixed with 1 bag of polyfoam, the mixture would cost $3.20 a bag. What is the cost of each type of rubber?

36. If 4 liters of peanut oil were blended with 1 liter of corn oil, the blend would sell for $3.88 a liter, and if 7 liters of peanut oil were blended with 3 liters of corn oil, the blend would sell for $4.07 a liter. How much is a liter of pure peanut oil and how much is a liter of pure corn oil?

12.3
SOLVING TWO EQUATIONS BY GRAPHING

To solve a pair of (nonequivalent consistent) linear equations in x and y we must find their common root. Since the graph of an equation in x and y is the set of points in the plane corresponding to the roots of the equation, the common root of a pair of equations must correspond to a common point in the graphs of the two equations. That is, if two linear equations in x and y have a unique common root,

1. their graphs, which are straight lines, must cross, and
2. the point of intersection must correspond to their common root.

For example, the pair of equations $x = y + 1$ and $2x - y = 3$ can be solved graphically by graphing both equations on the same set of axes:

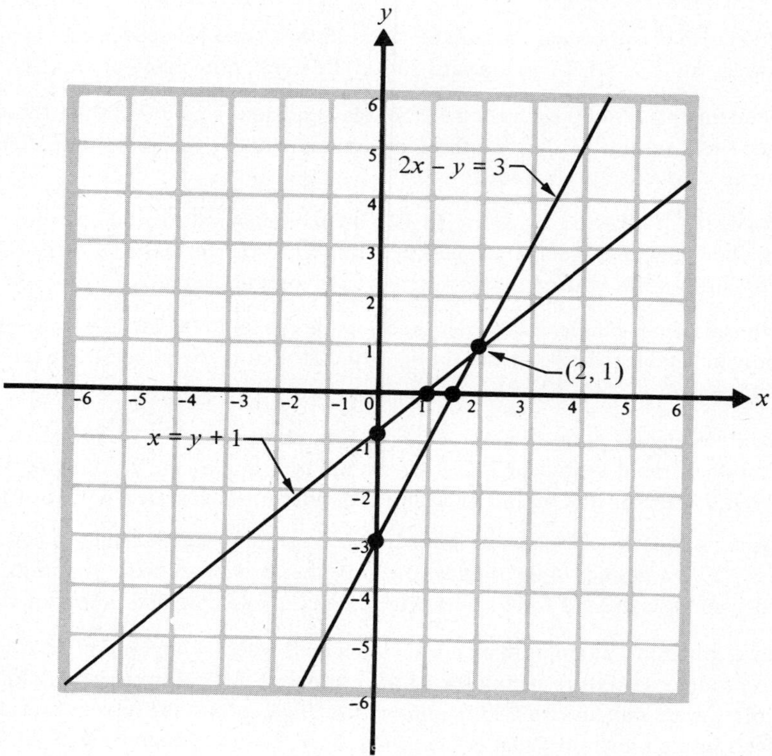

The graphs of these two equations intersect at the point (2, 1), which corresponds to the common root, because both equations are satisfied when x is 2 and y is 1.

Estimating Graphical Solutions

Solving a pair of equations algebraically, by the Substitution Method, yields a precise solution. When we solve a pair of equations graphically, on the other hand, we can only estimate the coordinates of the point of intersection. For example, we can estimate the coordinates of points in the plane as follows:

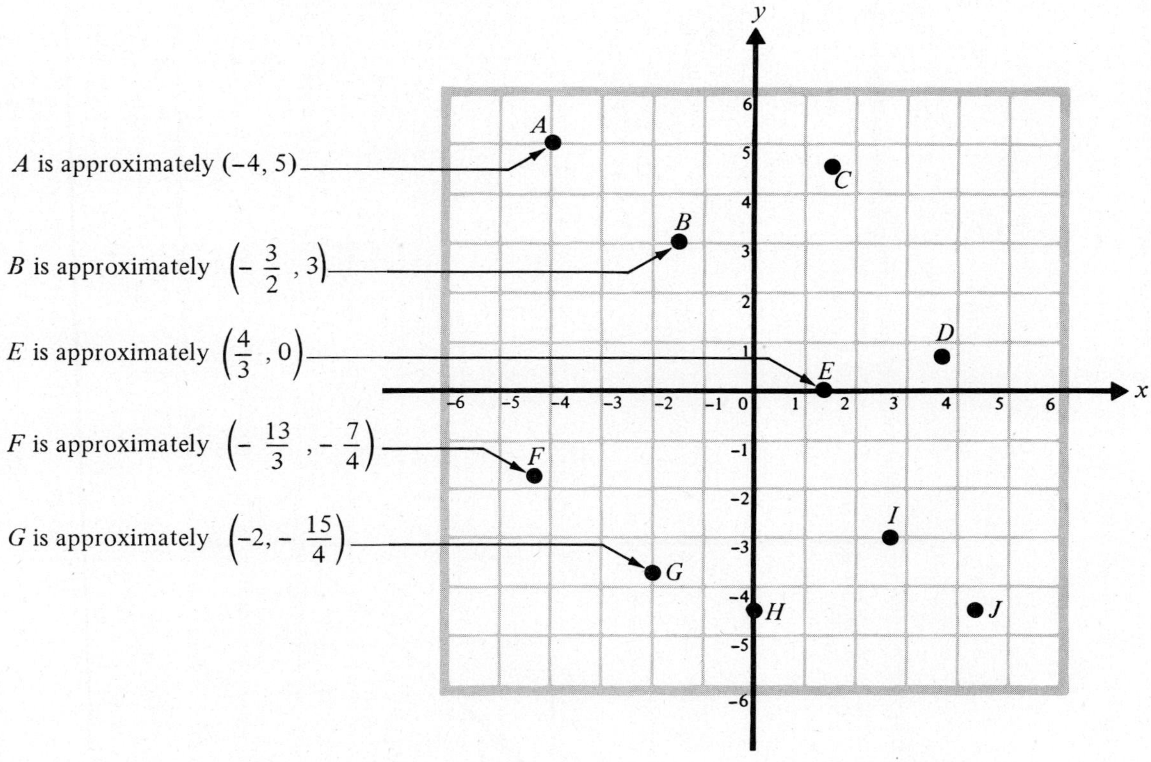

Hence, solving a pair of equations graphically yields only an approximate solution.

To simplify the task of estimating the coordinates of the point of intersection (and to increase the accuracy of our estimates), we will only consider pairs of equations in this section with integer-valued solutions.

6. Estimate the coordinates of the points C, D, H, I, J.

Look at the following example of *solving two equations by graphing.*

TYPICAL PROBLEM

$x - y = 5$ and $7x + 3y = 15$

↓

We first form a table of values for the first equation and graph it:

$x - y = 5$

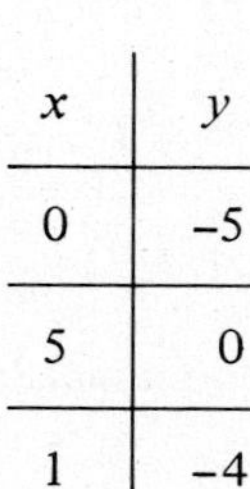

x	y
0	-5
5	0
1	-4

↓

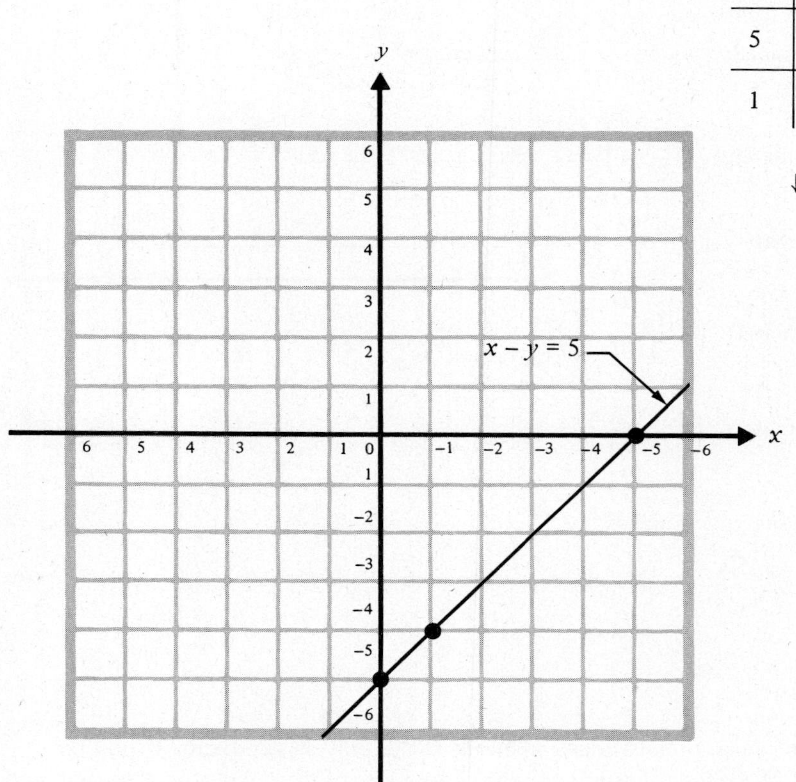

$7x + 3y = 15$

x	y
0	5
$\frac{15}{7}$	0
3	-2

↓

Then we form a table of values for the second equation and we graph it along with the first equation on the same set of axes:

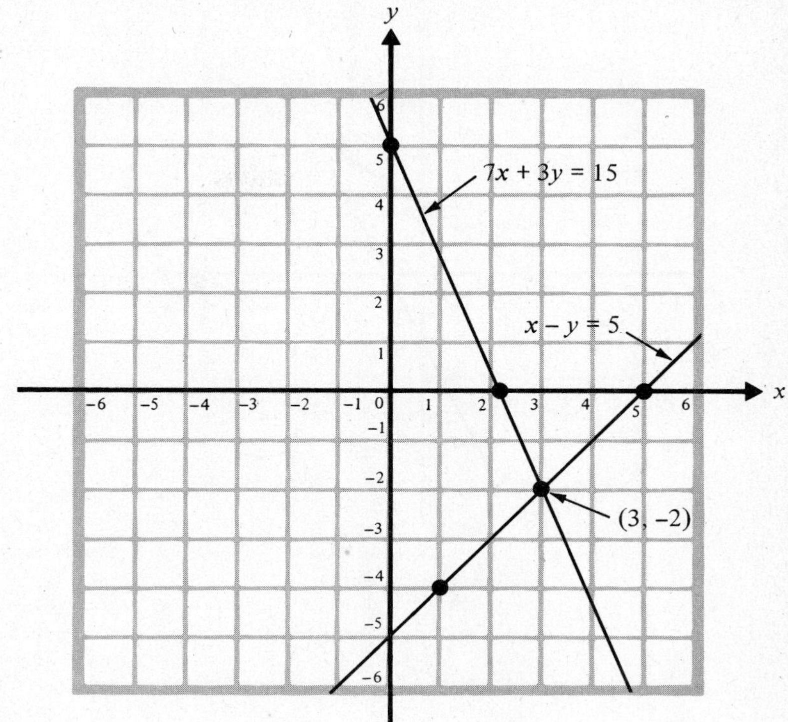

We then estimate the coordinates of the point where the two lines intersect. This point corresponds to the common root of the two equations:

The point of intersection is (3, − 2).

Hence, $x - y = 5$ and $7x + 3y = 15 \Rightarrow x = 3$ and $y = -2$.

Each equation is graphed by the method in Section 11.4.

EXAMPLES

Solve the following pairs of equations by graphing and estimate the points of intersection.

1. $4x = 2y + 4$ and $2x + 6 = 3y$
2. $6x + y = -6$ and $2y - 6 = -3x$
3. $5x = 10 - 2y$ and $3x + 8 = -4y$
4. $5x = 3y - 9$ and $2x = 7y + 8$

Solutions

1. $4x = 2y + 4$

x	y
0	−2
1	0
2	2

$2x + 6 = 3y$

x	y
−3	0
0	2
−6	−2

↓

2. $6x + y = -6$

x	y
−1	0
0	−6
−2	6

$2y - 6 = -3x$

x	y
2	0
0	3
1	$\frac{3}{2}$

↓

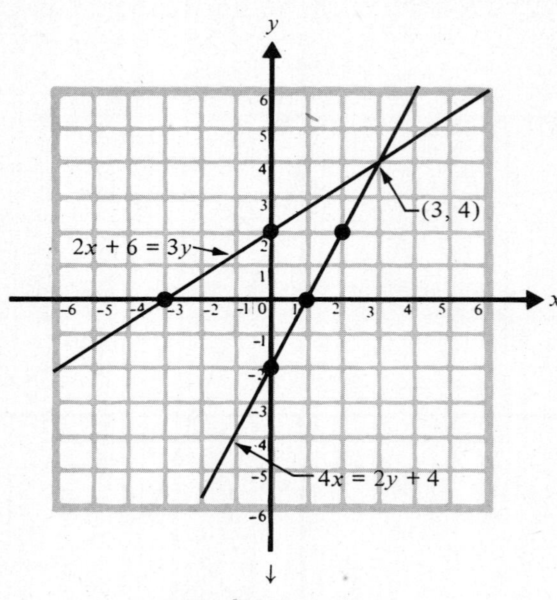

↓

$x = 3$ and $y = 4$

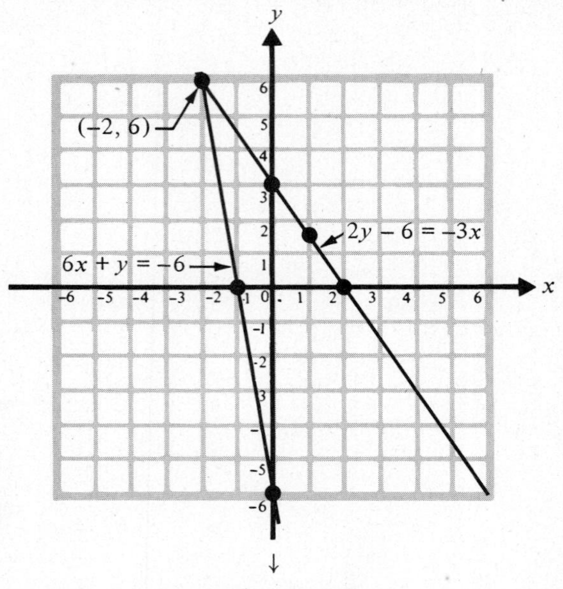

↓

$x = -2$ and $y = 6$

3. $5x = 10 - 2y$ $\quad$ $3x + 8 = -4y$

x	y
2	0
0	5
1	$\frac{5}{2}$

x	y
0	-2
$\frac{-8}{3}$	0
2	$\frac{-7}{2}$

↓

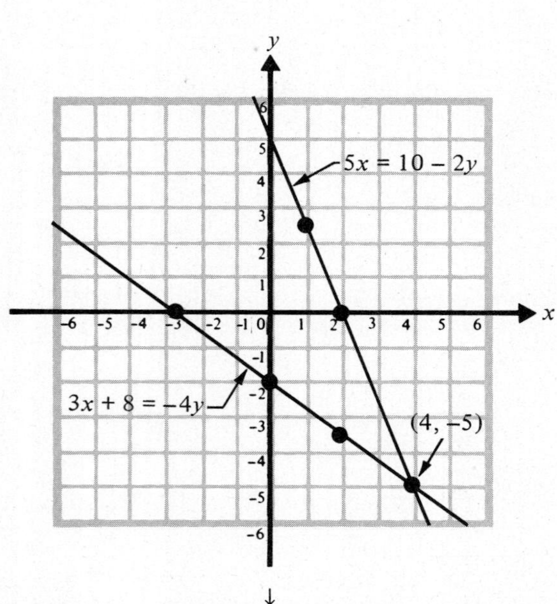

↓

$x = 4$ and $y = -5$

4. $5x = 3y - 9$ $\quad$ $2x = 7y + 8$

x	y
0	3
$\frac{-9}{5}$	0
-1	$\frac{4}{3}$

x	y
0	$\frac{-8}{7}$
4	0
2	$\frac{-4}{7}$

↓

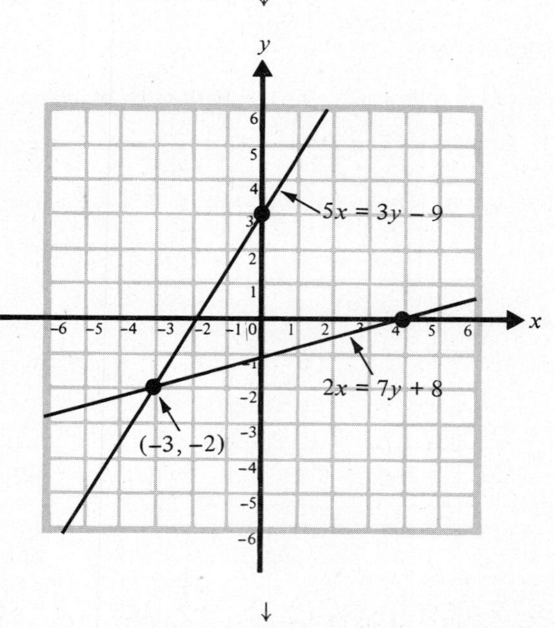

↓

$x = -3$ and $y = -2$

SPECIAL CASES

When solving a pair of equations graphically, if one of the equations has the form

$$ax = c \quad \text{or} \quad by = c$$

where a, b, and c represent numbers with a and b nonzero, then we treat it as an equation in x and y of the form

$$ax + 0y = c \quad \text{or} \quad 0x + by = c$$

for the purposes of graphing. For example, to solve the pair of equations

$$3x - 2y = -6 \quad \text{and} \quad x = -4$$

graphically, we determine the tables of values and graph as follows:

$3x - 2y = -6$

x	y
0	3
−2	0
−4	−3

$x + 0y = -4$

x	y
−4	5
−4	0
−4	−3

$2x + 5y = -10$

x	y
0	−2
−5	0
5	−4

$0x + y = -4$

x	y
5	−4
0	−4
−3	−4

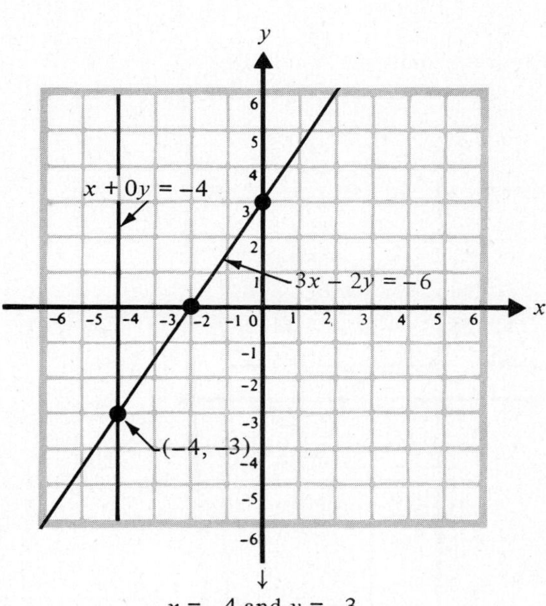

$x = -4$ and $y = -3$

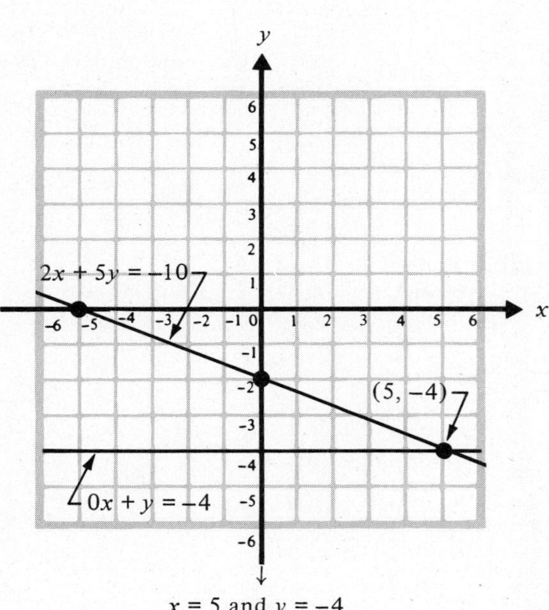

$x = 5$ and $y = -4$

Note that the graph of $x = -4$ is a *vertical* line, whereas the graph of $y = -4$ is a *horizontal* line.

7. Graph
   a. $x = 2$ and $y = -3$
   b. $x = -4$ and $x = y$
   c. $x + y = 2$ and $y = 2$

OBSERVATIONS AND REFLECTIONS

Solving a pair of linear equations algebraically using the Substitution Method yields a solution of the form $x = x$-value and $y = y$-value, whereas solving graphically yields the point of intersection (x value, y value) in the plane, but for the same x and y values. We can, if we wish, take the algebraic solution and, treating it as we would any other pair of equations, find its common root graphically.

For example, the algebraic solution to $x = y + 1$ and $2x - y = 3$ is $x = 2$ and $y = 1$; we saw in the introduction to this section that the graphical solution is

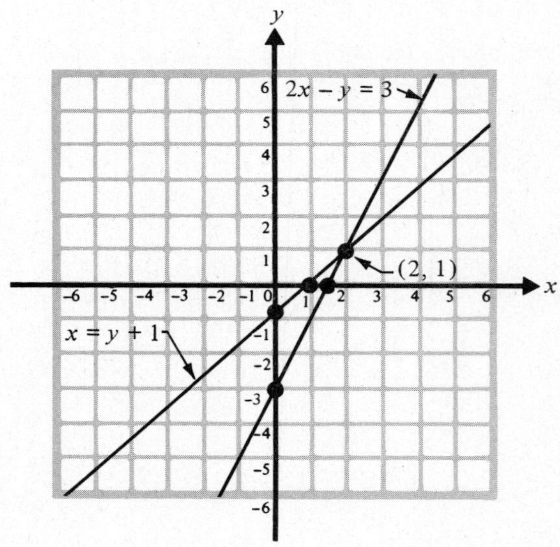

If we now graph the pair of equations $x = 2$ and $y = 1$, we obtain

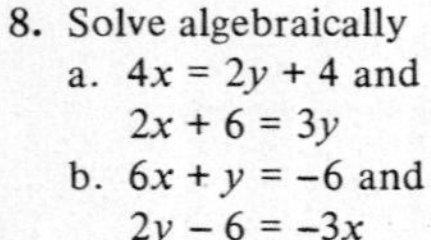
8. Solve algebraically
a. $4x = 2y + 4$ and $2x + 6 = 3y$
b. $6x + y = -6$ and $2y - 6 = -3x$

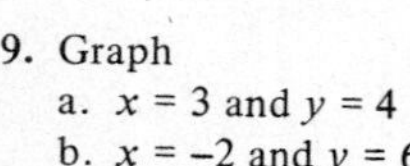
9. Graph
a. $x = 3$ and $y = 4$
b. $x = -2$ and $y = 6$

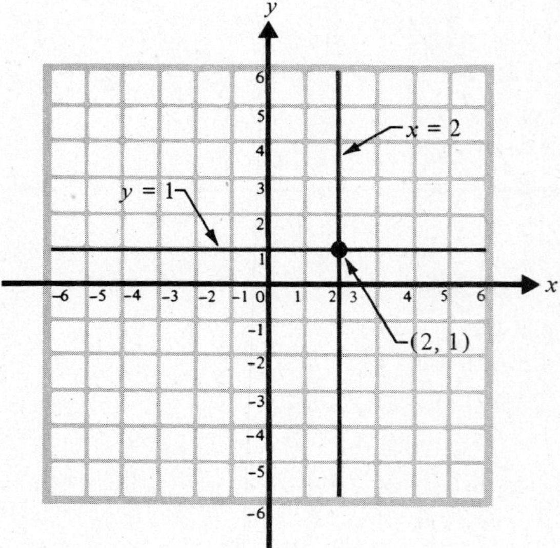

Observe that the graphs of $x = 2$ and $y = 1$ intersect at the same point as the graphs of $x = y + 1$ and $2x - y = 3$. Given any pair of linear equations with a unique solution, the graphs of their solution are a horizontal and vertical line that cross at the same point as the graphs of the two equations.

EXERCISES

Graph each of the following pairs of equations and estimate the value of the common root. In these exercises, all pairs of equations have common roots with integer values.

1. $3x - 5y = -15$ and $x - y = -1$
2. $x - 6y = 6$ and $x - y = -4$
3. $2x + y = 2$ and $y = 3 - x$
4. $-3x + y = 3$ and $y = -x - 5$
5. $4y = 5x + 20$ and $3x = -4y - 12$
6. $3y = 4x + 12$ and $5x = -3y - 15$
7. $y = 2x - 6$ and $x = 2$
8. $y = -3x + 6$ and $y = -3$
9. $2x - y = 4$ and $5y = 2x + 4$
10. $2x + 3y = 6$ and $3x = -y - 5$
11. $4x + 12 = -3y$ and $5x + 6 = -6y$
12. $3x + 6 = -2y$ and $5x + 12 = -3y$
13. $5x = 2y + 10$ and $5y = -6x - 25$
14. $5x = 4y - 20$ and $8y = -5x - 20$
15. $y = -1$ and $7x + 17 = 4y$
16. $x = 4$ and $7x - 23 = 5y$
17. $8x - 3y = 12$ and $4x - 5y = -8$
18. $5x - 3y = 10$ and $4x - 5y = -5$
19. $5x = 4y - 24$ and $-3y = x + 1$
20. $3x = 4y - 15$ and $7y = -3x + 18$
21. $20 = 5x - 6y$ and $4x + 9y = 16$
22. $15 = -4x + 3y$ and $9x + 4y = 20$
23. $5y = 5 - 3x$ and $6x + 5y = 20$
24. $2y = -8 - 3x$ and $7x + 6y = -12$

ANSWERS TO MARGIN EXERCISES

1. a. $7y - 3$
b. $6 - 2y$
c. $7y - 3 = 6 - 2y$

2. a. $x = 2$
b. $x = 2$

3. a, b, and d are dependent.

4. a. $5x = 15$
b. $-4y = 12$

5. a. $y + 7 = 2y - 9$
b. $x - 5 = 12 + 2x$

6. C is approximately $\left(\frac{3}{2}, \frac{9}{2}\right)$
D is approximately $\left(\frac{15}{4}, \frac{3}{4}\right)$
H is approximately $\left(0, \frac{-9}{2}\right)$
I is approximately $\left(\frac{11}{4}, -3\right)$
J is approximately $\left(\frac{13}{3}, \frac{-9}{2}\right)$

8. a. $x = 3$ and $y = 4$
b. $x = -2$ and $y = 6$

9. Compare your answers with the graphical solutions of Examples 1 and 2.

CHAPTER REVIEW

SECTION 12.1

Solve for x and y.

1. $8x + 5y = -9$ and $y = 3$
2. $7x = 3y + 46$ and $x = 4$
3. $2x = -6y - 24$ and $y = -5$
4. $-7x + 24 = -9y - 25$ and $y = 0$
5. $x + 4y = 24$ and $y = 2x - 3$
6. $3x = y - 11$ and $x = 5 - 4y$
7. $7y + 16 = 2x$ and $y = 2x + 8$
8. $-4x + 13 = -7y - 31$ and $x = -11 - y$
9. $5x + 3y = 46$ and $x = 4y$
10. $-9x - 16 = -y$ and $y = 5x - 4$
11. $8y = -3x - 18$ and $x = -4y - 2$
12. $9y - 17 = -4x + 1$ and $x = 6y - 12$
13. $3x - 4y = -5$ and $x = 3$
14. $5x = 7 - 2y$ and $y = \frac{-2}{3}$
15. $7y - 1 = 3x$ and $x = \frac{5}{6}$
16. $-8 + 2y = -9 + 8x$ and $y = \frac{-7}{6}$
17. $3x - 5y = -9$ and $x = 4y - 6$
18. $5x = -2 - 4y$ and $y = -3 - 5x$
19. $3 - 2x = 10y$ and $x = 7y - 3$
20. $3x - 4 = -4y - 5$ and $y = -7x - 9$
21. $-2x + 3y = -5$ and $x = -3y - 1$
22. $-5y + 2 = -4x$ and $y = -4x + 6$
23. $-4x = -7y - 13$ and $y = -4x - 7$
24. $-4x - 3 = 2y - 10$ and $x = -8y + 3$

SECTION 12.2

Solve for x and y.

25. $2x + 3y = -1$ and $2x + y = -7$
26. $-3x - 5y = -23$ and $-4y = x - 3$
27. $4x = 7y + 10$ and $x - 4y = -2$
28. $-5y = -4x - 32$ and $-8y = -x - 8$
29. $11 - 4x = 5y$ and $5x + y = -2$
30. $3x = 1 - 11y$ and $5y = 3 - x$
31. $4x - 14 = 18 - 4y$ and $x + 6y = -7$
32. $7 + 6x = 4y - 9$ and $-4x - 15 = 4 - y$
33. $7x + 3 = 5y + 20$ and $3x - 7 = y - 10$
34. $2x + 14 = 15 - 11y$ and $9 - x = 7 + 4y$
35. $9x = 8y - 1$ and $11 - 2x = 5 - y$
36. $4x + 7 = 8y - 5$ and $x - 5 = -3y - 8$
37. $3x + 4y = -24$ and $5x + 3y = -7$
38. $7x + 5y = -17$ and $3x = 17 - 7y$
39. $2x - 20 = 3y$ and $3x + 8y = 5$
40. $-5y - 9 = -6x$ and $-3y = 1 - 8x$
41. $2x - 3 = 9y - 11$ and $5x - 7y = 11$
42. $3x + 7 = 10 - 3y$ and $-2x = 5y - 23$
43. $7x + 4y = 4$ and $x + 7y = -2$
44. $4x + 5y = 11$ and $5x = 5 - y$

45. $5x + 9 = -3y$ and $3y = -6 - x$

46. $2y - 1 = -3x$ and $-6y = x + 3$

47. $4x + 7 = 7y + 1$ and $2x + y = 3$

48. $-3x + 8 = 5 - 2y$ and $-3y = 2 - x$

SECTION 12.3

Graph and estimate the common root.

49. $x - 3y = -6$ and $x - y = 2$

50. $2x - y = 6$ and $y = -x - 3$

51. $5y = 4x + 20$ and $2x = -5y - 10$

52. $y = -2x - 6$ and $x = -2$

53. $x + 2y = 6$ and $3x = -y - 2$

54. $-3x + 12 = -4y$ and $-6x + 6 = -5y$

55. $3x = 5y + 15$ and $8y = 25 - 5x$

56. $y = 2$ and $3x + 7 = 8y$

57. $7x - 4y = -20$ and $x - 3y = 2$

58. $5x = 2y - 8$ and $-y = 7x + 15$

59. $12 = -7x + 4y$ and $6x + 5y = 15$

60. $4y = 8 - 7x$ and $9x + 4y = 16$

13 TRANSLATING WORD PROBLEMS

The exercises in Chapters 10 and 12 contain many word problems that, *once translated into equations*, can be solved using the methods of those chapters. Translating from word problems to equations, however, is not always an easy task. In this chapter we will study this part of the process of solving word problems.

The word problems in Chapters 10 and 12 were grouped according to the methods used to solve their corresponding equations. Here we will only consider the process of translating word problems into equations, and we will consider number problems, consecutive integer problems, mixture problems, and so forth in separate sections.

Using Literals to Represent Numbers

Word problems talk about numerical quantities without specifying all of their values. The usual task in solving a word problem is to determine the unknown quantities using the information given. The algebraic approach to solving word problems is to use literals to *name* unknown quantities. We can then express the information given about the unknown quantities in algebraic equations whose roots are the quantities named by the literals.

The word problems in this chapter all translate into *linear equations* such as those in Chapters 10 and 12. Hence, once we have translated these problems into equations, we can use the methods of those chapters to determine the roots.

13.1 NUMBER PROBLEMS WITH ONE UNKNOWN

Number problems with one unknown are a class of word problems that can be identified as follows:

1. They involve a single unknown quantity.

2. The information given about the quantity is in the form of a numerical relation. (The unknown quantity may be mentioned more than once in the relation.)

For example, the word problem

> Six times a certain number is 20 more than the number. Find the number.

has a single unknown quantity (in this case the number 4) and states a numerical relation of this number (in this case that 6 times 4 is 20 more than 4).

Translating

To translate such a word problem into an equation we must do three things:

1. Name the unknown quantity with a literal.
2. Identify the parts of the stated numerical relation. We must identify the two members of the equation and the equality word(s).
3. Translate each member into an algebraic expression, and separate the two members by the equals sign, =.

For example, to translate the word problem given, we first name the unknown with a literal, say, x. We then identify the parts of the relation:

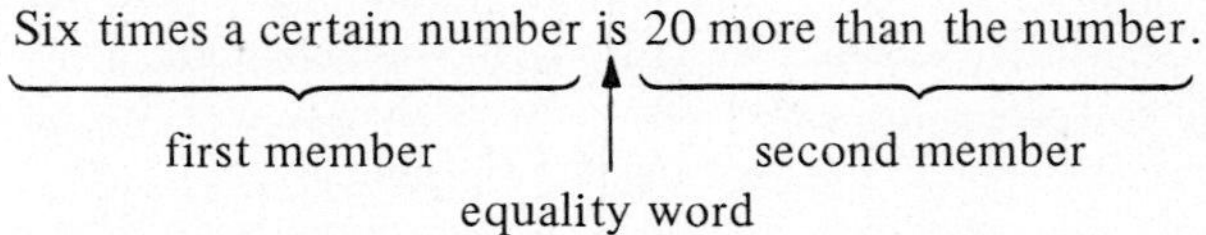

Finally, we translate each member:

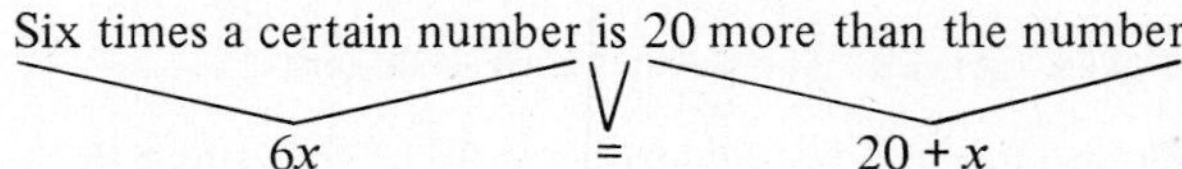

We learned how to translate word phrases into algebraic expressions in earlier chapters, so to translate these sentences we need only learn to identify the equality word or words. There are a number of ways to express equality in English. Let A and B represent word phrases and look at the following table to see how $A = B$ can be represented:

1. Translate
   a. The number is 6.
   b. Twice the number equals 8.
   c. Adding 1 to the number yields 3.
   d. Increasing the number by 4 results in twice the number.
   e. When 5 is subtracted from a number, the result is 4.
   f. If twice a number is added to 6, the result is 24.

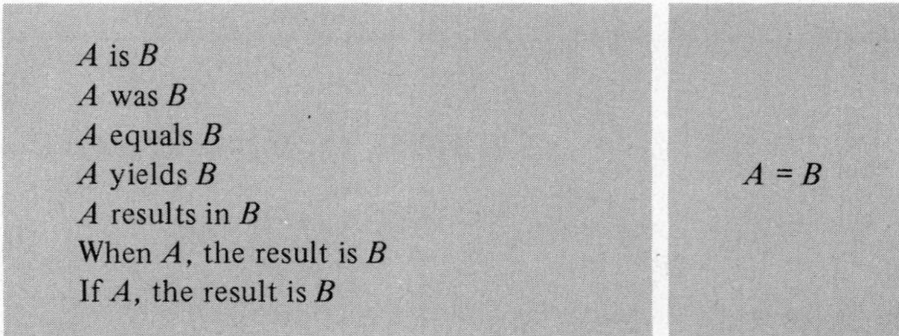

A is B A was B A equals B A yields B A results in B When A, the result is B If A, the result is B	$A = B$

Look at the following example *of translating number problems in one unknown into equations.*

TYPICAL PROBLEM

The product of 12 and the sum of 6 and a certain number is 16 less than the number. Find the number.

↓

First we read the problem and double underline the equality word(s):

The product of 12 and the sum of 6 and a certain number <u>is</u> 16 less than the number. Find the number.

Then we translate the phrase on each side of the equality word(s) and form the equation. The sentence Find the number becomes Find the root of the equation:

↓

$$12(6 + x) = x - 16$$

Find the root.

Hence, the word problem translates into the equation $12(6 + x) = x - 16$, whose solution is $x = -8$. Therefore, -8 is the number specified in the word problem.

EXAMPLES

Translate the following number problems in one unknown into equations.

1. Three times a certain number is 12 more than the number. Find the number.
2. When 9 subtracted from a certain number is multiplied by 7, the result is 28. Find the number.
3. Reducing 7 times a certain number by 11 yields 41 decreased by the number. Find the number.
4. If 6 added to a certain number is divided by 5, the result is the number increased by 2. Find the number.

Solutions

1. Three times a certain number <u>is</u> 12 more than the number.

↓

$$3x = 12 + x \qquad (\Rightarrow x = 6)$$

2. When 9 subtracted from a certain number is multiplied by 7, <u>the result is</u> 28.

↓

$$(x - 9)7 = 28 \qquad (\Rightarrow x = 13)$$

3. Reducing 7 times a certain number by 11 <u>yields</u> 41 decreased by the number.

↓

$$7x - 11 = 41 - x \qquad \left(\Rightarrow x = \frac{13}{2}\right)$$

4. If 6 added to a certain number is divided by 5, <u>the result is</u> the number increased by 2.

↓

$$\frac{6 + x}{5} = x + 2 \qquad (\Rightarrow x = -1)$$

2. Translate
  a. The number is reduced by 5.
  b. The number is 11 reduced by 5.
  c. Four times the number is increased by 12.
  d. Four times the number is the number increased by 12.

SPECIAL CASES

The equality word in a problem is frequently *is*. However, *is* can appear in a word problem without being the equality word. Consider, for example,

When 3 is added to a number, the result is 5.

Note that the word *is* appears twice in this sentence. Its first appearance is in the phrase "is added to," which indicates the operation of addition. The word *is* also appears in the phrase "the result is," where it represents the equals sign.

When the word *is* appears in a phrase such as "a is subtracted from b" or "a is multiplied by b," it is part of a phrase indicating an operation. In this case it does not represent the equals sign.

EXERCISES

Translate each of the following word problems into an equation and then solve it. These problems can be solved by the methods in Chapter 10.

1. The sum of 12 and a certain number equals 21 decreased by twice the number. Find the number.
2. The sum of a certain number and 30 is 7 times the number. What is the number?
3. If 28 plus a certain number equals the product of −6 and the number, what is the number?
4. If 18 less than a certain number equals 3 times the number, what is the number?
5. If the product of a certain number and 9 equals 12 reduced by the number, what is the number?
6. A certain number multiplied by 21 is 15 more than the number. Find the number.
7. Fourteen is equal to 18 subtracted from the sum of a certain number and 43. What is the number?
8. Eight times the sum of a certain number and −11 is 20. Find the number.
9. When the sum of 24 and a certain number is multiplied by 4, the result is 16. Find the number.
10. If 5 less than a certain number is multiplied by 3, the result is −75. What is the number?
11. When 9 times a certain number is subtracted from 61, the result is 19. What is the number?
12. Thirty-seven times a certain number is 8 less than the product of the number and 25. Find the number.
13. If 15 less than a certain number is multiplied by 12, the result is twice the number. What is the number?
14. If the sum of a certain number and 31 is multiplied by 7, the result is 89 less the number. Find the number.
15. The product of 76 and a certain number is 43 times the sum of the number and −1. What is the number?
16. When 18 times a certain number is decreased by 45, the result is the number multiplied by −7. What is the number?

17. Fifty-eight decreased by 4 times a certain number is the sum of 8 and 6 times the number. Find the number.

18. When the product of a certain number and 7 is reduced by 12, the result is 78 decreased by 11 times the number. What is the number?

19. If 8 more than a certain number is multiplied by −6, the result is 74 less than the number minus 52. What is the number?

20. When the sum of 43 and a certain number is subtracted from 69, the result is the same as when 12 more than the number is multiplied by 4. Find the number.

These problems can be solved by the methods in Section 10.4.

21. The sum of 12 and a certain number is $\frac{3}{5}$ the number. What is the number?

22. When a certain number is divided by 9, the result is 15 diminished by the number. What is the number?

23. A certain number increased by 21 equals $\frac{5}{6}$ the number. Find the number.

24. When a certain number is subtracted from 74, the result is the number divided by 5. What is the number?

25. Fifteen increased by $\frac{3}{4}$ a certain number is $\frac{3}{10}$. What is the number?

26. Nine times a certain number divided by 11 is $\frac{3}{4}$. Find the number.

27. If the sum of a certain number and 6 is divided by 8, the result is $\frac{2}{5}$. Find the number.

28. The product of $\frac{9}{16}$ and a certain number equals 10 reduced by $\frac{1}{6}$ the number. What is the number?

29. A certain number times $\frac{7}{9}$ equals 2 subtracted from $\frac{3}{5}$ the number. What is the number?

30. A certain number divided by 27 equals $\frac{1}{9}$ the sum of the number and 4. Find the number.

13.2 CONSECUTIVE INTEGER PROBLEMS

Consecutive integer problems are a special class of number problems with one unknown. To understand how to deal with these problems we must look at some properties of integers.

Consecutive Integers

Remember that the integers are the numbers

$$\ldots -4, -3, -2, -1, 0, +1, +2, +3, +4, \ldots$$

Two integers are called *consecutive integers* if one is 1 more than the other. For example, the pairs of integers

$$1,2 \qquad 7,8 \qquad -6,-5 \qquad -1,0$$

are each consecutive. Any two or more integers are called *consecutive* if they can be arranged so that each is 1 more than the preceding one. For example,

$$0, 1, 2 \quad \text{and} \quad 7, 8, 9, 10 \quad \text{and} \quad -8, -7, -6$$

are consecutive, whereas

$$0, 2, 3 \quad \text{and} \quad -2, -1, 1, 2 \quad \text{and} \quad 6, 8, 10$$

are not consecutive.

Note that two consecutive integers are nearest neighbors in that there are no other *integers* between them. (Of course, an unlimited number of *fractions* lie between any two consecutive integers.)

When talking about consecutive integers, we use some specialized words. The smallest integer is called the *first integer* or the *first consecutive integer.* The next larger integer is called the *second integer* or the *second consecutive integer.* Successively larger integers are referred to as the *third, fourth,* and so forth integers. The largest integer is also called the *last integer* or the *last consecutive integer*. For example, in the consecutive integers 3, 4, 5, the first integer is 3, the second is 4, and the last is 5.

3. Which of these sets of integers are consecutive?
   a. 2 4
   b. 5, 6
   c. 3 4. 5
   d. 3, 5, 7

Translating into Symbols

To translate a consecutive integer problem into an equation, the first step is to name the consecutive integers using a literal (by tradition the letter n). We proceed as follows:

The first integer	is named	n
The second integer	is named	$n + 1$
The third integer	is named	$n + 2$
The fourth integer	is named	$n + 3$
and so forth.		

When we name each integer, the number added to n is the *difference* between that integer and the first integer. Hence, for example, the third integer is named $n + 2$ because the third integer is 2 more than the first. (Remember that this is true in general. In any set of consecutive integers, the second integer is 1 more than the first, the third integer is 2 more than the first, the fourth integer is 3 more than the first, and so forth.)

4. Translate
   a. The sum of two consecutive integers
   b. Twice the second consecutive integer

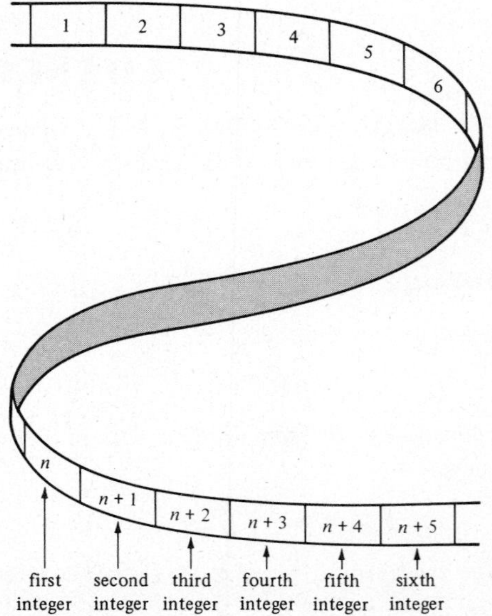

Consecutive Even or Odd Integers

Remember that the integer multiples of 2 are called the *even integers*:

$$\ldots -8, -6, -4, -2, 0, +2, +4, +6, +8, \ldots$$

The other integers, of course, are called the *odd integers*:

$$\ldots -9, -7, -5, -3, -1, +1, +3, +5, +7, +9, \ldots$$

Any two or more *even* integers are called *consecutive even integers* if they can be arranged so that each is two more than the preceding one. For example,

$$0, 2, 4 \quad \text{and} \quad -8, -6 \quad \text{and} \quad 4, 6, 8, 10$$

are consecutive even integers.

Any two or more *odd* integers are called *consecutive odd integers* if they can be arranged so that each is two more than the preceding one. For example,

$$-1, 1 \quad \text{and} \quad 3, 5, 7 \quad \text{and} \quad 15, 17, 19, 21$$

are consecutive odd integers.

When we talk about consecutive even or odd integers, we use the same specialized words that are used in talking about consecutive integers. That is, the smallest integer is called the *first integer* or the *first consecutive even (or odd) integer*. The next larger integer is called the *second integer* or the *second consecutive even (or odd) integer.* Successively larger even (or odd) integers are referred to as the *third, fourth,* and so forth integers. The largest integer is also called the *last integer* or the *last consecutive even (or odd) integer.* For example, in the consecutive even integers 4, 6, 8, the first integer is 4, the second is 6, and the last is 8. Similarly, in the consecutive odd integers 3, 5, 7, the first integer is 3, the second is 5, and the last is 7.

The first step in translating a consecutive even or odd integer problem into an equation is to name the integers, using a literal (again by tradition the letter n). **This is done the same way for both even and odd integers**. The first integer is always named n. Each successive integer is named as n plus the *difference* between that integer and the first integer. Hence, for example, the second integer is named $n + 2$ because the second integer is 2 more than the first. (Remember that the difference between any two consecutive *even* integers is 2, and the difference between any two consecutive *odd* integers is also 2.) Similarly, the third integer is named $n + 4$ because the third integer is 4 more than the first. Summarizing, we proceed as follows:

5. Translate
   a. The sum of two consecutive even integers
   b. Twice the second consecutive odd integer

The first even (or odd) integer	is named	n
The second even (or odd) integer	is named	$n + 2$
The third even (or odd) integer	is named	$n + 4$
The fourth even (or odd) integer	is named	$n + 6$
and so forth.		

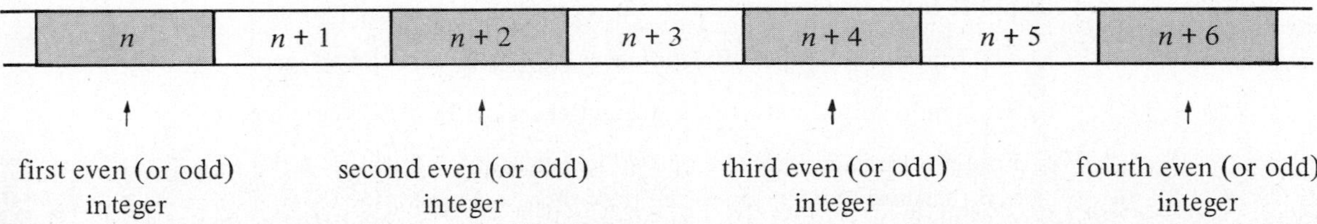

Look at the following examples of *translating consecutive integer problems into equations.*

TYPICAL PROBLEMS

1. The sum of two consecutive integers is 105. Find the integers.

↓

First we read the problem and double underline the equality word(s):

The sum of two consecutive integers is 105. Find the integers.

Then we translate the phrase on each side of the equality word(s) and form the equation. The sentence Find the integers becomes Find n and $n + 1$.

↓

$n + n + 1 = 105$

Find n and $n + 1$.

Hence, the word problem translates into the equation $n + n + 1 = 105$, whose solution is $n = 52$. Therefore, the consecutive integers are 52 and 53.

2. If the sum of two consecutive odd integers is 56, what are the integers?

↓

First we read the problem and double underline the equality word(s):

The sum of two consecutive odd integers is 56. What are the integers?

Then we translate the phrase on each side of the equality word(s) and form the equation. The clause What are the integers becomes Find n and $n + 2$:

↓

$n + n + 2 = 56$

Find n and $n + 2$.

Hence, the word problem translates into the equation $n + n + 2 = 56$, whose solution is $n = 27$. Therefore, the consecutive odd integers are 27 and 29.

3. The sum of two consecutive even integers is –22. Find the integers.

↓

First we read the problem and double underline the equality word(s):

The sum of two consecutive even integers is –22. Find the integers.

Then we translate the phrase on each side of the equality word(s) and form the equation. The sentence Find the integers becomes Find n and $n + 2$.

↓

$n + n + 2 = -22$

Find n and $n + 2$.

Hence, the word problem translates into the equation $n + n + 2 = -22$, whose solution is $n = -12$. Therefore, the consecutive even integers are –12 and –10.

EXAMPLES

Translate the following consecutive integer problems into equations.

1. The sum of four consecutive integers is 54. What are the integers?
2. The sum of three consecutive even integers is 138. Find the integers.
3. Find the three consecutive odd integers for which 6 less than the sum of the first and the third integers is 3 more than the second integer.

4. Find the two consecutive integers for which one-third of the first integer is 12 less than half the second integer.

Solutions

1. The sum of four consecutive integers <u>is</u> 54.

$$n + n + 1 + n + 2 + n + 3 = 54 \qquad (\Rightarrow n = 12 \rightarrow 12, 13, 14, 15)$$

2. The sum of three consecutive even integers <u>is</u> 138.

$$n + n + 2 + n + 4 = 138 \qquad (\Rightarrow n = 44 \rightarrow 44, 46, 48)$$

3. 6 less than the sum of the first and the third (odd) integers <u>is</u> 3 more than the second (odd) integer.

$$(n + n + 4) - 6 = 3 + n + 2 \qquad (\Rightarrow n = 7 \rightarrow 7, 9, 11)$$

4. One-third of the first integer <u>is</u> 12 less than half the second integer.

$$\frac{n}{3} = \left(\frac{n+1}{2}\right) - 12 \qquad (\Rightarrow n = 69 \rightarrow 69, 70)$$

OBSERVATIONS AND REFLECTIONS

When translating from a consecutive integer or a consecutive even (or odd) integer problem into an equation, keep the following points in mind:

1. The *word problem* specifies that the unknown quantities are integers, whereas the corresponding *equation* does not. The root of the equation may or may not be an integer. If the root is a fraction rather than an integer, then the word problem has no solution.
2. In a consecutive even (or odd) integer problem, the *word problem* specifies whether the unknown integers are even or odd, whereas the corresponding *equation* does not. Changing a consecutive *even* integer problem into a consecutive *odd* integer problem by changing the word *even* to *odd* throughout does not change the corresponding equation. If the word problem specifies even integers but the root of the equation is odd, then the problem has no solution. Similarly, the problem has no solution if it specifies odd integers and the root is even.

EXERCISES

Translate each of the following word problems into an equation and then solve it. These problems can be solved by the methods in Chapter 10.

1. The sum of two consecutive integers is 95. What are the integers?
2. The sum of two consecutive integers is 169. What are the integers?
3. The sum of two consecutive odd integers is 120. Find the integers.

4. The sum of two consecutive even integers is 54. What are the integers?

5. The sum of three consecutive integers is 39. What are the integers?

6. The sum of four consecutive integers is 134. Find the integers.

7. If the sum of three consecutive even integers is 168, what are the integers?

8. The sum of three consecutive odd integers is -189. Find the integers.

9. If the sum of three consecutive even integers is 12 more than twice the largest integer, what are the integers?

10. The sum of four consecutive odd integers is -17 added to 5 times the second integer. What are the integers?

11. When the sum of three consecutive integers is multiplied by 4, the result is 88 less than 20 times the second integer. Find the integers.

12. Find the three consecutive odd integers such that twice the sum of the second and third integers is 135 more than the first integer.

13. Find the four consecutive integers such that 6 times the second integer is 70 more than the sum of the other three integers.

14. Find the four consecutive even integers such that 50 less than the sum of the first two integers is 36 more than the third integer.

15. Find the three consecutive integers such that 10 times the third integer is 171 more than the sum of the first and second integers.

These problems can be solved by the methods in Section 10.4.

16. One-third the sum of two consecutive integers is 33. What are the integers?

17. If the sum of three consecutive integers is divided by 9, the result is 15. What are the integers?

18. If $\frac{1}{10}$ the second of three consecutive integers is added to the third integer, the result is 34. Find the three consecutive integers.

19. Find the two consecutive integers such that if $\frac{2}{3}$ times the first integer is added to $\frac{1}{6}$ the second integer, the result is 11.

20. Find the two consecutive integers such that the sum of $\frac{3}{8}$ the first integer and $\frac{1}{5}$ the second integer is 83.

21. Find the three consecutive integers such that $\frac{10}{7}$ the first integer equals the third integer added to half the second integer.

22. If the sum of the second and fourth integers is divided by 4, the result is 6 less than the third integer. What are the four consecutive integers?

23. One-fifth the sum of the second and third integers is $\frac{5}{14}$ the first integer. What are the three consecutive integers?

24. Find the three consecutive integers so that if the second integer divided by 11 is added to the third integer divided by 2, the result is $\frac{5}{8}$ the first integer.

25. If $\frac{4}{15}$ the first of four consecutive integers is subtracted from the third integer, the result is 29 less than the fourth integer. What are the integers?

13.3 NUMBER PROBLEMS WITH TWO UNKNOWNS

In the last two sections we have studied number problems in one unknown. In this section we will work with *number problems with two unknowns.* These problems can be identified as follows:

1. They involve two unknown quantities.
2. The information given about the unknown quantities is in the form of two numerical relations.

For example, the word problem

One number is 3 more than 4 times the other, and their sum is 8. Find the numbers.

has two unknown quantities (in this case the numbers 7 and 1) and states two numerical relations involving these numbers (in this case that 7 is 3 more than 4 times 1, and 7 plus 1 is 8).

To translate such a word problem into equations we must do three things:

1. Name the unknown quantities with literals.
2. Identify the two numerical relations within the word problem.
3. Translate each relation into an equation.

For example, to translate the word problem given, we first name the two unknown numbers with literals, say x and y. Then we identify the numerical relations and translate them into equations:

One number is 3 more than 4 times the other $\Rightarrow x = 3 + 4y$

and

their sum is 8 $\Rightarrow x + y = 8$

Once we have translated the word problem into a pair of equations, we can find the unknown numbers by solving the two equations using the methods in Chapter 12.

6. Translate
   a. The sum of two numbers is −15.
   b. The smaller of two numbers is 7 less than the larger number.
   c. Twice the smaller of two numbers less the larger number is 27.
   d. The first of two numbers increased by 11 is 3 times the second number.

Look at the following examples of *translating number problems with two unknowns into equations.*

TYPICAL PROBLEMS

1. The larger of two numbers is 8 more than 3 times the smaller number. If the sum of 5 times the smaller number and 4 times the larger number is 15, what are the numbers?

↓

First we read the problem and double underline the equality word(s) for each relation:

The larger of two numbers <u>is</u> 8 more than 3 times the smaller number. The sum of 5 times the smaller and 4 times the larger number <u>is</u> 15. What are the numbers?

↓

Then we translate the phrase on each side of the equality word(s) and form an equation for each relation. The clause What are the numbers becomes Find the common root of the two equations:

$$l = 8 + 3s$$
and
$$5s + 4l = 15$$
Find the common root.

Hence, the word problem translates into the equations $l = 8 + 3s$ and $5s + 4l = 15$, whose solution is $s = -1$ and $l = 5$. Therefore, -1 and 5 are the numbers specified by the word problem.

2. The first of two numbers increased by 12 is 9 times the sum of the second number and 7. If the product of 3 and the first number is diminished by 57, the result is 3 times the second number. Find the numbers.

↓

First we read the problem and double underline the equality word(s) for each relation:

The first of two numbers increased by 12 <u>is</u> 9 times the sum of the second number and 7. If the product of 3 and the first number is diminished by 57, <u>the result is</u> 3 times the second number. Find the numbers.

↓

Then we translate the phrase on each side of the equality word(s) and form an equation for each relation. The sentence Find the numbers becomes Find the common root of the two equations:

$$f + 12 = 9(s + 7)$$
and
$$3f - 57 = 3s$$
Find the common root.

Hence, the word problem translates into the equations $f + 12 = 9(s + 7)$ and $3f - 57 = 3s$, whose solution is $f = 15$ and $s = -4$. Therefore, 15 and -4 are the numbers specified by the word problem.

EXAMPLES

Translate the following word problems into equations.

1. The sum of two numbers is 22. If one number is 4 more than twice the other number, what are the numbers?

2. The larger of two numbers is 11 more than the smaller number. When 5 times the smaller number is added to 7 times the larger number, the result is –163. Find the numbers.

3. Fourteen more than the smaller of two numbers is 6 less than the larger number. The smaller number plus 8 times the larger number is 52. Find the numbers.

4. Twice the smaller of two numbers equals 7 subtracted from the larger number. If the sum of 5 times the smaller number and –3 times the larger number is –30, what are the numbers?

Solutions

1. The sum of two numbers is 22. One number is 4 more than twice the other number.

$$\downarrow$$

$$x + y = 22 \quad \text{and} \quad x = 4 + 2y \quad (\Rightarrow x = 16 \text{ and } y = 6)$$

2. The larger of two numbers is 11 more than the smaller number. When 5 times the smaller number is added to 7 times the larger number, the result is –163.

$$\downarrow$$

$$l = 11 + s \quad \text{and} \quad 5s + 7l = -163 \quad (\Rightarrow s = -20 \text{ and } l = -9)$$

3. Fourteen more than the smaller of two numbers is 6 less than the larger number. The smaller number plus 8 times the larger number is 52.

$$\downarrow$$

$$14 + s = l - 6 \quad \text{and} \quad s + 8l = 52 \quad (\Rightarrow s = -12 \text{ and } l = 8)$$

4. Twice the smaller of two numbers equals 7 subtracted from the larger number. The sum of 5 times the smaller number and –3 times the larger number is –30.

$$\downarrow$$

$$2s = l - 7 \quad \text{and} \quad 5s - 3l = -30 \quad (\Rightarrow s = 9 \text{ and } l = 25)$$

EXERCISES

Translate each of the following word problems into equations and then solve them. These problems can be solved by the methods in Chapter 12.

1. The larger of two numbers equals the sum of 9 and the smaller number. If the smaller number equals 29 reduced by the larger number, what are the numbers?
2. The first of two numbers is 7 times the second. If the first number equals 24 decreased by the second number, what are the numbers?
3. The smaller of two numbers equals the larger number reduced by 6, and the larger number equals the smaller number multiplied by 4. Find the numbers.
4. The sum of two numbers is 7. If one number is equal to 19 added to the other number, what are the numbers?
5. The product of the first of two numbers and –8 equals the second number. If the second number is 27 less than the first number, what are the numbers?
6. One of two numbers is 8 less than the second number. What are the numbers if the product of the second number and 5 equals the first number?
7. The first of two numbers is 8 increased by the second number. If the sum of the two numbers is 50, what are the numbers?

8. The first of two numbers decreased by the second number is −33. If the first number is 12 times as great as the second number, what are the numbers?

9. The sum of two numbers is 31. Find the numbers if the smaller number decreased by the larger number is −19.

10. Subtracting 20 from one of two numbers gives the other number. Find the numbers if adding one number to the other number yields 8.

11. The first of two numbers is 25 more than the second number. If the sum of the second number and the product of 4 and the first number is 75, what are the numbers?

12. The larger of two numbers diminished by the smaller number is 15, and the larger number subtracted from 8 times the smaller number is 13. Find the numbers.

13. When the product of 3 and the first of two numbers is increased by the second number, the result is 5. If the sum of the first number and 3 times the second number equals 47, what are the numbers?

14. The smaller of two numbers increased by the larger number equals 21. If the product of 4 and the smaller number is subtracted from the larger number, the result is −19. What are the numbers?

15. If the first of two numbers subtracted from 25 is the second number, and if −52 increased by the second number multiplied by 6 is the first number, what are the numbers?

16. The first of two numbers less 5 times the second number is 28. If the product of 3 and the first number is reduced by 8 times the second number, the result is 70. Find the numbers.

17. The first of two numbers is the sum of 2 and 6 times the second number. If the second number added to 12 times the first number is −49, what are the numbers?

18. When 3 times the first of two numbers is increased by the product of 5 and the second number, the result is 69. What are the numbers if 2 times the second number is 27 minus the first number?

19. When the product of 4 and a certain number is subtracted from 7 times another number, the result is 28. What are the numbers if the sum of the numbers is 15?

20. If the product of 11 and the first of two numbers is added to the product of 3 and the second number, the result is 30. The first number subtracted from 4 times the second number is 40. Find the numbers.

13.4 PERIMETER PROBLEMS

Perimeter problems are a special class of number problems. In this section we will deal with problems about perimeters of rectangles.

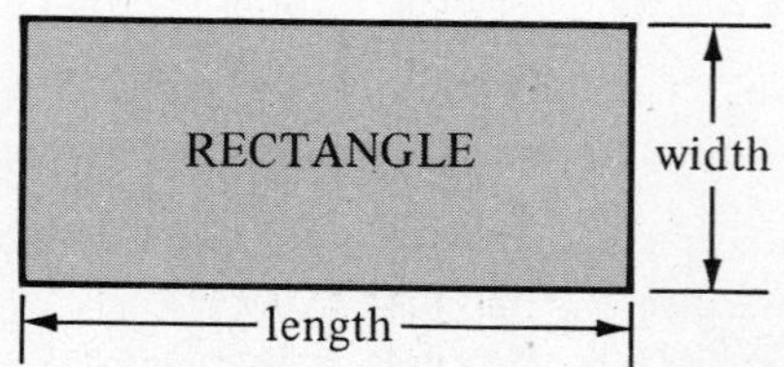

Remember that the *perimeter* of a rectangle is the sum of the lengths of its sides. If we let l stand for the length of a rectangle and w stand for its width (as is traditional), we write the perimeter of a rectangle as

$$\text{perimeter} = l + w + l + w = 2l + 2w$$

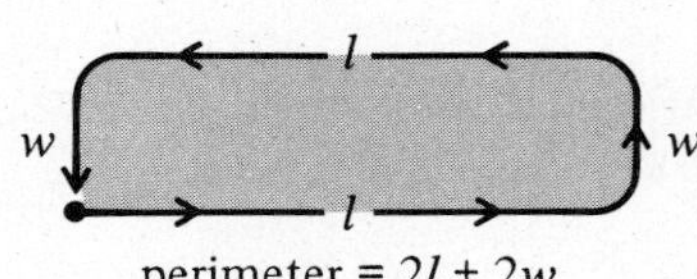

perimeter = $2l + 2w$

7. Determine the perimeter of a rectangle whose length and width are
   a. l = 9cm and w = 5cm
   b. l = 8m and w = 7m
   c. l = 3km and w = 1km
   d. l = 28mm and w = 15mm

(*Peri-* is a Greek prefix, meaning around or encircling. The suffix *-meter* is Latin (derived from Greek), meaning measure.)

In any word problem about a rectangle, mention of the perimeter implies the equation $2l + 2w$ = number. For example, the perimeter is 22 meters implies the equation $2l + 2w = 22$.

Look at the following examples of *translating perimeter problems into equations.*

TYPICAL PROBLEMS

1. The width of a rectangle is 4 meters less than its length. If the perimeter of the rectangle is 36 meters, what are its dimensions?

First we read the problem and double underline the equality word(s) for each relation:

↓

The width of a rectangle is 4 meters less than its length. The perimeter of the rectangle is 36 meters. What are its dimensions?

Then we translate the phrase on each side of the equality word(s) and form an equation for each relation. The clause What are its dimensions becomes Find the common root of the two equations:

↓

$w = l - 4$
and
$2l + 2w = 36$
Find the common root.

Hence, the word problem translates into the equations $w = l - 4$ and $2l + 2w = 36$, whose solution is $w = 7$ and $l = 11$. Therefore, the dimensions of the rectangle are 7 by 11 meters.

2. The length of a rectangle diminished by 6 centimeters is 3 times its width, and the perimeter is 108 centimeters. Find its dimensions.

First we read the problem and double underline the equality word(s) for each relation:

↓

The length of a rectangle diminished by 6 centimeters is 3 times its width. The perimeter is 108 centimeters. Find its dimensions.

Then we translate the phrase on each side of the equality word(s) and form an equation for each relation. The sentence Find its dimensions becomes Find the common root of the two equations:

↓

$l - 6 = 3w$
and
$2l + 2w = 108$
Find the common root.

Hence, the word problem translates into the equations $l - 6 = 3w$ and $2l + 2w = 108$, whose solution is $w = 12$ and $l = 42$. Therefore, the dimensions of the rectangle are 12 by 42 centimeters.

EXAMPLES

Translate the following perimeter problems into equations.

1. What are the dimensions of a rectangle whose length is 7 times its width, and whose perimeter is 176 meters?
2. The perimeter of a rectangle is 84 centimeters, and its length is 13 centimeters less than 10 times its width. What are its dimensions?
3. The length of a rectangle increased by 28 kilometers equals 12 times its width. If the perimeter is 178 kilometers, find its dimensions.
4. The perimeter of a rectangle is 18 millimeters. Eight millimeters more than its width is 6 millimeters more than its length. What are its dimensions?

Solutions

1. The length (of a rectangle) is 7 times its width. The perimeter is 176 meters.

$$\downarrow$$

$$l = 7w \quad \text{and} \quad 2l + 2w = 176 \qquad (\Rightarrow w = 11 \text{ and } l = 77)$$

2. The perimeter of a rectangle is 84 centimeters. Its length is 13 centimeters less than 10 times its width.

$$\downarrow$$

$$2l + 2w = 84 \quad \text{and} \quad l = 10w - 13 \qquad (\Rightarrow w = 5 \text{ and } l = 37)$$

3. The length of a rectangle increased by 28 kilometers equals 12 times its width. The perimeter is 178 kilometers.

$$\downarrow$$

$$l + 28 = 12w \quad \text{and} \quad 2l + 2w = 178 \qquad (\Rightarrow w = 9 \text{ and } l = 80)$$

4. The perimeter of a rectangle is 18 millimeters. Eight millimeters more than its width is 6 millimeters more than its length.

$$\downarrow$$

$$2l + 2w = 18 \quad \text{and} \quad 8 + w = 6 + l \qquad \left(\Rightarrow w = \frac{7}{2} \text{ and } l = \frac{11}{2}\right)$$

EXERCISES

Translate each of the following word problems into equations and then solve them. These problems can be solved by the methods in Chapter 12.

1. The length of a rectangle is 12 meters more than its width. What are the dimensions of the rectangle if its perimeter is 84 meters?
2. What are the dimensions of a rectangle whose width is 7 meters less than its length, and whose perimeter is 90 meters?
3. The length of a rectangle is 9 times its width, and its perimeter is 100 centimeters. Find its dimensions.
4. The width of a rectangle added to 10 meters equals its length. Find its dimensions if its perimeter is 28 meters.
5. The length of a rectangle decreased by 18 millimeters equals its width, and its perimeter is 64 millimeters. What are its dimensions?

6. The length of a rectangle equals 4 meters increased by 4 times its width, and its perimeter is 98 meters. What are its dimensions?
7. Find the dimensions of a rectangle whose width is 45 centimeters less than 3 times its length, and whose perimeter is 70 centimeters.
8. What are the dimensions of a rectangle whose length is 6 times its width decreased by 100 millimeters and whose perimeter is 500 millimeters?
9. If the length of a rectangle multiplied by 4 is subtracted from its width, the result is −101 meters. What are its dimensions if its perimeter is 108 meters?
10. Twelve times the width of a rectangle added to 7 centimeters equals its length. What are its dimensions if its perimeter is 92 centimeters?
11. What are the dimensions of a rectangle whose perimeter is 208 meters and whose length increased by 14 meters equals its width added to 26 meters?
12. If the perimeter of a rectangle is 46 kilometers, and its length plus 7 times its width is 47 kilometers, what are its dimensions?
13. If the length of a rectangle is subtracted from 5 times its width, the result is 62 centimeters. What are its dimensions if its perimeter is 68 centimeters?
14. Find the dimensions of a rectangle whose perimeter is 30 kilometers and whose length subtracted from 12 kilometers less than its width is −17 kilometers.
15. The perimeter of a rectangle is 276 millimeters. If 30 millimeters more than its width decreased by its length is −2 millimeters, what are its dimensions?

13.5 COIN PROBLEMS

Mixture problems are a large class of number problems that involve both the number and value of various types of objects. We will first consider a special class of mixture problems called *coin problems.* In these problems the values of the objects considered (coins) are well known, whereas in other mixture problems the values must either be given in the problem or solved for. We will consider problems about two types of coins. For example, we will see problems about pennies and nickels, dimes and quarters, nickels and half-dollars, and so forth. To understand how to translate these problems we must be aware of the following facts:

1. When we *count* coins, the coins do not all have to be the same type. When we count nickels and dimes, for example, the *number* of coins is just the number of nickels plus the number of dimes.
2. When we determine the *dollar value* of a collection of coins, we must replace each coin with its *penny* equivalent. A *nickel* is worth 5 pennies, a *dime* is worth 10 pennies, a *quarter* is worth 25 pennies, and a *half-dollar* is worth 50 pennies.

For example, if we have 7 nickels and 8 dimes, the *number* of coins is $7 + 8 = 15$, whereas the *value* of the coins is $7(5¢)$ plus $8(10¢) = 35¢ + 80¢$, for a total of \$1.15 (i.e., $115¢$). More generally, if we have n nickels and d dimes (where n and d stand for positive integers), we have $n + d$ coins whose value is $(5n + 10d)¢$.

8. What is the value (penny equivalent) of
   a. 9 quarters
   b. 7 dimes
   c. 13 nickels
   d. 11 half-dollars
   e. 3 dimes and 7 nickels
   f. 5 quarters and 8 dimes

	number	·	unit value	=	total value
pennies	p		1¢		p¢
nickels	n		5¢		$5n$¢
dimes	d		10¢		$10d$¢
quarters	q		25¢		$25q$¢
half-dollars	h		50¢		$50h$¢

Coin problems almost always contain two relations: one about the number of coins and one about the value of the coins. If we are told, for example, that we have $1.35 worth of nickels and dimes, and that we have 21 coins in all, then

1. The *numerical* relation is that the number of nickels n plus the number of dimes d is 21. That is, $n + d = 21$.
2. The *value* relation is that the value of the nickels (in cents) $5n$ plus the value of the dimes (in cents) $10d$ is 135 cents. That is, $5n + 10d = 135$.

(We can solve the equations $n + d = 21$ and $5n + 10d = 135$ to determine that we have 11 nickels and 8 dimes.)

The Equality Words in Coin Problems

Generally, the words that represent the equals sign in a verbal equation separate the two members of the equation. However, coin problems have exceptions such as

Brian has 5 more dimes than nickels.

and

Ann has 10 times as many pennies as quarters.

When we translate these sentences, the word *has* represents the equals sign. The *first* coin mentioned is the *left* member of the equation, and the rest of the sentence makes up the right member. For example,

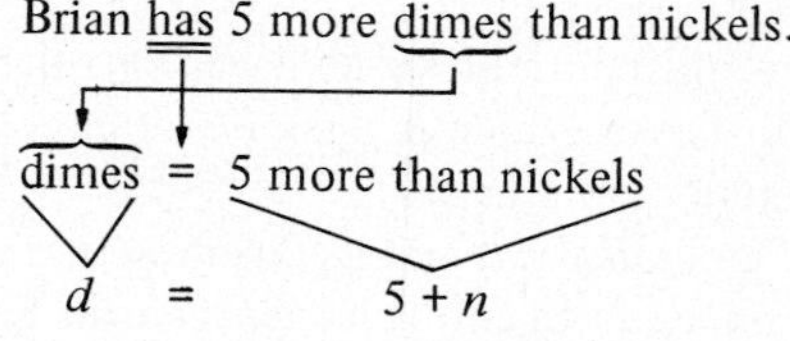

and

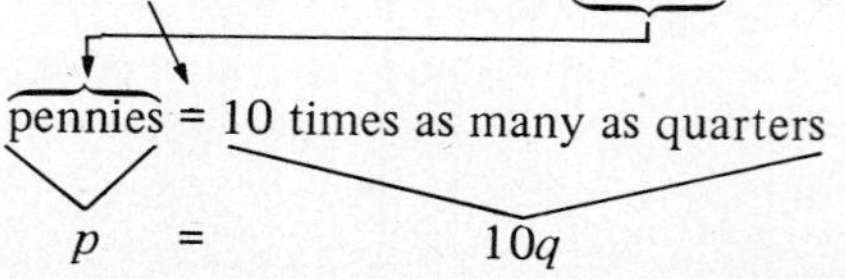

9. Translate
   a. Ben has 3 fewer quarters than dimes.
   b. Eva has twice as many nickels as pennies.

13.5

Look at the following examples of *translating coin problems into equations.*

TYPICAL PROBLEMS

1. Janice has 12 more dimes than nickels. If the coins are worth \$2.25, how many coins does she have?

We read the problem, identifying the numerical and the value relations:

We consider the numerical relation first and double underline the equality word(s). If the equality word(s) come first we indicate with an arrow that the first coin mentioned belongs to the left of the equality word(s):

↓

Janice has 12 more dimes than nickels.

↓

Then we form the equation:

$$d = 12 + n$$

and

We next consider the value relation and form the equation representing the total value of the coins in cents:

$$5n + 10d = 225$$

Hence, the coin problem translates into the equations $d = 12 + n$ and $5n + 10d = 225$, whose solution is $n = 7$ and $d = 19$. Therefore, Janice has 7 nickels and 19 dimes for a total of 26 coins.

2. Frank has some nickels and quarters worth \$2.05. If 8 times the number of quarters is 11 more than the number of nickels, how many nickels does Frank have?

We read the problem, identifying the numerical and the value relations:

↓

We consider the numerical relation first and double underline the equality word(s):

8 times the number of quarters is 11 more than the number of nickels.

↓

Then we form the equation:

$$8q = 11 + n$$

and

We next consider the value relation and form the equation representing the total value of the coins in cents:

$$5n + 25q = 205$$

Hence, the coin problem translates into the equations $8q = 11 + n$ and $5n + 25q = 205$, whose solution is $n = 21$ and $q = 4$. Therefore, Frank has 21 nickels.

EXAMPLES

Translate the following coin problems into equations.

1. Phil has 7 times as many dimes as quarters. If his coins are worth \$5.70, how many coins does he have?
2. Kath has 12 fewer nickels than pennies. If her nickels and pennies are worth \$1.50, how many nickels does she have?
3. Juanita has \$16.70 in dimes and half-dollars. If 5 times the number of dimes is 3 more than the number of half-dollars, how many half-dollars does Juanita have?
4. Peter has 250 coins worth \$6.10. If he has only quarters and pennies, how many of each coin does he have?

Solutions

1. Phil has 7 times as many dimes as quarters. (The dimes and quarters are worth $5.70.)

↓

$d = 7q$ and $10d + 25q = 570$ ($\Rightarrow d = 42$ and $q = 6$)

2. Kath has 12 fewer nickels than pennies. (The nickels and pennies are worth $1.50.)

↓

$n = p - 12$ and $p + 5n = 150$ ($\Rightarrow n = 23$ and $p = 35$)

3. Five times the number of dimes is 3 more than the number of half-dollars. (The dimes and half-dollars are worth $16.70.)

↓

$5d = 3 + h$ and $10d + 50h = 1670$ ($\Rightarrow d = 7$ and $h = 32$)

4. Peter has 250 coins. (The quarters and pennies are worth $6.10.)

↓

$q + p = 250$ and $p + 25q = 610$ ($\Rightarrow q = 15$ and $p = 235$)

EXERCISES

Translate each of the following problems into equations and then solve them. These problems can be solved by the methods in Chapter 12.

1. If Steve has 12 more dimes than quarters for a total of $5.75, how many coins does he have?
2. Claire has 20 fewer nickels than dimes. If the coins are worth $8.75, how many coins does she have?
3. Paul has nickels and quarters worth $8.10. If he has 58 coins, how many quarters does he have?
4. If Maria has 9 times as many dimes as pennies, and together they are worth $18.20, how many coins does she have?
5. Linda has $180 in change. If she has only nickels and pennies, and if she has 30 fewer nickels than pennies, how many coins does she have?
6. Aaron has 12 times as many quarters as half-dollars. If he has $17.50, how many coins does he have?
7. If Olga has 2 more half-dollars than pennies and the coins are worth $3.55, how many coins does she have?
8. Bruce saved dimes and half-dollars worth $6.80. If he had 32 coins, how many half-dollars did he have?
9. Jose has 4 times as many pennies as quarters, and together they are worth $6.96. How many coins does he have?
10. Nick saves only pennies and Lisa saves dimes. Together they have 560 coins worth $11.00. How many coins has each of them saved?

11. Joan has \$12.75. If she has 12 more dimes than 5 times her nickels, how many coins does she have?

12. If Alan has 31 fewer nickels than quarters and together the coins are worth \$12.55, how much are his nickels worth?

13. Mario counted his coins and found that 20 times the number of dimes was 7 more than the number of quarters. If the coins were worth \$8.45, how many coins did he have?

14. If 12 times the number of Judy's nickels is added to 4 times her pennies, the result is 64. How many coins does she have if they are worth 22¢ ?

15. Carl has 8 fewer quarters than half-dollars. If the coins are worth \$19.00, how many coins does Carl have?

16. Fourteen more than the number of Kerry's dimes is 9 times the number of her nickels. If the dimes and nickels are worth \$3.35, how many coins does she have?

17. Julie has nickels and quarters worth \$4.00. If 3 times the number of quarters is 8 less than the number of nickels, how many quarters does Julie have?

18. If Eve adds 10 times the number of dimes she has to 4 times her pennies, she gets 112. How many coins does she have if they are worth 73¢ ?

19. If twice the number of Tony's quarters is subtracted from 5 times his pennies, the result is 158. How many coins does Tony have if they are worth \$5.65?

20. Susan has quarters and dimes worth \$8.30. If the number of quarters is subtracted from 93, the result equals 5 times the number of dimes. How many quarters does Susan have?

13.6 MIXTURE PROBLEMS

Mixture problems are a large class of problems that involve the number and values of various types of objects. For example, a mixture problem can involve the *number of pounds* (or kilograms) of each ingredient in a mixture and the *cost per pound* (or kilogram) of each ingredient (and of the mixture itself). Alternatively, a mixture problem might involve the *number* and *cost* of tickets for various locations in a concert hall. For example, consider the problem If thirty pounds of cashews worth \$1.90 a pound are mixed with eighty pounds of peanuts worth 80¢ a pound, how much is the mixture worth per pound.

To solve such a problem we must use the following three facts:

> 1. The *total cost* of each *ingredient* can be determined by multiplying the *amount* of the ingredient by its *unit value* (cost per unit). Similarly, the *total cost* of the *mixture* is the product of the *total amount* and the *unit cost* of the mixture.

For example,

pounds of cashews • cost per pound = cost of cashews
pounds of peanuts • cost per pound = cost of peanuts
pounds of mixture • cost per pound = cost of mixture

2. The *amount* of the *mixture* can be determined by adding the *amounts* of its *ingredients.*

   For example,

   pounds of cashews + pounds of peanuts = pounds of mixture

3. The *total cost* of the *mixture* can be determined by adding the *total costs* of its *ingredients.*

   For example,

   cost of cashews + cost of peanuts = cost of mixture

To display these relations, so that we can use them more effectively, we can set up a table.

	Number of pounds	·	Cost per pound (Unit value) in cents	=	Total cost in cents
cashews	30	·	190	=	30(190)
peanuts	80	·	80	=	80(80)
mixture	(30 + 80)	·	x	=	$(30 + 80)x$

Observe that the three rows state the three equations specified in 1. The first column gives the number of pounds of each item, and we have used 2 to determine the number of pounds of the mixture. We can read the first column vertically as an equation because the sum of the top two entries equals the bottom entry.

Now, using 3 we can read the third column vertically to form an equation again because the sum of the top two entries in this column equals the bottom entry. We obtain

$$30(190) + 80(80) = (30 + 80)x$$

(Solving this equation yields a cost per pound of $1.10 for the mixture.)

Look at the following examples of *translating mixture problems to equations.*

TYPICAL PROBLEMS

1. If 12 kilograms of sunflower seeds worth \$1.05 a kilogram are mixed with 8 kilograms of peanuts worth 75¢ a kilogram, how much is the mixture worth per kilogram?

↓

First we read the problem and set up a table to display the relations:

	Number of kilograms	·	Cost per kilogram in cents	=	Total value in cents
sunflower seeds	12	·	105	=	12(105)
peanuts	8	·	75	=	8(75)
mixture	12 + 8	·	x	=	$(12 + 8)x$

Then we form the equation from the third column in the table (and simplify each term of the equation):

$$12(105) + 8(75) = (12 + 8)x$$
$$\downarrow$$
$$1{,}260 + 600 = 20x$$

Hence, the mixture problem translates into the equation $1{,}260 + 600 = 20x$, whose solution is $x = 93$. Therefore, the mixture is worth 93¢ per kilogram.

2. How many pounds of corn meal costing 25¢ a pound should be mixed with 15 pounds of flour costing 16¢ a pound to produce a mixture worth 22¢ a pound?

↓

First we read the problem and set up a table to display the relations:

	Number of pounds	·	Cost per pound in cents	=	Total value in cents
corn meal	x	·	25	=	$x(25)$
flour	15	·	16	=	15(16)
mixture	$(x + 15)$	·	22	=	$(x + 15)22$

Then we form the equation from the third column in the table (and simplify the terms of the equation):

$$x(25) + 15(16) = (x + 15)22$$
$$\downarrow$$
$$25x + 240 = 22x + 330$$

Hence, the mixture problem translates into the equation $25x + 240 = 22x + 330$, whose solution is $x = 30$. Therefore, 30 pounds of corn meal should be mixed with the flour.

3. Alcohol costs 26¢ a bottle and peroxide costs 32¢ a bottle. Dr. Flint's order of alcohol and peroxide costs a total of \$8.60. If there are 13 more bottles of alcohol than peroxide, how many bottles of alcohol did Dr. Flint order?

First we read the problem and set up a table to display the relations:

↓

	Number of bottles	·	Cost per bottle in cents	=	Total value in cents
alcohol	a	·	26	=	$26a$
peroxide	p	·	32	=	$32p$
Total Order	*		—		860

*This entry is omitted because the numerical relation does not involve the sum of the unknowns.

Then we form an equation from the third column in the table, and we form the numerical relation:

↓

$26a + 32p = 860$ and $a = 13 + p$

Hence, the mixture problem translates into the equations $26a + 32p = 860$ and $a = 13 + p$, whose solution is $a = 22$ and $p = 9$. Therefore, there were 22 bottles of alcohol in Dr. Flint's order.

4. The school bookstore has in stock a few elementary algebra review books costing \$3.50 each and a few elementary algebra textbooks costing \$11.00 each. If it has a total of 20 books worth \$130.00, how many of each book is in stock?

First we read the problem and set up a table to display the relations:

↓

	Number of books	·	Cost per book in cents	=	Total value in cents
review books	x	·	350	=	$350x$
textbook	y	·	1,100	=	$1{,}100y$
both books	20		—		13,000

Then we form the equations from the first column and from the third column:

↓

$x + y = 20$ and $350x + 1{,}100y = 13{,}000$

Hence, the mixture problem translates into the equations $x + y = 20$ and $350x + 1{,}100y = 13{,}000$, whose solution is $x = 12$ and $y = 8$. Therefore, there are 12 review books and 8 textbooks in stock.

5. If 3 kilograms of raisins are mixed with 6 kilograms of peanuts, the mixture is worth 80¢ a kilogram. On the other hand, if 1 kilogram of raisins is mixed with 4 kilograms of peanuts, the mixture is worth 78¢ a kilogram. How much does a kilogram of raisins cost, and how much does a kilogram of peanuts cost?

First we read the problem and set up a table for each mixture to display the relations:

↓

	Number of kilograms	·	Cost per kilogram in cents	=	Total value in cents
raisins	3	·	x	=	$3x$
peanuts	6	·	y	=	$6y$
mixture	3 + 6	·	80	=	$(3 + 6) \cdot 80$

and

	Number of kilograms	·	Cost per kilogram in cents	=	Total value in cents
raisins	1	·	x	=	x
peanuts	4	·	y	=	$4y$
mixture	1 + 4	·	78	=	$(1 + 4) \cdot 78$

↓

Then we form an equation from the third column of each table (and simplify the terms of each equation):

$3x + 6y = (3 + 6) \cdot 80$ and $x + 4y = (1 + 4) \cdot 78$

↓

$3x + 6y = 720$ and $x + 4y = 390$

Hence, the mixture problem translates into the equations $3x + 6y = 720$ and $x + 4y = 390$ whose solutions are $x = 90$ and $y = 75$. Therefore, a kilogram of raisins costs 90¢ and a kilogram of peanuts costs 75¢.

EXAMPLES

Translate the following mixture problems into equations.

1. If 30 pounds of sirloin worth \$1.90 a pound are ground together with 40 pounds of chuck worth \$1.20 a pound to produce hamburger, how much should the hamburger cost per pound?
2. How many gallons of paint costing \$9.30 a gallon should be mixed with 10 gallons of paint costing \$12.00 a gallon to produce a blend worth \$10.20 a gallon?

3. Black tea costs \$2.40 a kilogram and red tea costs \$1.86 a kilogram. If there is twice as much red tea as black tea in a special blend and if this mixture costs a total of \$61.20, how many kilograms of each type of tea are in the mixture?

4. Eileen and Gail worked a total of 66 hours between them last week and earned \$282.90. If Eileen earns \$3.55 an hour and Gail earns \$4.90 an hour, how many hours did each work?

5. Seven kilograms of candy and 30 kilograms of nuts sold for \$27.30. Later, 9 kilograms of the same candy and 1 kilogram of the same nuts sold for \$8.80. How much does 1 kilogram of candy cost? How much does 1 kilogram of nuts cost?

Solutions

1.

	Number of pounds	·	Cost per pound in cents	=	Total value in cents
sirloin	30	·	190	=	30(190)
chuck	40	·	120	=	40(120)
hamburger	30 + 40	·	x	=	$(30 + 40)x$

↓

$$30(190) + 40(120) = (30 + 40)x$$

↓

$$5{,}700 + 4{,}800 = 70x \qquad (\Rightarrow x = 150)$$

2.

	Number of gallons	·	Cost per gallon in cents	=	Total value in cents
cheap paint	10	·	1,200	=	10(1,200)
expensive paint	x	·	930	=	$x(930)$
blend	$10 + x$	·	1,020	=	$(10 + x)1{,}020$

↓

$$10(1{,}200) + x(930) = (10 + x)1{,}020$$

↓

$$12{,}000 + 930x = 10{,}200 + 1{,}020x \qquad (\Rightarrow x = 20)$$

3.

	Number of kilograms	·	Cost per kilogram in cents	=	Total value in cents
black tea	x	·	240	=	$240x$
red tea	y	·	186	=	$186y$
blend	*		–		6,120

*This entry is omitted because the numerical relation does not involve the sum of the unknowns.

↓

$$y = 2x \quad \text{and} \quad 240x + 186y = 6{,}120 \qquad (\Rightarrow x = 10 \text{ and } y = 20)$$

4.

	Number of hours	•	Salary per hour in cents	=	Total salary in cents
Eileen	x	•	355	=	$355x$
Gail	y	•	490	=	$490y$
together	66	•	—	=	28,290

↓

$x + y = 66$ and $355x + 490y = 28{,}290$ ($\Rightarrow x = 30$ and $y = 36$)

5.

	Number of kilograms	•	Cost per kilogram in cents	=	Total value in cents
candy	7	•	x	=	$7x$
nuts	30	•	y	=	$30y$
1st mixture	7 + 30		—		2,730

and

	Number of kilograms	•	Cost per kilogram in cents	=	Total value in cents
candy	9	•	x	=	$9x$
nuts	1	•	y	=	y
2nd mixture	9 + 1		—		880

↓

$7x + 30y = 2{,}730$ and $9x + y = 880$ ($\Rightarrow x = 90$ and $y = 70$)

EXERCISES

Translate each of the following word problems into equations and then solve them. These problems can be solved by the methods in Chapter 10.

1. If 20 kilograms of grass seed worth \$1.20 a kilogram are mixed with 15 kilograms of seed worth \$1.90 a kilogram, how much is a kilogram of the mixture worth?
2. If 4 pounds of ground chuck at \$1.59 a pound are mixed with 7 pounds of ground sirloin at \$1.92 a pound, how much should a butcher charge for a pound of the mixture?
3. If 9 pounds of candy worth 67¢ a pound were mixed with 5 pounds of candy worth 95¢ a pound, how much would a pound of the mixture be worth?

4. A manufacturer wishes to mix 40 kilograms of tea at $3.44 a kilogram with 15 kilograms of tea at $5.64 a kilogram. How much will a kilogram of the mixture be worth?

5. If 24 pounds of currants worth 86¢ a pound are mixed with 13 pounds of raisins worth $1.23 a pound, how much is the mixture worth?

6. How much should be charged per kilogram for a tea produced from 17 kilograms of China tea at $2.08 a kilogram and 23 kilograms of Ceylon tea at $2.88 a kilogram?

7. How many pounds of peanuts worth 72¢ a pound should be mixed with 10 pounds of cashews worth $2.07 a pound to produce a mixture worth $1.02 a pound?

8. How many bags of sand costing $4.00 a bag must be mixed with 47 bags of cement costing $5.40 a bag to produce concrete worth $4.94 a bag?

9. A fabric manufacturer wants to produce fine cotton-dacron material to sell at a cost of 98¢ a meter. If the cotton costs 70¢ a meter and the dacron fiber costs $1.30 a meter, how many meters of cotton must be blended with 14 meters of dacron fiber?

10. How many kilograms of Colombian coffee at $8.00 a kilogram should be mixed with 55 kilograms of Brazilian coffee at $6.00 a kilogram to make a blend worth $6.90 a kilogram?

11. How many liters of wine vinegar at $3.50 a liter should be mixed with 25 liters of white vinegar at $1.60 a liter to produce a blend worth $3.00 a liter?

12. How much wheat flour costing 20¢ a pound should be mixed with 5 pounds of soya flour costing 62¢ a pound to produce a flour worth 30¢ a pound?

These problems can be solved by the methods in Chapter 12.

13. A furniture company sells desks and chairs. The desks sell for $250.00 each and the chairs for $65.00 each. If a total of 19 chairs and desks were sold for $2,715.00, how many desks were sold?

14. A carpenter ordered 45 pounds of nails and screws. The nails cost 85¢ a pound and the screws cost $1.07 a pound. How many pounds of screws did he order if the order cost him $44.19?

15. Eleven hundred tickets were sold at a hockey game. Adults paid $4.50 each and senior citizens paid $2.90. If the total receipts for the game were $4,470.00, how many adult tickets were sold?

16. Twenty-seven plumbers earn a total of $9,330.00 a week. If apprentices earn $240.00 a week and master plumbers earn $430.00 a week, how many apprentices are there?

17. Students paid $1.20 each for a school play and adults paid $2.00. If 435 tickets were sold for a total of $610.00, how many student tickets were sold and how many adult tickets were sold?

18. Thirty-nine books costing $485.00 were delivered to a school bookstore. If the books were mathematics books costing $11.00 each and chemistry books costing $15.00, how many chemistry books were delivered?

19. Both the customer elevator and the freight elevator bring people to the twentieth floor. The customer elevator can carry 16 fewer people than the freight elevator. If the customer elevator makes 10 trips and the freight elevator makes 7 trips, 520 people can get to the twentieth floor. How many people can each elevator carry per trip?

20. Stewart has 35¢ special delivery stamps and 15¢ air mail stamps. He has 11 more air mail stamps than special delivery stamps. How many of each does he have if together the stamps are worth $5.65?

21. The larger of two pick-up trucks can carry 5 more tons of sand than the smaller truck. If the larger truck makes 7 full-capacity deliveries and the smaller truck makes 8 full-capacity deliveries, they will deliver 245 tons of sand. What is the capacity of each truck?

22. Esther has 6 more 5¢ stamps than 8¢ stamps. The stamps are worth $3.16. How many of each stamp does she have?

23. Warren has foreign coins worth $17.37. The French coins are worth 33¢ each and the German coins are worth 29¢ each. If he has 15 fewer French coins than German coins, how many of each type of coin does Warren have?

24. A bricklayer can lay 20 more bricks than his helper can in an hour. If the bricklayer works for 5 hours and the helper works for 4 hours, they can lay 262 bricks. How many bricks can each man lay per hour?

25. One customer purchased 6 chairs and 1 table for $459.00. Another customer bought 3 tables and 10 chairs for $945.00. What is the price of one table and one chair?

26. Alice bought 3 pounds of apples and 2 pounds of pears for $2.85. At the same store her friend paid $2.56 for 1 pound of apples and 3 pounds of pears. What is the cost of a pound of apples and what is the cost of a pound of pears?

27. Five pounds of chocolate are mixed with 1 pound of nuts to make a candy selling for $1.70 a pound. If 3 pounds of chocolate were mixed with 2 pounds of nuts, the candy would sell for $1.49 a pound. What is the cost of a pound of chocolate?

28. Five kilograms of Colombian coffee are mixed with 2 kilograms of Brazilian coffee to produce a mixture selling for $8.52 a kilogram. If 6 kilograms of Colombian coffee were mixed with 1 kilogram of Brazilian coffee, the mixture would sell for $8.76 a kilogram. What is the value of one kilogram of each type of coffee?

29. Eight bags of foam rubber are mixed with 1 bag of polyfoam to produce a mixture costing $3.47 a bag. If 14 bags of foam rubber were mixed with 4 bags of polyfoam, the mixture would cost $3.44 a bag. What is the cost of each type of rubber?

30. If 5 liters of peanut oil were blended with 3 liters of corn oil, the blend would sell for $4.05 a liter, whereas if 3 liters of peanut oil were blended with 1 liter of corn oil, the blend would sell for $4.00 a liter. How much is a liter of pure peanut oil and how much is a liter of pure corn oil?

ANSWERS TO MARGIN EXERCISES

1. a. $x = 6$
b. $2x = 8$
c. $1 + x = 3$
d. $x + 4 = 2x$
e. $x - 5 = 4$
f. $2x + 6 = 24$

2. a. $x - 5$
b. $x = 11 - 5$
c. $4x + 12$
d. $4x = x + 12$

3. b and c are consecutive

4. a. $n + n + 1$
b. $2(n + 1)$

5. a. $n + n + 2$
b. $2(n + 2)$

6. a. $x + y = -15$
b. $s = l - 7$
c. $2s - l = 27$
d. $f + 11 = 3s$

7. a. 28cm
b. 30m
c. 8km
d. 86mm

8. a. 225 cents
b. 70 cents
c. 65 cents
d. 550 cents
e. 65 cents
f. 205 cents

9. a. $q = d - 3$
b. $n = 2p$

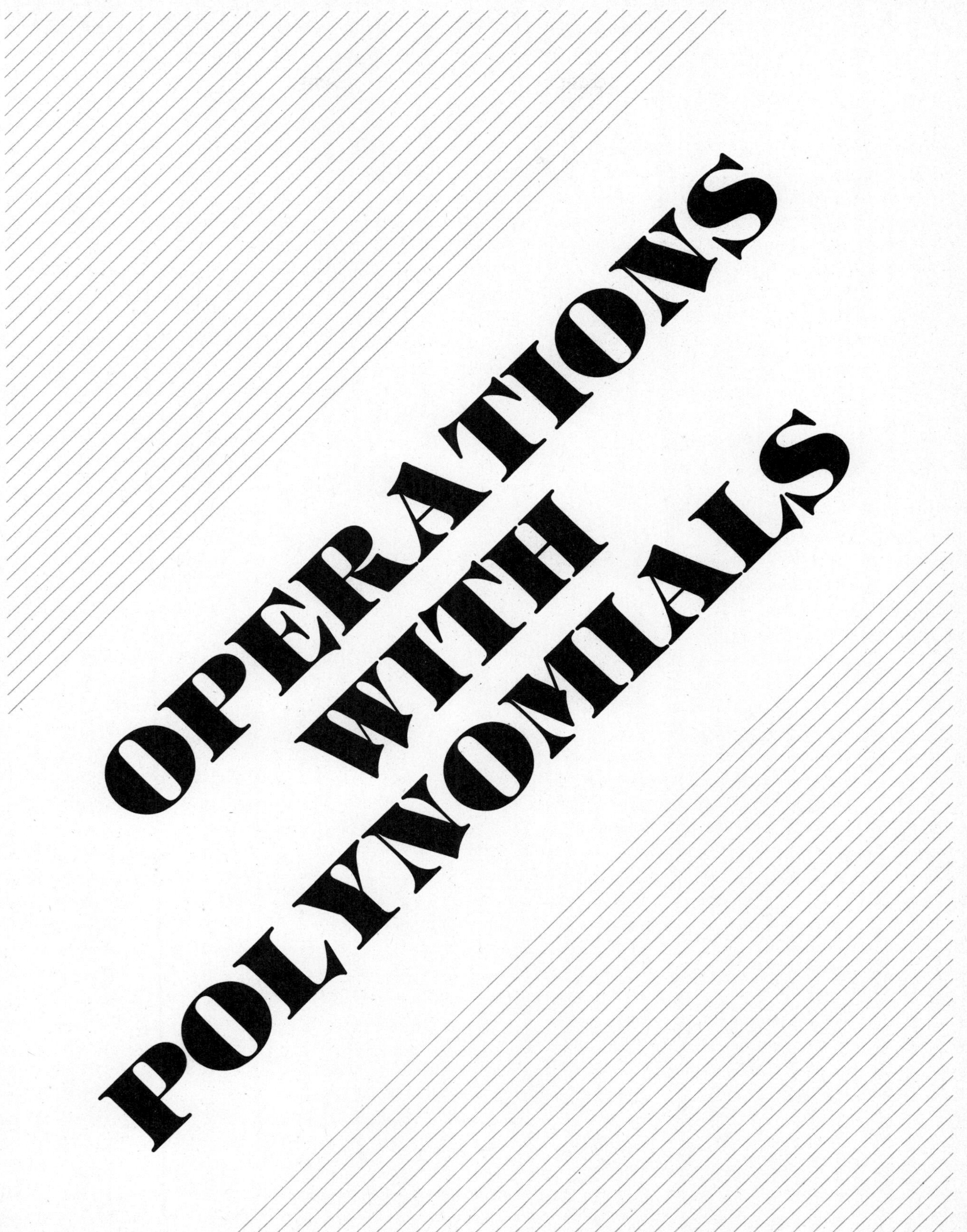
OPERATIONS
WITH
POLYNOMIALS

14 FACTORING MONOMIALS FROM MULTINOMIALS

Factoring is the reverse of multiplication. In particular, factoring a monomial from a multinomial is the reverse of distributing a monomial over a multinomial. (Remember that a multinomial is a polynomial with two or more terms.)

As we saw in Chapter 4, the product of a monomial and a multinomial is another multinomial whose terms are products of the monomial and the corresponding terms in the first multinomial. For example, the product of A and $(B + C)$ is $AB + AC$. The indicated product $A(B + C)$ is said to be in *factored form.*

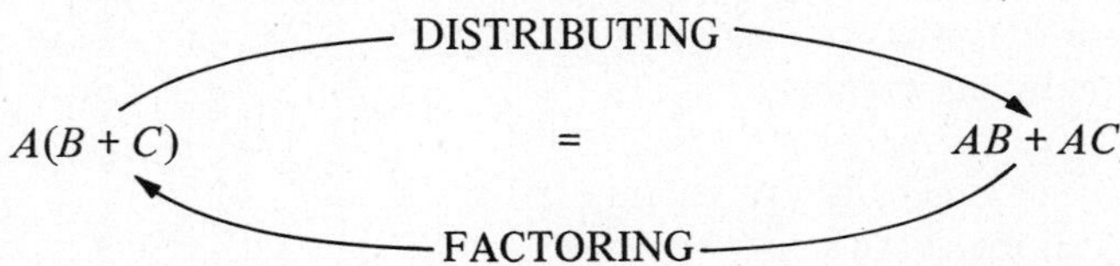

When factoring a monomial from a multinomial, we want to factor out the biggest monomial possible. We want to factor $4x - 12$ as $4(x - 3)$, for example, rather than as $2(2x - 6)$. We will say that a multinomial (with integer coefficients) is in *lowest terms* if no monomial except 1 or -1 can be factored out. For example, $x - 3$ is in lowest terms, whereas $2x - 6$ and $4x - 12$ are not. Factoring the biggest possible monomial from a multinomial then yields the indicated product of a monomial and a multinomial in lowest terms.

14.1 FACTORING OUT INTEGERS

We will begin the study of factoring monomials from multinomials by considering integer factors of multinomials. Remember that the product of an integer and a multinomial is called an *integer multiple* of the multinomial. For example, $4x - 12$ is an integer multiple of $x - 3$. So $4x - 12$ can be expressed as the product of 4 and $(x - 3)$.

1. Compute
a. $2(x+1)$
b. $2(x-1)$
c. $-2(-x+1)$
d. $-2(-x-1)$

2. Factor
a. $4x+2$
$6x+3$
b. $4x-2$
$6x-3$
c. $-4x+2$
$-6x+3$
d. $-4x-2$
$-6x-3$

Look at the pattern in the following examples and try to see how to factor out an integer.

$$x+2=1(x+2) \qquad x-2=1(x-2)$$
$$2x+4=2(x+2) \qquad 2x-4=2(x-2)$$
$$3x+6=3(x+2) \qquad 3x-6=3(x-2)$$
$$4x+8=4(x+2) \qquad 4x-8=4(x-2)$$

$$-x+2=1(-x+2) \qquad -x-2=1(-x-2)$$
$$-2x+4=2(-x+2) \qquad -2x-4=2(-x-2)$$
$$-3x+6=3(-x+2) \qquad -3x-6=3(-x-2)$$
$$-4x+8=4(-x+2) \qquad -4x-8=4(-x-2)$$

Did you notice that in each case the integer factored out divides both terms of the original binomial? Can you see how to determine which integer to factor out?

Our goal is to factor out the largest possible integer and thereby reduce the multinomial to lowest terms. However, we do not need to determine this integer before we begin factoring. We will factor out one integer after another until we reach the indicated product we want. For example, rather than factor $4x - 12$ directly to $4(x-3)$, we can factor it in two steps: first to $2(2x-6)$ and then to $2 \cdot 2(x-3) = 4(x-3)$.

In this section, we only discuss the process of factoring an integer out of a multinomial, and we will only encounter expressions that are integer multiples of multinomials in lowest terms.

Look at the following example of *factoring out an integer.*

TYPICAL PROBLEM

$$8x^3 - 72y^3 + 56z^3$$

First we set up our format by drawing a line down to the left of the multinomial:

$$\begin{array}{l|l} & 8x^3 - 72y^3 + 56z^3 \end{array}$$

We then find a common divisor of the terms. We write this divisor to the left of the line, and then divide each term by this number:

$$\begin{array}{r|l} 2 & 8x^3 - 72y^3 + 56z^3 \\ & 4x^3 - 36y^3 + 28z^3 \end{array}$$

Then we find a common divisor of the resulting multinomial and repeat the above step. (We repeat this step until the multinomial is in lowest terms.) Then we draw a line under the reduced multinomial:

$$\begin{array}{r|l} 2 & 8x^3 - 72y^3 + 56z^3 \\ 4 & 4x^3 - 36y^3 + 28z^3 \\ & x^3 - 9y^3 + 7z^3 \\ \hline \end{array}$$

Every multinomial can be factored as the product of a monomial and a multinomial in lowest terms.

Finally we place the reduced multinomial in parentheses and attach the product of the divisors:

$$8(x^3 - 9y^3 + 7z^3)$$

Hence, $8x^3 - 72y^3 + 56z^3 = 8(x^3 - 9y^3 + 7z^3)$.

EXAMPLES

Factor the following multinomials.

1. $5x^2 - 20$
2. $18x + 162$
3. $245x - 140y$
4. $-216x + 96y$
5. $99x^2 + 66$
6. $9x^2 + 24xy - 15y^2$

Solutions

1. $5 \,|\, 5x^2 - 20$; $x^2 - 4$; $5(x^2 - 4)$

2. $2 \,|\, 18x + 162$; $9 \,|\, 9x + 81$; $x + 9$; $18(x + 9)$

3. $5 \,|\, 245x - 140y$; $7 \,|\, 49x - 28y$; $7x - 4y$; $35(7x - 4y)$

4. $4 \,|\, -216x + 96y$; $6 \,|\, -54x + 24y$; $-9x + 4y$; $24(-9x + 4y)$ or $-24(9x - 4y)$

5. $3 \,|\, 99x^2 + 66$; $11 \,|\, 33x^2 + 22$; $3x^2 + 2$; $33(3x^2 + 2)$

6. $3 \,|\, 9x^2 + 24xy - 15y^2$; $3x^2 + 8xy - 5y^2$; $3(3x^2 + 8xy - 5y^2)$

SPECIAL CASES

1. Because -1 is always a divisor of every term of a multinomial, -1 can be factored from *any* multinomial. Every multinomial equals -1 times the negative of the multinomial. For example,

$$-9x + 4y = -1(9x - 4y)$$

2. The negative of any positive common divisor of the terms of a multinomial is also a common divisor of the terms and can always be factored out instead of the positive integer. The product of a positive integer and a multinomial equals the product of the negative of the integer and the negative of the multinomial. For example,

$$\begin{aligned} -x + 2 &= 1(-x + 2) = -1(x - 2) \\ -2x + 4 &= 2(-x + 2) = -2(x - 2) \\ -3x + 6 &= 3(-x + 2) = -3(x - 2) \\ -4x + 8 &= 4(-x + 2) = -4(x - 2) \end{aligned}$$

3. Factor -1 from
a. $5x + 2$
b. $5x - 2$
c. $-5x + 2$
d. $-5x - 2$

4. Factor a negative integer from
a. $3x + 9y$
b. $9x - 3y$
c. $-3y + 9x$
d. $-9y - 3x$

Checking

Because *distributing* and *factoring* a monomial are reverse operations, we can check the accuracy of a factorization by distributing. For example, we can check the factorization of $4x - 12$ into $4(x - 3)$ by the computation

$$4(x - 3) = 4(x) + 4(-3) = 4x - 12$$

Because we get the original expression, the factorization is correct. (This check does not guarantee, however, that the factorization is complete. The multinomial factor may not be in lowest terms, and we may still be able to factor out another monomial.)

5. Check the solutions to examples 1 to 6 by distributing.

OBSERVATIONS AND REFLECTIONS

1. We can also factor out an integer by the Method of Primes. We can factor the coefficient of every term in a multinomial into a product of primes and then factor out every common prime factor. For example,

$$42x^2 - 28y^2 \quad \text{can be written as} \quad 2 \cdot 3 \cdot 7 \cdot x^2 - 2 \cdot 2 \cdot 7 \cdot y^2$$

The common prime factors are 2 and 7, so their product can be factored out as

$$2 \cdot 3 \cdot 7 \cdot x^2 - 2 \cdot 2 \cdot 7 \cdot y^2 = 2 \cdot 7(3 \cdot x^2 - 2 \cdot y^2) = 14(3x^2 - 2y^2)$$

6. Factor each of these by the Method of Primes
a. $12x - 30$
b. $54y^3 + 90x^2$

Compare this use of the Method of Primes with its use in reducing simple fractions (Section 6.1) and in determining the least common multiple (LCM) of two integers (Section 7.4).

★ 2. Any common divisor of the terms of a multinomial can be factored out. For example, $4x - 12$ equals $2(2x - 6)$. However, we must factor out the largest divisor to reduce the multinomial to lowest terms. This largest divisor is the greatest common divisor (GCD) of the terms of the multinomial.

7. Determine the GCD of
a. $12x^2$ and 20
b. $18x$ and $30y$

★ 3. To reduce a multinomial to lowest terms we can factor out the GCD of its terms in either one step or in several steps. For example, we can reduce $4x - 12$ to lowest terms by factoring 4 out in a single step or by factoring 2 out first and then factoring $2x - 6$ (i.e., $4x - 12 = 2(2x - 6) = 2 \cdot 2(x - 3)$). If we factor a multinomial in several steps, the product of the integers we factor out is always the greatest common divisor (GCD) of the terms of the multinomial (provided, of course, that the multinomial has no literal factors). The result of factoring the multinomial in several steps is always the same as the result of factoring the GCD out in a single step.

8. Factor out the GCD of the terms of
a. $12x^2 + 20$
b. $18x - 30y$

9. Factor in two steps, first factoring out 2:
a. $12x^2 + 20$
b. $18x - 30y$

$$P \div (a \cdot b) = (P \div a) \div b$$
(where P represents a polynomial and a and b represent nonzero monomials)

★ 4. We can factor out any common divisor of the terms of a multinomial, because any common divisor of two or more quantities is a divisor of their sum.

If a is a divisor of b, and a is a divisor of c, then a is a divisor of (b plus c)

$$a|b \quad \text{and} \quad a|c \quad \Rightarrow \quad a|(b+c)$$

(where a, b, and c represent monomials)

Hence, any monomial that divides every term of a polynomial is a divisor of the polynomial. For example, 3 is a divisor of $(9x + 15)$, because 3 is a divisor of both $9x$ and 15.

The converse of this statement is not always true. A number can be a divisor of a sum without being a divisor of its terms. For example, the numerical sum $9 + 15$ equals 24, which is divisible by 2, although neither 9 nor 15 is divisible by 2. (In Chapter 16 we will see trinomials with binomial divisors, although no binomial can divide the individual terms of a trinomial.)

10. Determine the positive divisors of
a. 48
b. 180
c. $48x + 180$

EXERCISES

Factor each of the following expressions completely.

1. $6x^2 - 18$	2. $25x^3 - 15y^2$	3. $-30x + 24y$
4. $52x^2 + 13$	5. $18 - 30x$	6. $42x - 35y^3$
7. $-32x + 18y$	8. $16y^2 - 32x^2$	9. $14x^4 + 7$
10. $40x - 120$	11. $18y + 81$	12. $108x^2 - 84y^2$
13. $-144x + 72y$	14. $169y^3 - 130x$	15. $81y + 54z - 18$

16. $3xy - 12xz + 6yz$ 17. $-40x - 60y + 80z$ 18. $12x + 30y - 18x^2y$

19. $-8w^2x - 14x^2y + 10y^2z$ 20. $-30x^3 - 45y^2 - 15$

21. $18k^2 + 27r^2 + 45k^2r$ 22. $72x^2 + 81xy - 99y$

23. $15x^4 - 105x^2y^2 - 75$ 24. $144a^3 + 156ab^2 - 168b^3$

14.2 FACTORING OUT MONOMIALS

To understand how to factor a monomial from a multinomial, we must consider the special case of factoring out a single power. Look at the pattern in the following examples:

$$x + x^2 = x(1 + x) \qquad x^2 + x^3 = x^2(1 + x)$$
$$x + x^3 = x(1 + x^2) \qquad x^2 + x^4 = x^2(1 + x^2)$$
$$x + x^4 = x(1 + x^3) \qquad x^2 + x^5 = x^2(1 + x^3)$$

$$x^3 + x^4 = x^3(1 + x) \qquad x^4 + x^5 = x^4(1 + x)$$
$$x^3 + x^5 = x^3(1 + x^2) \qquad x^4 + x^6 = x^4(1 + x^2)$$
$$x^3 + x^6 = x^3(1 + x^3) \qquad x^4 + x^7 = x^4(1 + x^3)$$

11. Factor
a. $x^3 + x^7$
b. $x^{12} + x^6$
c. $x + x^9$
d. $x^8 + x^5$

Can you see which power of x to factor out?

In the previous section we saw how to factor an integer from a multinomial. To factor a monomial from a multinomial we first factor out the largest possible integer and then the literals common to all the terms, one base at a time. Look at the pattern in the following examples to see how to do this.

$$2x^2 + 2x^5 = 2x^2(1 + x^3) \qquad w^2x^2 + w^2x^5 = w^2x^2(1 + x^3)$$
$$4x^2 + 2x^5 = 2x^2(2 + x^3) \qquad w^4x^2 + w^2x^5 = w^2x^2(w^2 + x^3)$$
$$6x^2 + 2x^5 = 2x^2(3 + x^3) \qquad w^6x^2 + w^2x^5 = w^2x^2(w^4 + x^3)$$
$$8x^2 + 2x^5 = 2x^2(4 + x^3) \qquad w^8x^2 + w^2x^5 = w^2x^2(w^6 + x^3)$$

12. Factor
a. $x^3y^2 + x^5y^3$
b. $x^6y^3 + x^2y^9$
c. $3x^4 + 3x^{12}$
d. $9x^6 + 6x^9$

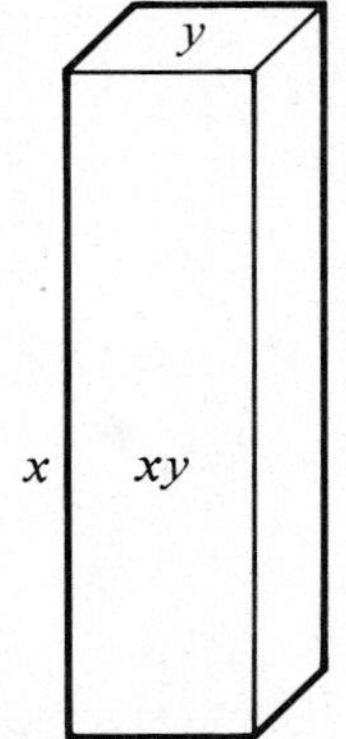

+

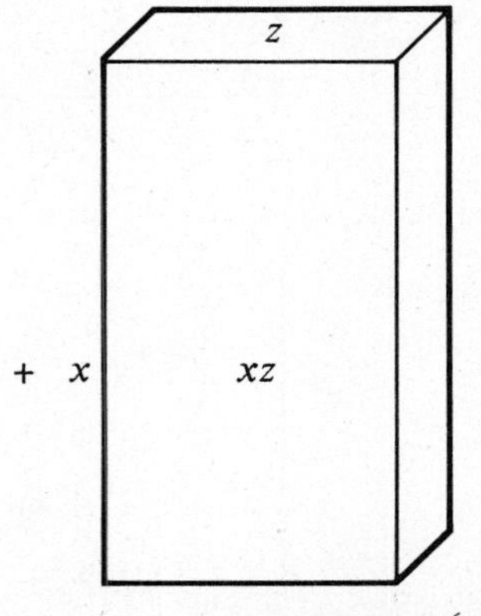

=

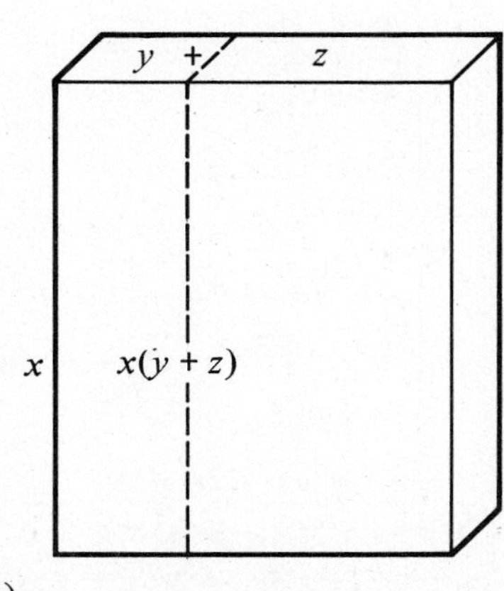

$xy + xz = x(y + z)$

Look at the following example of *factoring out a monomial.*

TYPICAL PROBLEM

$$24x^5y + 6x^4y^2 - 18x^3y^5$$

First we set up our format by drawing a line down to the left of the multinomial:

$$\left|\ 24x^5y + 6x^4y^2 - 18x^3y^5\right.$$

We then factor out the largest possible integer. Then we factor out the common literals, one base at a time. We then draw a line under the reduced multinomial:

$$\begin{array}{r|l} & 24x^5y + 6x^4y^2 - 18x^3y^5 \\ 6 & 4x^5y + x^4y^2 - 3x^3y^5 \\ x^3 & 4x^2y + xy^2 - 3y^5 \\ y & 4x^2 + xy - 3y^4 \\ \hline \end{array}$$

To determine the numerical coefficient of the monomial factor we use the method in the previous section. The format for these two sections is the same.

Finally we place the reduced multinomial in parentheses and attach the product of the divisors:

$$6x^3y(4x^2 + xy - 3y^4)$$

Hence, $24x^5y + 6x^4y^2 - 18x^3y^5 = 6x^3y(4x^2 + xy - 3y^4)$.

EXAMPLES

Every multinomial can be factored as the product of a monomial and a multinomial in lowest terms.

Factor the following multinomials.

1. $24x^3 - 36x^2$
2. $40ab^2c - 8b^2c$
3. $52s^2t - 39st^2$
4. $x^4y^6 - 9x^5y^4z^3 + 6x^2y^3z^4$
5. $-7r^3s + 35r^4s^2t - 14r^5s^3t^2$
6. $52x^3y^2z^4 + 52x^3y^5z^3 - 52x^4y^8z^4$

Solutions

1.
$$\begin{array}{r|l} & 24x^3 - 36x^2 \\ 12 & 2x^3 - 3x^2 \\ x^2 & 2x - 3 \\ \hline \end{array}$$
$12x^2(2x - 3)$

2.
$$\begin{array}{r|l} & 40ab^2c - 8b^2c \\ 8 & 5ab^2c - b^2c \\ b^2 & 5ac - c \\ c & 5a - 1 \\ \hline \end{array}$$
$8b^2c(5a - 1)$

3.
$$\begin{array}{r|l} & 52s^2t - 39st^2 \\ 13 & 4s^2t - 3st^2 \\ s & 4st - 3t^2 \\ t & 4s - 3t \\ \hline \end{array}$$
$13st(4s - 3t)$

4.
$$\begin{array}{r|l} & x^4y^6 - 9x^5y^4z^3 + 6x^2y^3z^4 \\ x^2 & x^2y^6 - 9x^3y^4z^3 + 6y^3z^4 \\ y^3 & x^2y^3 - 9x^3yz^3 + 6z^4 \\ \hline \end{array}$$
$x^2y^3(x^2y^3 - 9x^3yz^3 + 6z^4)$

5.
$$\begin{array}{r|l} & -7r^3s + 35r^4s^2t - 14r^5s^3t^2 \\ 7 & -r^3s + 5r^4s^2t - 2r^5s^3t^2 \\ r^3 & -s + 5rs^2t - 2r^2s^3t^2 \\ s & -1 + 5rst - 2r^2s^2t^2 \\ \hline \end{array}$$
$7r^3s(-1 + 5rst - 2r^2s^2t^2)$
or $-7r^3s(1 - 5rst + 2r^2s^2t^2)$

6.
$$\begin{array}{r|l} & 52x^3y^2z^4 + 52x^3y^5z^3 - 52x^4y^8z^4 \\ 52 & x^3y^2z^4 + x^3y^5z^3 - x^4y^8z^4 \\ x^3 & y^2z^4 + y^5z^3 - xy^8z^4 \\ y^2 & z^4 + y^3z^3 - xy^6z^4 \\ z^3 & z + y^3 - xy^6z \\ \hline \end{array}$$
$52x^3y^2z^3(z + y^3 - xy^6z)$

SPECIAL CASES

The product of a monomial and a multinomial equals the product of the negative of the monomial and the negative of the multinomial. For example,

$$2x^2(-9x + 4y) = -2x^2(9x - 4y)$$

When factoring a multinomial, it is sometimes preferable to factor out a negative monomial if the first term of the multinomial is negative. Hence, for example, $-18x^3 + 8x^2y$ is frequently factored as $-2x^2(9x - 4y)$, rather than as $2x^2(-9x + 4y)$.

13. Factor a negative monomial from
a. $-x^4 + 4x^3$
b. $-4x^4 - x^3$
c. $-6x^3y^5 + x^2y$
d. $-x^3y^5 - 6x^2y$

OBSERVATIONS AND REFLECTIONS

1. Factoring provides the theoretical foundation for our method of adding like terms in Section 3.1. For example, we simplify the sum $3x + 5x$ to $8x$ by adding the coefficients and attaching the common literal. This method is equivalent to factoring out the common literal component and then combining the numerical coefficients. For example,

$$3x + 5x \;=\; x(3 + 5) \;=\; x(8) \;=\; 8x$$

2. Factoring also provides the theoretical foundation for our method of adding fractions with common denominators, in Sections 7.1 and 9.1. For example, we simplify the sum $\frac{3}{x} + \frac{5}{x}$ to $\frac{8}{x}$ by adding the numerators and placing this sum over the common denominator. This method is equivalent to factoring out the common denominator (which is a common factor) and then combining the numerators. For example,

$$\frac{3}{x} + \frac{5}{x} = 3 \cdot \frac{1}{x} + 5 \cdot \frac{1}{x} = \frac{1}{x}(3 + 5) = \frac{1}{x}(8) = \frac{8}{x}$$

14. Similarly, add these fractions by factoring
a. $\frac{7}{3x} + \frac{4}{3x}$
b. $\frac{5}{x} - \frac{9}{x}$
c. $\frac{4x}{5} + \frac{9}{5}$

★ 3. As we saw in Chapter 8, the common divisor of two monomials with the largest positive numerical coefficient and the largest exponents is called their *greatest common divisor* (GCD). For example, the GCD of $8w^2x^5$ and $12w^3x^2$ is $4w^2x^2$. Hence, the biggest monomial that can be factored out of a multinomial is the GCD of its terms. For example, $4w^2x^2$ is the biggest monomial that can be factored from $8w^2x^5 + 12w^3x^2$.

To reduce a multinomial to lowest terms we must factor out the GCD of its terms. For example, $8w^2x^5 + 12w^3x^2 = 4w^2x^2(2x^3 + 3w)$ and $2x^3 + 3w$ is in lowest terms.

15. Determine the GCD of
a. $9x^3y^2$ and $21x^2y^6$
b. $8x^3y^2$ and $24x^2y$

★ 4. Factoring a monomial from a binomial and reducing a monomial fraction are closely related. Consider the two problems:

Factor $8x + 12$ Reduce $\frac{8x}{12}$

The quantities $8x$ and 12 are both multiples of 4. We can factor $8x + 12$ into $4(2x + 3)$, and we can reduce $\frac{8x}{12}$ to $\frac{2x}{3}$. In both problems we divide $8x$ and 12 by 4. When *factoring* we divide to get the reduced binomial factor of the indicated product $4(2x + 3)$. When *reducing* we divide to get the numerator and denominator of the reduced fraction $\frac{2x}{3}$.

16. Factor
a. $9x^3y^2 + 21x^2y^6$
b. $8x^3y^2 + 24x^2y$

Despite the similarity of these computations, however, there is a significant difference in our goals in solving these two problems. When we *factor* we must find an indicated product with the same *value* as the original polynomial. Hence, the quantity that we factor from the polynomial must be a factor of the resulting expression. For example, $8x + 12 = 4(2x + 3)$, and not $(2x + 3)$. On the other hand, when we *reduce a fraction* we must find a reduced fraction with a numerator and denominator in the same *ratio* as those in the original fraction. Hence, we must divide the numerator and denominator of the original fraction by the *same* quantity to preserve their ratio, but we need not preserve this quantity. For example,

17. Reduce

a. $\dfrac{9x^3y^2}{21x^2y^6}$

b. $\dfrac{8x^3y^2}{24x^2y}$

$$\frac{8x}{12} = \frac{2x \cdot 4}{12 \cdot 4} = \frac{2x}{12} \cdot \frac{4}{4} = \frac{2x}{3}$$

(and $\frac{4}{4}$ can be reduced to 1 and forgotten).

EXERCISES

Factor each of the following expressions completely.

1. $7x^3y - 15xy$
2. $12x^5y^2 - y^2$
3. $9x^6y^3 - 14x^2y^2$
4. $21x^5 - 44x^4$
5. $-4x^7y^5z^2 + x^4y^5z^6$
6. $x^6y^4z^2 - 2x^6y^5z$
7. $7x^2y + 42y^2$
8. $16rs^2t^3 - 24st^2$
9. $12xy^2z - 8xz^2$
10. $24x^2y - 12xy^2$
11. $14xy^3 + 21xy^4$
12. $26x^2yz^2 - 13x^2z$
13. $15xy^3 - 75x^2y^4$
14. $8xz^4 - 64z^3$
15. $52x^2y + 65xy^3$
16. $64r^2s^4 - 96rs^4t^2$
17. $-28xy^3z + 28y^3$
18. $144x^7y^4 + 156x^6y^6$
19. $12x^2y^4z + 18x^2y^3 - 9y^2$
20. $30x^4y^6z^2 + 30x^3y^6z^3 + 30x^4y^5z^3$
21. $27x^3y^2 + 36x^2y^3z - 18x^2z^3$
22. $15a^2b^3c^2 - 20a^5b^6c^5 + 25a^8b^9c^8$
23. $110x^9 + 165x^6 - 143x^3$
24. $60x^7y^2 + 90x^4y^9 - 60x^2y^4$

ANSWERS TO MARGIN EXERCISES

1. a. $2x+2$
b. $2x-2$
c. $2x-2$
d. $2x+2$

2. a. $2(2x+1)$
$3(2x+1)$
b. $2(2x-1)$
$3(2x-1)$
c. $-2(2x-1)$
$-3(2x-1)$
d. $-2(2x+1)$
$-3(2x+1)$

3. a. $-1(-5x-2)$
b. $-1(-5x+2)$
c. $-1(5x-2)$
d. $-1(5x+2)$

4. a. $-3(-x-3y)$
b. $-3(-3x+y)$
c. $-3(y-3x)$
d. $-3(3y+x)$

6. a. $2 \cdot 3(2 \cdot x - 5) = 6(2x-5)$
b. $2 \cdot 3 \cdot 3(3 \cdot y^3 + 5 \cdot x^2) = 18(3y^3 + 5x^2)$

7. a. 4
b. 6

8. a. $4(3x^2+5)$
b. $6(3x-5y)$

9. a. $2(6x^2+10) = 4(3x^2+5)$
b. $2(9x-15y) = 6(3x-5y)$

10. a. 1, 2, 3, 4, 6, 8, 12, 16, 24, 48
b. 1, 2, 3, 4, 5, 6, 9, 10, 12, 15, 18, 20, 30, 36, 45, 60, 90, 180
c. 1, 2, 3, 4, 6, 12

11. a. $x^3(1+x^4)$
b. $x^6(x^6+1)$
c. $x(1+x^8)$
d. $x^5(x^3+1)$

12. a. $x^3y^2(1+x^2y)$
b. $x^2y^3(x^4+y^6)$
c. $3x^4(1+x^8)$
d. $3x^6(3+2x^3)$

13. a. $-x^3(x-4)$
b. $-x^3(4x+1)$
c. $-x^2y(6xy^4-1)$
d. $-x^2y(xy^4+6)$

14. a. $\frac{1}{3x}(7+4) = \frac{1}{3x}(11) = \frac{11}{3x}$
b. $\frac{1}{x}(5-9) = \frac{1}{x}(-4) = \frac{-4}{x}$
c. $\frac{1}{5}(4x+9) = \frac{4x+9}{5}$

15. a. $3x^2y^2$
b. $8x^2y$

16. a. $3x^2y^2(3x+7y^4)$
b. $8x^2y(xy+3)$

17. a. $\frac{3x}{7y^4}$
b. $\frac{xy}{3}$

CHAPTER REVIEW

SECTION 14.1 Factor completely.

1. $12x + 48$
2. $20 - 48x^2$
3. $28y + 32z$
4. $8x^2 - 24z^2$
5. $-15x + 10y$
6. $54x + 9$
7. $18y^2 - 9y + 27$
8. $-21x - 42y + 35z$
9. $25x^2 - 75y + 100$
10. $32y^2 + 64x - 144$
11. $-12x^2 - 36y^2 - 72$
12. $21x + 35y^2 + 49$

SECTION 14.2 Factor completely.

13. $12x^3 - 41x^2$
14. $35x^7y^4 - 36x^6y^5$
15. $4x^7y^5 - 5x^6y^6$
16. $27x^5 - 36x^4$
17. $48a^2b^3c^4 - 36b^3c$
18. $15x^6 - 20x^7$
19. $18wxy + 22xyz$
20. $96x^8 - 108x^9$
21. $36km^4 - 9k$
22. $-6xy + 24x^2y^2z - 30wxyz^2$
23. $-9k^2m + 18km^3n - 27km^2n^3$
24. $42x^7y^4 - 84x^9y^8 + 126x^9y^{11}$

CUMULATIVE TEST

The following problems test your understanding of this chapter and of related subject matter in Chapter 4. Before taking this test, thoroughly review Sections 4.2, 4.3, 14.1, and 14.2.

Once you have finished the test, compare your answers with the answers provided at the back of the book. Note the section number of each problem missed, and thoroughly review those sections again.

Factor or multiply.

1. $16x + 64$
2. $13x^6 - 36x^4$
3. $56x - 77 - 21x^3$
4. $6(5x + 9)$
5. $42a^2b^4c - 56a^3b$
6. $9y - 45z$
7. $s^2t(s^4t - 1)$
8. $27wxy + 48xyz$
9. $28 - 14x^2$
10. $6x(-9 + 4x - x^2)$
11. $12x^4 + 25x^3$
12. $63x - 28y^2$
13. $8(-4x + y - 6z)$
14. $-6k^3n^2 + 30k^2m^4n^2 - 42k^2m^2n^4$
15. $-20x + 12y - 4$
16. $-uv^3(4u^4v^3 - 9uv^6)$
17. $14s^4t^7 - 5s^8t^6$
18. $18x^2 - 24y^2 + 36xy$
19. $-5(6 - 8A)$
20. $-8xy + 40x^3y^2z - 24wx^2y^3z^4$

15 MULTIPLICATION OF POLYNOMIALS

In *algebraic products* of polynomials such as $(3x + 4)(2x - 7)$, $(8x)(x^2 - 3x + 7)(2x^2 - 5x)$, and $(3a - 2b)^3$, each quantity is called *a factor* of the product. For example, in the product $(8x)(x^2 - 3x + 7)(2x^2 - 5x)$, the factors are $8x$, $x^2 - 3x + 7$, and $2x^2 - 5x$.

Polynomial multiplication is indicated in any of the following ways:

1. $(3x + 4)(2x - 7)$	with both multinomials in parentheses.
2. $(3a - 2b)^3$	using exponential notation; the multinomial is inside the parentheses and the exponent is outside the parentheses.
3. $-9y^2(3y - 4)(2y + 1)$	with the monomial first and the other factors in parentheses.
4. $(8x)(x^2 - 3x + 7)(2x^2 - 5x)$	with every factor inside parentheses.

Every multinomial must be in parentheses, but a monomial factor, if it is first, need not be in parentheses. A dot can be placed between the parentheses in an indicated product.

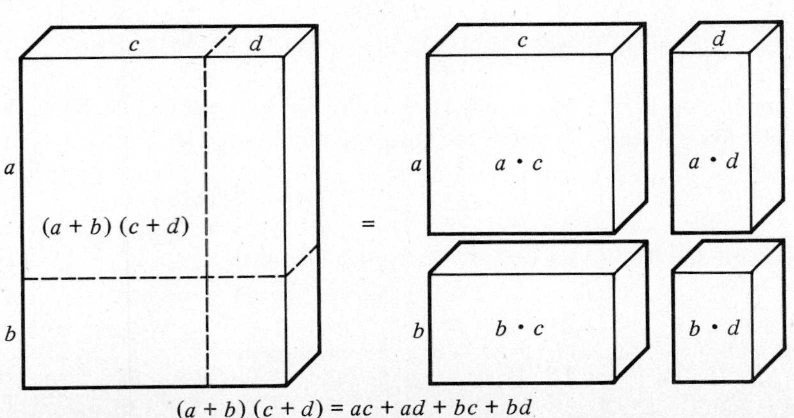

15.1
MULTIPLYING TWO BINOMIALS

In this section we will begin the study of multiplication of polynomials by considering products of two binomials, such as $(3x - 2)(4x + 5)$.

To see how to multiply two binomials, look at the pattern in the following numerical computations. In the first row of each computation, the expression on the left is an indicated product of sums, and the expression on the right is its multiplied-out form.

$$(1 - 1)(3 + 4) = (1)(3) + (1)(4) + (-1)(3) + (-1)(4) = 3 + 4 - 3 - 4$$
$$(0) \cdot (7) = 0$$

$$(2 - 1)(3 + 4) = (2)(3) + (2)(4) + (-1)(3) + (-1)(4) = 6 + 8 - 3 - 4$$
$$(1) \cdot (7) = 7$$

$$(3 - 1)(3 + 4) = (3)(3) + (3)(4) + (-1)(3) + (-1)(4) = 9 + 12 - 3 - 4$$
$$(2) \cdot (7) = 14$$

$$(4 - 1)(3 + 4) = (4)(3) + (4)(4) + (-1)(3) + (-1)(4) = 12 + 16 - 3 - 4$$
$$(3) \cdot (7) = 21$$

1. Form a similar diagram for
   a. $(2 - 3)(2 + 5)$
   b. $(3 - 3)(2 + 5)$

In each example we have added in going from the first row to the second, and in each row we have multiplied in going from left to right. Can you guess from these examples how to multiply two binomials?

The product of two binomials is a sum of four terms. Each term in this sum is itself a product of two monomials, one from each binomial factor. That is,

$$(a + b)(c + d) = ac + ad + bc + bd$$

2. Similarly, compute the products
   a. $(a + b)(c - d)$
   b. $(a - b)(c - d)$

This sum simplifies to a trinomial (three terms) when two of the four terms are *like* terms (or to a binomial if these two like terms are negatives of each other). For example, the product $(3x - 2)(4x + 5)$ equals the indicated sum of products:

$$(3x)(4x) + (3x)(+5) + (-2)(4x) + (-2)(+5)$$

which simplifies to $12x^2 + 15x - 8x - 10 = 12x^2 + 7x - 10$, a trinomial. Similarly, the product $(3x - 2)(3x + 2)$ equals the indicated sum of products:

$$(3x)(3x) + (3x)(+2) + (-2)(3x) + (-2)(+2)$$

which simplifies to $9x^2 + 6x - 6x - 4 = 9x^2 - 4$, a binomial.

3. Simplify
   a. $(x)(3x) + (x)(+2) + (+4)(3x) + (+4)(+2)$
   b. $(5x)(6x) + (5x)(-1) + (-3)(6x) + (-3)(-1)$
   c. $(2x)(2x) + (2x)(+9) + (-9)(2x) + (-9)(+9)$

When multiplying polynomials, we use a vertical format.

Look at the following examples of *multiplying two binomials.*

TYPICAL PROBLEMS

1. $(3x - 2)(4x + 5)$

We first change to vertical format, writing one factor under the other and dropping parentheses:

$$\begin{array}{r} 4x + 5 \\ \underline{3x - 2} \end{array}$$

We then multiply each term of the top binomial by the first term of the bottom binomial. We multiply from left to right:

$$\begin{array}{r} 4x + 5 \\ \underline{3x - 2} \\ 12x^2 + 15x \end{array}$$

Then we multiply each term of the top binomial by the second term of the bottom binomial. We again multiply from left to right and we line up like terms as we write this second product:

$$\begin{array}{rrr} & 4x & + 5 \\ & \underline{3x} & \underline{- 2} \\ 12x^2 & + 15x & \\ & - 8x & - 10 \end{array}$$

Finally we add the two products:

$$\begin{array}{rrr} & 4x & + 5 \\ & \underline{3x} & \underline{- 2} \\ 12x^2 & + 15x & \\ & \underline{- 8x} & \underline{- 10} \\ 12x^2 & + 7x & - 10 \end{array}$$

Hence, $(3x - 2)(4x + 5) = 12x^2 + 7x - 10$.

2. $(3x - 5)(7x^2 - 8)$

We first change to vertical format, writing one factor under the other and dropping parentheses:

$$\begin{array}{r} 3x - 5 \\ \underline{7x^2 - 8} \end{array}$$

Then we multiply each term of the top binomial by the first term of the bottom binomial. We multiply from left to right:

$$\begin{array}{r} 3x - 5 \\ \underline{7x^2 - 8} \\ 21x^3 - 35x^2 \end{array}$$

We then multiply each term of the top binomial by the second term of the bottom binomial. Since there are no like terms, we write this product below and to the right of the previous product:

$$\begin{array}{rrrr} & 3x - 5 & & \\ & \underline{7x^2 - 8} & & \\ 21x^3 & - 35x^2 & & \\ & & - 24x & + 40 \end{array}$$

Finally we add the two products by bringing down the terms:

$$\begin{array}{rrrr} & 3x - 5 & & \\ & \underline{7x^2 - 8} & & \\ 21x^3 & - 35x^2 & & \\ & & \underline{- 24x} & \underline{+ 40} \\ 21x^3 & - 35x^2 & - 24x & + 40 \end{array}$$

To multiply each pair of terms we use the method in Section 2.3.

Hence, $(3x - 5)(7x^2 - 8) = 21x^3 - 35x^2 - 24x + 40$.

WARNING: Do not put one term under another unless they are like terms.

EXAMPLES

Every product of polynomials can be multiplied, yielding a polynomial.

Multiply the following pairs of binomials.

1. $(x + 4)(3x - 7)$
2. $(5x - 9)(x - 4)$
3. $(8x + 3)(6x + 5)$
4. $(7x - 9y)(4x - 5y)$
5. $(3x - 4)(x - 2y)$
6. $(4x^2 - 5x)(3x + 2)$

Solutions

1. $$\begin{array}{r} x + 4 \\ 3x - 7 \\ \hline 3x^2 + 12x \\ - 7x - 28 \\ \hline 3x^2 + 5x - 28 \end{array}$$

2. $$\begin{array}{r} 5x - 9 \\ x - 4 \\ \hline 5x^2 - 9x \\ - 20x + 36 \\ \hline 5x^2 - 29x + 36 \end{array}$$

3. $$\begin{array}{r} 8x + 3 \\ 6x + 5 \\ \hline 48x^2 + 18x \\ 40x + 15 \\ \hline 48x^2 + 58x + 15 \end{array}$$

4. $$\begin{array}{r} 7x - 9y \\ 4x - 5y \\ \hline 28x^2 - 36xy \\ - 35xy + 45y^2 \\ \hline 28x^2 - 71xy + 45y^2 \end{array}$$

5. $$\begin{array}{r} 3x - 4 \\ x - 2y \\ \hline 3x^2 - 4x \\ - 6xy + 8y \\ \hline 3x^2 - 4x - 6xy + 8y \end{array}$$

6. $$\begin{array}{r} 4x^2 - 5x \\ 3x + 2 \\ \hline 12x^3 - 15x^2 \\ 8x^2 - 10x \\ \hline 12x^3 - 7x^2 - 10x \end{array}$$

SPECIAL CASES

When multiplying two binomials that differ *only* in the sign of their second terms, we obtain another binomial. To see this, consider the product of $(a + b)(a - b)$:

4. Compute
a. $(x - y)(x + y)$
b. $(x^2 + 1)(x^2 - 1)$

$$\begin{array}{r} a + b \\ a - b \\ \hline a^2 + ab \\ - ab - b^2 \\ \hline a^2 - b^2 \end{array}$$

You have already seen several multiplication tables. Here is a table showing products of binomials. Verify the entries to check your understanding of multiplication.

times	$x - 3$	$x - 2$	$x - 1$	x	$x + 1$	$x + 2$	$x + 3$
$x + 3$							
$x + 2$		$x^2 - 4$	$x^2 + x - 2$	$x^2 + 2x$	$x^2 + 3x + 2$	$x^2 + 4x + 4$	
$x + 1$		$x^2 - x - 2$	$x^2 - 1$	$x^2 + x$	$x^2 + 2x + 1$	$x^2 + 3x + 2$	
x		$x^2 - 2x$	$x^2 - x$	x^2	$x^2 + x$	$x^2 + 2x$	
$x - 1$		$x^2 - 3x + 2$	$x^2 - 2x + 1$	$x^2 - x$	$x^2 - 1$	$x^2 + x - 2$	
$x - 2$		$x^2 - 4x + 4$	$x^2 - 3x + 2$	$x^2 - 2x$	$x^2 - x - 2$	$x^2 - 4$	
$x - 3$							

5. Complete the table.

OBSERVATIONS AND REFLECTIONS

1. In place of our procedure for multiplying two binomials, which uses a vertical format, an alternative method using a horizontal format, called the *FOIL Method*, is also commonly used.

FOIL METHOD: The factors are multiplied in the following order.

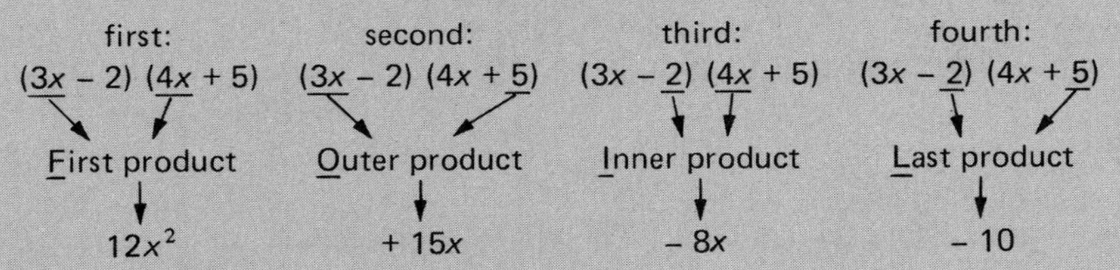

6. Use the FOIL Method to compute
   a. $(3x - 4)(2x + 5y)$
   b. $(x^2 - 6)(7x - 3)$

2. When changing a product of two binomials to vertical format, we do not specify which binomial is to be written above the other. Because multiplication is both commutative and associative, polynomials can be written and multiplied in any order. For example,

$$(3x - 2)(4x + 5) \quad = \quad (4x + 5)(3x + 2)$$

$$\begin{array}{r} 3x - 2 \\ \underline{4x + 5} \\ 12x^2 - 8x \qquad \\ \underline{15x - 10} \\ 12x^2 + 7x - 10 \end{array} \qquad \qquad \begin{array}{r} 4x + 5 \qquad \\ \underline{3x - 2 \qquad} \\ 12x^2 + 15x \qquad \\ \underline{- 8x - 10} \\ 12x^2 + 7x - 10 \end{array}$$

7. Similarly, compute $(2x + 5)(x + 4)$ in two ways.

★ 3. In each computation in 2, two partial products are formed and then combined. Each partial product can be viewed as a distribution. Consider again the computations:

$$\begin{array}{rcr} & & 3x - 2 \\ & & \underline{4x + 5} \\ \mathbf{4x(3x - 2)} & = & 12x^2 - 8x \qquad \\ \underline{\mathbf{5(3x - 2)}} & = & \underline{15x - 10} \\ \mathbf{4x(3x - 2) + 5(3x - 2)} & = & 12x^2 + 7x - 10 \end{array} \quad \text{and} \quad \begin{array}{rcl} 4x + 5 \qquad & & \\ \underline{3x - 2 \qquad} & & \\ 12x^2 + 15x \qquad & = & \mathbf{3x(4x + 5)} \\ \underline{- 8x - 10} & = & \underline{\mathbf{-2(4x + 5)}} \\ 12x^2 + 7x - 10 & = & \mathbf{3x(4x + 5) - 2(4x + 5)} \end{array}$$

Both $4x(3x - 2) + 5(3x - 2)$ and $3x(4x + 5) - 2(4x + 5)$ can be viewed as indicated distributions of one binomial over the other.

$$(a + b)(c + d) = (a + b)c + (a + b)d = ac + bc + ad + bd$$

and

$$(a + b)(c + d) = a(c + d) + b(c + d) = ac + ad + bc + bd$$

(where *a*, *b*, *c*, and *d* represent monomials)

8. Distribute
a. $(x + 1)$ over $(3x + 2)$
b. $(4x - 3)$ over $(2x + 8)$
c. $(2x - 5)$ over $(3x - 7)$

Distribution can be used to multiply a multinomial by any other expression simply by treating the other expression as a monomial.

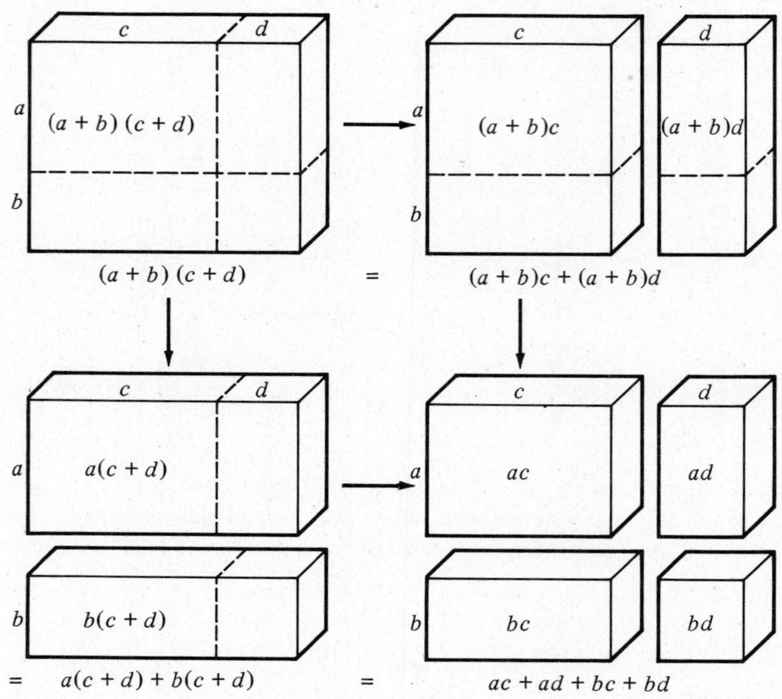

EXERCISES

Multiply to convert each of the following expressions to a polynomial.

1. $(x - 3)(x + 6)$
2. $(x + 8)(x + 4)$
3. $(3x + 1)(4x + 5)$
4. $(4x + 3)(6x + 1)$
5. $(9x - 1)(3x - 2)$
6. $(4x + 5)(x - 4)$
7. $(3x - 4y)(6x + 3y)$
8. $(7x + 2y)(4x - 9y)$
9. $(8y - x)(6y + 5x)$
10. $(6x - 9y)(3x - y)$
11. $(x - 4)(x^2 - 3)$
12. $(2x - 9)(3x^2 - 2)$
13. $(4x - 5)(x - 3y)$
14. $(3x + 5)(4x + 5y)$
15. $(8x - 1)(x^2 - 3x)$
16. $(3x + 5)(3x^2 + 5x)$
17. $(6x^2 - x)(x - 6)$
18. $(3x^2 - 6x)(4x^5 - 2x^4)$
19. $(3x + 2)(3x - 2)$
20. $(2x + 5)(2x - 5)$
21. $(6x^2 - y)(6x^2 + y)$
22. $(5x^3 - y^2)(5x^3 + y^2)$
23. $(5x^3 + 4y^5)(5x^3 - 4y^5)$
24. $(6xy^5 - 1)(6xy^5 + 1)$
25. $(5x + 1)(5x + 1)$
26. $(8x - 1)(8x - 1)$
27. $(3x - 7y)(3x - 7y)$
28. $(6x + 5y)(6x + 5y)$
29. $(2x^2 - 9y^3)(2x^2 - 9y^3)$
30. $(8x^4 - 3y^3)(8x^4 - 3y^3)$

15.2 MULTIPLYING MONOMIALS AND BINOMIALS

When computing the product of a monomial and two or more multinomials, we multiply the multinomials first. Then we multiply the resulting sum by the monomial. This last step, of course, involves distribution. We shall see that the vertical format used in multiplying multinomials can also be used in multiplying a multinomial by a monomial.

9. Distribute
a. $2x^2(5x + 7)$
b. $-4(x - 5)$

Look at the following example of *multiplying monomials and binomials.*

TYPICAL PROBLEM

$$9a^2(3a - 4)(2a + 1)$$

First we multiply the two binomials:

$$\begin{array}{r} 3a - 4 \\ 2a + 1 \\ \hline 6a^2 - 8a \qquad \\ 3a - 4 \\ \hline 6a^2 - 5a - 4 \end{array}$$

Then we multiply the product by the monomial:

$$\begin{array}{l} 6a^2 - 5a - 4 \\ 9a^2 \\ \hline 54a^4 - 45a^3 - 36a^2 \end{array}$$

Hence, $9a^2(3a - 4)(2a + 1) = 54a^4 - 45a^3 - 36a^2$.

To multiply two binomials we use the method in the previous section. Then to multiply the product of the binomials by the monomial we use the same vertical format.

EXAMPLES

Multiply the following monomials and binomials.

Every product of polynomials can be multiplied, yielding a polynomial.

1. $5x(5x - 1)(5x + 1)$
2. $4x^2(x - 4y)(7x + 3y)$
3. $-3x^3(6x + 5)(3x - 4)$
4. $4(x - 4)^2$
5. $x^3(x - y)^2$
6. $16x(8x - 2)^2$

Solutions

1.
$$\begin{array}{l} 5x - 1 \\ 5x + 1 \\ \hline 25x^2 - 5x \\ \qquad\quad 5x - 1 \\ \hline 25x^2 \qquad - 1 \\ \quad\downarrow \\ 25x^2 \qquad - 1 \\ 5x \\ \hline 125x^3 - 5x \end{array}$$

2.
$$\begin{array}{l} x - 4y \\ 7x + 3y \\ \hline 7x^2 - 28xy \\ \qquad\quad 3xy - 12y^2 \\ \hline 7x^2 - 25xy - 12y^2 \\ \quad\downarrow \\ 7x^2 - 25xy - 12y^2 \\ 4x^2 \\ \hline 28x^4 - 100x^3y - 48x^2y^2 \end{array}$$

3. $6x + 5$
$3x - 4$
$18x^2 + 15x$
$- 24x - 20$
$18x^2 - 9x - 20$
↓
$18x^2 - 9x - 20$
$-3x^3$
$-54x^5 + 27x^4 + 60x^3$

4. $x - 4$
$x - 4$
$x^2 - 4x$
$- 4x + 16$
$x^2 - 8x + 16$
↓
$x^2 - 8x + 16$
4
$4x^2 - 32x + 64$

5. $x - y$
$x - y$
$x^2 - xy$
$- xy + y^2$
$x^2 - 2xy + y^2$
↓
$x^2 - 2xy + y^2$
x^3
$x^5 - 2x^4y + x^3y^2$

6. $8x - 2$
$8x - 2$
$64x^2 - 16x$
$- 16x + 4$
$64x^2 - 32x + 4$
↓
$64x^2 - 32x + 4$
$16x$
$1024x^3 - 512x^2 + 64x$

SPECIAL CASES

10. Write without exponential notation
a. $(x + 3)^2$
b. $(2x - 5)^2$
c. $(3x + 4y)^2$

Exponential notation is a shorthand for multiplication. Therefore,

$$(x - 4)^2 = (x - 4)(x - 4) \qquad (8x - 2)^2 = (8x - 2)(8x - 2)$$

OBSERVATIONS AND REFLECTIONS

★ When computing a product of polynomials, we ordinarily multiply two factors at a time. We can compute the product $m(x + 1)(y - 1)$, for example, in any of the following ways:

11. Similarly, compute $2x(x - 3)(x + 4)$ in three ways.

$m(x + 1)(y - 1)$	or	$(x + 1)\,m\,(y - 1)$	or	$(x + 1)(y - 1)m$
$(mx + m)(y - 1)$		$(x + 1)(my - m)$		$(xy - x + y - 1)m$
$mxy - mx + my - m$		$mxy - mx + my - m$		$mxy - mx + my - m$ (This is the preferred way.)

Each computation involves one product of two binomials and one distribution of the monomial over a multinomial.

WARNING: The monomial m in the above example *cannot* be distributed over *both* binomials before they are multiplied together.

This is an illegal step:

$$m(x + 1)(y - 1) \neq (mx + m)(my - m)$$

EXERCISES

Multiply to convert each of the following expressions to a polynomial.

1. $4x(x + 3)(x - 2)$
2. $6(x - 5)(x - 8)$
3. $-4y(3y - 2)(y - 7)$
4. $x^7(x - 4)(3x - 2)$
5. $3xy(2x - 3y)(x + 2y)$
6. $4x^2(2x - 1)(2x + 3)$
7. $-4x^2y(3y - 2)(4y + 1)$
8. $-6(x - 8)(4x - 1)$
9. $10x^3(x + 4)(4x + 3)$
10. $-4y^4(2y + 5)(3y - 1)$
11. $x^6(x^3 + x^2)(x - 1)$
12. $-3x^4(x^5 - 3x^4)(x^3 - 2x^2)$
13. $4x^2(4x^2y - 5xy)(3x + 2)$
14. $3xy^3(x - 4y)(xy^2 - 2y^3)$
15. $-(x^2 + 3)(x^2 + 6)$
16. $4y^2(3x^2 - 4y)(x^2 + 4y)$
17. $-2x(3x - x^3)(2x - x^3)$
18. $6x^5(x^3 - 4x)(3x^7 + 7x^5)$
19. $5(x - 3)^2$
20. $-4x(x - 4)^2$
21. $12x(4x - 3)^2$
22. $-8x^2(2x - 4)^2$
23. $x^2y^4(3x + y^2)^2$
24. $-4xy(2x + 8y)^2$

15.3 MULTIPLYING POLYNOMIALS

In the previous sections, we have considered products of monomials and binomials. Here we will generalize to products of any number of polynomials with any number of terms. We will see such products as $(x^2 - 3x + 7)(2x^2 - 5x)$ and $(3a - 2b)^3$.

Remember that exponential notation is a shorthand. Just as x^2 means $x \cdot x$, x^3 means $x \cdot x \cdot x$, and x^4 means $x \cdot x \cdot x \cdot x$, so

$(3a - 2b)^2$	means	$(3a - 2b)(3a - 2b)$
$(3a - 2b)^3$	means	$(3a - 2b)(3a - 2b)(3a - 2b)$
$(3a - 2b)^4$	means	$(3a - 2b)(3a - 2b)(3a - 2b)(3a - 2b)$

Note that $(3a - 2b)^4$ is also $(3a - 2b)^2 \cdot (3a - 2b)^2$.

12. Compute
a. $x \cdot x^3$
b. $x^2 \cdot x^2$
c. $x^2 \cdot x^4$
d. $x^3 \cdot x^3$

Look at the following example of *multiplying polynomials.*

TYPICAL PROBLEM

$$(x^2 - 3x + 7)(2x^2 - 5x)$$

$$\downarrow$$

We first change to vertical format, writing one factor under the other and dropping parentheses:

$$\begin{array}{r} x^2 - 3x + 7 \\ 2x^2 - 5x \\ \hline \end{array}$$

$$\downarrow$$

We then multiply each term of the top polynomial by the first term of the bottom polynomial. We multiply from left to right:

$$\begin{array}{rrrr} & x^2 & -3x & +7 \\ & 2x^2 & -5x & \\ \hline 2x^4 & -6x^3 & +14x^2 & \end{array}$$

$$\downarrow$$

Then we multiply each term of the top polynomial by the next term of the bottom polynomial. We again multiply from left to right and line up like terms (if any). This step is repeated for each term of the bottom polynomial:

$$\begin{array}{rrrr} x^2 & -3x & +7 & \\ 2x^2 & -5x & & \\ \hline 2x^4 & -6x^3 & +14x^2 & \\ & -5x^3 & +15x^2 & -35x \end{array}$$

$$\downarrow$$

This method generalizes the methods in the previous two sections.

Finally we add the products:

$$\begin{array}{rrrr} x^2 & -3x & +7 & \\ 2x^2 & -5x & & \\ \hline 2x^4 & -6x^3 & +14x^2 & \\ & -5x^3 & +15x^2 & -35x \\ \hline 2x^4 & -11x^3 & +29x^2 & -35x \end{array}$$

Hence, $(x^2 - 3x + 7)(2x^2 - 5x) = 2x^4 - 11x^3 + 29x^2 - 35x$.

WARNING: Do not put one term under another unless they are like terms.

EXAMPLES

Every product of polynomials can be multiplied, yielding a polynomial.

Multiply the following polynomials.

1. $(2x^2 - 6x + 1)(x + 5)$
2. $(2a^2 + a - 5)^2$
3. $(x - 2)(x^3 - 2x^2 + 4x - 8)$
4. $(x^2 + 2)(x^2 - 2x + 4)$
5. $(3x + 2)^3$
6. $(x - y)^4$

Solutions

1.
$$\begin{array}{rrrr} 2x^2 & -6x & +1 & \\ & & x & +5 \\ \hline 2x^3 & -6x^2 & +x & \\ & 10x^2 & -30x & +5 \\ \hline 2x^3 & +4x^2 & -29x & +5 \end{array}$$

2.
$$\begin{array}{rrrrr} 2a^2 & +a & -5 & & \\ 2a^2 & +a & -5 & & \\ \hline 4a^4 & +2a^3 & -10a^2 & & \\ & 2a^3 & +a^2 & -5a & \\ & & -10a^2 & -5a & +25 \\ \hline 4a^4 & +4a^3 & -19a^2 & -10a & +25 \end{array}$$

3.
$$\begin{array}{l} x^3 - 2x^2 + 4x - 8 \\ \underline{x - 2} \\ x^4 - 2x^3 + 4x^2 - 8x \\ \underline{\quad - 2x^3 + 4x^2 - 8x + 16} \\ x^4 - 4x^3 + 8x^2 - 16x + 16 \end{array}$$

4.
$$\begin{array}{l} x^2 - 2x + 4 \\ \underline{x^2 + 2} \\ x^4 - 2x^3 + 4x^2 \\ \underline{\qquad\qquad 2x^2 - 4x + 8} \\ x^4 - 2x^3 + 6x^2 - 4x + 8 \end{array}$$

5.
$$\begin{array}{l} 3x + 2 \\ \underline{3x + 2} \\ 9x^2 + 6x \\ \underline{\qquad 6x + 4} \\ 9x^2 + 12x + 4 \\ \quad\downarrow \\ 9x^2 + 12x + 4 \\ \underline{3x + 2} \\ 27x^3 + 36x^2 + 12x \\ \underline{\qquad 18x^2 + 24x + 8} \\ 27x^3 + 54x^2 + 36x + 8 \end{array}$$

6. $x - y$ square the binomial
$$\begin{array}{l} \underline{x - y} \\ x^2 - xy \\ \underline{\quad - xy + y^2} \\ x^2 - 2xy + y^2 \\ \quad\downarrow \\ x^2 - 2xy + y^2 \text{ square the trinomial} \\ \underline{x^2 - 2xy + y^2} \\ x^4 - 2x^3y + x^2y^2 \\ \quad - 2x^3y + 4x^2y^2 - 2xy^3 \\ \underline{\qquad\qquad x^2y^2 - 2xy^3 + y^4} \\ x^4 - 4x^3y + 6x^2y^2 - 4xy^3 + y^4 \\ [(x - y)^2 \cdot (x - y)^2 = (x - y)^4] \end{array}$$

OBSERVATIONS AND REFLECTIONS

Consider the powers of the binomial $(x + y)$:

$$(x + y)^2 = x^2 + 2xy + y^2$$
$$(x + y)^3 = x^3 + 3x^2y + 3xy^2 + y^3$$
$$(x + y)^4 = x^4 + 4x^3y + 6x^2y^2 + 4xy^3 + y^4$$

and so forth. Hence, $(x + y)^2$ equals a sum of three terms, $(x + y)^3$ equals a sum of four terms, $(x + y)^4$ equals a sum of five terms, and so forth.

WARNING: A power of a sum is not a sum of powers.

This is an illegal step.

$$(x + y)^n \neq x^n + y^n$$

(where n represents a numerical value larger than 1)

13. Compute
a. $(x + y)(x^2 - xy + y^2)$
b. $(x - y)(x^2 + xy + y^2)$
c. $(x + y)(x - y)(x^2 + y^2)$
d. $(x + y)(x^4 - x^3y + x^2y^2 - xy^3 + y^4)$
e. $(x - y)(x^4 + x^3y + x^2y^2 + xy^3 + y^4)$

EXERCISES

Multiply to convert each of the following expressions to a polynomial.

1. $(x + 3)(x^2 + 2x + 4)$
2. $(x + 1)(3x^2 - 9x + 7)$
3. $(2x - 1)(4x^2 - 3x + 2)$
4. $(5x + 3)(5x^2 - x - 3)$
5. $(2x - y)(x^2 - 5xy + y^2)$
6. $(3x^2 - 4xy + 3y^2)(x - 3y)$

7. $(x^2 + 3)(x^2 + 3x + 9)$
8. $(x^2 + 4)(2x^2 + 7x + 3)$
9. $(x^2 + 2y^2)(x^2 + 3xy + 2y^2)$
10. $(x^2 + y^2)(4x^2 + 9xy + 12y^2)$
11. $(x + 2)(x^3 + 3x^2 + x + 3)$
12. $(2x - 1)(5x^3 - 3x^2 + x + 6)$
13. $(4x + 3)(x^3 + 8x^2 + x + 8)$
14. $(6x^3 - 3x^2 + 3x - 1)(x - 3)$
15. $(x^2 + 2x + 3)(x^2 + 3x + 2)$
16. $(3x^2 + 4x - 6)(5x^2 - 2x + 3)$
17. $(8x^2 - 4x + 2)(4x^2 - 2x + 1)$
18. $(2x^2 - 5xy - 3y^2)(6x^2 + 4xy - y^2)$
19. $(x + 2)^3$
20. $(3x - 1)^3$
21. $(-4x + 7)^3$
22. $(-9x + 8)^3$
23. $(x - 3y)^3$
24. $(8x - y)^3$
25. $(x^2 + y^2)^3$
26. $(3x^2 - 2y^2)^3$
27. $(4x^2 - 1)^4$
28. $(5x - 3)^4$
29. $(x - 2y)^4$
30. $(x^2 - y^2)^4$
31. $(x^2 - 2x + 1)^2$
32. $(x^2 - 5x + 1)^2$
33. $(x^4 - 3x^2 - 9)^2$
34. $(x^4 + 8x^2 + 4)^2$
35. $(2x^2 - 5xy + y^2)^2$
36. $(4x^2 + xy + 4y^2)^2$
37. $(-3x^2 - 2xy + y^2)^2$
38. $(6x^2 + 3xy + y^2)^2$
39. $(4x - 7y + 1)^2$
40. $(4x + 3y - 2)^2$

ANSWERS TO MARGIN EXERCISES

1. a. $(2 - 3)(2 + 5) = (2)(2) + (2)(5) + (-3)(2) + (-3)(5) = 4 + 10 - 6 - 15$
$(-1) \cdot (7) \quad = \quad -7$
b. $(3 - 3)(2 + 5) = (3)(2) + (3)(5) + (-3)(2) + (-3)(5) = 6 + 15 - 6 - 15$
$(0) \cdot (7) \quad = \quad 0$

2. a. $ac - ad + bc - bd$
b. $ac - ad - bc + bd$

3. a. $3x^2 + 14x + 8$
b. $30x^2 - 23x + 3$
c. $4x^2 - 81$

4. a. $x^2 - y^2$
b. $x^4 - 1$

6. a. $6x^2 + 15xy - 8x - 20y$
b. $7x^3 - 3x^2 - 42x + 18$

7. $2x^2 + 13x + 20$

8. a. $(x + 1)(3x) + (x + 1)(2)$
b. $(4x - 3)(2x) + (4x - 3)(8)$
c. $(2x - 5)(3x) + (2x - 5)(-7)$

9. a. $10x^3 + 14x^2$
b. $-4x + 20$

10. a. $(x + 3)(x + 3)$
b. $(2x - 5)(2x - 5)$
c. $(3x + 4y)(3x + 4y)$

11. $2x^3 + 2x^2 - 24x$

12. a. x^4
b. x^4
c. x^6
d. x^6

13. a. $x^3 + y^3$
b. $x^3 - y^3$
c. $x^4 - y^4$
d. $x^5 + y^5$
e. $x^5 - y^5$

CHAPTER REVIEW

SECTION 15.1

Multiply to convert to a polynomial.

1. $(x - 4)(x - 5)$
2. $(2x + 5)(x + 8)$
3. $(7x + 1)(8x - 1)$
4. $(5x - y)(2x + 5y)$
5. $(4y + 3x)(y - 8x)$
6. $(5x^2 + 1)(x - 4)$
7. $(7x - 2y)(x - 2)$
8. $(5x^2 + 2x)(x - 2)$
9. $(x^3 - 4x^2)(x^2 - 5x)$
10. $(9x + 4)(9x - 4)$
11. $(4x^4 - y)(4x^4 + y)$
12. $(5x^6y - 1)(5x^6y + 1)$
13. $(6x + 1)(6x + 1)$
14. $(8x - 3y)(8x - 3y)$
15. $(3x^3 - 10y^4)(3x^3 - 10y^4)$

SECTION 15.2

Multiply to convert to a polynomial.

16. $3x^2(3x - 1)(x - 3)$
17. $7x^2(7x - 1)(7x + 1)$
18. $5x^3(x - 6)(6x - 1)$
19. $x^3y^2(2x - y)(4x + y)$
20. $4x^3(5x^3 - x^2)(3x^4 - x^3)$
21. $-x^2y(x + 3y)^2$
22. $xy^2(4x + 6y)^2$
23. $-2x^5(4x - 5)(4x - 1)$
24. $4x(3x - 1)(3x + 1)$
25. $8x(3x^2 - 2x)(2x - 1)$
26. $-3y^3(y - 6)^2$
27. $3xy^2(9x - y)^2$

SECTION 15.3

Multiply to convert to a polynomial.

28. $(x - 4)(x^2 + 5x + 1)$
29. $(5x + 2)(2x^2 - 4x - 6)$
30. $(4x - 3y)(3x^2 - 4xy - y^2)$
31. $(x^2 - 1)(x^2 - x - 1)$
32. $(7x^2 - y^2)(2x^2 + 5xy + y^2)$
33. $(x + 4)(x^3 - 4x^2 - 3x - 1)$
34. $(2x^3 - 3x^2 + 4x - 5)(4x - 1)$
35. $(x^2 - 4x + 2)(x^2 + 4x - 2)$
36. $(x^2 - xy + y^2)(x^2 + 3xy - 9y^2)$
37. $(x - 3)^3$
38. $(-3x - 3)^3$
39. $(2x + 5y)^3$
40. $(x^2 - 3y)^3$
41. $(2x^2 + 2)^4$
42. $(3x + y^2)^4$
43. $(x^2 + 3x + 2)^2$
44. $(x^6 + 2x^3 + 8)^2$
45. $(3x^2 - 2xy + 6y^2)^2$
46. $(3x^2 + 7xy + 4y^2)^2$
47. $(3x + 2 - y)^2$

CUMULATIVE TEST

The following problems test your understanding of this chapter and of related subject matter in previous chapters. Before taking this test, thoroughly review Sections 2.3, 4.3, 15.1, 15.2, and 15.3.

Once you have finished the test, compare your answers with the answers provided at the back of the book. Note the section number of each problem missed, and thoroughly review those sections again.

Multiply.

1. $(3x + 8)(x + 9)$
2. $6x^2(6x - 1)(6x + 1)$
3. $(x - 3)(x^2 + 6x + 1)$
4. $3x^3y^2(6y^4 - 7x^2 - 2y^2)$
5. $8x^2(5x)(-3)$
6. $(3x - 1)^4$
7. $10x^3(x^2 - 4)(x^2 + 6)$
8. $-x^3(-4x + 1)$
9. $(x^2 - 6x + 4)(x^2 + 6x - 4)$
10. $-4y^2(-3x)(+7y)$
11. $3x(4x^3 - x^2)(6x^4 - x^3)$
12. $-4x^5(x^3 - 9x^{12})$
13. $(6x^2 + 1)(x - 8)$
14. $3x^3y^4(6x^2 + 5y^2)^2$
15. $(4x^2 - 7xy + 9y^2)^2$
16. $-2x^3y^2(xy^4 - x^2y^3 + x^3)$
17. $(7x - y)(3x + 2y)$
18. $(-5x^3 + x^2 + 4x)^2$
19. $10x(3x - 8)^2$
20. $5y^3(8y^6 - 7y^4 + 3y^2)$
21. $(6x + 5)^3$
22. $(3x^2 + 4x)(x - 6)$
23. $4x^3(x - 8)(8x - 1)$
24. $(x + 12)(x^3 - 2x^2 + 3x - 1)$
25. $(8x^3y^2)(x^4 - 8y^5)$
26. $5y^3(x - 5y)^2$
27. $-4xy^2(2x^2)(-9x)$
28. $(8x - 5)(3x^2 - 11x - 4)$
29. $(x - 4y)(x + 4y)$
30. $2x^4(6x - 1)(8x - 5)$

Quadratic Trinomials

We will first look at a special class of polynomials, called *quadratic trinomials.*

A *quadratic trinomial* is a trinomial of the form

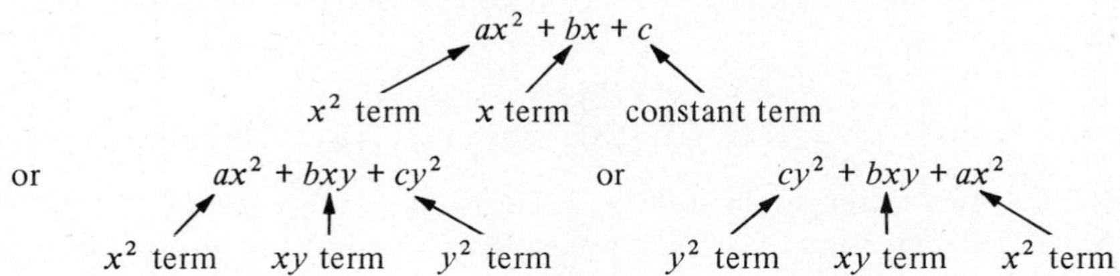

where a, b, and c represent nonzero (*signed*) integers. For example,

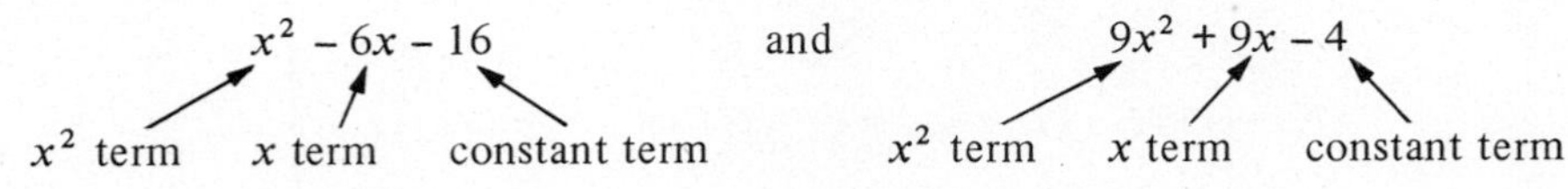

are quadratic trinomials in x, and

$$x^2 + 12xy - 45y^2 \quad \text{and} \quad y^2 - 5xy + 6x^2$$

(x^2 term, xy term, y^2 term) and (y^2 term, xy term, x^2 term)

are quadratic trinomials in x and y.

Simple and General (Quadratic) Trinomials

A quadratic trinomial is called a *simple trinomial* if the coefficient of x^2 (or y^2) is +1. For example,

$$x^2 - 6x - 16, \quad x^2 + 12xy - 45y^2, \quad \text{and} \quad y^2 - 5xy + 6x^2$$

are simple trinomials. All other quadratic trinomials are called *general trinomials.* For example,

$$9x^2 + 9x - 4 \quad \text{and} \quad 6x^2 - xy - 15y^2$$

are general trinomials. (If the coefficient of x^2 is -1, as, for example, in $-x^2 + 6x + 16$, then technically the trinomial is a product of a simple trinomial and -1.)

We will first study the factorization of simple trinomials and then look at general trinomials.

As we shall see, many quadratic trinomials cannot be factored into a product of monomials and binomials with integer coefficients. We shall say that such trinomials are *not factorable in integers*, or *not factorable* (for short).

16.1 COMPLEMENTARY FACTORS OF AN INTEGER

An essential step in factoring trinomials is to determine all the ways that an integer can be expressed as the indicated product of two integers. Two integers whose product is a given number are called a *complementary pair of factors* of the number, or a *pair of factors* (for short).

Consider, for example, the integer −12. The complementary pairs of factors of −12 are

$$\begin{array}{lll} (+1)(-12) & (+2)(-6) & (+3)(-4) \\ (-1)(+12) & (-2)(+6) & (-3)(+4) \end{array}$$

Note that every factor is a divisor of −12. Furthermore, every divisor of −12 appears in exactly one pair of factors. If we know the divisors of an integer, we should be able to determine the complementary pairs of factors of the integer. To see how this can be done, look at the following computations:

$$\begin{array}{lll} -12 \div (+1) = -12 & -12 \div (+2) = -6 & -12 \div (+3) = -4 \\ -12 = (+1)(-12) & -12 = (+2)(-6) & -12 = (+3)(-4) \\ -12 \div (-1) = +12 & -12 \div (-2) = +6 & -12 \div (-3) = +4 \\ -12 = (-1)(+12) & -12 = (-2)(+6) & -12 = (-3)(+4) \end{array}$$

1. Divide −12 by
   a. 4
   b. 6
   c. 12
   d. −4
   e. −6
   f. −12

As a second example, consider the integer +45 (whose divisors are 1, 3, 5, 9, 15, 45, and their negatives). To see how to find the complementary pairs of factors of +45, look at the following computations:

$$\begin{array}{lll} +45 \div (+1) = +45 & +45 \div (+3) = +15 & +45 \div (+5) = +9 \\ +45 = (+1)(+45) & +45 = (+3)(+15) & +45 = (+5)(+9) \end{array}$$

$+45 \div (-1) = -45$ $\quad$ $+45 \div (-3) = -15$ $\quad$ $+45 \div (-5) = -9$

$+45 = (-1)(-45)$ $\quad$ $+45 = (-3)(-15)$ $\quad$ $+45 = (-5)(-9)$

Thus, we see that the complementary factors of +45 are

$$(+1)(+45) \quad (+3)(+15) \quad (+5)(+9)$$
$$(-1)(-45) \quad (-3)(-15) \quad (-5)(-9)$$

Every divisor of +45 appears in exactly one pair of factors.

Did you note that the complementary factors of a negative integer differ in sign, whereas a pair of factors of a positive integer have the same sign? Did you also note that changing the sign of *each* integer in a complementary pair of factors yields another pair of factors?

2. Divide +45 by
a. 9
b. 15
c. 45
d. −9
e. −15
f. −45

3. Determine the complementary factors of
a. +2
b. −2

Look at the following example of *listing all pairs of factors of an integer.*

TYPICAL PROBLEM

List all pairs of factors of −24.

We first draw a line under the integer:

−24

Then we factor the integer as 1 times itself, and as −1 times its negative:

−24

(+1) (−24)
(−1) (+24)

We then find the next smallest positive factor. (We try the numbers 2, 3, 4, 5, . . . in increasing order.) We list this next factor with its complement and then we list their negatives:

−24

(+1) (−24)
(−1) (+24)

(+2) (−12)
(−2) (+12)

We continue to find larger and larger factors. We list each factor we find with its complement and then we list their negatives. We stop when we find a factor that is larger than its complement (this pair of factors has already been listed in reverse order):

−24

(+1) (−24)
(−1) (+24)

(+2) (−12)
(−2) (+12)

(+3) (−8)
(−3) (+8)

(+4) (−6)
(−4) (+6)

To determine the complement of each factor we divide the integer by the factor.

EXAMPLES

List all pairs of factors of the following integers.

1. +6
2. −35
3. +11
4. −48
5. +36
6. −98

Solutions

1. $+6$: $(+1)(+6)$, $(-1)(-6)$, $(+2)(+3)$, $(-2)(-3)$
2. -35: $(+1)(-35)$, $(-1)(+35)$, $(+5)(-7)$, $(-5)(+7)$
3. $+11$: $(+1)(+11)$, $(-1)(-11)$
4. -48: $(+1)(-48)$, $(-1)(+48)$, $(+2)(-24)$, $(-2)(+24)$, $(+3)(-16)$, $(-3)(+16)$, $(+4)(-12)$, $(-4)(+12)$, $(+6)(-8)$, $(-6)(+8)$
5. $+36$: $(+1)(+36)$, $(-1)(-36)$, $(+2)(+18)$, $(-2)(-18)$, $(+3)(+12)$, $(-3)(-12)$, $(+4)(+9)$, $(-4)(-9)$, $(+6)(+6)$, $(-6)(-6)$
6. -98: $(+1)(-98)$, $(-1)(+98)$, $(+2)(-49)$, $(-2)(+49)$, $(+7)(-14)$, $(-7)(+14)$

SPECIAL CASES

1. A prime number or the negative of a prime number has only two pairs of complementary factors: itself times +1 and its negative times −1. For example, the pairs of factors of −3 and +7 are

-3	and	$+7$
$(+1)(-3)$		$(+1)(+7)$
$(-1)(+3)$		$(-1)(-7)$

Similarly the number +1 has only two pairs of complementary factors:

$$(+1)(+1) \quad \text{and} \quad (-1)(-1)$$

2. Negating each factor in a complementary pair may not yield a distinct pair of factors. For example, negating each factor in $(+2)(-2)$ yields $(-2)(+2)$, which is the same pair of factors written in reverse order.

   Hence, in particular, the number −1 has only one pair of complementary factors, $(+1)(-1)$.

4. List all pairs of factors of
a. −5
b. +17
c. −19
d. +29

5. List all pairs of factors of
a. −9
b. −16
c. −25

OBSERVATIONS AND REFLECTIONS

★ Remember that an integer is an integer multiple of each of its divisors. Hence, dividing an integer by one of its divisors must yield another *integer*. For example, 2 is a divisor of −12, and $\frac{-12}{2} = -6$. Furthermore, the integer obtained is always another divisor of the integer. For example, $\frac{-12}{2} = -6$ and −6 is another divisor of 12.

If a is a divisor of b, then there is an integer c such that b equals (a times c).

$$a | b \Rightarrow b = a \cdot c \quad \text{and} \quad c = b \div a$$

(where a, b, and c represent integers)

6. Find the complementary factor of 120, given the factor
a. 6
b. −10
c. 24
d. −60

EXERCISES

List all complementary pairs of factors of each of the following integers.

1. +6	2. −15	3. −10
4. +13	5. +21	6. −25
7. −8	8. +27	9. +42
10. −63	11. −16	12. +100
13. +81	14. −60	15. −36
16. +80	17. +120	18. −176

16.2 FACTORING SIMPLE TRINOMIALS IN ONE LITERAL

We will begin the study of factoring quadratic trinomials by considering *simple trinomials* in x.

A *simple trinomial* in x is a trinomial of the form

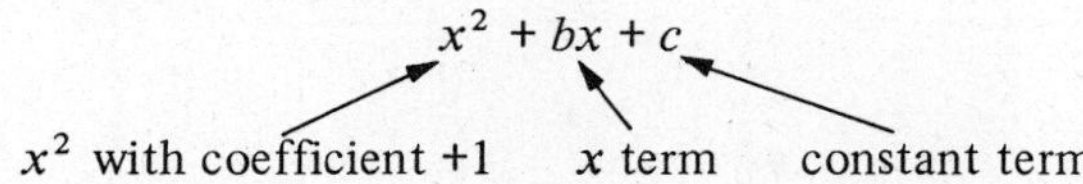

where b and c represent nonzero integers.

The process of factoring a simple trinomial in integers is the reverse of multiplication. Hence, to understand factoring we must look at each term in the product of two binomials.

Consider the following products of two *simple binomials* (x + integer):

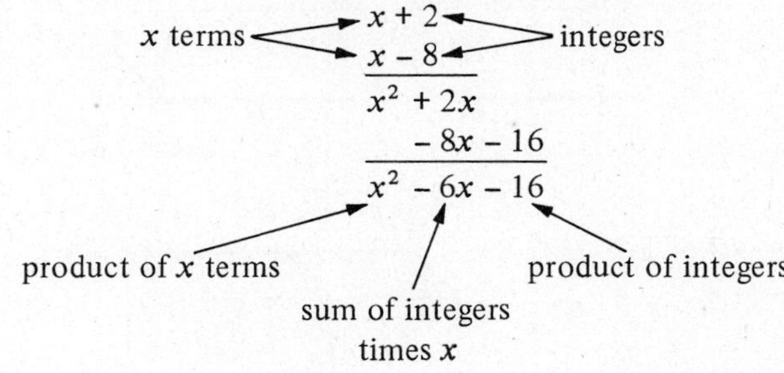

and

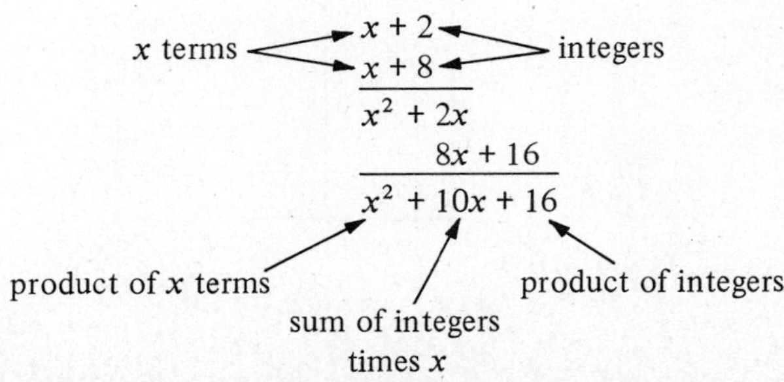

7. Compute
   a. $(x + 3)(x + 5)$
   b. $(x - 3)(x + 5)$
   c. $(x + 3)(x - 5)$
   d. $(x - 3)(x - 5)$

As these examples show, the product of two *simple binomials* has the following properties:

1. The *product* of their constant terms (integers) is the *constant term* of their product.
2. The *sum* of their constant terms is the *coefficient of* x in their product, and the integer with the larger numerical value has the same sign as this coefficient.

Can you see how to use these two properties to factor a simple trinomial in x? Check your ideas by looking at the following examples.

8. Match trinomials with binomial products.
   a. $x^2 + 9x + 14$
   b. $x^2 + 5x - 14$
   c. $x^2 - 5x - 14$
   d. $x^2 - 9x + 14$
   e. $(x - 2)(x + 7)$
   f. $(x - 2)(x - 7)$
   g. $(x + 2)(x - 7)$
   h. $(x + 2)(x + 7)$

$+3 + 5$ $(+3)(+5)$

$x^2 + 8x + 15 = (x + 3)(x + 5)$

these signs agree

$-3 + 5$ $(-3)(+5)$

$x^2 + 2x - 15 = (x - 3)(x + 5)$

these signs agree

$+3 - 5$ $(+3)(-5)$

$x^2 - 2x - 15 = (x + 3)(x - 5)$

these signs agree

$-3 - 5$ $(-3)(-5)$

$x^2 - 8x + 15 = (x - 3)(x - 5)$

these signs agree

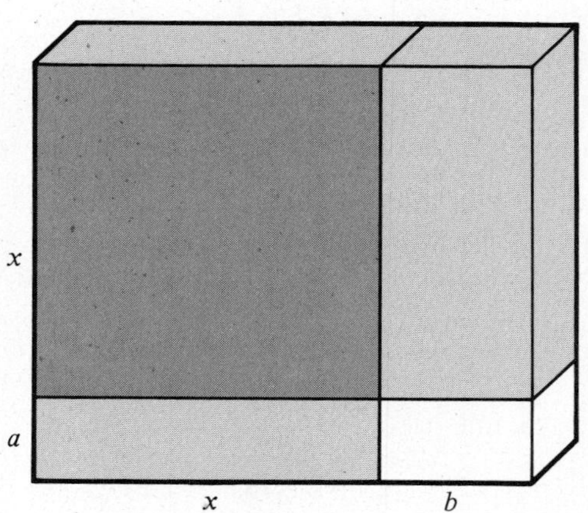

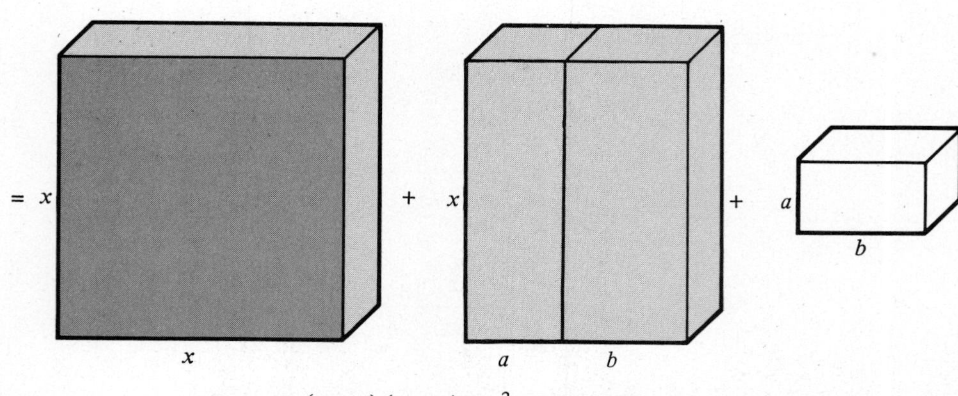

$$(x + a)(x + b) = x^2 + x(a + b) + ab$$

Look at the following example of *factoring simple trinomials in x.*

TYPICAL PROBLEM

$x^2 - 6x - 16$

We first set up our format, leaving space in each pair of parentheses for a constant term:

$(x \quad)(x \quad)$

Then we list pairs of factors of the constant term, and beside each pair we write their sum. (We only list pairs whose sum has the same sign as the x term of the trinomial.):

Factors of −16		*Sum of factors*
(+1) (−16)	→	+1 − 16 = − 15
(+2) (−8)	→	+2 − 8 = −6★
(+4) (−4)	→	+4 − 4 = 0

Then we select the pair of factors (if any) whose sum is the coefficient of the x term (starred above):

(+2) (−8)

Finally we write one factor in each pair of parentheses in the format:

$(x + 2)(x - 8)$

To list the pairs of factors of the constant term we use the method in the previous section.

Hence, $x^2 - 6x - 16 = (x + 2)(x - 8)$.

WARNING The trinomial is not factorable in integers unless one of the sums of factors is the coefficient of the x term.

EXAMPLES

Factor the following simple trinomials in x.

1. $x^2 + 7x + 12$
2. $x^2 - 2x - 35$
3. $x^2 - 9x + 8$
4. $x^2 + 5x - 50$
5. $x^2 + 11x + 30$
6. $x^2 - 3x + 7$

Solutions

1.

+12	Sum
(+1) (+12)	+1 + 12 = +13
(+2) (+6)	+2 + 6 = +8
(+3) (+4)	+3 + 4 = +7★

↓

$(x + 3)(x + 4)$

2.

−35	Sum
(+1) (−35)	+1 − 35 = −34
(+5) (−7)	+5 − 7 = −2★

↓

$(x + 5)(x - 7)$

3.

+8	Sum
(−1) (−8)	−1 − 8 = −9★
(−2) (−4)	−2 − 4 = −6

↓

$(x - 1)(x - 8)$

4.

−50	Sum
(−1) (+50)	−1 + 50 = +49
(−2) (+25)	−2 + 25 = +23
(−5) (+10)	−5 + 10 = +5★

↓

$(x - 5)(x + 10)$

5.

+30	Sum
(+1) (+30)	+1 + 30 = +31
(+2) (+15)	+2 + 15 = +17
(+3) (+10)	+3 + 10 = +13
(+5) (+6)	+5 + 6 = +11★

↓

$(x+5)(x+6)$

6.

+7	Sum
(−1) (−7)	−1 − 7 = −8

↓

none

↓

$x^2 - 3x + 7$ is not factorable in integers.

Checking

Because multiplying and factoring are reverse operations, we can check the accuracy of a factorization by multiplying. For example, we can check the factorization of $x^2 - 6x - 16$ into $(x+2)(x-8)$ by the computation

$$\begin{array}{r} x + 2 \\ x - 8 \\ \hline x^2 + 2x \qquad \\ - 8x - 16 \\ \hline x^2 - 6x - 16 \end{array}$$

Because we get back the original trinomial, the factorization is correct.

9. Check the solutions to examples 1 through 5 by multiplying.

OBSERVATIONS AND REFLECTIONS

1. Because multiplication is commutative, the two binomial factors of a trinomial may be written in either order. For example,

$$(x+2)(x-8) = (x-8)(x+2)$$

Hence, in factoring a simple trinomial we need not specify which integer to place in which factor in the form $(x \quad)(x \quad)$.

10. Pair equal products:
   a. $(x+3)(x-7)$
   b. $(x-7)(x-3)$
   c. $(x-3)(x+7)$
   d. $(x-3)(x-7)$
   e. $(x+7)(x-3)$
   f. $(x-7)(x+3)$

★ 2. Consider the products

$$\begin{array}{ccc} (x+2)(x-8) & \text{and} & (-x-2)(-x+8) \\ \downarrow & & \downarrow \\ \begin{array}{r} x + 2 \\ x - 8 \\ \hline x^2 + 2x \qquad \\ - 8x - 16 \\ \hline x^2 - 6x - 16 \end{array} & = & \begin{array}{r} -x - 2 \\ -x + 8 \\ \hline x^2 + 2x \qquad \\ - 8x - 16 \\ \hline x^2 - 6x - 16 \end{array} \end{array}$$

These computations show that the indicated products $(x+2)(x-8)$ and $(-x-2)(-x+8)$ both equal $x^2 - 6x - 16$. Hence, $x^2 - 6x - 16$ can be factored either as $(x+2)(x-8)$ or as $(-x-2)(-x+8)$. The products $(x+2)(x-8)$ and $(-x-2)(-x+8)$ are equal, because

$$(-1)(x+2) = (-x-2) \quad \text{and} \quad (-1)(x-8) = (-x+8)$$

That is,

$$(-x-2)(-x+8) = (-1)(x+2)(-1)(x-8) = (x+2)(x-8)$$

because $(-1)(-1)$ is +1.

11. Pair equal products.
   a. $(x-6)(x+8)$
   b. $(-x+8)(-x+6)$
   c. $(-x+6)(-x-8)$
   d. $(x-8)(x+6)$
   e. $(x-6)(x-8)$
   f. $(-x-6)(-x+8)$

EXERCISES

Factor each of the following expressions completely.

1. $x^2 + 2x + 1$
2. $x^2 + 3x + 2$
3. $x^2 - 3x - 4$
4. $x^2 - 4x + 4$
5. $x^2 + 5x + 6$
6. $x^2 - 3x - 10$
7. $x^2 + 4x - 12$
8. $x^2 + 6x - 7$
9. $x^2 + 5x - 14$
10. $x^2 - 9x + 8$
11. $x^2 + 4x - 5$
12. $x^2 + 4x + 5$
13. $x^2 - 3x + 28$
14. $x^2 + 5x - 24$
15. $x^2 + 11x + 28$
16. $x^2 - 12x + 32$
17. $x^2 - 6x - 27$
18. $x^2 - 7x - 44$
19. $x^2 - 5x + 8$
20. $x^2 - 16x + 64$
21. $x^2 + 21x - 22$
22. $x^2 - 12x + 20$
23. $x^2 + 15x - 54$
24. $x^2 + 20x + 64$
25. $x^2 + 13x - 90$
26. $x^2 - 17x - 84$
27. $x^2 - 21x - 72$
28. $x^2 + 10x + 21$
29. $x^2 + 7x - 30$
30. $x^2 - 35x + 36$

16.3 FACTORING SIMPLE TRINOMIALS IN TWO LITERALS

A *simple trinomial* in x and y has one of the two forms

$$x^2 + bxy + cy^2 \quad \text{or} \quad y^2 + bxy + cx^2$$

where b and c represent nonzero integers. For example,

$$x^2 + 12xy - 45y^2 \quad \text{and} \quad y^2 - 5xy + 6x^2$$

are simple trinomials in x and y. We will say that x is the *leading literal* in the first trinomial and y is the *leading literal* in the second one.

A trinomial in x and y with x leading behaves like a trinomial in x. For example,

$$x^2 + 12xy - 45y^2 = (x - 3y)(x + 15y)$$
$$x^2 + 12x - 45 = (x - 3)(x + 15)$$

Similarly, a trinomial in x and y with y leading behaves like a trinomial in y. For example,

$$y^2 - 5xy + 6x^2 = (y - 2x)(y - 3x)$$
$$y^2 - 5y + 6 = (y - 2)(y - 3)$$

Hence, we can factor a trinomial in x and y in much the same way as we factor a trinomial in one literal.

If a simple trinomial in x and y is factorable in integers, it is the product of two simple binomials in x and y. (In this section we will see a few trinomials that are not factorable in integers.)

12. Form the trinomial in one literal corresponding to
a. $x^2 - 3xy + 2y^2$
b. $y^2 - 2xy - 8x^2$
c. $x^2 + 5xy - 56y^2$
d. $y^2 + 11xy + 24x^2$

Look at the following example of *factoring simple trinomials in* x *and* y.

To factor a simple trinomial in x *and* y *we extend the method in the previous section.*

TYPICAL PROBLEM

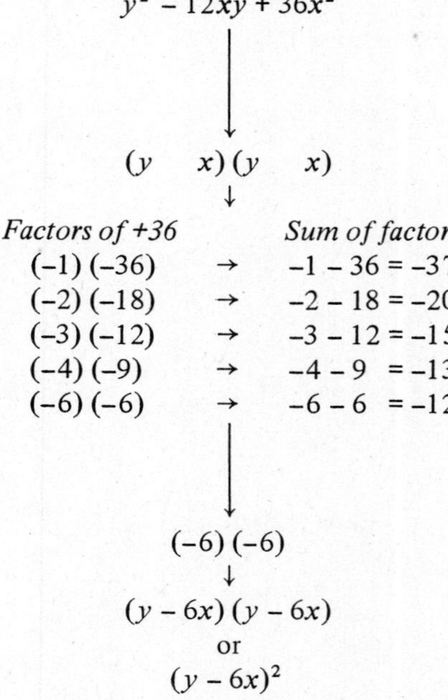

We set up our format with the leading literal first. We leave space in each pair of parentheses for the coefficient of the second term:

Then we list pairs of factors of the coefficient of the third term, and beside each pair we write their sum. (We only list pairs whose sum has the same sign as the xy term of the trinomial.):

Then we select the pair of factors (if any) whose sum is the coefficient of the xy term (starred above):

Finally we write one factor as the coefficient of the second literal in each pair of parentheses in the format:

Hence, $y^2 - 12xy + 36x^2 = (y - 6x)^2$.

WARNING The trinomial is not factorable in integers unless one of the sums of factors is the coefficient of the *xy* term.

EXAMPLES

Factor the following simple trinomials.

1. $x^2 - 3xy + 2y^2$
2. $y^2 - 2xy - 8x^2$
3. $x^2 + 5xy - 56y^2$
4. $y^2 + 11xy + 24x^2$

Solutions

1. $(x \quad y)(x \quad y)$

$\downarrow$

+2	sum
$(-1)(-2)$	$-1 - 2 = -3\star$

$\downarrow$

$(x - 1y)(x - 2y)$

or

$(x - y)(x - 2y)$

2. $(y \quad x)(y \quad x)$

$\downarrow$

−8	sum
$(+1)(-8)$	$+1 - 8 = -7$
$(+2)(-4)$	$+2 - 4 = -2\star$

$\downarrow$

$(y + 2x)(y - 4x)$

3. $(x \quad y)(x \quad y)$

↓

−56	sum
(−1) (+56)	−1 + 56 = +55
(−2) (+28)	−2 + 28 = +26
(−4) (+14)	−4 + 14 = +10
(−7) (+8)	−7 + 8 = +1

↓

none

↓

$x^2 + 5xy - 56y^2$ is not factorable in integers

4. $(y \quad x)(y \quad x)$

↓

+24	sum
(+1) (+24)	+1 + 24 = +25
(+2) (+12)	+2 + 12 = +14
(+3) (+8)	+3 + 8 = +11★
(+4) (+6)	+4 + 6 = +10

↓

$(y + 3x)(y + 8x)$

OBSERVATIONS AND REFLECTIONS

★ 1. When factoring a simple trinomial in x and y, we factor a trinomial with x leading into two binomials with the x term first and a trinomial with y leading into two binomials with the y term first. For example,

$$x^2 + 12xy - 45y^2 = (x - 3y)(x + 15y)$$

and

$$y^2 - 5xy + 6x^2 = (y - 2x)(y - 3x)$$

However, because addition is commutative, each factor could be written with the other term first. For example,

$$(x - 3y)(x + 15y) = (-3y + x)(15y + x)$$

and

$$(y - 2x)(y - 3x) = (-2x + y)(-3x + y)$$

★ 2. Note that $(2x - y)(3x - y) = (-2x + y)(-3x + y)$, because $(2x - y) = (-1)(-2x + y)$ and $(3x - y) = (-1)(-3x + y)$. Multiplying two factors in a product by −1 (or, equivalently, negating two factors) yields an expression with the same value. Hence, $(y - 2x)(y - 3x) = (2x - y)(3x - y)$.

13. Pair equal products.
a. $(y - 7x)(y + 4x)$
b. $(7x + y)(-4x + y)$
c. $(-7x + y)(-4x + y)$
d. $(y - 7x)(y - 4x)$
e. $(y + 7x)(y - 4x)$
f. $(-7x + y)(4x + y)$

14. Pair equal products.
a. $(x - 5y)(x - 9y)$
b. $(x + 5y)(x - 9y)$
c. $(5y - x)(-9y - x)$
d. $(9y - x)(-5y - x)$
e. $(5y - x)(9y - x)$
f. $(x + 9y)(x - 5y)$

EXERCISES

Factor each of the following expressions completely.

1. $x^2 + 8xy + 7y^2$
2. $x^2 - 6xy + 8y^2$
3. $y^2 + 5xy - 24x^2$
4. $x^2 - 4xy - 5y^2$
5. $y^2 + 9xy + 8x^2$
6. $x^2 - 11xy + 18y^2$
7. $y^2 + 7xy - 30x^2$
8. $x^2 - 8xy - 48y^2$
9. $y^2 + 12xy - 35x^2$
10. $x^2 - 10xy + 25y^2$
11. $x^2 + 8xy - 14y^2$
12. $y^2 - 16xy - 39x^2$
13. $y^2 + 15xy + 50x^2$
14. $y^2 - 18xy + 72x^2$
15. $x^2 + 9xy - 36y^2$
16. $x^2 - 11xy - 64y^2$
17. $y^2 + 16xy + 48x^2$
18. $x^2 - 14xy + 24y^2$

16.4
FACTORING PRODUCTS OF MONOMIALS AND SIMPLE TRINOMIALS

In Chapter 14 we learned how to factor a monomial from a polynomial. Combining this with what we have learned in this chapter, we can factor a new class of trinomials: products of monomials and simple trinomials. For example, consider

$$2x\,(x-3)\,(x+8) \qquad \text{and} \qquad y^3(x+2y)\,(x-5y)$$

$$\begin{array}{rrr} & x & -\;3 \\ & x & +\;8 \\ \hline x^2 & -\;3x & \\ & +\;8x & -\;24 \\ \hline x^2 & +\;5x & -\;24 \\ 2x & & \\ \hline 2x^3 & +\;10x^2 & -\;48x \end{array} \qquad\qquad \begin{array}{rrr} & x & +\;2y \\ & x & -\;5y \\ \hline x^2 & +\;2xy & \\ & -\;5xy & -\;10y^2 \\ \hline x^2 & -\;3xy & -\;10y^2 \\ y^3 & & \\ \hline x^2y^3 & -\;3xy^4 & -\;10y^5 \end{array}$$

When we factor a trinomial, we always factor out the monomial first, leaving a trinomial in *lowest terms.* Then we factor the reduced trinomial into a product of binomials if possible. The resulting indicated product is called the *completely factored form* of the original trinomial.

15. Factor
a. $2x^3 + 10x^2 - 48x$
b. $x^2y^3 - 3xy^4 - 10y^5$

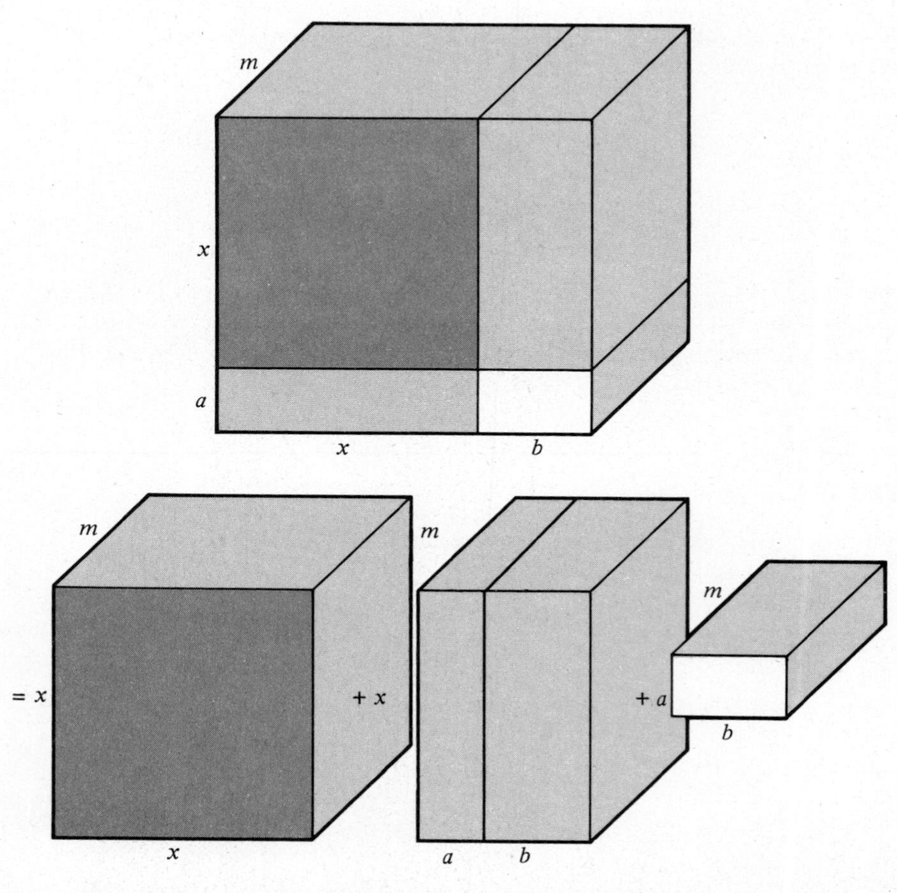

$m(x+a)\,(x+b) = mx^2 + mx(a+b) + mab$

16.4

Look at the following example of *factoring products of monomials and simple trinomials.*

TYPICAL PROBLEM

$$2x^3 + 10x^2 - 48x$$

We first set up our format and factor out the monomial. (We used this format in Chapter 14.):

$$2x \,\Big|\, \begin{array}{c} 2x^3 + 10x^2 - 48x \\ x^2 + 5x - 24 \end{array}$$

Then we factor the simple trinomial:

$$2x \,\Big|\, \begin{array}{c} 2x^3 + 10x^2 - 48x \\ x^2 + 5x - 24 \\ \downarrow \\ (x-3)(x+8) \end{array}$$

Finally, we attach the monomial factor to the binomial factors:

$$2x(x-3)(x+8)$$

Hence, $2x^3 + 10x^2 - 48x = 2x(x-3)(x+8)$.

To factor the simple trinomial we use the methods in the previous two sections.

EXAMPLES

Factor the following products of monomials and simple trinomials.

1. $-10x^2 - 30x + 280$
2. $-3x^2y^2 - 15x^2y + 72x^2$
3. $4x^2y + 12xy^2 - 72y^3$
4. $3x^4y^3 - 6x^5y^2 - 105x^6y$
5. $4x^4y^6 + 4x^5y^5 - 20x^6y^4$
6. $3x^3y^7 - 24x^4y^6 + 48x^5y^5$

Solutions

1. $-10 \,\Big|\, \begin{array}{c} x^2 + 3x - 28 \\ \downarrow \\ (x-4)(x+7) \end{array}$

   $-10(x-4)(x+7)$

2. $-3x^2 \,\Big|\, \begin{array}{c} y^2 + 5y - 24 \\ \downarrow \\ (y-3)(y+8) \end{array}$

   $-3x^2(y-3)(y+8)$

3. $4y \,\Big|\, \begin{array}{c} x^2 + 3xy - 18y^2 \\ \downarrow \\ (x-3y)(x+6y) \end{array}$

   $4y(x-3y)(x+6y)$

4. $3x^4y \,\Big|\, \begin{array}{c} y^2 - 2xy - 35x^2 \\ \downarrow \\ (y+5x)(y-7x) \end{array}$

   $3x^4y(y+5x)(y-7x)$

5. $4x^4y^4 \,\Big|\, \begin{array}{c} y^2 + xy - 5x^2 \\ \downarrow \\ \text{not factorable} \\ \text{in integers} \end{array}$

   $4x^4y^4(y^2 + xy - 5x^2)$

6. $3x^3y^5 \,\Big|\, \begin{array}{c} y^2 - 8xy + 16x^2 \\ \downarrow \\ (y-4x)(y-4x) \end{array}$

   $3x^3y^5(y-4x)(y-4x)$

   or $3x^3y^5(y-4x)^2$

EXERCISES

Factor each of the following expressions completely. These trinomials are products of monomials and simple trinomials (which may or may not be factorable in integers).

1. $6x^2 - 48x + 96$
2. $5y^2 - 10y - 120$
3. $8x^2 - 40x + 32$
4. $15x^2 - 45x + 30$
5. $9x^3 - 54x^2 - 144x$
6. $3x^3y + 21x^2y + 36xy$
7. $5x^3y^2 - 25x^3y + 30x^3$
8. $x^2y^2 - x^2y - 20x^2$
9. $2x^2y + 18xy - 16y$
10. $5x^2y^3 + 10x^2y^2 + 15x^2y$
11. $-3x^2 - 36x - 108$
12. $-x^4 - 15x^3 + 16x^2$
13. $-x^2y^2 + 7xy^2 + 18y^2$
14. $-3xy^2 + 27xy - 60x$
15. $-x^4y^4 - 10x^3y^4 - 24x^2y^4$
16. $-2x^6y^4 - 36x^5y^4 - 32x^4y^4$
17. $-x^5y^5 + 14x^5y^4 - 24x^5y^3$
18. $-3x^5y^2 + 36x^4y^2 + 84x^3y^2$
19. $4x^2 + 36xy + 32y^2$
20. $8x^2 - 32xy + 24y^2$
21. $11y^2 + 33xy - 44x^2$
22. $x^3y - x^2y^2 - 42xy^3$
23. $2y^4 + 12xy^3 + 10x^2y^2$
24. $8x^3 - 16x^2y + 8xy^2$
25. $2x^3y^2 + 10x^4y - 28x^5$
26. $6x^3y^2 - 24x^2y^3 - 30xy^4$
27. $4x^5y^4 + 36x^6y^3 + 96x^7y^2$
28. $3x^6y^6 - 18x^7y^5 + 27x^8y^4$
29. $5x^4y^7 + 15x^5y^6 - 50x^6y^5$
30. $40x^4y^2 - 40x^3y^3 - 1200x^2y^4$
31. $-x^4 - 8x^3y - 7x^2y^2$
32. $-3x^2y^2 + 18xy^3 - 24y^4$
33. $-8xy^3 - 32x^2y^2 + 48x^3y$
34. $-5x^3y^4 + 15x^4y^3 + 60x^5y^2$
35. $-4x^4y^8 - 40x^5y^7 - 64x^6y^6$
36. $-9x^8y^2 + 54x^7y^3 - 81x^6y^4$

16.5 FACTORING SIMPLE TRINOMIALS NOT IN STANDARD FORM

Because addition is commutative, the terms of a trinomial may be written in any order. A trinomial that is factorable in integers is factorable regardless of the order in which its terms appear. For example,

$$6x^2 - 5xy + y^2 = y^2 - 5xy + 6x^2 = (y - 2x)(y - 3x)$$

However, before we can use the methods we have developed in the last two sections, we must rearrange the terms into *standard form.*

1. The *standard form* of a simple trinomial in x is

$$x^2 + bx + c$$

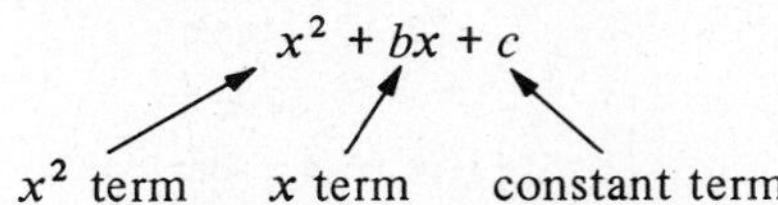

where b and c represent integers.

The terms are in decreasing order of x: the x^2 term is first with coefficient +1, the x term is second, and the constant term is last.

2. The *standard form* of a simple trinomial in x and y *with* x *leading* is

$$x^2 + bxy + cy^2$$

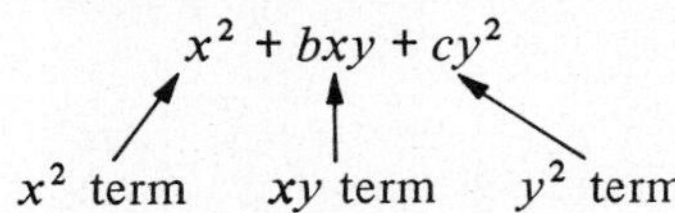

The terms are in decreasing order of x: the x^2 term is first with coefficient +1, the xy term is second, and the y^2 term is last.

3. The *standard form* of a simple trinomial in x and y *with* y *leading* is

$$y^2 + bxy + cx^2$$

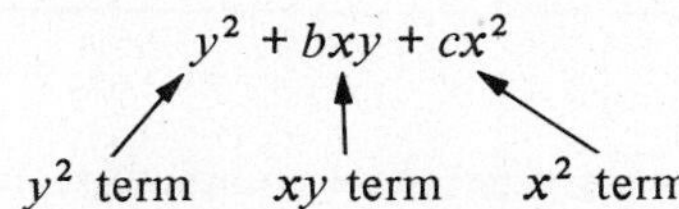

The terms are in decreasing order of y: the y^2 term is first with coefficient +1, the xy term is second, and the x^2 term is last.

In this section we will look at two types of quadratic trinomials:

1. Simple trinomials in one or two literals, not in standard form.
2. Negatives of simple trinomials in one or two literals, not in standard form. (The negative of a simple trinomial is easily converted to a product of −1 and the simple trinomial.)

These two types of trinomials look very much alike when they are not written in standard form.

Look at the following example of *factoring simple trinomials that are not in standard form.*

16. Place in standard form.
 a. $-3x + x^2 - 10$
 b. $24 - 11x + x^2$
 c. $3x - 28 + x^2$

17. Place in standard form with x leading.
 a. $12xy + 27y^2 + x^2$
 b. $40y^2 + x^2 - 14xy$
 c. $-72y^2 - 6xy + x^2$

18. Place in standard form with y leading.
 a. $-33x^2 - 8xy + y^2$
 b. $-13xy + 12x^2 + y^2$
 c. $26xy + y^2 - 56x^2$

19. Factor −1 from
 a. $-x^2 + x + 30$
 b. $-x^2 + 8xy - 12y^2$
 c. $-y^2 - 9xy + 36x^2$

TYPICAL PROBLEM

$$18x^2 + 7xy - y^2$$
$$\downarrow$$

We first arrange the terms in standard form: $-y^2 + 7xy + 18x^2$

$$\downarrow$$

We must factor out −1 if the coefficient of the first term is −1: $-1(y^2 - 7xy - 18x^2)$

$$\downarrow$$

Finally we factor the simple trinomial: $-1(y + 2x)(y - 9x)$

Hence, $18x^2 + 7xy - y^2 = -1(y + 2x)(y - 9x)$.

To factor the simple trinomial we use the methods in Sections 16.2 and 16.3.

EXAMPLES

Place the following trinomials in standard form and then factor them.

1. $x - 12 + x^2$
2. $x - x^2 + 30$
3. $4xy - 32y^2 + x^2$
4. $-12y^2 + 8xy - x^2$
5. $36x^2 - y^2 - 9xy$
6. $30x^2 + 11xy + y^2$

Solutions

1. $x^2 + x - 12$
$\downarrow$
$(x - 3)(x + 4)$

2. $-x^2 + x + 30$
$\downarrow$
$-1(x^2 - x - 30)$
$\downarrow$
$-1(x + 5)(x - 6)$
or
$-(x + 5)(x - 6)$

3. $x^2 + 4xy - 32y^2$
$\downarrow$
$(x - 4y)(x + 8y)$

4. $-x^2 + 8xy - 12y^2$
$\downarrow$
$-1(x^2 - 8xy + 12y^2)$
$\downarrow$
$-1(x - 2y)(x - 6y)$
or
$-(x - 2y)(x - 6y)$

5. $-y^2 - 9xy + 36x^2$
$\downarrow$
$-1(y^2 + 9xy - 36x^2)$
$\downarrow$
$-1(y - 3x)(y + 12x)$
or
$-(y - 3x)(y + 12x)$

6. $y^2 + 11xy + 30x^2$
$\downarrow$
$(y + 5x)(y + 6x)$

EXERCISES

Factor each of the following expressions completely, first placing the terms in standard form.

1. $8x - 9 + x^2$
2. $-5x + x^2 - 24$
3. $11x - 24 - x^2$
4. $9x - x^2 - 18$
5. $-x^2 + 33 + 8x$
6. $6x - x^2 + 40$
7. $x^2 - 30 + 7x$
8. $-24 + x^2 + 10x$
9. $13x + x^2 + 30$
10. $12x - 36 - x^2$
11. $4x - 4 - x^2$
12. $14x - 49 - x^2$
13. $-9xy - y^2 + 10x^2$
14. $3xy + 28x^2 - y^2$
15. $6y^2 + 5xy + x^2$
16. $-6xy + x^2 + 9y^2$
17. $-6x^2 - 7xy - y^2$
18. $9xy - x^2 - 14y^2$
19. $54x^2 - 3xy - y^2$
20. $2xy + x^2 - 48y^2$
21. $40y^2 - x^2 + 3xy$
22. $9xy - x^2 - 18y^2$
23. $-40x^2 - 6xy + y^2$
24. $11xy + 80x^2 - y^2$

16.6 FACTORING THE DIFFERENCE OF TWO SQUARES

Perfect Squares

Multiplying any integer by itself yields what is called a *perfect square*. The following table lists the first twenty numerical perfect squares:

$1 = 1 \cdot 1$	$36 = 6 \cdot 6$	$121 = 11 \cdot 11$	$256 = 16 \cdot 16$
$4 = 2 \cdot 2$	$49 = 7 \cdot 7$	$144 = 12 \cdot 12$	$289 = 17 \cdot 17$
$9 = 3 \cdot 3$	$64 = 8 \cdot 8$	$169 = 13 \cdot 13$	$324 = 18 \cdot 18$
$16 = 4 \cdot 4$	$81 = 9 \cdot 9$	$196 = 14 \cdot 14$	$361 = 19 \cdot 19$
$25 = 5 \cdot 5$	$100 = 10 \cdot 10$	$225 = 15 \cdot 15$	$400 = 20 \cdot 20$

Similarly, multiplying any monomial by itself yields a *perfect square* monomial. Remember that a monomial is the product of its numerical coefficient and its literal component. The following table gives some examples of literal perfect squares:

$x^2 = x \cdot x$	$x^4 = x^2 \cdot x^2$	$x^6 = x^3 \cdot x^3$	$x^8 = x^4 \cdot x^4$
$x^2y^2 = xy \cdot xy$	$x^4y^2 = x^2y \cdot x^2y$	$x^6y^2 = x^3y \cdot x^3y$	$x^8y^2 = x^4y \cdot x^4y$
$x^2y^4 = xy^2 \cdot xy^2$	$x^4y^4 = x^2y^2 \cdot x^2y^2$	$x^6y^4 = x^3y^2 \cdot x^3y^2$	$x^8y^4 = x^4y^2 \cdot x^4y^2$
$x^2y^6 = xy^3 \cdot xy^3$	$x^4y^6 = x^2y^3 \cdot x^2y^3$	$x^6y^6 = x^3y^3 \cdot x^3y^3$	$x^8y^6 = x^4y^3 \cdot x^4y^3$
$x^2y^8 = xy^4 \cdot xy^4$	$x^4y^8 = x^2y^4 \cdot x^2y^4$	$x^6y^8 = x^3y^4 \cdot x^3y^4$	$x^8y^8 = x^4y^4 \cdot x^4y^4$

20. Express as the product of two equal factors.
a. x^{12}
b. x^2y^{10}
c. $x^{14}y^8$

Note that only *even exponents* (2, 4, 6, 8, . . .) appear in these literal perfect squares, because only a power with an even exponent can be expressed as the indicated product of two equal powers, each with half the original exponent.

$$x^{2n} = x^n \cdot x^n \quad \text{and} \quad x^{2n}y^{2m} = x^ny^m \cdot x^ny^m$$

(where n and m represent numerical values)

21. Which of the following are perfect squares?
a. x^{24}
b. x^6y^{18}
c. $x^{25}y^{10}$
d. $x^{49}y^7$

For a monomial to be a perfect square its numerical coefficient must be a numerical perfect square and its literal component must be a literal perfect square. Look at the following examples of monomial perfect squares.

$4x^2 = 2x \cdot 2x$	$36x^4 = 6x^2 \cdot 6x^2$	$100x^6 = 10x^3 \cdot 10x^3$
$9x^4 = 3x^2 \cdot 3x^2$	$49x^8 = 7x^4 \cdot 7x^4$	$121x^{12} = 11x^6 \cdot 11x^6$
$16x^6 = 4x^3 \cdot 4x^3$	$64x^{12} = 8x^6 \cdot 8x^6$	$144x^{18} = 12x^9 \cdot 12x^9$
$25x^8 = 5x^4 \cdot 5x^4$	$81x^{16} = 9x^8 \cdot 9x^8$	$169x^{24} = 13x^{12} \cdot 13x^{12}$
$4x^2y^2 = 2xy \cdot 2xy$	$36x^4y^4 = 6x^2y^2 \cdot 6x^2y^2$	$100x^6y^6 = 10x^3y^3 \cdot 10x^3y^3$
$9x^4y^2 = 3x^2y \cdot 3x^2y$	$49x^8y^6 = 7x^4y^3 \cdot 7x^4y^3$	$121x^{12}y^{10} = 11x^6y^5 \cdot 11x^6y^5$
$16x^6y^2 = 4x^3y \cdot 4x^3y$	$64x^{12}y^8 = 8x^6y^4 \cdot 8x^6y^4$	$144x^{18}y^{14} = 12x^9y^7 \cdot 12x^9y^7$
$25x^8y^2 = 5x^4y \cdot 5x^4y$	$81x^{16}y^{10} = 9x^8y^5 \cdot 9x^8y^5$	$169x^{24}y^{18} = 13x^{12}y^9 \cdot 13x^{12}y^9$

22. Express as the product of two equal factors.
a. $144x^2$
b. $49x^8$
c. $64x^4y^2$
d. $81x^6y^{12}$

Note that the perfect squares in the above examples are positive. (Remember that a monomial is said to be positive if its coefficient is positive.) In fact, every perfect square is positive, because multiplying any expression by itself yields a positive quantity. For example, $(3x^2) \cdot (3x^2) = 9x^4$ and $(-3x^2) \cdot (-3x^2) = 9x^4$.

The *negative* of a perfect square, such as $-9x^4$, can be expressed as the product of two quantities that differ only in sign. For example, $-9x^4 = (3x^2) \cdot (-3x^2)$.

23. Express as the product of two equal negative factors.
a. 16
b. $100x^6$
c. x^8y^4
d. $81x^{10}y^2$

$$b^2 = (b)\,(b) = (-b)\,(-b)$$

and

$$-b^2 = -(b)\,(b) = (-b)\,(b) = (b)\,(-b)$$

(where b represents a signless monomial)

24. Express as the product of two factors that differ only in sign.
a. $-144x^2$
b. $-49y^8$
c. $-64x^4y^2$
d. $-81x^6y^{12}$

The Difference of Two Squares

The *difference of two squares* is a sum of two terms, the first of which is a perfect square, and the second of which is the negative of a perfect square. As we saw in Chapter 15, the product of two binomials that differ only in the sign of the second term is another binomial. For example, the product of $x + 4$ and $x - 4$ is

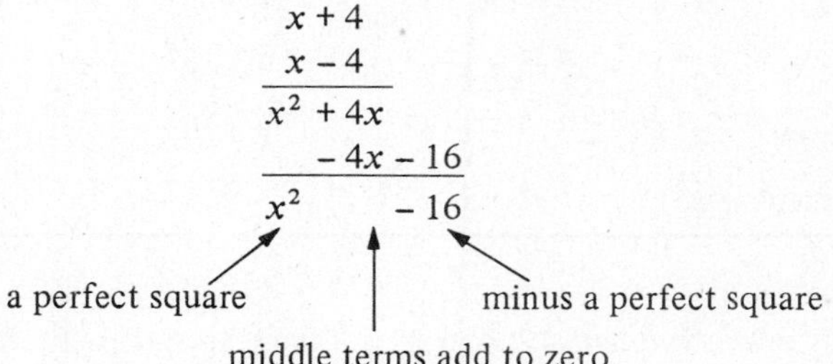

25. Multiply
a. $(8x + 3)(8x - 3)$
b. $(x^3 + 2y)(x^3 - 2y)$
c. $(5 - y^5)(5 + y^3)$

Because the sign of a product is + if the signs of the two factors agree, and – if they differ, the product of two binomials that differ only in the sign of the second term is always a difference of two squares. Further, every difference of two squares can be factored into the product of two binomials that differ only in the sign of the second term.

26. Reduce to lowest terms
a. $8x^2 - 18$
b. $75x^2 - 147y^2$
c. $16x^6 - 100x^4$
d. $9x^8y^3 - 144x^2y^9$

$$a^2 - b^2 = (a + b)(a - b)$$

(where a and b represent signless monomials)

In this section we will also consider expressions that are products of monomials and the difference of two squares. When factoring polynomials, we always factor out any monomial factor before looking for binomial factors.

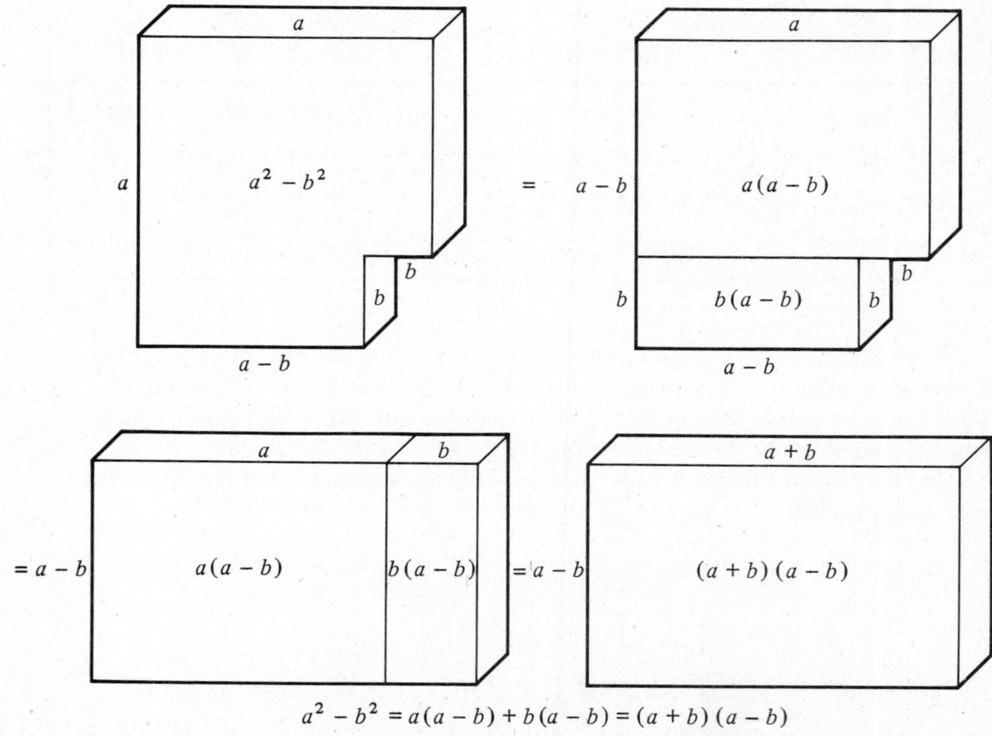

$$a^2 - b^2 = a(a - b) + b(a - b) = (a + b)(a - b)$$

Look at the following examples of *factoring the difference of two squares.*

TYPICAL PROBLEMS

1. $16x^2 - 25y^6$

We first set up our format, leaving space in each set of parentheses for a term on each side of the sign:

$$16x^2 - 25y^6$$
$$\downarrow$$
$$(\quad + \quad)(\quad - \quad)$$

We then find two equal factors of the first term (if possible), and place them in the format in both pairs of parentheses:

$$16x^2 = 4x \cdot 4x$$
$$\downarrow$$
$$(4x + \quad)(4x - \quad)$$

Then we find two equal factors of the second term (if possible), ignoring its minus sign, and place them in the format in both pairs of parentheses:

$$25y^6 = 5y^3 \cdot 5y^3$$
$$\downarrow$$
$$(4x + 5y^3)(4x - 5y^3)$$

Hence, $16x^2 - 25y^6 = (4x + 5y^3)(4x - 5y^3)$.

2. $18x^3 - 98x$

We first set up our format and factor out the monomial. (We used this format in Chapter 14.):

$$2x \;\Big|\; \begin{array}{l} 18x^3 - 98x \\ 9x^2 - 49 \end{array}$$

Then we factor the difference of two squares:

$$2x \;\Big|\; \begin{array}{l} 18x^3 - 98x \\ 9x^2 - 49 \\ \downarrow \\ (3x + 7)(3x - 7) \end{array}$$

Finally we attach the monomial factor to the two binomial factors:

$$2x(3x + 7)(3x - 7)$$

Hence, $18x^3 - 98x = 2x(3x + 7)(3x - 7)$.

EXAMPLES

Factor the following differences of two squares.

1. $4x^2 - 25$
2. $25x^2 - 16y^2$
3. $100x^2y^2 - 81$
4. $36x^4 - 49y^6$
5. $150x^3 - 96x$
6. $98x^5 - 72x^3$

Solutions

1. $(\quad + \quad)(\quad - \quad)$

$$\downarrow$$
$$4x^2 = 2x \cdot 2x$$
and
$$25 = 5 \cdot 5$$
$$\downarrow$$
$$(2x + 5)(2x - 5)$$

2. $(\quad + \quad)(\quad - \quad)$

$$\downarrow$$
$$25x^2 = 5x \cdot 5x$$
and
$$16y^2 = 4y \cdot 4y$$
$$\downarrow$$
$$(5x + 4y)(5x - 4y)$$

3. $100x^2y^2 = 10xy \cdot 10xy$

and

$81 = 9 \cdot 9$

$\downarrow$

$(10xy + 9)(10xy - 9)$

4. $36x^4 = 6x^2 \cdot 6x^2$

and

$49y^6 = 7y^3 \cdot 7y^3$

$\downarrow$

$(6x^2 + 7y^3)(6x^2 - 7y^3)$

5. $6x$ | $25x^2 - 16$

$\downarrow$

$(5x + 4)(5x - 4)$

$6x(5x + 4)(5x - 4)$

6. $2x^3$ | $49x^2 - 36$

$\downarrow$

$(7x + 6)(7x - 6)$

$2x^3(7x + 6)(7x - 6)$

EXERCISES

Factor each of the following expressions completely.

1. $49x^2 - 16$
2. $64x^2 - 81$
3. $9x^2 - 49$
4. $36x^2 - 25$
5. $100x^2 - 81y^2$
6. $16x^2 - 121y^2$
7. $64x^2y^2 - 9$
8. $144x^2y^2 - 1$
9. $100x^6 - 49y^8$
10. $4x^4 - y^{10}$
11. $y^8 - 36x^6$
12. $81y^8 - 4x^{12}$
13. $x^3 - 81x$
14. $8x^{11} - 242x^7$
15. $400y^8 - 9y^4$
16. $72x^7 - 2x^3$
17. $128x^3 - 18x^9$
18. $25x^7y^8 - xy^2$
19. $363x^4y^4 - 12y^4$
20. $72x^2y^2 - 98x^2y^8$
21. $162xy - 200x^7y^5$
22. $144x^8y^9 - 225x^{10}y^5$
23. $32x^8y^6 - 72x^2y^8$
24. $49x^{16}y^9 - 169x^6y^{15}$

16.7 FACTORING GENERAL TRINOMIALS IN X

In this section we will consider *general trinomials* in x. A *general trinomial* is a quadratic trinomial of the form

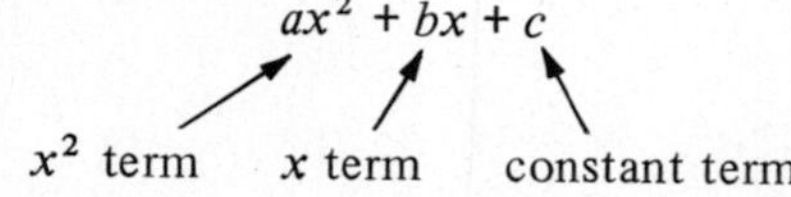

where a, b, and c represent nonzero (*signed*) integers and $a \neq 1$. For example,

$$9x^2 + 9x - 4 \quad \text{and} \quad 8x^2 + 22x + 12$$

are general trinomials.

Remember that a general trinomial is in *lowest terms* if its terms have no common divisors except 1 and –1. For example, $9x^2 + 9x - 4$ is in lowest terms, whereas $8x^2 + 22x + 12$ is not (its terms are all divisible by 2). When factoring a general trinomial, we always reduce it to lowest terms (by factoring out the largest integer possible) before determining its binomial factors (if any).

Factoring a general trinomial of the form $ax^2 + bx + c$ is a more involved process than factoring a simple trinomial of the form $x^2 + bx + c$ because we must consider the coefficient a of the x^2 term (since $a \neq 1$) as well as the numbers b and c.

27. Reduce to lowest terms.
a. $18x^2 - 64x + 30$
b. $24x^2 - 40x - 224$
c. $36x^2 + 51x - 120$
d. $50x^2 + 560x + 600$

Method A (Sum of Products Method)

In this section we will present two methods for factoring general trinomials. The first method is the more traditional one. To understand it, we must look at the product of two general binomials (constant $\cdot$ x + constant). For example, consider the product of $(5x + 4)$ and $(3x - 8)$:

$$\begin{array}{r} 5x + 4 \\ 3x - 8 \\ \hline 15x^2 + 12x \qquad \\ -40x - 32 \\ \hline 15x^2 - 28x - 32 \end{array}$$

x^2 term $\quad$ x term $\quad$ constant term

The product of these two general binomials is a general trinomial with the following properties:

28. Multiply
a. $(2x + 3)(4x + 5)$
b. $(2x - 3)(4x + 5)$
c. $(2x + 3)(4x - 5)$
d. $(2x - 3)(4x - 5)$

1. Its x^2 term, $15x^2$, is the product of the x terms of the binomials.

$$(5x)(3x)$$
$$15x^2 - 28x - 32 = (5x + 4)(3x - 8)$$

2. Its constant term, -32, is the product of the constant terms of the binomials.

$$(+4)(-8)$$
$$15x^2 - 28x - 32 = (5x + 4)(3x - 8)$$

3. Its x term, $-28x$, is the sum of two terms, each the product of the x term of one binomial and the constant term of the other.

$$15x^2 - 28x - 32 = (5x + 4)(3x - 8) \quad \text{and} \quad \begin{array}{r} (5x)(-8) = -40x \\ (3x)(+4) = +12x \\ \hline -28x \end{array}$$

Can you see how these properties might be used to factor a general trinomial? Check your ideas by looking at the following example.

$$15x^2 - 4x - 32 = (3x + 4)(5x - 8) \quad \text{and} \quad \begin{array}{r} (3x)(-8) = -24x \\ (5x)(+4) = +20x \\ \hline -4x \end{array}$$

To determine the binomial factors of a general trinomial we first factor its x^2 term and constant term. For example, to determine that $15x^2 - 28x - 32$ equals $(5x + 4)(3x - 8)$, we first observe that $15x^2 = 3x \cdot 5x$ and $-32 = (+4)(-8)$. Then we

29. List two binomial factorizations using the four terms
a. $2x, 5x, 3, 7$
b. $x, 9x, 8, 11$

form the two binomials $(5x + 4)$ and $(3x - 8)$ containing these four factors. However, the same four factors can also be arranged into the indicated product $(3x + 4)(5x - 8)$, which equals $15x^2 - 4x - 32$. Since we cannot see by inspection which of the products $(3x + 4)(5x - 8)$ or $(5x + 4)(3x - 8)$ equals $15x^2 - 28x - 32$, we must multiply them both to see which (if either) is the correct factorization.

Look at the following example of *factoring a general trinomial.*

TYPICAL PROBLEM (Method A) $\qquad 4x^2 - 11x + 6$

$\downarrow$

We first set up our format, leaving space in each pair of parentheses for a binomial factor: $\qquad (\quad)(\quad)$

$\downarrow$

Then we list all pairs of positive factors of the x^2 term. (We always factor x^2 as $x \cdot x$; never as $x^2 \cdot 1$):

$4x^2$
$(x)(4x)$
$(2x)(2x)$

and

We next list all pairs of factors in the constant term:

$+6$
$(+1)(+6)$
$(-1)(-6)$
$(+2)(+3)$
$(-2)(-3)$

$\downarrow$

Then we form a binomial factorization. To do this we substitute a pair of factors of the x^2 term into the left sides of the parentheses and a pair of factors of the constant term into the right sides of the parentheses in the format: $\qquad (x + 1)(4x + 6)$

$\downarrow$

We check this possible factorization by multiplying: $\qquad 4x^2 + 10x + 6$

$\downarrow$

Next we interchange the constant terms in the two binomials and check the new factorization by multiplying: $\qquad (x + 6)(4x + 1) = 4x^2 + 25x + 6$

$\cdot$
$\cdot$
$\cdot$

We repeat the above three steps with each possible set of factors until one set multiplies to the original trinomial. (If we try every possible set of factors with no success, the trinomial is not factorable in integers): $\qquad (x - 2)(4x - 3) = 4x^2 - 11x + 6$

Hence, $4x^2 - 11x + 6 = (x - 2)(4x - 3)$.

When factoring a general trinomial we always factor out any numerical factor first, using the method in Section 14.1. To multiply binomials in order to check each possible factorization we use the method in Section 15.1.

EXAMPLES

Factor the following general trinomials in x.

1. $9x^2 - 32x + 15$
2. $3x^2 - 5x - 28$
3. $12x^2 + 17x - 40$
4. $5x^2 + 56x + 60$
5. $36x^2 + 42x - 120$
6. $192x^2 - 280x + 72$

Solutions (Method A)

1. $9x^2 - 32x + 15$

↓

( )( )

↓

Positive factors of $9x^2$	Factors of +15
$(x)(9x)$	$(+1)(+15)$
$(3x)(3x)$	$(-1)(-15)$
	$(+3)(+5)$
	$(-3)(-5)$

↓

$(x + 1)(9x + 15) = 9x^2 + 24x + 15$
$(x + 15)(9x + 1) = 9x^2 + 136x + 15$

$(x - 1)(9x - 15) = 9x^2 - 24x + 15$
$(x - 15)(9x - 1) = 9x^2 - 136x + 15$

$(x + 3)(9x + 5) = 9x^2 + 32x + 15$
$(x + 5)(9x + 3) = 9x^2 + 48x + 15$

$(x - 3)(9x - 5) = 9x^2 - 32x + 15$★
$(x - 5)(9x - 3) = 9x^2 - 48x + 15$

$(3x + 1)(3x + 15) = 9x^2 + 48x + 15$
$(3x - 1)(3x - 15) = 9x^2 - 48x + 15$

$(3x + 3)(3x + 5) = 9x^2 + 24x + 15$
$(3x - 3)(3x - 5) = 9x^2 - 24x + 15$

↓

$9x^2 - 32x + 15 = (x - 3)(9x - 5)$

2. $3x^2 - 5x - 28$

↓

( )( )

↓

Positive factors of $3x^2$	Factors of −28
$(x)(3x)$	$(+1)(-28)$
	$(-1)(+28)$
	$(+2)(-14)$
	$(-2)(+14)$
	$(+4)(-7)$
	$(-4)(+7)$

↓

$(x + 1)(3x - 28) = 3x^2 - 25x - 28$
$(x - 28)(3x + 1) = 3x^2 - 83x - 28$

$(x - 1)(3x + 28) = 3x^2 + 25x - 28$
$(x + 28)(3x - 1) = 3x^2 + 83x - 28$

$(x + 2)(3x - 14) = 3x^2 - 8x - 28$
$(x - 14)(3x + 2) = 3x^2 - 40x - 28$

$(x - 2)(3x + 14) = 3x^2 + 8x - 28$
$(x + 14)(3x - 2) = 3x^2 + 40x - 28$

$(x + 4)(3x - 7) = 3x^2 + 5x - 28$
$(x - 7)(3x + 4) = 3x^2 - 17x - 28$

$(x - 4)(3x + 7) = 3x^2 - 5x - 28$★
$(x + 7)(3x - 4) = 3x^2 + 17x - 28$

↓

$3x^2 - 5x - 28 = (x - 4)(3x + 7)$

3. $12x^2 + 17x - 40$

↓

( )( )

↓

Positive factors of $12x^2$	Factors of −40
$(x)(12x)$	$(+1)(-40)$
$(2x)(6x)$	$(-1)(+40)$
$(3x)(4x)$	$(+2)(-20)$
	$(-2)(+20)$
	$(+4)(-10)$
	$(-4)(+10)$
	$(+5)(-8)$
	$(-5)(+8)$

↓

48 possible factorizations

↓

$12x^2 + 17x - 40 = (3x + 8)(4x - 5)$

4. $5x^2 + 56x + 60$

↓

( )( )

↓

Positive factors of $5x^2$	Factors of +60
$(x)(5x)$	$(+1)(+60)$
	$(-1)(-60)$
	$(+2)(+30)$
	$(-2)(-30)$
	$(+3)(+20)$
	$(-3)(-20)$
	$(+4)(+15)$
	$(-4)(-15)$
	$(+5)(+12)$
	$(-5)(-12)$
	$(+6)(+10)$
	$(-6)(-10)$

↓

24 possible factorizations

↓

$5x^2 + 56x + 60 = (x + 10)(5x + 6)$

5.

	$36x^2 + 42x - 120$
2	
	$18x^2 + 21x - 60$
3	
	$6x^2 + 7x - 20$
	↓
	24 possible factorizations
	↓
	$(2x + 5)(3x - 4)$
$6(2x + 5)(3x - 4)$	

6.

	$192x^2 - 280x + 72$
2	
	$96x^2 - 140x + 36$
2	
	$48x^2 - 10x + 18$
2	
	$24x^2 - 35x + 9$
	↓
	24 possible factorizations
	↓
	$(3x - 1)(8x - 9)$
$8(3x - 1)(8x - 9)$	

Method B
(Corresponding Simple Trinomial Method)

An alternative method for factoring a general trinomial makes use of a (not obvious) relationship between general trinomials and simple trinomials. This relationship relies on two facts:

1. Any simple trinomial can be converted to a special type of general trinomial by replacing x by an integer times x. For example, replacing x by $8x$ in the simple trinomial $x^2 + 2x - 120$ yields the general trinomial

30. Convert to a general trinomial by replacing
a. x by $5x$ in $x^2 - 3x - 10$
b. x by $4x$ in $x^2 + 15x + 44$

$$x^2 + 2x - 120$$
$$(8x)^2 + 2(8x) - 120 = 64x^2 + 16x - 120$$

Note that the first term, $64x^2$, is the square of $8x$, and the second term, $16x$, is a multiple of $8x$.

2. Although most general trinomials are not of this special type, any general trinomial can be converted simply by multiplying it by the coefficient of its x^2 term. Multiplying the general trinomial $8x^2 + 2x - 15$ by 8, for example, yields

31. Multiply
a. $5(5x^2 - 3x - 2)$
b. $4(4x^2 + 15x + 11)$

$$8(8x^2 + 2x - 15) = (8x)^2 + 2(8x) - (8 \cdot 15)$$

which, as we just saw above, corresponds to the simple trinomial $x^2 + 2x - 120$.

Our second method for factoring a general trinomial such as $8x^2 + 2x - 15$ makes use of these two facts as follows. We first factor

$$8(8x^2 + 2x - 15) = (8x)^2 + 2(8x) - 120$$

much as we would the simple trinomial $x^2 + 2x - 120$. Then we divide by 8 to get the factorization of $8x^2 + 2x - 15$.

Although its theoretical basis is not elementary, this method has two practical advantages. It is the quickest way of factoring most general trinomials, and if the general trinomial cannot be factored in integers (which is frequently the case), then neither can the corresponding simple trinomial.

Look at the following example of *factoring general trinomials.*

TYPICAL PROBLEM (Method B)

We first multiply the trinomial by the coefficient of the x^2 term:

Then we factor $(4x)^2 - 11(4x) + 24$ as we would a simple trinomial. To do this we first set up our format, but with $4x$ in each pair of parentheses in place of x. We leave space in each pair of parentheses for a constant term:

Then we list pairs of factors of the new constant term, and beside each pair we write their sum. (We only list pairs whose sum has the same sign as the x term of the trinomial.):

And then we select the pair of factors (if any) whose sum is the coefficient of the x term (starred above). We write one factor in each pair of parentheses in the format. (This is the factorization of 4 times $4x^2 - 11x + 6$.):

Finally we reduce each binomial to lowest terms and discard the factored integer to get the factorization of $4x^2 - 11x + 6$:

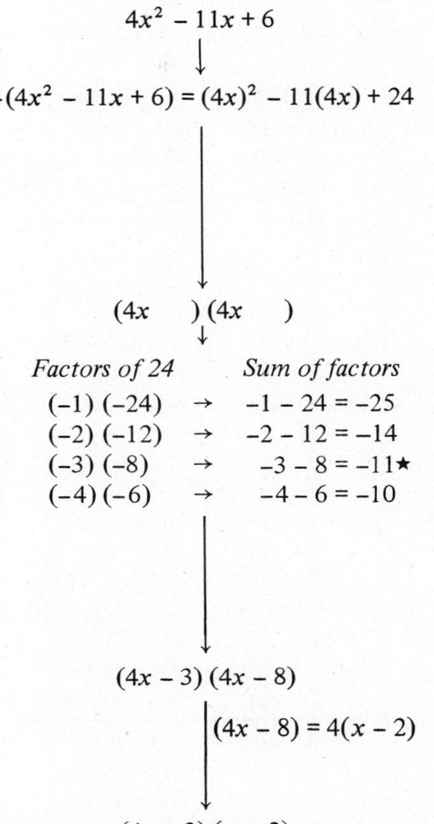

When factoring a general trinomial we first factor out any numerical factor using the method in Section 14.1. To reduce each binomial to lowest terms in the last step we again use the method in Section 14.1, before discarding the integer(s) factored out.

Hence, $4x^2 - 11x + 6 = (4x - 3)(x - 2)$.

WARNING: The trinomial is not factorable in integers unless one of the sums of factors is the coefficient of the x term.

EXAMPLES

Factor the following general trinomials in x.

1. $9x^2 - 32x + 15$
2. $3x^2 - 5x - 28$
3. $12x^2 + 17x - 40$
4. $5x^2 + 56x + 60$
5. $36x^2 + 42x - 120$
6. $192x^2 - 280x + 72$

Solutions (Method B)

1.

$$9x^2 - 32x + 15$$
$$\downarrow$$
$$9(9x^2 - 32x + 15) = (9x)^2 - 32(9x) + 135$$
$$\downarrow$$

+135	sum
(−1) (−135)	−1 − 135 = −136
(−3) (−45)	−3 − 45 = −48
(−5) (−27)	−5 − 27 = −32★
(−9) (−15)	−9 − 15 = −24

$$\downarrow$$
$$(9x - 5)(9x - 27) = 9(9x^2 - 32x + 15)$$
$$\downarrow$$
$$(9x - 5)(x - 3) = 9x^2 - 32x + 15$$

2.

$$3x^2 - 5x - 28$$
$$\downarrow$$
$$3(3x^2 - 5x - 28) = (3x)^2 - 5(3x) - 84$$
$$\downarrow$$

−84	sum
(+1) (−84)	+1 − 84 = −83
(+2) (−42)	+2 − 42 = −40
(+3) (−28)	+3 − 28 = −25
(+4) (−21)	+4 − 21 = −17
(+6) (−14)	+6 − 14 = −8
(+7) (−12)	+7 − 12 = −5★

$$\downarrow$$
$$(3x + 7)(3x - 12) = 3(3x^2 - 5x - 28)$$
$$\downarrow$$
$$(3x + 7)(x - 4) = 3x^2 - 5x - 28$$

3.

$$12x^2 + 17x - 40$$
$$\downarrow$$
$$12(12x^2 + 17x - 40)$$
$$= (12x)^2 + 17(12x) - 480$$
$$\downarrow$$

12 possible factorizations

−480	sum
(−15) (+32)	−15 + 32 = +17★

$$\downarrow$$
$$(12x - 15)(12x + 32)$$
$$= 12(12x^2 + 17x - 40)$$
$$\downarrow$$
$$(4x - 5)(3x + 8) = 12x^2 + 17x - 40$$

4.

$$5x^2 + 56x + 60$$
$$\downarrow$$
$$5(5x^2 + 56x + 60)$$
$$= (5x)^2 + 56(5x) + 300$$
$$\downarrow$$

9 possible factorizations

+300	sum
(+6) (+50)	+6 + 50 = +56★

$$\downarrow$$
$$(5x + 6)(5x + 50)$$
$$= 5(5x^2 + 56x + 60)$$
$$\downarrow$$
$$(5x + 6)(x + 10) = 5x^2 + 56x + 60$$

5.

$$36x^2 + 42x - 120$$

2

$$18x^2 + 21x - 60$$

3

$$6x^2 + 7x - 20$$
$$\downarrow$$
$$6(6x^2 + 7x - 20)$$
$$= (6x)^2 + 7(6x) - 120$$
$$\downarrow$$

8 possible factorizations

$$\downarrow$$
$$(6x - 8)(6x + 15)$$
$$= 6(6x^2 + 7x - 20)$$
$$\downarrow$$
$$(3x - 4)(2x + 5)$$
$$= 6x^2 + 7x - 20$$

$$6(3x - 4)(2x + 5)$$

6.

$$192x^2 - 280x + 72$$

2

$$96x^2 - 140x + 36$$

2

$$48x^2 - 70x + 18$$

2

$$24x^2 - 35x + 9$$
$$\downarrow$$
$$24(24x^2 - 35x + 9)$$
$$= (24x)^2 - 35(24x) + 216$$
$$\downarrow$$

8 possible factorizations

$$\downarrow$$
$$(24x - 8)(24x - 27)$$
$$= 24(24x^2 - 35x + 9)$$
$$\downarrow$$
$$(3x - 1)(8x - 9)$$
$$= 24x^2 - 35x + 9$$

$$8(3x - 1)(8x - 9)$$

OBSERVATIONS AND REFLECTIONS

1. If a general trinomial is in lowest terms, its binomial factors are also in lowest terms, because any divisor of a binomial also divides any multiple of the binomial. For example, because 3 is a divisor of $3x - 6$, 3 also divides $(3x - 6)(2x + 5) = 6x^2 + 3x - 30$. Similarly, 3 divides every other multiple of $3x - 6$. Hence, no multiple of $3x - 6$ is in lowest terms. In other words, no polynomial in lowest terms has $3x - 6$ as a divisor, precisely because $3x - 6$ is not itself in lowest terms.

   In general, if P is a polynomial in lowest terms, every multinomial divisor of P is also in lowest terms. No *monomial* can divide a divisor of the polynomial P unless it also divides P.

> If P is a divisor of Q and Q is a divisor of R, P is a divisor of R
>
> $P|Q$ and $Q|R \Rightarrow P|R$
>
> (where P, Q, and R represent polynomials)

2. Using the sum of products method (method A) for factoring a general trinomial in lowest terms, we find every indicated product of two binomials that yields the x^2 term and constant term of the general trinomial. We must then multiply out each product to determine which one yields the correct x term. Frequently, however, possible factorizations can be rejected without multiplying because a binomial factor is not in lowest terms. From paragraph 1, we know that a binomial factor of a general trinomial in lowest terms must also be in lowest terms. For example, using method A to factor $9x^2 - 22x - 15$, we consider the following factorizations and their products:

$(x + 1)(9x - 15) = 9x^2 - 6x - 15$	$(x - 1)(9x + 15) = 9x^2 + 6x - 15$
★$(x - 15)(9x + 1) = 9x^2 - 134x - 15$	★$(x + 15)(9x - 1) = 9x^2 + 134x - 15$
★$(x + 3)(9x - 5) = 9x^2 + 22x - 15$	★$(x - 3)(9x + 5) = 9x^2 - 22x - 15$
$(x - 5)(9x + 3) = 9x^2 - 42x - 15$	$(x + 5)(9x - 3) = 9x^2 + 42x - 15$
$(3x + 1)(3x - 15) = 9x^2 - 42x - 15$	$(3x - 1)(3x + 15) = 9x^2 + 42x - 15$
$(3x - 3)(3x + 5) = 9x^2 + 6x - 15$	$(3x + 3)(3x - 5) = 9x^2 - 6x - 15$

   However, only the starred factorizations are real contenders, because the others contain a binomial divisible by 3.

   This situation arises whenever the coefficient of the x^2 term and the constant term have a common divisor larger than 1. For example, in $9x^2 - 22x - 15$, 9 and -15 are both divisible by 3.

3. Look again at the above list of factorizations and their products, and compare the two factorizations in each row. In the first row, for example, look at

$$(x + 1)(9x - 15) = 9x^2 - 6x - 15 \quad \text{vs.} \quad (x - 1)(9x + 15) = 9x^2 + 6x - 15$$

   Note that the binomial factors on the left differ from the binomial factors on the right only in the signs of the constant factors. Similarly, the trinomial on the left differs from the trinomial on the right only in the sign of the x term. This sign relationship holds in general: If we take the factorization of any general trinomial $ax^2 + bx + c$ (for any signed integers a, b, and c) and change the signs of the constant terms of both factors, we obtain the factorization of $ax^2 - bx + c$. Hence, to check the possible factorizations of a general trinomial, we need only multiply half of them and look for a product whose x term has a coefficient with the same numerical value as the original trinomial's x term. For example, to check the list of possible factorizations of $9x^2 - 22x - 15$, we need only multiply the factorizations in the left column. Every possible factorization yields a different x term, and only $(x + 3)(9x - 5)$ on the left yields an x term whose coefficient has the cor-

32. List the possible factorizations in lowest terms of
a. $4x^2 + 17x + 4$
b. $2x^2 - 7x + 6$
c. $5x^2 + 49x - 10$

33. Factor
a. $12x^2 + 11x - 15$, given $12x^2 - 11x - 15 = (3x - 5)(4x + 3)$
b. $35x^2 + 38x + 8$, given $35x^2 - 38x + 8 = (7x - 2)(5x - 4)$
c. $63x^2 - 92x - 11$, given $63x^2 + 92x - 11 = (7x + 11)(9x - 1)$

rect numerical value. Because it yields an x term with the wrong sign, we negate the two constant terms to get the correct factorization of $9x^2 - 22x - 15$, namely $(x - 3)(9x + 5)$.

EXERCISES

Factor each of the following expressions completely.
In these exercises, complete the factorization.

1. $25x^2 - 45x + 18$
   $(\quad x - 6)(\quad x - 3)$
2. $21x^2 - 19x - 12$
   $(\quad x - 4)(\quad x + 3)$
3. $22x^2 + 105x - 25$
   $(\quad x + 5)(\quad x - 5)$
4. $15x^2 + 22x - 48$
   $(\quad x - 6)(\quad x + 8)$
5. $6x^2 + 43x - 15$
   $(\quad x + 15)(\quad x - 1)$
6. $22x^2 + 25x + 7$
   $(\quad x + 7)(\quad x + 1)$
7. $14x^2 + x - 3$
   $(7x \quad)(2x \quad)$
8. $8x^2 - 5x - 13$
   $(8x \quad)(x \quad)$
9. $10x^2 + x - 9$
   $(x \quad)(10x \quad)$
10. $10x^2 + 29x + 21$
    $(5x \quad)(2x \quad)$
11. $27x^2 - 21x + 4$
    $(9x \quad)(3x \quad)$
12. $20x^2 + 13x - 33$
    $(x \quad)(20x \quad)$

In these exercises, all trinomials can be factored in integers.

13. $3x^2 + 22x + 7$
14. $5x^2 + 3x - 2$
15. $11x^2 - 16x + 5$
16. $5x^2 - 34x - 7$
17. $3x^2 + 40x + 13$
18. $2x^2 + 21x - 11$
19. $6x^2 - 7x + 2$
20. $6x^2 + 13x + 5$
21. $12x^2 + 4x - 1$
22. $14x^2 + 33x - 5$
23. $7x^2 - 24x + 9$
24. $12x^2 + 41x - 11$
25. $15x^2 - 17x + 4$
26. $22x^2 + 51x - 10$
27. $6x^2 - 5x - 25$
28. $14x^2 + 39x + 10$
29. $9x^2 - 38x + 33$
30. $6x^2 - 25x + 21$
31. $6x^2 + 23x + 20$
32. $10x^2 + 11x - 8$
33. $18x^2 - 41x + 21$
34. $20x^2 - 3x - 9$
35. $4x^2 + x - 18$
36. $14x^2 - 29x + 12$
37. $9x^2 + 17x - 30$
38. $8x^2 + 27x + 22$
39. $27x^2 - 33x + 10$
40. $6x^2 - 79x - 27$
41. $21x^2 + 17x - 30$
42. $14x^2 + 23x + 8$
43. $18x^2 + 35x - 25$
44. $33x^2 - 31x - 40$
45. $35x^2 - 48x + 16$
46. $12x^2 + 77x + 30$
47. $27x^2 + 15x - 28$
48. $12x^2 - 23x + 10$

In these exercises, all trinomials have a monomial factor, but not necessarily binomial factors with integer conefficients.

49. $6x^2 + 15x - 9$
50. $12x^2 + 64x + 20$
51. $15x^2 + 110x + 35$
52. $9x^2 - 42x + 45$
53. $72x^2 - 36x - 27$
54. $18x^2 - 54x - 48$
55. $80x^2 - 30x - 50$
56. $60x^2 - 51x + 9$
57. $54x^2 - 207x + 90$
58. $60x^2 + 60x - 225$
59. $50x^2 + 365x - 75$
60. $54x^2 - 114x + 66$

ANSWERS TO MARGIN EXERCISES

1. a. -3
b. -2
c. -1
d. $+3$
e. $+2$
f. $+1$

2. a. $+5$
b. $+3$
c. $+1$
d. -5
e. -3
f. -1

3. a. $(+1)(+2)$
$(-1)(-2)$
b. $(+1)(-2)$
$(-1)(+2)$

4. a. $(+1)(-5)$
$(-1)(+5)$
b. $(+1)(+17)$
$(-1)(-17)$
c. $(+1)(-19)$
$(-1)(+19)$
d. $(+1)(+29)$
$(-1)(-29)$

5. a. $(+1)(-9)$
$(-1)(+9)$
$(+3)(-3)$
b. $(+1)(-16)$
$(-1)(+16)$
$(+2)(-8)$
$(-2)(+8)$
$(+4)(-4)$
c. $(+1)(-25)$
$(-1)(+25)$
$(+5)(-5)$

6. a. $+20$
b. -12
c. $+5$
d. -2

7. a. $x^2 + 8x + 15$
b. $x^2 + 2x - 15$
c. $x^2 - 2x - 15$
d. $x^2 - 8x + 15$

8. a and h
b and e
c and g
d and f

10. a and f
b and d
c and e

11. a and c
b and e
d and f

12. a. $x^2 - 3x + 2$
b. $y^2 - 2y - 8$
c. $x^2 + 5x - 56$
d. $y^2 + 11y + 24$

13. a and f
b and e
c and d

14. a and e
b and d
c and f

15. a. $2x(x - 3)(x + 8)$
b. $y^3(x + 2y)(x - 5y)$

16. a. $x^2 - 3x - 10$
b. $x^2 - 11x + 24$
c. $x^2 + 3x - 28$

17. a. $x^2 + 12xy + 27y^2$
b. $x^2 - 14xy + 40y^2$
c. $x^2 - 6xy - 72y^2$

18. a. $y^2 - 8xy - 33x^2$
b. $y^2 - 13xy + 12x^2$
c. $y^2 + 26xy - 56x^2$

19. a. $-1(x^2 - x - 30)$
b. $-1(x^2 - 8xy + 12y^2)$
c. $-1(y^2 + 9xy - 36x^2)$

20. a. $x^6 \cdot x^6$
b. $xy^5 \cdot xy^5$
c. $x^7y^4 \cdot x^7y^4$

21. a and b

22. a. $12x \cdot 12x$
b. $7x^4 \cdot 7x^4$
c. $8x^2y \cdot 8x^2y$
d. $9x^3y^6 \cdot 9x^3y^6$

23. a. $(-4)(-4)$
b. $(-10x^3)(-10x^3)$
c. $(-x^4y^2)(-x^4y^2)$
d. $(-9x^5y)(-9x^5y)$

24. a. $(12x)(-12x)$
b. $(7y^4)(-7y^4)$
c. $(8x^2y)(-8x^2y)$
d. $(9x^3y^6)(-9x^3y^6)$

25. a. $64x^2 - 9$
b. $x^6 - 4y^2$
c. $25 - y^{10}$

26. a. $2(x^2 - 9)$
b. $3(25x^2 - 49y^2)$
c. $4x^4(4x^2 - 25)$
d. $9x^2y^3(x^6 - 16y^6)$

27. a. $2(9x^2 - 32x + 15)$
b. $8(3x^2 - 5x - 28)$
c. $3(12x^2 + 17x - 40)$
d. $10(5x^2 + 56x + 60)$

28. a. $8x^2 + 22x + 15$
b. $8x^2 - 2x - 15$
c. $8x^2 + 2x - 15$
d. $8x^2 - 22x + 15$

29. a. $(2x + 3)(5x + 7)$
$(2x + 7)(5x + 3)$
b. $(x + 8)(9x + 11)$
$(x + 11)(9x + 8)$

31. a. $25x^2 - 15x - 10$
b. $16x^2 + 60x + 44$

30. a. $(5x)^2 - 3(5x) - 10$
$= 25x^2 - 15x - 10$
b. $(4x)^2 + 15(4x) + 44$
$= 16x^2 + 60x + 44$

32. a. $(x + 4)(4x + 1)$
$(x - 4)(4x - 1)$
b. $(x + 2)(2x + 3)$
$(x - 2)(2x - 3)$
$(x + 6)(2x + 1)$
$(x - 6)(2x - 1)$
c. $(x + 5)(5x - 2)$
$(x - 5)(5x + 2)$
$(x + 10)(5x - 1)$
$(x - 10)(5x + 1)$

33. a. $(3x + 5)(4x - 3)$
b. $(7x + 2)(5x + 4)$
c. $(7x - 11)(9x + 1)$

CHAPTER REVIEW

SECTION 16.1 List all complementary pairs of factors.

1. +4
2. −9
3. +11
4. −12
5. +105
6. −90
7. +126
8. −72
9. +270

SECTION 16.2 Factor completely.

10. $x^2 + 5x + 6$
11. $x^2 + 2x - 15$
12. $x^2 - 19x + 48$
13. $x^2 - 8x - 9$
14. $x^2 - 15x - 54$
15. $x^2 + 11x - 90$
16. $x^2 + 5x - 24$
17. $x^2 + 12x + 36$
18. $x^2 - 21x + 108$
19. $x^2 - 7x - 18$
20. $x^2 + 30x + 96$
21. $x^2 + 16x - 80$
22. $x^2 + 10x - 48$
23. $x^2 + 16x + 64$
24. $x^2 - 18x + 56$

SECTION 16.3 Factor completely.

25. $x^2 + 11xy + 18y^2$
26. $y^2 - 9xy - 36x^2$
27. $y^2 + 11xy - 56x^2$
28. $x^2 - 18xy + 81y^2$
29. $x^2 - 15xy - 72y^2$
30. $y^2 + 19xy + 88x^2$
31. $y^2 + 34xy + 64x^2$
32. $x^2 - 29xy - 96y^2$
33. $y^2 + 29xy + 120x^2$

SECTION 16.4 Factor completely.

34. $4x^2 + 28x - 32$
35. $30y^2 + 150y + 120$
36. $4x^2y - 16xy - 128y$
37. $x^5 + 3x^4 - 18x^3$
38. $7x^6 - 49x^5 + 63x^4$
39. $-5x^2 + 40x - 75$
40. $-2y^3 - 32y^2 - 160y$
41. $-x^2y^8 + 11x^2y^7 + 60x^2y^6$
42. $-2x^7y^6 - 34x^6y^6 + 36x^5y^6$
43. $12x^2 - 180xy + 168y^2$
44. $x^8y^2 + 15x^7y^3 + 26x^6y^4$
45. $-6x^5 + 36x^4y + 72x^3y^2$
46. $-9y^2 - 63xy + 162x^2$
47. $x^4y^6 + x^5y^5 - 42x^6y^4$
48. $4x^2y^5 - 36x^3y^4 + 32x^4y^3$
49. $16x^2 + 96xy + 144y^2$
50. $-x^9y^3 + 10x^{10}y^2 + 33x^{11}y$
51. $8x^3y^4 + 48x^4y^3 - 128x^5y^2$

SECTION 16.5 Factor completely, first placing in standard form.

52. $18x - x^2 - 32$
53. $44 + 20x - x^2$
54. $3x - 18 + x^2$
55. $22x - 121 - x^2$
56. $52 - x^2 - 9x$
57. $19xy + 48y^2 + x^2$
58. $17xy - 66x^2 - y^2$
59. $5xy - x^2 - 6y^2$
60. $17xy - x^2 - 60y^2$

61. $35y^2 + x^2 + 12xy$
62. $-10x + 24 - x^2$
63. $16x^2 + 6xy - y^2$

SECTION 16.6

Factor completely.

64. $9x^2 - 1$
65. $36x^2 - 49$
66. $81x^2 - 64y^2$
67. $49x^2y^2 - 4$
68. $144x^6 - 169y^4$
69. $225y^8 - 81x^{10}$
70. $18x^4 - 32x^2$
71. $72x^6 - 98x^4$
72. $27x^6 - 48x^4$
73. $32x^3 - 72xy^2$
74. $98x^5y^4 - 2x^{11}$
75. $144x^5y^3 - 225x^3y$

SECTION 16.7

Factor completely.

76. $6x^2 - 41x - 30$
    $(\ \ x + 2)(\ \ x - 15)$
77. $55x^2 - 84x + 32$
    $(\ \ x - 4)(\ \ x - 8)$
78. $14x^2 - 53x - 45$
    $(\ \ x + 5)(\ \ x - 9)$
79. $180x^2 - 123x + 20$
    $(12x \quad)(15x \quad)$
80. $42x^2 - 39x - 36$
    $(2x \quad)(21x \quad)$
81. $24x^2 + 154x + 225$
    $(4x \quad)(6x \quad)$
82. $3x^2 + 11x - 4$
83. $7x^2 - 18x + 11$
84. $6x^2 + 13x - 15$
85. $14x^2 + 45x + 25$
86. $35x^2 + 9x - 26$
87. $6x^2 - 77x + 242$
88. $9x^2 - x - 38$
89. $8x^2 + 22x + 15$
90. $27x^2 + 160x - 12$
91. $12x^2 - 28x + 15$
92. $24x^2 + 10x - 25$
93. $18x^2 + 105x + 98$
94. $16x^2 + 6x - 27$
95. $7x^2 - 88x + 81$
96. $120x^2 + 26x - 3$
97. $24x^2 - 95x - 125$
98. $112x^2 + 150x - 27$
99. $120x^2 - 252x + 108$
100. $30x^2 - 48x + 18$
101. $40x^2 + 92x - 20$
102. $49x^2 - 280x - 84$
103. $40x^2 - 136x + 160$
104. $20x^2 + 146x + 42$
105. $48x^2 + 108x - 72$

CUMULATIVE TEST

The following problems test your understanding of this chapter and of related subject matter in Chapter 14. Before taking this test, thoroughly review Sections 14.1, 14.2, 16.2, 16.3, 16.4, 16.6, and 16.7.

Once you have finished the test, compare your answers with the answers provided at the back of the book. Note the section number of each problem missed, and thoroughly review those sections again.

Factor completely.

1. $x^2 - 2x - 48$
2. $x^2 + 18xy + 72y^2$
3. $4x^2 - 132x - 432$
4. $27x^3 - 9$
5. $6x^2 + 7x - 20$
6. $49x^2 - 25$
7. $x^2 - 35x + 96$
8. $50x^2 - 128$
9. $10y^2 + 460y - 960$
10. $x^2 + 7xy - 98y^2$
11. $36x^5y^3 - 25x^3y^4$
12. $56x^2 - 70x - 21$
13. $4y^4 - 81x^2$
14. $12y^2 + 168xy - 864x^2$
15. $28x^3 + 63x + 56$
16. $x^2 + 19x + 84$
17. $242x^7y^3 - 72x^5y^5$
18. $16x^2 - 29x - 6$
19. $-16x^2 - 48y^2 - 64z^2$
20. $14x^2 + 65x - 25$
21. $x^3 - 144x$
22. $x^2 + 25xy + 156y^2$
23. $6x^2y^3 + 18x^3y^2 - 1080x^4y$
24. $4x^3y + x^4y^3 - 4x^4y^5$
25. $21x^2 - 106x - 32$
26. $x^2 - 28x + 96$
27. $9x^6 - 100y^8$
28. $x^8y^3 + 48x^7y^4 + 135x^6y^5$
29. $96x^2 - 12x - 108$
30. $36y^3 - 54x^2 - 42z$
31. $y^2 + 19xy - 150x^2$
32. $35x^4y^2 - 70x^5y^4 + 105x^5y^6$
33. $144y^5 - 100y^3$
34. $18x^2 - 43x + 25$
35. $-3x^4 + 45x^3 + 378x^2$
36. $121y^4 - 16x^6$
37. $x^2 - 31x - 180$
38. $36x^2 - 120x - 125$
39. $15x^3y^6 + 16x^8y^2 - x^3y^2$
40. $y^2 + 22xy + 96x^2$

An algebraic fraction whose numerator and/or denominator is a multinomial (two or more terms) is called a *multinomial fraction*. For example,

$$\frac{15x^3}{25x^2 - 30x^5}, \quad \frac{2a^3 - 10a^2}{6a^4 - 30a^3}, \quad \text{and} \quad \frac{2x^4 + 2x^3 - 84x^2}{4x^3 + 20x^2 - 56x}$$

are multinomial fractions.

Algebraic fractions were introduced in Chapter 8 with the study of monomial fractions, and, as you would expect, all the definitions and ideas introduced there apply equally to multinomial fractions.

Multiplying the numerator and denominator of a fraction by any nonzero polynomial results in an *equivalent fraction*. For example,

$$\frac{3x - 4}{7x + 9} = \frac{8(3x - 4)}{8(7x + 9)} = \frac{5x^2(3x - 4)}{5x^2(7x + 9)} = \frac{(5x^2 + 8)(3x - 4)}{(5x^2 + 8)(7x + 9)}$$

Similarly, dividing the numerator and denominator of a fraction by any common divisor (monomial or multinomial) results in an equivalent fraction. For example,

$$\frac{(5x^2 + 8)(3x - 4)}{(5x^2 + 8)(7x + 9)} = \frac{\overset{1}{\cancel{(5x^2 + 8)}}(3x - 4)}{\underset{1}{\cancel{(5x^2 + 8)}}(7x + 9)} = \frac{(3x - 4)}{(7x + 9)}$$

A multinomial fraction is *reduced* if its numerator and denominator have no common divisors except 1 (and –1). To reduce a multinomial fraction we must learn how to identify and cancel both monomial and multinomial common divisors.

17.1 FACTORING OUT MONOMIALS TO REDUCE FRACTIONS

A multinomial fraction can be reduced by cancelling any common divisors of its numerator and denominator. In this section, we will see how to find and cancel common *monomial divisors.*

The best way to find the common monomial divisors of two polynomials is to reduce each polynomial to lowest terms by factoring out the largest monomial possible. Then we can look at the monomial factors of the two factored expressions to find any common divisors.

1. Reduce
   a. $\frac{27x^2}{9x}$
   b. $\frac{16x^3}{64x^2}$
   c. $\frac{18x^3y^5}{27x^7y^9}$
   d. $\frac{40x^4y^4}{5x^5y^2}$

Remember that the terms of a multinomial in a fraction are not themselves factors of the fraction. For example, in the fraction $\frac{3x+7}{12x}$ we cannot reduce by cancelling $3x$ top and bottom.

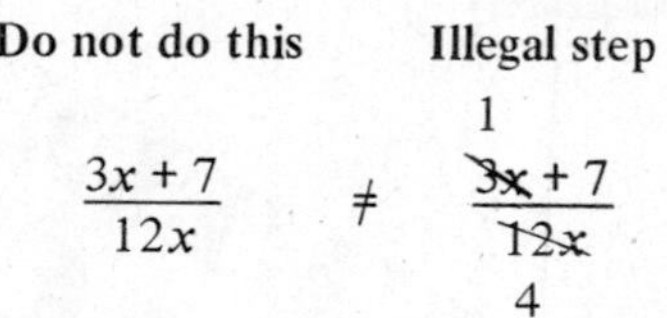

Do not do this **Illegal step**

$$\frac{3x+7}{12x} \neq \frac{\overset{1}{\cancel{3x}}+7}{\underset{4}{\cancel{12x}}}$$

Multinomial and monomial factors in a fraction cannot be reduced together.

Look at the following example of *factoring out monomials to reduce fractions.*

TYPICAL PROBLEM

$$\frac{15x^3}{25x^2 - 30x^5}$$

$\downarrow$

First we completely factor the binomial:

$$\frac{15x^3}{5x^2\,(5 - 6x^3)}$$

$\downarrow$

Then we reduce the monomial factors. (Remember that monomial terms within the binomial cannot be reduced.):

$$\frac{3x}{5 - 6x^3}$$

To factor the binomial we use the method in Section 14.2. Then to reduce the monomial factors we use the method in Section 8.3.

Hence, $\frac{15x^3}{25x^2 - 30x^5} = \frac{3x}{5 - 6x^3}$

WARNING: Binomial factors can only be reduced with binomial factors.

EXAMPLES

Every algebraic fraction can be reduced to a fraction in lowest terms.

Factor and reduce the following fractions.

1. $\frac{81x^2 - 27x^3}{9x}$
2. $\frac{16x^3}{64x^3 - 128x^2}$
3. $\frac{36x^5y^7 - 54x^3y^5}{27x^7y^9}$
4. $\frac{40x^4y^4}{25x^7y^2 - 55x^5y^4}$

Solutions

1. $\frac{27x^2(3-x)}{9x}$

   $\downarrow$

   $3x(3-x)$

   or

   $9x - 3x^2$

2. $\frac{16x^3}{64x^2(x-2)}$

   $\downarrow$

   $\frac{x}{4(x-2)}$

   or

   $\frac{x}{4x-8}$

3. $\dfrac{18x^3y^5(2x^2y^2 - 3)}{27x^7y^9}$
↓
$\dfrac{2(2x^2y^2 - 3)}{3x^4y^4}$
or
$\dfrac{4x^2y^2 - 6}{3x^4y^4}$

4. $\dfrac{40x^4y^4}{5x^5y^2(5x^2 - 11y^2)}$
↓
$\dfrac{8y^2}{x(5x^2 - 11y^2)}$
or
$\dfrac{8y^2}{5x^3 - 11xy^2}$

OBSERVATIONS AND REFLECTIONS

1. Remember that a polynomial fraction is *reduced* if its numerator and denominator have no common divisors except 1 and −1. The two polynomials in a fraction have no common *monomial* divisors if no monomial is a divisor of *every* term of *both* polynomials. For example, $\dfrac{8x - 27}{9x}$ is reduced even though $9x$ shares a common divisor with each of $8x$ and -27.

2. Note, however, that because the denominator of this fraction is a *monomial*, we can split it into the sum of two monomial fractions and reduce the two monomial fractions separately:

$$\frac{8x - 27}{9x} = \frac{8x}{9x} + \frac{-27}{9x} = \frac{8}{9} + \frac{-3}{x}$$

This is the reverse of the process of adding monomial fractions presented in Chapter 9.

2. Which fractions are reduced?

a. $\dfrac{12x^2}{15x + 24}$

b. $\dfrac{21x + 20}{18x}$

c. $\dfrac{40}{60x - 48}$

d. $\dfrac{35x - 90}{21}$

3. Convert to a sum of fractions and reduce

a. $\dfrac{5 - 6x^3}{3x}$

b. $\dfrac{12x - 15}{5x}$

c. $\dfrac{54x^3 - 49}{21x^2}$

d. $\dfrac{16x^2 + 9}{12}$

EXERCISES

Reduce each of the following expressions to a fraction in lowest terms.

1. $\dfrac{8x + 24}{24}$

2. $\dfrac{24x}{12x - 48}$

3. $\dfrac{42 + 12x}{48xy^2}$

4. $\dfrac{28x^3}{24x - 40}$

5. $\dfrac{35x^7y^2}{30 + 25y}$

6. $\dfrac{72x - 45}{30x^4}$

7. $\dfrac{7x^4 + 4x^3}{x^7}$

8. $\dfrac{x^3}{9x^3 - 8x^2}$

9. $\dfrac{9x^5 + 5x^6}{x^3}$

10. $\dfrac{4x^2y^2 + xy^2}{x^3y^2}$

11. $\dfrac{-2x^3y + 9x^4y}{x^2y^4}$

12. $\dfrac{x^5y^3}{5x^2y^5 + 6x^2y^6}$

13. $\dfrac{45x^6 + 15x^5}{3x^2}$

14. $\dfrac{40x^8}{-64x^6 + 16x^7}$

15. $\dfrac{24x^7 - 18x^5}{16x^2}$

16. $\dfrac{49x^2y^2 - 14xy^2}{49x^5y}$

17. $\dfrac{48x^6y}{72x^3y + 45x^4y}$

18. $\dfrac{48x^4y^2 + 72x^5y^2}{32x^2y^6}$

17.2
REDUCING BINOMIAL FRACTIONS

4. Reduce $\dfrac{3(4x-9)}{5x(4x-9)}$

In this section we will look at ratios of binomials with common monomial and binomial divisors. To reduce a binomial fraction we must identify and cancel any common monomial or binomial divisors. To do this, we must first completely factor the numerator and denominator.

Look at the following example of *reducing binomial fractions.*

TYPICAL PROBLEM

$$\frac{2a^3 - 10a^2}{6a^4 - 30a^3}$$

$$\downarrow$$

We first completely factor each binomial:

$$\frac{2a^2(a-5)}{6a^3(a-5)}$$

$$\downarrow$$

Then we reduce the monomial factors:

$$\frac{1(a-5)}{3a(a-5)}$$

$$\downarrow$$

And then we reduce the binomial factors:

$$\frac{1}{3a}$$

To reduce binomial fractions we extend the method in the previous section.

Hence, $\dfrac{2a^3 - 10a^2}{6a^4 - 30a^3} = \dfrac{1}{3a}$.

EXAMPLES

Every algebraic fraction can be reduced to a fraction in lowest terms.

Factor and reduce the following binomial fractions.

1. $\dfrac{6x^4 + 9x^3}{8x^2 + 12x}$
2. $\dfrac{15x - 12y}{35x - 28y}$
3. $\dfrac{4x^3 - 16x^4}{24x^9 + 88x^8}$
4. $\dfrac{18x^5 - 36x^3}{72x^8 - 144x^6}$
5. $\dfrac{x^3 - 5x^7}{x^9 + 5x^6}$
6. $\dfrac{6y^3 - 14xy}{28x - 12y^2}$

Solutions

1. $\dfrac{3x^3(2x+3)}{4x(2x+3)} \downarrow \dfrac{3x^2}{4}$

2. $\dfrac{3(5x-4y)}{7(5x-4y)} \downarrow \dfrac{3}{7}$

3. $\dfrac{4x^3(1-4x)}{8x^8(3x+11)} \downarrow \dfrac{1-4x}{2x^5(3x+11)}$ or $\dfrac{1-4x}{6x^6+22x^5}$

4. $\dfrac{18x^3(x^2-2)}{72x^6(x^2-2)} \downarrow \dfrac{1}{4x^3}$

5. $\dfrac{x^3(1-5x^4)}{x^6(x^3+5)} \downarrow \dfrac{1-5x^4}{x^3(x^3+5)}$ or $\dfrac{1-5x^4}{x^6+5x^3}$

6. $\dfrac{2y(3y^2-7x)}{4(7x-3y^2)} \downarrow \dfrac{-y}{2}$

SPECIAL CASES

A *multinomial* is a common divisor of the numerator and denominator of a fraction in factored form if

1. It appears as a factor of both the numerator and denominator (although the terms of the multinomial need not appear in the same order). For example, $(x + y)$ is a common divisor of the numerator and denominator of $\frac{3x(x+y)}{(x+y)(x-y)}$ and of $\frac{4xy(y+x)}{6x(x+y)(3y-x)}$. Hence,

$$\frac{3x\overset{1}{\cancel{(x+y)}}}{\underset{1}{\cancel{(x+y)}}(x-y)} = \frac{3x}{(x-y)} \quad \text{and} \quad \frac{4xy\overset{1}{\cancel{(y+x)}}}{6x\underset{1}{\cancel{(x+y)}}(3y-x)} = \frac{2y}{3(3y-x)}$$

2. It appears as a factor of the numerator, and its negative appears as a factor of the denominator. Remember that two multinomials are negatives of each other if they contain the same terms with opposite signs (although corresponding terms need not appear in the same order). For example, $(x - y)$ and $(y - x)$ are negatives, so $(x - y)$ is a common divisor of the numerator and denominator of $\frac{3x(-x+y)}{(x+y)(x-y)}$ and of $\frac{4xy(y-x)}{6x(x-y)(3y-x)}$. Hence,

$$\frac{3x\overset{-1}{\cancel{(-x+y)}}}{(x+y)\underset{1}{\cancel{(x-y)}}} = \frac{-3x}{(x-y)} \quad \text{and} \quad \frac{4xy\overset{-1}{\cancel{(y-x)}}}{6x\underset{1}{\cancel{(x-y)}}(3y-x)} = \frac{-2y}{3(3y-x)}$$

When a binomial factor is in both the numerator and denominator of a fraction, reduce it to 1 in both places. When a binomial is in the numerator and its negative is in the denominator of a fraction, reduce the binomial in the numerator to −1 and its negative in the denominator to 1.

5. Reduce

a. $\frac{9x^2(2x^2 - 5y)}{(2x^2 - 5y)(9x^2 - 5y)}$

b. $\frac{2x(5x + 4y)}{9y(4y + 5x)}$

6. Reduce

a. $\frac{9x^2(2x^2 - 5y)}{(-2x^2 + 5y)(9x^2 - 5y)}$

b. $\frac{2x(4y - 5x)}{9y(5x - 4y)}$

EXERCISES

Reduce each of the following expressions to a fraction in lowest terms.

1. $\frac{8x - 16}{24x - 48}$
2. $\frac{60x + 90}{48x + 72}$
3. $\frac{14x - 21}{35x - 42}$
4. $\frac{15x - 45}{30x - 85}$
5. $\frac{32x + 24x^2}{27x + 36}$
6. $\frac{20x + 80x^2}{120x + 30}$
7. $\frac{x^6 + 5x^5}{5x^2 + x^3}$
8. $\frac{x^7 + 7x^6}{7x^2 + x^3}$
9. $\frac{28x - 42}{28x - 84}$
10. $\frac{3x^7 - 12x^6}{12x^9 - 16x^8}$
11. $\frac{6x^4y - 9x^4}{12x^7y - 18x^6}$
12. $\frac{12x^7y^5 - 20x^7y^4}{36x^{10}y^2 - 60x^9y^2}$
13. $\frac{24x^5y^7 - 48x^5y^6}{-48x^3y^8 + 24x^3y^9}$
14. $\frac{16x^9y^6 + 48x^8y^5}{18x^3y^7 + 54x^4y^8}$
15. $\frac{30x^2y^9 - 12xy^9}{20xy^5 - 8y^5}$
16. $\frac{2x^6y^7 - 3x^5y^7}{10x^9y^5 - 15x^8y^5}$
17. $\frac{36x^9y^6 - 72x^8y^5}{81x^4y^8 - 27x^4y^7}$
18. $\frac{180x^2y^9 - 72xy^9}{120x^7y^5 - 48x^6y^5}$

These fractions can be reduced by cancelling negative binomials.

19. $\dfrac{20 - 20x}{16x - 16}$ 20. $\dfrac{24x - 36}{126 - 84x}$ 21. $\dfrac{27x + 9}{-24x - 72x^2}$

22. $\dfrac{32y^2 - 80y}{120xy - 48x}$ 23. $\dfrac{15x^5 - 4x^4}{4x^3 - 15x^4}$ 24. $\dfrac{9x^6 - 8x^5y}{8y^3 - 9xy^2}$

25. $\dfrac{7x^5y^4 - 15x^5y^3}{15x^2y^7 - 7x^2y^8}$ 26. $\dfrac{12x^6y - 6x^5y^2}{12x^6y - 24x^7}$ 27. $\dfrac{16x^4y - 32x^3y^2}{72y^6 - 36xy^5}$

28. $\dfrac{120xy^4 - 30y^5}{20xy^4 - 80x^2y^3}$ 29. $\dfrac{72x^3y^4 - 36x^3y^3}{45x^8 - 90x^7y}$ 30. $\dfrac{32x^5y^4 - 48x^4y^5}{60y^2 - 40xy}$

17.3 REDUCING MULTINOMIAL FRACTIONS

In this section we will look at multinomial fractions containing factorable binomials and trinomials. We can rewrite these fractions in factored form using the methods in Chapter 16, and then reduce them using the methods in the previous section.

Look at the following example of *reducing fractions containing binomials and trinomials.*

TYPICAL PROBLEM

$$\frac{2x^4 + 2x^3 - 84x^2}{4x^3 + 20x^2 - 56x}$$

$\downarrow$

We first completely factor the numerator and denominator of the fraction:

$$\frac{2x^2(x - 6)(x + 7)}{4x(x + 7)(x - 2)}$$

$\downarrow$

Then we reduce the fraction. We first reduce the monomial factors, and then we reduce the binomial factors:

$$\frac{x(x - 6)}{2(x - 2)}$$

To reduce multinomial fractions we generalize the methods in the previous two sections.

Hence, $\dfrac{2x^4 + 2x^3 - 84x^2}{4x^3 + 20x^2 - 56x} = \dfrac{x(x - 6)}{2(x - 2)}$.

EXAMPLES

Every algebraic fraction can be reduced to a fraction in lowest terms.

Factor and reduce the following fractions.

1. $\dfrac{4x - 8}{x^2 - 10x + 16}$ 2. $\dfrac{x^2 + 9x + 14}{3x^3 + 21x^2}$

3. $\dfrac{x^2 + 6x + 8}{x^2 - x - 6}$ 4. $\dfrac{x^2 - 36}{x^2 - 3x - 18}$

5. $\dfrac{3x^3 - 3x^2 - 216x}{2x^2 + 8x - 64}$ 6. $\dfrac{9x^2 + 27x - 36}{3x^2 - 48}$

Solutions

1. $\dfrac{4(x-2)}{(x-8)(x-2)} \downarrow \dfrac{4}{x-8}$

2. $\dfrac{(x+7)(x+2)}{3x^2(x+7)} \downarrow \dfrac{x+2}{3x^2}$

3. $\dfrac{(x+4)(x+2)}{(x+2)(x-3)} \downarrow \dfrac{x+4}{x-3}$

4. $\dfrac{(x+6)(x-6)}{(x-6)(x+3)} \downarrow \dfrac{x+6}{x+3}$

5. $\dfrac{3x(x+8)(x-9)}{2(x+8)(x-4)} \downarrow \dfrac{3x(x-9)}{2(x-4)}$ or $\dfrac{3x^2-27x}{2x-8}$

6. $\dfrac{9(x+4)(x-1)}{3(x+4)(x-4)} \downarrow \dfrac{3(x-1)}{x-4}$ or $\dfrac{3x-3}{x-4}$

EXERCISES

Reduce each of the following expressions to a fraction in lowest terms.

1. $\dfrac{x-3}{x^2-x-6}$
2. $\dfrac{x+4}{x^2+2x-8}$
3. $\dfrac{x+9}{x^2+8x-9}$
4. $\dfrac{x^2+9x+20}{x+5}$
5. $\dfrac{x^2-x-56}{4x+28}$
6. $\dfrac{6x+42}{x^2+5x-14}$
7. $\dfrac{4x+28}{x^2-49}$
8. $\dfrac{9x-36}{x^2-16}$
9. $\dfrac{x^5+5x^4}{x^2+x-20}$
10. $\dfrac{x^2+2x-48}{x^8+8x^7}$
11. $\dfrac{6x^5-60x^4}{x^2-8x-20}$
12. $\dfrac{x^2-14x+33}{5x^4-15x^3}$
13. $\dfrac{7x-35}{4x^2-36x+80}$
14. $\dfrac{6x^2+30x-84}{9x+63}$
15. $\dfrac{6x^3+90x^2+264x}{8x^4+32x^3}$
16. $\dfrac{10x^2+30x-400}{15x^2-75x}$
17. $\dfrac{x^2-x-12}{x^2-3x-4}$
18. $\dfrac{x^2-3x-10}{x^2-4x-12}$
19. $\dfrac{x^2-3x-18}{x^2-8x+12}$
20. $\dfrac{x^2-11x+28}{x^2-13x+36}$
21. $\dfrac{x^2-15x+44}{x^2-17x+66}$
22. $\dfrac{x^2-23x+120}{x^2-17x+72}$
23. $\dfrac{x^2-20x+99}{x^2-21x+108}$
24. $\dfrac{x^2+22x+105}{x^2+7x-120}$
25. $\dfrac{x^2+x-20}{x^2-16}$
26. $\dfrac{x^2-49}{x^2+5x-14}$
27. $\dfrac{x^2-25}{x^2-11x+30}$
28. $\dfrac{x^2-13x+22}{x^2-121}$
29. $\dfrac{3x^2-18x-48}{4x^2+4x-8}$
30. $\dfrac{6x^2+18x-168}{2x^2+20x+42}$
31. $\dfrac{8x^2-8x-336}{12x^4+12x^3-672x^2}$
32. $\dfrac{18x^6-108x^5-486x^4}{4x^9-8x^8-60x^7}$

33. $\dfrac{3x^2 - 192}{5x^2 - 10x - 400}$ 34. $\dfrac{7x^2 + 70x - 525}{6x^2 - 150}$

35. $\dfrac{14x^3 - 56x}{7x^4 + 63x^3 + 98x^2}$ 36. $\dfrac{8x^2 + 80x - 448}{16x^5 - 256x^3}$

17.4 PRODUCTS AND QUOTIENTS OF MULTINOMIAL FRACTIONS

7. Simplify

a. $\dfrac{2x^3y^2}{9x^2y^7} \cdot \dfrac{6y^8}{8x^6y^2}$

b. $\dfrac{8x^2y^5}{6x^7y} \div \dfrac{10xy^8}{12x^5y^5}$

We will now consider products and quotients of multinomial fractions. These behave very much like the products and quotients of monomial fractions we saw in Chapter 8. By combining the methods in Chapter 8 with those in the previous sections we can simplify any product or quotient of fractions to a single reduced fraction. (Remember that dividing one fraction by a second is equivalent to multiplying the first fraction by the reciprocal of the second fraction.)

Look at the following example of *simplifying products and quotients of polynomial fractions.*

TYPICAL PROBLEM

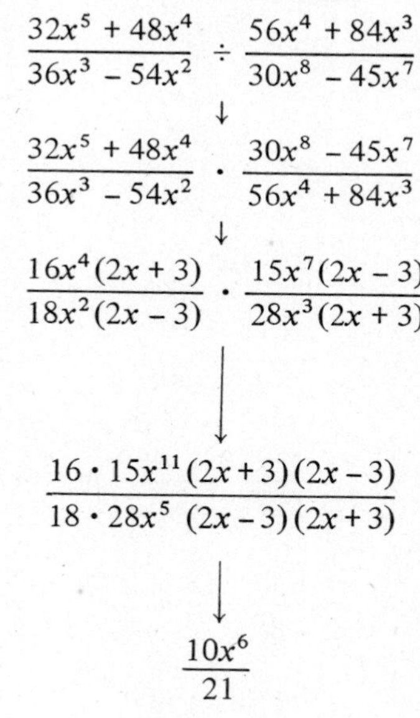

$$\frac{32x^5 + 48x^4}{36x^3 - 54x^2} \div \frac{56x^4 + 84x^3}{30x^8 - 45x^7}$$

We first invert the fraction after the division sign and change the operation to multiplication:

$$\frac{32x^5 + 48x^4}{36x^3 - 54x^2} \cdot \frac{30x^8 - 45x^7}{56x^4 + 84x^3}$$

Then we completely factor each numerator and denominator:

$$\frac{16x^4(2x+3)}{18x^2(2x-3)} \cdot \frac{15x^7(2x-3)}{28x^3(2x+3)}$$

We then form a single fraction. We write integers over integers, like powers over like powers, and binomials over binomials:

$$\frac{16 \cdot 15x^{11}(2x+3)(2x-3)}{18 \cdot 28x^5\,(2x-3)(2x+3)}$$

Finally we reduce numbers with numbers, like powers with like powers, and binomials with binomials:

$$\frac{10x^6}{21}$$

Hence, $\dfrac{32x^5 + 48x^4}{36x^3 - 54x^2} \div \dfrac{56x^4 + 84x^3}{30x^8 - 45x^7} = \dfrac{10x^6}{21}$

To simplify products and quotients of multinomial fractions we parallel the methods in Sections 8.4 and 8.5. To reduce the single fraction that we form we use the methods in the previous two sections.

EXAMPLES

Every product or quotient of algebraic fractions can be simplified to a single reduced fraction.

Simplify the following products and quotients of polynomial fractions.

1. $\dfrac{14x^2}{24x^2 - 64x} \cdot \dfrac{36x^6 - 96x^5}{21x^7}$ 2. $\dfrac{40x^4 + 12x^3}{48x^2 + 18x} \div \dfrac{28x^6}{56x^3 + 21x^2}$

3. $\dfrac{6x^3y^3 - 4x^4y^2}{27x^3y^7 - 18x^2y^8} \div \dfrac{8x^6y^2}{18xy^8 - 12y^9}$

4. $\dfrac{16x^3y^5 - 56x^2y^5}{12x^8y + 42x^7y} \cdot \dfrac{24x^6y^5 + 84x^5y^5}{20xy^9 - 30xy^8}$

5. $\dfrac{x^2 + 4x - 32}{x^2 - 13x - 48} \cdot \dfrac{x^2 - 20x + 64}{x + 8}$

6. $\dfrac{x^2 + x - 156}{x^2 + 6x - 91} \div \dfrac{x^2 - 17x + 60}{x^2 - 49}$

Solutions

1. $\dfrac{14x^2}{8x(3x - 8)} \cdot \dfrac{12x^5(3x - 8)}{21x^7}$

$\downarrow$

$\dfrac{14 \cdot 12x^7(3x - 8)}{8 \cdot 21x^8(3x - 8)}$

$\downarrow$

$\dfrac{1}{x}$

2. $\dfrac{40x^4 + 12x^3}{48x^2 + 18x} \cdot \dfrac{56x^3 + 21x^2}{28x^6}$

$\downarrow$

$\dfrac{4x^3(10x + 3)}{6x(8x + 3)} \cdot \dfrac{7x^2(8x + 3)}{28x^6}$

$\downarrow$

$\dfrac{4 \cdot 7x^5(10x + 3)(8x + 3)}{6 \cdot 28x^7(8x + 3)}$

$\downarrow$

$\dfrac{10x + 3}{6x^2}$

3. $\dfrac{6x^3y^3 - 4x^4y^2}{27x^3y^7 - 18x^2y^8} \cdot \dfrac{18xy^8 - 12y^9}{8x^6y^2}$

$\downarrow$

$\dfrac{2x^3y^2(3y - 2x)}{9x^2y^7(3x - 2y)} \cdot \dfrac{6y^8(3x - 2y)}{8x^6y^2}$

$\downarrow$

$\dfrac{2 \cdot 6x^3y^{10}(3y - 2x)(3x - 2y)}{9 \cdot 8x^8y^9(3x - 2y)}$

$\downarrow$

$\dfrac{y(3y - 2x)}{x^5}$

or

$\dfrac{3y^2 - 2xy}{6x^5}$

4. $\dfrac{8x^2y^5(2x - 7)}{6x^7y(2x + 7)} \cdot \dfrac{12x^5y^5(2x + 7)}{10xy^8(2y - 3)}$

$\downarrow$

$\dfrac{8 \cdot 12x^7y^{10}(2x - 7)(2x + 7)}{6 \cdot 10x^8y^9(2x + 7)(2y - 3)}$

$\downarrow$

$\dfrac{8y(2x - 7)}{5x(2y - 3)}$

or

$\dfrac{16xy - 56y}{10xy - 15x}$

5. $\dfrac{(x - 4)(x + 8)}{(x + 3)(x - 16)} \cdot \dfrac{(x - 4)(x - 16)}{(x + 8)}$

$\downarrow$

$\dfrac{(x - 4)(x - 4)(x + 8)(x - 16)}{(x + 3)(x + 8)(x - 16)}$

$\downarrow$

$\dfrac{(x - 4)(x - 4)}{x + 3}$

or

$\dfrac{x^2 - 8x + 16}{x + 3}$

6. $\dfrac{x^2 + x - 156}{x^2 + 6x - 91} \cdot \dfrac{x^2 - 49}{x^2 - 17x + 60}$

$\downarrow$

$\dfrac{(x - 12)(x + 13)}{(x - 7)(x + 13)} \cdot \dfrac{(x + 7)(x - 7)}{(x - 5)(x - 12)}$

$\downarrow$

$\dfrac{(x + 7)(x - 7)(x - 12)(x + 13)}{(x - 5)(x - 7)(x - 12)(x + 13)}$

$\downarrow$

$\dfrac{x + 7}{x - 5}$

EXERCISES

Simplify each of the following expressions to a reduced fraction.

1. $\dfrac{6x^3}{3x^2 - 12x} \cdot \dfrac{7x - 28}{40x^2}$

2. $\dfrac{8x^4}{3x^3 - 15x^2} \cdot \dfrac{8x - 40}{28x^2}$

3. $\dfrac{12x^3 + 28x^2}{8x^4} \cdot \dfrac{18x^2}{9x^5 + 21x^4}$

4. $\dfrac{28x^4 - 20x^3}{20x^3} \cdot \dfrac{18x}{63x^2 - 45x}$

5. $\dfrac{16x^2 + 48x}{36x^7} \div \dfrac{32x^4 + 144x^3}{81x^2}$

6. $\dfrac{28x^4 - 84x^3}{32x^2} \div \dfrac{42x^2 - 98x}{56x^8}$

7. $\dfrac{15x - 10}{35x^4} \cdot \dfrac{42x + 54}{18x^3 - 12x^2}$

8. $\dfrac{45x^4 + 30x^3}{80x^2 + 60x} \cdot \dfrac{35x^4}{63x^9 + 42x^8}$

9. $\dfrac{147x^9 + 42x^8}{18x^3} \div \dfrac{28x^3 + 8x^2}{40x^2 - 16x}$

10. $\dfrac{70x^5 - 84x^4}{15x^7} \div \dfrac{40x^2 - 48x}{75x^3 - 100x^2}$

11. $\dfrac{52x^7 - 78x^6}{72x^2 - 36x} \div \dfrac{39x^4}{16x^3 - 8x^2}$

12. $\dfrac{44x^5 + 154x^4}{72x^4 + 180x^3} \div \dfrac{55x^3}{60x^2 + 150x}$

13. $\dfrac{12x^3 - 33x^2}{165x^7 - 75x^6} \cdot \dfrac{15x^9 - 55x^8}{32x^6 - 88x^5}$

14. $\dfrac{45x^2 - 10x}{36x - 16} \cdot \dfrac{42x^4 - 12x^3}{63x^4 - 14x^3}$

15. $\dfrac{15x - 3}{32x^2 - 56x} \cdot \dfrac{24x^3 - 42x^2}{25x^5 - 5x^4}$

16. $\dfrac{5x - 20}{28x^3 - 35x^2} \cdot \dfrac{24x^7 - 30x^6}{4x^4 - 16x^3}$

17. $\dfrac{20x^3 + 32x^2}{27x^6 + 72x^5} \div \dfrac{18x^2 + 42x}{21x^3 + 56x^2}$

18. $\dfrac{28x^3 + 35x^2}{48x^2 + 36x} \div \dfrac{28x + 42}{20x^5 + 15x^4}$

19. $\dfrac{12x^2y^2 - 20x^2y}{45x^6y - 60x^6} \cdot \dfrac{15x^2y^4 - 20x^2y^3}{6xy^4}$

20. $\dfrac{27x^4y^4 + 90x^3y^5}{36xy^7 + 120x^2y^6} \cdot \dfrac{44y^5}{45x^5 + 150x^4y}$

21. $\dfrac{11x^5y}{48x^4y^4 - 108x^3y^4} \div \dfrac{88x^6y^5 - 198x^5y^5}{32x^2y^3 - 72xy^3}$

22. $\dfrac{5x^6y}{60x^3y^3 - 48x^2y^3} \div \dfrac{75x^2y^4 - 60xy^4}{45x^7y^2 - 36x^6y^2}$

23. $\dfrac{72xy^4 + 16y^5}{54x^6y^7 + 12x^5y^6} \cdot \dfrac{63x^3y^2 + 18x^2y^3}{27x^4y^3 + 6x^3y^4}$

24. $\dfrac{63x^9y^2 - 7x^8y^3}{12x^6y + 44x^5y^2} \cdot \dfrac{15x^5y^6 + 55x^4y^7}{27xy^7 - 3y^8}$

25. $\dfrac{x^2 + 3x - 28}{x + 9} \div \dfrac{x - 4}{x^2 + 16x + 63}$

26. $\dfrac{x^2 - 19x + 88}{x + 6} \div \dfrac{x - 11}{x^2 - 2x - 48}$

27. $\dfrac{x - 6}{x^2 + 24x - 52} \cdot \dfrac{x - 2}{x^2 - 24x + 108}$

28. $\dfrac{x - 3}{x^2 + 5x - 150} \cdot \dfrac{x - 10}{x^2 - 27x + 72}$

29. $\dfrac{x^2 - 22x + 112}{x^2 + 10x - 56} \cdot \dfrac{x^2 - 8x + 16}{x - 14}$

30. $\dfrac{x^2 - 22x + 72}{x + 16} \cdot \dfrac{x^2 + 10x - 96}{x^2 - 12x - 108}$

31. $\dfrac{x + 7}{x^2 - 20x + 99} \div \dfrac{x^2 + 19x + 84}{x^2 - 21x + 108}$

32. $\dfrac{x + 3}{x^2 - 28x - 128} \div \dfrac{x^2 - 21x - 72}{x^2 + 28x + 96}$

33. $\dfrac{x^2 - 17x + 60}{x^2 - 24x + 135} \cdot \dfrac{x^2 - 2x - 63}{x^2 - 5x - 84}$

34. $\dfrac{x^2 - 21x + 104}{x^2 - 24x + 80} \cdot \dfrac{x^2 - 14x - 120}{x^2 - 7x - 78}$

35. $\dfrac{x^2 - 5x - 36}{x^2 + 14x + 49} \div \dfrac{x^2 - 23x + 126}{x^2 - 49}$

36. $\dfrac{x^2 - 2x - 120}{x^2 + 9x - 90} \div \dfrac{x^2 - 7x - 60}{x^2 - 225}$

17.5 LONG DIVISION

We have considered the problem of reducing ratios and quotients of polynomials to lowest terms. Now we will take a different approach to simplifying quotients of polynomials with a different goal in mind. We will look at the quotient of a polynomial in x divided by a linear binomial in x, such as $(4x^2 + 4x - 9) \div (2x - 1)$. Our goal will be to express this quotient of polynomials as the sum of another polynomial and a fraction with a numerical numerator. For example, we will express

$$(4x^2 + 4x - 9) \div (2x - 1) \quad \text{as} \quad 2x + 3 + \frac{-6}{2x - 1}.$$

$$\text{Original Polynomial} \div \text{Binomial} = \text{Second Polynomial} + \frac{\text{Integer}}{\text{Binomial}}$$

To accomplish this we will use a procedure called *long division*, which is an extension of arithmetic long division. The long-division format for $(4x^2 + 4x - 9) \div (2x - 1)$, for example, is

8. Compute by long division.
a. $602 \div 7$
b. $324 \div 12$
c. $506 \div 11$

$$2x - 1 \overline{\big) 4x^2 + 4x - 9}$$

binomial — $2x - 1$; polynomial — $4x^2 + 4x - 9$

The quotient $(4x^2 + 4x - 9) \div (2x - 1)$ equals the sum $2x + 3 + \dfrac{-6}{2x - 1}$, and this sum, obtained by long division, is also called the *quotient*. (That is, when we compute a quotient using long division, we also call the result of the computation a quotient.) Each term in this sum is called a *partial quotient*, and the numerator of the fraction is called the *remainder.*

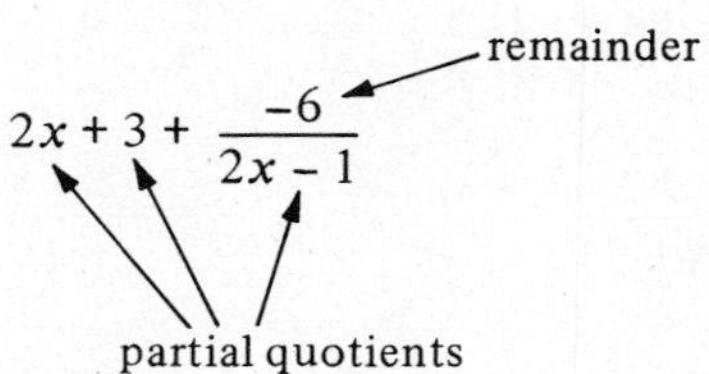

Long division is a three-step cycle of division, multiplication, and subtraction. In the subtraction step, we compute the difference of two polynomials. To subtract a polynomial we form its negative (by changing the sign of each term) and then add.

Look at the following table to see some examples of subtracting one polynomial from another.

	SUBTRACT THIS						
subtract → from ↓	$-5x - 3$	$-5x - 2$	$-5x - 1$	$-5x$	$-5x + 1$	$-5x + 2$	$-5x + 3$
FROM THIS $-5x + 3$							
$-5x + 2$		4	3	2	1	0	
$-5x + 1$		3	2	1	0	−1	
$-5x$		2	1	0	−1	−2	
$-5x - 1$		1	0	−1	−2	−3	
$-5x - 2$		0	−1	−2	−3	−4	
$-5x - 3$							

9. Complete the table.

Look at the following example of *dividing a polynomial by a binomial.*

TYPICAL PROBLEM

$(4x^2 + 4x - 9) \div (2x - 1)$

First we change to the long-division format:

$$2x - 1 \overline{)4x^2 + 4x - 9}$$

Then we divide the x^2 term of the polynomial by the x term of the divisor to get the first partial quotient:

$$\frac{4x^2}{2x} \to 2x$$

We write this partial quotient above the line:

$$\begin{array}{r} 2x \\ 2x - 1 \overline{)4x^2 + 4x - 9} \end{array}$$

Then we multiply the binomial by this partial quotient:

$$2x(2x - 1) \to 4x^2 - 2x$$

And we subtract this product from the polynomial. (To subtract this product we form its negative and add.):

$$\begin{array}{r} 2x \\ 2x - 1 \overline{)4x^2 + 4x - 9} \\ \underline{\overset{-}{\oplus} 4x^2 \overset{+}{\ominus} 2x} \\ 6x - 9 \end{array}$$

Then we divide the x term of this difference by the x term of the binomial to get the next partial quotient:

$$\frac{6x}{2x} \to 3$$

Then we multiply the binomial by the second partial quotient:

$$3(2x - 1) \to 6x - 3$$

And we subtract this product from the previous difference. (We form its negative and add.):

$$\begin{array}{r} 2x + 3 \\ 2x - 1 \overline{)4x^2 + 4x - 9} \\ \underline{-4x^2 + 2x} \\ 6x - 9 \\ \underline{\overset{-}{\oplus} 6x \overset{+}{\ominus} 3} \\ -6 \end{array}$$

Then if this new difference is not zero, we form a fraction by placing it over the binomial and add this fraction to the other partial quotients:

$$2x + 3 + \frac{-6}{2x - 1}$$

Hence, $(4x^2 + 4x - 9) \div (2x - 1) = 2x + 3 + \dfrac{-6}{2x - 1}$

The above quotient can also be written as $2x + 3$, R(−6), where R stands for *remainder.*

To determine each partial quotient we divide using the method in Section 8.3.

EXAMPLES

Simplify the following quotients using long division.

1. $(x^2 + 3x - 9) \div (x + 5)$

2. $(4x^2 - 57) \div (x - 6)$

Every quotient of a multinomial in x divided by a linear binomial in x can be simplified, using long division, to the sum of a polynomial and a fraction with a numerical numerator.

3. $(4x^2 - 8x - 5) \div (2x + 1)$

4. $(8x^2 + 4x - 28) \div (2x - 3)$

5. $(x^3 - 13x + 11) \div (x + 4)$

6. $(6x^3 - 7x^2 + 4x) \div (2x - 1)$

Solutions

1.
$$x + 5 \overline{\big)\, x^2 + 3x - 9}\quad\text{quotient: } x - 2$$
$$\overset{-}{\oplus} x^2 \overset{-}{\oplus} 5x$$
$$-2x - 9$$
$$\overset{+}{\ominus} 2x \overset{+}{\ominus} 10$$
$$1$$
$$\downarrow$$
$$x - 2 + \frac{1}{x + 5}$$
or
$$x - 2,\ \mathrm{R}(1)$$

2.
$$x - 6 \overline{\big)\, 4x^2 \qquad - 57}\quad\text{quotient: } 4x + 24$$
$$\overset{-}{\oplus} 4x^2 \overset{+}{\ominus} 24x$$
$$24x - 57$$
$$\overset{-}{\oplus} 24x \overset{+}{\ominus} 144$$
$$87$$
$$\downarrow$$
$$4x + 24 + \frac{87}{x - 6}$$
or
$$4x + 24,\ \mathrm{R}(87)$$

3.
$$2x + 1 \overline{\big)\, 4x^2 - 8x - 5}\quad\text{quotient: } 2x - 5$$
$$\overset{-}{\oplus} 4x^2 \overset{-}{\oplus} 2x$$
$$-10x - 5$$
$$\overset{+}{\ominus} 10x \overset{+}{\ominus} 5$$
$$0$$
$$\downarrow$$
$$2x - 5$$

4.
$$2x - 3 \overline{\big)\, 8x^2 + 4x - 28}\quad\text{quotient: } 4x + 8$$
$$\overset{-}{\oplus} 8x^2 \overset{+}{\ominus} 12x$$
$$16x - 28$$
$$\overset{-}{\oplus} 16x \overset{+}{\ominus} 24$$
$$-4$$
$$\downarrow$$
$$4x + 8 + \frac{-4}{2x + 3}$$
or
$$4x + 8,\ \mathrm{R}(-4)$$

5.
$$x + 4 \overline{\big)\, x^3 \qquad - 13x + 11}\quad\text{quotient: } x^2 - 4x + 3$$
$$\overset{-}{\oplus} x^3 \overset{-}{\oplus} 4x^2$$
$$-4x^2 - 13x + 11$$
$$\overset{+}{\ominus} 4x^2 \overset{+}{\ominus} 16x$$
$$3x + 11$$
$$\overset{-}{\oplus} 3x \overset{-}{\oplus} 12$$
$$-1$$
$$\downarrow$$
$$x^2 - 4x + 3 + \frac{-1}{x + 4}$$
or
$$x^2 - 4x + 3,\ \mathrm{R}(-1)$$

6.
$$2x - 1 \overline{\big)\, 6x^3 - 7x^2 + 4x}\quad\text{quotient: } 3x^2 - 2x + 1$$
$$\overset{-}{\oplus} 6x^3 \overset{+}{\ominus} 3x^2$$
$$-4x^2 + 4x$$
$$\overset{+}{\ominus} 4x^2 \overset{-}{\oplus} 2x$$
$$2x$$
$$\overset{-}{\oplus} 2x \overset{+}{\ominus} 1$$
$$1$$
$$\downarrow$$
$$3x^2 - 2x + 1 + \frac{1}{2x - 1}$$
or
$$3x^2 - 2x + 1,\ \mathrm{R}(1)$$

SPECIAL CASES

Remember that only like terms can be combined, and different powers of x are unlike terms. Hence, in a polynomial such as $x^3 - 7x + 5$, which has a missing x^2 term, we should leave space for this missing term when setting up our long-division format. For example, $(x^3 - 7x + 5) \div (x - 8)$ should be set up as $x - 8\overline{)x^3 \qquad - 7x + 5}$ rather than as $x - 8\overline{)x^3 - 7x + 5}$.

10. Rewrite each of these quotients, leaving space where needed.

a. $x - 2\overline{)x^2 + 1}$

b. $x + 5\overline{)x^3 - 4x^2 + 6}$

c. $x - 1\overline{)x^3 - 1}$

Checking

When we perform long division, we can make computational errors and arrive at the wrong quotient. We can check the accuracy of a computation by multiplying. For example, we can check example 1, $(x^2 + 3x - 9) \div (x + 5) = x - 2 + \frac{1}{x+5}$, as follows:

$$(x^2 + 3x - 9) \div (x + 5) = x - 2 + \frac{1}{x+5}$$

$$(x^2 + 3x - 9) \stackrel{?}{=} (x + 5)\left(x - 2 + \frac{1}{x+5}\right)$$

$$\stackrel{?}{=} (x + 5)(x - 2) + (x + 5)\left(\frac{1}{x+5}\right)$$

$$\stackrel{?}{=} x^2 + 3x - 10 + 1$$

$$= x^2 + 3x - 9$$

Because we get the original polynomial back, our quotient is correct.

11. Compute

a. $(x - 4)\left(x + 9 + \frac{32}{x-4}\right)$

b. $(2x + 3)\left(4x + \frac{-9}{2x+3}\right)$

OBSERVATIONS AND REFLECTIONS

1. Algebraic long division is similar to arithmetic long division. Compare the following two long divisions:

$$\begin{array}{r} 3t + 2 + \frac{5}{t+2} \\ t + 2\overline{)3t^2 + 8t + 9} \\ \underline{-3t^2 - 6t} \\ 2t + 9 \\ \underline{-2t - 4} \\ 5 \end{array} \quad \text{and} \quad \begin{array}{r} 32 + \frac{5}{12} \\ 12\overline{)389} \\ \underline{-36} \\ 29 \\ \underline{-24} \\ 5 \end{array}$$

Note that the numerical coefficients in the computation on the left are exactly the same as the numerals in the computation on the right. These two computations are equivalent when $t = 10$.

12. Compute

a. $17\overline{)189}$

b. $t + 7\overline{)t^2 + 8t + 9}$

c. $11\overline{)685}$

d. $t + 1\overline{)6t^2 + 8t + 5}$

e. $32\overline{)677}$

f. $3t + 2\overline{)6t^2 + 7t + 7}$

13. Which of these polynomials are multiples of $x - 3$?
a. $3x^2 - 2x - 21$
b. $2x^2 + 11x + 15$
c. $8x^2 - 25x + 3$

14. Which of these polynomials are multiples of $5x + 2$?
a. $10x^2 - x - 2$
b. $20x^2 + 23x + 6$
c. $35x^2 - x + 6$

★ 2. Consider the binomial $ax + b$, where a and b represent integers. We can determine whether a given polynomial in x is a multiple of this binomial $ax + b$ by long division. If the polynomial is in fact a multiple of $ax + b$, we will get a remainder of zero when we divide the polynomial by $ax + b$. For example, we can demonstrate that $2x^2 - x - 6$ is a multiple of $x - 2$ by performing the computation

$$(2x^2 - x - 6) \div (x - 2) \rightarrow \begin{array}{r} 2x + 3 \\ x - 2 \overline{\smash{)}\, 2x^2 - x - 6} \\ -\, 2x^2 + 4x \\ \hline 3x - 6 \\ -\, 3x + 6 \\ \hline 0 \end{array}$$, with a remainder of zero.

Hence, $(x - 2)(2x + 3) = 2x^2 - x - 6$, and $2x^2 - x - 6$ is a multiple of $x - 2$.

EXERCISES

Simplify each of the following expressions by long division.

1. $(x^2 + 6x - 5) \div (x - 3)$
2. $(x^2 + 4x - 1) \div (x - 2)$
3. $(x^2 - 5x - 6) \div (x + 4)$
4. $(x^2 + 6x + 7) \div (x + 3)$
5. $(3x^2 - 5x + 1) \div (x - 4)$
6. $(3x^2 - 16x + 16) \div (x - 4)$
7. $(6x^2 + 15x - 17) \div (x + 4)$
8. $(4x^2 + 11x - 14) \div (x + 7)$
9. $(3x^2 + 21x - 4) \div (x + 7)$
10. $(2x^2 + 18x - 45) \div (x + 10)$
11. $(11x^2 + 40x - 26) \div (x + 4)$
12. $(3x^2 - 8x - 4) \div (x - 8)$
13. $(x^2 + 5) \div (x + 5)$
14. $(x^2 + 4) \div (x - 6)$
15. $(x^2 - 9) \div (x + 3)$
16. $(x^2 + 4x) \div (x + 4)$
17. $(3x^2 - 7) \div (x - 4)$
18. $(7x^2 + 3x) \div (x - 3)$
19. $(6x^2 - 24) \div (x + 2)$
20. $(3x^2 - 75) \div (x - 5)$
21. $(x^3 - 3x^2 + 4) \div (x - 1)$
22. $(x^3 - 9x^2 + 4x) \div (x - 3)$
23. $(x^3 - 4x + 8) \div (x - 4)$
24. $(x^3 - 3x + 9) \div (x - 3)$
25. $(4x^2 + 4x - 7) \div (2x - 1)$
26. $(6x^2 - 11x - 11) \div (3x - 1)$
27. $(12x^2 + 11x + 18) \div (4x + 1)$
28. $(6x^2 - 19x + 19) \div (3x - 2)$
29. $(6x^2 - 19x + 4) \div (2x - 3)$
30. $(6x^2 - x + 24) \div (2x + 1)$
31. $(8x^2 + 15x - 9) \div (8x - 1)$
32. $(9x^2 - 9x + 1) \div (3x + 2)$
33. $(12x^2 + x - 4) \div (3x - 5)$
34. $(12x^2 + 13x + 12) \div (4x + 3)$
35. $(12x^2 - 17x + 1) \div (4x - 3)$
36. $(99x^2 - 23x - 1) \div (9x + 2)$

37. $(4x^2 - 7) \div (2x - 1)$

38. $(12x^2 + 5) \div (2x + 5)$

39. $(18x^2 + 4) \div (3x + 1)$

40. $(32x^2 - 16) \div (4x + 3)$

41. $(6x^2 + 5x) \div (2x - 3)$

42. $(12x^2 + 2x) \div (3x - 1)$

43. $(8x^3 - 15) \div (2x + 5)$

44. $(25x^3 + x - 4) \div (5x - 3)$

45. $(16x^3 - 9x + 13) \div (4x - 7)$

46. $(20x^3 - 25x - 4) \div (2x + 3)$

47. $(15x^3 + 31x^2 + 7) \div (5x + 2)$

48. $(12x^3 + 11x^2 - 18) \div (3x - 4)$

ANSWERS TO MARGIN EXERCISES

1. a. $3x$
b. $\frac{x}{4}$
c. $\frac{2}{3x^4y^4}$
d. $\frac{8y^2}{x}$

2. b and d

3. a. $\frac{5}{3x} - 2x^2$
b. $\frac{12}{5} - \frac{3}{x}$
c. $\frac{18x}{7} - \frac{7}{3x^2}$
d. $\frac{4x^2}{3} + \frac{3}{4}$

4. $\frac{3}{5x}$

5. a. $\frac{9x^2}{9x^2 - 5y}$
b. $\frac{2x}{9y}$

6. a. $\frac{-9x^2}{9x^2 - 5y}$
b. $\frac{-2x}{9y}$

7. a. $\frac{y}{6x^5}$
b. $\frac{8y}{5x}$

8. a. 86
b. 27
c. 46

10. a. $x - 2\overline{)x^2 \quad + 1}$
b. $x + 5\overline{)x^3 - 4x^2 \quad + 6}$
c. $x - 1\overline{)x^3 \quad - 1}$

11. a. $x^2 + 5x - 4$
b. $8x^2 + 12x - 9$

12. a. $11 + \frac{2}{17}$
b. $t + 1 + \frac{2}{t + 7}$
c. $62 + \frac{3}{11}$
d. $6t + 2 + \frac{3}{t + 1}$
e. $21 + \frac{5}{32}$
f. $2t + 1 + \frac{5}{3t + 2}$

13. a and c

14. a and b

CHAPTER REVIEW

SECTION 17.1 Reduce to lowest terms.

1. $\dfrac{14x + 21}{28}$
2. $\dfrac{-18x}{30x - 60}$
3. $\dfrac{6x^5}{4x^3 - 8x^2}$
4. $\dfrac{9x^5}{12x^3 - 6x^2}$
5. $\dfrac{-4x^3 + 18x^4}{6x^2}$
6. $\dfrac{18x^2y^4}{36x^3y^2 + 45x^4y}$
7. $\dfrac{36x^4y^3 + 60x^5y^2}{28x^2y^6}$
8. $\dfrac{x^2y^3}{5x^4y - 3x^3y^2}$
9. $\dfrac{2x^6y^4 + 9x^5y^3}{x^6y^2}$

SECTION 17.2 Reduce to lowest terms.

10. $\dfrac{18x - 54}{36x - 108}$
11. $\dfrac{x^9 + 6x^8}{6x^4 + 24x^3}$
12. $\dfrac{36x^{12} + 63x^{11}}{105x^3 + 60x^4}$
13. $\dfrac{9x^8y^2 - 15x^8y}{-15x^3y^5 + 9x^3y^6}$
14. $\dfrac{8x^{10} - 14x^9}{35x^2 + 20x^3}$
15. $\dfrac{35x^2 + 63x}{36x^8 + 20x^9}$
16. $\dfrac{18x^4y^3 - 30x^3y^4}{-30x^7y^8 + 18x^8y^7}$
17. $\dfrac{60x^2y^9 - 24xy^9}{40x^7y^5 - 16x^6y^5}$
18. $\dfrac{48x^7y^5 - 80x^7y^4}{144x^{10}y^2 - 32x^9y^2}$
19. $\dfrac{18 - 18x}{24x - 24}$
20. $\dfrac{16x^6 - 9x^5}{9x^2 - 16x^3}$
21. $\dfrac{18x^6y - 9x^7y^2}{18x^8y - 36x^9}$
22. $\dfrac{24x^5y - 36x^4y^2}{144y^7 - 96xy^6}$
23. $\dfrac{54xy^4 - 6y^5}{4xy^4 - 36x^2y^3}$
24. $\dfrac{36x^6 - 27x^5y}{18y^6 + 24xy^5}$

SECTION 17.3 Reduce to lowest terms.

25. $\dfrac{x^2 + x - 20}{x + 5}$
26. $\dfrac{x - 6}{x^2 - 3x - 18}$
27. $\dfrac{x^2 - 2x - 15}{8x + 24}$
28. $\dfrac{x^2 - 10x + 21}{6x - 18}$
29. $\dfrac{3x^3 - 21x^2}{x^2 - 49}$
30. $\dfrac{4x^9 + 48x^8}{x^2 + 20x + 96}$
31. $\dfrac{15x + 135}{12x^3 + 72x^2 - 324x}$
32. $\dfrac{4x^4 + 24x^3}{4x^4 + 64x^3 + 240x^2}$
33. $\dfrac{x^2 + 11x + 28}{x^2 + 15x + 56}$
34. $\dfrac{x^2 - 9x + 8}{x^2 - 2x - 48}$
35. $\dfrac{x^2 - 8x - 84}{x^2 + 13x + 42}$
36. $\dfrac{x^2 + 22x + 112}{x^2 + 24x + 140}$
37. $\dfrac{x^2 + 5x - 24}{x^2 - 9}$
38. $\dfrac{x^2 - 64}{x^2 - 16x + 64}$
39. $\dfrac{4x^3 + 28x^2 - 72x}{6x^2 + 78x + 216}$
40. $\dfrac{3x^4 + 15x^3 - 18x^2}{12x^2 - 72x - 864}$
41. $\dfrac{8x^2 - 648}{16x^5 - 80x^4 - 576x^3}$
42. $\dfrac{15x^2 - 735}{18x^4 + 270x^3 + 1008x^2}$

SECTION 17.4 Simplify to a reduced fraction.

43. $\dfrac{75x^6}{15x^3 - 120x^2} \cdot \dfrac{4x - 32}{40x^2}$
44. $\dfrac{20x^6 - 45x^5}{75x} \cdot \dfrac{24x^2}{24x^2 - 54x}$

45. $\dfrac{36x^6 + 84x^5}{99x^3} \div \dfrac{48x^3 + 120x^2}{18}$

46. $\dfrac{36x^4 - 126x^3}{36x^4 - 54x^3} \cdot \dfrac{21x}{28x^7 - 98x^6}$

47. $\dfrac{160x^4 + 60x^3}{45x^9} \div \dfrac{128x^2 + 48x}{90x^5 + 54x^4}$

48. $\dfrac{51x^4 - 102x^3}{48x^2 - 120x} \div \dfrac{51x^5}{18x^3 - 45x^2}$

49. $\dfrac{35x^3 - 20x^2}{36x^5 - 48x^4} \cdot \dfrac{56x^2 - 48x}{21x^6 - 12x^5}$

50. $\dfrac{66x - 11}{15x^3 - 40x^2} \cdot \dfrac{27x^5 - 72x^4}{48x^4 - 8x^3}$

51. $\dfrac{45x^2 + 75x}{16x^5 + 40x^4} \div \dfrac{50x^7 + 75x^6}{6x^3 + 15x^2}$

52. $\dfrac{15x^3y^2 - 42x^3y}{24x^8y - 42x^8} \cdot \dfrac{32x^3y^2 - 56x^3y}{10xy^3}$

53. $\dfrac{7x^4y}{70x^2y^4 - 30xy^4} \div \dfrac{147x^2y^5 - 63xy^5}{42x^8y^3 - 18x^7y^3}$

54. $\dfrac{30xy^4 + 80y^5}{36x^5y^2 + 48x^4y} \cdot \dfrac{108x^6y^4 + 81x^5y^5}{27x^6y^2 + 72x^5y^3}$

55. $\dfrac{x^2 + 4x - 45}{x + 12} \div \dfrac{x - 5}{x^2 + 21x + 108}$

56. $\dfrac{x - 7}{x^2 + 26x - 120} \cdot \dfrac{x - 4}{x^2 - 19x + 84}$

57. $\dfrac{x^2 - 20x + 96}{x^2 + 6x - 72} \cdot \dfrac{x^2 - 12x + 36}{x - 12}$

58. $\dfrac{x + 8}{x^2 + 10x - 96} \div \dfrac{x^2 + 22x + 112}{x^2 - 20x + 84}$

59. $\dfrac{x^2 - 10x - 144}{x^2 + 4x - 192} \cdot \dfrac{x^2 + 13x - 48}{x^2 - 21x + 54}$

60. $\dfrac{x^2 - 26x + 144}{x^2 + 18x + 81} \div \dfrac{x^2 - 22x + 72}{x^2 - 81}$

SECTION 17.5

Simplify by long division.

61. $(x^2 - 7x - 8) \div (x - 1)$

62. $(4x^2 + 2x + 1) \div (x - 2)$

63. $(8x^2 - 2x + 7) \div (x + 3)$

64. $(4x^2 - 13x + 1) \div (x - 1)$

65. $(3x^2 - 17x - 15) \div (x - 3)$

66. $(7x^2 + 22x + 3) \div (x + 3)$

67. $(x^2 - 4x) \div (x + 2)$

68. $(x^2 - 5x) \div (x + 5)$

69. $(x^3 - 4x + 7) \div (x + 2)$

70. $(x^3 + 6x^2 - 12x) \div (x + 1)$

71. $(6x^3 - x^2 + 7) \div (x + 1)$

72. $(4x^3 - 12x^2 + 16) \div (x - 3)$

73. $(8x^2 + 10x + 7) \div (2x + 3)$

74. $(8x^2 - 26x - 15) \div (2x - 5)$

75. $(12x^2 - 5x - 12) \div (3x + 1)$

76. $(12x^2 + 29x - 17) \div (3x + 5)$

77. $(8x^2 + 26x + 13) \div (2x + 5)$

78. $(8x^2 - 10x + 5) \div (2x - 3)$

79. $(9x^2 - 1) \div (3x - 2)$

80. $(25x^2) \div (5x + 6)$

81. $(18x^2 + 15x) \div (6x + 1)$

82. $(18x^3 - 11x - 5) \div (3x - 4)$

83. $(27x^3 + 500) \div (3x + 8)$

84. $(28x^3 + 19x^2 + 25) \div (4x + 5)$

CUMULATIVE TEST

The following problems test your understanding of this chapter and Chapter 8. Before taking this test, thoroughly review Sections 8.3, 8.4, 8.5, 17.1, 17.2, 17.3, and 17.4.

Once you have finished the test, compare your answers with the answers provided at the back of the book. Note the section number of each problem missed, and thoroughly review those sections again.

Simplify.

1. $\dfrac{16x + 24}{32}$

2. $\dfrac{9x - 36}{12x - 48}$

3. $\dfrac{x - 7}{x^2 - 15x + 56}$

4. $\dfrac{24x - 44}{12x} \cdot \dfrac{9x^2}{4}$

5. $\dfrac{-6x^3}{18x^7}$

6. $\dfrac{x^2 - 8x - 48}{5x + 20}$

7. $\dfrac{x^3}{15} \div \dfrac{x^5}{45}$

8. $x^3y^2 \cdot \dfrac{x^4}{y^3}$

9. $\dfrac{24x^3y^2 - 30xy^2}{18x^4y}$

10. $\dfrac{18y^6}{64x^3} \div \dfrac{21y^8}{40x^6}$

11. $\dfrac{28x^5 - 63x^4}{35x^2} \cdot \dfrac{40x^6}{48x^2 - 108x}$

12. $\dfrac{48x^3y^4}{42xy^6}$

13. $\dfrac{16x^8 - 28x^9}{20x^3 + 32x^4}$

14. $\dfrac{80x^4 - 32x^3}{112x^3 - 32x^2} \cdot \dfrac{27x}{90x^5 - 36x^4}$

15. $\dfrac{-1}{8x^2} \cdot \dfrac{16}{x^5}$

16. $\dfrac{27x^8}{15x^3 - 6x^2}$

17. $\dfrac{8xy^3}{15z^4} \div \dfrac{x^3y}{20z^5}$

18. $\dfrac{35x^4y^2 - 63x^4y}{56x^2y - 98x^2} \cdot \dfrac{32x^3y^8 - 56x^3y^7}{48xy^6}$

19. $\dfrac{x^2 + 8x - 48}{x^2 - 16}$

20. $\dfrac{26y^4 - 39y^3}{18y^7 - 27y^6}$

21. $\dfrac{36x^4 - 84x^3}{49x^8} \div \dfrac{72x^6 - 120x^5}{154x^2}$

22. $\dfrac{x^2 - 13x - 48}{x^2 - 29x - 96}$

23. $\dfrac{x^2 - 18x - 144}{x^2 + 14x - 147} \cdot \dfrac{x^2 + 17x - 84}{x^2 - 28x + 96}$

24. $\dfrac{-26x^6y^4z^5}{-39x^8y^4z^3}$

25. $\dfrac{5x^3}{6}\left(-\dfrac{7x}{30}\right)$

26. $\dfrac{35x^3y}{6x + 25y} \cdot \dfrac{6y^3}{49x^2}$

27. $\dfrac{3x^8 - 5x^7}{6x^4 - 10x^3}$

28. $\dfrac{12x^2}{5y^4} \cdot \dfrac{18x^3}{48y^4}$

29. $\dfrac{-4x^4 + 14x^3}{12x^2}$

30. $\dfrac{36x^5y^2}{56x^3y^5 - 24x^2y^6} \div \dfrac{63x^8y^4 - 27x^7y^5}{175x^3y - 75x^2y^2}$

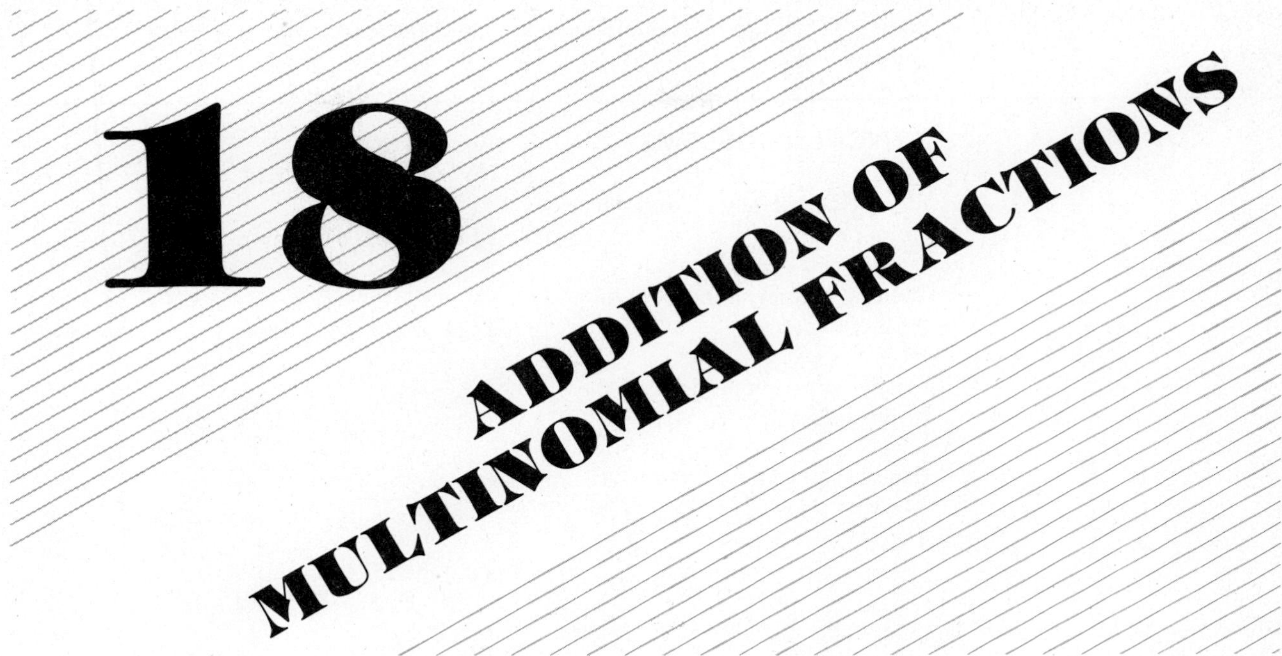

18 ADDITION OF MULTINOMIAL FRACTIONS

In this chapter we will study addition of multinomial fractions. In preparation, we must first learn how to determine the least common multiple (LCM) of two polynomials. Then we will look at sums of fractions with polynomial denominators. We will also see complex fractions and sums of complex fractions. (The multinomial denominators in this chapter can be factored using the methods in Chapters 14 and 16.)

18.1 DETERMINING THE LEAST COMMON MULTIPLE OF POLYNOMIALS

Remember that the least common multiple (LCM) of two monomials is the product of the LCM of their numerical coefficients and the LCM of their literal components. For example, the LCM of 6 and 8 is 24 and the LCM of x^2y and y^4 is x^2y^4, so the LCM of $6x^2y$ and $8y^4$ is $24x^2y^4$. Similarly, as we shall see, the LCM of two polynomials in completely factored form is the product of the LCM of their monomial factors and the LCM of their multinomial factors.

1. Determine the LCM of
   a. $16x^9$ and $24x^4$
   b. $3x^3y^2$ and $9x^6y^2$

When a polynomial is completely factored, each multinomial factor is divisible only by itself and 1 (and their negatives). To see how to determine the LCM of two completely factored products of multinomials, look at the pattern in the following tables.

LCM	$(x+1)$	$(x+1)^2$	$(x+1)^3$
$(x+1)$	$(x+1)$	$(x+1)^2$	
$(x+1)^2$	$(x+1)^2$	$(x+1)^2$	
$(x+1)^3$			

and

LCM	$(x+1)$	$(x+1)^2$	$(x+1)^3$
$(x-2)$	$(x+1)(x-2)$	$(x+1)^2(x-2)$	
$(x-2)^2$	$(x+1)(x-2)^2$	$(x+1)^2(x-2)^2$	
$(x-2)^3$			

2. Complete the tables.

Compare these tables with those in Section 9.3. Can you see the relationship between the LCM of multinomials and the LCM of literals?

Look at the following examples of determining the *least common multiple (LCM) of two polynomials.*

TYPICAL PROBLEMS

1. Find the LCM of
$42x^4y^2 - 105x^3y^2$ and $14x^3y^3 - 35x^2y^3$

First we completely factor each multinomial:

$21x^3y^2(2x - 5)$ and $7x^2y^3(2x - 5)$

Then we determine the LCM of the monomials:

LCM of $21x^3y^2$ and $7x^2y^3 = 21x^3y^3$

Finally we attach the LCM of the binomial factors:

$21x^3y^3(2x - 5)$

Hence, the LCM of $42x^4y^2 - 105x^3y^2$ and $14x^3y^3 - 35x^2y^3$ is $21x^3y^3(2x - 5)$.

2. Find the LCM of
$x^2 + 2x - 15$ and $x^2 - 10x + 21$

To factor each multinomial we use the methods in Chapters 14 and 16. To determine the LCM of the monomials we use the method in Section 9.3. When the LCM contains binomials, it is usually left in factored form.

First we completely factor each multinomial:

$(x - 3)(x + 5)$ and $(x - 3)(x - 7)$

Then we form the LCM of the binomial factors:

$(x - 3)(x + 5)(x - 7)$

Hence, the LCM of $x^2 + 2x - 15$ and $x^2 - 10x + 21$ is $(x - 3)(x + 5)(x - 7)$.

EXAMPLES

Every pair of polynomials has two LCM's that are negatives of each other.

Find the LCM of each pair of polynomials.

1. $21x^4y^2 - 45x^3y^3$ and $9x^6y^2$
2. $84x^2y^4 - 70xy^5$ and $42x^6y^3 - 49x^5y^4$
3. $12x^3 + 9x^2$ and $4x^5 + 3x^4$
4. $60x^5y^6 + 36x^4y^7$ and $225x^4y^3 + 135x^3y^4$
5. $x^2 + x - 12$ and $x^2 - 16x + 39$
6. $x^2 - 13x + 36$ and $x^2 - x - 72$

Solutions

1. $3x^3y^2(7x - 15y)$ and $9x^6y^2$
$\downarrow$
LCM $= 9x^6y^2(7x - 15y)$

2. $14xy^4(6x - 5y)$ and $7x^5y^3(6x - 7y)$
$\downarrow$
LCM $= 14x^5y^4(6x - 5y)(6x - 7y)$

3. $3x^2(4x + 3)$ and $x^4(4x + 3)$
$\downarrow$
LCM $= 3x^4(4x + 3)$

4. $12x^4y^6(5x + 3y)$ and $45x^3y^3(5x + 3y)$
$\downarrow$
LCM $= 180x^4y^6(5x + 3y)$

5. $(x + 4)(x - 3)$ and $(x - 13)(x - 3)$
$\downarrow$
LCM $= (x + 4)(x - 13)(x - 3)$

6. $(x - 9)(x - 4)$ and $(x + 8)(x - 9)$
$\downarrow$
LCM $= (x - 4)(x + 8)(x - 9)$

SPECIAL CASES

If two binomials are negatives of each other, then either may be selected to form the LCM. The negative of the LCM of two expressions is also considered the LCM. For example, the LCM of $3x(x - y)$ and $2y(y - x)$ is either $6xy(x - y)$ or $6xy(y - x)$.

3. Determine the LCM of
a. $3x - 5y$ and $5y - 3x$
b. $4x^2y(x - 4y)$ and $6xy^3(4y - x)$
c. $(2x - 7y)(y - 5x)$ and $(2y - 7x)(5x - y)$
d. $(3y + 8x)(4x - 9y)$ and $(9y - 4x)(8x + 3y)$

EXERCISES

Find the LCM (Least Common Multiple). Leave the LCM in factored form.

1. $6x^2, 3x + 2$
2. $10x^3, 2x + 5$
3. $8x^3, 12x^2 - 8x$
4. $9x^2, 18x^3 - 15x^2$
5. $15xy^2, 6x^2y - 21xy^2$
6. $16x^2y, 24x^2y + 40xy^2$
7. $6x - 15, 2x - 5$
8. $9x - 12, 3x - 4$
9. $12x^2 - 27x, 5x - 9$
10. $16x^2 + 20x, 4x + 7$
11. $48x^5 - 88x^4, 54x^3 - 99x^2$
12. $96x^6 + 36x^5, 56x^3 + 21x^2$
13. $72x^3 + 60x^2, 30x^6 - 25x^5$
14. $27x^5 + 63x^4, 24x^8 - 56x^7$
15. $6x^4y - 20x^3y^2, 15x^2y^3 - 50xy^4$
16. $72x^5y - 30x^4y^2, 84x^3y^3 - 35x^2y^4$
17. $9x^5y - 21x^4y^2, 36x^2y^3 - 63xy^4$
18. $36x^4y^2 + 8x^3y^3, 56x^3y^4 + 16x^2y^5$
19. $x^2 + 6x + 5, x^2 + 2x - 15$
20. $x^2 - 14x + 24, x^2 - 9x + 14$
21. $x^2 - 8x - 65, x^2 - 9x - 52$
22. $x^2 + 16x + 63, x^2 + 7x - 18$
23. $x^2 - 20x + 36, x^2 - 11x + 18$
24. $x^2 + 13x - 48, x^2 + 4x - 21$
25. $x^2 + 18x + 80, x^2 + 20x + 96$
26. $x^2 - 24x + 144, x^2 + 4x - 192$
27. $x^2 - x - 20, x^2 - 16$
28. $x^2 - 18x + 72, x^2 - 36$
29. $x^2 + 14x + 49, x^2 - 49$
30. $x^2 + 28x - 60, x^2 - 4$

18.2 ADDING MONOMIAL AND BINOMIAL FRACTIONS

Sums of multinomial fractions behave very much like sums of monomial fractions, and they are combined in much the same way. The major difference lies in the need to factor the denominators of multinomial fractions in order to determine the LCM. Once this is done, we proceed as usual to convert the sum to a sum with common denominators and then combine the numerators.

4. Simplify
a. $\frac{1}{6x} - \frac{7}{2x}$
b. $\frac{-3}{40x^2y^2} + \frac{5y}{8x^2}$

Look at the following example of *adding fractions containing monomial and binomial denominators.*

TYPICAL PROBLEM

$$\frac{11x-9}{42x-56}-\frac{3x-5}{105x-140}$$

We first completely factor the binomial denominators:

$$\frac{11x-9}{14(3x-4)}-\frac{3x-5}{35(3x-4)}$$

Then to avoid sign errors we move all minus signs to the numerators (so that no minus signs are before the fraction bars).

$$\frac{11x-9}{14(3x-4)}+\frac{-3x+5}{35(3x-4)}$$

We then determine the LCM of the denominators:

$$\text{LCM} = 70(3x-4)$$

Then we convert the sum to a sum with like denominators. To do this, we first multiply each term by the LCM to determine the new numerators:

$$70(3x-4)\left(\frac{11x-9}{14(3x-4)}\right)+70(3x-4)\left(\frac{-3x+5}{35(3x-4)}\right)$$

And then we form new fractions by placing each new numerator over the LCM:

$$\frac{\overset{5}{\cancel{70(3x-4)}}\left(\frac{11x-9}{\cancel{14(3x-4)}}\right)}{70(3x-4)}+\frac{\overset{2}{\cancel{70(3x-4)}}\left(\frac{-3x+5}{\cancel{35(3x-4)}}\right)}{70(3x-4)}$$

At this point we have a sum of fractions with common denominators:

$$\frac{55x-45}{70(3x-4)}+\frac{-6x+10}{70(3x-4)}$$

Now we add by forming a single fraction:

$$\frac{55x-45-6x+10}{70(3x-4)}$$

And then we simplify the numerator:

$$\frac{49x-35}{70(3x-4)}$$

Finally we reduce the fraction to lowest terms:

$$\frac{7(7x-5)}{70(3x-4)}$$

$$\frac{7x-5}{10(3x-4)}$$

Hence, $\dfrac{11x-9}{42x-56}-\dfrac{3x-5}{105x-140}=\dfrac{7x-5}{10(3x-4)}$

To add fractions with monomial and binomial denominators, we parallel the method in Section 9.4 for adding fractions with monomial denominators. To determine the LCM of the denominators we use the method in the previous section. To reduce the fraction in the last step we use the methods in Chapter 17.

EXAMPLES

Add the following fractions containing monomial and binomial denominators.

Every sum of algebraic fractions can be simplified to a single reduced fraction.

1. $\dfrac{1}{6x} - \dfrac{7}{6x^2 - 14x}$

2. $\dfrac{-3}{40x^2y^2} - \dfrac{2x - 5y}{16x^3y^2 + 24x^2y^3}$

3. $\dfrac{-2x}{27x - 45} + \dfrac{3x}{36x - 60}$

4. $\dfrac{-x}{4x + 1} + \dfrac{x - 2}{3x - 5}$

5. $\dfrac{10x - 9}{60x - 24} - \dfrac{4x - 15}{40x - 16}$

6. $\dfrac{5x}{8x - 12} - \dfrac{8}{21 - 14x}$

Solutions

1. $\text{LCM} = 6x(3x - 7)$

$$\downarrow$$

$$\frac{\overset{(3x-7)}{\cancel{6x(3x-7)}}\left(\dfrac{1}{\cancel{6x}}\right)}{6x(3x - 7)} + \frac{\overset{3}{\cancel{6x(3x-7)}}\left(\dfrac{-7}{\cancel{2x(3x-7)}}\right)}{6x(3x - 7)}$$

$$\downarrow$$

$$\frac{3x - 7}{6x(3x - 7)} + \frac{-21}{6x(3x - 7)}$$

$$\downarrow$$

$$\frac{3x - 7 - 21}{6x(3x - 7)}$$

$$\downarrow$$

$$\frac{3x - 28}{6x(3x - 7)}$$

2. $\text{LCM} = 40x^2y^2(2x + 3y)$

$$\downarrow$$

$$\frac{\overset{(2x+3y)}{\cancel{40x^2y^2(2x+3y)}}\left(\dfrac{-3}{\cancel{40x^2y^2}}\right)}{40x^2y^2(2x + 3y)} + \frac{\overset{5}{\cancel{40x^2y^2(2x+3y)}}\left(\dfrac{-2x + 5y}{\cancel{8x^2y^2(2x+3y)}}\right)}{40x^2y^2(2x + 3y)}$$

$$\downarrow$$

$$\frac{-6x - 9y}{40x^2y^2(2x + 3y)} + \frac{-10x + 25y}{40x^2y^2(2x + 3y)}$$

$$\downarrow$$

$$\frac{-6x - 9y - 10x + 25y}{40x^2y^2(2x + 3y)}$$

$$\downarrow$$

$$\frac{-16x + 16y}{40x^2y^2(2x + 3y)}$$

$$\downarrow$$

$$\frac{-2x + 2y}{5x^2y^2(2x + 3y)} \text{ or } -\frac{2x - 2y}{5x^2y^2(2x + 3y)}$$

3. LCM = 36(3x − 5)

$$\downarrow$$

$$\frac{\overset{4}{\cancel{36(3x-5)}}\left(\dfrac{-2x}{\cancel{9(3x-5)}}\right)}{36(3x-5)} + \frac{\overset{3}{\cancel{36(3x-5)}}\left(\dfrac{3x}{\cancel{12(3x-5)}}\right)}{36(3x-5)}$$

$$\downarrow$$

$$\frac{-8x}{36(3x-5)} + \frac{9x}{36(3x-5)}$$

$$\downarrow$$

$$\frac{-8x+9x}{36(3x-5)}$$

$$\downarrow$$

$$\frac{x}{36(3x-5)}$$

4. LCM = (4x + 1) (3x − 5)

$$\downarrow$$

$$\frac{\overset{(3x-5)}{\cancel{(4x+1)}(3x-5)}\left(\dfrac{-x}{\cancel{4x+1}}\right)}{(4x+1)(3x-5)} + \frac{\overset{(4x+1)}{(4x+1)\cancel{(3x-5)}}\left(\dfrac{x-2}{\cancel{3x-5}}\right)}{(4x+1)(3x-5)}$$

$$\downarrow$$

$$\frac{-3x^2+5x}{(4x+1)(3x-5)} + \frac{4x^2-7x-2}{(4x+1)(3x-5)}$$

$$\downarrow$$

$$\frac{-3x^2+5x+4x^2-7x-2}{(4x+1)(3x-5)}$$

$$\downarrow$$

$$\frac{x^2-2x-2}{(4x+1)(3x-5)}$$

5. LCM = 24(5x − 2)

$$\downarrow$$

$$\frac{\overset{2}{\cancel{24(5x-2)}}\left(\dfrac{10x-9}{\cancel{12(5x-2)}}\right)}{24(5x-2)} + \frac{\overset{3}{\cancel{24(5x-2)}}\left(\dfrac{-4x+15}{\cancel{8(5x-2)}}\right)}{24(5x-2)}$$

$$\downarrow$$

$$\frac{20x-18}{24(5x-2)} + \frac{-12x+45}{24(5x-2)}$$

$$\downarrow$$

$$\frac{20x-18-12x+45}{24(5x-2)}$$

$$\downarrow$$

$$\frac{8x+27}{24(5x-2)}$$

6.

$$\text{LCM} = 28(2x-3)$$

$$\downarrow$$

$$\frac{\overset{7}{\cancel{28(2x-3)}}\left(\frac{5x}{\cancel{4(2x-3)}}\right)}{28(2x-3)} + \frac{\overset{4}{\cancel{28(2x-3)}}\left(\frac{8}{\cancel{7(2x-3)}}\right)}{28(2x-3)}$$

$$\downarrow$$

$$\frac{35x}{28(2x-3)} + \frac{32}{28(2x-3)}$$

$$\downarrow$$

$$\frac{35x+32}{28(2x-3)}$$

SPECIAL CASES

Given a sum of polynomial fractions whose denominators are in completely factored form, we will say that the sum is in *standard form* if no two denominators contain factors that are negatives of each other. A sum can be put in standard form by negating factors in the denominators and at the same time negating corresponding numerators. For example, the sum $\frac{5}{2x-3y} + \frac{7x}{y(3y-2x)}$ is not in standard form, because $2x - 3y$ and $3y - 2x$ are negatives of each other. But it can be put in standard form by changing it to either

$$\frac{5}{2x-3y} + \frac{-7x}{y(2x-3y)} \quad \text{or} \quad \frac{-5}{3y-2x} + \frac{7x}{y(3y-2x)}$$

When combining multinomial fractions, first put them in standard form to avoid confusing the LCM of the denominators with the product of the denominators. For example, the LCM of $3x(x-y)$ and $2y(y-x)$ is either $6xy(x-y)$ or $6xy(y-x)$, but not $6xy(x-y)(y-x)$.

5. Convert each sum to standard form.

a. $\frac{1}{x-3y} + \frac{1}{3y-x}$

b. $\frac{1}{x(x-y)} - \frac{1}{y(-x+y)}$

c. $\frac{-1}{y^2+4x} + \frac{1}{-4x-y^2}$

EXERCISES

Simplify each of the following expressions to a single reduced fraction.

1. $\frac{x}{2} + \frac{1}{x+2}$
2. $\frac{3}{4} - \frac{4x}{3x+y}$
3. $\frac{-1}{3x} + \frac{x-5}{4x+1}$
4. $\frac{7x+4}{6x+5} + \frac{-5}{8x}$
5. $\frac{2x+3}{x+3} + \frac{4x}{3}$
6. $-\frac{4x-1}{6} + \frac{8}{2x-9}$
7. $\frac{-5}{6x^2-9x} + \frac{7}{6x^2}$
8. $\frac{-9}{14x} + \frac{7}{2x^2-14x}$
9. $\frac{x}{30x^3} - \frac{1}{15x^4+30x^3}$
10. $-\frac{5x}{32xy+12y} + \frac{-3x}{40y}$
11. $\frac{x}{40y^2} - \frac{x}{5xy^2+40y^2}$
12. $-\frac{4}{15x^3-60x^2} + \frac{x}{120x^2}$
13. $-\frac{8x-7y}{14x^3y-16x^2y} + \frac{7}{16x^2y}$
14. $-\frac{2x-5y}{18x^3y^3-21x^2y^4} + \frac{11}{45x^2y^3}$
15. $\frac{-15}{36x-8} + \frac{2x-3}{36}$

16. $\dfrac{18}{15x^8 + 25x^7} - \dfrac{4x - 5}{30x^7}$

17. $\dfrac{3x - 1}{30x^5} + \dfrac{7}{36x^4 - 6x^3}$

18. $\dfrac{8x + 3y}{32y^2} + \dfrac{15}{64x - 24y}$

19. $\dfrac{2x}{2x + 5} + \dfrac{3}{x + 3}$

20. $\dfrac{-8}{7x - 4} + \dfrac{-12x}{11x - 6}$

21. $\dfrac{5x - 1}{2x - 1} + \dfrac{-4}{x + 3}$

22. $\dfrac{2x + 7}{3x - 1} + \dfrac{-6}{x + 6}$

23. $\dfrac{7x - 2y}{2x - 7y} - \dfrac{3x + 7y}{2x + 7y}$

24. $\dfrac{3x - y}{2x + 9y} - \dfrac{x + 9y}{8x - 7y}$

25. $\dfrac{5x}{24x + 60} - \dfrac{7}{12x + 36}$

26. $\dfrac{6x}{63x - 14} + \dfrac{5}{21x - 70}$

27. $\dfrac{5x - 1}{30x - 15} - \dfrac{4}{15x - 60}$

28. $\dfrac{2x + 7}{21x - 7} - \dfrac{6x}{7x + 42}$

29. $\dfrac{x + 1}{12x - 24} + \dfrac{x + 6}{12x + 24}$

30. $\dfrac{5x + 1}{30x - 75} + \dfrac{3x + 1}{45x + 60}$

In these exercises, the denominators have a common binomial factor.

31. $\dfrac{1}{x + 1} + \dfrac{1}{5x + 5}$

32. $\dfrac{x}{3x + 4} + \dfrac{3x}{15x + 20}$

33. $\dfrac{2}{3x + 9} - \dfrac{3}{4x + 12}$

34. $\dfrac{7x}{3x - 15} - \dfrac{5x}{6x - 30}$

35. $\dfrac{3}{3x + 1} - \dfrac{x + 3}{9x + 3}$

36. $\dfrac{2x - 1}{6x - 4} - \dfrac{5x}{18x - 12}$

37. $\dfrac{x - 5}{24x + 20} + \dfrac{4x}{18x + 15}$

38. $\dfrac{3}{28x - 21} + \dfrac{3x + 7}{36x - 27}$

39. $\dfrac{7x + 15}{5x - 2} + \dfrac{x - 16}{35x - 14}$

40. $\dfrac{8x - 11}{8x + 28} + \dfrac{8x + 15}{2x + 7}$

41. $\dfrac{8x - 15}{81x - 36} - \dfrac{6x - 5}{54x - 24}$

42. $\dfrac{4x - 5}{18x - 48} - \dfrac{9x - 28}{96x - 256}$

In these exercises, the denominators have negative binomial factors.

43. $\dfrac{3}{8y - 2x} + \dfrac{2}{x - 4y}$

44. $\dfrac{8y}{10x - 25y} + \dfrac{14y}{5y - 2x}$

45. $\dfrac{15y}{16x - 10y} - \dfrac{5x}{35y - 56x}$

46. $\dfrac{15}{16x - 36} - \dfrac{2}{45 - 20x}$

47. $-\dfrac{11y}{27x - 126y} + \dfrac{7y}{84y - 18x}$

48. $\dfrac{12x}{45y - 180x} + \dfrac{8y}{300x - 75y}$

49. $\dfrac{x - y}{x - 9y} + \dfrac{5x}{18y - 2x}$

50. $\dfrac{4x}{40x - 5} + \dfrac{2x + 1}{1 - 8x}$

51. $\dfrac{5x - 2y}{35y - 14x} - \dfrac{y - 2x}{24x - 60y}$

52. $\dfrac{3x + 1}{40x - 65} - \dfrac{-3}{26 - 16x}$

53. $\dfrac{8y + x}{96x - 126y} - \dfrac{9y}{210y - 160x}$

54. $\dfrac{5x + 4y}{270x - 225y} - \dfrac{7y - 2x}{135y - 162x}$

18.3 ADDING FRACTIONS WITH MULTINOMIAL DENOMINATORS

In this section we will look at sums of fractions with denominators that are simple binomials and trinomials. After we factor the denominators using the methods in Chapter 16, we can combine these fractions as we did in the previous section.

Look at the following example of *adding fractions with polynomial denominators*.

TYPICAL PROBLEM

$$\frac{12}{x^2 + 2x - 15} - \frac{3x}{x^2 - 5x + 6}$$

$$\downarrow$$

We first completely factor the denominators:

$$\frac{12}{(x+5)(x-3)} - \frac{3x}{(x-3)(x-2)}$$

$$\downarrow$$

Then to avoid sign errors we move all minus signs to the numerators (so that no minus signs are before the fraction bars):

$$\frac{12}{(x+5)(x-3)} + \frac{-3x}{(x-3)(x-2)}$$

$$\downarrow$$

We then determine the LCM of the denominators:

$$\text{LCM} = (x+5)(x-3)(x-2)$$

$$\downarrow$$

Then we convert the sum to a sum with like denominators. To do this, we first multiply each term by the LCM to determine the new numerators:

$$(x+5)(x-3)(x-2)\left(\frac{12}{(x+5)(x-3)}\right) + (x+5)(x-3)(x-2)\left(\frac{-3x}{(x-3)(x-2)}\right)$$

$$\downarrow$$

And then we form new fractions by placing each new numerator over the LCM:

$$\frac{\overset{(x-2)}{\cancel{(x+5)(x-3)}(x-2)}\left(\frac{12}{\cancel{(x+5)(x-3)}}\right)}{(x+5)(x-3)(x-2)} + \frac{\overset{(x+5)}{(x+5)\cancel{(x-3)(x-2)}}\left(\frac{-3x}{\cancel{(x-3)(x-2)}}\right)}{(x+5)(x-3)(x-2)}$$

$$\downarrow$$

At this point we have a sum of fractions with common denominators:

$$\frac{12x - 24}{(x+5)(x-3)(x-2)} + \frac{-3x^2 - 15x}{(x+5)(x-3)(x-2)}$$

$$\downarrow$$

Now we add by forming a single fraction:

$$\frac{12x - 24 - 3x^2 - 15x}{(x+5)(x-3)(x-2)}$$

$$\downarrow$$

And then we simplify the numerator:

$$\frac{-3x^2 - 3x - 24}{(x+5)(x-3)(x-2)}$$

Hence, $\frac{12}{x^2 + 2x - 15} - \frac{3x}{x^2 - 5x + 6} = \frac{-3x^2 - 3x - 24}{(x+5)(x-3)(x-2)}$

This method is the same as in the previous section. To determine the LCM of the denominators we use the method in Section 18.1.

EXAMPLES

Every sum of algebraic fractions can be simplified to a single reduced fraction.

Add the following fractions with polynomial denominators.

1. $\dfrac{7}{x+6} + \dfrac{3x-5}{x^2+4x-12}$

2. $\dfrac{4}{x^2-x-2} + \dfrac{6}{x^2-4x+4}$

3. $\dfrac{2x}{x^2-4x+3} - \dfrac{5}{x^2-x-6}$

4. $\dfrac{1}{x^2-4} - \dfrac{18}{x^2+2x-8}$

Solutions

1.

$$\text{LCM} = (x+6)(x-2)$$
$$\downarrow$$
$$\frac{\overset{(x-2)}{\cancel{(x+6)(x-2)}}\left(\dfrac{7}{\cancel{x+6}}\right)}{(x+6)(x-2)} + \frac{\overset{1}{\cancel{(x+6)(x-2)}}\left(\dfrac{3x-5}{\cancel{(x+6)(x-2)}}\right)}{(x+6)(x-2)}$$
$$\downarrow$$
$$\frac{7x-14}{(x+6)(x-2)} + \frac{3x-5}{(x+6)(x-2)}$$
$$\downarrow$$
$$\frac{7x-14+3x-5}{(x+6)(x-2)}$$
$$\downarrow$$
$$\frac{10x-19}{(x+6)(x-2)}$$

2.

$$\text{LCM} = (x+1)(x-2)^2$$
$$\downarrow$$
$$\frac{\overset{(x-2)}{\cancel{(x+1)(x-2)^2}}\left(\dfrac{4}{\cancel{(x+1)(x-2)}}\right)}{(x+1)(x-2)^2} + \frac{\overset{(x+1)}{\cancel{(x+1)(x-2)^2}}\left(\dfrac{6}{\cancel{(x-2)(x-2)}}\right)}{(x+1)(x-2)^2}$$
$$\downarrow$$
$$\frac{4x-8}{(x+1)(x-2)^2} + \frac{6x+6}{(x+1)(x-2)^2}$$
$$\downarrow$$
$$\frac{4x-8+6x+6}{(x+1)(x-2)^2}$$
$$\downarrow$$
$$\frac{10x-2}{(x+1)(x-2)^2}$$

3. $$\text{LCM} = (x-3)(x-1)(x+2)$$

$$\downarrow$$

$$\frac{\overset{(x-2)}{\cancel{(x-3)(x-1)(x+2)}}\left(\dfrac{2x}{\cancel{(x-3)(x-1)}}\right)}{(x-3)(x-1)(x+2)} + \frac{\overset{(x-1)}{\cancel{(x-3)(x-1)(x+2)}}\left(\dfrac{-5}{\cancel{(x-3)(x+2)}}\right)}{(x-3)(x-1)(x+2)}$$

$$\downarrow$$

$$\frac{2x^2-4x}{(x-3)(x-1)(x+2)} + \frac{-5x+5}{(x-3)(x-1)(x+2)}$$

$$\downarrow$$

$$\frac{2x^2-4x-5x+5}{(x-3)(x-1)(x+2)}$$

$$\downarrow$$

$$\frac{2x^2-9x+5}{(x-3)(x-1)(x+2)}$$

4. $$\text{LCM} = (x+2)(x-2)(x+4)$$

$$\downarrow$$

$$\frac{\overset{(x+4)}{\cancel{(x+2)(x-2)(x+4)}}\left(\dfrac{1}{\cancel{(x+2)(x-2)}}\right)}{(x+2)(x-2)(x+4)} + \frac{\overset{(x+2)}{\cancel{(x+2)(x-2)(x+4)}}\left(\dfrac{-18}{\cancel{(x-2)(x+4)}}\right)}{(x+2)(x-2)(x+4)}$$

$$\downarrow$$

$$\frac{x+4}{(x+2)(x-2)(x+4)} + \frac{-18x-36}{(x+2)(x-2)(x+4)}$$

$$\downarrow$$

$$\frac{x+4-18x-36}{(x+2)(x-2)(x+4)}$$

$$\downarrow$$

$$\frac{-17x-32}{(x+2)(x-2)(x+4)}$$

EXERCISES

Simplify each of the following expressions to a single reduced fraction.

1. $\dfrac{1}{x^2-x-6} + \dfrac{1}{x-3}$

2. $\dfrac{3}{x-4} + \dfrac{8}{x^2-3x-4}$

3. $\dfrac{9x}{x^2+7x+12} - \dfrac{4}{x+3}$

4. $\dfrac{7}{x+2} - \dfrac{15x}{x^2-5x-14}$

5. $\dfrac{6x+1}{x^2+7x-30} + \dfrac{4}{x+10}$

6. $\dfrac{7}{x+12} - \dfrac{7x-2}{x^2+16x+48}$

7. $\dfrac{5}{x^2-5x-24} + \dfrac{-2}{x^2-6x-27}$

8. $\dfrac{3}{x^2+11x+28} + \dfrac{6}{x^2-7x-44}$

9. $\dfrac{4x}{x^2-5x-36} - \dfrac{6}{x^2-10x+9}$

10. $\dfrac{9}{x^2+10x+16} - \dfrac{2x}{x^2-5x-14}$

11. $\dfrac{-8x}{x^2-20x+96} + \dfrac{12}{x^2-22x+112}$

12. $\dfrac{-8}{8x-1} - \dfrac{2x-1}{64x^2-1}$

13. $\dfrac{4x-1}{4x^2-25} + \dfrac{5}{2x+5}$

14. $\dfrac{8x-7}{49x^2-64} + \dfrac{4}{7x-8}$

15. $\dfrac{1}{x^2-10x+16} - \dfrac{1}{x^2-64}$

16. $\dfrac{4}{x^2-225} - \dfrac{7}{x^2+17x+30}$

17. $\dfrac{3}{x^2 - 1} + \dfrac{16}{x^2 + 27x - 28}$

18. $\dfrac{8}{x^2 - 144} + \dfrac{9x}{x^2 - 8x - 48}$

19. $\dfrac{4}{x^2 + 30x - 64} - \dfrac{5x}{x^2 - 4}$

20. $\dfrac{13}{x^2 - 64} - \dfrac{21x}{x^2 - 16x + 64}$

18.4 ALGEBRAIC COMPLEX FRACTIONS

We looked at numerical complex fractions in Chapters 6 and 7 and the simplest algebraic complex fractions in Chapter 8. Here we will look at complex fractions and sums of complex fractions such as

$$\frac{6 + \frac{4}{x}}{1 + \frac{3x}{2}}, \quad \frac{\frac{4x}{x+3}}{\frac{2}{x} - \frac{1}{x+3}}, \text{ and } \frac{\frac{3}{2x-3}}{2} + \frac{1}{2}$$

6. Simplify

a. $\dfrac{\frac{7}{2} - \frac{4}{3}}{\frac{7}{6} - \frac{3}{4}}$

b. $\dfrac{\frac{1}{7}}{\frac{2}{3} - \frac{4}{7}} + \dfrac{1}{2}$

These complex fractions contain sums of monomials and binomials.

Every such expression can be simplified to a single reduced polynomial fraction. This resulting fraction is occasionally a monomial fraction, although this is not necessarily the case even when the original complex fraction contains only monomials.

Look at the following example of *simplifying algebraic complex fractions.*

TYPICAL PROBLEM

$$\frac{y - \frac{1}{y}}{y - 1} - 1$$

We first change each complex fraction to the division format and place parentheses around the numerator and the denominator:

$$\left[\left(y - \frac{1}{y}\right) \div (y - 1)\right] - 1$$

Then we simplify inside each pair of parentheses. *At this point, each complex fraction has been converted to a quotient of fractions:*

$$\left[\left(\frac{y^2 - 1}{y}\right) \div (y - 1)\right] - 1$$

We then invert the fraction after each division sign and change the operation to multiplication:

$$\left[\frac{(y + 1)(y - 1)}{y} \cdot \frac{1}{y - 1}\right] - 1$$

Then we simplify each product. *At this point, each complex fraction has been converted to a single fraction:*

$$\frac{y + 1}{y} - 1$$

To simplify each expression we use the methods in Chapters 8 and 9 for simplifying monomial fractions, and the methods in Chapters 17 and 18 for simplifying multinomial fractions.

We then simplify further if necessary:

$$\frac{1}{y}$$

Hence, $\dfrac{y - \frac{1}{y}}{y - 1} - 1 = \dfrac{1}{y}$

Every algebraic complex fraction and every sum of algebraic complex fractions can be simplified to a single reduced fraction.

EXAMPLES

Simplify the following algebraic complex fractions.

1. $\dfrac{6+\dfrac{4}{x}}{1+\dfrac{3x}{2}}$

2. $\dfrac{6x-\dfrac{24}{x}}{3\left(\dfrac{2}{x}+1\right)}$

3. $\dfrac{\dfrac{7}{2}-\dfrac{4}{3x}}{\dfrac{7}{6}-\dfrac{3}{4x}}$

4. $\dfrac{\dfrac{1}{7}}{\dfrac{2x}{3}-\dfrac{4x}{7}}+\dfrac{1}{2}$

5. $\dfrac{\dfrac{4x}{x+3}}{\dfrac{2}{x}-\dfrac{1}{x+3}}$

6. $\dfrac{\dfrac{x+4}{x+1}}{\dfrac{3}{x+1}-\dfrac{4}{x}}$

7. $\dfrac{\dfrac{3}{2x-3}}{2}+\dfrac{1}{2}$

8. $\dfrac{1+\dfrac{4}{x}+\dfrac{x+4}{3}}{x+3}$

Solutions

1. $\left(6+\dfrac{4}{x}\right)\div\left(1+\dfrac{3x}{2}\right)$
$\downarrow$
$\left(\dfrac{6x+4}{x}\right)\div\left(\dfrac{2+3x}{2}\right)$
$\downarrow$
$\dfrac{6x+4}{x}\cdot\dfrac{2}{2+3x}$
$\downarrow$
$\dfrac{4}{x}$

2. $\left(6x-\dfrac{24}{x}\right)\div\left(3\left(\dfrac{2}{x}+1\right)\right)$
$\downarrow$
$\left(\dfrac{6x^2-24}{x}\right)\div\left(\dfrac{6+3x}{x}\right)$
$\downarrow$
$\dfrac{6x^2-24}{x}\cdot\dfrac{x}{6+3x}$
$\downarrow$
$2(x-2)$

3. $\left(\dfrac{7}{2}-\dfrac{4}{3x}\right)\div\left(\dfrac{7}{6}-\dfrac{3}{4x}\right)$
$\downarrow$
$\left(\dfrac{21x-8}{6x}\right)\div\left(\dfrac{14x-9}{12x}\right)$
$\downarrow$
$\dfrac{21x-8}{6x}\cdot\dfrac{12x}{14x-9}$
$\downarrow$
$\dfrac{2(21x-8)}{14x-9}$

4. $\left[\left(\dfrac{1}{7}\right)\div\left(\dfrac{2x}{3}-\dfrac{4x}{7}\right)\right]+\dfrac{1}{2}$
$\downarrow$
$\left[\dfrac{1}{7}\div\dfrac{2x}{21}\right]+\dfrac{1}{2}$
$\downarrow$
$\left[\dfrac{1}{7}\cdot\dfrac{21}{2x}\right]+\dfrac{1}{2}$
$\downarrow$
$\dfrac{3}{2x}+\dfrac{1}{2}$
$\downarrow$
$\dfrac{3+x}{2x}$

5. $\left(\dfrac{4x}{x+3}\right)\div\left(\dfrac{2}{x}-\dfrac{1}{x+3}\right)$
$\downarrow$
$\left(\dfrac{4x}{x+3}\right)\div\left(\dfrac{x+6}{x(x+3)}\right)$
$\downarrow$
$\dfrac{4x}{x+3}\cdot\dfrac{x(x+3)}{x+6}$
$\downarrow$
$\dfrac{4x^2}{x+6}$

6. $\left(\dfrac{x+4}{x+1}\right)\div\left(\dfrac{3}{x+1}-\dfrac{4}{x}\right)$
$\downarrow$
$\left(\dfrac{x+4}{x+1}\right)\div\left(\dfrac{-x-4}{x(x+1)}\right)$
$\downarrow$
$\dfrac{x+4}{x+1}\cdot\dfrac{x(x+1)}{-x-4}$
$\downarrow$
$-x$

7. $\left[\left(\frac{3}{2x-3}\right) \div (2)\right] + \frac{1}{2}$

$\downarrow$

$\left[\frac{3}{2x-3} \cdot \frac{1}{2}\right] + \frac{1}{2}$

$\downarrow$

$\frac{3}{2(2x-3)} + \frac{1}{2}$

$\downarrow$

$\frac{x}{2x-3}$

8. $\left(1 + \frac{4}{x} + \frac{x+4}{3}\right) \div (x+3)$

$\downarrow$

$\frac{x^2+7x+12}{3x} \div (x+3)$

$\downarrow$

$\frac{x^2+7x+12}{3x} \cdot \frac{1}{x+3}$

$\downarrow$

$\frac{x+4}{3x}$

OBSERVATIONS AND REFLECTIONS

To simplify a complex fraction with a numerator and denominator that are both sums of fractions, such as

$$\frac{\frac{7}{2} - \frac{4}{3x}}{\frac{7}{6} - \frac{3}{4x}},$$

we can adopt the following alternative method.

We first determine the LCM of all four denominators: LCM of 2, $3x$, 6, and $4x = 12x$

$\downarrow$

Then we multiply the numerator and denominator of the complex fraction by the LCM:

$$\frac{12x\left(\frac{7}{2} - \frac{4}{3x}\right)}{12x\left(\frac{7}{6} - \frac{3}{4x}\right)}$$

$\downarrow$

And then we simplify:

$$\frac{\overset{6x}{\cancel{12x}}\left(\frac{7}{\cancel{2}}\right) + \overset{4}{\cancel{12x}}\left(\frac{-4}{\cancel{3x}}\right)}{\overset{2x}{\cancel{12x}}\left(\frac{7}{\cancel{6}}\right) + \overset{3}{\cancel{12x}}\left(\frac{-3}{\cancel{4x}}\right)}$$

(with 1 under each cancelled denominator)

$\downarrow$

$$\frac{42x - 16}{14x - 9}$$

This method is usually more efficient for simplifying such complex fractions, except when the LCM of the four denominators is difficult to compute or is too large. (Compare the above computation with the solution to example 3.)

---

7. Use this method to compute

a. $\dfrac{\frac{3}{4} - \frac{2}{3}}{\frac{7}{8} + \frac{5}{3}}$

b. $\dfrac{\frac{7}{x} - \frac{3}{5}}{\frac{3}{4x} - 2}$

EXERCISES

Simplify each of the following expressions to a reduced fraction.

1. $\dfrac{2x - \dfrac{3x}{y}}{\dfrac{x}{y}}$

2. $\dfrac{\dfrac{4y}{x} - 6y}{4xy^2}$

3. $\dfrac{\dfrac{y}{3} - \dfrac{x}{2y}}{\dfrac{x}{5} - \dfrac{3x^2}{10y^2}}$

4. $\dfrac{4x - \dfrac{1}{9x}}{6 + \dfrac{1}{x}}$

5. $\dfrac{\dfrac{3}{2} - \dfrac{4}{x}}{\dfrac{9x}{8} - 3}$

6. $\dfrac{\dfrac{2x}{y} - 7}{4y - \dfrac{8x}{7}}$

7. $\dfrac{1 + \dfrac{3}{x}}{\dfrac{x}{3} - \dfrac{3}{x}}$

8. $\dfrac{\dfrac{3x}{8} - \dfrac{6}{x}}{18 - \dfrac{9x}{2}}$

9. $\dfrac{\dfrac{9}{2} + \dfrac{18}{5x}}{\dfrac{35}{2} + \dfrac{14}{x}}$

10. $\dfrac{\dfrac{24}{x} - \dfrac{32}{3}}{\dfrac{50}{9} - \dfrac{25}{2x}}$

11. $\dfrac{\dfrac{3}{5}}{\dfrac{2x}{3} - \dfrac{8x}{5}} + \dfrac{1}{4}$

12. $\dfrac{\dfrac{7}{8}}{\dfrac{3x}{4} - \dfrac{15x}{16}} + \dfrac{5}{9}$

13. $\dfrac{x + \dfrac{3}{4}}{\dfrac{2}{5}} - \dfrac{x - \dfrac{1}{8}}{\dfrac{3}{5}}$

14. $\dfrac{3x - \dfrac{1}{3}}{\dfrac{3}{4}} + \dfrac{2x + \dfrac{1}{4}}{\dfrac{3}{2}}$

15. $\dfrac{6x + \dfrac{1}{5}}{\dfrac{3}{8}} - \dfrac{4x + \dfrac{1}{3}}{\dfrac{1}{4}}$

16. $\dfrac{4}{\dfrac{2}{3} - \dfrac{x}{4}} - \dfrac{x}{\dfrac{4}{9} - \dfrac{x}{6}}$

17. $\dfrac{3}{\dfrac{1}{2} - \dfrac{2x}{9}} - \dfrac{x}{\dfrac{3}{8} - \dfrac{x}{6}}$

18. $\dfrac{x - \dfrac{3}{5}}{\dfrac{6}{25}} - \dfrac{7x - 4}{\dfrac{8}{5}}$

19. $\dfrac{1}{1 - \dfrac{1}{1 - x}}$

20. $\dfrac{6}{2 + \dfrac{4}{x - 2}}$

21. $\dfrac{14}{\dfrac{6}{7x - 4} + 5}$

22. $\dfrac{10}{\dfrac{8}{5y - 3} + 1}$

23. $\dfrac{\dfrac{9x}{x + 4}}{\dfrac{3}{x} - \dfrac{6}{x + 4}}$

24. $\dfrac{\dfrac{8y}{y - 2}}{\dfrac{4}{y - 2} - \dfrac{8}{y}}$

25. $\dfrac{\dfrac{x - 2}{x - 8}}{\dfrac{4}{x} + \dfrac{12}{x - 8}}$

26. $\dfrac{\dfrac{3x - 2}{2x - 1}}{\dfrac{3}{2x - 1} - \dfrac{6}{x}}$

27. $\dfrac{\dfrac{15y - 25}{2y - 5}}{\dfrac{14}{2y - 5} + \dfrac{2}{y}}$

28. $\dfrac{\dfrac{8x - 96}{3x - 8}}{\dfrac{12}{x} - \dfrac{28}{3x - 8}}$

29. $\dfrac{\dfrac{5}{x - 5}}{4} + \dfrac{5}{4}$

30. $\dfrac{\dfrac{2}{6y - 1}}{3} + \dfrac{2}{3}$

31. $\dfrac{\dfrac{3}{4x-9}}{4} + \dfrac{5}{12}$

32. $\dfrac{\dfrac{4}{5x+4}}{5} + \dfrac{7}{15}$

33. $\dfrac{\dfrac{5}{y-3} - \dfrac{2}{y+3}}{\dfrac{6}{y^2-9}}$

34. $\dfrac{\dfrac{14}{2x-5} + \dfrac{6}{2x+5}}{\dfrac{5x}{4x^2-25}}$

35. $\dfrac{1 - \dfrac{2}{x} + \dfrac{x-2}{5}}{x+5}$

36. $\dfrac{1 - \dfrac{6}{x} + \dfrac{x-6}{4}}{x+4}$

ANSWERS TO MARGIN EXERCISES

1. a. $48x^9$
b. $9x^6y^2$

3. a. $3x - 5y$
b. $12x^2y^3(x - 4y)$
c. $(2x - 7y)(2y - 7x)(5x - y)$
d. $(3y + 8x)(4x - 9y)$

4. a. $\dfrac{-10}{3x}$
b. $\dfrac{-3 + 25y^3}{40x^2y^2}$

5. a. $\dfrac{1}{x - 3y} + \dfrac{-1}{x - 3y}$
b. $\dfrac{1}{x(x - y)} + \dfrac{1}{y(x - y)}$
c. $\dfrac{-1}{4x + y^2} + \dfrac{-1}{4x + y^2}$

6. a. $\dfrac{26}{5}$
b. 2

7. a. $\dfrac{2}{61}$
b. $\dfrac{140 - 12x}{15 - 40x}$

CHAPTER REVIEW

SECTION 18.1

Find the LCM (Least Common Multiple). Leave the LCM in factored form.

1. $15x^2, 5x + 3$
2. $12x^3, 18x^2 - 12x$
3. $18x^2y^2, 42x^2y + 24xy^2$
4. $6x - 14, 3x - 7$
5. $18x^2 - 30x, 4x - 5$
6. $80x^7 - 144x^6, 75x^4 - 135x^3$
7. $54x^4 + 30x^3, 99x^6 - 55x^5$
8. $39x^7y^2 + 52x^6y^3, 27x^4y^6 + 36x^3y^7$
9. $18x^4y^3 + 30x^5y^4, 48x^2y^4 + 60xy^5$
10. $x^2 + 5x - 36, x^2 + 4x - 45$
11. $x^2 - 17x - 60, x^2 - 30x + 200$
12. $x^2 + 8x - 20, x^2 + 2x - 80$
13. $x^2 + 14x + 40, x^2 - 3x - 28$
14. $x^2 + 23x + 126, x^2 - 81$
15. $x^2 - 24x + 144, x^2 - 144$

SECTION 18.2

Simplify to a single reduced fraction.

16. $\frac{2x}{5} - \frac{3}{2x+5}$
17. $\frac{3x}{5} - \frac{4x}{3x-4}$
18. $\frac{9x-1}{6} + \frac{6x}{x-9}$
19. $\frac{-7}{8x^3} - \frac{9}{8x^3 + 14x^2}$
20. $\frac{11x}{70} + \frac{3}{28x^2 + 70x}$
21. $\frac{13x}{30y} + \frac{25x}{18xy - 24y}$
22. $\frac{-15x - 8}{30x^4 - 75x^3} - \frac{3}{20x^3}$
23. $\frac{2x}{6x^2 + 27x} - \frac{8x+1}{48x^2}$
24. $\frac{-4x - 15}{15x^8} + \frac{7}{9x^9 - 3x^8}$
25. $\frac{5x}{8x-7} + \frac{-10}{6x+7}$
26. $\frac{9x}{2x-5} - \frac{2x-5}{4x-5}$
27. $\frac{6x-5}{3x+5} + \frac{2x-5}{2x+5}$
28. $\frac{5}{9x-36} - \frac{4x}{27x-72}$
29. $\frac{9x}{24x-40} - \frac{2x-5}{32x-40}$
30. $\frac{6x-5}{54x+90} - \frac{4x-5}{36x+90}$
31. $\frac{5}{12x+30} + \frac{3}{4x+10}$
32. $\frac{9x}{16x-8} + \frac{7x}{10x-5}$
33. $\frac{7}{15x+25} + \frac{4x+1}{3x+5}$
34. $\frac{7x-1}{35x+10} - \frac{15x}{21x+6}$
35. $\frac{18x-5}{12x-16} - \frac{11x-9}{18x-24}$
36. $\frac{22x-3}{10x-60} - \frac{7x-32}{25x-150}$
37. $\frac{2x}{y-7x} + \frac{7y}{28x-4y}$
38. $\frac{11x}{24y-22x} - \frac{14x}{33x-36y}$
39. $\frac{-11}{72x-156y} + \frac{-7}{104y-48x}$
40. $-\frac{7x-3y}{12x-18y} + \frac{y-2x}{3y-2x}$
41. $-\frac{13y}{42x-90y} - \frac{7y-x}{105y-49x}$
42. $\frac{-11x}{72y-104x} - \frac{8x+y}{156x-108y}$

SECTION 18.3 Simplify to a single reduced fraction.

43. $\frac{6}{x-5}+\frac{4}{x^2-7x+10}$

44. $\frac{7}{x+6}+\frac{3x-5}{x^2+4x-12}$

45. $\frac{8}{x-2}+\frac{28}{x^2-11x+18}$

46. $\frac{2x}{x^2-4x+3}-\frac{5}{x^2-x-6}$

47. $\frac{7}{x^2+16x-36}-\frac{3}{x^2+6x-16}$

48. $\frac{9x}{49x^2-16}-\frac{5}{7x-4}$

49. $\frac{x-4}{121x^2-4}+\frac{2}{11x+2}$

50. $\frac{1}{x^2-4}-\frac{18}{x^2+2x-8}$

51. $\frac{5x}{x^2-6x-27}+\frac{16}{x^2-81}$

52. $\frac{4}{x^2-144}+\frac{1}{x^2-9x-36}$

SECTION 18.4 Simplify to a reduced fraction.

53. $\dfrac{6xy-\frac{x}{2y}}{\frac{3x}{y}}$

54. $\dfrac{y-\frac{3x}{y}}{\frac{2x}{y}-\frac{2y}{3}}$

55. $\dfrac{\frac{3x}{5}-\frac{3}{2}}{\frac{4}{5}-\frac{2}{x}}$

56. $\dfrac{\frac{21}{x}-6}{\frac{x}{7}-\frac{7}{4x}}$

57. $\dfrac{8+\frac{6}{x}}{1+\frac{4x}{3}}$

58. $\dfrac{\frac{4}{5}-\frac{x}{6}}{\frac{5}{56}-\frac{3}{7x}}$

59. $\dfrac{\frac{65}{9}}{\frac{10x}{3}-\frac{25x}{27}}-\frac{4}{5}$

60. $\dfrac{x-\frac{3}{4}}{\frac{9}{28}}-\dfrac{x-\frac{2}{3}}{\frac{2}{7}}$

61. $\dfrac{2x}{\frac{4x}{35}-\frac{1}{21}}+\dfrac{7}{\frac{1}{6x}-\frac{2}{5}}$

62. $\dfrac{12}{9+\frac{18}{y-3}}$

63. $\dfrac{16}{\frac{3}{8x-7}+5}$

64. $\dfrac{\frac{15x}{x+5}}{\frac{9}{x}-\frac{24}{x+5}}$

65. $\dfrac{\frac{9y-7}{3y-5}}{\frac{14}{y}-\frac{6}{3y-5}}$

66. $\dfrac{\frac{60x+24}{5x+3}}{\frac{75}{5x+3}+\frac{30}{x}}$

67. $\dfrac{\frac{9}{2x-7}}{4}+\frac{3}{4}$

68. $\dfrac{\frac{6}{8x+3}}{2}+\frac{11}{16}$

69. $\dfrac{\frac{11}{7x-1}-\frac{10}{7x+1}}{\frac{7}{49x^2-1}}$

70. $\dfrac{1+\frac{8}{x}+\frac{x+8}{7}}{x+7}$

CUMULATIVE TEST

The following problems test your understanding of this chapter and related subject matter of previous chapters. Before taking this test, thoroughly review Sections 9.4, 17.4, 18.2, and 18.3.

Once you have finished the test compare your answers with the answers provided at the back of the book. Note the section number of each problem missed, and thoroughly review those sections again.

Simplify.

1. $\dfrac{5}{x+8} - \dfrac{3}{8}$

2. $\dfrac{14x}{x^2+2x-48} - \dfrac{3}{x+8}$

3. $\dfrac{7x+5}{25} + \dfrac{12x-5}{15x}$

4. $\dfrac{36x^2+84x}{80x^3} \cdot \dfrac{45x}{27x+63}$

5. $\dfrac{6}{x-3y} - \dfrac{8}{6x-5y}$

6. $\dfrac{14}{5y^2} + \dfrac{17}{20xy}$

7. $\dfrac{7}{18x-12} + \dfrac{11}{30x-6}$

8. $-\dfrac{8x-9}{64x^2-25} + \dfrac{5}{8x+5}$

9. $\dfrac{60x^8}{45x^5-120x^4} \cdot \dfrac{12x-32}{32x^5}$

10. $\dfrac{-3x+4y}{15xy} - \dfrac{9x-y}{21xy}$

11. $-\dfrac{9x+4}{6} + \dfrac{8x}{4x-3}$

12. $\dfrac{-5x+2}{4x} + \dfrac{5}{6}$

13. $\dfrac{7}{x^2+9x-36} + \dfrac{8}{x^2-17x+42}$

14. $\dfrac{11x}{21y} - \dfrac{9x}{28xy-35y}$

15. $\dfrac{-8}{9xy} \div \dfrac{28y^2}{36x-45y}$

16. $\dfrac{-7}{84x^2y^5} + \dfrac{8x-1}{36x^2y^4+24xy^4}$

17. $-\dfrac{8x-3}{12} - \dfrac{3x-7}{20}$

18. $\dfrac{15x}{x^2-3x-108} - \dfrac{11}{x^2-144}$

19. $\dfrac{5}{8x^2} + \dfrac{7}{12x}$

20. $\dfrac{9x}{14x-42} - \dfrac{5x}{12x-36}$

21. $-\dfrac{4x}{x^2+6x-16} - \dfrac{3}{x^2-10x+16}$

22. $\dfrac{x^2-22x+96}{x^2+7x-144} \div \dfrac{x-16}{x^2-18x+81}$

23. $\dfrac{8x-11}{11} - \dfrac{16x-7}{11x}$

24. $\dfrac{4x}{9} + \dfrac{3x}{4x-7}$

25. $\dfrac{91x-26}{21x^3+56x^2} \cdot \dfrac{45x^7+120x^6}{77x^3-22x^2}$

26. $-\dfrac{2y}{4x+7y} + \dfrac{12x}{8x-9y}$

27. $\dfrac{6}{x-4} + \dfrac{20}{x^2 - 22x + 72}$

28. $\dfrac{4}{9} - \dfrac{3x-8}{9x}$

29. $\dfrac{24x^8 - 64x^7y}{6x^3y^8 + 90x^2y^9} \div \dfrac{18x^8y^2 - 48x^7y^3}{15xy^6 + 225y^7}$

30. $\dfrac{3}{8x} - \dfrac{2x-5}{8x+1}$

NONLINEAR
EQUATIONS

19 NONLINEAR FRACTIONAL EQUATIONS

Nonlinear Equations and Derived Equations

A reduced fraction is *linear* only if its numerator is linear and its denominator is a nonzero integer. A reduced fraction is *nonlinear* if its numerator is nonlinear, or if its denominator contains any *literal expression* (even if the literal expression is linear). For example, $\frac{x}{4}$ and $\frac{3x-2}{9}$ are linear fractions, whereas $\frac{4}{x}$ and $\frac{9}{3x-2}$ are nonlinear fractions. Hence, an equation in which the unknown appears in the denominator is nonlinear. For example, $\frac{4x-3}{6x} - \frac{1}{4} = \frac{1}{2x}$ and $\frac{5}{2x-6} - \frac{7x}{3x-9} = -\frac{1}{4}$ are nonlinear equations.

Our first step in solving such equations is to clear denominators. To do so we must multiply by an expression containing x. Remember that transformation rule 4 in Chapter 10 states that multiplying an equation by any nonzero *number* yields an equivalent equation. However, when we multiply by a literal expression containing x, we may occasionally get an equation that is not equivalent to the original one. That is, we may get an equation with a root that is not a root of the original equation. This is why we call the equation we obtain when clearing the denominators of a nonlinear fractional equation a *derived equation.* Keep in mind that a nonlinear fractional equation is not always equivalent to its derived equation.

If the derived equation is linear, as is sometimes the case, it is called the *derived linear equation.* For example, the derived equation of $\frac{4}{x} = 1$ is the linear equation $4 = x$, so $4 = x$ is the derived linear equation of $\frac{4}{x} = 1$.

Extraneous Roots

Every root of the original fractional equation is a root of its derived equation. But a root of the derived equation need not be a root of the original equation. When it is not, this root is called an *extraneous root*. Consider, for example, the equation

$$\frac{x}{x-3} = \frac{3}{x-3}$$

Multiplying this equation by $x - 3$ yields $x = 3$. But evaluating the original equation when $x = 3$ yields the nonsense equation

$$\frac{3}{0} = \frac{3}{0}$$

which has no meaning. (Fractions with a denominator of zero are not defined.) Hence, 3 is a root of the derived equation but not of the original equation. Therefore, the original equation has no roots.

Evaluating a fractional equation for an extraneous root, such as the number 3 in the example, always yields a nonsense equation with at least one denominator equal to zero.

Testing for Extraneous Roots

To determine whether a root of the derived equation is also a root of the original equation, it is sufficient to evaluate the nonconstant denominators of the original equation for this value of x. If none of these denominators reduces to zero, the root is good. If any of these denominators reduces to zero, however, the root is an extraneous root. For example, in the fractional equation $\frac{x}{x-3} = \frac{3}{x-3}$, the denominator $x - 3$ reduces to zero when $x = 3$.

19.1 SOLVING EQUATIONS WITH MONOMIAL DENOMINATORS

The simplest nonlinear fractional equations are those in which every denominator is an integer or an integer times the unknown. We will consider equations of this type separately in this section.

The equations in this section have derived linear equations without an extraneous root. Hence, for these equations, the solution to the derived equation is also the solution to the fractional equation. (The second typical problem on the next page is the only exception to this rule.)

1. Which fractions are linear?
   a. $\frac{3+2x}{5}$
   b. $\frac{3}{x}$
   c. $\frac{5x-1}{3x}$

Look at the following examples of *solving equations containing fractions with monomial denominators, whose derived equations are linear.*

TYPICAL PROBLEMS

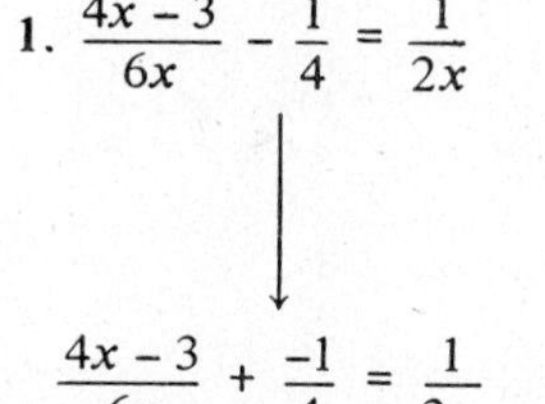

1. $\dfrac{4x-3}{6x} - \dfrac{1}{4} = \dfrac{1}{2x}$

2. $\dfrac{4x+3}{6x} - \dfrac{1}{4} = \dfrac{1}{2x}$

As we do when adding fractions, we first move all minus signs to the numerators to avoid sign errors:

1. $\dfrac{4x-3}{6x} + \dfrac{-1}{4} = \dfrac{1}{2x}$

2. $\dfrac{4x+3}{6x} + \dfrac{-1}{4} = \dfrac{1}{2x}$

Then we clear fractions. To do this, we first determine the LCM of the denominators:

1. LCM = $12x$

2. LCM = $12x$

And then we multiply each term by the LCM:

1. $\overset{2}{\cancel{12x}}\left(\dfrac{4x-3}{\cancel{6x}}\right) + \overset{3x}{\cancel{12x}}\left(\dfrac{-1}{\cancel{4}}\right) = \overset{6}{\cancel{12x}}\left(\dfrac{1}{\cancel{2x}}\right)$

2. $\overset{2}{\cancel{12x}}\left(\dfrac{4x+3}{\cancel{6x}}\right) + \overset{3x}{\cancel{12x}}\left(\dfrac{-1}{\cancel{4}}\right) = \overset{6}{\cancel{12x}}\left(\dfrac{1}{\cancel{2x}}\right)$

At this point each member of the equation is a sum of monomials with no fractions and no parentheses.

1. $8x - 6 - 3x = 6$

2. $8x + 6 - 3x = 6$

Finally we solve:

1. $5x = 12$

2. $5x = 0$

The solution must not make any denominator zero, so x must not be zero:

1. $x = \dfrac{12}{5}$ → O.K.

2. $x = 0$ → extraneous root

Hence, $\dfrac{4x-3}{6x} - \dfrac{1}{4} = \dfrac{1}{2x} \Rightarrow x = \dfrac{12}{5}$,

whereas $\dfrac{4x+3}{6x} - \dfrac{1}{4} = \dfrac{1}{2x}$ has no solution.

To solve equations with monomial denominators we essentially use the method in Section 10.4. To determine the LCM of the denominators we use the method in Section 9.3.

EXAMPLES

Solve the following equations with monomial denominators and linear derived equations.

1. $\dfrac{7}{x} + \dfrac{9}{2} = 1 + \dfrac{4}{x}$

2. $\dfrac{2}{5} + \dfrac{5}{4x} = \dfrac{8}{5x} - \dfrac{11}{4}$

3. $\dfrac{7-4x}{10x} = \dfrac{3}{10} + \dfrac{7}{15x}$

4. $-\dfrac{9-7x}{5x} - \dfrac{3}{4} - \dfrac{1}{4x} = \dfrac{2x-7}{10x}$

Solutions

1.
$$\text{LCM} = 2x$$
$$\downarrow$$
$$\overset{2}{\cancel{2x}}\left(\frac{7}{\cancel{x}}\right) + \overset{x}{\cancel{2x}}\left(\frac{9}{\cancel{2}}\right) = 2x(1) + \overset{2}{\cancel{2x}}\left(\frac{4}{\cancel{x}}\right)$$
$$\downarrow$$
$$14 + 9x = 2x + 8$$
$$\downarrow$$
$$7x = -6$$
$$\downarrow$$
$$x = \frac{-6}{7}$$

2.
$$\text{LCM} = 20x$$
$$\downarrow$$
$$\overset{4x}{\cancel{20x}}\left(\frac{2}{\cancel{5}}\right) + \overset{5}{\cancel{20x}}\left(\frac{5}{\cancel{4x}}\right) = \overset{4}{\cancel{20x}}\left(\frac{8}{\cancel{5x}}\right) + \overset{5x}{\cancel{20x}}\left(\frac{-11}{\cancel{4}}\right)$$
$$\downarrow$$
$$8x + 25 = 32 - 55x$$
$$\downarrow$$
$$63x = 7$$
$$\downarrow$$
$$x = \frac{1}{9}$$

3.
$$\text{LCM} = 30x$$
$$\downarrow$$
$$\overset{3}{\cancel{30x}}\left(\frac{7 - 4x}{\cancel{10x}}\right) = \overset{3x}{\cancel{30x}}\left(\frac{3}{\cancel{10}}\right) + \overset{2}{\cancel{30x}}\left(\frac{7}{\cancel{15x}}\right)$$
$$\downarrow$$
$$21 - 12x = 9x + 14$$
$$\downarrow$$
$$-21x = -7$$
$$\downarrow$$
$$x = \frac{1}{3}$$

4.
$$\text{LCM} = 20x$$
$$\downarrow$$
$$\overset{4}{\cancel{20x}}\left(\frac{-9 + 7x}{\cancel{5x}}\right) + \overset{5x}{\cancel{20x}}\left(\frac{-3}{\cancel{4}}\right) + \overset{5}{\cancel{20x}}\left(\frac{-1}{\cancel{4x}}\right) = \overset{2}{\cancel{20x}}\left(\frac{2x - 7}{\cancel{10x}}\right)$$
$$\downarrow$$
$$-36 + 28x - 15x - 5 = 4x - 14$$
$$\downarrow$$
$$9x = 27$$
$$\downarrow$$
$$x = 3$$

EXERCISES

Solve each of the following equations. In these exercises, the derived equations are linear and the original equations have solutions.

1. $\frac{2}{x} = \frac{3}{4}$
2. $\frac{9}{x} = \frac{3}{4}$
3. $\frac{5}{3x} = \frac{7}{2}$
4. $\frac{8}{5x} = \frac{6}{7}$
5. $\frac{12}{13x} = \frac{3}{5}$
6. $\frac{1}{x} + \frac{5}{3} = 1 + \frac{3}{x}$
7. $\frac{3}{x} + \frac{7}{4} = 2 - \frac{7}{8}$
8. $\frac{5}{x} - 3 = \frac{2}{x} + \frac{3}{5}$
9. $\frac{5}{x} - 4 = \frac{5}{8} - \frac{3}{8x}$
10. $\frac{3}{5x} - 2 = \frac{1}{x} - \frac{4}{5}$
11. $2 - \frac{7}{3x} = \frac{3}{x} + \frac{2}{3}$
12. $\frac{1}{3} + \frac{3}{2x} = \frac{2}{3x} - \frac{1}{2}$
13. $\frac{3}{4} + \frac{7}{6x} = \frac{5}{6} - \frac{1}{4x}$
14. $\frac{7}{8x} - \frac{2}{3} = \frac{1}{8} + \frac{5}{3x}$
15. $\frac{2 - 3x}{4x} = \frac{1}{4} + \frac{5}{6x}$
16. $\frac{3 - 7x}{9x} = \frac{2}{9} - \frac{4}{15x}$
17. $\frac{9}{16} - \frac{3x - 5}{6x} = \frac{-1}{6x}$
18. $\frac{1}{3} - \frac{4x - 9}{4x} = \frac{2}{3x}$
19. $\frac{3x - 1}{4x} - \frac{2}{3} = \frac{5x - 4}{3x} + \frac{1}{4}$
20. $\frac{5x - 2}{4x} + \frac{1}{6} = \frac{x - 9}{6x} - \frac{3}{4x}$
21. $\frac{7x - 2}{16x} - \frac{1}{2} = \frac{3x - 4}{2x} + \frac{5}{16x}$
22. $\frac{2x - 3}{6x} + \frac{11}{4x} = \frac{5x - 1}{12x} - \frac{2}{3}$
23. $\frac{3x - 8}{30x} + \frac{4}{5x} = \frac{2x - 1}{6x} - \frac{7}{10}$
24. $\frac{4x - 9}{12x} - \frac{5}{8} = \frac{3 - 10x}{6x} - \frac{13}{24x}$

19.2 SOLVING EQUATIONS WITH BINOMIAL DENOMINATORS

In this section we will look at equations containing terms with binomial denominators. We will again consider only equations with derived linear equations that have no extraneous root. Hence, for these equations, the solution to the derived equation is also the solution to the fractional equation.

Look at the following example of *solving equations containing fractions with binomial denominators, whose derived equations are linear.*

TYPICAL PROBLEM

$$\frac{5}{2x-6} - \frac{7x}{3x-9} = -\frac{1}{4}$$

We first move all minus signs to the numerators to avoid sign errors:

$$\frac{5}{2x-6} + \frac{-7x}{3x-9} = \frac{-1}{4}$$

Then we clear fractions. To do this, we first factor the denominators and then determine the LCM of the denominators:

$$\text{LCM} = 12(x-3)$$

And then we multiply each term by the LCM:

$$\overset{6}{\cancel{12(x-3)}}\left(\frac{5}{\cancel{2(x-3)}}\right) + \overset{4}{\cancel{12(x-3)}}\left(\frac{-7x}{\cancel{3(x-3)}}\right) = \overset{3(x-3)}{\cancel{12(x-3)}}\left(\frac{-1}{\cancel{4}}\right)$$

At this point each member of the equation is a sum of monomials with no fractions and no parentheses.

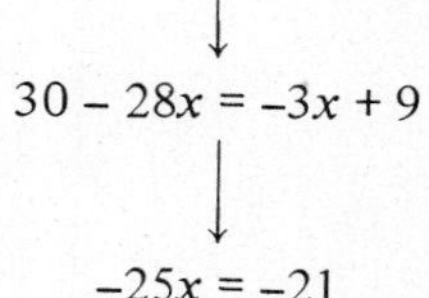

$$30 - 28x = -3x + 9$$

Having cleared fractions, we now solve the derived equation:

$$-25x = -21$$

$$\boxed{x = \frac{21}{25}}$$

Then we evaluate the nonconstant denominators. The solution must not make any of them zero:

$2x - 6$ and $3x - 9$, when $x = \frac{21}{25}$

$$2\left(\frac{21}{25}\right) - 6 \qquad 3\left(\frac{21}{25}\right) - 9$$

$$\frac{42}{25} - 6 \qquad \frac{63}{25} - 9$$

$$\frac{-108}{25} \neq 0 \qquad \frac{-162}{25} \neq 0$$

O.K.   O.K.

To solve equations containing binomial denominators we parallel the method in the previous section. To determine the LCM of the denominators we use the method in Section 18.1.

Hence, $\frac{5}{2x-6} - \frac{7x}{3x-9} = -\frac{1}{4} \Rightarrow x = \frac{21}{25}$.

EXAMPLES

Solve the following equations containing binomial denominators and linear derived equations.

1. $\dfrac{6x+5}{3x+8} - \dfrac{3}{7} = \dfrac{3x+5}{3x+8}$

2. $\dfrac{4}{3x-2} = \dfrac{-3}{2x-7}$

3. $\dfrac{3x}{5x-15} + \dfrac{11}{4x-12} = -\dfrac{1}{2}$

4. $\dfrac{8}{6x-9} - \dfrac{7x}{12-8x} = \dfrac{5}{6}$

Solutions

1.

$$\text{LCM} = 7(3x+8)$$
$$\downarrow$$
$$\overset{7}{\cancel{7(3x+8)}}\left(\frac{6x+5}{\cancel{3x+8}}\right) + \overset{(3x+8)}{\cancel{7(3x+8)}}\left(\frac{-3}{\cancel{7}}\right) = \overset{7}{\cancel{7(3x+8)}}\left(\frac{3x+5}{\cancel{3x+8}}\right)$$
$$\downarrow$$
$$42x + 35 - 9x - 24 = 21x + 35$$
$$\downarrow$$
$$12x = 24$$
$$\downarrow$$
$$x = 2$$

2.

$$\text{LCM} = (3x-2)(2x-7)$$
$$\downarrow$$
$$\overset{(2x-7)}{\cancel{(3x-2)(2x-7)}}\left(\frac{4}{\cancel{3x-2}}\right) = \overset{(3x-2)}{\cancel{(3x-2)(2x-7)}}\left(\frac{-3}{\cancel{2x-7}}\right)$$
$$\downarrow$$
$$8x - 28 = -9x + 6$$
$$\downarrow$$
$$17x = 34$$
$$\downarrow$$
$$x = 2$$

3.

$$\text{LCM} = 20(x-3)$$
$$\downarrow$$
$$\overset{4}{\cancel{20(x-3)}}\left(\frac{3x}{\cancel{5(x-3)}}\right) + \overset{5}{\cancel{20(x-3)}}\left(\frac{11}{\cancel{4(x-3)}}\right) = \overset{10(x-3)}{\cancel{20(x-3)}}\left(\frac{-1}{\cancel{2}}\right)$$
$$\downarrow$$
$$12x + 55 = -10x + 30$$
$$\downarrow$$
$$22x = -25$$
$$\downarrow$$
$$x = \frac{-25}{22}$$

4.

$$\text{LCM} = 12(2x-3)$$
$$\downarrow$$
$$\overset{4}{\cancel{12(2x-3)}}\left(\frac{8}{\cancel{3(2x-3)}}\right) + \overset{3}{\cancel{12(2x-3)}}\left(\frac{7x}{\cancel{4(2x-3)}}\right) = \overset{2(2x-3)}{\cancel{12(2x-3)}}\left(\frac{5}{\cancel{6}}\right)$$
$$\downarrow$$
$$32 + 21x = 20x - 30$$
$$\downarrow$$
$$x = -62$$

OBSERVATIONS AND REFLECTIONS

To check whether a root of the derived equation is an extraneous root, we check each nonconstant denominator of the original fractional equation. For example, in the typical problem $\frac{5}{2x-6} - \frac{7}{3x-9} = -\frac{1}{4}$, we checked the denominators $2x - 6$ and $3x - 9$ for the root of the derived equation. However, rather than check each nonconstant denominator separately we could have checked the LCM of the denominators. If any denominator reduces to zero, the LCM of the denominators will also reduce to zero. For example, the LCM of $2x - 6$ and $3x - 9$ is $12(x - 3)$ and all three expressions reduce to zero when $x = 3$. Hence, rather than check both denominators we need only check $12(x - 3)$ to see if the root is extraneous.

Thus, we can determine whether a root is extraneous by checking a single expression rather than each denominator separately.

2. Check the solutions to examples 1 to 4 for extraneous roots by evaluating each LCM.

EXERCISES

Solve each of the following equations. In these exercises, the derived equations are linear and the original equations have solutions.

1. $\frac{3}{x+2} = \frac{1}{4}$
2. $\frac{3}{4x-1} = \frac{5}{9}$
3. $\frac{2}{3x+4} = \frac{1}{2x}$
4. $\frac{2}{3x-8} = \frac{4}{9x}$
5. $\frac{2}{x-5} + \frac{5}{6} = \frac{3}{x-5}$
6. $\frac{1}{2x-9} - \frac{3}{4} = \frac{4}{2x-9}$
7. $\frac{6}{x-6} + \frac{3}{2} = \frac{1}{x-6} - 1$
8. $\frac{17}{9x-5} - 2 = \frac{7}{9x-5} - \frac{4}{7}$
9. $\frac{3x}{4x-5} + \frac{1}{6} = \frac{4}{4x-5} + 1$
10. $\frac{3}{6x-5} - \frac{2}{3} = \frac{-3x}{6x-5} - 2$
11. $\frac{3x-10}{2x-7} = 1$
12. $\frac{16x+1}{8x-5} = 3$
13. $\frac{2x-9}{5x+2} = \frac{3}{4}$
14. $\frac{8x+4}{12x-7} = \frac{4}{3}$
15. $\frac{8x-7}{2x-1} - \frac{3x-1}{2x-1} = \frac{3}{4}$
16. $\frac{8x-3}{5x-12} - \frac{3}{4} = \frac{3x+1}{5x-12}$
17. $\frac{7x-4}{3x-2} - \frac{1}{3} - \frac{5x-8}{3x-2} = 2$
18. $\frac{4x-3}{8x-11} + 2 = \frac{5x-10}{8x-11} - \frac{3}{2}$
19. $\frac{3}{x+1} = \frac{2}{x-3}$
20. $\frac{2}{x-4} = \frac{3}{x-5}$
21. $\frac{6}{3x-1} = \frac{8}{3x+1}$
22. $\frac{16}{8x-1} = \frac{-9}{x-7}$

23. $\frac{2}{4x-1} - \frac{3}{x-2} = 0$

24. $\frac{6}{10x-1} - \frac{9}{x-5} = 0$

25. $\frac{3x}{x-3} + 2 = \frac{1}{4x-12}$

26. $\frac{5}{2x-5} - 3 = \frac{2x}{6x-15}$

27. $\frac{8}{5x-2} - 1 = \frac{-x}{25x-10}$

28. $\frac{-18}{6x-14} = \frac{7x}{3x-7} + 4$

29. $\frac{-9}{6x-14} = 3 - \frac{5x}{9x-21}$

30. $\frac{5}{20x-36} + 1 = \frac{-x}{30x-54}$

31. $\frac{x}{2x+10} + \frac{1}{3} = \frac{25}{6x+30}$

32. $\frac{17x}{12x-12} - \frac{1}{4} = \frac{13}{3x-3}$

33. $\frac{3x}{6x-2} + \frac{1}{2} = \frac{22}{12x-4}$

34. $\frac{13}{40x+30} - \frac{4}{5} = \frac{9x}{8x+6}$

35. $\frac{-4}{90x-18} + \frac{5}{6} = \frac{16x}{15x-3}$

36. $\frac{9}{4x-4} = \frac{-23x}{32x-32} + \frac{1}{8}$

37. $\frac{3}{42x+63} + \frac{3}{7} = \frac{2x}{18x+27}$

38. $\frac{3x}{40x-20} = \frac{3}{24x-12} - \frac{1}{3}$

39. $\frac{9x}{30x-36} - \frac{5}{8} = \frac{-5}{80x-96}$

40. $\frac{-9}{40x-25} = \frac{21x}{24x-15} - \frac{2}{5}$

41. $\frac{-10}{48x-12} = \frac{5}{6} - \frac{x}{40x-10}$

42. $\frac{25x}{28x-42} - \frac{6}{7} = \frac{27}{24x-36}$

19.3 SOLVING FORMULA-TYPE EQUATIONS

In this section we will study equations such as $\frac{CD}{B^2} = \frac{AB}{CD}$ and $3B^2 - 7C^3 = 2AB$. Such equations arise in many applications of mathematics, and are called *formula-type* equations. In a *formula-type equation* each literal usually represents an independently measurable quantity. For example, in the formula for the perimeter of a rectangle, $P = 2l + 2w$, the length, width, and perimeter can each be measured directly in any given rectangle.

Most of the equations in this section are not linear, because the terms contain products of literals and literals appear in the denominators. In general, such equations can be solved the same way as the equations we have seen already. That is, if the equation contains fractions, we first clear denominators. Then we solve for the literal that we want by isolation and division.

However, because these equations contain products of literals and literal denominators, we will have to multiply and divide by literal expressions. Keep in mind that multiplying or dividing by a literal expression does not always yield an equivalent equation because of subtle difficulties related to multiplying and dividing by zero.

3. Solve for x.
a. $2x + y = 5y + 7$
b. $9 - 3x + 4y = 7 + x$
c. $\frac{5x-12}{8} = \frac{4x+9}{6}$

Look at the following examples of *solving formula-type equations.*

TYPICAL PROBLEMS Solve for A: 1. $$\frac{CD}{B^2} = \frac{AB}{CD}$$

We first clear fractions. To do this, we determine the LCM (least common multiple) of the denominators:

$$\text{LCM} = B^2CD$$

And then we multiply each term by the LCM:

$$B^2CD\left(\frac{CD}{B^2}\right) = B^2CD\left(\frac{AB}{CD}\right)$$

$$\overset{CD}{\cancel{B^2CD}}\left(\frac{CD}{\cancel{B^2}}\right) = \overset{B^2}{\cancel{B^2CD}}\left(\frac{AB}{\cancel{CD}}\right)$$

At this point each member of the equation is a sum of monomials with no fractions and no parentheses:

$$C^2D^2 = AB^3$$

Then we isolate the A term on the left side by shifting terms:

$$-AB^3 = -C^2D^2$$

We then negate both sides if the coefficient of the A term is negative:

$$AB^3 = C^2D^2$$

Finally we divide by the other factors of the A term:

$$\frac{AB^3}{B^3} = \frac{C^2D^2}{B^3}$$

And we reduce:

$$A = \frac{C^2D^2}{B^3}$$

Hence, $\dfrac{CD}{B^2} = \dfrac{AB}{CD} \Rightarrow A = \dfrac{C^2D^2}{B^3}$

Solve for A: 2. $$3B^2 - 7C^3 = 2AB$$

We isolate the A term on the left side by shifting terms:

$$-2AB = -3B^2 + 7C^3$$

We then negate both sides if the coefficient of the A term is negative:

$$2AB = 3B^2 - 7C^3$$

Finally we divide by the other factors of the A term:

$$\frac{2AB}{2B} = \frac{3B^2 - 7C^3}{2B}$$

And we reduce:

$$A = \frac{3B^2 - 7C^3}{2B}$$

To solve formula-type equations we parallel the methods in the previous sections.

Hence, $3B^2 - 7C^3 = 2AB \Rightarrow A = \dfrac{3B^2 - 7C^3}{2B}$

EXAMPLES

Solve the following formula-type equations for A.

1. $AB = \dfrac{C}{D}$
2. $\dfrac{9A}{4B} = \dfrac{2B^3}{5C}$
3. $3x - 4Ay = 12$
4. $8B = 4B^2 - 5A$
5. $\dfrac{2A}{7} - \dfrac{3B}{2} = \dfrac{9C}{14}$
6. $\dfrac{2A+5}{3} = \dfrac{B-4}{2}$

Solutions

1. LCM $= D$

$$\downarrow$$
$$D(AB) = \overset{1}{\cancel{D}}\left(\frac{C}{\cancel{D}}\right)$$
$$\downarrow$$
$$ABD = C$$
$$\downarrow$$
$$A = \frac{C}{BD}$$

2. LCM $= 20BC$

$$\downarrow$$
$$\overset{5C}{\cancel{20BC}}\left(\frac{9A}{\cancel{4B}}\right) = \overset{4B}{\cancel{20BC}}\left(\frac{2B^3}{\cancel{5C}}\right)$$
$$\downarrow$$
$$45AC = 8B^4$$
$$\downarrow$$
$$A = \frac{8B^4}{45C}$$

3. $-4Ay = 12 - 3x$

$$\downarrow$$
$$4Ay = -12 + 3x$$
$$\downarrow$$
$$A = \frac{-12 + 3x}{4y}$$

4. $5A = 4B^2 - 8B$

$$\downarrow$$
$$A = \frac{4B^2 - 8B}{5}$$

5. LCM $= 14$

$$\downarrow$$
$$\overset{2}{\cancel{14}}\left(\frac{2A}{\cancel{7}}\right) + \overset{7}{\cancel{14}}\left(\frac{-3B}{\cancel{2}}\right) = \overset{1}{\cancel{14}}\left(\frac{9C}{\cancel{14}}\right)$$
$$\downarrow$$
$$4A - 21B = 9C$$
$$\downarrow$$
$$4A = 9C + 21B$$
$$\downarrow$$
$$A = \frac{9C + 21B}{4}$$

6. LCM $= 6$

$$\downarrow$$
$$\overset{2}{\cancel{6}}\left(\frac{2A+5}{\cancel{3}}\right) = \overset{3}{\cancel{6}}\left(\frac{B-4}{\cancel{2}}\right)$$
$$\downarrow$$
$$4A + 10 = 3B - 12$$
$$\downarrow$$
$$4A = 3B - 22$$
$$\downarrow$$
$$A = \frac{3B - 22}{4}$$

OBSERVATIONS AND REFLECTIONS

When each member of a formula-type equation is a *single fraction*, the LCM of the denominators is not necessarily the best multiplier to use when clearing denominators. Instead, we can use the *product* of the denominators to clear the fractions and multiply each numerator by the denominator of the other member. This procedure is called *cross multiplication.*

4. Solve for A using cross multiplication

a. $\frac{5A}{6B} = \frac{7B^2}{8C}$

b. $\frac{2A-9}{4} = \frac{7+B}{10}$

CROSS MULTIPLICATION

$$\frac{A}{B} = \frac{C}{D} \Rightarrow A \cdot D = B \cdot C$$

(where A, B, C, and D represent polynomials)

Keep in mind, however, that cross multiplication can only be used when each member of the equation is a *single* fraction.

EXERCISES

Solve for A in each of the following equations.

1. $AB = C$
2. $AKL = M^2$
3. $4AB = C$
4. $9B^2 = 3AD$
5. $\frac{AK}{4} = \frac{K^5}{2}$
6. $\frac{-BC}{9} = \frac{2AC}{15}$
7. $\frac{-K}{2} = \frac{4AR^2}{9}$
8. $\frac{6AL^2}{5} = \frac{L^4M^6}{15}$
9. $\frac{AB^3}{C} = B^2C$
10. $C^2E^3 = \frac{AE^2}{C^3}$
11. $\frac{BE^4}{D} = \frac{ACD}{E}$
12. $\frac{B^3}{C^3} = \frac{AB^2}{C^3}$
13. $AB = C + D$
14. $Ax^2 = 2x - 5$
15. $3A = 4x + y$
16. $Ar^2 - 2Br = D$
17. $Ab - b^3 = 9$
18. $4Ax + 5y = 6z$
19. $9x - 5Ay = 40$
20. $R^3T^2 = T^3 - AR$
21. $2x^2 - 5x + A = 0$
22. $y = 9x^2 + 8Ax$
23. $S^2 + T = AST^3 - 1$
24. $Bx^2 - 5Ax + 3 = C$
25. $\frac{A}{F} = B + \frac{E}{B}$
26. $\frac{A}{H} = \frac{B}{D} - E$
27. $\frac{AG}{H} = R + \frac{HG}{R}$
28. $\frac{AB}{D} = \frac{D}{E} + BE$
29. $AF + \frac{G}{F} = \frac{B}{G}$
30. $\frac{C}{L} - AM = \frac{L}{M}$
31. $\frac{BK}{D} = AG - \frac{B}{K}$
32. $\frac{GM}{P} = \frac{H}{G} + AB$

33. $\frac{2A}{5} + \frac{B}{4} = \frac{7C}{10}$

34. $\frac{9A}{4} - \frac{5B}{6} = \frac{4C}{3}$

35. $\frac{6x}{5} - \frac{2A}{3} = \frac{xy}{5}$

36. $\frac{4T}{3} - \frac{3A}{4} = \frac{TW}{6}$

37. $\frac{A+4}{8} = \frac{B-5}{6}$

38. $\frac{A+9}{6} = \frac{B-3}{10}$

39. $\frac{3A-8}{12} = \frac{5B+2}{9}$

40. $\frac{6A-8}{25} = \frac{2B-9}{15}$

41. $\frac{3A-4}{14} = \frac{B}{7} - \frac{3C-1}{4}$

42. $\frac{5A+7}{18} = \frac{B}{9} - \frac{5C-2}{6}$

43. $\frac{3F-5}{16} = \frac{4H}{3} - \frac{3A-1}{12}$

44. $\frac{9G-4}{20} = \frac{3E}{8} - \frac{6A-5}{10}$

45. $\frac{B}{2} = \frac{Ax^2+y^2}{3BD}$

46. $\frac{E^2}{3} = \frac{Aw^3-4v}{4E}$

47. $\frac{4B^2}{5} = \frac{AR^2-R}{2B}$

48. $\frac{9K}{4} = \frac{AS-T^2}{KT}$

ANSWERS TO MARGIN EXERCISES

1. a

3. a. $x = \frac{4y+7}{2}$

b. $x = \frac{2y+1}{2}$

c. $x = -72$

4. a. $A = \frac{21B^3}{20C}$

b. $A = \frac{2B+59}{10}$

CHAPTER REVIEW

SECTION 19.1 Solve.

1. $\frac{6}{x} = \frac{8}{5}$
2. $\frac{4}{15x} = \frac{16}{3}$
3. $\frac{5}{x} = \frac{10}{7}$
4. $\frac{1}{x} + \frac{3}{4} = 2 + \frac{6}{x}$
5. $\frac{3}{x} - 5 = \frac{4}{9} - \frac{1}{9x}$
6. $\frac{1}{4} + \frac{2}{5x} = \frac{7}{4x} - \frac{4}{5}$
7. $\frac{1}{6} - \frac{7x-5}{2x} = \frac{5}{6x}$
8. $\frac{4-3x}{8x} = \frac{3}{8} + \frac{5}{12x}$
9. $\frac{4x-1}{10x} = \frac{9}{5} + \frac{1}{4x}$
10. $\frac{1}{4} - \frac{7x-11}{8x} = \frac{3}{4x}$
11. $\frac{4x-7}{10x} - \frac{1}{3} = \frac{8x-1}{3x} + \frac{7}{10}$
12. $\frac{4x-9}{8x} + \frac{7}{6x} = \frac{9x-11}{24x} - \frac{3}{4}$

SECTION 19.2 Solve.

13. $\frac{6}{2x-7} = \frac{2}{3}$
14. $\frac{4}{2x-5} = \frac{6}{5x}$
15. $\frac{5}{6x-1} - \frac{1}{3} = \frac{4}{6x-1}$
16. $\frac{-6}{4x+3} - \frac{3}{4} = \frac{5}{4x+3} - 2$
17. $\frac{4}{2x-3} + 2 = \frac{-x}{2x-3} - \frac{1}{4}$
18. $\frac{7x+3}{5x-3} = 2$
19. $\frac{5x-1}{8x-3} = \frac{5}{6}$
20. $\frac{3x+1}{9x-8} = \frac{1}{6} - \frac{4x+5}{9x-8}$
21. $\frac{2x+7}{4x-7} + \frac{3}{5} = \frac{1-2x}{4x-7} - 1$
22. $\frac{2}{2x+3} = \frac{5}{2x+5}$
23. $\frac{12}{3x-8} = \frac{-3}{x+2}$
24. $\frac{8}{3x-7} + \frac{12}{6x-1} = 0$
25. $\frac{28}{16x+4} = \frac{9x}{4x+1} - 5$
26. $\frac{-12}{20x-15} - 1 = \frac{11x}{4x-3}$
27. $\frac{4x}{28x-21} = \frac{-5}{20x-15} - 1$
28. $\frac{4x}{14x+7} = \frac{-22}{70x+35} - \frac{2}{5}$
29. $\frac{7x}{9x-4} = \frac{13}{36x-16} - \frac{3}{4}$
30. $\frac{2}{12x-36} - \frac{1}{4} = \frac{2x}{3x-9}$
31. $\frac{-3}{50x-50} + \frac{3}{5} = \frac{7x}{30x-30}$
32. $\frac{-7x}{12x-24} - \frac{3}{4} = \frac{25}{15x-30}$
33. $\frac{7x}{48x-48} = \frac{1}{54x-54} + \frac{2}{9}$

SECTION 19.3 Solve for A.

34. $AR^2 = T$
35. $AC = 7B$
36. $\frac{R^2}{6} = \frac{AM}{4}$

37. $\frac{5AB^3}{6} = \frac{B^5C^2}{8}$

38. $R^2T^3 = \frac{A}{RT}$

39. $\frac{B}{E} = \frac{AB}{CD}$

40. $Ak^2 = 5k - 9$

41. $AT^2 + 3BT = K$

42. $5x - 12Ay = 20$

43. $R^2S = T^2 - AR$

44. $3y^3 - 8y^2 + A = 0$

45. $Ax^2 - 11Bx - 21 = C$

46. $\frac{A}{K} = D - \frac{M}{D}$

47. $\frac{AH}{E} = HP + \frac{E}{P}$

48. $AB - \frac{K}{B} = \frac{L}{K}$

49. $\frac{RW}{V} = AU + \frac{W}{R}$

50. $\frac{5A}{9} + \frac{B}{6} = \frac{3C}{4}$

51. $\frac{5x}{8} - \frac{7A}{9} = \frac{xz}{12}$

52. $\frac{A+5}{15} = \frac{B-6}{9}$

53. $\frac{2A-9}{14} = \frac{6B+7}{21}$

54. $\frac{2A-7}{12} = \frac{B}{6} - \frac{5C-1}{8}$

55. $\frac{8E-3}{18} = \frac{3M}{4} - \frac{2A-9}{9}$

56. $\frac{B}{4} = \frac{Ax - 2x^2}{5By}$

57. $\frac{6T^2}{5} = \frac{Ax - T}{4T}$

CUMULATIVE TEST

The following problems test your understanding of this chapter and related subject matter of previous chapters. Before taking this test, thoroughly review Sections 10.2, 10.4, 11.2, 19.1, 19.2, and 19.3.

Once you have finished the test, compare your answers with the answers provided at the back of the book. Note the section number of each problem missed, and thoroughly review those sections again.

Solve for x or A.

1. $5x = 7y - 11$

2. $Y^3Z^4 = \frac{A}{YZ}$

3. $\frac{8}{x} = \frac{12}{7}$

4. $\frac{4}{2x-3} = \frac{2}{3}$

5. $a^2 + b^2 = Aab - 25$

6. $12x = -3$

7. $\frac{11}{12} - \frac{5x-16}{24x} = \frac{19}{12x}$

8. $3x - \frac{7}{12} = 4 + \frac{7x}{4}$

9. $\frac{x}{3x-5} = \frac{32}{18x-30} - \frac{5}{6}$

10. $\frac{4A}{5B} = \frac{12B^3}{75C}$

11. $\frac{2x}{5} - 1 = 3x$

12. $\frac{6x-1}{12x} = \frac{1}{6} + \frac{11}{4x}$

13. $8x - 15y = 6y + 21z$

14. $2 - \frac{x}{5} = \frac{3}{4}$

15. $\frac{3A-5}{18} = \frac{3B}{4} - \frac{2B-5}{12}$

16. $\frac{8x+10}{6x-1} = 2$

17. $8y - 13x - 5z = 7y - 5z$

18. $\frac{5x+1}{4} = 2 + \frac{x}{5}$

19. $\frac{3x+16}{20x} - \frac{5}{28} = \frac{13}{20} - \frac{3x+16}{28x}$

20. $\frac{12}{12x-4} = \frac{x}{3x-1} - 3$

21. $12y - 3x = 16z - 1$

22. $11 + 9x = 19 - 4x$

23. $\frac{5}{8} + \frac{x}{3} = \frac{7x}{12} + 1$

24. $\frac{4}{x} - 3 = \frac{5}{6} + \frac{1}{6x}$

25. $At - t^2 = 2$

26. $18y - 12x + 15z = 10x - y$

27. $\frac{2x+5}{9x-4} = \frac{3}{5} - \frac{4x+1}{9x-4}$

28. $\frac{4}{x} = \frac{16}{9}$

29. $14x = 8y - 10z$

30. $\frac{x-6}{4} = \frac{1-6x}{6}$

31. $\frac{3}{2x-7} = \frac{8}{2x+3}$

32. $\frac{7x}{5} - \frac{8x-3}{6} = \frac{4}{15} + \frac{17x}{10}$

33. $\frac{4A}{9} + \frac{5B}{6} = \frac{7C}{12}$

34. $-9x = 4y - 2x + 10z$

35. $\frac{1}{4} + \frac{2}{9x} = \frac{7}{4x} - \frac{2}{3}$

36. $\frac{2}{5x-2} + 3 = \frac{-3x}{5x-2} - \frac{2}{3}$

37. $9 - 8x = 7$

38. $\frac{A}{w} = B - \frac{C}{v}$

39. $\frac{11}{8} - \frac{3x-10}{4x} = \frac{5}{8x}$

40. $\frac{11x}{24x+36} - \frac{3}{4} = \frac{2}{6x+9}$

20 QUADRATIC EQUATIONS

Quadratic Equations

In Chapter 16 we studied quadratic trinomials. In this chapter we will look at equations containing quadratic trinomials in x.

An equation of the form

$$ax^2 + bx + c = dx^2 + ex + f$$

where a through f represent integers with a and d not equal, is called a *quadratic equation.* For example, $-15x - 98 = x^2 - 36x$ and $2x^2 + 5x - 8 = x^2 + 6x + 4$ are quadratic equations.

An equation of the form

$$ax^2 + bx + c = 0$$

where a, b, and c represent integers with a greater than 0, is called a *quadratic equation in general form.* For example, $x^2 = 0$, $x^2 + 1 = 0$, $5x^2 - 2x = 0$, and $3x^2 + 2x - 8 = 0$ are quadratic equations in general form. Note that every quadratic equation can be simplified to a quadratic equation in general form using the methods in Chapter 11.

Roots of a Quadratic Equation

A quadratic equation is a nonlinear equation, because x^2 is not linear. A quadratic equation may have *no* roots, a *single* root, or *two* roots. For example,

$x^2 = -1$	has no roots, because any signed number raised to the second power is nonnegative
$x^2 = 0$	has one root, 0
$x^2 = x$	has two roots, 0 and 1

All the equations we will study here have at least one root.

Quadratic equations cannot be solved using the methods in Chapter 10 because of the x^2 terms. The method we will use in this chapter relies on factoring.

20.1
SOLVING FACTORED EQUATIONS

In this section we will look at equations in which the left member is completely factored and the right member is zero, such as $6x(x - 4) = 0$ or $(11x + 1)(x + 13)^2 = 0$.

1. Evaluate $3x(x + 2)$, when
   a. $x = 0$
   b. $x = -2$

2. Evaluate $(x - 5)(x + 7)$, when
   a. $x = 5$
   b. $x = -7$

3. Evaluate $-x^2(4x - 1)$, when
   a. $x = 0$
   b. $x = \frac{1}{4}$

4. Evaluate $(3x - 5)(5x + 3)$, when
   a. $x = \frac{5}{3}$
   b. $x = \frac{-3}{5}$

Roots

What can we tell about the roots of such an equation? Because the right member is 0, the roots are the values of x that make the left member zero. Certainly any value of x that makes one of the factors on the left zero makes the left member zero, because zero times any expression is zero. For example, $6x(x - 4)$, when $x = 4$ is

$$6x(x - 4) = 6(4)(4 - 4) = 6(4)(0) = 0$$

Hence, 4 is a root of $6x(x - 4) = 0$.

Conversely, any value of x that makes the left member zero makes at least one of its factors zero. This can be seen from the following property of numbers:

LAW OF ZERO PRODUCTS

If the product of two factors is zero, then at least one of the two factors is itself zero.

$A \cdot B = 0 \Rightarrow A = 0$, *or* $B = 0$, *or* $A = 0$ and $B = 0$

$n \cdot A = 0 \Rightarrow A = 0$ $\qquad$ $A^2 = 0 \Rightarrow A = 0$

(where A and B represent polynomials and n represents a nonzero number)

5. Solve
   a. $4x = 0$
   b. $9(x - 1) = 0$
   c. $-7(x + 2) = 0$

6. Solve
   a. $(x - 1)^2 = 0$
   b. $(x + 2)^2 = 0$

It follows that if one of two factors of zero is a nonzero constant, the other factor must be zero. For example, if $6x = 0$ then $x = 0$. Similarly, if $-6(x - 4) = 0$ then $(x - 4) = 0$. Further, if any power of a factor is zero, then the factor is zero. For example, if $x^2 = 0$ then $x = 0$. Similarly, if $(x - 4)^2 = 0$ then $(x - 4) = 0$.

Derived Equations

7. Form the derived equations of
   a. $8x(x + 1) = 0$
   b. $-4x = 0$
   c. $(x - 2)(4x + 1) = 0$
   d. $x^2(3x - 5) = 0$

We can see that to find the roots we must look at the distinct nonconstant factors of the left member. With each of them we will form a *derived equation* by setting it equal to zero. For example, the equation $6x(x - 4) = 0$ yields the derived equations $x = 0$ and $(x - 4) = 0$. Similarly the equation $(11x + 1)(x + 13)^2 = 0$ yields the derived equations $(11x + 1) = 0$ and $(x + 13) = 0$.

The root of each derived equation is a root of the original equation, and every root of the original equation is a root of one of its derived equations. To emphasize that an equation with two or more roots does not completely specify the value of the unknown, we must always indicate that its solutions (of the form "literal = root") are *alternative solutions.* For example, because $6x(x - 4)$ has the roots 0 and 4, we will say that the solution to this equation is $x = 0$ *or* $x = 4$.

Look at the following examples of *solving factored equations in which the right member is zero.*

TYPICAL PROBLEMS

1. $$6x(x-4)=0$$

We first form the derived equations by setting each distinct nonconstant factor of the left member equal to zero:

$$\downarrow$$
$$x=0\ ;\ x-4=0$$

Then we solve each derived equation. We separate the solutions by the word "or":

$$\downarrow$$
$$x=0 \text{ or } x=4$$

Hence, $6x(x-4)=0 \quad \Rightarrow \quad x=0$ or $x=4$.

2. $$(11x+1)(x+13)^2=0$$

We first form the derived equations by setting each distinct nonconstant factor of the left member equal to zero:

$$\downarrow$$
$$11x+1=0\ ;\ x+13=0$$

Then we solve each derived equation. We separate the solutions by the word "or":

$$\downarrow$$
$$x=\frac{-1}{11} \text{ or } x=-13$$

Hence, $(11x+1)(x+13)^2=0 \quad \Rightarrow \quad x=\frac{-1}{11}$ or $x=-13$.

To solve each derived equation we use the methods in Chapter 10.

EXAMPLES

Solve the following factored equations.

1. $x(x-8)=0$
2. $x(2x-1)^2=0$
3. $-(x+2)(x-9)=0$
4. $(2x+8)(9x-2)=0$
5. $7(3x-5)^2=0$
6. $-x(x+7)(x-7)=0$
7. $8x(8x-1)^2=0$
8. $-6(x-2)(x-3)=0$

Solutions

1. $x=0\ ;\ (x-8)=0$
   $\downarrow$
   $x=0$ or $x=8$

2. $x=0\ ;\ (2x-1)=0$
   $\downarrow$
   $x=0$ or $x=\frac{1}{2}$

3. $(x+2)=0\ ;\ (x-9)=0$
   $\downarrow$
   $x=-2$ or $x=9$

4. $(2x+8)=0\ ;\ (9x-2)=0$
   $\downarrow$
   $x=-4$ or $x=\frac{2}{9}$

5. $(3x - 5) = 0$
$\downarrow$
$x = \frac{5}{3}$

6. $x = 0$; $(x + 7) = 0$; $(x - 7) = 0$
$\downarrow$
$x = 0$ or $x = -7$ or $x = 7$

7. $x = 0$; $(8x - 1) = 0$
$\downarrow$
$x = 0$ or $x = \frac{1}{8}$

8. $(x - 2) = 0$; $(x - 3) = 0$
$\downarrow$
$x = 2$ or $x = 3$

OBSERVATIONS AND REFLECTIONS

In the introduction to this section we observed that if the product of two expressions is 0, then at least one of them must itself be 0. To put this another way, the product of two *nonzero* expressions is *not* 0.

$$P \neq 0 \text{ and } Q \neq 0 \quad \Rightarrow \quad P \cdot Q \neq 0$$

(where P and Q represent polynomials)

Note that this relationship is not true for addition. It is easy to find two nonzero expressions whose sum is zero (i.e., two expressions that are negatives of each other). For example, the sum of -1 and 1 is 0.

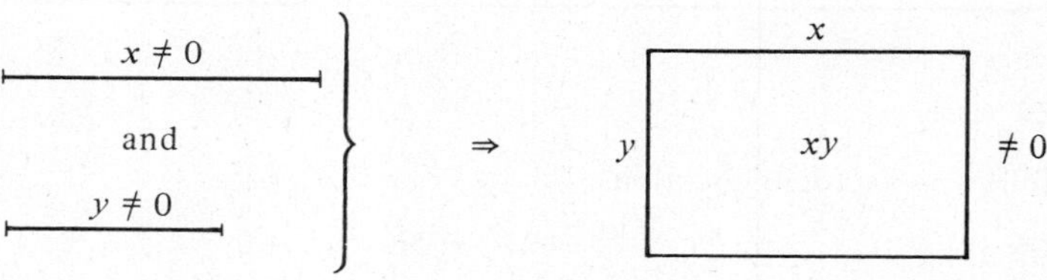

EXERCISES

Determine all solutions of each of the following equations.

1. $x(x - 1) = 0$
2. $x(x + 4) = 0$
3. $(x - 4)(x + 5) = 0$
4. $(x + 3)(x - 8) = 0$
5. $-x(3x - 1) = 0$
6. $-x(5x + 2) = 0$
7. $-(2x + 5)(3x - 4) = 0$
8. $-(4x - 1)(8x + 3) = 0$
9. $3x(x - 5) = 0$
10. $5x(x - 2) = 0$
11. $6x(2x - 1) = 0$
12. $-4x(7x - 2) = 0$
13. $3x(x - 5)(2x + 5) = 0$
14. $5x(3x - 7)(x + 3) = 0$
15. $x^2(x - 6) = 0$
16. $-x(x - 4)^2 = 0$
17. $x(3x - 2)^2 = 0$
18. $x^2(4x + 5) = 0$

20.2 SOLVING QUADRATIC EQUATIONS IN GENERAL FORM

In this section we will look at quadratic equations in *general form.* Remember that an equation of the form

$$ax^2 + bx + c = 0$$

where a, b, and c represent integers with a greater than 0, is called a *quadratic equation in general form.* In this chapter we will only consider equations in which the left member can be factored in integers using the methods of Chapters 14 and 16. Hence, with these equations we will always be able to factor the left member into a product of linear monomials and binomials, and then solve them using the method of the previous section.

We will also discuss a class of equations (in general form) in which the left member is x times a quadratic trinomial in x, such as $x^3 + 3x^2 + 2x = 0$. These latter equations can be solved in the same manner as quadratic equations.

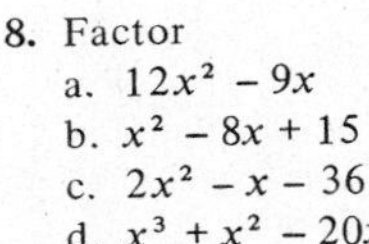

8. Factor
   a. $12x^2 - 9x$
   b. $x^2 - 8x + 15$
   c. $2x^2 - x - 36$
   d. $x^3 + x^2 - 20x$

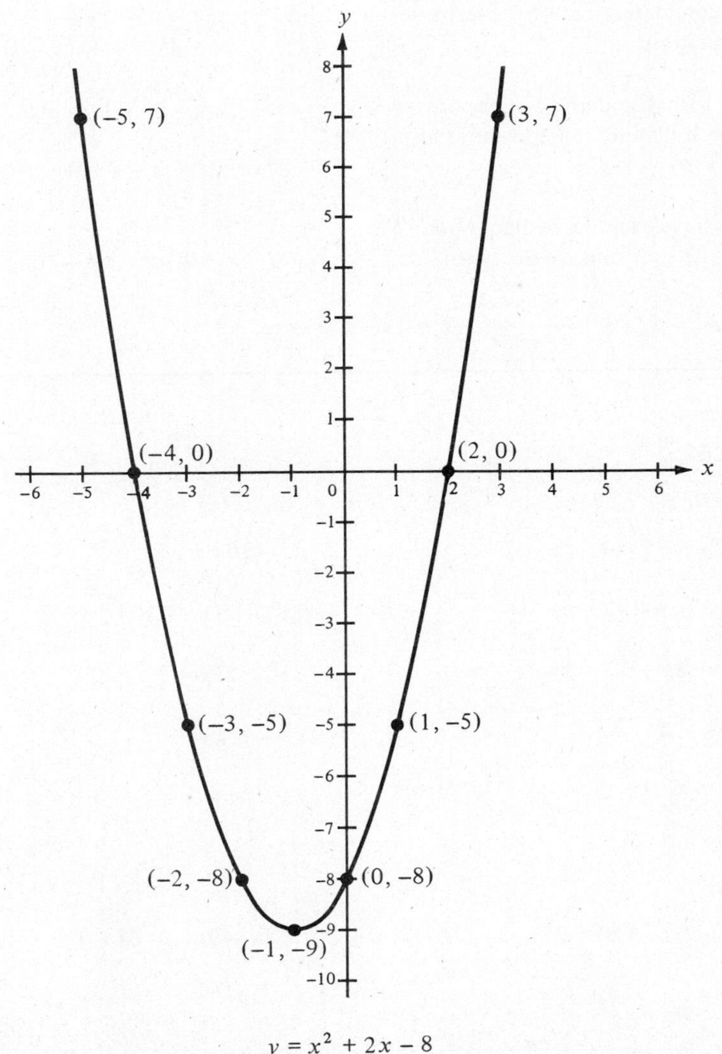

$y = x^2 + 2x - 8$

Look at the following examples of *solving equations in general form.*

TYPICAL PROBLEMS

1. $x^2 - 3x - 10 = 0$

We first completely factor the left side of the equation: $(x - 5)(x + 2) = 0$

Then we form the derived equations by setting each distinct nonconstant factor equal to zero: $x - 5 = 0 \ ; \ x + 2 = 0$

Then we solve each derived equation. We separate the solutions by the word "or": $x = 5$ or $x = -2$

Hence, $x^2 - 3x - 10 = 0 \Rightarrow x = 5$ or $x = -2$.

2. $4x^3 - 52x^2 + 144x = 0$

We first completely factor the left side of the equation: $4x(x - 4)(x - 9)$

Then we form the derived equations by setting each distinct nonconstant factor equal to zero: $x = 0 \ ; \ x - 4 = 0 \ ; \ x - 9 = 0$

Then we solve each derived equation. We separate the solutions by the word "or": $x = 0$ or $x = 4$ or $x = 9$

Hence, $4x^2 - 52x + 144 \Rightarrow x = 0$ or $x = 4$ or $x = 9$.

To factor the left member of the equation we use the methods in Chapter 16. To solve the derived equation we use the method in the previous section.

EXAMPLES

Solve the following equations in general form by factoring.

1. $x^2 - 2x - 35 = 0$
2. $x^2 - 15x + 54 = 0$
3. $6x^2 + 72x + 192 = 0$
4. $7x^3 - 77x^2 - 294x = 0$
5. $21x^2 - 68x + 32 = 0$
6. $6x^2 + 29x - 120 = 0$
7. $x^2 - 36 = 0$
8. $x^2 - 10x = 0$
9. $3x^2 - 75 = 0$
10. $12x^2 - 18x = 0$

Solutions

1. $(x - 7)(x + 5) = 0$
   $\downarrow$
   $(x - 7) = 0 \ ; \ (x + 5) = 0$
   $\downarrow$
   $x = 7$ or $x = -5$

2. $(x - 9)(x - 6) = 0$
   $\downarrow$
   $(x - 9) = 0 \ ; \ (x - 6) = 0$
   $\downarrow$
   $x = 9$ or $x = 6$

3. $6(x+8)(x+4)=0$

$\downarrow$

$(x+8)=0 \quad ; \quad (x+4)=0$

$\downarrow$

$x=-8$ or $x=-4$

4. $7x(x-14)(x+3)=0$

$\downarrow$

$x=0 \quad ; \quad (x-14)=0 \quad ; \quad (x+3)=0$

$\downarrow$

$x=0$ or $x=14$ or $x=-3$

5. $(3x-8)(7x-4)=0$

$\downarrow$

$(3x-8)=0 \quad ; \quad (7x-4)=0$

$\downarrow$

$3x=8 \quad ; \quad 7x=4$

$\downarrow$

$x=\frac{8}{3}$ or $x=\frac{4}{7}$

6. $(2x+15)(3x-8)=0$

$\downarrow$

$(2x+15)=0 \quad ; \quad (3x-8)=0$

$\downarrow$

$2x=-15 \quad ; \quad 3x=8$

$\downarrow$

$x=\frac{-15}{2}$ or $x=\frac{8}{3}$

7. $(x+6)(x-6)=0$

$\downarrow$

$(x+6)=0 \quad ; \quad (x-6)=0$

$\downarrow$

$x=-6$ or $x=6$

8. $x(x-10)=0$

$\downarrow$

$x=0 \quad ; \quad (x-10)=0$

$\downarrow$

$x=0$ or $x=10$

9. $3(x^2-25)=0$

$\downarrow$

$3(x+5)(x-5)=0$

$\downarrow$

$(x+5)=0 \quad ; \quad (x-5)=0$

$\downarrow$

$x=-5$ or $x=5$

10. $6x(2x-3)=0$

$\downarrow$

$x=0 \quad ; \quad (2x-3)=0$

$\downarrow$

$x=0$ or $x=\frac{3}{2}$

SPECIAL CASES

If the left member of a quadratic equation in general form is the square of a binomial, the equation has only one root. For example,

$$x^2+2x+1=0 \quad \Rightarrow \quad (x+1)^2=0 \quad \Rightarrow \quad (x+1)=0 \quad \Rightarrow \quad x=-1$$

so -1 is the only root of x^2+2x+1.

9. Factor and solve
a. $x^2-2x+1=0$
b. $x^2+4x+4=0$
c. $x^2-10x+25=0$
d. $4x^2-12x+9=0$

OBSERVATIONS AND REFLECTIONS

The *general form* of the quadratic equation is the only form in which factoring can be used to solve the equation. If a product equals zero, one of the factors must equal zero. But if a product equals any other number, nothing can be said about either factor in isolation. For example, if $a \cdot b = 0$, then $a=0$ or $b=0$, or both are zero. But if, on the other hand, $a \cdot b = 3$, then nothing can be said about a or b alone (except, of course, that neither equals zero). (For example, if $a \cdot b = 3$, then a could be 3 and b could be 1, or a could be -6 and b could be $-\frac{1}{2}$, and so forth.)

EXERCISES

Determine all solutions of each of the following equations.

In these exercises, all trinomials are factorable into binomials.

1. $x^2 + 10x + 21 = 0$
2. $x^2 + 6x - 55 = 0$
3. $x^2 - 14x - 15 = 0$
4. $x^2 + 16x + 39 = 0$
5. $x^2 - 23x + 42 = 0$
6. $x^2 + 13x - 30 = 0$
7. $x^2 + 29x - 30 = 0$
8. $x^2 - 38x + 105 = 0$
9. $x^2 + 36x + 99 = 0$
10. $x^2 + 20x - 125 = 0$
11. $x^2 - x - 56 = 0$
12. $x^2 + 21x + 90 = 0$

In these exercises, all trinomials are factorable into monomials and simple binomials.

13. $3x^2 - 6x - 45 = 0$
14. $12x^2 - 60x + 48 = 0$
15. $-18x^2 - 108x - 144 = 0$
16. $-3x^2 - 45x + 102 = 0$
17. $-x^3 + 6x^2 + 16x = 0$
18. $x^3 + 11x^2 + 30x = 0$
19. $2x^3 - 60x^2 + 378x = 0$
20. $-3x^3 + 123x^2 + 126x = 0$
21. $-2x^3 + 18x^2 + 324x = 0$
22. $7x^2 - 217x + 1470 = 0$
23. $4x^2 + 180x + 1800 = 0$
24. $-6x^3 - 348x^2 + 720x = 0$

In these exercises, all trinomials are factorable general trinomials.

25. $5x^2 + 12x + 7 = 0$
26. $3x^2 - 20x - 7 = 0$
27. $6x^2 + 13x - 5 = 0$
28. $5x^2 + 28x + 15 = 0$
29. $14x^2 - 41x + 15 = 0$
30. $6x^2 - 41x - 22 = 0$
31. $35x^2 + x - 12 = 0$
32. $25x^2 - 150x + 176 = 0$
33. $21x^2 + 82x + 40 = 0$
34. $25x^2 + 5x - 42 = 0$

In these exercises, all binomials are factorable into products of monomials and linear binomials.

35. $x^2 + 5x = 0$
36. $x^2 - 8x = 0$
37. $3x^2 + 18x = 0$
38. $7x^2 - 42x = 0$
39. $6x^2 + 27x = 0$
40. $12x^2 - 15x = 0$
41. $x^2 - 1 = 0$
42. $x^2 - 4 = 0$
43. $x^2 - 81 = 0$
44. $x^2 - 16 = 0$
45. $4x^2 - 49 = 0$
46. $64x^2 - 1 = 0$
47. $36x^2 - 121 = 0$
48. $144x^2 - 49 = 0$

20.3
SOLVING QUADRATIC EQUATIONS

We will now discuss a large class of equations, which are not in general form but are equivalent to the equations in general form that we saw in the previous section. Such equations must be converted to general form by the methods in Chapter 11 before they can be solved by factoring.

Look at the following examples of *solving quadratic equations.*

TYPICAL PROBLEMS

1. $2x^2 + 5x - 8 = x^2 + 6x + 4$

We first shift all terms to the left side to place the equation in general form:

$$\downarrow$$
$$x^2 - x - 12 = 0$$

We then completely factor the left side of the equation:

$$\downarrow$$
$$(x - 4)(x + 3) = 0$$

Then we form the derived equations by setting each distinct nonconstant factor equal to zero:

$$\downarrow$$
$$x - 4 = 0 \quad ; \quad x + 3 = 0$$

Finally we solve each derived equation. We separate the solutions by the word "or":

$$\downarrow$$
$$x = 4 \quad \text{or} \quad x = -3$$

Hence, $2x^2 + 5x - 8 = x^2 + 6x + 4 \quad \Rightarrow \quad x = 4$ or $x = -3$.

2. $7x^3 + 54x^2 - 189x = 4x^3 + 15x^2 - 45x$

We first shift all terms to the left side to place the equation in general form:

$$\downarrow$$
$$3x^3 + 39x^2 - 144x = 0$$

We then completely factor the left side of the equation:

$$\downarrow$$
$$3x(x + 16)(x - 3) = 0$$

Then we form the derived equations by setting each distinct nonconstant factor equal to zero:

$$\downarrow$$
$$x = 0 \quad ; \quad x + 16 = 0 \quad ; \quad x - 3 = 0$$

Finally we solve each derived equation. We separate the solutions by the word "or":

$$\downarrow$$
$$x = 0 \quad \text{or} \quad x = -16 \quad \text{or} \quad x = 3$$

Hence, $7x^3 + 54x^2 - 189x = 4x^3 + 15x^2 - 45x \quad \Rightarrow \quad x = 0$ or $x = -16$ or $x = 3$

To solve quadratic equations we extend the method in the previous section.

EXAMPLES

Solve the following quadratic equations by shifting terms and factoring.

1. $-98 - 15x = x^2 - 36x$
2. $8x^2 + 11x - 29 = 9x^2 + 31x - 125$
3. $3x^2 + 51x - 113 = 9x^2 - 33x + 175$
4. $7x^3 - 114x^2 + 325x = 3x^3 - 22x^2 - 155x$
5. $24x - 108 = 30x^2 + 35x - 136$
6. $99x^2 - 37x + 146 = 15x^2 - 137x + 121$

Solutions

1.
$$-x^2 + 21x - 98 = 0$$
$$\downarrow$$
$$-(x - 14)(x - 7) = 0$$
$$\downarrow$$
$$(x - 14) = 0 \;;\; (x - 7) = 0$$
$$\downarrow$$
$$x = 14 \text{ or } x = 7$$

2.
$$-x^2 - 20x + 96 = 0$$
$$\downarrow$$
$$-(x + 24)(x - 4) = 0$$
$$\downarrow$$
$$(x + 24) = 0 \;;\; (x - 4) = 0$$
$$\downarrow$$
$$x = -24 \text{ or } x = 4$$

3.
$$-6x^2 + 84x - 288 = 0$$
$$\downarrow$$
$$-6(x - 6)(x - 8) = 0$$
$$\downarrow$$
$$(x - 6) = 0 \;;\; (x - 8) = 0$$
$$\downarrow$$
$$x = 6 \text{ or } x = 8$$

4.
$$4x^3 - 92x^2 + 480x = 0$$
$$\downarrow$$
$$4x(x - 15)(x - 8) = 0$$
$$\downarrow$$
$$x = 0 \;;\; (x - 15) = 0 \;;\; (x - 8) = 0$$
$$\downarrow$$
$$x = 0 \text{ or } x = 15 \text{ or } x = 8$$

5.
$$-30x^2 - 11x + 28 = 0$$
$$\downarrow$$
$$-(5x - 4)(6x + 7) = 0$$
$$\downarrow$$
$$(5x - 4) = 0 \;;\; (6x + 7) = 0$$
$$\downarrow$$
$$5x = 4 \;;\; 6x = -7$$
$$\downarrow$$
$$x = \frac{4}{5} \text{ or } x = \frac{-7}{6}$$

6.
$$84x^2 + 100x + 25 = 0$$
$$\downarrow$$
$$(6x + 5)(14x + 5) = 0$$
$$\downarrow$$
$$(6x + 5) = 0 \;;\; (14x + 5) = 0$$
$$\downarrow$$
$$6x = -5 \;;\; 14x = -5$$
$$\downarrow$$
$$x = \frac{-5}{6} \text{ or } x = \frac{-5}{14}$$

EXERCISES

Determine all solutions of each of the following equations.

In these exercises, all equations, when placed in general form, contain factorable trinomials.

1. $x^2 - x = 12$
2. $x^2 + 5x = 24$
3. $x^2 + 24 = 10x$
4. $x^2 + 44 = -15x$
5. $x^2 = 3x + 70$
6. $x^2 = 17x + 60$
7. $x^2 + x + 72 = -17x$
8. $x^2 + 22x + 120 = 56x$

9. $x^2 - 10x - 100 = 6x + 125$

10. $x^2 + 15x + 60 = 14x + 300$

11. $10x + 36 = x^2 - 25x + 300$

12. $10x + 11 = x^2 + 101x + 101$

13. $3x^2 - 5x + 110 = 2x^2 - 3x + 334$

14. $8x^2 - 11x + 60 = 9x^2 - 89x - 100$

In these exercises, all equations, when placed in general form, contain trinomials which are factorable into monomials and simple trinomials.

15. $6x^2 - 48x + 74 = 18x - 94$

16. $5x^2 - 14x + 235 = -84x + 70$

17. $x^3 - 17x^2 - 58x = -18x^2 + 14x$

18. $-x^3 - 43x^2 + 192x = -28x^2 + 66x$

19. $5x^3 + 17x^2 + 143x = 3x^3 - 23x^2 - 49x$

20. $8x^3 - 43x^2 + 592x = 6x^3 + 37x^2 - 158x$

21. $5x^3 - 58x^2 - 172x = 7x^3 + 94x^2 - 812x$

22. $5x^3 + 186x^2 - 44x = 17x^3 + 54x^2 - 188x$

In these exercises, all equations, when placed in general form, contain factorable general trinomials.

23. $14x^2 + 75x = 11$

24. $17x^2 = 49x + 6$

25. $6x^2 + 38x + 11 = 19x - 4$

26. $33x - 62 = 55x^2 - 54x - 36$

27. $64x^2 + 38x + 60 = 46x^2 + 23x + 85$

28. $73x^2 + 13x + 63 = 53x^2 + x + 36$

29. $56x^2 + 89x - 94 = 80x^2 + 15x - 49$

30. $37x^2 - 83x - 144 = 67x^2 + 14x - 84$

ANSWERS TO MARGIN EXERCISES

1. a. 0
b. 0

2. a. 0
b. 0

3. a. 0
b. 0

4. a. 0
b. 0

5. a. $x = 0$
b. $x = 1$
c. $x = -2$

6. a. $x = 1$
b. $x = -2$

7. a. $x = 0$; $x + 1 = 0$
b. $x = 0$
c. $x - 2 = 0$; $4x + 1 = 0$
d. $x = 0$; $3x - 5 = 0$

8. a. $3x(4x - 3)$
b. $(x - 3)(x - 5)$
c. $(2x - 9)(x + 4)$
d. $x(x - 4)(x + 5)$

9. a. $x = 1$
b. $x = -2$
c. $x = 5$
d. $x = \frac{3}{2}$

CHAPTER REVIEW

SECTION 20.1 Determine all solutions.

1. $x(x+9)=0$
2. $-(x-7)(x+10)=0$
3. $x(4x-3)=0$
4. $(9x-2)(3x-8)=0$
5. $-7x(x+5)=0$
6. $8x(8x-1)=0$
7. $-12x(4x-9)(x+9)=0$
8. $x^2(x+8)=0$
9. $-x(6x-11)^2=0$

SECTION 20.2 Determine all solutions.

10. $x^2-11x+10=0$
11. $x^2+4x-77=0$
12. $x^2+21x+110=0$
13. $x^2-10x-75=0$
14. $x^2-51x+98=0$
15. $x^2+4x-60=0$
16. $5x^2+20x-385=0$
17. $6x^2-72x+162=0$
18. $-x^3-10x^2+875x=0$
19. $3x^3+252x^2+729x=0$
20. $-10x^3-130x^2+900x=0$
21. $-3x^2+72x-432=0$
22. $11x^2-14x+3=0$
23. $6x^2-7x-10=0$
24. $14x^2+29x+15=0$
25. $18x^2-45x-50=0$
26. $16x^2+54x-81=0$
27. $x^2+7x=0$
28. $4x^2+64x=0$
29. $15x^2+21x=0$
30. $x^2-25=0$
31. $x^2-36=0$
32. $81x^2-25=0$
33. $25x^2-169=0$

SECTION 20.3 Determine all solutions.

34. $x^2+9x=-14$
35. $x^2-45=4x$
36. $x^2=20x-75$
37. $x^2+3x-54=-12x$
38. $x^2+20x+80=2x+15$
39. $8x+120=x^2-42x-480$
40. $5x^2-40x+280=4x^2+16x-63$
41. $-9x^2-84x+324=-21x+162$
42. $x^3-11x^2+42x=8x^2-18x$
43. $4x^3+52x^2+314x=9x^3-78x^2-646x$
44. $x^2-59x-362=4x^2-36x-250$
45. $15x^2+3=14x$
46. $20x+34=22x^2+13x+25$
47. $48x^2-16x-35=28x^2+21x-17$
48. $96x^2+58x+134=8x^2+75x+140$

CUMULATIVE TEST

The following problems test your understanding of this chapter and related subject matter of previous chapters. Before taking this test, thoroughly review Sections 10.2, 11.2, 20.1, 20.2, and 20.3.

Once you have finished the test, compare your answers with the answers provided at the back of the book. Note the section number of each problem missed, and thoroughly review those sections again.

Solve for x.

1. $x(3x - 8) = 0$
2. $x^2 - 19x + 48 = 0$
3. $x^2 - 18x = -56$
4. $16x^2 - 49 = 0$
5. $x^3 - 32x^2 - 33x = 23x^2 + 23x$
6. $10 - 7x + 14 - 5x = 0$
7. $-16x(4x - 7)(x + 7) = 0$
8. $x^2 - 5x - 126 = 3x + 54$
9. $8x^3 + 15x^2 - 27x = 0$
10. $6y - 5x = 8y - 16$
11. $14x^2 + 27x + 9 = 0$
12. $30x^2 - 35 = 17x$
13. $3x(3x - 1) = 0$
14. $3 - 4x - 15 = 5x$
15. $8x^2 + 56x = 0$
16. $8x - 12y + 3z = 4x - 2z$
17. $x^2(x - 14) = 0$
18. $-8x - 3x - x - 1 = 0$
19. $(3x - 5)(6x + 1) = 0$
20. $x^2 = 2x + 48$
21. $18x^2 - 42x + 20 = 0$
22. $8y - 11x = 16y - 4$
23. $4x^2 + 68x - 72 = 0$
24. $-9x(x + 10) = 0$
25. $6 - 3x = 8x - 6$
26. $7x^2 - 12x - 14 = 6x^2 + 14x - 62$
27. $x^2 - 21x - 28 = 7x^2 - 38x - 16$
28. $x^2 + 2x - 80 = 0$
29. $3x - 18 = 2 + 15y$
30. $4x^2 - 18x - 68 = 3x^2 + 15x + 40$

In the previous chapter we saw how to solve quadratic equations such as $x^2 - 4 = 0$ by factoring. However, many quadratic equations are not factorable in integers and cannot be solved in this manner. For example, $x^2 - 2 = 0$ cannot be solved by factoring, because 2 is not a perfect square. Nevertheless, many of these equations do have roots, and we will see how to solve them using *square roots* at the end of this chapter.

Powers and Roots

Remember that multiplying an expression by itself any number of times yields a *power* of the expression. For example, the powers of 2 are 2, 4, 8, 16 . . . . An expression is said to be a *root* of each of its powers. In particular, an expression is the *square root* of its square. For example,

0 is the square root of 0 because $0 \cdot 0 = 0$
1 is the square root of 1 because $1 \cdot 1 = 1$
2 is the square root of 4 because $2 \cdot 2 = 4$
3 is the square root of 9 because $3 \cdot 3 = 9$

and so forth.

To talk about square roots we will use the radical sign $\sqrt{}$, which is read the square root of. (The sign $\sqrt{}$ is a modified form of the letter *r*, which stands for *radix*, a Latin word meaning root.) For example,

$$\sqrt{0} = 0 \qquad \sqrt{1} = 1 \qquad \sqrt{4} = 2 \qquad \sqrt{9} = 3 \qquad \sqrt{16} = 4$$

and so forth. Note that both $2 \cdot 2$ and $(-2)(-2)$ equal 4, so 2 and -2 are both square roots of 4. The radical sign is used exclusively for non-negative roots, so $\sqrt{4} = 2$ and $-\sqrt{4} = -2$.

In this chapter will study the square roots of non-negative integers, fractions, and literal numbers. We will not discuss square roots of negative numbers (*imaginary numbers*), because the square of every signed number is positive.

Most square roots of integers and fractions do not equal an integer or a fraction and, hence, are called *irrational numbers.* For example, $\sqrt{2}$ $\sqrt{3}$, $\sqrt{5}$, $\sqrt{6}$, and $\sqrt{7}$ are irrational, whereas $\sqrt{1}$, $\sqrt{4}$, and $\sqrt{9}$ are not. (Integers and fractions are called *rational numbers* to distinguish them from the irrational numbers.)

Locating Square Roots on the Number Line

In Chapter 6 we saw that integers and fractions can be represented on the number line. Square roots of non-negative numbers can also be located on the number line:

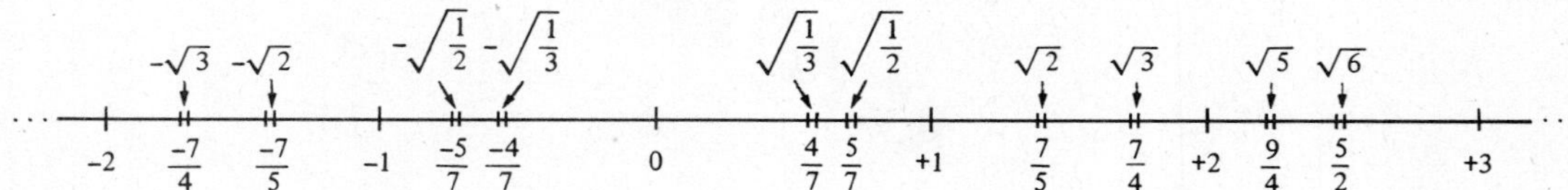

Note that $\sqrt{2}$ is very close to $\frac{7}{5}$ [after all, $\left(\frac{7}{5}\right)^2 = \frac{49}{25}$, which is very close to 2].

Similarly, $\sqrt{3}$ is close to $\frac{7}{4}$, and $\sqrt{5}$ is close to $\frac{9}{4}$. In fact, all square roots can be approximated (as closely as desired) by fractions.

Square Roots of Literals

As we saw in Section 16.6 when studying the difference of two squares, every literal to an even power is a perfect square. We will define the *principal square root* of an even power of a literal to be the literal with half the exponent. That is, $\sqrt{x^2} = x$, $\sqrt{x^4} = x^2$, $\sqrt{x^6} = x^3$, and so forth.

21.1 REDUCING SQUARE ROOTS TO SIMPLEST FORM

One number is the *square root* of another if multiplying it by itself yields the second number. For example, 2 and –2 are square roots of 4, and 3 and –3 are square roots of 9. The non-negative square root of a number is *indicated* by placing the number under the *radical sign* $\sqrt{}$. For example, $2 = \sqrt{4}$, and $3 = \sqrt{9}$.

The product of a square root and an integer or fraction is indicated by writing the integer or fraction before the radical sign, as in $2\sqrt{3}$, $\frac{3}{4}\sqrt{7}$, and $-8\sqrt{12}$.

Equivalent Expressions

Just as different fractions can represent the same quantity $\left(\text{for example, } \frac{1}{2} = \frac{2}{4} = \frac{3}{6}\right)$, different expressions containing square roots can represent the same quantity. We have already seen that $\sqrt{4} = 2$, $\sqrt{9} = 3$, and so forth. Similarly, different products of integers and square roots can be equal. For example, look at the following computations. We are using perfect squares under the radical signs so that you can check the results.

$$2\sqrt{4} = \sqrt{16} \qquad 2\sqrt{9} = \sqrt{36} \qquad 2\sqrt{16} = \sqrt{64}$$

$$2 \cdot 2 = 4 \qquad 2 \cdot 3 = 6 \qquad 2 \cdot 4 = 8$$

$3\sqrt{4} = \sqrt{36}$ $\quad 3 \cdot 2 = 6$

$3\sqrt{9} = \sqrt{81}$ $\quad 3 \cdot 3 = 9$

$3\sqrt{16} = \sqrt{144}$ $\quad 3 \cdot 4 = 12$

$4\sqrt{4} = \sqrt{64}$ $\quad 4 \cdot 2 = 8$

$4\sqrt{9} = \sqrt{144}$ $\quad 4 \cdot 3 = 12$

$4\sqrt{16} = \sqrt{256}$ $\quad 4 \cdot 4 = 16$

Can you see how to convert the product of an integer and a square root to a square root? Look again at the computations above and then compare them with these computations.

$2\sqrt{4} = \sqrt{16}$ $\quad 4 \cdot 4 = 16$

$2\sqrt{9} = \sqrt{36}$ $\quad 4 \cdot 9 = 36$

$2\sqrt{16} = \sqrt{64}$ $\quad 4 \cdot 16 = 64$

$3\sqrt{4} = \sqrt{36}$ $\quad 9 \cdot 4 = 36$

$3\sqrt{9} = \sqrt{81}$ $\quad 9 \cdot 9 = 81$

$3\sqrt{16} = \sqrt{144}$ $\quad 9 \cdot 16 = 144$

$4\sqrt{4} = \sqrt{64}$ $\quad 16 \cdot 4 = 64$

$4\sqrt{9} = \sqrt{144}$ $\quad 16 \cdot 9 = 144$

$4\sqrt{16} = \sqrt{256}$ $\quad 16 \cdot 16 = 256$

In each example we have squared each number in going from the first row to the second. [Note how easy it is to square an indicated square root: $(\sqrt{4}\,)^2 = 4$, $(\sqrt{9}\,)^2 = 9$, $(\sqrt{16}\,)^2 = 16$, and so forth.] As these examples show, we can bring a factor under the radical sign by squaring it. For example, $3\sqrt{4} = \sqrt{3^2 \cdot 4} = \sqrt{36}$.

1. Find the square of
   a. $\sqrt{5}$
   b. $\sqrt{6}$
   c. $\sqrt{7}$

2. Bring the integer factor under the radical sign.
   a. $2\sqrt{5}$
   b. $5\sqrt{3}$
   c. $6\sqrt{2}$

$$a\sqrt{b} = \sqrt{a^2 b}$$

(where a and b represent non-negative numbers)

We can reverse this process and bring factors out from under the radical sign. To do so we must factor the number under the radical sign into two factors, one of which is a perfect square. For example, $12 = 4 \cdot 3$, so $\sqrt{12} = \sqrt{4 \cdot 3} = 2\sqrt{3}$.

Reducing a Square Root to Simplest Form

A square root, or the product of a number and a square root, is said to be in *simplest form* if the number under the radical sign is not divisible by any perfect square (except 1). (We call such numbers *square-free*.) For example, $\sqrt{2}$, $3\sqrt{6}$, and $-4\sqrt{30}$ are in simplest form, whereas $\sqrt{18}$, $3\sqrt{24}$, and $-2\sqrt{90}$ are not.

To reduce a square root to simplest form we must determine the largest perfect square that divides the number under the radical sign. To do this we can use the Method of Primes and write the prime factorization of the number under the radical sign. Then we can separate the factors into a product of primes to even powers and a square-free product. For example,

3. Pair equivalent expressions.
   a. $\sqrt{12}$
   b. $3\sqrt{2}$
   c. $\sqrt{27}$
   d. $3\sqrt{3}$
   e. $\sqrt{18}$
   f. $2\sqrt{3}$

$200 = 2^3 \cdot 5^2 = (2^2 \cdot 5^2)(2)$, so $\sqrt{200} = \sqrt{(2^2 \cdot 5^2)(2)} = 2 \cdot 5\sqrt{2} = 10\sqrt{2}$

and

$96 = 2^5 \cdot 3 = (2^4)(2 \cdot 3)$, so $\sqrt{96} = \sqrt{(2^4)(2 \cdot 3)} = 2^2\sqrt{2 \cdot 3} = 4\sqrt{6}$

Literal Roots

We treat radical expressions containing literals the same way we treat numerical expressions. We define $\sqrt{x}$ the square root of x to be the non-negative quantity whose square is x, which we assume is also non-negative. It follows that $\sqrt{x^2} = x$, $\sqrt{x^3} = x\sqrt{x}$, $\sqrt{x^4} = x^2$, $\sqrt{x^5} = x^2\sqrt{x}$, $\sqrt{x^6} = x^3, \ldots$.

Just as with numerical square roots, a monomial square root is said to be in simplest form if the monomial under the radical sign is square-free; that is, if it is not divisible by any perfect square monomial (except 1). (See the examples of monomial perfect squares in Section 16.6.) For example, $\sqrt{7x}$, $5x^2y\sqrt{xy}$, and $-4x^3\sqrt{10}$ are in simplest form, whereas $\sqrt{7x^3}$, $3x\sqrt{5xy^2}$, and $-9x\sqrt{12x^5}$ are not. Note that in a monomial square root the radical factor is written last. For example, $4x^3\sqrt{10}$ is never written as $4\sqrt{10}\,x^3$ (which looks too much like $4\sqrt{10x^3}$).

Look at the following examples of *reducing square roots to simplest form.*

TYPICAL PROBLEMS

1.

We first factor the number under the radical sign into a product of primes:

Then we separate the primes into two factors: one a perfect square and the other square-free:

We then factor out the perfect square and multiply the square-free factors together:

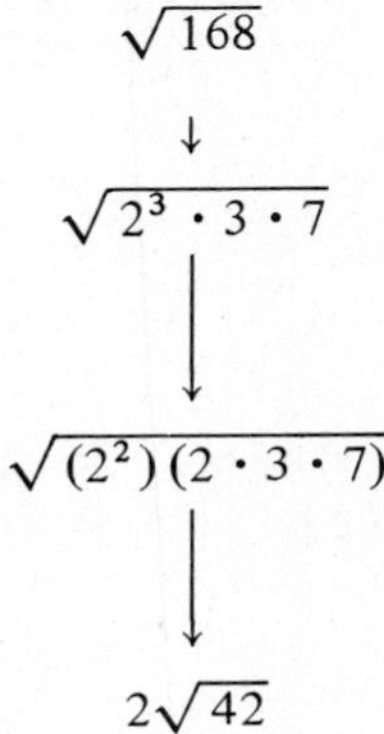

Hence, $\sqrt{168} = 2\sqrt{42}$.

2.

We first factor the numerical coefficient under the radical sign into a product of primes:

Then we separate the product under the radical sign into two factors: one a perfect square and the other square-free:

We then factor out the perfect square:

And we simplify further if necessary:

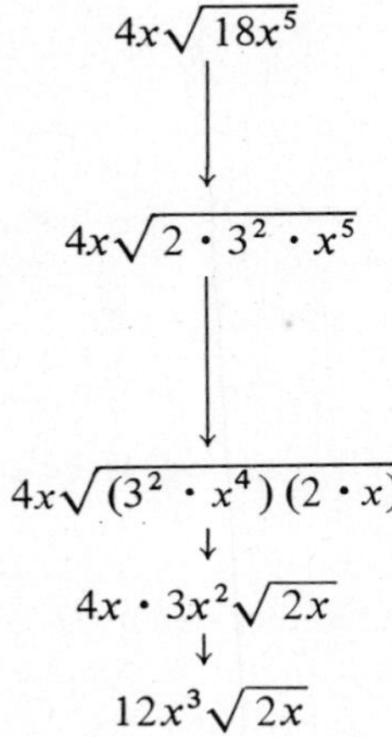

A table of prime factorizations of the numbers from 1 to 200 appears on the inside front cover.

Hence, $4x\sqrt{18x^5} = 12x^3\sqrt{2x}$.

EXAMPLES

Every square root can be reduced to a square root in simplest form.

Simplify the following expressions containing square roots using the Method of Primes.

1. $\sqrt{24}$
2. $9\sqrt{176}$

3. $\sqrt{81x^4y^3}$

4. $3x^5y\sqrt{30x^3y^6}$

Solutions

1. $\sqrt{2^3 \cdot 3} \rightarrow \sqrt{(2^2)(2 \cdot 3)} \rightarrow 2\sqrt{6}$

2. $9\sqrt{2^4 \cdot 11} \rightarrow 9(2^2)\sqrt{(11)} \rightarrow 36\sqrt{11}$

3. $\sqrt{3^4x^4y^3} \rightarrow \sqrt{(3^4x^4y^2)(y)} \rightarrow 3^2x^2y\sqrt{y} \rightarrow 9x^2y\sqrt{y}$

4. $3x^5y\sqrt{2 \cdot 3 \cdot 5 \cdot x^3y^6} \rightarrow 3x^5y\sqrt{(x^2y^6)(2 \cdot 3 \cdot 5 \cdot x)} \rightarrow 3x^5y(xy^3)\sqrt{2 \cdot 3 \cdot 5 \cdot x} \rightarrow 3x^6y^4\sqrt{30x}$

OBSERVATIONS AND REFLECTIONS

1. The square root of any positive number can be approximated as closely as desired by fractions. For example $\frac{7}{5}$, $\frac{99}{70}$, and $\frac{19{,}601}{13{,}860}$ are successively closer approximations of $\sqrt{2}$. $\left[\text{In fact, } \left(\frac{19{,}601}{13{,}860}\right)^2 = 2 + \frac{1}{192{,}099{,}600}.\right]$

The following procedure can be used to approximate square roots. Given a square root such as $\sqrt{5}$, pick a reasonably close approximation, such as 2. Next divide the number under the radical sign, 5, by this number. Then take their average (the average of 2 and $5 \div 2$) to get a better approximation:

$$\frac{2 + \frac{5}{2}}{2} = \frac{9}{4}$$

Repeating this process using $\frac{9}{4}$ instead of 2 gives the still better approximation

$$\frac{\frac{9}{4} + \left(5 \div \frac{9}{4}\right)}{2} = \frac{\frac{9}{4} + \frac{20}{9}}{2} = \frac{161}{72}$$

$\left[\text{In fact, } \left(\frac{161}{72}\right)^2 = 5 + \frac{1}{5{,}184}.\right]$

In general, if the fraction a is an approximation of $\sqrt{b}$, where a and b represent positive numbers, then

$$\frac{a + \frac{b}{a}}{2} = \frac{a^2 + b}{2a}$$

is a closer approximation of $\sqrt{b}$.

4. Find a closer approximation than

a. $\frac{7}{4}$ of $\sqrt{3}$

b. $\frac{5}{2}$ of $\sqrt{6}$

2. We can represent square roots of positive integers geometrically using the *Pythagorean theorem*. This theorem states that in a right triangle the square of the length of the longest side (the side opposite the right angle) equals the sum of the squares of the lengths of the other two sides. For example, the most familiar right triangle is the 3, 4, 5 triangle and $3^2 + 4^2 = 5^2$. Similarly, we can construct the 1, 1, $\sqrt{2}$ and 1, 2, $\sqrt{5}$ right triangles (because $1^2 + 1^2 = 2$ and $1^2 + 2^2 = 5$):

5. Determine the longest side of the right triangle whose other sides are
a. 2 and 3
b. 5 and 12

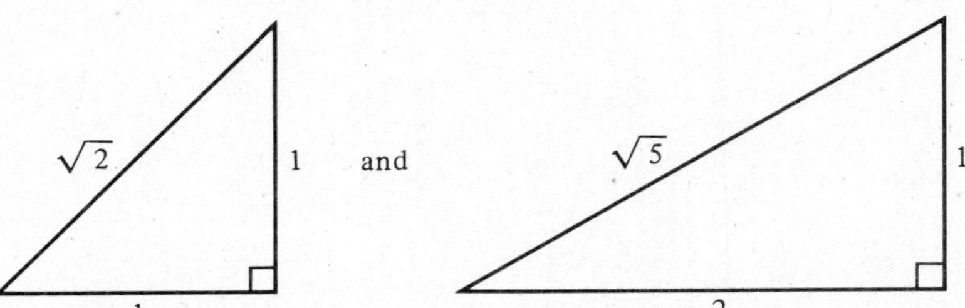

PYTHAGOREAN THEOREM

In an a, b, c right triangle,

$$a^2 + b^2 = c^2$$

(where a, b, and c represent positive numbers)

Note that $a^2 + b^2 = c^2$ implies that

$$a = \sqrt{c^2 - b^2} \qquad b = \sqrt{c^2 - a^2} \qquad c = \sqrt{a^2 + b^2}$$

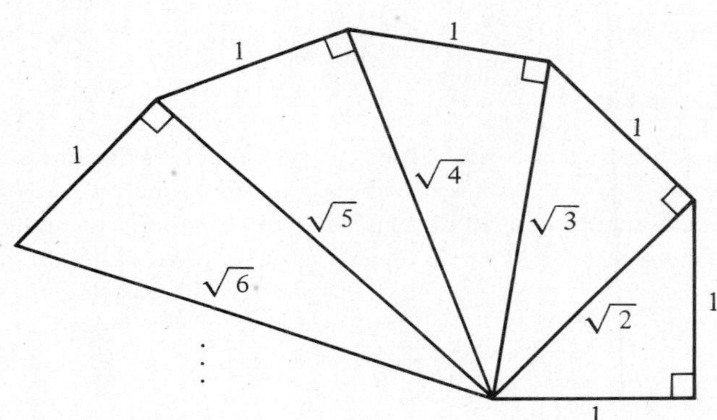

EXERCISES

Reduce each of the following expressions to simplest form.

1. $\sqrt{12}$	2. $\sqrt{20}$	3. $\sqrt{40}$
4. $\sqrt{49}$	5. $\sqrt{68}$	6. $\sqrt{63}$
7. $\sqrt{121}$	8. $\sqrt{90}$	9. $\sqrt{112}$
10. $\sqrt{96}$	11. $\sqrt{132}$	12. $\sqrt{156}$
13. $6\sqrt{54}$	14. $7\sqrt{48}$	15. $9\sqrt{98}$

16. $3\sqrt{88}$
17. $16\sqrt{126}$
18. $18\sqrt{150}$
19. $\sqrt{x^4y^3}$
20. $\sqrt{x^2y^7}$
21. $\sqrt{x^3y^6z^3}$
22. $\sqrt{x^8y^5z^7}$
23. $\sqrt{32x^5}$
24. $\sqrt{36y^9}$
25. $\sqrt{104x^{16}}$
26. $\sqrt{84x^{11}}$
27. $\sqrt{72x^2y^4}$
28. $\sqrt{64xy^7}$
29. $\sqrt{80x^5y^2}$
30. $\sqrt{198x^{10}y^5}$
31. $4x^2\sqrt{99x^4}$
32. $7y^3\sqrt{76y^8}$
33. $12xy^2\sqrt{144xy^2}$
34. $20x^4y\sqrt{81x^5y^4}$
35. $x^3y^2\sqrt{140y^6}$
36. $3xy^8\sqrt{192x^9y^4}$

21.2 PRODUCTS CONTAINING SQUARE ROOTS

The Product of Two Square Roots

To see how to simplify the product of two square roots, look at the following computations. We are using perfect squares under the radical signs so that you can check the results.

$$\sqrt{4}\sqrt{9} = 2 \cdot 3 = 6 = \sqrt{36}$$
$$\sqrt{4}\sqrt{25} = 2 \cdot 5 = 10 = \sqrt{100}$$
$$\sqrt{9}\sqrt{25} = 3 \cdot 5 = 15 = \sqrt{225}$$

As these examples show, a product of square roots equals the square root of a product.

$$\sqrt{a}\sqrt{b} = \sqrt{ab}$$

(where a and b represent positive quantities)

Hence, to simplify a product of square roots we multiply the quantities under the radical signs and reduce to simplest form. For example,

$$\sqrt{6} \cdot \sqrt{10} = \sqrt{60} = \sqrt{4 \cdot 15} = 2\sqrt{15}$$

6. Multiply
   a. $\sqrt{2}\sqrt{11}$
   b. $\sqrt{5}\sqrt{6}$
   c. $\sqrt{3}\sqrt{7}$
   d. $\sqrt{10}\sqrt{13}$

Like Square Roots

We call two numerical square roots in simplest form *like square roots* if the numbers under the radical signs are the same. For example, $\sqrt{6}$ and $3\sqrt{6}$ are like square roots, as are $5\sqrt{2}$ and $-3\sqrt{2}$, whereas $2\sqrt{3}$ and $3\sqrt{2}$ are not like square roots. Like square roots can be combined into a single term in the same way as like monomials. For example,

$x + x = 2x$	$\sqrt{3} + \sqrt{3} = 2\sqrt{3}$
$3x + 4x = 7x$	$3\sqrt{3} + 4\sqrt{3} = 7\sqrt{3}$
$5x - 9x = -4x$	$5\sqrt{3} - 9\sqrt{3} = -4\sqrt{3}$

On the other hand, a sum of unlike square roots in simplest form cannot be simplified. For example, $\sqrt{2} + \sqrt{3}$, $3\sqrt{2} + 4\sqrt{3}$, and $5\sqrt{2} - 9\sqrt{3}$ cannot be simplified.

7. Combine like square roots.
   a. $2\sqrt{3} + 4\sqrt{3}$
   b. $5\sqrt{2} - \sqrt{2}$
   c. $-3\sqrt{5} - 3\sqrt{5}$
   d. $-6\sqrt{7} - 3\sqrt{7}$

8. Reduce to simplest form and add.
   a. $8\sqrt{2} - 5\sqrt{8}$
   b. $\sqrt{27} + \sqrt{12}$
   c. $-\sqrt{45} - \sqrt{20}$
   d. $-2\sqrt{28} + \sqrt{63}$

In defining like square roots we require that the square roots be in simplest form so that we can easily determine whether they can be combined. For example, the sum of $\sqrt{2}$ and $\sqrt{18}$ can be simplified because $\sqrt{18} = 3\sqrt{2}$, and $\sqrt{2}$ and $3\sqrt{2}$ are like square roots.

$$\sqrt{2} + \sqrt{18} = \sqrt{2} + 3\sqrt{2} = 4\sqrt{2}$$

When we work with expressions containing square roots, we always reduce the square roots to simplest form before we perform any other operations.

Products of Sums Containing Square Roots

Sums containing square roots can be multiplied in the same way as the sums of polynomials we saw in Chapters 4 and 15.

Look at the following examples of *multiplying expressions containing square roots.*

TYPICAL PROBLEMS

1. $2\sqrt{42xy}(x\sqrt{30x})$

We first form a product, writing the monomials together and the product of the square roots as the square root of a product:

$$(2 \cdot x)(\sqrt{42xy \cdot 30x})$$

Then we factor each number under the radical sign into a product of primes:

$$2x\sqrt{2 \cdot 3 \cdot 7 \cdot 2 \cdot 3 \cdot 5 \cdot x^2 y}$$

We then separate the product under the radical sign into two factors: one a perfect square and the other square-free:

$$2x\sqrt{(2^2 \cdot 3^2 \cdot x^2)(5 \cdot 7 \cdot y)}$$

Then we factor out the perfect square:

$$2x(2 \cdot 3 \cdot x)\sqrt{5 \cdot 7 \cdot y}$$

Finally, we simplify:

$$12x^2\sqrt{35y}$$

Hence, $2\sqrt{42xy}(x\sqrt{30x}) = 12x^2\sqrt{35y}$

2. $(4 - \sqrt{6})(\sqrt{2} - \sqrt{3})$

We first change to vertical format, writing one factor under the other and dropping the parentheses:

$$\begin{array}{r} 4 - \sqrt{6} \\ \sqrt{2} - \sqrt{3} \\ \hline \end{array}$$

Then we multiply the two binomials:

$$\begin{array}{l} 4\sqrt{2} - \sqrt{2 \cdot 6} \\ \qquad\qquad -4\sqrt{3} - \sqrt{3 \cdot 6} \\ \hline 4\sqrt{2} - \sqrt{2 \cdot 6} - 4\sqrt{3} - \sqrt{3 \cdot 6} \end{array}$$

To multiply expressions containing sums we parallel the methods in Chapter 15. To reduce each square root to simplest form we use the method in the previous section.

We then reduce each term to simplest form and combine like square roots:

$$4\sqrt{2} - 2\sqrt{3} - 4\sqrt{3} - 3\sqrt{2}$$

$$\sqrt{2} - 6\sqrt{3}$$

Hence, $(4 - \sqrt{6})(\sqrt{2} - \sqrt{3}) = \sqrt{2} - 6\sqrt{3}$.

EXAMPLES

Multiply the following expressions containing square roots.

Every product containing square roots can be multiplied to a square root or a sum of unlike square roots in simplest form.

1. $(5\sqrt{39})(-4\sqrt{6})$
2. $x^2\sqrt{y}(\sqrt{34xy})$
3. $\sqrt{14}(7-3\sqrt{14})$
4. $5\sqrt{3}(\sqrt{39}+6\sqrt{13})$
5. $(4+\sqrt{10})(\sqrt{3}-8\sqrt{30})$
6. $(x+6+\sqrt{22})(x+6-\sqrt{22})$

Solutions

1. $5(-4)\sqrt{39 \cdot 6}$
$\downarrow$
$-20\sqrt{3 \cdot 13 \cdot 2 \cdot 3}$
$\downarrow$
$-20\sqrt{(3^2)(2 \cdot 13)}$
$\downarrow$
$-20(3)\sqrt{2 \cdot 13}$
$\downarrow$
$-60\sqrt{26}$

2. $x^2\sqrt{y \cdot 34xy}$
$\downarrow$
$x^2\sqrt{2 \cdot 17 \cdot xy^2}$
$\downarrow$
$x^2\sqrt{(y^2)(2 \cdot 17 \cdot x)}$
$\downarrow$
$x^2(y)\sqrt{2 \cdot 17 \cdot x}$
$\downarrow$
$x^2y\sqrt{34x}$

3. $7\sqrt{14} - 3(14)$
$\downarrow$
$7\sqrt{14} - 42$

4. $5\sqrt{3 \cdot 39} + 30\sqrt{3 \cdot 13}$
$\downarrow$
$15\sqrt{13} + 30\sqrt{39}$

5. $4+\sqrt{10}$
$\sqrt{3} - 8\sqrt{30}$
$4\sqrt{3} + \sqrt{3 \cdot 10}$
$- 32\sqrt{30} - 8\sqrt{10 \cdot 30}$
$4\sqrt{3} + \sqrt{3 \cdot 10} - 32\sqrt{30} - 8\sqrt{10 \cdot 30}$
$\downarrow$
$4\sqrt{3} + \sqrt{30} - 32\sqrt{30} - 80\sqrt{3}$
$\downarrow$
$-76\sqrt{3} - 31\sqrt{30}$

6. $x+6+\sqrt{22}$
$x+6-\sqrt{22}$
$x^2 + 6x + x\sqrt{22}$
$6x + 36 + 6\sqrt{22}$
$- x\sqrt{22} - 6\sqrt{22} - 22$
$x^2 + 12x + 36 - 22$
$\downarrow$
$x^2 + 12x + 14$

SPECIAL CASES

1. The product of two identical square roots equals the quantity under the radical sign. Thus,

$$\sqrt{5} \cdot \sqrt{5} = 5 \quad \text{and} \quad (3\sqrt{5})(4\sqrt{5}) = 3 \cdot 4 \cdot 5 = 60$$

$$a\sqrt{c} \cdot b\sqrt{c} = abc$$

(where *a*, *b*, and *c* represent numbers with *c* non-negative)

9. Multiply
a. $3\sqrt{11} \cdot 2\sqrt{11}$
b. $4\sqrt{xy} \cdot 9\sqrt{xy}$
c. $5x^2\sqrt{2x^3}\ \sqrt{2x^3}$

2. Two binomials that differ only in the sign of their second terms are called *conjugate binomials* when they contain square roots. For example,

$$(4+\sqrt{6}) \text{ and } (4-\sqrt{6}) \text{ are conjugate binomials}$$
$$(\sqrt{3}+\sqrt{2}) \text{ and } (\sqrt{3}-\sqrt{2}) \text{ are conjugate binomials}$$

Remember that the product of two binomials that differ only in the sign of their second terms is a difference of two squares. In particular, a product of conjugate binomials is a difference of squares. However, squaring a square root yields the number under the radical sign. Hence, the product of two conjugate binomials equals an expression without any radical signs. For example,

$$(4+\sqrt{6})(4-\sqrt{6}) \quad \text{and} \quad (\sqrt{2}+\sqrt{3})(\sqrt{2}-\sqrt{3})$$

$$\begin{array}{rrr} & 4 & +\sqrt{6} \\ & 4 & -\sqrt{6} \\ \hline & 16 & +4\sqrt{6} \\ & & -4\sqrt{6} \;-6 \\ \hline & 16 & -6 \\ & \downarrow & \\ & 10 & \end{array} \qquad \begin{array}{rrr} & \sqrt{2} & +\sqrt{3} \\ & \sqrt{2} & -\sqrt{3} \\ \hline & 2 & +\sqrt{6} \\ & & -\sqrt{6} \;-3 \\ \hline & 2 & -3 \\ & \downarrow & \\ & -1 & \end{array}$$

10. Multiply
a. $(3+\sqrt{7})(3-\sqrt{7})$
b. $(\sqrt{5}-\sqrt{10})(\sqrt{5}+\sqrt{10})$

Here are two multiplication tables for square roots. Verify the entries to check your understanding of multiplication.

times	$\sqrt{2}$	$\sqrt{3}$	$\sqrt{5}$	$\sqrt{6}$	$\sqrt{10}$
$\sqrt{2}$	2	$\sqrt{6}$	$\sqrt{10}$	$2\sqrt{3}$	
$\sqrt{3}$	$\sqrt{6}$	3	$\sqrt{15}$	$3\sqrt{2}$	
$\sqrt{5}$	$\sqrt{10}$	$\sqrt{15}$	5	$\sqrt{30}$	
$\sqrt{6}$	$2\sqrt{3}$	$3\sqrt{2}$	$\sqrt{30}$	6	
$\sqrt{10}$					

and

times	$\sqrt{x}$	x	$x\sqrt{x}$	x^2	$x^2\sqrt{x}$
$\sqrt{x}$	x	$x\sqrt{x}$	x^2		
x	$x\sqrt{x}$	x^2	$x^2\sqrt{x}$		
$x\sqrt{x}$	x^2	$x^2\sqrt{x}$	x^3		
x^2					
$x^2\sqrt{x}$					

11. Complete the tables.

OBSERVATIONS AND REFLECTIONS

1. In the previous section we used the fact that $a\sqrt{b}=\sqrt{a^2b}$ in reducing square roots to simplest form. This fact can be derived from the two properties

$$\sqrt{a}\,\sqrt{b}=\sqrt{ab} \quad \text{and} \quad \sqrt{a^2}=a$$

where a and b represent non-negative quantities, because

$$a\sqrt{b}=\sqrt{a^2}\,\sqrt{b}=\sqrt{a^2b}$$

For example, $\sqrt{12}=\sqrt{4\cdot 3}=\sqrt{4}\,\sqrt{3}=2\sqrt{3}$.

2. Two like square roots can be combined because the like square root is a common factor of both terms and can be factored out. For example,

$$\sqrt{3}+\sqrt{3}=1\sqrt{3}+1\sqrt{3}=(1+1)\sqrt{3}=2\sqrt{3}$$
$$3\sqrt{3}+4\sqrt{3}=(3+4)\sqrt{3}=7\sqrt{3}$$
$$5\sqrt{3}-9\sqrt{3}=(5-9)\sqrt{3}=-4\sqrt{3}$$

3. A sum of two unlike square roots can frequently be factored. For example, 3 is a common divisor of 6 and 15, so

$$\sqrt{6}+\sqrt{15}=\sqrt{3}\,\sqrt{2}+\sqrt{3}\,\sqrt{5}=\sqrt{3}\,(\sqrt{2}+\sqrt{5})$$

Similarly,

$$\sqrt{3}+\sqrt{15}=1\sqrt{3}+\sqrt{3}\,\sqrt{5}=\sqrt{3}\,(1+\sqrt{5})$$

4. Look at the product of the trinomials $(x-5+\sqrt{3})$ and $(x-5-\sqrt{3})$:

$$\begin{array}{llll} x-5+\sqrt{3} & & & \\ x-5-\sqrt{3} & & & \\ \hline x^2-5x+x\sqrt{3} & & & \\ \quad -5x & +25-5\sqrt{3} & & \\ \qquad -x\sqrt{3} & & +5\sqrt{3} & -3 \\ \hline x^2-10x & +25 & & -3 \\ \quad\downarrow & & & \\ x^2-10x+22 & & & \end{array}$$

Note that the numerical terms in the trinomials, namely $-5+\sqrt{3}$ and $-5-\sqrt{3}$, are conjugate binomials. Adding x to each of two conjugate binomials of the form $a+\sqrt{b}$ and $a-\sqrt{b}$, where a and b represent integers with b positive, and then multiplying the resulting trinomials always yields a quadratic trinomial in x with integer coefficients.

12. Factor out common square roots.

a. $\sqrt{10}+\sqrt{6}$

b. $\sqrt{10}-\sqrt{15}$

c. $-\sqrt{14}+2\sqrt{7}$

d. $-\sqrt{22}-\sqrt{11}$

EXERCISES

Multiply to convert each of the following expressions to a radical expression in simplest form.

1. $\sqrt{21}\sqrt{35}$
2. $\sqrt{33}\sqrt{55}$
3. $\sqrt{154}\sqrt{42}$
4. $\sqrt{165}\sqrt{30}$
5. $(\sqrt{91})(\sqrt{105})$
6. $(\sqrt{66})(\sqrt{102})$
7. $3\sqrt{22}\cdot\sqrt{143}$
8. $5\sqrt{65}\cdot\sqrt{91}$
9. $\sqrt{85}\,(2\sqrt{10})$
10. $\sqrt{187}\,(6\sqrt{34})$
11. $(3\sqrt{26})(4\sqrt{13})$
12. $(6\sqrt{11})(\sqrt{77})$
13. $\sqrt{2x}\sqrt{xy}$
14. $\sqrt{5y}\sqrt{5xy}$
15. $x^2\sqrt{x}\,(\sqrt{6x})$
16. $xy^3\sqrt{xy}\,(\sqrt{26x})$
17. $(3xy\sqrt{14y})(2y^4\sqrt{7x})$
18. $(7x^5\sqrt{38x})(xy^6\sqrt{19y})$
19. $\sqrt{5}\,(6-5\sqrt{5})$
20. $\sqrt{7}\,(2\sqrt{7}-9)$
21. $3\sqrt{6}\,(\sqrt{3}+4\sqrt{6})$
22. $2\sqrt{15}\,(\sqrt{5}+6\sqrt{15})$
23. $\sqrt{21}\,(2\sqrt{42}-3\sqrt{21})$
24. $\sqrt{22}\,(8\sqrt{110}-5\sqrt{22})$
25. $\sqrt{33}\,(\sqrt{11}+\sqrt{3})$
26. $\sqrt{26}\,(\sqrt{13}+\sqrt{2})$
27. $5\sqrt{10}\,(3\sqrt{2}-8\sqrt{5})$
28. $6\sqrt{6}\,(9\sqrt{3}-6\sqrt{2})$

29. $\sqrt{14}\,(-\sqrt{35}+2\sqrt{10})$

30. $\sqrt{21}\,(-3\sqrt{77}+\sqrt{6})$

31. $(6+\sqrt{7})(3-\sqrt{7})$

32. $(8-\sqrt{11})(7+\sqrt{11})$

33. $(4+3\sqrt{10})(2\sqrt{10}+\sqrt{5})$

34. $(6+\sqrt{14})(8\sqrt{7}+3\sqrt{14})$

35. $(3\sqrt{3}-5)(4\sqrt{6}-2\sqrt{2})$

36. $(8\sqrt{2}-1)(9\sqrt{10}-3\sqrt{5})$

37. $(-4-\sqrt{21})(-3+\sqrt{14})$

38. $(-6+\sqrt{26})(-5-\sqrt{6})$

39. $(\sqrt{2}-\sqrt{7})(\sqrt{3}+\sqrt{2})$

40. $(\sqrt{5}+\sqrt{11})(\sqrt{7}-\sqrt{5})$

41. $(2\sqrt{3}+5\sqrt{6})(4\sqrt{2}-6\sqrt{3})$

42. $(5\sqrt{2}-2\sqrt{10})(3\sqrt{5}-4\sqrt{2})$

43. $(x+6+\sqrt{11})(x+6-\sqrt{11})$

44. $(x+9+\sqrt{17})(x+9-\sqrt{17})$

45. $(x-5+\sqrt{51})(x-5-\sqrt{51})$

46. $(x-7+\sqrt{95})(x-7-\sqrt{95})$

47. $(x+8+4\sqrt{10})(x+8-4\sqrt{10})$

48. $(x+3+6\sqrt{2})(x+3-6\sqrt{2})$

21.3 QUOTIENTS AND RATIOS CONTAINING SQUARE ROOTS

The Quotient of Two Square Roots

To see how to simplify quotients and ratios of square roots, look at the following computations. We are using perfect squares under the radical signs so that you can check the results.

$$\frac{\sqrt{4}}{\sqrt{9}}=\frac{2}{3}=\sqrt{\frac{4}{9}} \qquad \frac{\sqrt{4}}{\sqrt{25}}=\frac{2}{5}=\sqrt{\frac{4}{25}} \qquad \frac{\sqrt{9}}{\sqrt{25}}=\frac{3}{5}=\sqrt{\frac{9}{25}}$$

As these examples show, a ratio of square roots equals the square root of a fraction.

$$\frac{\sqrt{a}}{\sqrt{b}}=\sqrt{\frac{a}{b}}$$

(where a and b represent positive quantities)

13. Reduce

a. $\dfrac{\sqrt{6}}{\sqrt{3}}$

b. $\dfrac{\sqrt{30}}{\sqrt{10}}$

Hence, to simplify a ratio or quotient of square roots we form a fraction under the radical sign and reduce it. We then convert the square root of the reduced fraction to a ratio of square roots, because it is considered improper to leave a fraction under a radical sign. For example,

$$\frac{\sqrt{6}}{\sqrt{10}}=\sqrt{\frac{6}{10}}=\sqrt{\frac{3}{5}}=\frac{\sqrt{3}}{\sqrt{5}}$$

Standard Form—Rationalizing the Denominator

Fractional expressions containing square roots can have several equivalent forms. For example, $\dfrac{\sqrt{2}}{3}=\dfrac{2}{3\sqrt{2}}$, $\dfrac{x}{\sqrt{xy}}=\dfrac{\sqrt{xy}}{y}$, and $\dfrac{3\sqrt{3}}{2+\sqrt{2}}=\dfrac{3\sqrt{6}}{2\sqrt{2}+2}=\dfrac{6\sqrt{3}-3\sqrt{6}}{2}$.

A fraction containing square roots is in *standard form* if all square roots are in simplest form and there are no square roots in the denominator. For example, $\frac{\sqrt{2}}{3}$, $\frac{\sqrt{xy}}{y}$, and $\frac{6\sqrt{3}-3\sqrt{6}}{2}$ are in standard form, whereas $\frac{2}{3\sqrt{2}}$, $\frac{x}{\sqrt{xy}}$, and $\frac{3\sqrt{3}}{2+\sqrt{2}}$ are not. Every fraction containing square roots is equivalent to a fraction in standard form. Converting a fraction whose denominator contains square roots to standard form is called *rationalizing the denominator.*

To see how to convert a fraction with a *monomial denominator* to standard form, look at the following computations.

$$\frac{1}{\sqrt{2}} = \frac{\sqrt{2}}{2} \qquad \frac{1}{\sqrt{3}} = \frac{\sqrt{3}}{3}$$

$$\frac{2}{\sqrt{2}} = \frac{2\sqrt{2}}{2} = \sqrt{2} \qquad \frac{2}{\sqrt{3}} = \frac{2\sqrt{3}}{3}$$

$$\frac{3}{\sqrt{2}} = \frac{3\sqrt{2}}{2} \qquad \frac{3}{\sqrt{3}} = \frac{3\sqrt{3}}{3} = \sqrt{3}$$

$$\frac{1}{\sqrt{6}} = \frac{\sqrt{6}}{6} \qquad \frac{2}{\sqrt{6}} = \frac{2\sqrt{6}}{6} = \frac{\sqrt{6}}{3}$$

$$\frac{3}{\sqrt{6}} = \frac{3\sqrt{6}}{6} = \frac{\sqrt{6}}{2} \qquad \frac{4}{\sqrt{6}} = \frac{4\sqrt{6}}{6} = \frac{2\sqrt{6}}{3}$$

$$\frac{5}{\sqrt{6}} = \frac{5\sqrt{6}}{6} \qquad \frac{6}{\sqrt{6}} = \frac{6\sqrt{6}}{6} = \sqrt{6}$$

14. Pair equivalent fractions:

a. $\frac{3}{\sqrt{2}}$

b. $\frac{\sqrt{6}}{2}$

c. $\frac{3\sqrt{2}}{2}$

d. $\frac{\sqrt{2}}{\sqrt{3}}$

e. $\frac{\sqrt{3}}{\sqrt{2}}$

f. $\frac{\sqrt{6}}{3}$

Can you see what you should multiply the numerator and denominator by to rationalize the denominator?

A *binomial denominator* can also be rationalized by multiplying. In the last section we saw that two binomials containing square roots are called *conjugate binomials* if they differ only in the sign of their second terms. Further, we saw that the product of two conjugate binomials is a difference of squares and, hence, free of square roots. For example, $(1+\sqrt{3})(1-\sqrt{3}) = 1 - 3 = -2$, so

$$\frac{1}{1+\sqrt{3}} = \frac{(1-\sqrt{3})}{(1+\sqrt{3})(1-\sqrt{3})} = -\frac{1-\sqrt{3}}{2}$$

and

$$\frac{1}{1-\sqrt{3}} = \frac{(1+\sqrt{3})}{(1-\sqrt{3})(1+\sqrt{3})} = -\frac{1+\sqrt{3}}{2}$$

Similarly, $(\sqrt{3}+\sqrt{2})(\sqrt{3}-\sqrt{2}) = 3 - 2 = 1$, so

$$\frac{1}{\sqrt{3}+\sqrt{2}} = \frac{(\sqrt{3}-\sqrt{2})}{(\sqrt{3}+\sqrt{2})(\sqrt{3}-\sqrt{2})} = \sqrt{3}-\sqrt{2}$$

and

$$\frac{1}{\sqrt{3}-\sqrt{2}} = \frac{(\sqrt{3}+\sqrt{2})}{(\sqrt{3}-\sqrt{2})(\sqrt{3}+\sqrt{2})} = \sqrt{3}+\sqrt{2}$$

When reducing quotients or ratios containing square roots, we always rationalize the denominator.

15. Pair conjugate binomials.

a. $3-\sqrt{2}$

b. $\sqrt{3}+\sqrt{2}$

c. $\sqrt{3}+2$

d. $\sqrt{3}-\sqrt{2}$

e. $3+\sqrt{2}$

f. $\sqrt{3}-2$

Look at the following examples of *reducing quotients and ratios containing square roots.*

TYPICAL PROBLEMS

1. $8x\sqrt{15} \div 30y\sqrt{6y}$

We first form a fraction:

$$\frac{8x\sqrt{15}}{30y\sqrt{6y}}$$

Then we rewrite the fraction as the product of a fraction and the square root of a fraction:

$$\frac{8x}{30y}\sqrt{\frac{15}{6y}}$$

We then reduce each fraction:

$$\frac{4x}{15y}\sqrt{\frac{5}{2y}}$$

Next we rewrite the product as a single fraction, converting the square root to a ratio of square roots:

$$\frac{4x\sqrt{5}}{15y\sqrt{2y}}$$

We then multiply top and bottom by the square root in the denominator to rationalize the denominator:

$$\frac{4x\sqrt{5}\cdot\sqrt{2y}}{15y\sqrt{2y}\cdot\sqrt{2y}}$$

And then we simplify:

$$\frac{4x\sqrt{10y}}{30y^2}$$

Finally we reduce the fraction again if necessary:

$$\frac{2x\sqrt{10y}}{15y^2}$$

Hence, $8x\sqrt{15} \div 30y\sqrt{6y} = \dfrac{2x\sqrt{10y}}{15y^2}$.

2. $\dfrac{6\sqrt{5}}{\sqrt{5}-\sqrt{2}}$

We multiply top and bottom by the conjugate binomial of the denominator to rationalize the denominator:

$$\frac{6\sqrt{5}(\sqrt{5}+\sqrt{2})}{(\sqrt{5}-\sqrt{2})(\sqrt{5}+\sqrt{2})}$$

And then we simplify:

$$\frac{6\cdot 5+6\sqrt{10}}{5-2}$$

$$\frac{30+6\sqrt{10}}{3}$$

Finally we reduce if possible:

$$\frac{6(5+\sqrt{10})}{3}$$

$$2(5+\sqrt{10}) = 10+2\sqrt{10}$$

To multiply expressions containing square roots when rationalizing denominators we use the methods in the previous section.

Hence, $\dfrac{6\sqrt{5}}{\sqrt{5}-\sqrt{2}} = 10+2\sqrt{10}$.

EXAMPLES

Reduce the following quotients and fractions containing square roots to standard form.

Every quotient or ratio containing square roots can be simplified to a single reduced fraction in standard form.

1. $12\sqrt{130} \div 16\sqrt{195}$

2. $8x^3y^2\sqrt{35x} \div 28y^3\sqrt{5x}$

3. $\dfrac{8\sqrt{3} - 3\sqrt{26}}{2\sqrt{6}}$

4. $\dfrac{\sqrt{30}}{\sqrt{6} - \sqrt{10}}$

Solutions

1. $$\frac{12\sqrt{130}}{16\sqrt{195}} \to \frac{12}{16}\sqrt{\frac{130}{195}} \to \frac{3}{4}\sqrt{\frac{2}{3}} \to \frac{3\sqrt{2}\cdot\sqrt{3}}{4\sqrt{3}\cdot\sqrt{3}} \to \frac{3\sqrt{6}}{12} \to \frac{\sqrt{6}}{4}$$

2. $$\frac{8x^3y^2\sqrt{35x}}{28y^3\sqrt{5x}} \to \frac{8x^3y^2}{28y^3}\sqrt{\frac{35x}{5x}} \to \frac{2x^3}{7y}\sqrt{7} \to \frac{2x^3\sqrt{7}}{7y}$$

3. $$\frac{(8\sqrt{3} - 3\sqrt{26})\cdot\sqrt{6}}{2\sqrt{6}\cdot\sqrt{6}} \to \frac{8\sqrt{18} - 3\sqrt{156}}{2\cdot 6} \to \frac{8\cdot 3\sqrt{2} - 3\cdot 2\sqrt{39}}{12} \to \frac{24\sqrt{2} - 6\sqrt{39}}{12} \to \frac{6(4\sqrt{2} - \sqrt{39})}{12} \to \frac{4\sqrt{2} - \sqrt{39}}{2}$$

4. $$\frac{\sqrt{30}}{\sqrt{6} - \sqrt{10}} \to \frac{\sqrt{30}\,(\sqrt{6} + \sqrt{10})}{(\sqrt{6} - \sqrt{10})(\sqrt{6} + \sqrt{10})} \to \frac{\sqrt{30\cdot 6} + \sqrt{30\cdot 10}}{6 - 10} \to \frac{6\sqrt{5} + 10\sqrt{3}}{-4} \to -\frac{2(3\sqrt{5} + 5\sqrt{3})}{4} \to -\frac{3\sqrt{5} + 5\sqrt{3}}{2}$$

SPECIAL CASES

16. Reduce

a. $\dfrac{\sqrt{12}}{\sqrt{12}}$ b. $\dfrac{7\sqrt{3}}{\sqrt{3}}$

c. $\dfrac{-4\sqrt{7}}{6\sqrt{7}}$ d. $\dfrac{-9\sqrt{15}}{-18\sqrt{15}}$

The quotient or ratio of two identical square roots equals 1. Thus,

$$\frac{\sqrt{5}}{\sqrt{5}} = 1 \quad \text{and} \quad \frac{3\sqrt{5}}{4\sqrt{5}} = \frac{3}{4}$$

$$\frac{a\sqrt{c}}{b\sqrt{c}} = \frac{a}{b}$$

(where a, b, and c represent numbers with c non-negative)

Here are two tables of ratios of square roots with the entries in standard form. Verify the entries to check your understanding of the standard form of a fraction.

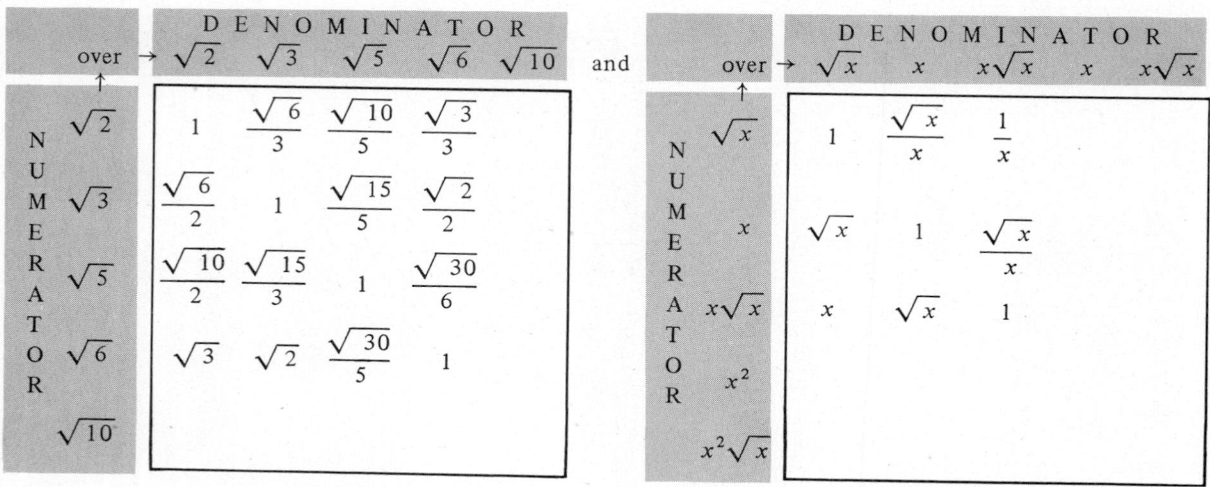

NUMERATOR over → DENOMINATOR	$\sqrt{2}$	$\sqrt{3}$	$\sqrt{5}$	$\sqrt{6}$	$\sqrt{10}$
$\sqrt{2}$	1	$\frac{\sqrt{6}}{3}$	$\frac{\sqrt{10}}{5}$	$\frac{\sqrt{3}}{3}$	
$\sqrt{3}$	$\frac{\sqrt{6}}{2}$	1	$\frac{\sqrt{15}}{5}$	$\frac{\sqrt{2}}{2}$	
$\sqrt{5}$	$\frac{\sqrt{10}}{2}$	$\frac{\sqrt{15}}{3}$	1	$\frac{\sqrt{30}}{6}$	
$\sqrt{6}$	$\sqrt{3}$	$\sqrt{2}$	$\frac{\sqrt{30}}{5}$	1	
$\sqrt{10}$					

and

NUMERATOR over → DENOMINATOR	$\sqrt{x}$	x	$x\sqrt{x}$	x	$x\sqrt{x}$
$\sqrt{x}$	1	$\frac{\sqrt{x}}{x}$	$\frac{1}{x}$		
x	$\sqrt{x}$	1	$\frac{\sqrt{x}}{x}$		
$x\sqrt{x}$	x	$\sqrt{x}$	1		
x^2					
$x^2\sqrt{x}$					

17. Complete the tables.

Here is a table of ratios with binomials containing square roots. Although it will involve some computations, verify the entries to check your understanding of the process of rationalizing binomial denominators. The entries are in standard form.

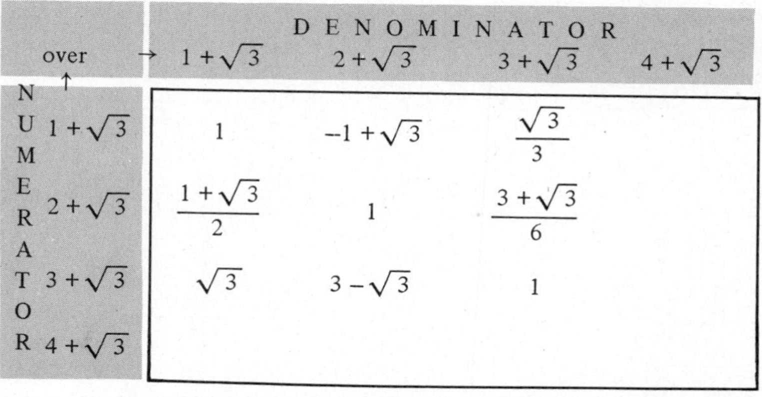

NUMERATOR over → DENOMINATOR	$1+\sqrt{3}$	$2+\sqrt{3}$	$3+\sqrt{3}$	$4+\sqrt{3}$
$1+\sqrt{3}$	1	$-1+\sqrt{3}$	$\frac{\sqrt{3}}{3}$	
$2+\sqrt{3}$	$\frac{1+\sqrt{3}}{2}$	1	$\frac{3+\sqrt{3}}{6}$	
$3+\sqrt{3}$	$\sqrt{3}$	$3-\sqrt{3}$	1	
$4+\sqrt{3}$				

18. Complete the table.

EXERCISES

Simplify each of the following expressions to a reduced fraction in standard form.

1. $\sqrt{35} \div \sqrt{42}$
2. $\sqrt{39} \div \sqrt{26}$
3. $6\sqrt{30} \div 8\sqrt{66}$

4. $18\sqrt{33} \div 12\sqrt{110}$ 5. $28\sqrt{170} \div 16\sqrt{119}$ 6. $40\sqrt{182} \div 24\sqrt{130}$

7. $\dfrac{\sqrt{65}}{\sqrt{70}}$ 8. $\dfrac{\sqrt{78}}{\sqrt{102}}$ 9. $\dfrac{10\sqrt{105}}{55\sqrt{30}}$

10. $\dfrac{35\sqrt{143}}{21\sqrt{55}}$ 11. $\dfrac{25\sqrt{19}}{8\sqrt{190}}$ 12. $\dfrac{9\sqrt{11}}{5\sqrt{165}}$

13. $x^2y\sqrt{xy} \div y^3\sqrt{x}$ 14. $x^6\sqrt{xy} \div x^2y^3\sqrt{y}$

15. $3x^3y^4\sqrt{10y} \div 4y^5\sqrt{15x}$ 16. $7xy^5\sqrt{6x} \div 5x^4\sqrt{14y}$

17. $\dfrac{27x^6y^3\sqrt{7x}}{6x^2y\sqrt{21xy}}$ 18. $\dfrac{32x^7y^2\sqrt{19y}}{40x^3y^6\sqrt{38xy}}$

19. $\dfrac{8-\sqrt{6}}{\sqrt{2}}$ 20. $\dfrac{5-\sqrt{10}}{\sqrt{5}}$

21. $\dfrac{3+2\sqrt{21}}{3\sqrt{6}}$ 22. $\dfrac{4+3\sqrt{14}}{8\sqrt{10}}$

23. $\dfrac{\sqrt{35}-\sqrt{77}}{\sqrt{21}}$ 24. $\dfrac{\sqrt{39}-\sqrt{65}}{\sqrt{26}}$

25. $\dfrac{4\sqrt{10}}{5+\sqrt{15}}$ 26. $\dfrac{7\sqrt{6}}{6+\sqrt{22}}$

27. $\dfrac{5\sqrt{14}}{3-2\sqrt{6}}$ 28. $\dfrac{8\sqrt{15}}{6-3\sqrt{10}}$

29. $\dfrac{3+\sqrt{5}}{5+\sqrt{5}}$ 30. $\dfrac{2+\sqrt{6}}{4+\sqrt{6}}$

31. $\dfrac{\sqrt{3}-2\sqrt{2}}{5+3\sqrt{6}}$ 32. $\dfrac{\sqrt{3}-5\sqrt{5}}{7+2\sqrt{15}}$

33. $\dfrac{2\sqrt{15}}{\sqrt{10}-\sqrt{6}}$ 34. $\dfrac{3\sqrt{35}}{\sqrt{21}-\sqrt{15}}$

35. $\dfrac{\sqrt{10}-4\sqrt{3}}{\sqrt{15}+3\sqrt{2}}$ 36. $\dfrac{\sqrt{15}-3\sqrt{7}}{\sqrt{35}+5\sqrt{3}}$

21.4
EVALUATING

In this section we extend the methods in Sections 8.6 and 9.5 to evaluate expressions containing square roots. We will consider expressions with one or two literals and evaluate them for both integer and fractional values of the literals.

To see some examples of evaluating expressions containing square roots, look at the following table.

evaluated at →	0	1	2	3	4
$2\sqrt{x}$	0	2	$2\sqrt{2}$	$2\sqrt{3}$	
$\sqrt{2x}$	0	$\sqrt{2}$	2	$\sqrt{6}$	
$\sqrt{\frac{2}{x}}$	*	$\sqrt{2}$	1	$\frac{\sqrt{6}}{3}$	
$\sqrt{\frac{x}{2}}$					

19. Complete the table.

*undefined

Look at the following examples of *evaluating expressions containing square roots.*

TYPICAL PROBLEMS

1. $\sqrt{3x} - 2\sqrt{11y-1}$, when $x = 8, y = 5$

We first substitute the numbers given using parentheses:

$$\sqrt{3(8)} - 2\sqrt{11(5)-1}$$

Then we simplify each expression under a radical sign:

$$\sqrt{24} - 2\sqrt{54}$$

We then reduce each square root to simplest form:

$$2\sqrt{6} - 6\sqrt{6}$$

Finally we combine like square roots:

$$-4\sqrt{6}$$

Hence, evaluating $\sqrt{3x} - 2\sqrt{11y-1}$, when $x = 8, y = 5$ yields $-4\sqrt{6}$.

2. $\dfrac{4\sqrt{x}}{\sqrt{3y}}$, when $x = \dfrac{3}{8}, y = \dfrac{9}{5}$

We first substitute the numbers given using parentheses:

$$\frac{4\sqrt{\left(\frac{3}{8}\right)}}{\sqrt{3\left(\frac{9}{5}\right)}}$$

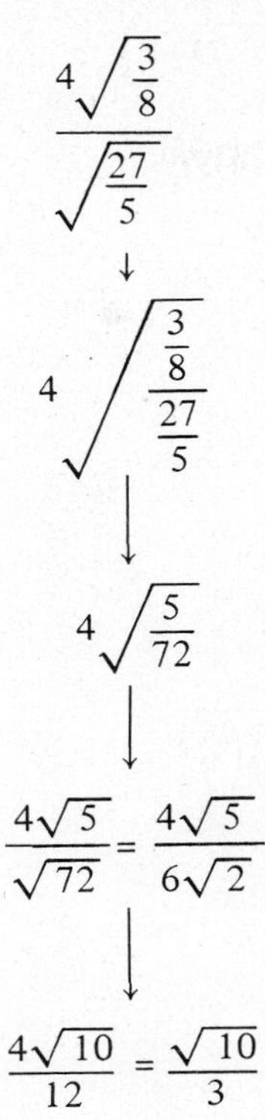

Then we simplify each expression under a radical sign:

$$\frac{4\sqrt{\frac{3}{8}}}{\sqrt{\frac{27}{5}}}$$

We then rewrite the fraction as the product of a fraction and the square root of a (complex) fraction:

$$4\sqrt{\frac{\frac{3}{8}}{\frac{27}{5}}}$$

And then we simplify the (complex) fraction under the radical sign:

$$4\sqrt{\frac{5}{72}}$$

Next we rewrite the product as a single fraction, converting the square root to a ratio of square roots. We reduce each square root to simplest form:

$$\frac{4\sqrt{5}}{\sqrt{72}} = \frac{4\sqrt{5}}{6\sqrt{2}}$$

And then we rationalize the denominator and reduce:

$$\frac{4\sqrt{10}}{12} = \frac{\sqrt{10}}{3}$$

Hence, evaluating $\frac{4\sqrt{x}}{\sqrt{3y}}$, when $x = \frac{3}{8}$, $y = \frac{9}{5}$ yields $\frac{\sqrt{10}}{3}$.

To evaluate expressions containing square roots we extend the methods in Sections 8.6 and 9.5. To simplify the numerical expressions we use the methods in the previous sections and the Order of Operations in Section 6.5.

EXAMPLES

Evaluate the following expressions containing square roots.

1. $2\sqrt{9 + 8x}$, when $x = 14$
2. $x\sqrt{10xy}$, when $x = 14, y = 21$
3. $\sqrt{x^2 - 4y}$, when $x = 6, y = -11$
4. $3\sqrt{10x}$, when $x = \frac{8}{15}$
5. $\sqrt{3 - xy}$, when $x = \frac{3}{4}$, $y = \frac{5}{2}$
6. $\frac{30 + \sqrt{2x}}{6}$, when $x = 10$

Solutions

1. $2\sqrt{9 + 8(14)}$

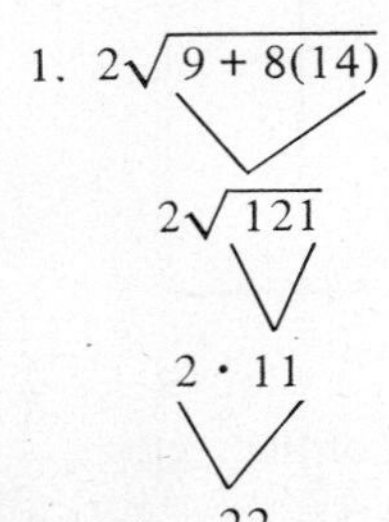

$2\sqrt{121}$

$2 \cdot 11$

22

2. $(14)\sqrt{10(14)(21)}$

$14\sqrt{2 \cdot 5 \cdot 2 \cdot 7 \cdot 3 \cdot 7}$

$14(2 \cdot 7)\sqrt{3 \cdot 5}$

$196\sqrt{15}$

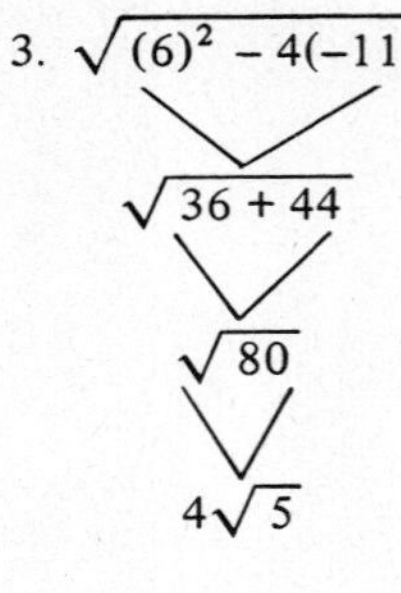

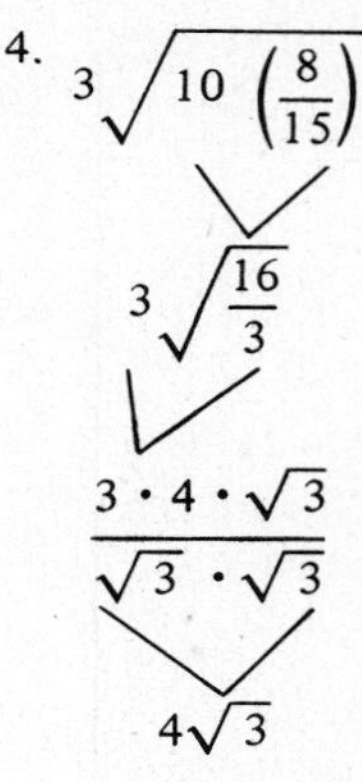

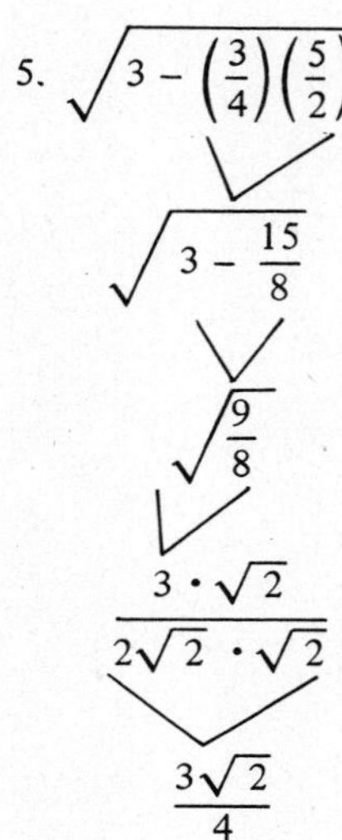

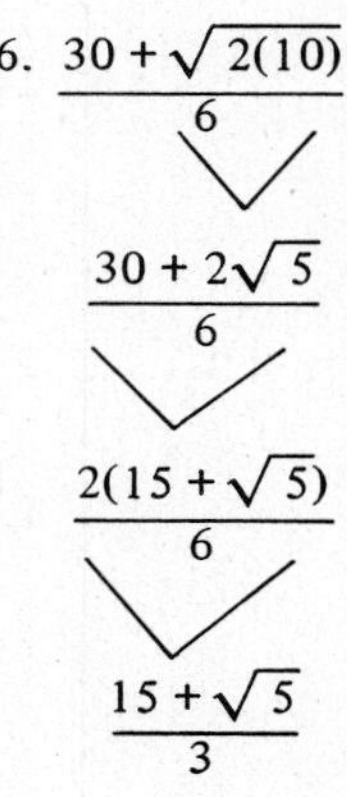

Here is a table of evaluations with fractional substitutions. Verify the entries to check your understanding of evaluations. The entries are in standard form.

evaluated at → ↓	$\frac{1}{2}$	$\frac{2}{3}$	$\frac{3}{4}$	$\frac{3}{2}$	$\frac{4}{3}$
$2\sqrt{x}$	$\sqrt{2}$	$\frac{2\sqrt{6}}{3}$	$\sqrt{3}$		
$\sqrt{2x}$	1	$\frac{2\sqrt{3}}{3}$	$\frac{\sqrt{6}}{2}$		
$\sqrt{\frac{2}{x}}$	2	$\sqrt{3}$	$\frac{2\sqrt{6}}{3}$		
$\sqrt{\frac{x}{2}}$					

20. Complete the table.

Compare this table with the table of evaluations given at the beginning of the section.

OBSERVATIONS AND REFLECTIONS

1. Remember that dividing any number by zero yields a meaningless or undefined answer. Hence, a fraction such as $\frac{1}{\sqrt{x}}$ or $\frac{3y}{\sqrt{2x}}$ cannot be evaluated when $x = 0$ because $\sqrt{0} = 0$.
2. An expression containing square roots cannot be evaluated for any value of a literal that makes the number under any radical sign negative. For example, in evaluating $\sqrt{3x}$ we require that x be non-negative, and in evaluating $\sqrt{5 - x}$ we require that x be no greater than 5. Similarly, in evaluating $\sqrt{x^2 - 4y}$ we require that $4y$ be no greater than x^2. Note that x or y can be negative here, because x^2 is always positive and $-4y$ is positive when y is negative.

EXERCISES

Evaluate each of the following expressions, simplifying completely. Convert all fractions to standard form.

1. $\sqrt{6x}$, when $x = 24$
2. $\sqrt{14x}$, when $x = 63$
3. $\sqrt{5xy}$, when $x = 8, y = 15$
4. $\sqrt{3xy}$, when $x = 75, y = 20$
5. $\sqrt{7x - 4}$, when $x = 8$
6. $\sqrt{5x - 2}$, when $x = 13$
7. $\sqrt{4x + 25}$, when $x = -4$
8. $\sqrt{3x + 27}$, when $x = -5$
9. $\sqrt{6x + 7y}$, when $x = 8, y = -3$
10. $\sqrt{9x + 5y}$, when $x = 9, y = -5$
11. $\sqrt{5x - 14y}$, when $x = -10, y = -9$
12. $\sqrt{6x - 18y}$, when $x = -3, y = -7$
13. $\sqrt{6x} - 2\sqrt{10}$, when $x = 15$
14. $\sqrt{14x} - 8\sqrt{7}$, when $x = 18$
15. $x\sqrt{22} + 5\sqrt{10x}$, when $x = 55$
16. $x\sqrt{15} + 2\sqrt{39x}$, when $x = 65$
17. $3\sqrt{5x} - 2\sqrt{2y}$, when $x = 98, y = 45$
18. $4\sqrt{7x} - 5\sqrt{3y}$, when $x = 75, y = 112$
19. $6\sqrt{6x} - 10\sqrt{14y}$, when $x = 28, y = 27$
20. $9\sqrt{10x} - 4\sqrt{15y}$, when $x = 12, y = 32$
21. $\sqrt{x^2 - 4y}$, when $x = -8, y = 4$
22. $\sqrt{x^2 - 4y}$, when $x = -6, y = 7$
23. $\sqrt{x^2 - 4y}$, when $x = 11, y = -8$
24. $\sqrt{x^2 - 4y}$, when $x = 13, y = -5$
25. $\frac{10 + \sqrt{3x}}{4}$, when $x = 8$
26. $\frac{12 + \sqrt{5x}}{6}$, when $x = 18$
27. $\frac{-10 - 2\sqrt{5x}}{20}$, when $x = 15$
28. $\frac{-14 - 7\sqrt{2x}}{28}$, when $x = 22$

29. $\sqrt{30x}$, when $x = \frac{5}{3}$

30. $\sqrt{42x}$, when $x = \frac{7}{2}$

31. $\sqrt{3xy}$, when $x = \frac{4}{15}, y = \frac{20}{7}$

32. $\sqrt{7xy}$, when $x = \frac{9}{14}, y = \frac{8}{3}$

33. $\sqrt{8x + 3}$, when $x = \frac{2}{3}$

34. $\sqrt{7x + 4}$, when $x = \frac{4}{9}$

35. $\sqrt{x + 3y}$, when $x = \frac{1}{9}, y = \frac{1}{5}$

36. $\sqrt{3x + y}$, when $x = \frac{3}{4}, y = \frac{3}{7}$

37. $\sqrt{3x - 2y}$, when $x = \frac{5}{27}, y = \frac{3}{14}$

38. $\sqrt{6x - 2y}$, when $x = \frac{1}{4}, y = \frac{12}{25}$

39. $4\sqrt{10} - 3\sqrt{x}$, when $x = \frac{2}{5}$

40. $5\sqrt{14} - 6\sqrt{x}$, when $x = \frac{2}{7}$

41. $3\sqrt{x} + 10\sqrt{y}$, when $x = \frac{3}{2}, y = \frac{2}{3}$

42. $9\sqrt{x} + 2\sqrt{y}$, when $x = \frac{10}{3}, y = \frac{3}{10}$

43. $2\sqrt{2x} - 3\sqrt{y}$, when $x = \frac{5}{4}, y = \frac{18}{5}$

44. $3\sqrt{3x} - 5\sqrt{y}$, when $x = \frac{7}{9}, y = \frac{12}{7}$

45. $\sqrt{x^2 - 4y}$, when $x = \frac{3}{10}, y = \frac{-2}{5}$

46. $\sqrt{x^2 - 4y}$, when $x = \frac{5}{12}, y = \frac{-1}{6}$

47. $\sqrt{x^2 - 4y}$, when $x = \frac{-10}{3}, y = \frac{5}{3}$

48. $\sqrt{x^2 - 4y}$, when $x = \frac{-12}{5}, y = \frac{6}{5}$

21.5 SOLVING QUADRATIC EQUATIONS BY COMPLETING THE SQUARE

In Chapter 19 we saw how to solve quadratic equations that can be factored in integers. In this chapter we have learned about square roots so that we can solve the quadratic equations that are not factorable in integers.

Remember that a quadratic equation may have no roots, one root, or two roots. If a quadratic equation with *two* roots is not factorable in integers, then its roots are irrational numbers *that can be represented using square roots.*

Solving Equations by Taking Square Roots

The simplest equations with irrational roots are equations such as $x^2 = 2$. To solve them we use the fact that if two quantities are equal, their square roots are equal. Hence, if $x^2 = 2$ then

$$x = \sqrt{2} \quad \text{or} \quad x = -\sqrt{2}$$

(Remember that $\sqrt{2}$ and $-\sqrt{2}$ are both square roots of 2.)

The idea of taking the square root of both sides of the equation can be used to solve all quadratic equations. To see how, we must first look at squares of binomials.

21. Solve by taking square roots.
a. $x^2 = 9$
b. $x^2 = 10$
c. $x^2 = 12$

Perfect Square Trinomials

In Section 16.6 we saw that a perfect square monomial is formed by multiplying a monomial by itself. Similarly, a perfect square trinomial is formed by multiplying a binomial by itself. Consider $(x + B)^2$, where B represents a monomial:

$$\begin{array}{r} x + B \\ x + B \\ \hline x^2 + Bx \quad\quad \\ Bx + B^2 \\ \hline x^2 + 2Bx + B^2 \end{array}$$

Note that the third term B^2 is a perfect square, and the middle term $2Bx$ is twice the product of the terms of the binomial $x + B$. Every trinomial of this form is the square of a binomial. For example,

$$x^2 + 6x + 9 = (x + 3)^2$$
$$x^2 - 14x + 49 = (x - 7)^2$$
$$x^2 + 18x + 81 = (x + 9)^2$$

22. Factor
a. $x^2 - 4x + 4$
b. $x^2 + 8x + 16$
c. $x^2 - 20x + 100$

Completing the Square

Look at the following computations to see how to solve (simple) quadratic equations by completing the square. (Remember that adding the same quantity to both sides of an equation yields an equivalent equation.)

$$\begin{array}{ccc} x^2 + 6x = 7 & \text{and} & x^2 - 14x = -34 \\ \downarrow & & \downarrow \\ x^2 + 6x + 9 = 7 + 9 & & x^2 - 14x + 49 = -34 + 49 \\ (x + 3)^2 = 16 & & (x - 7)^2 = 15 \\ \downarrow & & \downarrow \\ x + 3 = 4 \text{ or } x + 3 = -4 & & x - 7 = \sqrt{15} \text{ or } x - 7 = -\sqrt{15} \\ \downarrow & & \downarrow \\ x = 1 \text{ or } x = -7 & & x = 7 + \sqrt{15} \text{ or } x = 7 - \sqrt{15} \end{array}$$

Can you see what number to add to both members to form a perfect square trinomial?

23. Complete the square.
a. $x^2 + 10x$
b. $x^2 - 12x$
c. $x^2 - 16x$

Look at the following example of *solving quadratic equations by completing the square.*

TYPICAL PROBLEM

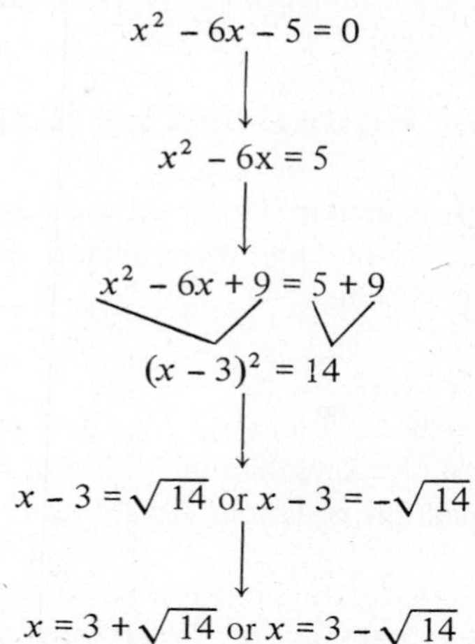

We first shift the constant term to the right member of the equation:

Then we square half the coefficient of the x term and add it to both members of the equation:

We then factor the perfect square trinomial on the left:

Next we take the square root of each member of the equation:

And then we solve the resulting equations:

Hence, $x^2 - 6x - 5 = 0 \Rightarrow x = 3 + \sqrt{14}$ or $x = 3 - \sqrt{14}$.

EXAMPLES

Solve the following equations by completing the square.

1. $x^2 + 14x + 46 = 0$
2. $x^2 - 8x - 56 = 0$
3. $x^2 - 16x + 63 = 0$
4. $x^2 - 28x + 196 = 0$

Solutions

1.
$$x^2 + 14x = -46$$
$$\downarrow$$
$$x^2 + 14x + 49 = -46 + 49$$
$$\downarrow$$
$$(x + 7)^2 = 3$$
$$\downarrow$$
$$x + 7 = \sqrt{3} \text{ or } x + 7 = -\sqrt{3}$$
$$\downarrow$$
$$x = -7 + \sqrt{3} \text{ or } x = -7 - \sqrt{3}$$

2.
$$x^2 - 8x = 56$$
$$\downarrow$$
$$x^2 - 8x + 16 = 56 + 16$$
$$\downarrow$$
$$(x - 4)^2 = 72$$
$$\downarrow$$
$$x - 4 = \sqrt{72} \text{ or } x - 4 = -\sqrt{72}$$
$$\downarrow$$
$$x = 4 + 6\sqrt{2} \text{ or } x = 4 - 6\sqrt{2}$$

3.
$$x^2 - 16x = -63$$
$$\downarrow$$
$$x^2 - 16x + 64 = -63 + 64$$
$$\downarrow$$
$$(x - 8)^2 = 1$$
$$\downarrow$$
$$x - 8 = 1 \text{ or } x - 8 = -1$$
$$\downarrow$$
$$x = 9 \text{ or } x = 7$$

4.
$$x^2 - 28x + 196 = 0$$
$$\downarrow$$
$$(x - 14)^2 = 0$$
$$\downarrow$$
$$x - 14 = 0$$
$$\downarrow$$
$$x = 14$$

SPECIAL CASES

If when we have shifted the constant term to the right and completed the square on the left both members of the equation become perfect squares, we could have solved the equation by factoring. For example,

$$x^2 - 2x - 15 = 0 \quad\text{or}\quad x^2 - 2x - 15 = 0$$

$$\downarrow \qquad\qquad \downarrow$$

$$x^2 - 2x + 1 = 15 + 1 \qquad (x + 3)(x - 5) = 0$$

$$(x - 1)^2 = 16 \qquad x + 3 = 0 \text{ or } x - 5 = 0$$

$$\downarrow \qquad\qquad \downarrow$$

$$x - 1 = 4 \text{ or } x - 1 = -4 \qquad x = -3 \text{ or } x = 5$$

$$\downarrow$$

$$x = 5 \text{ or } x = -3$$

EXERCISES

Solve each of the following equations by completing the square.

1. $x^2 + 4x - 17 = 0$
2. $x^2 + 6x - 21 = 0$
3. $x^2 - 10x + 7 = 0$
4. $x^2 - 8x + 6 = 0$
5. $x^2 + 16x - 57 = 0$
6. $x^2 + 20x - 69 = 0$
7. $x^2 - 12x - 18 = 0$
8. $x^2 - 16x - 8 = 0$
9. $x^2 + 22x + 72 = 0$
10. $x^2 + 26x + 105 = 0$
11. $x^2 - 30x + 78 = 0$
12. $x^2 - 32x + 96 = 0$
13. $x^2 + 28x - 2 = 0$
14. $x^2 + 24x - 18 = 0$
15. $x^2 - 36x + 155 = 0$
16. $x^2 - 40x + 204 = 0$
17. $x^2 + 32x + 56 = 0$
18. $x^2 + 28x + 70 = 0$

21.6 THE QUADRATIC FORMULA

The Solutions of a Simple Quadratic Equation

In the previous section we saw how to solve (simple) quadratic equations of the special form

$$x^2 + 2Bx + C = 0$$

where B and C represent (signed) integers by completing the square. If we solve

$x^2 + 2Bx + C = 0$ formally, we obtain

$$x^2 + 2Bx + C = 0$$
$$\downarrow$$
$$x^2 + 2Bx = -C$$
$$\downarrow$$
$$x^2 + 2Bx + B^2 = B^2 - C$$
$$\downarrow$$
$$(x + B)^2 = B^2 - C$$
$$\downarrow$$
$$x + B = \sqrt{B^2 - C} \text{ or } x + B = -\sqrt{B^2 - C}$$
$$\downarrow$$
$$x = -B + \sqrt{B^2 - C} \text{ or } x = -B - \sqrt{B^2 - C}$$

Note that the roots $-B + \sqrt{B^2 - C}$ and $-B - \sqrt{B^2 - C}$ are conjugate binomials. Hence, to express these two roots more compactly we introduce the symbol $\pm$, which is read *plus or minus.* Then we can rewrite the solution of $x^2 + Bx + C = 0$ as

THE SPECIAL FORMULA

$$x^2 + 2Bx + C = 0$$
$$\Downarrow$$
$$x = -B \pm \sqrt{B^2 - C}$$
$$\downarrow$$
$$x = -B + \sqrt{B^2 - C} \text{ or } x = -B - \sqrt{B^2 - C}$$

(where B and C represent signed numbers)

Hence, to solve an equation of the special form $x^2 + 2Bx + C = 0$ we can evaluate the special formula. For example,

$$x^2 + 6x - 7 = 0 \quad\text{and}\quad x^2 - 14x + 34 = 0$$
$$\downarrow \qquad\qquad \downarrow$$
$$B = 3 \text{ and } C = -7 \qquad B = -7 \text{ and } C = 34$$
$$\downarrow \qquad\qquad \downarrow$$
$$x = -(3) \pm \sqrt{(3)^2 - (-7)} \qquad x = -(-7) \pm \sqrt{(-7)^2 - (34)}$$
$$\downarrow \qquad\qquad \downarrow$$
$$x = -3 \pm \sqrt{16} = -3 \pm 4 \qquad x = 7 \pm \sqrt{15}$$
$$\downarrow \qquad\qquad \downarrow$$
$$x = 1 \text{ or } x = -7 \qquad x = 7 + \sqrt{15} \text{ or } x = 7 - \sqrt{15}$$

The Solutions of a General Quadratic Equation

We can develop a formula for the roots of a general quadratic equation of the form

$$ax^2 + bx + c = 0$$

where a, b, and c represent integers and $a \neq 0$ by using the (not obvious) fact that every equation of this form corresponds to an equation of the special form $x^2 + 2Bx + C = 0$: Multiplying both members of $ax^2 + bx + c = 0$ by $4a$ yields

$$4a(ax^2 + bx + c) = 4a(0)$$
$$\downarrow$$
$$4a^2x^2 + 4abx + 4ac = 0$$
$$\downarrow$$

$$(2ax)^2 + 2b(2ax) + 4ac = 0$$

$$x^2 + 2Bx + C = 0$$

$$\Downarrow$$

$$x = -B \pm \sqrt{B^2 - C}$$

$$2ax = -b \pm \sqrt{b^2 - 4ac}$$

$$\downarrow$$

$$x = \frac{-b \pm \sqrt{b^2 - 4ac}}{2a}$$

(Compare this with Method B for factoring general trinomials in Section 16.7.)

THE QUADRATIC FORMULA

$$ax^2 + bx + c = 0$$

$$\Downarrow$$

$$x = \frac{-b \pm \sqrt{b^2 - 4ac}}{2a}$$

$$\downarrow$$

$$x = \frac{-b + \sqrt{b^2 - 4ac}}{2a} \text{ or } x = \frac{-b - \sqrt{b^2 - 4ac}}{2a}$$

(where a, b, and c represent signed numbers with $a = 0$)

Hence, to solve an equation of the general form $ax^2 + bx + c = 0$ we can evaluate the quadratic formula.

Look at the following example of *solving quadratic equations using the quadratic formula.*

TYPICAL PROBLEM

$$3x^2 - 7x - 3 = 0$$

$$\downarrow$$

We first determine the values for a, b, and c:

$$a = 3, b = -7, c = -3$$

$$\downarrow$$

Then we evaluate the quadratic formula for these values, using parentheses when we substitute:

$$x = \frac{-b \pm \sqrt{b^2 - 4ac}}{2a}, \text{ when } a = 3, b = -7, c = -3$$

$$x = \frac{-(-7) \pm \sqrt{(-7)^2 - 4(3)(-3)}}{2(3)}$$

$$\downarrow$$

$$x = \frac{7 \pm \sqrt{49 + 36}}{6}$$

$$\downarrow$$

$$x = \frac{7 + \sqrt{85}}{6} \text{ or } x = \frac{7 - \sqrt{85}}{6}$$

Hence, $3x^2 - 7x - 3 = 0 \quad \Rightarrow \quad x = \frac{7 + \sqrt{85}}{6}$ or $x = \frac{7 - \sqrt{85}}{6}$.

To evaluate the quadratic formula we use the method in Section 21.4.

EXAMPLES

Solve the following equations by evaluating the quadratic formula.

1. $x^2 - 21x + 109 = 0$
2. $x^2 + 15x + 27 = 0$
3. $5x^2 - 6x - 2 = 0$
4. $2x^2 + 7x - 9 = 0$

Solutions

1.

$$a = 1, b = -21, c = 109$$
$$\downarrow$$
$$x = \frac{-(-21) \pm \sqrt{(-21)^2 - 4(1)(109)}}{2(1)}$$
$$\downarrow$$
$$x = \frac{21 \pm \sqrt{441 - 436}}{2}$$
$$\downarrow$$
$$x = \frac{21 \pm \sqrt{5}}{2}$$
$$\downarrow$$
$$x = \frac{21 + \sqrt{5}}{2} \text{ or } x = \frac{21 - \sqrt{5}}{2}$$

2.

$$a = 1, b = 15, c = 27$$
$$\downarrow$$
$$x = \frac{-(15) \pm \sqrt{(15)^2 - 4(1)(27)}}{2(1)}$$
$$\downarrow$$
$$x = \frac{-15 \pm \sqrt{225 - 108}}{2}$$
$$\downarrow$$
$$x = \frac{-15 \pm \sqrt{117}}{2}$$
$$\downarrow$$
$$x = \frac{-15 \pm 3\sqrt{13}}{2}$$
$$\downarrow$$
$$x = \frac{-15 + 3\sqrt{13}}{2} \text{ or } x = \frac{-15 - 3\sqrt{13}}{2}$$

3.

$$a = 5, b = -6, c = -2$$
$$\downarrow$$
$$x = \frac{-(-6) \pm \sqrt{(-6)^2 - 4(5)(-2)}}{2(5)}$$
$$\downarrow$$
$$x = \frac{6 \pm \sqrt{36 + 40}}{10}$$
$$\downarrow$$
$$x = \frac{6 \pm \sqrt{76}}{10}$$
$$\downarrow$$
$$x = \frac{6 \pm 2\sqrt{19}}{10}$$
$$\downarrow$$
$$x = \frac{3 \pm \sqrt{19}}{5}$$
$$\downarrow$$
$$x = \frac{3 + \sqrt{19}}{5} \text{ or } x = \frac{3 - \sqrt{19}}{5}$$

4.

$$a = 2, b = 7, c = -9$$
$$\downarrow$$
$$x = \frac{-(7) \pm \sqrt{(7)^2 - 4(2)(-9)}}{2(7)}$$
$$\downarrow$$
$$x = \frac{-7 \pm \sqrt{49 + 72}}{14}$$
$$\downarrow$$
$$x = \frac{-7 \pm \sqrt{121}}{14}$$
$$\downarrow$$
$$x = \frac{-7 \pm 11}{14}$$
$$\downarrow$$
$$x = \frac{-7 + 11}{14} \text{ or } x = \frac{-7 - 11}{14}$$
$$\downarrow$$
$$x = \frac{2}{7} \text{ or } x = \frac{-9}{7}$$

SPECIAL CASES

1. In the quadratic formula the quantity $b^2 - 4ac$ is called the *discriminant.* If the discriminant evaluates to a perfect square, then the roots of the quadratic equation are fractions (or integers) and the equation could have been solved by factoring. For example,

$$6x^2 + 11x - 10 = 0$$
$$\downarrow$$
$$a = 6, b = 11, c = -10$$
$$\downarrow$$
$$x = \frac{-(11) \pm \sqrt{(11)^2 - 4(6)(-10)}}{2(6)}$$
$$\downarrow$$
$$x = \frac{-11 \pm \sqrt{361}}{12}$$
$$\downarrow$$
$$x = \frac{-11 \pm 19}{12}$$
$$\downarrow$$
$$x = \frac{2}{3} \text{ or } x = \frac{-5}{2}$$

or

$$6x^2 + 11x - 10 = 0$$
$$\downarrow$$
factor using the methods in Section 16.7
$$\downarrow$$
$$(3x - 2)(2x + 5) = 0$$
$$\downarrow$$
$$3x - 2 = 0 \text{ or } 2x + 5 = 0$$
$$\downarrow$$
$$x = \frac{2}{3} \text{ or } x = \frac{-5}{2}$$

2. In particular, if the discriminant evaluates to 0, the equation has only one root, the fraction $\frac{-b}{2a}$. For example,

$$9x^2 - 12x + 4 = 0$$
$$\downarrow$$
$$a = 9, b = -12, c = 4$$
$$\downarrow$$
$$x = \frac{-(-12) \pm \sqrt{(-12)^2 - 4(9)(4)}}{2(9)}$$
$$\downarrow$$
$$x = \frac{12 \pm \sqrt{0}}{18}$$
$$\downarrow$$
$$x = \frac{2}{3}$$

or

$$9x^2 - 12x + 4$$
$$\downarrow$$
factor using the methods in Section 16.7
$$\downarrow$$
$$(3x - 2)^2 = 0$$
$$\downarrow$$
$$3x - 2 = 0$$
$$\downarrow$$
$$x = \frac{2}{3}$$

OBSERVATIONS AND REFLECTIONS

1. In Chapter 11 we saw how to graph linear equations in x and y. Similarly, we can graph quadratic equations of the form $y = ax^2 + bx + c$. Solving the equation

$ax^2 + bx + c = 0$ gives the x coordinates of the points where the graph of the equation $y = ax^2 + bx + c$ crosses the x axis. For example, look at the following diagram, which shows the graphs of

$$y = x^2 + 1 \qquad y = x^2 \qquad y = x^2 - 1$$

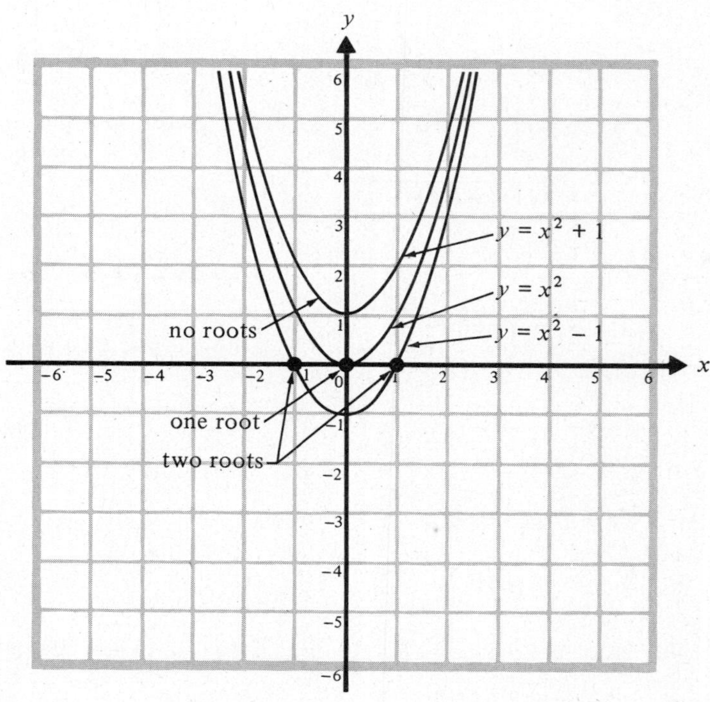

Observe that evaluating $x^2 + 1$ for any value of x yields a positive quantity no smaller than 1, so the graph of $y = x^2 + 1$ never crosses the x axis and $x^2 + 1 = 0$ has no roots. Similarly, x^2 is non-negative for all values of x, so the graph of $y = x^2$ touches the x axis only at the origin and $x^2 = 0$ has one root. On the other hand, evaluating $x^2 - 1$ yields both positive and negative values, so $y = x^2 - 1$ crosses the x axis twice and $x^2 - 1$ has two roots.

2. The three cases described in 1 can be distinguished by looking at the *discriminant* $b^2 - 4ac$ of the trinomial. If $b^2 - 4ac$ is positive, as it is for $x^2 - 1$, the graph of $y = ax^2 + bx + c$ crosses the x axis twice and $ax^2 + bx + c = 0$ has two roots. If $b^2 - 4ac$ is 0, as it is for x^2, the graph of $y = ax^2 + bx + c$ just touches the x axis and $ax^2 + bx + c = 0$ has one root. On the other hand, if $b^2 - 4ac$ is negative, as it is for $x^2 + 1$, the graph of $y = ax^2 + bx + c$ never crosses the x axis and $ax^2 + bx + c = 0$ has no roots.

24. Evaluate the discriminant to determine the number of roots of
a. $x^2 + 4x + 4 = 0$
b. $x^2 + 3x + 5 = 0$
c. $x^2 - 4x + 6 = 0$
d. $x^2 - 6x + 4 = 0$

EXERCISES

Solve each of the following equations using the quadratic formula.

1. $x^2 + 3x - 6 = 0$
2. $x^2 + 5x - 5 = 0$
3. $x^2 - 11x + 4 = 0$
4. $x^2 - 9x + 6 = 0$
5. $x^2 + 13x + 12 = 0$
6. $x^2 + 7x + 10 = 0$
7. $x^2 - 15x + 9 = 0$
8. $x^2 - 11x + 14 = 0$
9. $x^2 - 9x - 22 = 0$

10. $x^2 - 7x - 30 = 0$ 11. $x^2 + 17x + 34 = 0$ 12. $x^2 + 19x + 43 = 0$

13. $2x^2 - 5x - 5 = 0$ 14. $2x^2 - 7x - 7 = 0$ 15. $3x^2 + 4x - 7 = 0$

16. $3x^2 + 2x - 5 = 0$ 17. $5x^2 - 8x + 2 = 0$ 18. $5x^2 - 10x + 1 = 0$

19. $8x^2 + 9x - 1 = 0$ 20. $8x^2 + 7x - 3 = 0$ 21. $7x^2 + 6x - 5 = 0$

22. $7x^2 + 8x - 2 = 0$ 23. $4x^2 - 12x + 3 = 0$ 24. $4x^2 - 14x + 11 = 0$

25. $5x^2 + 11x + 2 = 0$ 26. $5x^2 + 13x + 6 = 0$ 27. $10x^2 - 2x - 3 = 0$

28. $10x^2 - 4x - 1 = 0$ 29. $4x^2 + 20x + 25 = 0$ 30. $9x^2 + 24x + 16 = 0$

ANSWERS TO MARGIN EXERCISES

1. a. 5
b. 6
c. 7

2. a. $\sqrt{20}$
b. $\sqrt{75}$
c. $\sqrt{72}$

3. a and f
b and e
c and d

4. a. $\frac{97}{56}$
b. $\frac{49}{20}$

5. a. $\sqrt{13}$
b. 13

6. a. $\sqrt{22}$
b. $\sqrt{30}$
c. $\sqrt{21}$
d. $\sqrt{130}$

7. a. $6\sqrt{3}$
b. $4\sqrt{2}$
c. $-6\sqrt{5}$
d. $-9\sqrt{7}$

8. a. $-2\sqrt{2}$
b. $5\sqrt{3}$
c. $-5\sqrt{5}$
d. $-\sqrt{7}$

9. a. 66
b. $36xy$
c. $10x^5$

10. a. 2
b. -5

12. a. $\sqrt{2}\,(\sqrt{5} + \sqrt{3})$
b. $\sqrt{5}\,(\sqrt{2} - \sqrt{3})$
c. $\sqrt{7}\,(-\sqrt{2} + 2)$
d. $\sqrt{11}\,(-\sqrt{2} - 1)$

13. a. $\sqrt{2}$
b. $\sqrt{3}$

14. a and c
b and e
d and f

15. a and e
b and d
c and f

16. a. 1
b. 7
c. $\frac{-2}{3}$
d. $\frac{1}{2}$

21. a. $x = 3$ or $x = -3$
b. $x = \sqrt{10}$ or $x = -\sqrt{10}$
c. $x = 2\sqrt{3}$ or $x = -2\sqrt{3}$

22. a. $(x - 2)^2$
b. $(x + 4)^2$
c. $(x - 10)^2$

23. a. $x^2 + 10x + 25$
b. $x^2 - 12x + 36$
c. $x^2 - 16x + 64$

24. a. one root
b. no roots
c. no roots
d. two roots

CHAPTER REVIEW

SECTION 21.1 Reduce to simplest form.

1. $\sqrt{28}$
2. $\sqrt{45}$
3. $\sqrt{75}$
4. $\sqrt{169}$
5. $\sqrt{108}$
6. $\sqrt{136}$
7. $8\sqrt{50}$
8. $6\sqrt{117}$
9. $15\sqrt{135}$
10. $\sqrt{x^3y^6}$
11. $\sqrt{x^9y^6z^4}$
12. $\sqrt{24x^7}$
13. $\sqrt{196y^{16}}$
14. $\sqrt{60x^6y^3}$
15. $\sqrt{200x^8y^{11}}$
16. $7x^2\sqrt{120y^{10}}$
17. $18xy^6\sqrt{100x^5y^{15}}$
18. $5x^4y^2\sqrt{189x^3y^{12}}$

SECTION 21.2 Multiply to convert to a radical expression in simplest form.

19. $\sqrt{15}\sqrt{35}$
20. $\sqrt{42}\sqrt{105}$
21. $\sqrt{95}\sqrt{114}$
22. $4\sqrt{51}\sqrt{119}$
23. $\sqrt{66}(9\sqrt{21})$
24. $(2\sqrt{57})(2\sqrt{19})$
25. $\sqrt{6x}\sqrt{xy}$
26. $y^4\sqrt{y}(\sqrt{13xy})$
27. $(6x^2y\sqrt{70y})(6xy^3\sqrt{10x})$
28. $\sqrt{3}(8-6\sqrt{3})$
29. $4\sqrt{10}(\sqrt{5}+5\sqrt{2})$
30. $\sqrt{35}(2\sqrt{105}-7\sqrt{35})$
31. $\sqrt{51}(\sqrt{17}+\sqrt{3})$
32. $2\sqrt{15}(3\sqrt{3}-15\sqrt{5})$
33. $\sqrt{22}(-\sqrt{33}+6\sqrt{14})$
34. $(5+\sqrt{14})(4-\sqrt{14})$
35. $(6+2\sqrt{30})(3\sqrt{30}+\sqrt{15})$
36. $(5\sqrt{7}-1)(3\sqrt{14}-8\sqrt{2})$
37. $(-2-\sqrt{10})(-6+\sqrt{35})$
38. $(\sqrt{2}-\sqrt{11})(\sqrt{5}+\sqrt{2})$
39. $(7\sqrt{2}+3\sqrt{14})(9\sqrt{2}-4\sqrt{7})$
40. $(x+8+\sqrt{19})(x+8-\sqrt{19})$
41. $(x-6+\sqrt{66})(x-6-\sqrt{66})$
42. $(x+11+5\sqrt{7})(x+11-5\sqrt{7})$

SECTION 21.3 Simplify to a reduced fraction in standard form.

43. $\sqrt{34}\div\sqrt{51}$
44. $18\sqrt{22}\div 14\sqrt{77}$
45. $14\sqrt{190}\div 8\sqrt{133}$
46. $\dfrac{\sqrt{85}}{\sqrt{95}}$
47. $\dfrac{21\sqrt{154}}{35\sqrt{42}}$
48. $\dfrac{45\sqrt{17}}{12\sqrt{170}}$
49. $x^3y\sqrt{xy}\div xy^4\sqrt{y}$
50. $5xy^6\sqrt{6x}\div 8x^3\sqrt{15y}$

51. $\dfrac{36x^5y^2\sqrt{5x}}{16x^2y\sqrt{30xy}}$

52. $\dfrac{6-\sqrt{14}}{\sqrt{7}}$

53. $\dfrac{15+2\sqrt{21}}{9\sqrt{15}}$

54. $\dfrac{\sqrt{42}-\sqrt{30}}{\sqrt{66}}$

55. $\dfrac{6\sqrt{15}}{4+\sqrt{6}}$

56. $\dfrac{7\sqrt{21}}{8-2\sqrt{35}}$

57. $\dfrac{5+\sqrt{7}}{6+\sqrt{7}}$

58. $\dfrac{\sqrt{2}-4\sqrt{5}}{4+3\sqrt{10}}$

59. $\dfrac{5\sqrt{21}}{\sqrt{35}-\sqrt{15}}$

60. $\dfrac{\sqrt{33}-9\sqrt{2}}{\sqrt{22}+2\sqrt{3}}$

Evaluate, simplifying completely. Convert fractions to standard form. SECTION 21.4

61. $\sqrt{10x}$, when $x = 45$

62. $\sqrt{6xy}$, when $x = 27, y = 18$

63. $\sqrt{11x-3}$, when $x = 5$

64. $\sqrt{5x+26}$, when $x = -2$

65. $\sqrt{7x+8y}$, when $x = 8, y = -3$

66. $\sqrt{4x-16y}$, when $x = -4, y = -8$

67. $\sqrt{6x} - 6\sqrt{14}$, when $x = 21$

68. $x\sqrt{21} + 6\sqrt{35x}$, when $x = 60$

69. $8\sqrt{7x} - 7\sqrt{2y}$, when $x = 72, y = 63$

70. $7\sqrt{21x} - 11\sqrt{35y}$, when $x = 45, y = 12$

71. $\sqrt{x^2-4y}$, when $x = -10, y = 7$

72. $\sqrt{x^2-4y}$, when $x = 9, y = -11$

73. $\dfrac{8+\sqrt{7x}}{6}$, when $x = 12$

74. $\dfrac{-12-4\sqrt{3x}}{24}$, when $x = 21$

75. $\sqrt{70x}$, when $x = \dfrac{7}{5}$

76. $\sqrt{5xy}$, when $x = \dfrac{9}{10}, y = \dfrac{18}{7}$

77. $\sqrt{6x+5}$, when $x = \dfrac{4}{5}$

78. $\sqrt{3x+y}$, when $x = \dfrac{1}{4}, y = \dfrac{3}{5}$

79. $\sqrt{5x-2y}$, when $x = \dfrac{1}{2}, y = \dfrac{5}{9}$

80. $2\sqrt{6} - 5\sqrt{x}$, when $x = \dfrac{2}{3}$

81. $9\sqrt{x} + 6\sqrt{y}$, when $x = \dfrac{3}{2}, y = \dfrac{8}{3}$

82. $5\sqrt{2x} - 3\sqrt{y}$, when $x = \dfrac{7}{4}, y = \dfrac{50}{7}$

83. $\sqrt{x^2-4y}$, when $x = \dfrac{11}{10}, y = \dfrac{-3}{16}$

84. $\sqrt{x^2 - 4y}$, when $x = \frac{-13}{4}$, $y = \frac{1}{4}$

SECTION 21.5 Solve by completing the square.

85. $x^2 + 2x - 19 = 0$

86. $x^2 - 8x + 10 = 0$

87. $x^2 + 18x - 63 = 0$

88. $x^2 - 14x - 50 = 0$

89. $x^2 + 24x + 108 = 0$

90. $x^2 - 34x + 97 = 0$

91. $x^2 + 26x - 20 = 0$

92. $x^2 - 38x + 161 = 0$

93. $x^2 + 30x + 93 = 0$

SECTION 21.6 Solve using the quadratic formula.

94. $x^2 + 7x - 7 = 0$

95. $x^2 - 11x + 9 = 0$

96. $x^2 + 9x + 14 = 0$

97. $x^2 - 13x + 11 = 0$

98. $x^2 - 5x - 24 = 0$

99. $x^2 + 15x + 45 = 0$

100. $2x^2 - 9x - 9 = 0$

101. $3x^2 + 8x - 11 = 0$

102. $5x^2 - 12x + 3 = 0$

103. $8x^2 + 5x - 2 = 0$

104. $7x^2 + 4x - 2 = 0$

105. $4x^2 - 16x + 5 = 0$

106. $5x^2 + 9x + 4 = 0$

107. $10x^2 - 6x - 1 = 0$

108. $4x^2 + 28x + 49 = 0$

ANSWERS

CHAPTER 1

1.0

1. −3
2. +1
3. −3
4. −2
5. +2
6. −1
7. +2
8. −2
9. −4
10. +4
11. −4
12. +5
13. 0
14. +6
15. −5
16. 0
17. −5
18. +3

1.1

1. −7
2. −10
3. −6
4. −3
5. 1
6. −14
7. 10
8. 0
9. −14
10. 0
11. −5
12. −6
13. −13
14. −1
15. −2
16. 8
17. −16
18. −77
19. 34
20. −53
21. 77
22. −26
23. 31
24. −8
25. −42
26. −96
27. −53
28. −21
29. −102
30. 133
31. −102
32. −38
33. −173
34. 275
35. 391
36. −417

1.1 Phrases

1. $8 + 7 = 15$
2. $9 - 13 = -4$
3. $16 - 5 = 11$
4. $17 + 4 = 21$
5. $11 - 13 = -2$
6. $4 - 6 = -2$
7. $5 + 31 = 36$
8. $12 - 15 = -3$
9. $11 + 2 = 13$
10. $1 - 5 = -4$
11. $14 + 8 = 22$
12. $16 - 19 = -3$

1.2

1. −10
2. −9
3. −45
4. 1
5. 15
6. −6

7. −14 8. 0 9. 4 10. 0 11. −12 12. 14

13. 8 14. −4 15. −5 16. −9 17. −13 18. −1

19. −1 20. 0 21. −3 22. −2 23. 8 24. −46

1.3

1. −7 2. 26 3. −35 4. 19 5. 31 6. 61

7. −14 8. −6 9. 1 10. −20 11. 81 12. −8

13. −6 14. −49 15. −2 16. 9 17. 9 18. −4

19. 8 20. 0 21. −12 22. −4 23. −16 24. 8

25. −59 26. 0 27. 15 28. −60 29. −15 30. 0

31. −121 32. 0 33. −27 34. −36 35. 146 36. −84

1.3 Phrases

1. $-9-8=-17$ 2. $-7-5=-12$ 3. $-27-(-16)=-11$ 4. $-3+(-11)=-14$

5. $-14-(-14)=0$ 6. $-3-(-2)=-1$ 7. $5-(-1)=6$ 8. $-15+(-4)=-19$

9. $-24-(-7)=-17$ 10. $-9-(-6)=-3$ 11. $-21-19=-40$ 12. $-13-22=-35$

1.4

1. −70 2. −148 3. 126 4. 72 5. −24 6. −294

7. −64 8. 24 9. −27 10. −70 11. 56 12. 0

13. 0 14. −60 15. −48 16. 0 17. −63 18. 64

19. 168 20. −6 21. −24 22. −120 23. 252 24. −288

25. 30 26. −72 27. 0 28. −60 29. 72 30. 0

31. −90 32. 0 33. 210 34. 0 35. 0 36. −126

1.4 Phrases

1. $4(-12)=-48$ 2. $3(-7)=-21$ 3. $2(-16)=-32$ 4. $(-5)(-9)=45$

5. $(-2)(-6)(3)=36$ 6. $8(-1)(5)=-40$ 7. $(-4)(-2)(-3)=-24$ 8. $2(9)(6)=108$

9. $(-10)(7)(-4)=280$ 10. $13(-1)(2)=-26$ 11. $2(-8)(10)=-160$ 12. $2(-15)(-4)=120$

1.5

1. 11 2. −15 3. 9 4. 15 5. 34 6. 9

7. −19 8. 30 9. −3 10. −13 11. 17 12. 30

13. −4 14. −29 15. −5 16. 27 17. 48 18. −84

19. −15 20. 16 21. 18 22. −36 23. 4 24. −19

25. −3 26. 10 27. 41 28. 0 29. 19 30. 8

31. 28   32. 1   33. 20   34. 35   35. −46   36. −52

37. −6   38. 44   39. 44   40. 0   41. −26   42. 5

43. 44   44. −25

1.5 Phrases

1. (5)(12) + 18 = 78   2. 8 + (7)(3) = 29   3. 15 + (17 + 6) = 38   4. 50 − (9)(8) = −22   5. 4(13 − 19) = −24

6. 19 − (12 + 9) = −2   7. 90 − (6)(7) = 48   8. (20 + 15)(−9) = −315   9. (15 + 3) − (−8) = 26   10. −56 + 2(27) = −2

11. 3(9 − 18) = −27   12. −72 + (8)(9) = 0   13. (4)(3) − 6 = 6   14. (43 − 18)(−3) = −75   15. −10 − (5)(6) = −40

16. −80 − (15)(2) = −110   17. 4(16) − 32 = 32   18. 2(16 − 24) = −16   19. 86 − (16)9 = −58   20. 25(−3 − 4) = −175

1.6 Phrases

1. 6 + 13 = 19   2. 5 − 11 = −6   3. 16 − 9 = 7   4. 23 − 8 = 15   5. 26 + 8 = 34

6. 21 − 25 = −4   7. −31 − (−13) = −18   8. −6 − 17 = −23   9. −10 − 7 = −17   10. −3 − 19 = −22

11. −25 − (−5) = −20   12. 18 − (−18) = 36   13. 2(−19) = −38   14. −8(−7) = 56   15. 6(3)(−5) = −90

16. 2(7)(5) = 70   17. (−2)(−3)(9) = 54   18. −7(2)(−7) = 98   19. −8(−4)(5) = 160   20. 2(−15)(4) = −120

21. (4 · 5) + 6 = 26   22. 9 + (10 · 3) = 39   23. 18 + (12 + 11) = 41   24. 63 − (6 · 10) = 3   25. 27 − (9 + 12) = 6

26. 14 − (12 + 13) = −11   27. 56 − (9 · 4) = 20   28. (8 + 22)(−9) = −270   29. (25 · 7) − 80 = 95   30. (42 + 11) + 17 = 70

31. (12)(9) − 99 = 9   32. (16 + 31) + (−62) = −15   33. −17 − (3 · 16) = −65

34. −18 − (14 · 5) = −88   35. −26 − (8)(7) = −82   36. (−11)(2 + 9) = −121

1 — Chapter Review

1. −1   2. −3   3. −5   4. +3   5. +2   6. 0

7. +5   8. −3   9. +6   10. −7   11. −6   12. 6

13. 7   14. 0   15. −5   16. −17   17. −5   18. −32

19. 18   20. 91   21. 34   22. −34   23. −29   24. −130

25. −59   26. 168   27. 347   28. 6   29. −9   30. 3

31. 16   32. 2   33. 4   34. 6   35. 4   36. 9

37. 4   38. 0   39. 0   40. 16   41. 13   42. −42

43. 40   44. 21   45. −20   46. −3   47. −6   48. 13

49. 4   50. −7   51. −30   52. −26   53. 0   54. −24

55. −17   56. −109   57. −35   58. 21   59. 60   60. −66

61. −60   62. 0   63. −84   64. −44   65. −44   66. −60

67. −288   68. 108   69. −144   70. −90   71. −72   72. 0

73. 42   74. 72   75. 30   76. −19   77. −15   78. 50

79. 26   80. −10   81. 0   82. 40   83. 31   84. 160

85. −1 86. −21 87. −8 88. 24 89. −6 90. 12
91. 3 92. −42 93. 9 94. −17 95. 0 96. −45
97. −48

1 — Cumulative Test

1. (1.1) 27 2. (1.2) 6 3. (1.3) −3 4. (1.4) −81 5. (1.1) −3 6. (1.2) 0
7. (1.4) −132 8. (1.2) −7 9. (1.1) −44 10. (1.3) 4 11. (1.2) −37 12. (1.4) 0
13. (1.1) 26 14. (1.4) 120 15. (1.3) 14 16. (1.2) −3 17. (1.4) −147 18. (1.1) −33
19. (1.2) 3 20. (1.4) −84 21. (1.3) 2 22. (1.1) −38 23. (1.2) −3 24. (1.4) 0
25. (1.2) −1 26. (1.3) −25 27. (1.4) −36 28. (1.2) −2 29. (1.3) −2 30. (1.4) 70

CHAPTER 2

2.1

1. $35xy$ 2. $27wxy$ 3. $6AB$ 4. $12mnp$ 5. $28xyz$ 6. $-36wx$
7. $48abcd$ 8. $-55xyz$ 9. $-54mn$ 10. $-28AB$ 11. $-36rst$ 12. $-32xy$
13. $27rs$ 14. $6wxyz$ 15. $-32wxyz$ 16. $21xyz$ 17. $36xyz$ 18. $24wxyz$
19. $-252KMN$ 20. $60abcd$ 21. $-63abcd$ 22. $18ABC$ 23. $-648tu$ 24. $-35x$

2.1 Phrases

1. $5a(-7b) = -35ab$ 2. $(-12x)(3y) = -36xy$ 3. $2(xy) = 2xy$
4. $(9c)(-6d) = -54cd$ 5. $4x(-10z)(3y) = -120xyz$ 6. $2(9x)(-y) = -18xy$
7. $-5w(-9x)(-2u) = -90uwx$ 8. $2(5abc)(-13de) = -130abcde$ 9. $abx(12w)(z) = 12abwxz$
10. $8ad(-2b)(-2c) = 32abcd$ 11. $-18a(-5b)(2cd) = 180abcd$ 12. $20xy(-10z)(-9w) = 1800wxyz$

2.2

1. x^5 2. n^7 3. r^4 4. c^5 5. x^4 6. w^5
7. K^6 8. x^{11} 9. n^{13} 10. b^{10} 11. w^{11} 12. Y^{15}
13. x^8 14. s^9 15. $-z^5$ 16. $-x^7$ 17. t^{12} 18. x^8
19. 81 20. 128 21. 256 22. 243 23. 1024 24. 216
25. 1024 26. 729 27. 625 28. 1 29. 4096 30. 256
31. −1 32. −1 33. 1 34. 16 35. 81 36. −512

2.2 Phrases

1. $a^3 \cdot a^4 = a^7$ 2. $m^9 \cdot m^2 = m^{11}$ 3. $(y)^2 = y^2$
4. $x^3(x^4 \cdot x^5) = x^{12}$ 5. $b^9(b^0 \cdot b^6) = b^{15}$ 6. $x(x^8 \cdot x^9) = x^{18}$

7. $v^4(v^7 \cdot v^3) = v^{14}$
8. $a^8(a^{10} \cdot a^{11}) = a^{29}$
9. $(c)^5 \cdot c = c^6$
10. $3^2 \cdot 3^3 = 3^5 = 243$
11. $7^2 \cdot 7 = 7^3 = 343$
12. $2(8)(8^2) = 1024$
13. $(-1)^3(-1)^3 = (-1)^6 = 1$
14. $5(5^2 \cdot 5^0) = 5^3 = 125$
15. $2(-3)^3(-3)^2 = -486$
16. $(-2)^2(-2)(-2)^4 = (-2)^7 = -128$
17. $4^2(4 \cdot 4^0) = 4^3 = 64$
18. $(-2)^2(-2)^4 = (-2)^6 = 64$

2.3

1. $a^2b^2c^2$
2. $-r^2s^2t^2$
3. a^3b^5
4. R^4S^4
5. k^6m^7
6. $-s^7t^7$
7. abr^3s^3
8. abx^7y^{10}
9. $-x^4y^4z^2$
10. A^5B^4C
11. $x^7y^5z^8$
12. $k^5r^6t^3$
13. $30A^5$
14. $-120x^7$
15. $-6r^5$
16. $18t^6$
17. $54A^3B^6$
18. $-24x^{11}y^2$
19. $-30A^3B^6$
20. $8x^3y^5$
21. $-44r^3s^3$
22. $81x^5y^5$
23. $-60k^3m^{12}$
24. $-216a^3b^9$
25. $64x^{13}$
26. $-64a^{10}$
27. $60y^6$
28. $-36x^5$
29. $-35m^9n^9$
30. $-240s^{10}t^{11}$
31. $45x^{11}y^{10}$
32. $28u^6v^8$
33. $3a^3b^8c^2$
34. $-10w^5x^4y^4$
35. $-126r^5s^6t^{10}$
36. $-72w^5x^3y^6$
37. $-t^{14}$
38. r^{17}
39. $-3k^9$
40. $-8x^9$
41. $5t^{13}$
42. $8y^{16}$

2.3 Phrases

1. $5\text{m}(7\text{m}^2) = 35\text{m}^3$
2. $9\text{cm}(12\text{cm}^2) = 108\text{cm}^3$
3. $2(30\text{mm}^3) = 60\text{mm}^3$
4. $4\text{km}(6\text{km}^2) = 24\text{km}^3$
5. $3x^2y(-4xyz)(7y^2z^4) = -84x^3y^4z^5$
6. $2(9x^3)(-9x^4z^5) = -162x^7z^5$
7. $-18a^3b^2(a^6b^8c^4)(-2a^5bc^5) = 36a^{14}b^{11}c^9$
8. $3abx(7a^2b^5x^3)(a^3bx) = 21a^6b^7x^5$
9. $-8ad^5(-3a^5d^3)(-5a^5d) = -120a^{11}d^9$
10. $2(4x^3y^2z^2)(25x^6y^8z^0) = 200x^9y^{10}z^2$
11. $r^{10}s^8t^3(r^6s^7t^4)(r^5s^2t) = r^{21}s^{17}t^8$
12. $-6xyz(2)(-4x^3y^0z^0) = 48x^4yz$

2.4

1. 20
2. 18
3. −28
4. −72
5. −27
6. −30
7. 16
8. 9
9. −108
10. −192
11. −36
12. −42
13. 0
14. 0
15. 400
16. 360
17. 36
18. 32
19. 48
20. 12
21. −147
22. −288
23. −270
24. −160
25. 60
26. 96
27. 240
28. 162
29. −600
30. −288

2 – Chapter Review

1. $12xy$
2. $60mn$
3. $16xyz$
4. $24qrst$
5. $3wxy$
6. $-28abc$
7. $60km$
8. $-35KLMN$
9. $6ABC$
10. $-216xyz$
11. $-8abwyz$
12. $-288rt$
13. b^4
14. K^6
15. w^5
16. t^5
17. x^{18}
18. p^{11}
19. r^{12}
20. b^8
21. $-A^{11}$
22. 125
23. 128
24. 512
25. 81
26. 128
27. 1
28. 1
29. −32
30. 625

31. wx^2y^2z 32. x^5y^5 33. $-x^7y^8$ 34. A^7B^8 35. $x^6y^4z^6$ 36. $r^{11}s^9t^{10}$

37. $27x^6$ 38. $-16t^8$ 39. $75m^2n^5$ 40. $-24x^5y^3$ 41. $-128a^3b^3$ 42. $-54x^4y^{12}$

43. $120x^{18}$ 44. $96t^{10}$ 45. $-96x^8y^8$ 46. $225y^9z^9$ 47. $9K^3L^{10}$ 48. $-160r^5s^{10}t^9$

49. $-A^{17}$ 50. $8k^{13}$ 51. $-14y^{24}$ 52. 56 53. −54 54. −32

55. 25 56. −84 57. −48 58. 0 59. 360 60. 108

61. 81 62. −300 63. −16 64. 144 65. 315 66. −4800

2 – Summary of Phrases

1. $(-5u)(-8x) = 40ux$
2. $(-7a)(4c) = -28ac$
3. $11b(-2a)(4c) = -88abc$
4. $2(-10xz)(5y) = -100xyz$
5. $7uv(7w)(z) = 49uvwz$
6. $(3ab)(-10d)(-6c) = 180abcd$
7. $x^4 \cdot x^2 = x^6$
8. $a^9(a)^{12} = a^{21}$
9. $u(u^3 \cdot u^8) = u^{12}$
10. $a^7(a)^3 = a^{10}$
11. $x(x^7 \cdot x^0) = x^8$
12. $5(5^2) = 125$
13. $2(3)(3)^2 = 54$
14. $(-2)(-2)^4 = -32$
15. $(-1)^9(-1)(-1)^0 = 1$
16. $(7\text{m})(11\text{m}) = 77\text{m}^2$
17. $(13\text{cm})(2\text{cm}^2) = 26\text{cm}^3$
18. $2a^2y(-9ayz)(3yz^2) = -54a^3y^3z^3$
19. $6axy(9a^2x^3)(x^4y^5) = 54a^3x^8y^6$
20. $2(14x^5y^3z^4)(3x^7y^0z) = 84x^{12}y^3z^5$
21. $70x^3y(2)(-4x^3y^{10}z^0) = -560x^6y^{11}$

CHAPTER 3

3.1

1. $-7y$ 2. $-5n$ 3. $-4x$ 4. $-12x$ 5. $-8y$ 6. $-15x$

7. $16yz$ 8. $13xy$ 9. $-10uv$ 10. $-14xz$ 11. $-14st$ 12. $-19wx$

13. $2x^2$ 14. $7x^2$ 15. $-3x^5$ 16. $-3y^2$ 17. $-7x^6$ 18. $-17x^3$

19. $-8xy^2$ 20. $-5v^2w$ 21. $-2s^4t^8$ 22. $-2m^3n^9$ 23. $12x^4y^3$ 24. $13u^3v^2$

3.1 Phrases

1. 9cm + 16cm = 25cm
2. 11km − 7km = 4km
3. 12g − 5g = 7g
4. 8L + 13L = 21L
5. 19m − 17m = 2m
6. 6mm − 4mm = 2mm
7. 19g + 31g = 50g
8. 18km − 15km = 3km
9. 2mL + 27mL = 29mL
10. 18g − 11g = 7g
11. 16cm + 6cm = 22cm
12. 25L − 24L = 1L

3.2

1. 0 2. $-5x$ 3. $-8y$ 4. $-7x$ 5. $-12n^2$ 6. $-18A^3$

7. z^5 8. $-8x^8$ 9. 0 10. $12n$ 11. $-8a$ 12. $2n$

13. $12r^3$ 14. $4x^6$ 15. $7t^7$ 16. $-y^2$ 17. $13t^2$ 18. $-4r^3$

19. $-6ab$ 20. $-4xy$ 21. 0 22. 0 23. $-31w^5x^2$ 24. $4s^7t^8$

3.3

1. $10x - 5$
2. $2x + 3xy$
3. $4A - 3A^2$
4. $-3w + 7w^2$
5. 12
6. $-3x^2 + 5x$
7. $-6w + 17w^2$
8. $x^2 + 6xy^2$
9. $14A^3 - 2A^2$
10. $-4x^3$
11. $13x^4 + 2x$
12. $6y^2 - 4xy^2$
13. $-3a^5 + 7b^4 + 13b^5$
14. $16T^3 - ST^3 - 9S^3$
15. $14x^3 - 4y^2 + 9y^3$
16. $15y^4 - 4z^2 + 7y^2$
17. $-4A^2 - 1 + 3A^3$
18. $u - 7v^3 + u^3$
19. $12x^2 - 8$
20. $-4n + 2m$
21. $-8k^3 + 1 + k$
22. $3y - 2xy - 5x$
23. $4x^2y - 4xy^3 - 9x^2y^3$
24. $-17A^2B^3 + A^3B + 11AB^3$

3.4

1. $-23x$
2. $-9A$
3. n
4. $10B$
5. $-7x$
6. $-51E$
7. $-7x^4$
8. $-11y^3$
9. $-21n^6$
10. $6x^2$
11. $-4k$
12. $-14xy$
13. $-10x^4$
14. x^2y
15. $-6x^2 - 14$
16. $-3x - 11y$
17. $a + 2b$
18. $-p - 2p^3$
19. $-2x - 4x^2$
20. $-10p - 3q$
21. $-11a^2$
22. $10y - 2y^2$
23. $-4x^2y^3$
24. $-30x^5y$
25. $-6x^9$
26. $-4w^7x^2$
27. $12x^6y^2 + 6x^7y^2$
28. $17x^4y^2$
29. $3x^4 - 7x^3$
30. $29w^6 + 6w^2$
31. $4x^4 + 5x^2$
32. $30x^7$
33. $36x^3 - 21x^6$
34. $12y^6 + 15y^9$
35. $-27t^{15}$
36. $-12w^7 + 16w^6$
37. $-10x^5 + 28x^7$
38. $6A^9$
39. $-4x$
40. $17y^2$
41. $30x^7$
42. $-6y^9$
43. $-6x^2y^4$
44. $-12xy^4$
45. $8x^6$
46. $15z^3$
47. $-3x^3 + 3y^3$
48. $8x^5 - 4x^3$
49. $-10x^4$
50. $27x^7$
51. $8x^3 - 6x^4$
52. $-3y^4 - 8y^5$
53. $-14x^4y^6$
54. x^6z^4

3.4 Phrases

1. $5(8\text{m}) + 15\text{m} = 55\text{m}$
2. $50\text{cm}^3 + (4\text{cm})(10\text{cm}^2) = 90\text{cm}^3$
3. $18\text{km} + (12\text{m} + 9\text{m}) = 18\text{km} + 21\text{m}$
4. $20\text{mm} - 8(2\text{mm}) = 4\text{mm}$
5. $12(18\text{g} - 6\text{g}) = 144\text{g}$
6. $8\text{m}^2 - 3\text{cm}(7\text{cm}) = 8\text{m}^2 - 21\text{cm}^2$
7. $17\text{m} - (3\text{cm} + 7\text{cm}) = 17\text{m} - 10\text{cm}$
8. $19\text{L} - (13\text{mL} + 9\text{mL}) = 19\text{L} - 22\text{mL}$
9. $90\text{cm}^3 - 6\text{cm}(7\text{cm}^2) = 48\text{cm}^3$
10. $(-20a + 15a)(-9b^2) = 45ab^2$
11. $(15w - 8w) - 10w^2 = 7w - 10w^2$
12. $56a^2 + 2(-27a^2) = 2a^2$
13. $3x^2y(9xy - 18xy) = -27x^3y^2$
14. $16c^2 + (8c)(-9c^2) = 16c^2 - 72c^3$
15. $(-7x)(-3y) - 6xy = 15xy$
16. $(-55a^2 + 25a^2) + 19a = -30a^2 + 19a$
17. $84x^3 - (0)(27x^3) = 84x^3$
18. $80w - (15w^2)(-2w) = 80w + 30w^3$
19. $88x^2y^3 - 2(44x^2y^3) = 0$
20. $(60k^3 - 56k^3) - 14k = 4k^3 - 14k$
21. $(-4x^2)(-16x) - 90x^2 = 64x^3 - 90x^2$
22. $2(73x^2 - 18x^2) = 110x^2$
23. $(6ab^2)(9ab) - 86a^2b^3 = -32a^2b^3$
24. $25x^5(-3x^2 + 7x^2) = 100x^7$

3.5

1. -2
2. -3
3. -1
4. -30
5. 18
6. 18
7. 15
8. 16
9. 8
10. 30
11. -1
12. -24
13. 45
14. 55
15. 3
16. -16
17. 51
18. 17

19. -88 20. -118 21. 14 22. 6 23. -77 24. -129

25. -5 26. 46 27. -54 28. -80 29. 66 30. 23

31. 600 32. 486 33. 9 34. 7 35. 33 36. 35

37. -21 38. -15 39. 12 40. -14 41. 83 42. 93

3 — Chapter Review

1. $-16y$ 2. $-7x$ 3. $-15z$ 4. $88wx$ 5. $-28xz$

6. $9st$ 7. $-33r^2$ 8. $-66A^5$ 9. $96x^7$ 10. $-38kt^2$

11. $-2A^3B^5$ 12. $14x^3z^6$ 13. $-2A$ 14. $-6x$ 15. $8s^4$

16. $-6y^9$ 17. $-x$ 18. A 19. $-9x^3$ 20. $5t^4$

21. $-5r^2$ 22. $-2xy$ 23. 0 24. $19m^3n^8$ 25. $20x - 8$

26. $2A - 3A^2$ 27. $-5m^3 + 6n^2$ 28. $7x^2 - 2xy^2 + 5y^2$ 29. $-A^2 - 11A^4$ 30. $8x^2y - 4y$

31. $-7A - 14B^2 + 2A^2$ 32. $-n^3 + 10m^4 + 2m^3$ 33. $-6x^3 - 6 + 7x^2$ 34. $4s^2 - 6 - 7s$ 35. $6mn + 6n$

36. $8x^2y - 17xy^3 - 6x^2y^3$ 37. $4x$ 38. $-18A$ 39. $-8x$ 40. $-5x^2$

41. $13x$ 42. $-6w$ 43. $-7mn^2$ 44. $-7r + 11t$ 45. $7y^2 - 2y$

46. $-4y^3 + 5y^2$ 47. $4x + 2x^2$ 48. $-6q^5r^8 - 6q^8r^5$ 49. 0 50. uvw

51. $-4x^5 - 13x^7$ 52. $-12r^3 - 10r^6$ 53. $-6t^9 + 20t^7$ 54. $-42b^5$ 55. $15x^9 - 32y^9$

56. $25x^2$ 57. $-9y^5$ 58. $-35y^3z^3$ 59. x^5 60. $3x^2y + 5xy^2$

61. $-11x^5$ 62. $4x^2y^3 - 6xy^3$ 63. $-x^7z^3$ 64. -3 65. -47

66. 29 67. 52 68. 66 69. -9 70. 40

71. -3 72. 204 73. -55 74. 42 75. -130

76. 5 77. -63 78. -13 79. 900 80. 440

81. 73 82. 49 83. 60 84. 118

3 — Summary of Phrases

1. 12km + 18km = 30km
2. 19mm − 2mm = 17mm
3. 7m − 4m = 3m
4. 25g − 14g = 11g
5. 35L − 23L = 12L
6. 11mL + 42mL = 53mL
7. (4g)(21) + 14g = 98g
8. 39mm − 5(7mm) = 4mm
9. 90cm² − (6cm)(9cm) = 36cm²
10. 31g − (24mg + 17g) = 14g − 24mg
11. $(-18a + 7a)(-8c^3) = 88ac^3$
12. $79a^2b + 2(-42a^2b) = -5a^2b$
13. $6ab^3(5ab - 17ab) = -72a^2b^4$
14. $13a^4 + (-9a^2)(2a) = 13a^4 - 18a^3$
15. $(-3w)(-6y) - 14wy = 4wy$
16. $(9xy)(-7x^3y^2) - 5x^3y^2 = -63x^4y^3 - 5x^3y^2$
17. $45a^4b^2 - (32a^3)(-ab^2) = 77a^4b^2$
18. $(-9x^5)(-11x^3) - 78x^{15} = 99x^8 - 78x^{15}$

3 — Cumulative Test

1. (2.1) $-20xy$
2. (3.1) $-34xy$
3. (2.3) $-90x^{11}$
4. (3.2) $9a$
5. (3.3) $2k - 4k^2$
6. (2.3) $-42x^{12}y^6$
7. (3.1) $-21r^3$
8. (2.1) $-24stxyz$
9. (3.2) $-12t^3$
10. (3.3) $18x^3 + 5y$
11. (2.3) $-72m^5n^5$
12. (3.2) $-8n^3$
13. (3.1) $-53st^4$
14. (2.3) $21y^4z^9$
15. (3.2) $-12AB$
16. (2.1) $6rst$
17. (3.2) 0
18. (3.3) $-m^2 + m^3$
19. (2.3) $-48x^4y^6$
20. (3.1) $2x^4$
21. (2.3) $216r^7s^6$
22. (3.3) $-15AC$
23. (3.2) $9b^8$
24. (2.1) $144abcd$
25. (3.2) $-12x$
26. (3.3) $3A - 3B$
27. (2.3) $-8x^{16}y^{10}$
28. (3.2) $-9yz^6$
29. (2.3) $144u^{10}v^6$
30. (3.3) $9mn - 13$

CHAPTER 4

4.1

1. $2i - 5$
2. $-7x + 2$
3. $-5r + s$
4. $27 - 8A^2$
5. $-18 - 9x$
6. $-3w + w^2$
7. $-3x^2 + 5x - 7$
8. $8a + 7b - 3c$
9. $7 - 8s - 9t$
10. $x + 13y$
11. $-6A + 2B$
12. $2 - 4w^3$
13. $7 + 7j^2$
14. $-6 + 14n$
15. $x^2 - 4x + 7$
16. $7y^2 - 4y - 9$
17. $-3 + 7w + 2w^2$
18. $-9 + x + 5x^2$

4.2

1. $10 + 2A$
2. $12k + 28$
3. $-9 + 54x$
4. $-24 + 12n$
5. $-18x + 30$
6. $-24a + 10b$
7. $-11A - 22B$
8. $120 - 160x$
9. $32 - 16A$
10. $63G - 9H$
11. $20x^2 - 24x$
12. $-36x^4 - 12y^2$
13. $-60x^3 + 4$
14. $49x - 42x^2$
15. $-27x^5 + 9x^4$
16. $15m^3 - 6m^2$
17. $80 + 8k^4$
18. $75x^2 + 125y^2$
19. $-20A - 55B + 5C$
20. $-240 - 170x + 150x^2$
21. $28 - 8a + 12b$
22. $-42x^2 - 14x + 35$
23. $-32y^2 - 56y - 40$
24. $16 - 32w - 64x$
25. $24 + 33y + 12y^2$
26. $40 - 50k + 30k^2$
27. $6a - 21b + 30$
28. $50x^2 - 20y^2 - 5z^2$
29. $-6 + 30x + 6x^2$
30. $-72x^2 + 63xy + 27y^2$
31. $-12x - 54$
32. $36m - 48n$
33. $-18x^3 + 18$
34. $-44r^2 - 4$
35. $2x^2 - 22x + 36$
36. $-9x^2 - 21xy + 3y^2$

4.2 Phrases

1. 8(3cm + 2m) = 24cm + 16m
2. 5(7km + 12m) = 35km + 60m
3. 6(4g − 15mg) = 24g − 90mg
4. 7(12L + 2mL) = 84L + 14mL
5. $-2(35ab^2 - 18a^2) = -70ab^2 + 36a^2$
6. $20(-8x^2 + 12xy) = -160x^2 + 240xy$
7. $-4(9a^2 - 11b) = -36a^2 + 44b$
8. $(-90x^2 + 20w^2)(-3) = 270x^2 - 60w^2$
9. $-9(7ab - 7ac) = -63ab + 63ac$
10. $15(8x - 5y) = 120x - 75y$
11. $-1(47x + 89y) = -47x - 89y$
12. $-11(11 - 8a) = -121 + 88a$

4.3

1. $3x^2 - 5x^3$
2. $7x^4 - 5x^3$
3. $-35y^8 - 63y^{12}$

4. $-6k^8 - 3k^{10}$
5. $-36a^6 - 8a^4$
6. $-48t^5 - 18t^3$
7. $-6x^5 + 42x^4 + 84x^3$
8. $-8z^7 - 40z^6 + 128z^5$
9. $-3x^6y^2 + x^4y^3$
10. $-6a^7b^3 + a^5b^8$
11. $-5s^7t^6 + st^{13}$
12. $-9u^5v^5 + uv^{11}$
13. $-18x^7y^7 - 48x^3y^4$
14. $-45A^{10}B^8 - 90A^3B^6$
15. $168m^8n - 72m^3n^5$
16. $100yz^9 - 60y^5z^7$
17. $15a^3b^3c^5 - 21ab^5c^5 + 27ab^3c^8$
18. $12x^5yz^3 - 64x^4y^9z^2 + 24x^3y^4z^3$

4.3 Phrases

1. 4g(2m + 5mm) = 8g m + 20g mm
2. 6m (8mg + 9g) = 48m mg + 54m g
3. 9cm^2(3g + 4mg) = 27cm^2g + 36cm^2 mg
4. 8mg ($6\text{m}^2 - 7\text{cm}^2$) = 48mg m^2 − 56mg cm^2
5. $-3x^2y^3(9x^2y - 4x^3y^2 = -27x^4y^4 + 12x^5y^5$
6. $11abc(7b - 3a) = 77ab^2c - 33a^2bc$
7. $10x^3(27x^4 + 19x^5) = 270x^7 + 190x^8$
8. $-xyz(x + y) = -x^2yz - xy^2z$
9. $-2a^2b(-7ab - 8ab^2) = 14a^3b^2 + 16a^3b^3$
10. $13a^2b^3c(a^3 - 2a^2) = 13a^5b^3c - 26a^4b^3c$
11. $-45xy^2(2x^3y^2 - x^5y^3) = -90x^4y^4 + 45x^6y^5$
12. $(100w^3 - 50x^4)(30w^3x^2) = 3000w^6x^2 - 1500w^3x^6$

4.4

1. $-7y + 10$
2. $68x + 30$
3. $-30x + 5$
4. $-34y - 4$
5. $-13y - 16$
6. $17x - 15$
7. $3x - 1$
8. $13x - 44$
9. $29y - 2$
10. $-y + 12$
11. $6x + 1$
12. $-47x - 52$
13. $-y - 30$
14. $-7y - 31$
15. $-10x + 51$
16. $5x + 1$
17. $14y - 26$
18. $-24y + 12$

4 — Chapter Review

1. $-4x - 9y$
2. $7x + 11y$
3. $14A - 82$
4. $25a + 11b - 8c$
5. $-4x^3 + 5x$
6. $3r - s^2$
7. $6A^2 + 7A$
8. $-2x^2 + 3xy - y^2$
9. $6 - 8t + 5t^2$
10. $28x + 35$
11. $-60n + 90$
12. $-12x + 15y$
13. $36x + 144$
14. $6 - 48A$
15. $36k^2 - 10k$
16. $-18 + 45n^3$
17. $-40x + 24y$
18. $90 + 42k^4$
19. $-8a - 40b + 32c$
20. $8 - 20w + 40w^2$
21. $-63x^2 - 3x - 90$
22. $48 + 10y - 24y^2$
23. $20p - 90q + 125$
24. $-11 + 88x + 77x^2$
25. $-24x - 64$
26. $-30y^3 + 270$
27. $2x^2 - 24xy + 30y^2$
28. $5x^4 - 2x^5$
29. $-24t^9 - 18t^8$
30. $-75k^7 - 15k^5$
31. $-8x^5 + 96x^4 + 160x^3$
32. $-5m^9n^5 + m^4n^8$
33. $-12x^9y^7 + xy^{14}$
34. $-88u^9v^4 - 16u^3v^2$
35. $121x^8y - 44x^4y^4$
36. $8K^7L^5M^2 - 36KL^7M^2 + 2KL^9M^5$
37. $-23x + 20$
38. $-22y + 30$
39. $-47x + 18$
40. $-12x - 22$
41. $26x - 24$
42. $y - 20$
43. -31
44. $-x - 71$
45. $-21y - 12$

4 — Summary of Phrases

1. 8(6cm + 9mm) = 48cm + 72mm
2. 12(5g + 8mg) = 60g + 96mg

3. $16(3x^3 - 4x^2) = 48x^3 - 64x^2$

4. $-6(7a^2 - 10a^5) = -42a^2 + 60a^5$

5. $25(4c - 4c^2) = 100c - 100c^2$

6. $-7(7b - 11) = -49b + 77$

7. $4g(6km + 21m) = 24g\ km + 84g\ m$

8. $3cm\ (7g + 8mg) = 21cm\ g + 24cm\ mg$

9. $7x^3(4x^2 - 13x) = 28x^5 - 91x^4$

10. $-abc(ab + ac) = -a^2b^2c - a^2bc^2$

11. $-5x^2y(-21x^2y + 16xy^2) = 105x^4y^2 - 80x^3y^3$

12. $20a^3b(6a^2b - 15ab^2) = 120a^5b^2 - 300a^4b^3$

4 — Cumulative Test

1. (4.1) $9A + 16$
2. (4.2) $40k - 96$
3. (3.4) $-39w$
4. (4.3) $2x^6y - 2x^2y^2$
5. (4.1) $2x^4 + 15y^2$
6. (4.2) $-60x^3 + 55x^2$
7. (4.3) $x^7y^5 + 17x^3y^8$
8. (3.4) $-18x^2$
9. (4.2) $36 + 54w - 90w^2$
10. (4.1) $-4x^3 + 8x^2 - x$
11. (4.3) $-s^2t^9 + s^5t^4 - s^{10}t^3$
12. (4.2) $42 - 14x^2$
13. (3.4) $2k - 6m$
14. (4.1) $-5x^2 + 8xy - 18y^2$
15. (4.3) $-36A + 12A^2 + 8A^3$
16. (4.2) $-77s^2t + 88st^2 + 33t^3$
17. (4.3) $7x^7y^2 - 7x^3y^4$
18. (3.4) $-16x + 11y$
19. (4.2) $54 - 63a$
20. (4.3) $-24x^9 + 56x^7 - 64x^5$

CHAPTER 5

5.1

1. $-12x - 2y$
2. $9 + 3x$
3. $-19A + 5B$
4. $-n - 7$
5. $x - 7$
6. $-6y - 4$
7. $4s + 11t$
8. $-12 + 4x^2$
9. $2y - 3x$
10. $-2i + 17j$
11. $a - 4b$
12. $-13p + 14q$
13. $-28m + 16n$
14. $x - 2$
15. $-6 - 16r$
16. $-3x^2y$
17. $7a - 13b$
18. $12K - 12L$
19. $6u - 11v$
20. $-3m - 12n$
21. $13x$
22. $-5A + 6B$
23. $-2y^2 - 5z$
24. $2x^2 - 17y$

5.2

1. $-10 - 2x$
2. $-25 - 3x$
3. $22A - 36$
4. $23B - 4$
5. $-23y - 5$
6. $-27x - 12$
7. $18 - 22k$
8. $20 - 35n$
9. $10 + 24x$
10. $21 + 21y$
11. $-5b - 17$
12. $-13a - 19$
13. $26x - 14y$
14. $-4k + 40$
15. $10x + 30y$
16. $-15u + 26v$
17. $-14a + b$
18. $9m - 76n$
19. $-53s - 59t$
20. $-40x + 30y$
21. $-76 + 58r$
22. $-36P$
23. $2a + 14b + 24c$
24. $-48x + 9z - 6y$
25. $11 + 30n - 28m$
26. $4 + 16x - 36x^2$
27. $-56t^2 + 85 - 36t$
28. $-44w - 23 + 26v$
29. $21x + 4y$
30. $-40s + 38t$
31. $24B + 30 - 16A$
32. $-21n + 35 + 21m$
33. $-30x^2 + 28x + 124$
34. $-94y^2 + 170y - 140$
35. $24x^2 - 102xy + 10y^2$
36. $-38x^2 + 42xy + 6y^2$

5.3

1. $-x + 7y$
2. $-13n + 7m$
3. $4A + 23B$
4. $11s - 3t$
5. $6x + 5$
6. $2n + 9m$
7. $6u + 11v$
8. $7x^2 - 2x$
9. $7A - 9B$
10. $6 - 18k$
11. $-2x + 4$
12. $3 - 13i$
13. $y - 1$
14. $-9 + 2A^2$
15. $12z^2 - 6z$
16. $30k - 60$
17. $52 + 56n$
18. $-8x + 32y$

19. $9w - 15$ 20. $24t + 80$ 21. $-60a + 134$ 22. $-39x - 35$ 23. $-23A + 12$ 24. $12x - 45$

25. $50i + 85$ 26. $-12n + 93$ 27. $10b$ 28. $36x - 12$ 29. $6T - 6$ 30. $-34x - 54y$

31. $7 - 4x$ 32. $-n - 12$ 33. $9A + 4$ 34. $6x + 9$ 35. $16z - 19$ 36. $2n + 15$

37. $-8v + 21w$ 38. $11 + 6A$ 39. $8t - 16$ 40. $-16 - 13i$ 41. $-16x^2 + 28$ 42. $4y^2 - 5$

5 — Chapter Review

1. $-30k - 14$ 2. $-3x - 4$ 3. $-5s + t$ 4. $7B - 15A$ 5. $-3K + 9$ 6. $-65u + 35v$

7. $2x + 11y$ 8. $-37m + 4n$ 9. $-3 - 6x^2$ 10. $6i - 3j$ 11. $-4A - 22B$ 12. $-2x - 8$

13. $-21 - 4x$ 14. $51A - 27$ 15. $-26y - 8$ 16. $25 - 23k$ 17. $3 + 36x$ 18. $-16a - 9$

19. $-11i - 57j$ 20. $20A - 56B$ 21. $83 - 36x$ 22. $-36r - 33t$ 23. $-12w - 8$ 24. $33y - 45 + 27z$

25. $48i - 96j + 64k$ 26. $12A - 18 - 14B$ 27. $-8x$ 28. $-22a - 12b$ 29. $-51x^2 + 30x + 93$ 30. $47x^2 - 104xy + 9y^2$

31. $-12A + 4B$ 32. $5x + 9x^2$ 33. $-K + L$ 34. $5x + 3y$ 35. $8u + 6v$ 36. $-7t - 2p$

37. $8A - 2B$ 38. $-12x + 44y$ 39. $-75a + 10b$ 40. $-6r + 30$ 41. $-38x + 20$ 42. $69 - 128y$

43. $-44p + 60$ 44. $-70K - 49L$ 45. $64u + 100v$ 46. $6k - 9$ 47. $10 + 4z$ 48. $4n + 14$

49. $-7 - 3t^2$ 50. 33 51. $9x + 3y$

5 — Cumulative Test

1. (5.1) $-39x - 13$ 2. (5.2) $-26 - 4x$ 3. (2.3) $-36y^9$ 4. (3.2) $-4s$

5. (3.4) $t^3 + 6t^2$ 6. (5.2) $-34x - 16y$ 7. (3.3) $-7x^2 - 8$ 8. (5.1) $4w - 6v$

9. (2.3) $12s^4t^2$ 10. (3.2) $-15x^4$ 11. (5.2) $-31k - 6$ 12. (3.3) $13yz^2 - 11z^2$

13. (3.4) $-27w^3x^4$ 14. (5.1) $-5A + 10$ 15. (2.3) $30a^6b^4$ 16. (4.2) $-60x + 4y - 104z$

17. (5.2) $10n - 35$ 18. (3.3) $-19x - 9y^2$ 19. (5.1) $-23x^2 - 5$ 20. (3.4) $-7rst$

21. (5.1) $-61A + 40B$ 22. (2.3) $126u^4v^5$ 23. (4.2) $6x + 36$ 24. (5.2) $27x - 62y$

25. (3.2) $-pq$ 26. (5.1) $5w - 8x$ 27. (3.4) $79s^3t^6$ 28. (2.3) $48w^4x^5$

29. (5.2) $12 + 37x$ 30. (4.2) $36r^2 + 72s^2 - 54t^2$ 31. (3.2) $14k^2$ 32. (3.4) $24x^8$

33. (5.1) $S - 13T$ 34. (2.3) $-16x^7y^6$ 35. (4.2) $-140x^2 + 10xy + 260y^2$ 36. (5.2) $-30m + 37$

37. (3.3) $-x + 10yz$ 38. (3.4) $-6r^3 - 60r^4$ 39. (5.1) $-22u$ 40. (5.2) $-46i + 17$

CHAPTER 6

6.1

1. $\frac{2}{5}$ 2. $\frac{3}{7}$ 3. $\frac{11}{5}$ 4. $\frac{5}{3}$ 5. $\frac{3}{8}$ 6. $\frac{2}{9}$

7. $\frac{3}{5}$ 8. $\frac{2}{7}$ 9. $\frac{15}{8}$ 10. $\frac{15}{4}$ 11. $\frac{5}{7}$ 12. $\frac{17}{19}$

13. $\frac{7}{11}$ 14. $\frac{5}{13}$ 15. $\frac{12}{13}$ 16. $\frac{7}{6}$ 17. $\frac{2}{3}$ 18. $\frac{2}{5}$

19. $\frac{5}{4}$ 20. $\frac{3}{5}$ 21. $\frac{4}{33}$ 22. $\frac{3}{10}$ 23. $\frac{23}{29}$ 24. $\frac{13}{11}$

6.2

1. $\frac{2}{9}$ 2. $\frac{10}{7}$ 3. $\frac{9}{14}$ 4. $\frac{1}{8}$ 5. 3 6. $\frac{1}{6}$

7. $\frac{5}{6}$ 8. $\frac{2}{13}$ 9. $\frac{2}{3}$ 10. $\frac{3}{4}$ 11. $\frac{-3}{14}$ 12. $\frac{-36}{43}$

13. $\frac{2}{5}$ 14. $\frac{-5}{16}$ 15. $\frac{-2}{3}$ 16. $\frac{1}{6}$ 17. $\frac{-18}{7}$ 18. $\frac{2}{9}$

19. $\frac{-3}{13}$ 20. $\frac{40}{11}$ 21. $\frac{3}{2}$ 22. $\frac{-3}{4}$ 23. $\frac{-8}{9}$ 24. $\frac{8}{15}$

6.3

1. $\frac{20}{3}$ 2. $\frac{1}{3}$ 3. 30 4. $\frac{28}{3}$ 5. $\frac{2}{9}$ 6. $\frac{2}{9}$

7. $\frac{-5}{12}$ 8. $\frac{-7}{8}$ 9. $\frac{-9}{4}$ 10. $\frac{3}{10}$ 11. $\frac{1}{9}$ 12. $\frac{-8}{7}$

13. $\frac{-3}{16}$ 14. $\frac{-1}{6}$ 15. $\frac{-3}{16}$ 16. $\frac{33}{4}$ 17. $\frac{6}{5}$ 18. 1

19. $\frac{1}{4}$ 20. 6 21. 2 22. $\frac{2}{5}$ 23. $\frac{10}{21}$ 24. −1

6.3 Phrases

1. $\frac{3}{7} \cdot \frac{5}{6} = \frac{5}{14}$ 2. $\frac{4}{5}\left(-\frac{15}{16}\right) = \frac{-3}{4}$ 3. $\frac{9}{10}\left(\frac{-5}{12}\right)(-3) = \frac{9}{8}$

4. $\left(\frac{1}{2} \cdot \frac{13}{14}\right)\left(-\frac{8}{7}\right) = \frac{-26}{49}$ 5. $\left(2 \cdot \frac{54}{25}\right)\left(-\frac{60}{63}\right) = \frac{-144}{35}$ 6. $\frac{-15}{35} \cdot \frac{1}{2}(-56) = 12$

7. $-5\left(\frac{-8}{25} \cdot \frac{21}{12}\right) = \frac{14}{5}$ 8. $\frac{-44}{30}\left(\frac{-52}{16} \cdot \frac{9}{-66}\right) = \frac{-13}{20}$ 9. $36\left(\frac{-13}{48} \cdot \frac{-2}{39}\right) = \frac{1}{2}$

10. $\frac{1}{3}\left(\frac{60}{77} \cdot \frac{22}{35}\right) = \frac{8}{49}$ 11. $\frac{7}{8}\left(\frac{4}{21} \cdot \frac{-6}{13}\right) = \frac{-1}{13}$ 12. $\frac{42}{15}\left(\frac{25}{14} \cdot \frac{-27}{12}\right) = \frac{-45}{4}$

6.4

1. $\frac{1}{6}$ 2. $\frac{2}{11}$ 3. 9 4. 10 5. $\frac{-2}{15}$ 6. $\frac{-2}{27}$

7. $\frac{-144}{5}$ 8. $\frac{-25}{4}$ 9. $\frac{4}{15}$ 10. $\frac{9}{2}$ 11. $\frac{-27}{10}$ 12. $\frac{13}{27}$

13. $\frac{1}{2}$ 14. 6 15. $\frac{1}{10}$ 16. $\frac{9}{49}$ 17. $\frac{-39}{50}$ 18. $\frac{5}{6}$

19. $\frac{27}{2}$ 20. 8 21. −8 22. −32 23. $\frac{35}{6}$ 24. $\frac{-1}{3}$

25. -1 26. $\frac{28}{27}$ 27. $\frac{-16}{21}$ 28. $\frac{-18}{49}$ 29. $\frac{1}{30}$ 30. $\frac{-180}{13}$

6.4 Phrases

1. $\dfrac{-18}{\frac{6}{5}} = -15$
2. $\frac{48}{64} \div (-144) = \frac{-1}{192}$
3. $\dfrac{125}{\frac{250}{800}} = 400$
4. $\frac{42}{50} \div (-84) = \frac{-1}{100}$
5. $\dfrac{\frac{5}{4}}{\frac{25}{-2}} = \frac{-1}{10}$
6. $45 \div \frac{-9}{10} = -50$
7. $\dfrac{\frac{15}{-27}}{\frac{-35}{-72}} = \frac{-8}{7}$
8. $-390 \div \left(-\frac{260}{3}\right) = \frac{9}{2}$
9. $\dfrac{\frac{35}{36}}{-45} = \frac{-7}{324}$
10. $\dfrac{-\frac{22}{35}}{121} = \frac{-2}{385}$
11. $\frac{-60}{7} \div \frac{70}{-3} = \frac{18}{49}$
12. $\frac{25}{16} \div \frac{-90}{48} = \frac{-5}{6}$

6.5

1. $\frac{1}{2}$ 2. $\frac{3}{4}$ 3. 3 4. $\frac{4}{13}$ 5. $\frac{3}{2}$ 6. $\frac{4}{3}$

7. 1 8. $-\frac{2}{5}$ 9. $\frac{1}{4}$ 10. $-\frac{2}{11}$ 11. $-\frac{1}{3}$ 12. $\frac{9}{4}$

13. $-\frac{5}{4}$ 14. $-\frac{4}{3}$ 15. $\frac{9}{2}$ 16. $\frac{15}{2}$ 17. 10 18. $\frac{4}{3}$

19. $\frac{27}{175}$ 20. 14 21. $\frac{9}{8}$ 22. $\frac{64}{25}$ 23. $\frac{3}{5}$ 24. $\frac{27}{25}$

25. $-\frac{1}{12}$ 26. 1 27. $-\frac{21}{2}$ 28. $-\frac{1}{20}$ 29. $\frac{2}{3}$ 30. $-\frac{25}{12}$

31. $-\frac{5}{8}$ 32. 1 33. 20 34. 6 35. $\frac{1}{3}$ 36. -9

37. $\frac{5}{3}$ 38. $-\frac{1}{5}$ 39. $-\frac{7}{108}$ 40. $-\frac{1}{6}$ 41. $\frac{5}{32}$ 42. $\frac{5}{8}$

43. $\frac{14}{3}$ 44. $\frac{135}{28}$ 45. $\frac{1}{4}$ 46. $\frac{9}{4}$ 47. $\frac{6}{25}$ 48. $\frac{147}{50}$

49. $\frac{3}{20}$ 50. $\frac{4}{15}$ 51. $\frac{28}{27}$ 52. $\frac{20}{7}$ 53. $\frac{4}{81}$ 54. $\frac{9}{100}$

55. $\frac{11}{3}$ 56. $\frac{13}{4}$ 57. $\frac{5}{108}$ 58. $\frac{5}{48}$ 59. 16 60. 36

6 – Chapter Review

1. $\frac{2}{7}$ 2. $\frac{13}{2}$ 3. $\frac{5}{16}$ 4. $\frac{3}{7}$ 5. $\frac{14}{3}$ 6. $\frac{11}{5}$

7. $\frac{3}{5}$ 8. $\frac{12}{5}$ 9. $\frac{4}{5}$ 10. $\frac{2}{3}$ 11. $\frac{4}{21}$ 12. $\frac{17}{31}$

13. $\frac{1}{2}$ 14. $\frac{2}{5}$ 15. $\frac{16}{3}$ 16. $\frac{5}{7}$ 17. $\frac{1}{4}$ 18. $-\frac{9}{16}$

19. $-\frac{7}{16}$ 20. $-\frac{1}{7}$ 21. $\frac{3}{16}$ 22. $\frac{5}{3}$ 23. $-\frac{3}{4}$ 24. $\frac{3}{14}$

25. 6 26. 21 27. $\frac{2}{7}$ 28. $-\frac{5}{24}$ 29. $-\frac{3}{5}$ 30. $\frac{1}{3}$

31. $-\frac{35}{13}$ 32. $\frac{8}{9}$ 33. $\frac{1}{40}$ 34. $\frac{7}{30}$ 35. $-\frac{3}{2}$ 36. $\frac{16}{9}$

37. $\frac{1}{12}$ 38. $\frac{108}{5}$ 39. $-\frac{1}{45}$ 40. $-\frac{147}{2}$ 41. $\frac{8}{3}$ 42. $-\frac{25}{28}$

43. 2 44. $\frac{3}{49}$ 45. $-\frac{39}{50}$ 46. 120 47. $-\frac{18}{5}$ 48. $\frac{20}{3}$

49. $-\frac{3}{8}$ 50. $-\frac{6}{55}$ 51. $-\frac{1}{12}$ 52. 4 53. $\frac{3}{2}$ 54. $\frac{3}{4}$

55. $\frac{6}{11}$ 56. 2 57. $\frac{1}{3}$ 58. $\frac{1}{2}$ 59. 84 60. $\frac{9}{4}$

61. $\frac{1}{5}$ 62. $\frac{15}{64}$ 63. $\frac{1}{2}$ 64. $\frac{4}{3}$ 65. $-\frac{8}{7}$ 66. 60

67. $\frac{1}{8}$ 68. 15 69. $\frac{4}{3}$ 70. $-\frac{1}{4}$ 71. $\frac{4}{3}$ 72. $\frac{14}{25}$

73. $\frac{22}{3}$ 74. $\frac{16}{9}$ 75. $\frac{4}{63}$ 76. $\frac{20}{63}$ 77. $\frac{16}{27}$ 78. $\frac{36}{25}$

79. $\frac{60}{7}$ 80. $\frac{1}{36}$ 81. 36

6 – Summary of Phrases

1. $-\frac{7}{8} \cdot \frac{16}{21} = \frac{-2}{3}$
2. $\left(\frac{2}{3}\right)\left(\frac{-6}{21} \cdot 7\right) = \frac{-4}{3}$
3. $\left(\frac{-18}{48}\right)\left(\frac{1}{2}(-64)\right) = 12$
4. $\left(\frac{-12}{51}\right)\left(\frac{17}{18} \cdot \frac{-27}{32}\right) = \frac{3}{16}$
5. $\frac{1}{2}\left(\frac{11}{3}\right)\left(-\frac{26}{55}\right) = \frac{-13}{15}$
6. $52\left(\frac{8}{26} \cdot \frac{10}{12}\right) = \frac{40}{3}$
7. $-\frac{18}{48} \div 72 = \frac{-1}{192}$
8. $\frac{25}{36} \div (-75) = \frac{-1}{108}$
9. $\dfrac{\frac{14}{33}}{\frac{42}{-55}} = \frac{-5}{9}$
10. $\dfrac{\frac{-72}{-25}}{\frac{-54}{35}} = \frac{-28}{15}$
11. $\dfrac{\frac{-24}{25}}{-48} = \frac{1}{50}$
12. $\frac{70}{-24} \div \frac{40}{16} = \frac{-7}{6}$

6 – Cumulative Test

1. (6.2) $\frac{4}{7}$ 2. (6.3) $\frac{5}{3}$ 3. (6.4) $\frac{3}{8}$ 4. (6.3) $\frac{5}{18}$ 5. (6.2) $\frac{-3}{11}$ 6. (6.4) $\frac{2}{63}$

7. (6.3) -1 8. (6.4) $\frac{21}{16}$ 9. (6.2) $\frac{1}{3}$ 10. (6.4) $\frac{-9}{32}$ 11. (6.3) $\frac{-3}{2}$ 12. (6.4) $\frac{-20}{3}$

13. (6.3) $\frac{9}{2}$ 14. (6.2) $\frac{-9}{4}$ 15. (6.4) $\frac{4}{27}$ 16. (6.3) $\frac{-33}{8}$ 17. (6.4) $\frac{28}{9}$ 18. (6.3) -2

19. (6.4) $\frac{-4}{27}$ 20. (6.3) $\frac{-9}{7}$

CHAPTER 7

7.1

1. 1	2. $\frac{3}{4}$	3. $\frac{8}{11}$	4. $\frac{7}{5}$	5. 0	6. 0
7. $-\frac{2}{5}$	8. $-\frac{1}{11}$	9. $-\frac{1}{2}$	10. $-\frac{2}{3}$	11. $-\frac{3}{2}$	12. 3
13. $-\frac{1}{3}$	14. $\frac{1}{2}$	15. $-\frac{4}{3}$	16. $-\frac{13}{9}$	17. $-\frac{17}{5}$	18. $-\frac{7}{3}$
19. $\frac{15}{8}$	20. $\frac{14}{9}$	21. $-\frac{4}{7}$	22. 0	23. $-\frac{8}{3}$	24. $-\frac{5}{11}$

7.1 Phrases

1. $\frac{2}{7}+\frac{4}{7}=\frac{6}{7}$	2. $\frac{7}{8}-\frac{3}{8}=\frac{1}{2}$	3. $\frac{19}{14}-\frac{13}{14}=\frac{3}{7}$	4. $\frac{-8}{5}+\frac{-4}{5}=\frac{-12}{5}$
5. $\frac{-12}{11}-\frac{10}{11}=-2$	6. $\frac{10}{3}-\frac{8}{3}=\frac{2}{3}$	7. $\frac{-5}{12}+\frac{3}{12}=\frac{-1}{6}$	8. $\frac{7}{25}-\frac{16}{25}=\frac{-9}{25}$
9. $\frac{8}{9}+\frac{-5}{9}=\frac{1}{3}$	10. $\frac{27}{35}-\frac{13}{35}=\frac{2}{5}$	11. $\frac{-33}{40}+\frac{8}{40}=\frac{-5}{8}$	12. $\frac{-3}{4}-\frac{15}{4}=\frac{-9}{2}$

7.2

1. $\frac{6}{10}$	2. $\frac{15}{21}$	3. $\frac{9}{24}$	4. $\frac{8}{18}$	5. $\frac{63}{18}$	6. $\frac{80}{24}$
7. $\frac{81}{45}$	8. $\frac{60}{55}$	9. $\frac{65}{13}$	10. $\frac{56}{7}$	11. $\frac{21}{35}$	12. $\frac{48}{36}$
13. $\frac{27}{63}$	14. $\frac{96}{60}$	15. $\frac{8}{4}$	16. $\frac{16}{4}$	17. $\frac{75}{15}$	18. $\frac{12}{2}$

7.3

1. $\frac{29}{6}$	2. $\frac{15}{4}$	3. $\frac{27}{8}$	4. $-\frac{1}{16}$	5. $\frac{5}{18}$	6. $-\frac{3}{20}$
7. $-\frac{11}{18}$	8. $\frac{1}{12}$	9. $\frac{47}{15}$	10. $\frac{4}{6}$	11. $\frac{13}{5}$	12. $\frac{37}{7}$
13. $\frac{3}{4}$	14. $\frac{61}{9}$	15. $-\frac{35}{8}$	16. $-\frac{17}{6}$	17. $\frac{211}{60}$	18. $\frac{55}{36}$
19. $\frac{67}{72}$	20. $\frac{23}{24}$	21. $-\frac{5}{20}$	22. $-\frac{11}{120}$	23. $\frac{11}{30}$	24. $\frac{9}{16}$

7.4

1. 12	2. 12	3. 18	4. 56	5. 60	6. 30
7. 36	8. 36	9. 70	10. 252	11. 144	12. 225
13. 140	14. 180	15. 22	16. 60	17. 72	18. 336

19. 60 20. 60 21. 210 22. 210 23. 540 24. 280

25. 7020 26. 1080 27. 780 28. 140 29. 504 30. 216

7.5

1. $\frac{5}{6}$ 2. $\frac{17}{21}$ 3. $-\frac{19}{36}$ 4. $-\frac{4}{15}$ 5. $-\frac{1}{4}$ 6. $-\frac{1}{6}$

7. $\frac{17}{12}$ 8. $\frac{15}{14}$ 9. $-\frac{19}{8}$ 10. $-\frac{16}{15}$ 11. $-\frac{1}{12}$ 12. $-\frac{1}{40}$

13. $\frac{19}{30}$ 14. $-\frac{11}{36}$ 15. $-\frac{17}{42}$ 16. $-\frac{11}{36}$ 17. $-\frac{34}{35}$ 18. $-\frac{5}{6}$

19. $-\frac{11}{48}$ 20. $\frac{1}{60}$ 21. $\frac{103}{48}$ 22. $\frac{11}{15}$ 23. $\frac{17}{72}$ 24. $-\frac{43}{18}$

25. $\frac{49}{40}$ 26. $\frac{1}{20}$ 27. $\frac{91}{15}$ 28. $\frac{67}{24}$ 29. $-\frac{31}{36}$ 30. $-\frac{13}{72}$

31. $-\frac{103}{90}$ 32. $\frac{29}{132}$ 33. $\frac{28}{15}$ 34. $\frac{56}{55}$ 35. $\frac{1}{3}$ 36. $\frac{3}{77}$

37. $\frac{55}{6}$ 38. $\frac{325}{24}$ 39. $\frac{57}{10}$ 40. $\frac{88}{45}$ 41. $\frac{13}{4}$ 42. $\frac{61}{8}$

7.5 Phrases

1. $\frac{-3}{5} + \frac{7}{4} = \frac{23}{20}$ 2. $\frac{2}{7} - \frac{2}{3} = \frac{-8}{21}$ 3. $\frac{-5}{9} + \frac{1}{2} = \frac{-1}{18}$ 4. $\frac{3}{8} - \frac{1}{6} = \frac{5}{24}$

5. $\frac{-7}{12} - \frac{7}{15} = \frac{-21}{20}$ 6. $\frac{-9}{28} + \frac{4}{35} = \frac{-29}{140}$ 7. $\frac{1}{42} - \frac{1}{54} = \frac{1}{189}$ 8. $\frac{-2}{9} - \frac{-3}{11} = \frac{5}{99}$

9. $\frac{-7}{18} + \frac{15}{72} = \frac{-13}{72}$ 10. $\frac{-9}{56} - \frac{5}{48} = \frac{-89}{336}$ 11. $\frac{-7}{30} - \frac{-3}{20} = \frac{-1}{12}$ 12. $\frac{3}{49} + \frac{-9}{77} = \frac{-30}{539}$

7.6

1. $\frac{3}{2}$ 2. $\frac{26}{9}$ 3. $\frac{8}{9}$ 4. $\frac{7}{4}$ 5. $-\frac{25}{6}$ 6. $-\frac{14}{9}$

7. 2 8. $-\frac{7}{2}$ 9. $-\frac{2}{9}$ 10. $\frac{15}{4}$ 11. $\frac{26}{9}$ 12. $\frac{5}{16}$

13. $-\frac{16}{25}$ 14. $\frac{6}{5}$ 15. $\frac{55}{9}$ 16. $\frac{42}{5}$ 17. $-\frac{50}{9}$ 18. 0

19. $\frac{1}{2}$ 20. $-\frac{9}{8}$ 21. $\frac{61}{36}$ 22. $\frac{1}{32}$ 23. $\frac{27}{25}$ 24. $\frac{5}{9}$

25. −2 26. −1 27. 37 28. 26 29. 18 30. 2

31. $-\frac{1}{5}$ 32. $\frac{11}{2}$ 33. $-\frac{11}{14}$ 34. $\frac{263}{30}$ 35. 44 36. 15

37. $\frac{17}{12}$ 38. $\frac{17}{35}$ 39. $\frac{13}{12}$ 40. $-\frac{7}{12}$ 41. $\frac{83}{36}$ 42. $-\frac{29}{10}$

43. $\frac{25}{27}$ 44. $\frac{15}{22}$ 45. $-\frac{52}{35}$ 46. $\frac{6}{25}$ 47. $-\frac{2}{49}$ 48. $\frac{12}{7}$

49. $\frac{220}{63}$ 50. $-\frac{3}{10}$ 51. $\frac{11}{15}$ 52. $\frac{16}{13}$ 53. $-\frac{1}{7}$ 54. $-\frac{13}{15}$

7.6 Phrases

1. $\left(\frac{1}{4}\right)\left(\frac{2}{7}\right)+\frac{1}{3}=\frac{17}{42}$
2. $\frac{9}{5}+\left(\frac{2}{3}\right)\left(\frac{-6}{7}\right)=\frac{43}{35}$
3. $\frac{-2}{7}+\left(\frac{3}{5}+\frac{-8}{35}\right)=\frac{3}{35}$
4. $\frac{-5}{12}-\left(\frac{7}{8}\right)\left(\frac{-5}{3}\right)=\frac{25}{24}$
5. $\frac{-3}{7}\left(\frac{1}{2}-\frac{5}{6}\right)=\frac{1}{7}$
6. $\frac{1}{12}-\left(\frac{4}{9}+\frac{5}{6}\right)=\frac{-43}{36}$
7. $\frac{7}{16}-\left(\frac{-3}{4}\right)\left(\frac{-10}{27}\right)=\frac{23}{144}$
8. $\left(\frac{2}{15}+\frac{3}{10}\right)\left(\frac{-20}{39}\right)=\frac{-2}{9}$
9. $\left(\frac{12}{7}+\frac{-9}{4}\right)-\frac{5}{8}=\frac{-65}{56}$
10. $\frac{-3}{11}+2\left(\frac{2}{7}\right)=\frac{23}{77}$
11. $\frac{-16}{7}\left(\frac{-3}{8}+\frac{-7}{12}\right)=\frac{46}{21}$
12. $\frac{-13}{36}+\left(\frac{-35}{16}\right)\left(\frac{-14}{45}\right)=\frac{23}{72}$
13. $\left(\frac{-35}{27}\right)\left(\frac{-18}{49}\right)-\frac{9}{14}=\frac{-1}{6}$
14. $\frac{-8}{35}+\left(\frac{12}{25}\right)\left(\frac{10}{21}\right)=0$
15. $\left(\frac{12}{49}\right)\left(\frac{25}{16}\right)-\frac{-15}{28}=\frac{45}{49}$
16. $\frac{5}{18}-\left(\frac{-7}{51}\right)\left(\frac{-17}{21}\right)=\frac{1}{6}$
17. $\frac{9}{25}-2\left(\frac{16}{15}\right)=\frac{-133}{75}$
18. $2\left(\frac{13}{27}+\frac{17}{18}\right)=\frac{77}{27}$

7 — Chapter Review

1. $\frac{35}{3}$ 2. $\frac{2}{3}$ 3. $\frac{15}{11}$ 4. $\frac{1}{2}$ 5. 12 6. $-\frac{1}{5}$

7. $-\frac{5}{7}$ 8. $-\frac{2}{3}$ 9. $-\frac{7}{5}$ 10. $\frac{14}{5}$ 11. 1 12. 0

13. $\frac{2}{8}$ 14. $\frac{20}{24}$ 15. $\frac{49}{28}$ 16. $\frac{135}{72}$ 17. $\frac{72}{12}$ 18. $\frac{65}{26}$

19. $\frac{24}{84}$ 20. $\frac{18}{6}$ 21. $\frac{100}{20}$ 22. $\frac{11}{10}$ 23. $\frac{47}{12}$ 24. $\frac{31}{10}$

25. $\frac{19}{24}$ 26. $\frac{71}{18}$ 27. $\frac{17}{3}$ 28. $-\frac{14}{5}$ 29. $-\frac{16}{3}$ 30. $\frac{50}{30}$

31. $-\frac{8}{15}$ 32. $\frac{11}{24}$ 33. $\frac{49}{28}$ 34. 72 35. 300 36. 28

37. 189 38. 252 39. 288 40. 54 41. 300 42. 864

43. 18 44. 84 45. 480 46. 2772 47. 330 48. 960

49. $\frac{2}{3}$ 50. $-\frac{1}{10}$ 51. $-\frac{4}{9}$ 52. $\frac{33}{32}$ 53. $-\frac{13}{14}$ 54. $\frac{13}{80}$

55. $\frac{1}{63}$ 56. $-\frac{119}{180}$ 57. $-\frac{23}{45}$ 58. $-\frac{13}{120}$ 59. $-\frac{1}{6}$ 60. $-\frac{47}{40}$

61. $-\frac{13}{40}$ 62. $\frac{253}{60}$ 63. $-\frac{17}{42}$ 64. $-\frac{41}{72}$ 65. $\frac{19}{10}$ 66. $\frac{25}{39}$

67. $\frac{251}{28}$ 68. $\frac{88}{45}$ 69. $-\frac{89}{21}$ 70. -1 71. $\frac{4}{3}$ 72. $-\frac{7}{5}$

73. $\frac{9}{8}$ 74. $\frac{21}{16}$ 75. $\frac{33}{16}$ 76. 3 77. $\frac{5}{12}$ 78. $-\frac{125}{18}$

79. $-\frac{3}{11}$ 80. 0 81. 6 82. $-\frac{4}{3}$ 83. -22 84. $\frac{7}{3}$

85. $\frac{5}{2}$ 86. $-\frac{10}{27}$ 87. 35 88. $\frac{43}{24}$ 89. $-\frac{19}{8}$ 90. $-\frac{25}{9}$

91. $\frac{9}{8}$ 92. $\frac{12}{31}$ 93. $\frac{3}{28}$ 94. $-\frac{5}{36}$ 95. $\frac{7}{32}$ 96. $-\frac{6}{5}$

7 – Summary of Phrases

1. $\frac{1}{8}+\frac{5}{8}=\frac{3}{4}$
2. $\frac{11}{12}-\frac{7}{12}=\frac{1}{3}$
3. $\frac{9}{5}-\frac{13}{5}=\frac{-4}{5}$
4. $\frac{-3}{11}+\frac{18}{11}=\frac{15}{11}$
5. $\frac{25}{36}-\frac{13}{36}=\frac{1}{3}$
6. $\frac{-29}{40}+\frac{19}{40}=\frac{-1}{4}$
7. $\frac{4}{5}-\frac{2}{3}=\frac{2}{15}$
8. $\frac{3}{8}-\frac{5}{6}=\frac{-11}{24}$
9. $\frac{3}{10}-\frac{1}{15}=\frac{7}{30}$
10. $\frac{9}{16}-\frac{2}{9}=\frac{49}{144}$
11. $\frac{-25}{36}+\frac{7}{24}=\frac{-29}{72}$
12. $\frac{5}{42}+\frac{-6}{35}=\frac{-11}{210}$
13. $\left(\frac{1}{3}\right)\left(\frac{2}{5}\right)+\frac{7}{15}=\frac{3}{5}$
14. $\frac{7}{4}-\left(\frac{8}{9}\right)\left(\frac{3}{4}\right)=\frac{13}{12}$
15. $\frac{-5}{12}-\left(\frac{-16}{21}\right)\left(\frac{35}{-24}\right)=\frac{-55}{36}$
16. $\left(\frac{4}{15}+\frac{3}{10}\right)-\frac{1}{10}=\frac{7}{15}$
17. $\left(\frac{45}{64}\right)\left(\frac{32}{54}\right)-\frac{8}{9}=\frac{-17}{36}$
18. $\left(\frac{18}{35}+\frac{-3}{14}\right)\left(\frac{30}{49}\right)=\frac{9}{49}$
19. $\left(\frac{26}{21}\right)\left(\frac{14}{39}\right)-\frac{11}{6}=\frac{-25}{18}$
20. $\frac{-11}{24}-\left(\frac{-9}{20}\right)\left(\frac{15}{-4}\right)=\frac{-103}{48}$
21. $\left(\frac{95}{-32}\right)\left(\frac{24}{75}\right)-\frac{3}{50}=\frac{-101}{100}$

7 – Cumulative Test

1. (7.1) $-\frac{1}{6}$ 2. (7.5) $\frac{5}{18}$ 3. (6.3) $-\frac{11}{40}$ 4. (6.4) $\frac{2}{63}$ 5. (7.1) $-\frac{8}{7}$ 6. (7.5) $-\frac{65}{72}$

7. (6.3) $\frac{4}{15}$ 8. (7.5) $-\frac{17}{4}$ 9. (6.4) $\frac{1}{2}$ 10. (7.5) $-\frac{7}{48}$ 11. (6.3) $\frac{3}{8}$ 12. (7.1) $-\frac{3}{7}$

13. (6.4) -1 14. (7.5) $-\frac{73}{36}$ 15. (6.3) $-\frac{27}{55}$ 16. (7.5) $-\frac{73}{30}$ 17. (6.4) $-\frac{3}{10}$ 18. (7.5) $-\frac{33}{28}$

19. (6.3) $-\frac{4}{5}$ 20. (7.5) $\frac{131}{60}$ 21. (7.1) $-\frac{11}{6}$ 22. (6.4) $-\frac{27}{2}$ 23. (7.5) $-\frac{11}{42}$ 24. (6.3) -1

25. (7.5) $-\frac{89}{72}$ 26. (6.4) $\frac{9}{20}$ 27. (6.3) $-\frac{1}{56}$ 28. (7.5) $\frac{13}{18}$ 29. (6.3) $-\frac{7}{15}$ 30. (7.5) $-\frac{19}{140}$

CHAPTER 8

8.1

1. 1 2. $\frac{1}{wy}$ 3. $\frac{n}{k}$ 4. p 5. $\frac{5}{8}$ 6. $\frac{7x}{3z}$

7. $\frac{3}{4}$ 8. $\frac{1}{4k}$ 9. $\frac{3}{2s}$ 10. 3 11. $\frac{y}{2}$ 12. $\frac{3s}{5}$

8.2

1. x^5 2. A 3. w^7 4. k^5 5. $\frac{1}{r}$ 6. 1

7. $\frac{1}{x^9}$ 8. $\frac{1}{R}$ 9. n^3 10. $\frac{1}{y^4}$ 11. 1 12. w

8.3

1. $\frac{x^6}{y^5}$ 2. $\frac{k^2}{m}$ 3. $-\frac{y^6}{x^4}$ 4. $\frac{1}{a^5 b^{10}}$ 5. $-\frac{LM}{K}$

6. $-\frac{1}{p^6}$ 7. $-\frac{5}{4x^3}$ 8. $-\frac{3}{2y}$ 9. $\frac{7A^2}{15}$ 10. $-\frac{t^4}{7}$

11. $\frac{2}{x^3}$ 12. $\frac{3k}{4}$ 13. $-\frac{3A^6B^2}{7}$ 14. $-\frac{9x^5}{11y^2}$ 15. $-\frac{2x^7}{9yz^2}$

16. $-\frac{x^7}{7y^4}$ 17. $\frac{4r}{9t}$ 18. $-\frac{5a^2}{14b^4c}$ 19. $-\frac{3}{4}$ 20. $-\frac{x}{3}$

21. $-\frac{8}{AB^5}$ 22. $-\frac{n^3}{5m^5}$ 23. $-8v$ 24. $\frac{1}{6x^2}$

8.4

1. $\frac{5}{x^9}$ 2. $\frac{7}{y^{11}}$ 3. $\frac{6}{7t^3}$ 4. $\frac{10}{3x^5}$ 5. $\frac{21x}{5y^7}$

6. $\frac{20u^4}{9v^7}$ 7. $\frac{2x^2}{21y^2}$ 8. $\frac{5a}{6b^3}$ 9. $\frac{A^7}{48B^2}$ 10. $\frac{5x^3}{12y^3}$

11. $\frac{1}{12i^4j^3}$ 12. $\frac{33}{104w^3x^3}$ 13. $\frac{w^3}{v^3x^2z^2}$ 14. $\frac{b^{10}c}{d^6}$ 15. $\frac{11x^2y^4}{10z}$

16. $\frac{4k^{20}}{49m^8}$ 17. $\frac{9A^2B^5}{16}$ 18. $-\frac{y^4}{12x^6}$ 19. $-\frac{45}{2kmn}$ 20. $-3w$

21. $\frac{6a^2}{b}$ 22. $\frac{3k^7t^2}{16}$ 23. $-\frac{3r^2s^4}{80t^7}$ 24. $\frac{v^{11}}{96w^3x^2}$

8.4 Phrases

1. $\left(\frac{6\text{g}}{\text{m}}\right)\left(\frac{5\text{m}}{\text{sec}}\right) = \frac{30\text{g}}{\text{sec}}$

2. $(9\text{mg})\left(\frac{3\text{m}}{\text{sec}^2}\right) = \frac{27\text{mg m}}{\text{sec}^2}$

3. $2\left(\frac{8\text{g}}{\text{m}^2}\right) = \frac{16\text{g}}{\text{m}^2}$

4. $\left(\frac{7\text{mg}}{5\text{m}^3}\right)\left(\frac{3\text{m}^3}{2\text{sec}^2}\right) = \frac{21\text{mg}}{10\text{sec}^2}$

5. $\left(\frac{5x^3}{-16}\right)\left(\frac{-4xy^2}{-6}\right)\left(\frac{27y^5}{-15x^4}\right) = \frac{3y^7}{8}$

6. $\frac{1}{2}\left(\frac{10A^2B^3}{9C^7}\right)\left(\frac{-21}{25}A^2C^3\right) = \frac{-7A^4B^3}{15C^4}$

7. $\left(\frac{x^9y^7z^2}{w^4}\right)\left(\frac{-w^8x^2}{y^2z}\right)\left(\frac{z^8}{wxy}\right) = -w^3x^{10}y^4z^9$

8. $\frac{1}{2}\left(\frac{9a^3b^2}{-8c^4d}\right)\left(\frac{-16cd^8}{15ab}\right) = \frac{3a^2bd^7}{5c^3}$

9. $\left(\frac{7A^3B^8}{8C^4}\right)\left(\frac{A^7C^2}{21B^2C}\right)\left(-4BC^5\right) = \frac{-A^{10}B^7C^2}{6}$

10. $\left(\frac{-6v^8z^4}{7w}\right)\left(\frac{10z^2}{-24w^3}\right)\left(\frac{-14w^9}{-5v^5}\right) = v^3w^5z^6$

11. $\left(\frac{A^8B^7C}{-D^3E^5}\right)\left(\frac{-B^2D^3E}{A^3C^4}\right)\left(\frac{C^3E^4}{A^5B^9}\right) = 1$

12. $\left(\frac{90x^3y^2}{18w^5}\right)\left(\frac{3w^2x^7}{-5y^2}\right)\left(\frac{-w^2}{21x^{10}y^2}\right) = \frac{1}{7wy^2}$

8.5

1. xy 2. x^2y^2 3. $-\frac{v}{u}$ 4. $-xyz^2$ 5. $\frac{x^2y^4}{3}$ 6. $\frac{9a^3b}{28}$

7. $\frac{1}{12y^3}$ 8. $\frac{121y^2}{5x^5}$ 9. $-\frac{13x^4}{21y}$ 10. $-\frac{4}{3w}$ 11. $\frac{4y^2z}{3x^2}$ 12. $-\frac{3x^2z}{2y^5}$

13. $\frac{4a}{b^4}$ 14. $-\frac{2z^4}{9y^2}$ 15. $\frac{v^6}{u^3}$ 16. $\frac{y^3}{xz^2}$ 17. $-\frac{r^2t^2}{s^2}$ 18. $\frac{x^9z^4}{y^2}$

19. $\frac{10y^2}{3x^2}$ 20. $-\frac{144m^3}{7k^2}$ 21. $\frac{8y^4}{15x^3}$ 22. $-\frac{10xy^2}{3}$ 23. $-\frac{3x^8}{7yz}$ 24. $\frac{4b^4}{5a}$

8.5 Phrases

1. $\frac{8\text{m}^3}{\frac{2\text{m}}{\text{sec}^2}} = 4\text{m}^2\ \text{sec}^2$

2. $\frac{\frac{5\text{g}}{\text{sec}^2}}{\frac{9\text{m}}{5\text{sec}^2}} = \frac{25\text{g}}{9\text{m}}$

3. $\frac{6\text{L}}{\text{sec}} \div 3\text{sec} = \frac{2\text{L}}{\text{sec}^2}$

4. $25\text{cm}^3 \div \frac{15\text{cm}^3}{\text{sec}} = \frac{5\text{sec}}{3}$

5. $\frac{\frac{-84y^8}{15x^6}}{\frac{63x^9}{12y^5}} = \frac{-16y^{13}}{15x^{15}}$

6. $\frac{72z^3}{-15x^4} \div \frac{-48z^5}{15x^7} = \frac{3x^3}{2z^2}$

7. $\frac{144x^{12}z^8}{\frac{-16x^5}{9z^8}} = -81x^7z^{16}$

8. $\frac{36b^3c^4}{-5a^9} \div (-96a^3b^2c) = \frac{3bc^3}{40a^{12}}$

9. $\frac{\frac{-38x^9y}{7z^4}}{19y^3z^4} = \frac{-2x^9}{7y^2z^8}$

10. $\frac{\frac{-A^{15}B^6}{C^8D^2}}{-A^{12}C^{10}E^3} = \frac{A^3B^6}{C^{18}D^2E^3}$

11. $\frac{-5a^3}{32b^7} \div \frac{-15a^4}{-8b^2} = \frac{-1}{12ab^5}$

12. $\frac{-17A^8B^7}{29C^3D} \div \frac{17A^8B^7}{29C^3D} = -1$

8.6

1. $\frac{4}{7}$ 2. $\frac{2}{5}$ 3. $-\frac{2}{25}$ 4. $-\frac{3}{16}$ 5. $\frac{4}{189}$ 6. $\frac{3}{175}$

7. $\frac{24}{7}$ 8. $\frac{36}{5}$ 9. $-\frac{15}{2}$ 10. -12 11. $-\frac{3}{32}$ 12. $-\frac{1}{81}$

13. $-\frac{7}{3}$ 14. $-\frac{11}{2}$ 15. $-\frac{5}{4}$ 16. $-\frac{2}{3}$ 17. 4 18. 8

19. $\frac{4}{175}$ 20. $\frac{1}{45}$ 21. $\frac{24}{7}$ 22. $\frac{8}{15}$ 23. 30 24. $-\frac{1}{27}$

25. $-\frac{5}{6}$ 26. $-\frac{27}{40}$ 27. $-\frac{16}{45}$ 28. $-\frac{16}{65}$ 29. $\frac{7}{2}$ 30. $\frac{7}{2}$

31. $-\frac{9}{4}$ 32. $-\frac{16}{3}$ 33. $-\frac{1}{3}$ 34. $\frac{21}{5}$ 35. $\frac{1}{10}$ 36. $-\frac{3}{28}$

37. $\frac{1}{5}$ 38. $\frac{3}{2}$ 39. $-\frac{20}{7}$ 40. $-\frac{28}{5}$ 41. $\frac{7}{24}$ 42. $\frac{35}{27}$

8 – Chapter Review

1. s 2. $\frac{a}{d}$ 3. $\frac{15}{4s}$ 4. $\frac{4t}{3}$ 5. $\frac{1}{2a}$ 6. 1

7. $\frac{1}{x^7}$ 8. 1 9. y^9 10. A 11. $\frac{1}{x}$ 12. $\frac{1}{z^2}$

13. $x^{12}y^{12}$ 14. $-\frac{s^2}{t^5}$ 15. $\frac{x^4}{y^4}$ 16. $-\frac{1}{3t^2}$ 17. $\frac{9}{16}$ 18. $\frac{2A^2}{3B}$

19. $-\frac{2x^2y^3}{3}$ 20. $\frac{1}{a^2b^3}$ 21. $\frac{3z^3}{4y^3}$ 22. $-\frac{11}{9R^4}$ 23. $\frac{2q^3}{5p^3}$ 24. $-\frac{2x^4}{5y}$

25. $\frac{6}{w^{11}}$ 26. $\frac{14}{5x^3}$ 27. $\frac{56}{3t^5u}$ 28. $\frac{2a^3}{15b^4}$ 29. $\frac{x^7}{6y^2}$ 30. $\frac{3}{4i^5j^4}$

31. $\frac{A^{12}}{BC^9D^4}$ 32. $\frac{k^{13}n}{44m^3}$ 33. $-\frac{v^2}{6u^2w}$ 34. $-\frac{16d}{9a^4b^7}$ 35. $-\frac{k^7}{r^9t^9}$ 36. $-\frac{8y^2z^2}{5x^6}$

37. x^2 38. $\frac{1}{x^3y^4}$ 39. $\frac{1}{s^4t^3}$ 40. $\frac{3}{a^2}$ 41. $\frac{7x^3}{8}$ 42. $\frac{3x^3y^2}{8}$

43. $\frac{3y}{4x^2}$ 44. $-\frac{m^4}{n}$ 45. $\frac{y}{x^3z}$ 46. $\frac{4y^5z^2}{3}$ 47. $\frac{21u^4}{80t^3}$ 48. $-\frac{9x^4}{28y^2}$

49. $\frac{3}{7}$ 50. $-\frac{4}{27}$ 51. $\frac{2}{81}$ 52. $\frac{30}{7}$ 53. $-\frac{12}{7}$ 54. $-\frac{1}{20}$

55. $-\frac{6}{5}$ 56. $-\frac{21}{10}$ 57. 12 58. $\frac{1}{147}$ 59. $\frac{2}{3}$ 60. $-\frac{1}{630}$

61. $-\frac{4}{9}$ 62. $-\frac{4}{35}$ 63. $\frac{7}{5}$ 64. $-\frac{8}{3}$ 65. $-\frac{11}{8}$ 66. $\frac{1}{36}$

67. $\frac{4}{5}$ 68. $-\frac{96}{5}$ 69. $\frac{35}{18}$

8 – Summary of Phrases

1. $18\text{cm}\left(\frac{5\text{g}}{9\text{cm}^2}\right) = \frac{10\text{g}}{\text{cm}}$

2. $2\left(\frac{27\text{L}}{4\text{sec}}\right) = \frac{27\text{L}}{2\text{sec}}$

3. $\frac{72x^3}{35}\left(\frac{1}{2}\cdot 21x^4\right) = \frac{108x^7}{5}$

4. $\left(\frac{22a^3b^2}{15c^4}\right)\left(\frac{-10ac^2}{33a^2}\right)\left(\frac{-a^3}{-2b^4}\right) = \frac{-2a^5}{9b^2c^2}$

5. $\frac{1}{2}\left(\frac{-12x^5y^2}{35y^3z}\right)\left(\frac{14xz}{-18xy^4}\right) = \frac{2x^5}{15y^5}$

6. $\left(\frac{8ab^2c}{-9a^2bc^2}\right)\left(\frac{-15a^2b^3}{16c^5}\right)(-24a^2b^3c^8) = -20a^3b^7c^2$

7. $\frac{25\text{km}^2}{\text{sec}} \div \frac{15\text{km}}{\text{sec}} = \frac{5\text{km}}{3}$

8. $\frac{\frac{18\text{g}}{7\text{cm}^2}}{\frac{9\text{sec}}{4\text{cm}^2}} = \frac{8\text{g}}{7\text{sec}}$

9. $\frac{-54x^8}{55y^3} \div \frac{81x^6}{15y^5} = \frac{-2x^2y^2}{11}$

10. $\frac{\frac{90a^8b^2}{-144c^8}}{\frac{-27b^5}{16ac^3}} = \frac{10a^9}{27b^3c^5}$

11. $\frac{49x^3y^2}{32z^9} \div \frac{42x^5y^2}{24z^9} = \frac{7}{8x^2}$

12. $120x^3y^5z^9 \div \frac{65x^2z^2}{26xy^8} = 48x^2y^{13}z^7$

8 – Cumulative Test

1. (8.1) $\frac{9}{14s}$
2. (8.3) $\frac{3B^2}{5A^2}$
3. (8.4) $\frac{x^2y^4}{2}$
4. (8.5) $\frac{1}{i^2j^2}$
5. (8.3) $-\frac{4}{9t^6}$
6. (8.5) $-\frac{81y^2}{2x^3}$
7. (8.1) $-3xz$
8. (8.4) $\frac{13}{5rs^3t}$
9. (8.3) $-\frac{4x^2y^5}{11}$
10. (8.5) $-\frac{3A^3}{4}$
11. (8.4) $\frac{5a^4}{3b^6}$
12. (8.3) $\frac{3}{8k}$
13. (8.5) $-\frac{x^{13}z^2}{y^{13}}$
14. (8.4) $-\frac{7x^4}{y^{11}z^3}$
15. (8.3) $-\frac{1}{2uv^4}$
16. (8.5) $\frac{t^5}{126s^8}$
17. (8.4) $-\frac{6s}{25t}$
18. (8.3) $\frac{7x^2z^6}{13y^4}$
19. (8.4) $-\frac{xy^2z^4}{3}$
20. (8.5) $\frac{64B^4}{125A^2}$

CHAPTER 9

9.1

1. $\frac{4}{x}$
2. $\frac{-1}{x^2}$
3. $\frac{9}{7y}$
4. $\frac{-40}{y^2z}$
5. $\frac{9u}{7v}$
6. $\frac{2y^3}{3x^2}$
7. $\frac{-17z}{x^2y}$
8. $\frac{3a^2}{b^3}$
9. $\frac{3s^2+5s^3}{8t}$
10. $\frac{5y^4-5y^2}{14x^2}$
11. $\frac{-16y+2w}{7x^3z}$
12. $\frac{-5x^3-11x^4}{8y^5}$
13. $\frac{7x+6}{6}$
14. $\frac{-3x+7}{7}$
15. $\frac{-5y-9}{2y}$
16. $\frac{3z+1}{9z^2}$
17. $\frac{7x-5}{5y}$
18. $\frac{5x-9}{12x}$
19. $\frac{5x+6}{3x}$
20. $\frac{y^2-8}{8y}$
21. -1
22. $\frac{-9}{x^2}$
23. $\frac{6x+2y}{3y^4}$
24. $\frac{21z-14}{18z}$

9.1 Phrases

1. $\frac{2\text{g}}{3\text{sec}} + \frac{8\text{g}}{3\text{sec}} = \frac{10\text{g}}{3\text{sec}}$
2. $\frac{12\text{L}}{5\text{sec}} - \frac{3\text{L}}{5\text{sec}} = \frac{9\text{L}}{5\text{sec}}$
3. $\frac{23\text{mg}}{3\text{cm}} - \frac{5\text{mg}}{3\text{cm}} = \frac{6\text{mg}}{\text{cm}}$
4. $\frac{5\text{m}^3}{8\text{sec}^2} + \frac{11\text{m}^3}{8\text{sec}^2} = \frac{2\text{m}^3}{\text{sec}^2}$
5. $\frac{-7a^3}{12b^2} - \frac{9a^3}{12b^2} = \frac{-4a^3}{3b^2}$
6. $\frac{7x}{-8w^2} - \frac{13x}{8w^2} = \frac{-5x}{2w^2}$
7. $\frac{2x-5}{18} + \frac{7x}{18} = \frac{9x-5}{18}$
8. $\frac{2-5a}{2a} - \frac{7}{2a} = \frac{-5-5a}{2a}$
9. $\frac{3x^2-2x}{5} + \frac{6x^2+4x}{5} = \frac{9x^2+2x}{5}$
10. $\frac{7a+3b}{12} - \frac{5a-12b}{12} = \frac{2a+15b}{12}$
11. $\frac{15x-19y}{10x} + \frac{6y-4x}{10x} = \frac{11x-13y}{10x}$
12. $\frac{17a-16b}{4c} - \frac{17a+16b}{4c} = \frac{-8b}{c}$

9.2

1. $\frac{8t-21}{54t}$ 2. $\frac{11x^2-15}{12x^2}$ 3. $\frac{8-7k^6}{18k^2}$ 4. $\frac{15-68A^3}{28A^2}$ 5. $\frac{2-7x^2}{14x^4}$ 6. $\frac{40-3B^3}{96B^6}$

7. $\frac{2r^4-15r^2s}{50s^2}$ 8. $\frac{25x^2-21x}{60y^3}$ 9. $\frac{36s^3-t^2}{9s^2t}$ 10. $\frac{175b^8+30a^2}{98a^2b^8}$ 11. $\frac{10R-21K^3}{36K^3R}$ 12. $\frac{8y^8+33x^8}{90x^6y^2}$

13. $\frac{12x-1}{20}$ 14. $\frac{12x+19}{36}$ 15. $\frac{30x-46y}{60}$ 16. $\frac{22z-21}{15}$ 17. $\frac{-3w-22}{28}$ 18. $\frac{-22x-27}{36}$

19. $\frac{27z-13}{15}$ 20. $\frac{13y+9}{12}$ 21. $\frac{18x-8}{9}$ 22. $\frac{30x+13}{30}$ 23. $\frac{11x-11}{90}$ 24. $\frac{z+6}{84}$

9.3

1. x^7 2. x^4 3. st^9 4. x^3y^3 5. u^4v 6. x^8y^9

7. $m^6n^7p^5$ 8. $a^8b^3c^3$ 9. $90x^4$ 10. $90y^3$ 11. $42k^4t^3$ 12. $36p^5q^7$

13. $224x^7y^2$ 14. $360u^4v^3$ 15. $480m^2n^4$ 16. $132x^7y^5$ 17. $420y^5z^3$ 18. $396x^7y^8$

19. $108x^7y^6z^4$ 20. $294a^8b^3c^3$ 21. $1400x^6y^2z^3$ 22. $110r^4s^3t^3$ 23. $704x^4y^6z^2$ 24. $952x^7y^8z^9$

9.4

1. $\frac{7}{6x^2}$ 2. $\frac{29}{20y^4}$ 3. $\frac{5}{84x^2}$ 4. $\frac{1}{63z}$ 5. $\frac{21x+25}{35x}$ 6. $\frac{-40-27z^3}{45z^3}$

7. $\frac{-55+30xy}{33xy}$ 8. $\frac{-105+16w}{120w}$ 9. $\frac{a+b}{ab}$ 10. $\frac{y^2-x^2y}{x^3}$ 11. $\frac{-x^2z^2+yz^5}{x^3y^2}$ 12. $\frac{-B^3+A^6}{A^2B^5}$

13. $\frac{-32y-27x^3}{12x^3y^2}$ 14. $\frac{-45s^3-12t^2}{12s^5t^3}$ 15. $\frac{15y+4x^2}{24x^2y}$ 16. $\frac{-16z+9y^3}{60y^4z}$ 17. $\frac{9y^4+6x^2}{16x^3y^2}$

18. $\frac{3m^6n^3+10n^2}{12m^8}$ 19. $\frac{5x+26}{10}$ 20. $\frac{-14y-45}{18}$ 21. $\frac{-14w+29}{28w}$ 22. $\frac{-63x+169}{33xy}$

23. $\frac{w+1}{4w}$ 24. $\frac{-43x-12}{12x}$ 25. $\frac{37x+1}{3x}$ 26. $\frac{5k-6}{75k}$ 27. $\frac{89y+19}{20y}$

28. $\frac{-56x-5}{30x^2}$ 29. $\frac{9n+80}{40}$ 30. $\frac{9z-47}{60}$ 31. $\frac{-37x+20}{60x}$ 32. $\frac{25x-9}{180y^4}$

33. $\frac{3w^2+5w+1}{2w}$ 34. $\frac{32y^2-25y+14}{28y}$ 35. $\frac{-9x^2-105x+88}{72x}$ 36. $\frac{-8t^2+9t+2}{4t}$

9.4 Phrases

1. $\frac{4\text{L}}{9\text{sec}}+\frac{7\text{L}}{15\text{sec}}=\frac{41\text{L}}{45\text{sec}}$ 2. $\frac{3\text{mg}}{8\text{cm}^2}-\frac{5\text{mg}}{21\text{cm}^2}=\frac{23\text{mg}}{168\text{cm}^2}$ 3. $\frac{8\text{m}}{3\text{sec}^2}+\frac{10\text{km}}{4\text{sec}^2}=\frac{32\text{m}+30\text{km}}{12\text{sec}^2}$

4. $\frac{7\text{cm}}{5\text{sec}}-\frac{9\text{mm}}{8\text{sec}}=\frac{56\text{cm}-45\text{mm}}{40\text{sec}}$ 5. $\frac{x+7}{2}-\frac{3x}{5}=\frac{-x+35}{10}$ 6. $-\frac{6a-7}{9}+\frac{8a}{21}=\frac{49-18a}{63}$

7. $\frac{4y+2}{7y}-\frac{13}{7y^2}=\frac{4y^2+2y-13}{7y^2}$ 8. $\frac{12a-7}{10}-\frac{16a+15}{15}=\frac{4a-51}{30}$

9. $\dfrac{25x-60y}{14}+\dfrac{8y-2x}{35}=\dfrac{121x-284y}{70}$

10. $\dfrac{8x-11}{6x}-\dfrac{12-15x}{33x^2}=\dfrac{88x^2-91x-24}{66x^2}$

11. $\dfrac{3x-7y}{5x}+\dfrac{7y-3x}{5y}=\dfrac{-3x^2+10xy-7y^2}{5xy}$

12. $\dfrac{-8a-5b}{12b}-\dfrac{10a+3b}{21a}=\dfrac{-56a^2-75ab-12b^2}{84ab}$

9.5

1. -1 2. -2 3. -5 4. -7 5. -9 6. -23 7. $\dfrac{1}{12}$

8. $\dfrac{1}{16}$ 9. -2 10. -1 11. $\dfrac{2}{3}$ 12. $\dfrac{9}{2}$ 13. $\dfrac{2}{3}$ 14. $\dfrac{3}{2}$

15. $\dfrac{15}{8}$ 16. $\dfrac{5}{9}$ 17. $\dfrac{-3}{10}$ 18. $\dfrac{-5}{6}$ 19. $\dfrac{1}{9}$ 20. $\dfrac{31}{8}$ 21. $\dfrac{-73}{10}$

22. $\dfrac{-43}{21}$ 23. $\dfrac{19}{8}$ 24. $\dfrac{13}{10}$ 25. $\dfrac{-1}{2}$ 26. $\dfrac{-2}{3}$ 27. $\dfrac{-21}{5}$ 28. $\dfrac{-16}{5}$

29. $\dfrac{1}{2}$ 30. $\dfrac{3}{8}$ 31. $\dfrac{3}{4}$ 32. $\dfrac{-2}{5}$ 33. $\dfrac{52}{9}$ 34. $\dfrac{58}{5}$ 35. $\dfrac{4}{5}$

36. $\dfrac{10}{7}$ 37. $\dfrac{8}{25}$ 38. $\dfrac{99}{8}$ 39. $\dfrac{-109}{8}$ 40. $\dfrac{-163}{27}$ 41. $\dfrac{55}{16}$ 42. $\dfrac{68}{27}$

43. $\dfrac{9}{20}$ 44. $\dfrac{5}{4}$ 45. $\dfrac{7}{36}$ 46. $\dfrac{-11}{72}$ 47. $\dfrac{-1}{10}$ 48. $\dfrac{117}{112}$

9 – Chapter Review

1. $\dfrac{-19x^2}{4y}$ 2. $\dfrac{-7y^2}{2z}$ 3. $\dfrac{11}{5t}$ 4. $\dfrac{-5}{12x^2}$ 5. $\dfrac{5a-21b}{12a^2b^2}$

6. $\dfrac{-3y^2-8x}{14xy^4}$ 7. $\dfrac{-3z-2}{8}$ 8. $\dfrac{-3x+3}{5xy}$ 9. $\dfrac{4y^2-11}{4z^2}$ 10. $\dfrac{11r-5}{5r^2}$

11. $\dfrac{2}{3v}$ 12. $\dfrac{7x-11}{7}$ 13. $\dfrac{9t-5}{12t}$ 14. $\dfrac{-10+7k^6}{12k^2}$ 15. $\dfrac{18-3x^2}{8x^4}$

16. $\dfrac{2r^4-9r^2s}{48s^2}$ 17. $\dfrac{-6rs^2+t^2}{3s^2t}$ 18. $\dfrac{27R-20K^3}{48K^3R}$ 19. $\dfrac{24x+57}{40}$ 20. $\dfrac{6w-25}{90}$

21. $\dfrac{-8x-61}{56}$ 22. $\dfrac{5z+9}{6}$ 23. $\dfrac{-21k+28}{50}$ 24. $\dfrac{-70y-19}{56}$ 25. t^7

26. x^3y^8 27. a^5b^{10} 28. $r^6s^5t^4$ 29. $32x^4$ 30. $36u^4v^3$

31. $45m^3n^2$ 32. $480x^4y^5$ 33. $105xy^2z$ 34. $400u^6v^8w^6$ 35. $135x^2y^5z^2$

36. $144a^4b^4c^6$ 37. $\dfrac{41}{35xy}$ 38. $\dfrac{-33}{80xy^3}$ 39. $\dfrac{35t^2-24}{42t^2}$ 40. $\dfrac{36+25y^2}{30y^2}$

41. $\dfrac{y^2+x}{x^2y}$ 42. $\dfrac{-x^2z^2-y^4z}{x^7y}$ 43. $\dfrac{22z^3-27}{18z^7}$ 44. $\dfrac{16+15x}{36x^2}$ 45. $\dfrac{21x^2-16x^2y}{28y^7}$

46. $\dfrac{-24t+73}{24}$ 47. $\dfrac{-24x-15}{40y}$ 48. $\dfrac{-w+2}{5w}$ 49. $\dfrac{13x+12}{24x}$ 50. $\dfrac{69x-28}{21y}$

51. $\frac{-32t + 13}{48}$ 52. $\frac{-77x + 73}{63xy}$ 53. $\frac{2w^2 - 25w + 7}{7w}$ 54. $\frac{8x^2 + 34x - 9}{30x}$ 55. $\frac{-9}{2}$ 56. -5

57. $\frac{-19}{2}$ 58. $\frac{5}{81}$ 59. -1 60. $\frac{1}{3}$ 61. 1 62. $\frac{29}{28}$

63. $\frac{-8}{3}$ 64. 6 65. $\frac{-57}{20}$ 66. $\frac{25}{12}$ 67. $\frac{-1}{3}$ 68. -5

69. $\frac{1}{2}$ 70. 2 71. $\frac{29}{2}$ 72. $\frac{12}{7}$ 73. $\frac{72}{7}$ 74. $\frac{-61}{16}$

75. $\frac{53}{15}$ 76. $\frac{49}{48}$ 77. $\frac{2}{225}$ 78. $\frac{1}{2}$

9 — Summary of Phrases

1. $\frac{10\text{L}}{3\text{sec}} + \frac{11\text{L}}{3\text{sec}} = \frac{7\text{L}}{\text{sec}}$

2. $\frac{96\text{g}}{5\text{cm}^2} - \frac{81\text{g}}{5\text{cm}^2} = \frac{3\text{g}}{\text{cm}^2}$

3. $\frac{-18a^2}{25b} + \frac{-17a}{25b} = \frac{-18a^2 - 17a}{25b}$

4. $\frac{17z}{16x^3y} - \frac{13z}{16x^3y} = \frac{z}{4x^3y}$

5. $\frac{6a + 7b}{9b} - \frac{4a - 3b}{9b} = \frac{2a + 10b}{9b}$

6. $\frac{27 - 14x}{13y} - \frac{-19 - 70x}{13y} = \frac{46 + 56x}{13y}$

7. $\frac{8\text{L}}{9\text{cm}^2} - \frac{5\text{L}}{6\text{cm}^2} = \frac{1\text{L}}{18\text{cm}^2}$

8. $\frac{4\text{km}}{3\text{sec}^2} + \frac{19\text{m}}{2\text{sec}^2} = \frac{8\text{km} + 57\text{m}}{6\text{sec}^2}$

9. $\frac{10x}{9y} - \frac{-7x}{8z} = \frac{80xz + 63xy}{72yz}$

10. $\frac{12 - a}{4b} + \frac{3c}{7b^3} = \frac{84b^2 - 7ab^2 + 12c}{28b^3}$

11. $\frac{3a - 8b}{14a^2b} + \frac{-5a + 2b}{12ab^2} = \frac{-35a^2 + 32ab - 48b^2}{84a^2b^2}$

12. $\frac{10x - 9y}{36y} - \frac{3x - 5y}{45x} = \frac{50x^2 - 57xy + 20y^2}{180xy}$

9 — Cumulative Test

1. (9.1) $\frac{-3x^3}{4y}$ 2. (9.4) $\frac{15x^2 - 14}{9x^2}$ 3. (8.4) $\frac{-3}{x^9}$ 4. (9.4) $\frac{117z^7 + 38x^2y^3}{144x^3yz^3}$

5. (8.5) $\frac{x^3}{y}$ 6. (9.1) $\frac{-x + 7}{7x}$ 7. (9.4) $\frac{-x^3z^3 - x^2y^5}{y^2z^4}$ 8. (8.3) $\frac{K^{11}}{LM^7N^7}$

9. (9.4) $\frac{12x + 25}{48}$ 10. (8.5) $\frac{6x^3y^5}{7}$ 11. (9.1) $\frac{14x - 21y^2}{9xy^2}$ 12. (8.4) $\frac{7}{9x^6y^4}$

13. (9.4) $\frac{15t + 6}{24t^2}$ 14. (8.5) $\frac{-n^2}{m^4}$ 15. (9.4) $\frac{64x^3 - 27x}{72}$ 16. (9.4) $\frac{12k - 29}{18k}$

17. (8.4) $\frac{-15u^2}{16t^6}$ 18. (9.1) $\frac{-14x + 13y}{18xy}$ 19. (9.4) $\frac{8A^2 - 35A^3B}{30B^3}$ 20. (8.5) $\frac{4A^{10}}{45B^5}$

21. (8.3) $\frac{9}{16x^3}$ 22. (9.4) $\frac{-16A + 17}{40}$ 23. (8.5) $\frac{3}{2t^3}$ 24. (9.4) $\frac{-38x + 39y}{144xy}$

25. (8.4) $\frac{x^5y^2}{7z^5}$ 26. (9.4) $\frac{-25x - 22}{14}$ 27. (8.5) $\frac{-21v^2}{2u^2w^2}$ 28. (9.4) $\frac{25y^2 - 6}{80y}$

29. (8.4) $\frac{-15}{4x}$ 30. (9.4) $\frac{17x^2 + 12}{30x^2}$

CHAPTER 10

10.1

1. $x = 6$ 2. $x = 9$ 3. $x = -10$ 4. $x = 1$ 5. $x = -4$ 6. $x = 8$

7. $x = 11$ 8. $x = -7$ 9. $x = 10$ 10. $x = 9$ 11. $x = 8$ 12. $x = -10$

13. $x = -3$ 14. $x = 4$ 15. $x = 8$ 16. $x = -6$ 17. $x = 4$ 18. $x = -10$

19. $x = 0$ 20. $x = 0$ 21. $x = -1$ 22. $x = 2$ 23. $x = 8$ 24. $x = -6$

10.1 Word Problems

1. $2x = 42 + x$, $x = 42$
2. $2x = 17 + x$, $x = 17$
3. $18 + x - 29 = 5$, $x = 16$
4. $51 - 2x = 121 - 3x$, $x = 70$
5. $5x = 4x - 88$, $x = -88$
6. $2x - 6 = x + 21$, $x = 27$
7. $7x - 78 = 6x$, $x = 78$
8. $70 + 10x = 42 + 9x$, $x = -28$
9. $17x - 5 = 19 + 16x$, $x = 24$
10. $18 - 12x = 11 - 13x$, $x = -7$

10.2

1. $x = -21$ 2. $x = 1$ 3. $x = 11$ 4. $x = 11$ 5. $x = 0$ 6. $x = 6$

7. $x = 9$ 8. $x = 3$ 9. $x = -3$ 10. $x = 6$ 11. $x = 1$ 12. $x = 10$

13. $x = 3$ 14. $x = -7$ 15. $x = -\frac{1}{2}$ 16. $x = \frac{1}{3}$ 17. $x = \frac{3}{2}$ 18. $x = -\frac{9}{2}$

19. $x = 2$ 20. $x = 1$ 21. $x = 7$ 22. $x = \frac{1}{4}$ 23. $x = \frac{1}{3}$ 24. $x = \frac{1}{2}$

25. $x = -\frac{1}{2}$ 26. $x = -\frac{1}{2}$ 27. $x = -\frac{2}{3}$ 28. $x = 2$ 29. $x = 4$ 30. $x = \frac{2}{3}$

31. $x = \frac{3}{2}$ 32. $x = -4$ 33. $x = -\frac{3}{2}$ 34. $x = -\frac{8}{3}$ 35. $x = -\frac{6}{5}$ 36. $x = \frac{3}{4}$

37. $x = -1$ 38. $x = 0$ 39. $x = \frac{19}{6}$ 40. $x = \frac{3}{4}$ 41. $x = 0$ 42. $x = 1$

43. $x = \frac{4}{5}$ 44. $x = \frac{1}{12}$ 45. $x = 3$ 46. $x = -\frac{3}{2}$ 47. $x = 1$ 48. $x = 2$

10.2 Word Problems

1. $10 = 30 - 4x$, $x = 5$
2. $9 + 7x = 37$, $x = 4$
3. $19 - 6x = 73$, $x = -9$
4. $15x - 12 = 30$, $x = \frac{14}{5}$
5. $31 - x = x - 13$, $x = 22$
6. $x + 55 = 11x$, $x = \frac{11}{2}$
7. $95 + x = 21x$, $x = \frac{19}{4}$
8. $x - 12 = 9x$, $x = -\frac{3}{2}$
9. $7x = 24 - x$, $x = 3$
10. $4x = 9x - 85$, $x = 17$
11. $17x - 24 = x + 40$, $x = 4$
12. $14x - 6 = 23x$, $x = -\frac{2}{3}$

10.3

1. $x = -3$ 2. $x = 1$ 3. $x = 2$ 4. $x = \frac{1}{5}$ 5. $x = -5$ 6. $x = 2$

7. $x = \frac{7}{3}$ 8. $x = -4$ 9. $x = \frac{11}{4}$ 10. $x = -\frac{1}{2}$ 11. $x = \frac{2}{3}$ 12. $x = -1$

13. $x = -\frac{13}{3}$ 14. $x = -\frac{1}{3}$ 15. $x = \frac{2}{3}$ 16. $x = -\frac{1}{2}$ 17. $x = 6$ 18. $x = \frac{3}{4}$

19. $x = 2$ 20. $x = 20$ 21. $x = -2$ 22. $x = 0$ 23. $x = 3$ 24. $x = \frac{1}{5}$

25. $x = \frac{23}{15}$ 26. $x = \frac{19}{4}$ 27. $x = \frac{9}{7}$ 28. $x = -\frac{5}{2}$ 29. $x = \frac{12}{5}$ 30. $x = \frac{4}{5}$

31. $x = -\frac{3}{2}$ 32. $x = 0$ 33. $x = -2$ 34. $x = 13$ 35. $x = -\frac{1}{4}$ 36. $x = \frac{18}{7}$

10.3 Word Problems

1. $6(x + 8) = 21$
   $x = -\frac{9}{2}$

2. $(27 + x)3 = 42$
   $x = -13$

3. $(x + 8)11 = 12 - x$
   $x = -\frac{19}{3}$

4. $9x = 20(x + 3)$
   $x = -\frac{60}{11}$

5. $40 + 7x = (8 + x) - 4$
   $x = -6$

6. $10(x + 18) = 4x$
   $x = -30$

7. $n + n + 1 = 39$
   19, 20

8. $n + n + 1 = 107$
   53, 54

9. $n + n + 2 = 84$
   41, 43

10. $n + n + 2 = 114$
    56, 58

11. $n + n + 1 + n + 2 = -81$
    $-28, -27, -26$

12. $n + n + 2 + n + 4 = 201$
    65, 67, 69

13. $n + n + 2 + n + 4 = 10 + 2(n + 4)$
    12, 14, 16

14. $n + n + 2 - 8 = 10 + n + 4$
    20, 22, 24, 26

15. $3(n + 2) = 6 + (n + n + 1)$
    1, 2, 3

16. $450(100) + 300(70) = 750x$
    $x = 88$
    88 cents per kilogram

17. $5(108) + 7(156) = 12x$
    $x = 136$
    $1.36 per pound

18. $10(70) + 30(90) = 40x$
    $x = 85$
    85 cents per pound

19. $28(240) + 42(380) = 70x$
    $x = 324$
    $3.24 per kilogram

20. $200(80) + 100(140) = 300x$
    $x = 100$
    $1.00 per pound

21. $80(595) + 120(650) = 200x$
    $x = 628$
    $6.28 per kilogram

22. $x(80) + 10(120) = (x + 10)96$
    $x = 15$
    15 pounds

23. $x(50) + 100(150) = (x + 100)130$
    $x = 25$
    25 bags

24. $x(300) + 40(250) = (x + 40)280$
    $x = 60$
    60 meters

25. $x(450) + 60(330) = (x + 60)402$
    $x = 90$
    90 kilograms

26. $x(200) + 20(125) = (x + 20)170$
    $x = 30$
    30 liters

27. $x(33) + 3(110) = (x + 3)54$
    $x = 8$
    8 pounds

10.4

1. $x = \frac{15}{4}$ 2. $x = \frac{10}{7}$ 3. $x = -\frac{5}{2}$ 4. $x = -6$ 5. $x = -\frac{5}{28}$ 6. $x = \frac{3}{22}$

7. $x = \frac{5}{8}$ 8. $x = \frac{48}{5}$ 9. $x = \frac{7}{4}$ 10. $x = \frac{21}{11}$ 11. $x = \frac{21}{20}$ 12. $x = -\frac{2}{7}$

13. $x = -\frac{1}{20}$ 14. $x = 1$ 15. $x = -\frac{29}{2}$ 16. $x = -\frac{3}{13}$ 17. $x = \frac{5}{2}$ 18. $x = -\frac{1}{3}$

19. $x = \frac{15}{4}$ 20. $x = -\frac{3}{28}$ 21. $x = -\frac{10}{9}$ 22. $x = \frac{54}{49}$ 23. $x = \frac{21}{10}$ 24. $x = \frac{3}{40}$

25. $x = \frac{56}{73}$ 26. $x = -2$ 27. $x = \frac{3}{16}$ 28. $x = -17$ 29. $x = \frac{47}{11}$ 30. $x = 3$

31. $x = -\frac{6}{5}$ 32. $x = \frac{14}{9}$ 33. $x = \frac{15}{19}$ 34. $x = -\frac{2}{5}$ 35. $x = \frac{6}{7}$ 36. $x = -\frac{5}{2}$

10.4 Word Problems

1. $\frac{8-x}{13} = 2$; $x = -18$
2. $3 + \frac{x}{7} = \frac{2}{3}$; $x = -\frac{49}{3}$
3. $7\left(\frac{4}{5}x\right) = \frac{10}{3}$; $x = \frac{25}{42}$
4. $\frac{1}{9}(x + 16) = 22$; $x = 182$
5. $\frac{3x}{4} = x - 17$; $x = 68$
6. $11 + x = \frac{4x}{7}$; $x = -\frac{77}{3}$
7. $\frac{x}{6} = 14 - x$; $x = 12$
8. $x + 21 = \frac{8x}{5}$; $x = 35$
9. $\frac{7x}{12} = \frac{9}{7} - \frac{x}{14}$; $x = \frac{108}{55}$
10. $\frac{3x}{10} = \frac{2}{15} - \frac{5x}{6}$; $x = \frac{2}{17}$
11. $\frac{1}{3} - x = \frac{8}{33} + \frac{x}{11}$; $x = \frac{1}{12}$
12. $\frac{x}{15} = \frac{2}{5}(x + 20)$; $x = -24$
13. $\frac{1}{3}(n + n + 1) = 11$; 16, 17
14. $\frac{n + n + 1 + n + 2}{9} = 9$; 26, 27, 28
15. $\frac{1}{4}(n + 1) + n + 2 = \frac{47}{2}$; 17, 18, 19
16. $\frac{1}{3}(n + 1 + n + 2) = \frac{7n}{12}$; −12, −11, −10
17. $\frac{n+1}{5} + \frac{n+2}{3} = \frac{n}{2}$; −26, −25, −24
18. $n + 2 - \frac{n}{3} = 18 + n + 3$; −57, −56, −55, −54

10.5

1. $x < 14$ 2. $x > -3$ 3. $x < -1$ 4. $x > 10$ 5. $x < -\frac{1}{8}$ 6. $x > -\frac{7}{3}$

7. $x < -\frac{5}{3}$ 8. $x > \frac{7}{3}$ 9. $x > -\frac{7}{3}$ 10. $x > \frac{3}{8}$ 11. $x < 2$ 12. $x < 0$

13. $x > \frac{5}{2}$ 14. $x > \frac{4}{3}$ 15. $x < 1$ 16. $x > -2$ 17. $x < 4$ 18. $x > -1$

19. $x < -2$ 20. $x < \frac{1}{2}$ 21. $x < 0$ 22. $x > 1$ 23. $x < -4$ 24. $x < 0$

10 — Chapter Review

1. $x = -9$ 2. $x = -5$ 3. $x = -7$ 4. $x = 0$ 5. $x = 11$ 6. $x = 2$

7. $x = -14$ 8. $x = 10$ 9. $x = 32$ 10. $x = 15$ 11. $x = 1$ 12. $x = 48$

13. $x = 6$ 14. $x = 8$ 15. $x = -5$ 16. $x = -8$ 17. $x = 5$ 18. $x = 3$

19. $x = 4$ 20. $x = -\frac{1}{3}$ 21. $x = -\frac{5}{2}$ 22. $x = \frac{1}{3}$ 23. $x = \frac{1}{4}$ 24. $x = \frac{1}{3}$

25. $x = \frac{5}{2}$ 26. $x = \frac{2}{3}$ 27. $x = \frac{7}{10}$ 28. $x = \frac{9}{10}$ 29. $x = -\frac{4}{9}$ 30. $x = 5$

31. $x = \frac{1}{5}$ 32. $x = -\frac{3}{2}$ 33. $x = 0$ 34. $x = -4$ 35. $x = -\frac{5}{4}$ 36. $x = 3$

37. $x = -7$ 38. $x = \frac{3}{2}$ 39. $x = \frac{5}{4}$ 40. $x = -5$ 41. $x = 10$ 42. $x = -\frac{6}{5}$

43. $x = \frac{9}{7}$ 44. $x = -\frac{4}{3}$ 45. $x = -\frac{7}{5}$ 46. $x = -3$ 47. $x = 0$ 48. $x = \frac{7}{2}$

49. $x = -\frac{17}{3}$ 50. $x = \frac{22}{7}$ 51. $x = \frac{5}{3}$ 52. $x = 6$ 53. $x = 8$ 54. $x = -\frac{9}{5}$

55. $x = \frac{9}{10}$ 56. $x = -\frac{13}{14}$ 57. $x = -\frac{11}{6}$ 58. $x = \frac{28}{3}$ 59. $x = \frac{7}{4}$ 60. $x = \frac{25}{8}$

61. $x = \frac{11}{60}$ 62. $x = -\frac{4}{9}$ 63. $x = -\frac{3}{2}$ 64. $x = \frac{4}{45}$ 65. $x = -\frac{15}{22}$ 66. $x = \frac{10}{3}$

67. $x = 3$ 68. $x = 23$ 69. $x = 9$ 70. $x = 5$ 71. $x = \frac{13}{12}$ 72. $x = -4$

73. $x < 5$ 74. $x < -15$ 75. $x < \frac{4}{3}$ 76. $x < -\frac{7}{3}$ 77. $x > \frac{4}{11}$ 78. $x < 3$

79. $x < -\frac{2}{3}$ 80. $x < 0$ 81. $x < -\frac{1}{3}$ 82. $x < \frac{3}{5}$ 83. $x > 1$ 84. $x < 0$

10 – Cumulative Test

1. (10.1) $x = 1$ 2. (10.2) $x = \frac{1}{2}$ 3. (10.3) $x = -\frac{13}{8}$ 4. (10.4) $x = -\frac{33}{5}$ 5. (10.5) $x > 2$

6. (10.3) $x = -\frac{44}{13}$ 7. (10.2) $x = \frac{1}{9}$ 8. (10.4) $x = \frac{8}{3}$ 9. (9.4) $\frac{35x - 12}{20}$ 10. (10.5) $x < 0$

11. (10.3) $x = 19$ 12. (10.4) $x = -\frac{11}{10}$ 13. (9.4) $\frac{-47x + 21}{42}$ 14. (10.1) $x = 27$ 15. (10.5) $x < -\frac{3}{2}$

16. (10.2) $x = 5$ 17. (10.2) $x = 1$ 18. (10.3) $x = 3$ 19. (10.4) $x = \frac{5}{12}$ 20. (10.3) $x = 3$

21. (10.5) $x > 1$ 22. (10.1) $x = 7$ 23. (10.4) $x = -\frac{15}{13}$ 24. (9.4) $\frac{141x + 98}{60}$ 25. (10.2) $x = \frac{1}{3}$

26. (10.3) $x = \frac{11}{6}$ 27. (10.2) $x = 4$ 28. (10.5) $x < \frac{7}{9}$ 29. (10.3) $x = \frac{3}{4}$ 30. (10.1) $x = -13$

31. (9.4) $\frac{-26x + 15}{80}$ 32. (10.4) $x = -\frac{11}{3}$ 33. (10.5) $x > \frac{3}{2}$ 34. (10.2) $x = -\frac{9}{2}$ 35. (10.3) $x = -\frac{14}{13}$

36. (10.4) $x = -\frac{1}{3}$ 37. (10.5) $x < -7$ 38. (10.2) $x = \frac{1}{3}$ 39. (10.4) $x = \frac{7}{6}$ 40. (10.5) $x > \frac{1}{2}$

CHAPTER 11

11.1

1. $x = 1 - 4y$
2. $x = 4y - 5$
3. $x = 5y$
4. $x = 13y$
5. $x = -11z - 6y$
6. $x = 3y - 4z$
7. $x = -4y$
8. $x = 5y$
9. $x = 9y - 3z$
10. $x = -5y + 7$
11. $x = -4y + 7z$
12. $x = 5y$
13. $x = -2y + 5z + 1$
14. $x = -2y + 9z$
15. $x = 9y - 4$
16. $x = -9z - 1 - 3y$
17. $x = 4y - 9 - z$
18. $x = -3z + 7 - y$

11.2

1. $x = \frac{7y - 3}{4}$
2. $x = \frac{6y + 2z}{11}$
3. $x = \frac{-y + 7z}{40}$
4. $x = \frac{2 + 9y}{5}$
5. $x = \frac{-7z + 20 + 3y}{4}$
6. $x = \frac{y + 9}{7}$
7. $x = \frac{-z + 3y + 4}{7}$
8. $x = \frac{7z - 8 + 9y}{3}$
9. $x = \frac{-3y + 1}{3}$
10. $x = \frac{11y + z}{3}$
11. $x = \frac{10z + 7y}{2}$
12. $x = \frac{-4z + 7y}{2}$
13. $x = \frac{-15 - 2y}{9}$
14. $x = \frac{z + 4y}{3}$
15. $x = \frac{-7z + 2y}{4}$
16. $x = \frac{-z + 7y}{7}$
17. $x = \frac{-7y - 2z}{2}$
18. $x = \frac{2y + 9z}{19}$
19. $x = \frac{-9z + 7y}{2}$
20. $x = \frac{-3y + 4z}{9}$
21. $x = \frac{1 + 8y + 2z}{13}$
22. $x = \frac{4 + 5y - 2z}{7}$
23. $x = \frac{5y + 3z}{7}$
24. $x = \frac{-14y + 5z}{14}$

11.3

1.

x	y
0	3
7	0
14	−3
−7	6
−28	15
28	−9

2.

x	y
0	5
4	0
12	−10
−12	20
−8	15
8	−5

3.

x	y
0	−2
3	0
12	6
15	8
−9	−8
−15	−12

4.

x	y
0	−3
5	0
15	6
25	12
−10	−9
−5	−6

5.

x	y
0	8
−14	0
7	12
−7	4
−4	$\frac{40}{7}$
$-\frac{49}{2}$	−6

6.

x	y
0	10
−12	0
18	25
6	15
−9	$\frac{5}{2}$
$\frac{12}{5}$	12

7.

x	y
0	−4
−5	0
10	−12
$-\frac{35}{2}$	10
−8	$\frac{12}{5}$
5	−8

8.

x	y
0	−3
−2	0
9	$-\frac{33}{2}$
−10	12
−8	9
$\frac{14}{3}$	−10

9.

x	y
0	$-\frac{24}{5}$
2	0
6	$\frac{48}{5}$
$\frac{23}{4}$	9
−8	−24
−3	−12

10.

x	y
0	−2
$\frac{18}{7}$	0
3	$\frac{1}{3}$
18	12
−12	$-\frac{34}{3}$
−9	−9

11.

x	y
0	3
$\frac{6}{5}$	0
5	$-\frac{19}{2}$
4	−7
−8	23
$\frac{24}{5}$	−9

12.

x	y
0	$\frac{10}{3}$
5	0
7	$-\frac{4}{3}$
−4	6
−10	10
$\frac{25}{2}$	−5

11.4

The graph is the line through

1. (1, 0), (0, −5) **2.** (2, 0), (0, 4) **3.** (−6, 0), (0, −2) **4.** (4, 0), (0, 1)

5. (−4, 0), (0, 5) **6.** (5, 0), (0, −6) **7.** (−4, 0), (0, −4) **8.** (3, 0), (0, 3)

9. (−6, 0), (0, 3) **10.** (6, 0), (0, −2) **11.** (5, 0), (0, 6) **12.** (3, 0), (0, 5)

13. (0, −4), (1, −1) **14.** (0, −6), (2, 2) **15.** (2, 0), (−1, −4) **16.** (1, 0), (−1, −3)

17. (0, −4), (−5, 2) **18.** (0, −6), (−2, −1) **19.** (3, 0), (1, 3) **20.** (2, 0), (−3, 6)

21. (3, 0), (−1, −1) **22.** (5, 0), (2, −1) **23.** (0, 3), (4, −2) **24.** (0, 1), (5, −2)

11 – Chapter Review

1. $x = 2 - 6y$ **2.** $x = -12y$ **3.** $x = 18y - 9z$ **4.** $x = 5y$ **5.** $x = 8y - 18 - 2z$

6. $x = -10y + 12$ **7.** $x = -4y + 9z$ **8.** $x = 18y - 8$ **9.** $x = -4y + 10z + 2$ **10.** $x = \frac{8y - 2}{3}$

11. $x = \frac{-14z + 10 + 6y}{8}$ **12.** $x = \frac{-9y + 3}{19}$ **13.** $x = \frac{22y + 4z}{9}$ **14.** $x = \frac{30 + 16y}{3}$ **15.** $x = \frac{-y + 6z}{18}$

16. $x = \frac{-8z + 14y}{7}$ **17.** $x = \frac{-2z + 10y}{5}$ **18.** $x = \frac{-19z + 14y}{4}$ **19.** $x = \frac{3 + 16y + 4z}{26}$ **20.** $x = \frac{12y + 7z}{14}$

21. $x = \frac{12y - 21z + 6}{10}$

22.

x	y
0	6
5	0
15	−12
−10	18
−5	12
10	−6

23.

x	y
0	−4
5	0
20	12
15	8
−10	−12
−20	−20

24.

x	y
0	21
−9	0
6	35
$-\frac{36}{7}$	9
−8	$\frac{7}{3}$
−15	−14

25.

x	y
0	−2
−3	0
12	−10
$-\frac{33}{2}$	9
−4	$\frac{2}{3}$
6	−6

26.

x	y
0	$-\frac{16}{7}$
2	0
9	8
$\frac{25}{2}$	12
−5	−8
$-\frac{27}{4}$	−10

27.

x	y
0	2
$\frac{14}{9}$	0
6	$-\frac{40}{7}$
$-\frac{49}{9}$	9
−7	11
7	−7

The graph is the line through

28. (1, 0), (0, −4) **29.** (5, 0), (0, 1) **30.** (4, 0), (0, −3) **31.** (−5, 0), (0, −5) **32.** (−6, 0), (0, 2) **33.** (4, 0), (0, 5)

34. (0, −3), (1, 2) **35.** (2, 0), (−1, −5) **36.** (0, −5), (−3, −1) **37.** (1, 0), (−1, 5) **38.** (4, 0), (1, −1) **39.** (0, 4), (4, 1)

CHAPTER 12

12.1

1. $x = 2 \;\&\; y = 3$
2. $x = -1 \;\&\; y = 2$
3. $x = 7 \;\&\; y = -4$
4. $x = 4 \;\&\; y = 1$
5. $x = -2 \;\&\; y = 5$
6. $x = -4 \;\&\; y = -2$
7. $x = -6 \;\&\; y = -5$
8. $x = 9 \;\&\; y = 0$
9. $x = 1 \;\&\; y = 2$
10. $x = 4 \;\&\; y = 2$
11. $x = 3 \;\&\; y = -1$
12. $x = -2 \;\&\; y = 4$
13. $x = -7 \;\&\; y = -2$
14. $x = -4 \;\&\; y = -5$
15. $x = 1 \;\&\; y = 5$
16. $x = -6 \;\&\; y = -7$
17. $x = 2 \;\&\; y = 3$
18. $x = 3 \;\&\; y = 1$
19. $x = 6 \;\&\; y = -1$
20. $x = 3 \;\&\; y = -8$
21. $x = -1 \;\&\; y = -9$
22. $x = -4 \;\&\; y = -5$
23. $x = -6 \;\&\; y = -9$
24. $x = 3 \;\&\; y = 9$
25. $x = -\frac{11}{3} \;\&\; y = 9$
26. $x = 4 \;\&\; y = -\frac{8}{5}$
27. $x = -\frac{3}{4} \;\&\; y = -\frac{7}{4}$
28. $x = \frac{7}{3} \;\&\; y = -\frac{5}{3}$
29. $x = -\frac{9}{8} \;\&\; y = \frac{3}{8}$
30. $x = \frac{4}{15} \;\&\; y = -\frac{1}{5}$
31. $x = -\frac{7}{10} \;\&\; y = -\frac{2}{5}$
32. $x = -\frac{2}{3} \;\&\; y = -\frac{11}{12}$
33. $x = \frac{1}{3} \;\&\; y = -\frac{1}{15}$
34. $x = \frac{6}{5} \;\&\; y = \frac{7}{5}$
35. $x = \frac{2}{3} \;\&\; y = \frac{1}{3}$
36. $x = -\frac{5}{7} \;\&\; y = \frac{1}{7}$
37. $x = -\frac{1}{3} \;\&\; y = \frac{1}{6}$
38. $x = -\frac{1}{3} \;\&\; y = -\frac{5}{6}$
39. $x = \frac{6}{5} \;\&\; y = -\frac{11}{5}$
40. $x = \frac{5}{6} \;\&\; y = \frac{7}{6}$
41. $x = \frac{10}{7} \;\&\; y = -\frac{6}{7}$
42. $x = \frac{8}{3} \;\&\; y = \frac{5}{3}$
43. $x = \frac{3}{4} \;\&\; y = \frac{3}{4}$
44. $x = -\frac{5}{3} \;\&\; y = \frac{5}{6}$
45. $x = -\frac{1}{5} \;\&\; y = \frac{1}{5}$
46. $x = -\frac{3}{8} \;\&\; y = -\frac{3}{8}$
47. $x = -\frac{1}{7} \;\&\; y = \frac{6}{7}$
48. $x = \frac{5}{7} \;\&\; y = \frac{8}{7}$

12.1 Word Problems

1. $l = 6 + s \;\&\; s = 10 - l$
   $l = 8 \;\&\; s = 2$
2. $f = 5s \;\&\; f = 12 - s$
   $f = 10 \;\&\; s = 2$
3. $s = l - 8 \;\&\; l = 3s$
   $s = 4 \;\&\; l = 12$
4. $x + y = 3 \;\&\; x = 7 + y$
   $x = 5 \;\&\; y = -2$
5. $2f = s \;\&\; s = f - 3$
   $f = -3 \;\&\; s = -6$
6. $f = s - 8 \;\&\; 5s = f$
   $f = -10 \;\&\; s = -2$
7. $f = 11 + s \;\&\; f + s = -3$
   $f = 4 \;\&\; s = -7$
8. $f - s = -12 \;\&\; f = 7s$
   $f = -14 \;\&\; s = -2$
9. $x - 15 = y \;\&\; x + y = -9$
   $x = 3 \;\&\; y = -12$
10. $f = 11 + s \;\&\; s + 4f = 39$
    $f = 10 \;\&\; s = -1$
11. $-16 - f = s \;\&\; -40 + 7s = f$
    $f = -19 \;\&\; s = 3$
12. $f = 21 + 6s \;\&\; s + 5f = -50$
    $f = -9 \;\&\; s = -5$
13. $l = 7 + w \;\&\; 2l + 2w = 62$
    $l = 19 \;\&\; w = 12$
    12 by 19 meters
14. $w = l - 3 \;\&\; 2l + 2w = 50$
    $l = 14 \;\&\; w = 11$
    11 by 14 centimeters
15. $l = 3w \;\&\; 2l + 2w = 120$
    $l = 45 \;\&\; w = 15$
    15 by 45 millimeters
16. $w + 12 = l \;\&\; 2l + 2w = 44$
    $l = 17 \;\&\; w = 5$
    5 by 17 meters
17. $l - 10 = w \;\&\; 2l + 2w = 80$
    $l = 25 \;\&\; w = 15$
    15 by 25 kilometers
18. $l = 24 + w \;\&\; 2l + 2w = 76$
    $l = 31 \;\&\; w = 7$
    7 by 31 millimeters
19. $w = l - 4 \;\&\; 2l + 2w = 28$
    $l = 9 \;\&\; w = 5$
    5 by 9 meters
20. $l = 4w \;\&\; 2l + 2w = 60$
    $l = 24 \;\&\; w = 6$
    6 by 24 centimeters
21. $7w = l \;\&\; 2l + 2w = 64$
    $l = 28 \;\&\; w = 4$
    4 by 28 kilometers
22. $d = 2 + q \;\&\; 10d + 25q = 195$
    $d = 7 \;\&\; q = 5$
    12 coins
23. $n = d - 7 \;\&\; 5n + 10d = 235$
    $n = 11 \;\&\; d = 18$
    29 coins
24. $d = 6p \;\&\; p + 10d = 915$
    $p = 15 \;\&\; d = 90$
    105 coins
25. $n = p - 140 \;\&\; p + 5n = 740$
    $p = 240 \;\&\; n = 100$
    340 coins
26. $q = 9h \;\&\; 25q + 50h = 825$
    $q = 27 \;\&\; h = 3$
    30 coins
27. $h = p + 3 \;\&\; p + 50h = 1{,}680$
    $p = 30 \;\&\; h = 33$
    63 coins
28. $p = 12q \;\&\; p + 25q = 222$
    $p = 72 \;\&\; q = 6$
    78 coins
29. $d = 5 + 3n \;\&\; 5n + 10d = 365$
    $n = 9 \;\&\; d = 32$
    41 coins
30. $n = q - 19 \;\&\; 5n + 25q = 775$
    $n = 10 \;\&\; q = 29$
    50 cents in nickels

31. $q = h - 13$ & $25q + 50h = 1{,}100$
$q = 6$ & $h = 19$
25 coins

32. $c = f - 20$ & $12c + 8f = 460$
$c = 15$ & $f = 35$
The customer elevator can carry 15 people and the freight elevator can carry 35 people

33. $a = s + 17$ & $55s + 15a = 1{,}305$
$s = 15$ & $a = 32$
15 special-delivery stamps and 32 air-mail stamps

34. $l = s + 3{,}000$ & $6l + 8s = 228{,}000$
$l = 18{,}000$ & $s = 15{,}000$
The larger truck can carry 18,000 pounds and the smaller truck can carry 15,000 pounds

35. $t = f + 26$ & $20t + 15f = 2{,}095$
$t = 71$ & $f = 45$
71 twenty-cent stamps and 45 fifteen-cent stamps

36. $9f + 22g = 844$ & $f = 18 + g$
$f = 40$ & $g = 22$
40 French coins and 22 German coins

37. $b = h + 12$ & $5b + 6h = 258$
$b = 30$ & $h = 18$
The bricklayer can lay 30 bricks and his helper 18 bricks per hour

12.2

1. $x = 2$ & $y = 3$
2. $x = 4$ & $y = 2$
3. $x = 3$ & $y = -4$
4. $x = -4$ & $y = -1$
5. $x = -5$ & $y = -3$
6. $x = 3$ & $y = 1$
7. $x = 2$ & $y = 1$
8. $x = 0$ & $y = -4$
9. $x = -3$ & $y = 4$
10. $x = -5$ & $y = -1$
11. $x = 3$ & $y = -6$
12. $x = 7$ & $y = -2$
13. $x = -8$ & $y = 1$
14. $x = -3$ & $y = 7$
15. $x = -4$ & $y = 0$
16. $x = 7$ & $y = 9$
17. $x = 3$ & $y = -5$
18. $x = 15$ & $y = 3$
19. $x = -4$ & $y = 1$
20. $x = 0$ & $y = -3$
21. $x = -3$ & $y = -8$
22. $x = 5$ & $y = 1$
23. $x = 2$ & $y = -9$
24. $x = 4$ & $y = -7$
25. $x = 5$ & $y = -3$
26. $x = -4$ & $y = 6$
27. $x = 7$ & $y = 2$
28. $x = 4$ & $y = 9$
29. $x = -3$ & $y = -6$
30. $x = -4$ & $y = -7$
31. $x = 0$ & $y = -9$
32. $x = 0$ & $y = -8$
33. $x = 3$ & $y = -4$
34. $x = 2$ & $y = -5$
35. $x = -1$ & $y = 4$
36. $x = -1$ & $y = 3$
37. $x = \frac{2}{5}$ & $y = -\frac{3}{5}$
38. $x = \frac{7}{3}$ & $y = -\frac{8}{3}$
39. $x = -\frac{4}{11}$ & $y = -\frac{5}{11}$
40. $x = \frac{7}{10}$ & $y = -\frac{4}{5}$
41. $x = -\frac{8}{7}$ & $y = \frac{5}{7}$
42. $x = \frac{5}{8}$ & $y = -\frac{3}{2}$
43. $x = \frac{1}{6}$ & $y = \frac{7}{6}$
44. $x = -\frac{9}{4}$ & $y = -\frac{5}{4}$
45. $x = \frac{5}{12}$ & $y = -\frac{1}{3}$
46. $x = -\frac{4}{5}$ & $y = \frac{7}{15}$
47. $x = \frac{9}{2}$ & $y = \frac{5}{6}$
48. $x = -\frac{4}{9}$ & $y = -\frac{1}{3}$

12.2 Word Problems

1. $s + l = -37$ & $s - l = 11$
$s = -24$ & $l = -13$

2. $l - s = 23$ & $4s - l = -38$
$s = -5$ & $l = 18$

3. $3f + s = -26$ & $f + 5s = 38$
$f = -12$ & $s = 10$

4. $s + l = 0$ & $l - 9s = 10$
$s = -1$ & $l = 1$

5. $6s - f = 27$ & $4f - 9s = -3$
$f = 15$ & $s = 7$

6. $-1f + 4s = 25$ & $2s = 35 + 3f$
$f = -9$ & $s = 4$

7. $14y - 4x = 48$ & $x + y = -3$
$x = -5$ & $y = 2$

8. $3f + 2s = 18$ & $5s - f = -74$
$f = 14$ & $s = -12$

9. $4l = 49 - w$ & $2l + 2w = 38$
$l = 10$ & $w = 9$
9 by 10 kilometers

10. $2l + 2w = 56$ & $l + 11 = w + 13$
$l = 15$ & $w = 13$
13 by 15 meters

11. $2l + 2w = 130$ & $l + 35 = 3w$
$l = 40$ & $w = 25$
25 by 40 centimeters

12. $11 - l = 6 - w$ & $2l + 2w = 58$
$l = 17$ & $w = 12$
12 by 17 meters

13. $2l + 2w = 234$ & $l - 16 = 37 + w$
$l = 85$ & $w = 32$
32 by 85 millimeters

14. $2l + 2w = 98$ & $81 + w = 4l$
$l = 26$ & $w = 23$
23 by 26 centimeters

15. $5n + 25q = 435$ & $n + q = 35$
$n = 22$ & $q = 13$
13 quarters

16. $10d + 50h = 770$ & $d + h = 29$
$d = 17$ & $h = 12$
12 half dollars

17. $p + d = 370$ & $p + 10d = 550$
$p = 350$ & $d = 20$
350 pennies, 20 dimes

18. $5d = q + 12$ & $10d + 25q = 780$
$d = 8$ & $q = 28$
36 coins

19. $3n + 12p = 189$ & $p + 5n = 87$
$p = 12$ & $n = 15$
27 coins

20. $21 + d = 7n$ & $5n + 10d = 240$
$n = 6$ & $d = 21$
27 coins

21. $5q = n - 8$ & $5n + 25q = 390$
$n = 43$ & $q = 7$
7 quarters

22. $13d + 2p = 128$ & $p + 10d = 71$
$d = 2$ & $p = 51$
53 coins

23. $15p - 2q = 10$ & $p + 25q = 1{,}006$
$p = 6$ & $q = 40$
46 coins

24. $q + d = 51$ & $10d + 25q = 975$
$d = 20$ & $q = 31$
31 quarters

25. $210d + 55c = 27{,}150$ & $d + c = 240$
$d = 90$ & $c = 150$
90 desks

26. $n + s = 195$ & $50n + 85s = 11{,}675$
$s = 55$ & $n = 140$
55 pounds

27. $750a + 600s = 690{,}000$ & $a + s = 950$
$a = 800$ & $s = 150$
800 adult tickets

28. $200a + 360m = 19{,}400$ & $a + m = 65$
$a = 25$ & $m = 40$
25 apprentices

29. $s + a = 1{,}260$ & $4s + 6a = 5{,}660$
$s = 950$ & $a = 310$
950 student tickets, 310 adult tickets

30. $12m + 14c = 6{,}120$ & $m + c = 480$
$m = 300$ & $c = 180$
180 chemistry books

31. $3c + t = 450$ & $2t + 10c = 1{,}140$
$c = 60$ & $t = 270$
$60 per chair, $270 per table

32. $3a + 5p = 355$ & $7a + 1p = 295$
$a = 35$ & $p = 50$
apples cost 35 cents and
pears cost 50 cents per pound

33. $9c + n = 1{,}620$ & $16c + 4n = 3{,}280$
$c = 160$ & $n = 180$
$1.60 per pound of chocolate

34. $2c + b = 1{,}620$ & $3c + 2b = 2{,}730$
$c = 510$ & $b = 600$
Colombian coffee costs $5.10 and
Brazilian coffee costs $6.00 per
kilogram

35. $7f + 3p = 3{,}240$ & $2f + p = 960$
$f = 360$ & $p = 240$
$3.60 per bag of foam-rubber and
$2.40 per bag of poly-form

36. $4p + 1c = 1{,}940$ & $7p + 3c = 4{,}070$
$p = 350$ & $c = 540$
peanut oil costs $3.50 and
corn oil costs $5.40 per liter

12.3

The lines intersect at the point

1. (5, 6) **2.** (−6, −2) **3.** (−1, 4) **4.** (−2, −3) **5.** (−4, 0) **6.** (−3, 0)

7. (2, −2) **8.** (3, −3) **9.** (3, 2) **10.** (−3, 4) **11.** (−6, 4) **12.** (−6, 6)

13. (0, −5) **14.** (−4, 0) **15.** (−3, −1) **16.** (4, 1) **17.** (3, 4) **18.** (5, 5)

19. (−4, 1) **20.** (−1, 3) **21.** (4, 0) **22.** (0, 5) **23.** (5, −2) **24.** (−6, 5)

12 – Chapter Review

1. $x = -3$ & $y = 3$ **2.** $x = 4$ & $y = -6$ **3.** $x = 3$ & $y = -5$ **4.** $x = 7$ & $y = 0$

5. $x = 4$ & $y = 5$ **6.** $x = -3$ & $y = 2$ **7.** $x = -6$ & $y = -4$ **8.** $x = -3$ & $y = -8$

9. $x = 8$ & $y = 2$ **10.** $x = -5$ & $y = -29$ **11.** $x = -14$ & $y = 3$ **12.** $x = 0$ & $y = 2$

13. $x = 3$ & $y = \frac{7}{2}$ **14.** $x = \frac{5}{3}$ & $y = -\frac{2}{3}$ **15.** $x = \frac{5}{6}$ & $y = \frac{1}{2}$ **16.** $x = -\frac{1}{6}$ & $y = -\frac{7}{6}$

17. $x = -\frac{6}{7}$ & $y = \frac{9}{7}$ **18.** $x = -\frac{2}{3}$ & $y = \frac{1}{3}$ **19.** $x = -\frac{3}{8}$ & $y = \frac{3}{8}$ **20.** $x = -\frac{7}{5}$ & $y = \frac{4}{5}$

21. $x = \frac{4}{3}$ & $y = -\frac{7}{9}$ **22.** $x = \frac{7}{6}$ & $y = \frac{4}{3}$ **23.** $x = -\frac{9}{8}$ & $y = -\frac{5}{2}$ **24.** $x = \frac{5}{3}$ & $y = \frac{1}{6}$

25. $x = -5$ & $y = 3$ **26.** $x = 11$ & $y = -2$ **27.** $x = 6$ & $y = 2$ **28.** $x = -8$ & $y = 0$

29. $x = -1$ & $y = 3$ **30.** $x = -7$ & $y = 2$ **31.** $x = 11$ & $y = -3$ **32.** $x = -6$ & $y = -5$

33. $x = -4$ & $y = -9$ **34.** $x = 6$ & $y = -1$ **35.** $x = 7$ & $y = 8$ **36.** $x = -3$ & $y = 0$

37. $x = 4$ & $y = -9$ **38.** $x = -6$ & $y = 5$ **39.** $x = 7$ & $y = -2$ **40.** $x = -1$ & $y = -3$

41. $x = 5$ & $y = 2$ **42.** $x = -6$ & $y = 7$ **43.** $x = \frac{4}{5}$ & $y = -\frac{2}{5}$ **44.** $x = \frac{2}{3}$ & $y = \frac{5}{3}$

45. $x = -\frac{3}{4}$ & $y = -\frac{7}{4}$ **46.** $x = \frac{3}{4}$ & $y = -\frac{5}{8}$ **47.** $x = \frac{5}{6}$ & $y = \frac{4}{3}$ **48.** $x = \frac{5}{7}$ & $y = -\frac{3}{7}$

The lines intersect at the point

49. (6, 4) **50.** (1, −4) **51.** (−5, 0) **52.** (−2, −2)

53. (−2, 4) **54.** (−4, −6) **55.** (5, 0) **56.** (3, 2)

57. (−4, −2) **58.** (−2, −1) **59.** (0, 3) **60.** (4, −5)

CHAPTER 13

13.1 Word Problems

1. $12 + x = 21 - 2x$, $x = 3$

2. $x + 30 = 7x$, $x = 5$

3. $28 + x = -6x$, $x = -4$

4. $x - 18 = 3x$, $x = -9$

5. $9x = 12 - x$, $x = \frac{6}{5}$

6. $21x = 15 + x$, $x = \frac{3}{4}$

7. $14 = (x + 43) - 18$, $x = -11$

8. $8(x - 11) = 20$, $x = \frac{27}{2}$

9. $(24 + x)4 = 16$, $x = -20$

10. $(x - 5)3 = -75$, $x = -20$

11. $61 - 9x = 19$, $x = \frac{14}{3}$

12. $37x = 25x - 8$, $x = -\frac{2}{3}$

13. $(x - 15)12 = 2x$, $x = 18$

14. $(x + 31)7 = 89 - x$, $x = -16$

15. $76x = 43(x - 1)$, $x = -\frac{43}{33}$

16. $18x - 45 = -7x$, $x = \frac{9}{5}$

17. $58 - 4x = 8 + 6x$, $x = 5$

18. $7x - 12 = 78 - 11x$, $x = 5$

19. $(8 + x)(-6) = (x - 52) - 74$, $x = \frac{78}{7}$

20. $69 - (43 + x) = (12 + x)4$, $x = -\frac{22}{5}$

21. $12 + x = \frac{3}{5}x$, $x = -30$

22. $\frac{x}{9} = 15 - x$, $x = \frac{27}{2}$

23. $x + 21 = \frac{5}{6}x$, $x = -126$

24. $74 - x = \frac{x}{5}$, $x = \frac{185}{3}$

25. $15 + \frac{3}{4}x = \frac{3}{10}$, $x = -\frac{98}{5}$

26. $9\left(\frac{x}{11}\right) = \frac{3}{4}$, $x = \frac{11}{12}$

27. $\frac{x + 6}{8} = \frac{2}{5}$, $x = -\frac{14}{5}$

28. $\frac{9}{16}x = 10 - \frac{1}{6}x$, $x = \frac{96}{7}$

29. $\frac{7}{9}x = \frac{3}{5}x - 2$, $x = -\frac{45}{4}$

30. $\frac{x}{27} = \frac{1}{9}(x + 4)$, $x = -6$

13.2 Word Problems

1. $n + n + 1 = 95$; 47, 48

2. $n + n + 1 = 169$; 84, 85

3. $n + n + 2 = 120$; 59, 61

4. $n + n + 2 = 54$; 26, 28

5. $n + n + 1 + n + 2 = 39$; 12, 13, 14

6. $n + n + 1 + n + 2 + n + 3 = 134$; 32, 33, 34, 35

7. $n + n + 2 + n + 4 = 168$; 54, 56, 58

8. $n + n + 2 + n + 4 = -189$; −65, −63, −61

9. $n + n + 2 + n + 4 = 12 + 2(n + 4)$; 14, 16, 18

10. $n + n + 2 + n + 4 + n + 6 = -17 + 5(n + 2)$; 19, 21, 23, 25

11. $4(n + n + 1 + n + 2) = 20(n + 1) - 88$; 10, 11, 12

12. $2(n + 2 + n + 4) = 135 + n$; 41, 43, 45

13. $6(n+1) = 70 + n + n + 2 + n + 3$
23, 24, 25, 26

14. $n + n + 2 - 50 = 36 + n + 4$
88, 90, 92, 94

15. $10(n+2) = 171 + n + n + 1$
19, 20, 21

16. $\frac{n + n + 1}{3} = 33$
49, 50

17. $\frac{n + n + 1 + n + 2}{9} = 15$
44, 45, 46

18. $\frac{n+1}{10} + n + 2 = 34$
29, 30, 31

19. $\frac{2}{3}n + \frac{n+1}{6} = 11$
13, 14

20. $\frac{3}{8}n + \frac{n+1}{5} = 83$
144, 145

21. $\frac{10}{7}n = n + 2 + \frac{n+1}{2}$
−35, −34, −33

22. $\frac{n + 1 + n + 3}{4} = n + 2 - 6$
10, 11, 12, 13

23. $\frac{n + 1 + n + 2}{5} = \frac{5}{14}n$
−14, −13, −12

24. $\frac{n+1}{11} + \frac{n+2}{2} = \frac{5}{8}n$
32, 33, 34

25. $n + 2 - \frac{4}{15}n = n + 3 - 29$
105, 106, 107, 108

13.3 Word Problems

1. $l = 9 + s$ & $s = 29 - l$
$s = 10$ & $l = 19$

2. $f = 7s$ & $f = 24 - s$
$s = 3$ & $f = 21$

3. $s = l - 6$ & $l = 4s$
$s = 2$ & $l = 8$

4. $x + y = 7$ & $x = 19 + y$
$x = 13$ & $y = -6$

5. $-8f = s$ & $s = f - 27$
$s = -24$ & $f = 3$

6. $x = y - 8$ & $5y = x$
$x = -10$ & $y = -2$

7. $f = 8 + s$ & $f + s = 50$
$f = 29$ & $s = 21$

8. $f - s = -33$ & $f = 12s$
$s = -3$ & $f = -36$

9. $s + l = 31$ & $s - l = -19$
$s = 6$ & $l = 25$

10. $x - 20 = y$ & $x + y = 8$
$x = 14$ & $y = -6$

11. $f = 25 + s$ & $s + 4f = 75$
$f = 20$ & $s = -5$

12. $l - s = 15$ & $8s - l = 13$
$l = 19$ & $s = 4$

13. $3f + s = 5$ & $f + 3s = 47$
$f = -4$ & $s = 17$

14. $s + l = 21$ & $l - 4s = -19$
$l = 13$ & $s = 8$

15. $25 - f = s$ & $-52 + 6s = f$
$f = 14$ & $s = 11$

16. $f - 5s = 28$ & $3f - 8s = 70$
$f = 18$ & $s = -2$

17. $f = 2 + 6s$ & $s + 12f = -49$
$f = -4$ & $s = -1$

18. $3f + 5s = 69$ & $2s = 27 - f$
$f = 3$ & $s = 12$

19. $7y - 4x = 28$ & $x + y = 15$
$x = 7$ & $y = 8$

20. $11f + 3s = 30$ & $4s - f = 40$
$f = 0$ & $s = 10$

13.4 Word Problems

1. $l = 12 + w$ & $2l + 2w = 84$
$l = 27$ & $w = 15$
15 by 27 meters

2. $w = l - 7$ & $2l + 2w = 90$
$l = 26$ & $w = 19$
19 by 26 meters

3. $l = 9w$ & $2l + 2w = 100$
$l = 45$ & $w = 5$
5 by 45 centimeters

4. $w + 10 = l$ & $2l + 2w = 28$
$l = 12$ & $w = 2$
2 by 12 meters

5. $l - 18 = w$ & $2l + 2w = 28$
$l = 25$ & $w = 7$
7 by 25 millimeters

6. $l = 4 + 4w$ & $2l + 2w = 98$
$l = 40$ & $w = 9$
9 by 40 meters

7. $w = 3l - 45$ & $2l + 2w = 70$
$l = 20$ & $w = 15$
15 by 20 centimeters

8. $l = 6w - 100$ & $2l + 2w = 500$
$l = 200$ & $w = 50$
50 by 200 millimeters

9. $w - 4l = -101$ & $2l + 2w = 108$
$l = 31$ & $w = 23$
23 by 31 meters

10. $12w + 7 = l$ & $2l + 2w = 92$
$l = 43$ & $w = 3$
3 by 43 centimeters

11. $l + 14 = w + 26$ & $2l + 2w = 208$
$l = 58$ & $w = 46$
46 by 58 meters

12. $2l + 2w = 46$ & $l + 7w = 47$
$l = 19$ & $w = 4$
4 by 19 kilometers

13. $5w - l = 62$ & $2l + 2w = 68$
$l = 18$ & $w = 16$
16 by 18 centimeters

14. $2l + 2w = 30$ & $(w - 12) - l = -17$
$l = 10$ & $w = 5$
5 by 10 kilometers

15. $2l + 2w = 276$ & $30 + (w - l) = -2$
$l = 85$ & $w = 53$
53 by 85 millimeters

13.5 Word Problems

1. $d = 12 + q$ & $10d + 25q = 575$
$d = 25$ & $q = 13$
38 coins

2. $n = d - 20$ & $5n + 10d = 875$
$n = 45$ & $d = 65$
110 coins

3. $n + q = 58$ & $5n + 25q = 810$
$n = 32$ & $q = 26$
26 quarters

4. $d = 9p$ & $p + 10d = 1{,}820$
$p = 20$ & $d = 180$
200 coins

5. $n = p - 30$ & $p + 5n = 180$
$p = 55$ & $n = 25$
80 coins

6. $q = 12h$ & $25q + 50h = 1{,}750$
$q = 60$ & $h = 5$
65 coins

7. $h = p + 2$ & $p + 50h = 355$
$p = 5$ & $h = 7$
12 coins

8. $d + h = 32$ & $10d + 50h = 680$
$d = 23$ & $h = 9$
9 half-dollars

9. $p = 4q$ & $p + 25q = 696$
$p = 96$ & $q = 24$
120 coins

10. $p + d = 560$ & $p + 10d = 1{,}100$
$p = 500$ & $d = 60$
500 pennies and 60 dimes

11. $d = 12 + 5n$ & $5n + 10d = 1{,}275$
$n = 21$ & $d = 117$
138 coins

12. $n = q - 31$ & $5n + 25q = 1{,}255$
$n = 16$ & $q = 47$
80 cents in nickels

13. $20d = 7 + q$ & $10d + 25q = 845$
$d = 2$ & $q = 33$
35 coins

14. $12n + 4p = 64$ & $p + 5n = 22$
$n = 3$ & $p = 7$
10 coins

15. $q = h - 8$ & $25q + 50h = 1{,}900$
$q = 20$ & $h = 28$
48 coins

16. $14 + d = 9n$ & $5n + 10d = 335$
$n = 5$ & $d = 31$
36 coins

17. $3q = n - 8$ & $5n + 25q = 400$
$n = 35$ & $q = 9$
9 quarters

18. $10d + 4p = 112$ & $p + 10d = 73$
$p = 13$ & $d = 6$
19 coins

19. $5p - 2q = 158$ & $p + 25q = 565$
$p = 40$ & $q = 21$
61 coins

20. $93 - q = 5d$ & $10d + 25q = 830$
$d = 13$ & $q = 28$
28 quarters

13.6 Word Problems

1. $20(120) + 15(190) = 35x$
$x = 150$
\$1.50 per kilogram

2. $4(159) + 7(192) = 11x$
$x = 180$
\$1.80 per pound

3. $9(67) + 5(95) = 14x$
$x = 77$
77 cents per pound

4. $40(344) + 15(564) = 55x$
$x = 404$
\$4.04 a kilogram

5. $24(86) + 13(123) = 37x$
$x = 99$
99 cents per pound

6. $17(208) + 23(288) = 40x$
$x = 254$
\$2.54 per kilogram

7. $72x + 10(207) = (x + 10)102$
$x = 35$
35 pounds

8. $400x + 47(540) = (x + 47)494$
$x = 23$
23 bags

9. $70x + 14(130) = (x + 14)98$
$x = 16$
16 meters

10. $800x + 55(600) = (x + 55)690$
$x = 45$
45 kilograms

11. $350x + 25(160) = (x + 25)300$
$x = 70$
70 liters

12. $20x + 5(62) = (x + 5)30$
$x = 16$
16 pounds

13. $250d + 65c = 2{,}715$ & $c + d = 19$
$d = 8$ & $c = 11$
8 desks

14. $85n + 107s = 4{,}419$ & $n + s = 45$
$n = 18$ & $s = 27$
27 pounds

15. $450a + 290s = 447{,}000$ & $a + s = 1{,}100$
$a = 800$ & $s = 300$
800 adult tickets

16. $240a + 430m = 9{,}330$ & $a + m = 27$
$a = 12$ & $m = 15$
12 apprentices

17. $120s + 200a = 61{,}000$ & $a + s = 435$
$s = 325$ & $a = 110$
325 student tickets & 110 adult tickets

18. $11m + 15c = 485$ & $m + c = 39$
$m = 25$ & $c = 14$
14 chemistry books

19. $10c + 7f = 520$ & $c = f - 16$
$c = 24$ & $f = 40$
the customer elevator carries 24 people and the freight elevator carries 40 people

20. $35s + 15a = 565$ & $a = 11 + s$
$s = 8$ & $a = 19$
8 special-delivery stamps and 19 air-mail stamps

21. $7l + 8s = 245$ & $l = 5 + s$
$l = 19$ & $s = 14$
the smaller truck can carry 14 tons and the larger truck can carry 19 tons

22. $5f + 8e = 316$ & $f = e + 6$
$f = 28$ & $e = 22$
28 five-cent stamps and
22 eight-cent stamps

23. $33f + 29g = 1{,}737$ & $f = g - 15$
$f = 21$ & $g = 36$
21 French coins and 36 German coins

24. $5b + 4h = 262$ & $b = 20 + h$
$b = 38$ & $h = 18$
a bricklayer lays 38 bricks per hour and a helper lays 18 bricks per hour

25. $6c + t = 459$ & $3t + 10c = 945$
$c = 54$ & $t = 135$
\$135 per table and \$54 per chair

26. $3a + 2p = 285$ & $a + 3p = 256$
$a = 49$ & $p = 69$
apples cost 49 cents and
pears cost 69 cents per pound

27. $5c + n = 6(170)$ & $3c + 2n = 5(149)$
$c = 185$ & $n = 95$
\$1.85 per pound of chocolate

28. $5c + 2b = 7(852)$ & $6c + b = 7(876)$
$c = 900$ & $b = 732$
Colombian coffee costs \$9.00 and Brazilian coffee costs \$7.32 per kilogram

29. $8f + p = 9(347)$ & $14f + 4p = 18(344)$
$f = 350$ & $p = 323$
\$3.50 per bag of foam-rubber and \$3.23 per bag of poly-foam

30. $5p + 3c = 8(405)$ & $3p + c = 4(400)$
$p = 390$ & $c = 430$
peanut oil costs \$3.90 and corn oil costs \$4.30 per liter

CHAPTER 14

14.1

1. $6(x^2 - 3)$
2. $5(5x^3 - 3y^2)$
3. $6(-5x + 4y)$
4. $13(4x^2 + 1)$
5. $6(3 - 5x)$
6. $7(6x - 5y^3)$
7. $2(-16x + 9y)$
8. $16(y^2 - 2x^2)$
9. $7(2x^4 + 1)$
10. $40(x - 3)$
11. $9(2y + 9)$
12. $12(9x^2 - 7y^2)$
13. $72(-2x + y)$
14. $13(13y^3 - 10x)$
15. $9(9y + 6z - 2)$
16. $3(xy - 4xz + 2yz)$
17. $20(-2x - 3y + 4z)$
18. $6(2x + 5y - 3x^2y)$
19. $2(-4w^2x - 7x^2y + 5y^2z)$
20. $15(-2x^3 - 3y^2 - 1)$
21. $9(2k^2 + 3r^2 + 5k^2r)$
22. $9(8x^2 + 9xy - 11y)$
23. $15(x^4 - 7x^2y^2 - 5)$
24. $12(12a^3 + 13ab^2 - 14b^3)$

14.2

1. $xy(7x^2 - 15)$
2. $y^2(12x^5 - 1)$
3. $x^2y^2(9x^4y - 14)$
4. $x^4(21x - 44)$
5. $x^4y^5z^2(-4x^3 + z^4)$
6. $x^6y^4z(z - 2y)$
7. $7y(x^2 + 6y)$
8. $8st^2(2rst - 3)$
9. $4xz(3y^2 - 2z)$
10. $12xy(2x - y)$
11. $7xy^3(2 + 3y)$
12. $13x^2z(2yz - 1)$
13. $15xy^3(1 - 5xy)$
14. $8z^3(xz - 8)$
15. $13xy(4x + 5y^2)$
16. $32rs^4(2r - 3t^2)$
17. $28y^3(-xz + 1)$
18. $12x^6y^4(12x + 13y^2)$
19. $3y^2(4x^2y^2z + 6x^2y - 3)$
20. $30x^3y^5z^2(xy + yz + xz)$
21. $9x^2(3xy^2 + 4y^3z - 2z^3)$
22. $5a^2b^3c^2(3 - 4a^3b^3c^3 + 5a^6b^6c^6)$
23. $11x^3(10x^6 + 15x^3 - 13)$
24. $30x^2y^2(2x^5 + 3x^2y^7 - 2y^2)$

14 – Chapter Review

1. $12(x + 4)$
2. $4(5 - 12x^2)$
3. $4(7y + 8z)$
4. $8(x^2 - 3z^2)$
5. $5(-3x + 2y)$
6. $9(6x + 1)$
7. $9(2y^2 - y + 3)$
8. $7(-3x - 6y + 5z)$
9. $25(x^2 - 3y + 4)$
10. $16(2y^2 + 4x - 9)$
11. $12(-x^2 - 3y^2 - 6)$
12. $7(3x + 5y^2 + 7)$
13. $x^2(12x - 41)$
14. $x^6y^4(35x - 36y)$
15. $x^6y^5(4x - 5y)$
16. $9x^4(3x - 4)$
17. $12b^3c(4a^2c^3 - 3)$
18. $5x^6(3 - 4x)$
19. $2xy(9w + 11z)$
20. $12x^8(8 - 9x)$
21. $9k(4m^4 - 1)$
22. $6xy(-1 + 4xyz - 5wz^2)$
23. $9km(-k + 2m^2n - 3mn^3)$
24. $42x^7y^4(1 - 2x^2y^4 + 3x^2y^7)$

14 — Cumulative Test

1. (14.1) $16(x+4)$
2. (14.2) $x^4(13x^2-36)$
3. (14.1) $7(8x-11-3x^3)$
4. (4.2) $30x+54$
5. (14.2) $14a^2b(3b^3c-4a)$
6. (14.1) $9(y-5z)$
7. (4.3) $s^6t^2-s^2t$
8. (14.2) $3xy(9w+16z)$
9. (14.1) $14(2-x^2)$
10. (4.3) $-54x+24x^2-6x^3$
11. (14.2) $x^3(12x+25)$
12. (14.1) $7(9x-4y^2)$
13. (4.2) $-32x+8y-48z$
14. (14.2) $6k^2n^2(-k+5m^4-7m^2n^2)$
15. (14.1) $4(-5x+3y-1)$
16. (4.3) $-4u^5v^6+9u^2v^9$
17. (14.2) $s^4t^6(14t-5s^4)$
18. (14.1) $6(3x^2-4y^2+6xy)$
19. (4.2) $-30+40A$
20. (14.2) $8xy(-1+5x^2yz-3wxy^2z^4)$

CHAPTER 15

15.1

1. $x^2+3x-18$
2. $x^2+12x+32$
3. $12x^2+19x+5$
4. $24x^2+22x+3$
5. $27x^2-21x+2$
6. $4x^2-11x-20$
7. $18x^2-15xy-12y^2$
8. $28x^2-55xy-18y^2$
9. $48y^2+34xy-5x^2$
10. $18x^2-33xy+9y^2$
11. $x^3-4x^2-3x+12$
12. $6x^3-27x^2-4x+18$
13. $4x^2-5x-12xy+15y$
14. $12x^2+20x+15xy+25y$
15. $8x^3-25x^2+3x$
16. $9x^3+30x^2+25x$
17. $6x^3-37x^2+6x$
18. $12x^7-30x^6+12x^5$
19. $9x^2-4$
20. $4x^2-25$
21. $36x^4-y^2$
22. $25x^6-y^4$
23. $25x^6-16y^{10}$
24. $36x^2y^{10}-1$
25. $25x^2+10x+1$
26. $64x^2-16x+1$
27. $9x^2-42xy+49y^2$
28. $36x^2+60xy+25y^2$
29. $4x^4-36x^2y^3+81y^6$
30. $64x^8-48x^4y^3+9y^6$

15.2

1. $4x^3+4x^2-24x$
2. $6x^2-78x+240$
3. $-12y^3+92y^2-56y$
4. $3x^9-14x^8+8x^7$
5. $6x^3y+3x^2y^2-18xy^3$
6. $16x^4+16x^3-12x^2$
7. $-48x^2y^3+20x^2y^2+8x^2y$
8. $-24x^2+198x-48$
9. $40x^5+190x^4+120x^3$
10. $-24y^6-52y^5+20y^4$
11. $x^{10}-x^8$
12. $-3x^{12}+15x^{11}-18x^{10}$
13. $48x^5y-28x^4y-40x^3y$
14. $3x^3y^5-18x^2y^6+24xy^7$
15. $-x^4-9x^2-18$
16. $12x^4y^2+32x^2y^3-64y^4$
17. $-12x^3+10x^5-2x^7$
18. $18x^{15}-30x^{13}-168x^{11}$
19. $5x^2-30x+45$
20. $-4x^3+32x^2-64x$
21. $192x^3-288x^2+108x$
22. $-32x^4+128x^3-128x^2$
23. $9x^4y^4+6x^3y^6+x^2y^8$
24. $-16x^3y-128x^2y^2-256xy^3$

15.3

1. $x^3+5x^2+10x+12$
2. $3x^3-6x^2-2x+7$
3. $8x^3-10x^2+7x-2$
4. $25x^3+10x^2-18x-9$
5. $2x^3-11x^2y+7xy^2-y^3$
6. $3x^3-13x^2y+15xy^2-9y^3$
7. $x^4+3x^3+12x^2+9x+27$
8. $2x^4+7x^3+11x^2+28x+12$
9. $x^4+3x^3y+4x^2y^2+6xy^3+4y^4$
10. $4x^4+9x^3y+16x^2y^2+9xy^3+12y^4$

11. $x^4 + 5x^3 + 7x^2 + 5x + 6$

12. $10x^4 - 11x^3 + 5x^2 + 11x - 6$

13. $4x^4 + 35x^3 + 28x^2 + 35x + 24$

14. $6x^4 - 21x^3 + 12x^2 - 10x + 3$

15. $x^4 + 5x^3 + 11x^2 + 13x + 6$

16. $15x^4 + 14x^3 - 29x^2 + 24x - 18$

17. $32x^4 - 32x^3 + 24x^2 - 8x + 2$

18. $12x^4 - 22x^3y - 40x^2y^2 - 7xy^3 + 3y^4$

19. $x^3 + 6x^2 + 12x + 8$

20. $27x^3 - 27x^2 + 9x - 1$

21. $-64x^3 + 336x^2 - 588x + 343$

22. $-729x^3 + 1944x^2 - 1728x + 512$

23. $x^3 - 9x^2y + 27xy^2 - 27y^3$

24. $512x^3 - 192x^2y + 24xy^2 - y^3$

25. $x^6 + 3x^4y^2 + 3x^2y^4 + y^6$

26. $27x^6 - 54x^4y^2 + 36x^2y^4 - 8y^6$

27. $256x^8 - 256x^6 + 96x^4 - 16x^2 + 1$

28. $625x^4 - 1500x^3 + 1350x^2 - 540x + 81$

29. $x^4 - 8x^3y + 24x^2y^2 - 32xy^3 + 16y^4$

30. $x^8 - 4x^6y^2 + 6x^4y^4 - 4x^2y^6 + y^8$

31. $x^4 - 4x^3 + 6x^2 - 4x + 1$

32. $x^4 - 10x^3 + 27x^2 - 10x + 1$

33. $x^8 - 6x^6 - 9x^4 + 54x^2 + 81$

34. $x^8 + 16x^6 + 72x^4 + 64x^2 + 16$

35. $4x^4 - 20x^3y + 29x^2y^2 - 10xy^3 + y^4$

36. $16x^4 + 8x^3y + 33x^2y^2 + 8xy^3 + 16y^4$

37. $9x^4 + 12x^3y - 2x^2y^2 - 4xy^3 + y^4$

38. $36x^4 + 36x^3y + 21x^2y^2 + 6xy^3 + y^4$

39. $16x^2 - 56xy + 8x + 49y^2 - 14y + 1$

40. $16x^2 + 24xy - 16x + 9y^2 - 12y + 4$

15 — Chapter Review

1. $x^2 - 9x + 20$

2. $2x^2 + 21x + 40$

3. $56x^2 + x - 1$

4. $10x^2 + 23xy - 5y^2$

5. $4y^2 - 29xy - 24x^2$

6. $5x^3 - 20x^2 + x - 4$

7. $7x^2 - 14x - 2xy + 4y$

8. $5x^3 - 8x^2 - 4x$

9. $x^5 - 9x^4 + 20x^3$

10. $81x^2 - 16$

11. $16x^8 - y^2$

12. $25x^{12}y^2 - 1$

13. $36x^2 + 12x + 1$

14. $64x^2 - 48xy + 9y^2$

15. $9x^6 - 60x^3y^4 + 100y^8$

16. $9x^4 - 30x^3 + 9x^2$

17. $343x^4 - 7x^2$

18. $30x^5 - 185x^4 + 30x^3$

19. $8x^5y^2 - 2x^4y^3 - x^3y^4$

20. $60x^{10} - 32x^9 + 4x^8$

21. $-x^4y - 6x^3y^2 - 9x^2y^3$

22. $16x^3y^2 + 48x^2y^3 + 36xy^4$

23. $-32x^7 + 48x^6 - 10x^5$

24. $36x^3 - 4x$

25. $48x^4 - 56x^3 + 16x^2$

26. $-3y^5 + 36y^4 - 108y^3$

27. $243x^3y^2 - 54x^2y^3 + 3xy^4$

28. $x^3 + x^2 - 19x - 4$

29. $10x^3 - 16x^2 - 38x - 12$

30. $12x^3 - 25x^2y + 8xy^2 + 3y^3$

31. $x^4 - x^3 - 2x^2 + x + 1$

32. $14x^4 + 35x^3y + 5x^2y^2 - 5xy^3 - y^4$

33. $x^4 - 19x^2 - 13x - 4$

34. $8x^4 - 14x^3 + 19x^2 - 24x + 5$

35. $x^4 - 16x^2 + 16x - 4$

36. $x^4 + 2x^3y - 11x^2y^2 + 12xy^3 - 9y^4$

37. $x^3 - 9x^2 + 27x - 27$

38. $-27x^3 - 81x^2 - 81x - 27$

39. $8x^3 + 60x^2y + 150xy^2 + 125y^3$

40. $x^6 - 9x^4y + 27x^2y^2 - 27y^3$

41. $16x^8 + 64x^6 + 96x^4 + 64x^2 + 16$

42. $81x^4 + 108x^3y^2 + 54x^2y^4 + 12xy^6 + y^8$

43. $x^4 + 6x^3 + 13x^2 + 12x + 4$

44. $x^{12} + 4x^9 + 20x^6 + 32x^3 + 64$

45. $9x^4 - 12x^3y + 40x^2y^2 - 24xy^3 + 36y^4$

46. $9x^4 + 42x^3y + 73x^2y^2 + 56xy^3 + 16y^4$

47. $9x^2 + 12x - 6xy + 4 - 4y + y^2$

15 — Cumulative Test

1. (15.1) $3x^2 + 35x + 72$

2. (15.2) $216x^4 - 6x^2$

3. (15.3) $x^3 + 3x^2 - 17x - 3$

4. (4.3) $18x^3y^6 - 21x^5y^2 - 6x^3y^4$

5. (2.3) $-120x^3$

6. (15.3) $81x^4 - 108x^3 + 54x^2 - 12x + 1$

7. (15.2) $10x^7 + 20x^5 - 240x^3$

8. (4.3) $4x^4 - x^3$

9. (15.3) $x^4 - 36x^2 + 48x - 16$

10. (2.3) $84xy^3$

11. (15.2) $72x^8 - 30x^7 + 3x^6$

12. (4.3) $-4x^8 + 36x^{17}$

13. (15.1) $6x^3 - 48x^2 + x - 8$

14. (15.2) $108x^7y^4 + 180x^5y^6 + 75x^3y^8$

15. (15.3) $16x^4 - 56x^3y + 121x^2y^2 - 126xy^3 + 81y^4$

16. (4.3) $-2x^4y^6 + 2x^5y^5 - 2x^6y^2$

17. (15.1) $21x^2 + 11xy - 2y^2$

18. (15.3) $25x^6 - 10x^5 - 39x^4 + 8x^3 + 16x^2$

19. (15.2) $90x^3 - 480x^2 + 640x$

20. (4.3) $40y^9 - 35y^7 + 15y^5$

21. (15.3) $216x^3 + 540x^2 + 450x + 125$

22. (15.1) $3x^3 - 14x^2 - 24x$

23. (15.2) $32x^5 - 260x^4 + 32x^3$

24. (15.3) $x^4 + 10x^3 - 21x^2 + 35x - 12$

25. (4.3) $8x^7y^2 - 64x^3y^7$

26. (15.2) $5x^2y^3 - 50xy^4 + 125y^5$

27. (2.3) $72x^4y^2$

28. (15.3) $24x^3 - 103x^2 + 23x + 20$

29. (15.1) $x^2 - 16y^2$

30. (15.2) $96x^6 - 76x^5 + 10x^4$

CHAPTER 16

16.1

1. (+1)(+6), (−1)(−6), (+2)(+3), (−2)(−3)

2. (1)(−15), (−1)(15), (3)(−5), (−3)(5)

3. (+1)(−10), (−1)(+10), (+2)(−5), (−2)(+5)

4. (1)(13), (−1)(−13)

5. (+1)(+21), (−1)(−21), (+3)(+7), (−3)(−7)

6. (1)(−25), (−1)(25), (5)(−5)

7. (+1)(−8), (−1)(+8), (+2)(−4), (−2)(+4)

8. (1)(27), (−1)(−27), (3)(9), (−3)(−9)

9. (+1)(+42), (−1)(−42), (+2)(+21), (−2)(−21), (+3)(+14), (−3)(−14), (+6)(+7), (−6)(−7)

10. (1)(−63), (−1)(63), (3)(−21), (−3)(21), (7)(−9), (−7)(9)

11. $(+1)(-16)$, $(-1)(+16)$, $(+2)(-8)$, $(-2)(+8)$, $(+4)(-4)$

12. $(1)(100)$, $(-1)(-100)$, $(2)(50)$, $(-2)(-50)$, $(4)(25)$, $(-4)(-25)$, $(5)(20)$, $(-5)(-20)$, $(10)(10)$, $(-10)(-10)$

13. $(+1)(+81)$, $(-1)(-81)$, $(+3)(+27)$, $(-3)(-27)$, $(+9)(+9)$, $(-9)(-9)$

14. $(1)(-60)$, $(-1)(60)$, $(2)(-30)$, $(-2)(30)$, $(3)(-20)$, $(-3)(20)$, $(4)(-15)$, $(-4)(15)$, $(5)(-12)$, $(-5)(12)$, $(6)(-10)$, $(-6)(10)$

15. $(+1)(-36)$, $(-1)(+36)$, $(+2)(-18)$, $(-2)(+18)$, $(+3)(-12)$, $(-3)(+12)$, $(+4)(-9)$, $(-4)(+9)$, $(+6)(-6)$

16. $(1)(80)$, $(-1)(-80)$, $(2)(40)$, $(-2)(-40)$, $(4)(20)$, $(-4)(-20)$, $(5)(16)$, $(-5)(-16)$, $(8)(10)$, $(-8)(-10)$

17. $(+1)(+120)$, $(-1)(-120)$, $(+2)(+60)$, $(-2)(-60)$, $(+3)(+40)$, $(-3)(-40)$, $(+4)(+30)$, $(-4)(-30)$, $(+5)(+24)$, $(-5)(-24)$, $(+6)(+20)$, $(-6)(-20)$, $(+8)(+15)$, $(-8)(-15)$, $(+10)(+12)$, $(-10)(-12)$

18. $(1)(-176)$, $(-1)(176)$, $(2)(-88)$, $(-2)(88)$, $(4)(-44)$, $(-4)(44)$, $(8)(-22)$, $(-8)(22)$, $(11)(-16)$, $(-11)(16)$

16.2

1. $(x+1)^2$ **2.** $(x+2)(x+1)$ **3.** $(x+1)(x-4)$ **4.** $(x-2)^2$ **5.** $(x+3)(x+2)$

6. $(x+2)(x-5)$ **7.** $(x+6)(x-2)$ **8.** $(x+7)(x-1)$ **9.** $(x-2)(x+7)$ **10.** $(x-8)(x-1)$

11. $(x+5)(x-1)$ **12.** N.F.* **13.** N.F. **14.** $(x+8)(x-3)$ **15.** $(x+4)(x+7)$

16. $(x-8)(x-4)$ **17.** $(x-9)(x+3)$ **18.** $(x+4)(x-11)$ **19.** N.F. **20.** $(x-8)^2$

21. $(x-1)(x+22)$ **22.** $(x-10)(x-2)$ **23.** $(x+18)(x-3)$ **24.** $(x+4)(x+16)$ **25.** $(x+18)(x-5)$

26. $(x-21)(x+4)$ **27.** $(x+3)(x-24)$ **28.** $(x+3)(x+17)$ **29.** $(x-3)(x+10)$ **30.** N.F.

16.3

1. $(x+7y)(x+y)$ **2.** $(x-2y)(x-4y)$ **3.** $(y-3x)(y+8x)$ **4.** $(x-5y)(x+y)$ **5.** $(y+8x)(y+x)$

6. $(x-2y)(x-9y)$ **7.** $(y+10x)(y-3x)$ **8.** $(x-12y)(x+4y)$ **9.** N.F.* **10.** $(x-5y)^2$

11. N.F. **12.** N.F. **13.** $(y+10x)(y+5x)$ **14.** $(y-12x)(y-6x)$ **15.** $(x-3y)(x+12y)$

16. N.F. **17.** $(y+12x)(y+4x)$ **18.** $(x-2y)(x-12y)$

16.4

1. $6(x-4)^2$ **2.** $5(y-6)(y+4)$ **3.** $8(x-4)(x-1)$

4. $15(x-1)(x-2)$ **5.** $9x(x-8)(x+2)$ **6.** $3xy(x+3)(x+4)$

7. $5x^3(y-3)(y-2)$ **8.** $x^2(y+4)(y-5)$ **9.** $2y(x^2+9x-8)$

10. $5x^2y(y^2+2y+3)$ **11.** $-3(x+6)^2$ **12.** $-x^2(x-1)(x+16)$

13. $-y^2(x-9)(x+2)$ **14.** $-3x(y-4)(y-5)$ **15.** $-x^2y^4(x+4)(x+6)$

16. $-2x^4y^4(x^2+18x+16)$ **17.** $-x^5y^3(y-12)(y-2)$ **18.** $-3x^3y^2(x+2)(x-14)$

19. $4(x+8y)(x+y)$ **20.** $8(x-3y)(x-y)$ **21.** $11(y-x)(y+4x)$

22. $xy(x-7y)(x+6y)$ **23.** $2y^2(y+5x)(y+x)$ **24.** $8x(x-y)^2$

*N.F. means not factorable in integers.

25. $2x^3(y+7x)(y-2x)$
26. $6xy^2(x-5y)(x+y)$
27. $4x^5y^2(y^2+9xy+24x^2)$
28. $3x^6y^4(y-3x)^2$
29. $5x^4y^5(y-2x)(y+5x)$
30. $40x^2y^2(x+5y)(x-6y)$
31. $-x^2(x+7y)(x+y)$
32. $-3y^2(x-4y)(x-2y)$
33. $-8xy(y^2+4xy-6x^2)$
34. $-5x^3y^2(y^2-3xy-12x^2)$
35. $-4x^4y^6(y+8x)(y+2x)$
36. $-9x^6y^2(x-3y)^2$

16.5

1. $(x-1)(x+9)$
2. $(x+3)(x-8)$
3. $-(x-3)(x-8)$
4. $-(x-3)(x-6)$
5. $-(x+3)(x-11)$
6. $-(x+4)(x-10)$
7. $(x-3)(x+10)$
8. $(x-2)(x+12)$
9. $(x+3)(x+10)$
10. $-(x-6)^2$
11. $-(x-2)^2$
12. $-(x-7)^2$
13. $-(y-x)(y+10x)$
14. $-(y+4x)(y-7x)$
15. $(x+2y)(x+3y)$
16. $(x-3y)^2$
17. $-(y+x)(y+6x)$
18. $-(x-2y)(x-7y)$
19. $-(y-6x)(y+9x)$
20. $(x-6y)(x+8y)$
21. $-(x+5y)(x-8y)$
22. $-(x-3y)(x-6y)$
23. $(y+4x)(y-10x)$
24. $-(y+5x)(y-16x)$

16.6

1. $(7x+4)(7x-4)$
2. $(8x+9)(8x-9)$
3. $(3x+7)(3x-7)$
4. $(6x+5)(6x-5)$
5. $(10x+9y)(10x-9y)$
6. $(4x+11y)(4x-11y)$
7. $(8xy+3)(8xy-3)$
8. $(12xy+1)(12xy-1)$
9. $(10x^3+7y^4)(10x^3-7y^4)$
10. $(2x^2+y^5)(2x^2-y^5)$
11. $(y^4+6x^3)(y^4-6x^3)$
12. $(9y^4+2x^6)(9y^4-2x^6)$
13. $x(x+9)(x-9)$
14. $2x^7(2x^2+11)(2x^2-11)$
15. $y^4(20y^2+3)(20y^2-3)$
16. $2x^3(6x^2+1)(6x^2-1)$
17. $2x^3(8+3x^3)(8-3x^3)$
18. $xy^2(5x^3y^3+1)(5x^3y^3-1)$
19. $3y^4(11x^2+2)(11x^2-2)$
20. $2x^2y^2(6+7y^3)(6-7y^3)$
21. $2xy(9+10x^3y^2)(9-10x^3y^2)$
22. $9x^8y^5(4y^2+5x)(4y^2-5x)$
23. $8x^2y^6(2x^3+3y)(2x^3-3y)$
24. $x^6y^9(7x^5+13y^3)(7x^5-13y^3)$

16.7

1. $(5x-6)(5x-3)$
2. $(3x-4)(7x+3)$
3. $(x+5)(22x-5)$
4. $(5x-6)(3x+8)$
5. $(2x+15)(3x-1)$
6. $(11x+7)(2x+1)$
7. $(7x-3)(2x+1)$
8. $(8x-13)(x+1)$
9. $(x+1)(10x-9)$
10. $(5x+7)(2x+3)$
11. $(9x-4)(3x-1)$
12. $(x-1)(20x+33)$
13. $(3x+1)(x+7)$
14. $(5x-2)(x+1)$
15. $(x-1)(11x-5)$
16. $(5x+1)(x-7)$
17. $(3x+1)(x+13)$
18. $(x+11)(2x-1)$
19. $(3x-2)(2x-1)$
20. $(3x+5)(2x+1)$
21. $(6x-1)(2x+1)$
22. $(2x+5)(7x-1)$
23. $(7x-3)(x-3)$
24. $(3x+11)(4x-1)$
25. $(5x-4)(3x-1)$
26. $(2x+5)(11x-2)$
27. $(3x+5)(2x-5)$
28. $(7x+2)(2x+5)$
29. $(9x-11)(x-3)$
30. $(6x-7)(x-3)$
31. $(3x+4)(2x+5)$
32. $(2x-1)(5x+8)$
33. $(2x-3)(9x-7)$
34. $(5x+3)(4x-3)$
35. $(4x+9)(x-2)$
36. $(7x-4)(2x-3)$
37. $(x+3)(9x-10)$
38. $(8x+11)(x+2)$
39. $(3x-2)(9x-5)$
40. $(3x+1)(2x-27)$
41. $(7x-6)(3x+5)$
42. $(2x+1)(7x+8)$
43. $(9x-5)(2x+5)$
44. $(3x-5)(11x+8)$
45. $(5x-4)(7x-4)$
46. $(12x+5)(x+6)$
47. $(3x+4)(9x-7)$
48. $(3x-2)(4x-5)$
49. $3(2x-1)(x+3)$
50. $4(3x+1)(x+5)$
51. $5(3x+1)(x+7)$
52. $3(3x-5)(x-3)$
53. $9(8x^2-4x-3)$
54. $6(3x^2-9x-8)$
55. $10(x-1)(8x+5)$
56. $3(4x-1)(5x-3)$
57. $9(2x-1)(3x-10)$
58. $15(2x-3)(2x+5)$
59. $5(2x+15)(5x-1)$
60. $6(9x^2-19x+11)$

16 – Chapter Review

1. (+1)(+4), (−1)(−4), (+2)(+2), (−2)(−2) 2. (+1)(−9), (−1)(+9), (+3)(−3) 3. (+1)(+11), (−1)(−11)

4. (+1)(−12), (−1)(+12), (+2)(−6), (−2)(+6), (+3)(−4), (−3)(+4)

5. (+1)(+105), (−1)(−105), (+3)(+35), (−3)(−35), (+5)(+21), (−5)(−21), (+7)(+15), (−7)(−15)

6. (+1)(−90), (−1)(+90), (+2)(−45), (−2)(+45), (+3)(−30), (−3)(+30), (+5)(−18), (−5)(+18), (+6)(−15), (−6)(+15), (+9)(−10), (−9)(+10)

7. (+1)(+126),(−1)(−126), (+2)(+63), (−2)(−63), (+3)(+42), (−3)(−42), (+6)(+21), (−6)(−21), (+7)(+18), (−7)(−18), (+9)(+14), (−9)(−14)

8. (+1)(−72), (−1)(+72), (+2)(−36), (−2)(+36), (+3)(−24), (−3)(+24), (+4)(−18), (−4)(+18), (+6)(−12), (−6)(+12), (+8)(−9), (−8)(+9)

9. (+1)(+270), (−1)(−270), (+2)(+135), (−2)(−135), (+3)(+90), (−3)(−90), (+5)(+54), (−5)(−54), (+6)(+45), (−6)(−45), (+9)(+30), (−9)(−30), (+10)(+27), (−10)(−27), (+15)(+18), (−15)(−18)

10. $(x+2)(x+3)$ 11. $(x-3)(x+5)$ 12. $(x-3)(x-16)$

13. $(x+1)(x-9)$ 14. $(x+3)(x-18)$ 15. N.F.*

16. $(x-3)(x+8)$ 17. $(x+6)^2$ 18. $(x-9)(x-12)$

19. $(x+2)(x-9)$ 20. N.F. 21. $(x-4)(x+20)$

22. N.F. 23. $(x+8)^2$ 24. $(x-4)(x-14)$

25. $(x+2y)(x+9y)$ 26. $(y+3x)(y-12x)$ 27. N.F.

28. $(x-9y)^2$ 29. N.F. 30. $(y+8x)(y+11x)$

31. $(y+2x)(y+32x)$ 32. $(x+3y)(x-32y)$ 33. $(y+5x)(y+24x)$

34. $4(x+8)(x-1)$ 35. $30(y+4)(y+1)$ 36. $4y(x-8)(x+4)$

37. $x^3(x+6)(x-3)$ 38. $7x^4(x^2-7x+9)$ 39. $-5(x-3)(x-5)$

40. $-2y(y^2+16y+80)$ 41. $-x^2y^6(y-15)(y+4)$ 42. $-2x^5y^6(x+18)(x-1)$

43. $12(x-y)(x-14y)$ 44. $x^6y^2(x+2y)(x+13y)$ 45. $-6x^3(x^2-6xy-12y^2)$

46. $-9(y-2x)(y+9x)$ 47. $x^4y^4(y-6x)(y+7x)$ 48. $4x^2y^3(y-x)(y-8x)$

49. $16(x+3y)^2$ 50. $-x^9y(y^2-10xy-33x^2)$ 51. $8x^3y^2(y-2x)(y+8x)$

52. $-(x-2)(x-16)$ 53. $-(x+2)(x-22)$ 54. $(x-3)(x+6)$

55. $-(x-11)^2$ 56. $-(x-4)(x+13)$ 57. $(x+3y)(x+16y)$

58. $-(y-6x)(y-11x)$ 59. $-(x-2y)(x-3y)$ 60. $-(x-5y)(x-12y)$

61. $(x+5y)(x+7y)$ 62. $-(x-2)(x+12)$ 63. $-(y+2x)(y-8x)$

64. $(3x+1)(3x-1)$ 65. $(6x+7)(6x-7)$ 66. $(9x+8y)(9x-8y)$

67. $(7xy+2)(7xy-2)$ 68. $(12x^3+13y^2)(12x^3-13y^2)$ 69. $(15y^4+9x^5)(15y^4-9x^5)$

70. $2x^2(3x+4)(3x-4)$ 71. $2x^4(6x+7)(6x-7)$ 72. $3x^4(3x+4)(3x-4)$

73. $8x(2x+3y)(2x-3y)$ 74. $2x^5(7y^2+x^3)(7y^2-x^3)$ 75. $9x^3y(4xy+5)(4xy-5)$

76. $(3x+2)(2x-15)$ 77. $(5x-4)(11x-8)$ 78. $(7x+5)(2x-9)$

*N.F. means not factorable in integers.

79. $(12x-5)(15x-4)$
80. $(2x-3)(21x+12)$
81. $(4x+9)(6x+25)$
82. $(3x-1)(x+4)$
83. $(7x-11)(x-1)$
84. $(6x-5)(x+3)$
85. $(7x+5)(2x+5)$
86. $(35x-26)(x+1)$
87. $(3x-22)(2x-11)$
88. $(9x-19)(x+2)$
89. $(4x+5)(2x+3)$
90. $(27x-2)(x+6)$
91. $(2x-3)(6x-5)$
92. $(4x+5)(6x-5)$
93. $(3x+14)(6x+7)$
94. $(8x-9)(2x+3)$
95. $(7x-81)(x-1)$
96. $(10x+3)(12x-1)$
97. $(24x+25)(x-5)$
98. $(56x-9)(2x+3)$
99. $12(2x-3)(5x-3)$
100. $6(x-1)(5x-3)$
101. $4(2x+5)(5x-1)$
102. $7(x-6)(7x+2)$
103. $8(5x^2-17x+20)$
104. $2(x+7)(10x+3)$
105. $12(4x^2+9x-6)$

16 – Cumulative Test

1. (16.2) $(x+6)(x-8)$
2. (16.3) $(x+6y)(x+12y)$
3. (16.4) $4(x+3)(x-36)$
4. (14.1) $9(3x^3-1)$
5. (16.7) $(3x-4)(2x+5)$
6. (16.6) $(7x+5)(7x-5)$
7. (16.2) $(x-3)(x-32)$
8. (16.6) $2(5x+8)(5x-8)$
9. (16.4) $10(y-2)(y+48)$
10. (16.3) $(x-7y)(x+14y)$
11. (14.2) $x^3y^3(36x^2-25y)$
12. (16.7) $7(2x-3)(4x+1)$
13. (16.6) $(2y^2+9x)(2y^2-9x)$
14. (16.4) $12(y-4x)(y+18x)$
15. (14.1) $7(4x^3+9x+8)$
16. (16.2) $(x+7)(x+12)$
17. (16.6) $2x^5y^3(11x+6y)(11x-6y)$
18. (16.7) $(16x+3)(x-2)$
19. (14.1) $-16(x^2+3y^2+4z^2)$
20. (16.7) $(14x-5)(x+5)$
21. (16.6) $x(x+12)(x-12)$
22. (16.3) $(x+12y)(x+13y)$
23. (16.4) $6x^2y(y-12x)(y+15x)$
24. (14.2) $x^3y(4+xy^2-4xy^4)$
25. (16.7) $(7x+2)(3x-16)$
26. (16.2) $(x-4)(x-24)$
27. (16.6) $(3x^3+10y^4)(3x^3-10y^4)$
28. (16.4) $x^6y^3(x+3y)(x+45y)$
29. (16.7) $12(x+1)(8x-9)$
30. (14.1) $6(6y^3-9x^2-7z)$
31. (16.3) $(y-6x)(y+25x)$
32. (14.2) $35x^4y^2(1-2x+3xy^4)$
33. (16.6) $4y^3(6y+5)(6y-5)$
34. (16.7) $(x-1)(18x-25)$
35. (16.4) $-3x^2(x+6)(x-21)$
36. (16.6) $(11y^2+4x^3)(11y^2-4x^3)$
37. (16.2) $(x+5)(x-36)$
38. (16.7) $(6x+5)(6x-25)$
39. (14.2) $x^3y^2(15y^4+16x^5-1)$
40. (16.3) $(y+6x)(y+16x)$

CHAPTER 17

17.1

1. $\frac{x+3}{3}$
2. $\frac{2x}{x-4}$
3. $\frac{7+2x}{8xy^2}$
4. $\frac{7x^3}{2(3x-5)}$
5. $\frac{7x^7y^2}{6+5y}$
6. $\frac{3(8x-5)}{10x^4}$
7. $\frac{7x+4}{x^4}$
8. $\frac{x}{9x-8}$
9. $9x^2+5x^3$
10. $\frac{4x+1}{x^2}$
11. $\frac{x(-2+9x)}{y^3}$
12. $\frac{x^3}{y^2(5+6y)}$
13. $15x^4+5x^3$
14. $\frac{5x^2}{2(-4+x)}$
15. $\frac{3x^3(4x^2-3)}{8}$
16. $\frac{y(7x-2)}{7x^4}$
17. $\frac{16x^3}{3(8+5x)}$
18. $\frac{3x^2(2+3x)}{4y^4}$

17.2

1. $\frac{1}{3}$
2. $\frac{5}{4}$
3. $\frac{2x-3}{5x-6}$
4. $\frac{3(x-3)}{6x-17}$
5. $\frac{8x}{9}$
6. $\frac{2x}{3}$
7. x^3
8. x^4
9. $\frac{2x-3}{2(x-3)}$
10. $\frac{3(x-4)}{4x^2(3x-4)}$
11. $\frac{2y-3}{2x^2(2xy-3)}$
12. $\frac{y^2(3y-5)}{3x^2(3x-5)}$
13. $\frac{x^2}{y^2}$
14. $\frac{8x^5(xy+3)}{9y^2(3xy+1)}$
15. $\frac{3xy^4}{2}$
16. $\frac{y^2}{5x^3}$
17. $\frac{4x^4(xy-2)}{3y^2(3y-1)}$
18. $\frac{3y^4}{2x^5}$
19. $\frac{-5}{4}$
20. $\frac{-2}{7}$
21. $\frac{-3}{8x}$
22. $\frac{2y(2y-5)}{3x(5y-2)}$
23. $-x$
24. $\frac{-x^5}{y^2}$
25. $\frac{-x^3}{y^4}$
26. $\frac{-y}{2x}$
27. $\frac{-4x^3}{9y^4}$
28. $\frac{-3y}{2x}$
29. $\frac{4y^3(2y-1)}{5x^4(x-2y)}$
30. $\frac{-4x^4y^3}{5}$

17.3

1. $\frac{1}{x+2}$
2. $\frac{1}{x-2}$
3. $\frac{1}{x-1}$
4. $x+4$
5. $\frac{x-8}{4}$
6. $\frac{6}{x-2}$
7. $\frac{4}{x-7}$
8. $\frac{9}{x+4}$
9. $\frac{x^4}{x-4}$
10. $\frac{x-6}{x^7}$
11. $\frac{6x^4}{x+2}$
12. $\frac{x-11}{5x^3}$
13. $\frac{7}{4(x-4)}$
14. $\frac{2(x-2)}{3}$
15. $\frac{3(x+11)}{4x^2}$
16. $\frac{2(x+8)}{3x}$
17. $\frac{x+3}{x+1}$
18. $\frac{x-5}{x-6}$
19. $\frac{x+3}{x-2}$
20. $\frac{x-7}{x-9}$
21. $\frac{x-4}{x-6}$
22. $\frac{x-15}{x-9}$
23. $\frac{x-11}{x-12}$
24. $\frac{x+7}{x-8}$
25. $\frac{x+5}{x+4}$
26. $\frac{x-7}{x-2}$
27. $\frac{x+5}{x-6}$
28. $\frac{x-2}{x+11}$
29. $\frac{3(x-8)}{4(x-1)}$
30. $\frac{3(x-4)}{x+3}$
31. $\frac{2(x+6)}{3x^2(x+8)}$
32. $\frac{9(x-9)}{2x^3(x-5)}$
33. $\frac{3(x-8)}{5(x-10)}$
34. $\frac{7(x+15)}{6(x+5)}$
35. $\frac{2(x-2)}{x(x+7)}$
36. $\frac{x+14}{2x^3(x+4)}$

17.4

1. $\frac{7}{20}$
2. $\frac{16}{21}$
3. $\frac{3}{x^4}$
4. $\frac{2}{5}$
5. $\frac{9(x+3)}{4x^7(2x+9)}$
6. $\frac{7x^8(x-3)}{2(3x-7)}$
7. $\frac{7x+9}{7x^6}$
8. $\frac{5}{4x^2(4x+3)}$
9. $\frac{7x^4(5x-2)}{3}$
10. $\frac{35(3x-4)}{12x^2}$
11. $\frac{4x^3(2x-3)}{27}$
12. $\frac{2x+7}{3x}$
13. $\frac{3x-11}{8x(11x-5)}$
14. $\frac{15x(7x-2)}{14(9x-4)}$
15. $\frac{9}{20x^3}$
16. $\frac{15x}{14}$
17. $\frac{14(5x+8)}{27x^2(3x+7)}$
18. $\frac{5x^5(4x+5)}{24(2x+3)}$
19. $\frac{2(3y-5)}{9x^3}$
20. $\frac{11y^3}{5x^2(3y+10x)}$

21. $\frac{1}{3x^2y^5(4x-9)}$
22. $\frac{x^9}{4y^4(5x-4)}$
23. $\frac{4(7x+2y)}{x^6y^3(9xy+2)}$
24. $\frac{35x^7}{12}$
25. $(x+7)^2$
26. $(x-8)^2$
27. $\frac{1}{(x-18)(x+26)}$
28. $\frac{1}{(x+15)(x-24)}$
29. $\frac{(x-4)(x-8)}{x+14}$
30. $\frac{(x-4)(x-6)}{x+6}$
31. $\frac{x-12}{(x-11)(x+12)}$
32. $\frac{x+24}{(x-24)(x-32)}$
33. $\frac{x-5}{x-15}$
34. $\frac{x-8}{x-4}$
35. $\frac{(x+4)(x-7)}{(x+7)(x-14)}$
36. $\frac{(x+10)(x-15)}{(x+5)(x-6)}$

17.5

1. $x+9$, R(22)*
2. $x+6$, R(11)
3. $x-9$, R(30)
4. $x+3$, R(−2)
5. $3x+7$, R(29)
6. $3x-4$
7. $6x-9$, R(19)
8. $4x-17$, R(105)
9. $3x$, R(−4)
10. $2x-2$, R(−25)
11. $11x-4$, R(−10)
12. $3x+16$, R(124)
13. $x-5$, R(30)
14. $x+6$, R(40)
15. $x-3$
16. x
17. $3x+12$, R(41)
18. $7x+24$, R(72)
19. $6x-12$
20. $3x+15$
21. x^2-2x-2, R(2)
22. $x^2-6x-14$, R(−42)
23. $x^2+4x+12$, R(56)
24. x^2+3x+6, R(27)
25. $2x+3$, R(−4)
26. $2x-3$, R(−14)
27. $3x+2$, R(16)
28. $2x-5$, R(9)
29. $3x-5$, R(−11)
30. $3x-2$, R(26)
31. $x+2$, R(−7)
32. $3x-5$, R(11)
33. $4x+7$, R(31)
34. $3x+1$, R(9)
35. $3x-2$, R(−5)
36. $11x-5$, R(9)
37. $2x+1$, R(−6)
38. $6x-15$, R(80)
39. $6x-2$, R(6)
40. $8x-6$, R(2)
41. $3x+7$, R(21)
42. $4x+2$, R(2)
43. $4x^2-10x+25$, R(−140)
44. $5x^2+3x+2$, R(2)
45. $4x^2+7x+10$, R(83)
46. $10x^2-15x+10$, R(−34)
47. $3x^2+5x-2$, R(11)
48. $4x^2+9x+12$, R(30)

17 – Chapter Review

1. $\frac{2x+3}{4}$
2. $\frac{-3x}{5(x-2)}$
3. $\frac{3x^3}{2(x-2)}$
4. $\frac{3x^3}{2(2x-1)}$
5. $\frac{x(-2+9x)}{3}$
6. $\frac{2y^3}{x(4y+5x)}$
7. $\frac{3x^2(3y+5x)}{7y^4}$
8. $\frac{y^2}{x(5x-3y)}$
9. $\frac{y(2xy+9)}{x}$
10. $\frac{1}{2}$
11. $\frac{x^5(x+6)}{6(x+4)}$
12. $\frac{3x^8}{5}$
13. $\frac{x^5}{y^4}$
14. $\frac{2x^7(4x-7)}{5(7+4x)}$
15. $\frac{7}{4x^7}$
16. $\frac{1}{x^4y^4}$
17. $\frac{3y^4}{2x^5}$
18. $\frac{y^2(3y-5)}{x^2(9x-2)}$
19. $\frac{-3}{4}$
20. $-x^3$
21. $\frac{y(2-xy)}{2x^2(y-2x)}$
22. $\frac{-x^4}{4y^5}$
23. $\frac{-3y}{2x}$
24. $\frac{3x^5(4x-3y)}{2y^5(3y+4x)}$
25. $x-4$
26. $\frac{1}{x+3}$
27. $\frac{x-5}{8}$
28. $\frac{x-7}{6}$

*R means Remainder.

29. $\frac{3x^2}{x+7}$ 30. $\frac{4x^8}{x+8}$ 31. $\frac{5}{4x(x-3)}$ 32. $\frac{x}{(x+10)}$

33. $\frac{x+4}{x+8}$ 34. $\frac{x-1}{x+6}$ 35. $\frac{x-14}{x+7}$ 36. $\frac{x+8}{x+10}$

37. $\frac{x+8}{x+3}$ 38. $\frac{x+8}{x-8}$ 39. $\frac{2x(x-2)}{3(x+4)}$ 40. $\frac{x^2(x-1)}{4(x-12)}$

41. $\frac{x+9}{2x^3(x+4)}$ 42. $\frac{5(x-7)}{6x^2(x+8)}$ 43. $\frac{x^2}{2}$ 44. $\frac{4x^5}{15}$

45. $\frac{3x+7}{11(2x+5)}$ 46. $\frac{3}{2x^5(2x-3)}$ 47. $\frac{5x+3}{2x^3}$ 48. $\frac{3(x-2)}{8x}$

49. $\frac{10(7x-6)}{9x^6(3x-4)}$ 50. $\frac{99}{40x}$ 51. $\frac{9(3x+5)}{40x^7(2x+3)}$ 52. $\frac{2(5y-14)}{5x^3y}$

53. $\frac{x^9}{5y^5}$ 54. $\frac{5y^5(4x+3y)}{2x^4(3xy+4)}$ 55. $(x+9)^2$ 56. $\frac{1}{(x-12)(x+30)}$

57. $\frac{(x-6)(x-8)}{x+12}$ 58. $\frac{x-14}{(x+14)(x+16)}$ 59. $\frac{x+8}{x-12}$ 60. $\frac{(x-8)(x-9)}{(x-4)(x+9)}$

61. $x-6$, R(−14)* 62. $4x+10$, R(21) 63. $8x-26$, R(85) 64. $4x-9$, R(−8)

65. $3x-8$, R(−39) 66. $7x+1$ 67. $x-6$, R(12) 68. $x-10$, R(50)

69. x^2-2x, R(7) 70. $x^2+5x-17$, R(17) 71. $6x^2-7x+7$ 72. $4x^2$, R(16)

73. $4x-1$, R(10) 74. $4x-3$, R(−30) 75. $4x-3$, R(−9) 76. $4x+3$, R(−32)

77. $4x+3$, R(−2) 78. $4x+1$, R(8) 79. $3x+2$, R(3) 80. $5x-6$, R(36)

81. $3x+2$, R(−2) 82. $6x^2+8x+7$, R(23) 83. $9x^2-24x+64$, R(−12) 84. $7x^2-4x+5$

17 – Cumulative Test

1. (17.1) $\frac{2x+3}{4}$ 2. (17.2) $\frac{3}{4}$ 3. (17.3) $\frac{1}{x-8}$ 4. (17.4) $\frac{3x(6x-11)}{4}$

5. (8.3) $\frac{-1}{3x^4}$ 6. (17.3) $\frac{x-12}{5}$ 7. (8.5) $\frac{3}{x^2}$ 8. (8.4) $\frac{x^7}{y}$

9. (17.1) $\frac{y(4x^2-5)}{3x^3}$ 10. (8.5) $\frac{15x^3}{28y^2}$ 11. (17.4) $\frac{2x^7}{3}$ 12. (8.3) $\frac{8x^2}{7y^2}$

13. (17.2) $\frac{x^5(4-7x)}{5+8x}$ 14. (17.4) $\frac{3}{2x^2(7x-2)}$ 15. (8.4) $\frac{-2}{x^7}$ 16. (17.1) $\frac{9x^6}{5x-2}$

17. (8.5) $\frac{32y^2z}{3x^2}$ 18. (17.4) $\frac{x^4y^2(5y-9)}{12}$ 19. (17.3) $\frac{x+12}{x+4}$ 20. (17.2) $\frac{13}{9y^3}$

21. (17.4) $\frac{11(3x-7)}{7x^8(3x-5)}$ 22. (17.3) $\frac{x-16}{x-32}$ 23. (17.4) $\frac{x+6}{x-7}$ 24. (8.3) $\frac{2z^2}{3x^2}$

25. (8.4) $\frac{-7x^4}{36}$ 26. (17.4) $\frac{30xy^4}{7(6x+25y)}$ 27. (17.2) $\frac{x^4}{2}$ 28. (8.4) $\frac{9x^5}{10y^8}$

29. (17.1) $\frac{x(-2x+7)}{6}$ 30. (17.4) $\frac{25}{2x^2y^6(7x-3y)}$

* R means Remainder.

CHAPTER 18

18.1

1. $6x^2(3x+2)$
2. $10x^3(2x+5)$
3. $8x^3(3x-2)$
4. $9x^2(6x-5)$
5. $15xy^2(2x-7y)$
6. $16x^2y(3x+5y)$
7. $3(2x-5)$
8. $3(3x-4)$
9. $3x(4x-9)(5x-9)$
10. $4x(4x+5)(4x+7)$
11. $72x^4(6x-11)$
12. $84x^5(8x+3)$
13. $60x^5(6x+5)(6x-5)$
14. $72x^7(3x+7)(3x-7)$
15. $10x^3y^3(3x-10y)$
16. $42x^4y^3(12x-5y)$
17. $9x^4y^3(3x-7y)(4x-7y)$
18. $8x^3y^4(9x+2y)(7x+2y)$
19. $(x+1)(x-3)(x+5)$
20. $(x-2)(x-7)(x-12)$
21. $(x+4)(x+5)(x-13)$
22. $(x-2)(x+7)(x+9)$
23. $(x-2)(x-9)(x-18)$
24. $(x-3)(x+7)(x+16)$
25. $(x+8)(x+10)(x+12)$
26. $(x-12)^2(x+16)$
27. $(x+4)(x-4)(x-5)$
28. $(x+6)(x-6)(x-12)$
29. $(x+7)^2(x-7)$
30. $(x+2)(x-2)(x+30)$

18.2

1. $\frac{x^2+2x+2}{2(x+2)}$
2. $\frac{-7x+3y}{4(3x+y)}$
3. $\frac{3x^2-19x-1}{3x(4x+1)}$
4. $\frac{56x^2+2x-25}{8x(6x+5)}$
5. $\frac{4x^2+18x+9}{3(x+3)}$
6. $\frac{-8x^2+38x+39}{6(2x-9)}$
7. $\frac{4x-21}{6x^2(2x-3)}$
8. $\frac{-9x+112}{14x(x-7)}$
9. $\frac{x^2+2x-2}{30x^3(x+2)}$
10. $\frac{-24x^2-59x}{40y(8x+3)}$
11. $\frac{x^2}{40y^2(x+8)}$
12. $\frac{x^2-4x-32}{120x^2(x-4)}$
13. $\frac{-15x+56y-56}{16x^2y(7x-8)}$
14. $\frac{36x-2y}{45x^2y^3(6x-7y)}$
15. $\frac{18x^2-31x-129}{36(9x-2)}$
16. $\frac{-12x^2-5x+133}{30x^7(3x+5)}$
17. $\frac{53x^2-9x+1}{30x^5(6x-1)}$
18. $\frac{64x^2+51y^2}{32y^2(8x-3y)}$
19. $\frac{2x^2+12x+15}{(2x+5)(x+3)}$
20. $\frac{-84x^2-40x+48}{(7x-4)(11x-6)}$
21. $\frac{5x^2+6x+1}{(2x-1)(x+3)}$
22. $\frac{2x^2+x+48}{(3x-1)(x+6)}$
23. $\frac{8x^2+52xy+35y^2}{(2x-7y)(2x+7y)}$
24. $\frac{22x^2-56xy-74y^2}{(2x+9y)(8x-7y)}$
25. $\frac{-5x^2+x-35}{12(2x+5)(x+3)}$
26. $\frac{18x^2-15x-10}{7(9x-2)(3x-10)}$
27. $\frac{5x^2-29x+8}{15(2x-1)(x-4)}$
28. $\frac{-16x^2+25x+42}{7(3x-1)(x+6)}$
29. $\frac{2x^2+7x-10}{12(x-2)(x+2)}$
30. $\frac{21x^2+10x-1}{15(2x-5)(3x+4)}$
31. $\frac{6}{5(x+1)}$
32. $\frac{8x}{5(3x+4)}$
33. $\frac{-1}{12(x+3)}$
34. $\frac{3x}{2(x-5)}$
35. $\frac{-x+6}{3(3x+1)}$
36. $\frac{x-3}{6(3x-2)}$
37. $\frac{19x-15}{12(6x+5)}$
38. $\frac{21x+76}{63(4x-3)}$
39. $\frac{50x+89}{7(5x-2)}$
40. $\frac{40x+49}{4(2x+7)}$
41. $\frac{-2x-15}{18(9x-4)}$
42. $\frac{37x+4}{96(3x-8)}$
43. $\frac{1}{2(x-4y)}$
44. $\frac{-62y}{5(2x-5y)}$
45. $\frac{10x+105y}{14(8x-5y)}$
46. $\frac{83}{20(4x-9)}$
47. $\frac{-43y}{18(3x-14y)}$
48. $\frac{-20x+8y}{75(4x-y)}$
49. $\frac{-3x-2y}{2(x-9y)}$
50. $\frac{-6x-5}{5(8x-1)}$
51. $\frac{-46x+17y}{84(2x-5y)}$
52. $\frac{6x-13}{10(8x-13)}$

53. $\dfrac{5x+67y}{30(16x-21y)}$ 54. $\dfrac{5x+47y}{135(6x-5)}$

18.3

1. $\dfrac{x+3}{(x+2)(x-3)}$ 2. $\dfrac{3x+11}{(x+1)(x-4)}$ 3. $\dfrac{5x-16}{(x+3)(x+4)}$ 4. $\dfrac{-8x-49}{(x+2)(x-7)}$

5. $\dfrac{10x-11}{(x-3)(x+10)}$ 6. $\dfrac{30}{(x+4)(x+12)}$ 7. $\dfrac{3x-29}{(x+3)(x-8)(x-9)}$ 8. $\dfrac{9x-9}{(x+4)(x+7)(x-11)}$

9. $\dfrac{4x^2-10x-24}{(x-1)(x+4)(x-9)}$ 10. $\dfrac{-2x^2-7x-63}{(x+2)(x-7)(x+8)}$ 11. $\dfrac{-8x^2+124x-144}{(x-8)(x-12)(x-14)}$ 12. $\dfrac{-66x-7}{(8x+1)(8x-1)}$

13. $\dfrac{14x-26}{(2x+5)(2x-5)}$ 14. $\dfrac{36x+25}{(7x+8)(7x-8)}$ 15. $\dfrac{10}{(x-2)(x+8)(x-8)}$ 16. $\dfrac{-3x+113}{(x+2)(x+15)(x-15)}$

17. $\dfrac{19x+100}{(x+1)(x-1)(x+28)}$ 18. $\dfrac{9x^2+116x+32}{(x+4)(x+12)(x-12)}$ 19. $\dfrac{-5x^2-156x+8}{(x+32)(x-2)(x+2)}$ 20. $\dfrac{-21x^2-155x-104}{(x+8)(x-8)^2}$

18.4

1. $2y-3$ 2. $\dfrac{2-3x}{2x^2y}$ 3. $\dfrac{5y}{3x}$ 4. $\dfrac{6x-1}{9}$ 5. $\dfrac{4}{3x}$

6. $\dfrac{-7}{4y}$ 7. $\dfrac{3}{x-3}$ 8. $-\dfrac{x+4}{12x}$ 9. $\dfrac{9}{35}$ 10. $\dfrac{-48}{25}$

11. $\dfrac{-18+7x}{28x}$ 12. $\dfrac{5x-42}{9x}$ 13. $\dfrac{5(2x+5)}{12}$ 14. $\dfrac{96x-5}{18}$ 15. $\dfrac{-4}{5}$

16. 6 17. 6 18. $\dfrac{-5x}{24}$ 19. $-\dfrac{1-x}{x}$ 20. $\dfrac{3(x-2)}{x}$

21. $\dfrac{2(7x-4)}{5x-2}$ 22. $\dfrac{2(5y-3)}{y+1}$ 23. $\dfrac{3x^2}{-x+4}$ 24. $\dfrac{2y^2}{-y+4}$ 25. $\dfrac{x}{16}$

26. $\dfrac{-x}{3}$ 27. $\dfrac{5y(3y-5)}{2(9y-5)}$ 28. x 29. $\dfrac{5(x-4)}{4(x-5)}$ 30. $\dfrac{4y}{6y-1}$

31. $\dfrac{5x-9}{3(4x-9)}$ 32. $\dfrac{7x+8}{3(5x+4)}$ 33. $\dfrac{y+7}{2}$ 34. $\dfrac{8(x+1)}{x}$ 35. $\dfrac{x-2}{5x}$

36. $\dfrac{x-6}{4x}$

18 — Chapter Review

1. $15x^2(5x+3)$ 2. $12x^3(3x-2)$ 3. $18x^2y^2(7x+4y)$ 4. $2(3x-7)$

5. $6x(3x-5)(4x-5)$ 6. $240x^6(5x-9)$ 7. $66x^5(9x+5)(9x-5)$ 8. $117x^6y^6(3x+4y)$

9. $12x^4y^4(3x+5y)(4x+5y)$ 10. $(x-4)(x-5)(x+9)$ 11. $(x+3)(x-10)(x-20)$ 12. $(x-2)(x-8)(x+10)$

13. $(x+4)(x-7)(x+10)$ 14. $(x+9)(x-9)(x+14)$ 15. $(x+12)(x-12)^2$ 16. $\dfrac{4x^2+10x-15}{5(2x+5)}$

17. $\dfrac{9x^2-32x}{5(3x-4)}$ 18. $\dfrac{9x^2-46x+9}{6(x-9)}$ 19. $\dfrac{-64x-49}{8x^3(4x+7)}$ 20. $\dfrac{22x^3+55x^2+15}{70x(2x+5)}$

21. $\frac{39x^2 + 73x}{30y(3x - 4)}$ **22.** $\frac{-78x + 13}{60x^3(2x - 5)}$ **23.** $\frac{16x^2 - 74x - 9}{48x^2(2x + 9)}$ **24.** $\frac{-12x^2 - 41x + 50}{15x^8(3x - 1)}$

25. $\frac{30x^2 - 45x + 70}{(8x - 7)(6x + 7)}$ **26.** $\frac{32x^2 - 25x - 25}{(2x - 5)(4x - 5)}$ **27.** $\frac{18x^2 + 15x - 50}{(3x + 5)(2x + 5)}$ **28.** $\frac{-4x^2 + 31x - 40}{9(x - 4)(3x - 8)}$

29. $\frac{30x^2 - 20x - 25}{8(3x - 5)(4x - 5)}$ **30.** $\frac{5x}{6(3x + 5)(2x + 5)}$ **31.** $\frac{7}{3(2x + 5)}$ **32.** $\frac{101x}{40(2x - 1)}$

33. $\frac{20x + 12}{5(3x + 5)}$ **34.** $\frac{-18x - 1}{5(7x + 2)}$ **35.** $\frac{32x + 3}{12(3x - 4)}$ **36.** $\frac{96x + 49}{50(x - 6)}$

37. $\frac{-8x + 7y}{4(7x - y)}$ **38.** $\frac{-61x}{6(11x - 12)}$ **39.** $\frac{-1}{24(6x - 13y)}$ **40.** $\frac{5x - 3y}{6(2x - 3y)}$

41. $\frac{-6x - 49y}{42(7x - 15y)}$ **42.** $\frac{17x - 2y}{24(13x - 9y)}$ **43.** $\frac{6x - 8}{(x - 2)(x - 5)}$ **44.** $\frac{10x - 19}{(x - 2)(x + 6)}$

45. $\frac{8x - 44}{(x - 2)(x - 9)}$ **46.** $\frac{2x^2 - x + 5}{(x - 1)(x + 2)(x - 3)}$ **47.** $\frac{4x + 2}{(x - 2)(x + 8)(x + 18)}$ **48.** $\frac{-26x - 20}{(7x + 4)(7x - 4)}$

49. $\frac{23x - 8}{(11x + 2)(11x - 2)}$ **50.** $\frac{-17x - 32}{(x + 2)(x - 2)(x + 4)}$ **51.** $\frac{5x^2 + 61x + 48}{(x + 3)(x + 9)(x - 9)}$ **52.** $\frac{5x + 24}{(x + 3)(x + 12)(x - 12)}$

53. $\frac{12y^2 - 1}{6}$ **54.** $\frac{-3}{2}$ **55.** $\frac{3x}{4}$ **56.** $\frac{-84}{2x + 7}$

57. $\frac{6}{x}$ **58.** $\frac{-28x}{15}$ **59.** $\frac{15 - 4x}{5x}$ **60.** $\frac{-7x}{18}$

61. 0 **62.** $\frac{4(y - 3)}{3(y - 1)}$ **63.** $\frac{2(8x - 7)}{5x - 4}$ **64.** $\frac{x^2}{-x + 3}$

65. $\frac{y(9y - 7)}{2(18y - 35)}$ **66.** $\frac{4x}{15}$ **67.** $\frac{3(x - 2)}{2(2x - 7)}$ **68.** $\frac{88x + 81}{16(8x + 3)}$

69. $x + 3$ **70.** $\frac{x + 8}{7x}$

18 – Cumulative Test

1. (18.2) $\frac{-3x + 16}{8(x + 8)}$ **2.** (18.3) $\frac{11x + 18}{(x - 6)(x + 8)}$ **3.** (9.4) $\frac{21x^2 + 75x - 25}{75x}$

4. (17.4) $\frac{3}{4x}$ **5.** (18.2) $\frac{28x - 6y}{(x - 3y)(6x - 5y)}$ **6.** (9.4) $\frac{56x + 17y}{20xy^2}$

7. (18.2) $\frac{68x - 29}{6(3x - 2)(5x - 1)}$ **8.** (18.3) $\frac{32x - 16}{(8x + 5)(8x - 5)}$ **9.** (17.4) $\frac{1}{2x}$

10. (9.4) $\frac{-22x + 11}{35xy}$ **11.** (18.2) $\frac{-36x^2 + 59x + 12}{6(4x - 3)}$ **12.** (9.4) $\frac{-5x + 6}{12x}$

13. (18.3) $\frac{15x - 2}{(x - 3)(x + 12)(x - 14)}$ **14.** (18.2) $\frac{44x^2 - 82x}{21y(4x - 5)}$ **15.** (17.4) $\frac{2(4x - 5y)}{7xy^3}$

16. (18.2) $\frac{-21x - 14 + 56x^2y - 7xy}{84x^2y^5(3x + 2)}$ **17.** (9.4) $\frac{-49x + 36}{60}$ **18.** (18.3) $\frac{15x^2 + 169x - 99}{(x + 9)(x + 12)(x - 12)}$

19. (9.4) $\dfrac{14x + 15}{24x^2}$

20. (18.2) $\dfrac{19x}{84(x - 3)}$

21. (18.3) $\dfrac{-4x^2 + 29x - 24}{(x - 2)(x + 8)(x - 8)}$

22. (17.4) $\dfrac{(x - 6)(x - 9)}{x + 16}$

23. (9.4) $\dfrac{8x^2 - 27x + 7}{11x}$

24. (18.2) $\dfrac{16x^2 - x}{9(4x - 7)}$

25. (17.4) $\dfrac{195x^2}{77}$

26. (18.2) $\dfrac{48x^2 + 68xy + 18y^2}{(4x + 7y)(8x - 9y)}$

27. (18.3) $\dfrac{6x - 88}{(x - 4)(x - 18)}$

28. (9.4) $\dfrac{x + 8}{9x}$

29. (17.4) $\dfrac{10}{3x^2y^4}$

30. (18.2) $\dfrac{-16x^2 + 64x + 3}{8x(8x + 1)}$

CHAPTER 19

19.1

1. $x = \dfrac{8}{3}$
2. $x = 12$
3. $x = \dfrac{10}{21}$
4. $x = \dfrac{28}{15}$
5. $x = \dfrac{20}{13}$
6. $x = 3$
7. $x = \dfrac{-24}{5}$
8. $x = \dfrac{5}{6}$
9. $x = \dfrac{43}{37}$
10. $x = \dfrac{-1}{3}$
11. $x = 4$
12. $x = -1$
13. $x = 17$
14. $x = -1$
15. $x = \dfrac{-1}{3}$
16. $x = \dfrac{3}{5}$
17. $x = -16$
18. $x = \dfrac{19}{8}$
19. $x = \dfrac{13}{22}$
20. $x = \dfrac{-7}{5}$
21. $x = 1$
22. $x = -4$
23. $x = \dfrac{-3}{2}$
24. $x = \dfrac{17}{33}$

19.2

1. $x = 10$
2. $x = \dfrac{8}{5}$
3. $x = 4$
4. $x = \dfrac{-16}{3}$
5. $x = \dfrac{31}{5}$
6. $x = \dfrac{5}{2}$
7. $x = 4$
8. $x = \dfrac{4}{3}$
9. $x = \dfrac{1}{2}$
10. $x = \dfrac{1}{3}$
11. $x = 3$
12. $x = 2$
13. $x = -6$
14. $x = \dfrac{5}{3}$
15. $x = \dfrac{3}{2}$
16. $x = -4$
17. $x = \dfrac{26}{15}$
18. $x = \dfrac{7}{6}$
19. $x = 11$
20. $x = 2$
21. $x = \dfrac{7}{3}$
22. $x = \dfrac{11}{8}$
23. $x = \dfrac{-1}{10}$
24. $x = \dfrac{-1}{4}$
25. $x = \dfrac{5}{4}$
26. $x = 3$
27. $x = \dfrac{25}{12}$
28. $x = 1$
29. $x = \dfrac{9}{4}$
30. $x = \dfrac{3}{2}$
31. $x = 3$
32. $x = \dfrac{7}{2}$
33. $x = 2$
34. $x = \dfrac{-1}{7}$
35. $x = \dfrac{-19}{21}$
36. $x = -4$
37. $x = \dfrac{-9}{4}$
38. $x = \dfrac{5}{7}$
39. $x = \dfrac{5}{2}$
40. $x = -1$
41. $x = 0$
42. $x = \dfrac{-9}{2}$

19.3

1. $A = \dfrac{C}{B}$
2. $A = \dfrac{M^2}{KL}$
3. $A = \dfrac{C}{4B}$
4. $A = \dfrac{3B^2}{D}$
5. $A = 2K^4$
6. $A = \dfrac{-5B}{6}$
7. $A = \dfrac{-9K}{8R^2}$
8. $A = \dfrac{L^2M^6}{18}$
9. $A = \dfrac{C^2}{B}$
10. $A = C^5E$
11. $A = \dfrac{BE^5}{CD^2}$
12. $A = B$
13. $A = \dfrac{C + D}{B}$
14. $A = \dfrac{2x - 5}{x^2}$
15. $A = \dfrac{4x + y}{3}$
16. $A = \dfrac{2Br + D}{r^2}$

17. $A = \frac{b^3 + 9}{b}$ 18. $A = \frac{-5y + 6z}{4x}$ 19. $A = \frac{9x - 40}{5y}$ 20. $A = \frac{T^3 - R^3T^2}{R}$

21. $A = -2x^2 + 5x$ 22. $A = \frac{y - 9x^2}{8x}$ 23. $A = \frac{S^2 + T + 1}{ST^3}$ 24. $A = \frac{Bx^2 + 3 - C}{5x}$

25. $A = \frac{B^2F + EF}{B}$ 26. $A = \frac{BH - DEH}{D}$ 27. $A = \frac{R^2H + GH^2}{GR}$ 28. $A = \frac{D^2 + BDE^2}{BE}$

29. $A = \frac{BF - G^2}{F^2G}$ 30. $A = \frac{CM - L^2}{LM^2}$ 31. $A = \frac{BD + BK^2}{DGK}$ 32. $A = \frac{G^2M - HP}{BGP}$

33. $A = \frac{-5B + 14C}{8}$ 34. $A = \frac{10B + 16C}{27}$ 35. $A = \frac{18x - 3xy}{10}$ 36. $A = \frac{16T - 2TW}{9}$

37. $A = \frac{4B - 32}{3}$ 38. $A = \frac{3B - 54}{5}$ 39. $A = \frac{20B + 32}{9}$ 40. $A = \frac{10B - 21}{18}$

41. $A = \frac{4B - 21C + 15}{6}$ 42. $A = \frac{2B - 15C - 1}{5}$ 43. $A = \frac{-9F + 64H + 19}{12}$ 44. $A = \frac{15E - 18G + 28}{24}$

45. $A = \frac{3B^2D - 2y^2}{2x^2}$ 46. $A = \frac{4E^3 + 12v}{3w^3}$ 47. $A = \frac{8B^3 + 5R}{5R^2}$ 48. $A = \frac{9K^2T + 4T^2}{4S}$

19 — Chapter Review

1. $x = \frac{15}{4}$ 2. $x = \frac{1}{20}$ 3. $x = \frac{7}{2}$ 4. $x = -4$

5. $x = \frac{4}{7}$ 6. $x = \frac{9}{7}$ 7. $x = \frac{1}{2}$ 8. $x = \frac{1}{9}$

9. $x = \frac{-1}{4}$ 10. $x = 1$ 11. $x = \frac{-1}{9}$ 12. $x = \frac{-4}{7}$

13. $x = 8$ 14. $x = \frac{-15}{4}$ 15. $x = \frac{2}{3}$ 16. $x = \frac{29}{20}$

17. $x = \frac{1}{2}$ 18. $x = 3$ 19. $x = \frac{9}{10}$ 20. $x = \frac{-4}{3}$

21. $x = \frac{1}{2}$ 22. $x = \frac{-5}{6}$ 23. $x = 0$ 24. $x = \frac{23}{21}$

25. $x = \frac{-12}{11}$ 26. $x = \frac{1}{25}$ 27. $x = \frac{7}{16}$ 28. $x = \frac{-3}{4}$

29. $x = \frac{5}{11}$ 30. $x = 1$ 31. $x = \frac{9}{5}$ 32. $x = \frac{-1}{8}$

33. $x = \frac{8}{3}$ 34. $A = \frac{T}{R^2}$ 35. $A = \frac{7B}{C}$ 36. $A = \frac{2R^2}{3M}$

37. $A = \frac{3B^2C^2}{20}$ 38. $A = R^3T^4$ 39. $A = \frac{CD}{E}$ 40. $A = \frac{5k - 9}{k^2}$

41. $A = \frac{K - 3BT}{T^2}$ 42. $A = \frac{5x - 20}{12y}$ 43. $A = \frac{-R^2S + T^2}{R}$ 44. $A = -3y^3 + 8y^2$

45. $A = \dfrac{11Bx + C + 21}{x^2}$ 46. $A = \dfrac{D^2K - KM}{D}$ 47. $A = \dfrac{EHP^2 + E^2}{HP}$ 48. $A = \dfrac{BL + K^2}{B^2K}$

49. $A = \dfrac{R^2W - VW}{RUV}$ 50. $A = \dfrac{-6B + 27C}{20}$ 51. $A = \dfrac{45x - 6xz}{56}$ 52. $A = \dfrac{5B - 45}{3}$

53. $A = \dfrac{12B + 41}{6}$ 54. $A = \dfrac{4B - 15C + 17}{4}$ 55. $A = \dfrac{-16E + 27M + 42}{8}$ 56. $A = \dfrac{5B^2y + 8x^2}{4x}$

57. $A = \dfrac{24T^3 + 5T}{5x}$

19 – Cumulative Test

1. (11.2) $x = \dfrac{7y - 11}{5}$ 2. (19.3) $A = Y^4Z^5$ 3. (19.1) $x = \dfrac{14}{3}$ 4. (19.2) $x = \dfrac{9}{2}$

5. (19.3) $A = \dfrac{a^2 + b^2 + 25}{ab}$ 6. (10.2) $x = \dfrac{-1}{4}$ 7. (19.1) $x = \dfrac{22}{17}$ 8. (10.4) $x = \dfrac{11}{3}$

9. (19.2) $x = \dfrac{19}{7}$ 10. (19.3) $A = \dfrac{B^4}{5C}$ 11. (10.4) $x = \dfrac{-5}{13}$ 12. (19.1) $x = \dfrac{17}{2}$

13. (11.2) $x = \dfrac{21y + 21z}{8}$ 14. (10.4) $x = \dfrac{25}{4}$ 15. (19.3) $A = \dfrac{21B + 25}{6}$ 16. (19.2) $x = 3$

17. (11.2) $x = \dfrac{y}{13}$ 18. (10.4) $x = \dfrac{5}{3}$ 19. (19.1) $x = \dfrac{12}{5}$ 20. (19.2) $x = 0$

21. (11.2) $x = \dfrac{12y - 16z + 1}{3}$ 22. (10.2) $x = \dfrac{8}{13}$ 23. (10.4) $x = \dfrac{-3}{2}$ 24. (19.1) $x = 1$

25. (19.3) $A = \dfrac{t^2 + 2}{t}$ 26. (11.2) $x = \dfrac{19y + 15z}{22}$ 27. (19.2) $x = -14$ 28. (19.1) $x = \dfrac{9}{4}$

29. (11.2) $x = \dfrac{4y - 5z}{7}$ 30. (10.4) $x = \dfrac{4}{3}$ 31. (19.2) $x = \dfrac{13}{2}$ 32. (10.4) $x = \dfrac{1}{7}$

33. (19.3) $A = \dfrac{-30B + 21C}{16}$ 34. (11.2) $x = \dfrac{-4y - 10z}{7}$ 35. (19.1) $x = \dfrac{5}{3}$ 36. (19.2) $x = \dfrac{1}{4}$

37. (10.2) $x = \dfrac{1}{4}$ 38. (19.3) $A = \dfrac{Bvw - Cw}{v}$ 39. (19.1) $x = -3$ 40. (19.2) $x = -5$

CHAPTER 20

20.1

1. $x = = 0$ or $x = 1$ 2. $x = 0$ or $x = -4$ 3. $x = 4$ or $x = -5$

4. $x = -3$ or $x = 8$ 5. $x = 0$ or $x = \dfrac{1}{3}$ 6. $x = 0$ or $x = \dfrac{-2}{5}$

7. $x = \dfrac{-5}{2}$ or $x = \dfrac{4}{3}$ 8. $x = \dfrac{1}{4}$ or $x = \dfrac{-3}{8}$ 9. $x = 0$ or $x = 5$

10. $x = 0$ or $x = 2$ 11. $x = 0$ or $x = \dfrac{1}{2}$ 12. $x = 0$ or $x = \dfrac{2}{7}$

13. $x = 0$ or $x = 5$ or $x = \frac{-5}{2}$

14. $x = 0$ or $x = \frac{7}{3}$ or $x = -3$

15. $x = 0$ or $x = 6$

16. $x = 0$ or $x = 4$

17. $x = 0$ or $x = \frac{2}{3}$

18. $x = 0$ or $x = \frac{-5}{4}$

20.2

1. $x = -3$ or $x = -7$
2. $x = 5$ or $x = -11$
3. $x = -1$ or $x = 15$
4. $x = -3$ or $x = -13$
5. $x = 2$ or $x = 21$
6. $x = 2$ or $x = -15$
7. $x = 1$ or $x = -30$
8. $x = 3$ or $x = 35$
9. $x = -3$ or $x = -33$
10. $x = 5$ or $x = -25$
11. $x = -7$ or $x = 8$
12. $x = -6$ or $x = -15$
13. $x = -3$ or $x = 5$
14. $x = 1$ or $x = 4$
15. $x = -2$ or $x = -4$
16. $x = 2$ or $x = -17$
17. $x = 0$ or $x = -2$ or $x = 8$
18. $x = 0$ or $x = -5$ or $x = -6$
19. $x = 0$ or $x = 9$ or $x = 21$
20. $x = 0$ or $x = -1$ or $x = 42$
21. $x = 0$ or $x = -9$ or $x = 18$
22. $x = 10$ or $x = 21$
23. $x = -15$ or $x = -30$
24. $x = 0$ or $x = 2$ or $x = -60$
25. $x = -1$ or $x = \frac{-7}{5}$
26. $x = 7$ or $x = \frac{-1}{3}$
27. $x = \frac{1}{3}$ or $x = \frac{-5}{2}$
28. $x = -5$ or $x = \frac{-3}{5}$
29. $x = \frac{5}{2}$ or $x = \frac{3}{7}$
30. $x = \frac{-1}{2}$ or $x = \frac{22}{3}$
31. $x = \frac{-3}{5}$ or $x = \frac{4}{7}$
32. $x = \frac{8}{5}$ or $x = \frac{22}{5}$
33. $x = \frac{-4}{7}$ or $x = \frac{-10}{3}$
34. $x = \frac{-7}{5}$ or $x = \frac{6}{5}$
35. $x = 0$ or $x = -5$
36. $x = 0$ or $x = 8$
37. $x = 0$ or $x = -6$
38. $x = 0$ or $x = 6$
39. $x = 0$ or $x = \frac{-9}{2}$
40. $x = 0$ or $x = \frac{5}{4}$
41. $x = 1$ or $x = -1$
42. $x = 2$ or $x = -2$
43. $x = 9$ or $x = -9$
44. $x = 4$ or $x = -4$
45. $x = \frac{7}{2}$ or $x = \frac{-7}{2}$
46. $x = \frac{1}{8}$ or $x = \frac{-1}{8}$
47. $x = \frac{11}{6}$ or $x = \frac{-11}{6}$
48. $x = \frac{7}{12}$ or $x = \frac{-7}{12}$

20.3

1. $x = -3$ or $x = 4$
2. $x = 3$ or $x = -8$
3. $x = 4$ or $x = 6$
4. $x = -4$ or $x = -11$
5. $x = -7$ or $x = 10$
6. $x = -3$ or $x = 20$
7. $x = -6$ or $x = -12$
8. $x = 4$ or $x = 30$
9. $x = -9$ or $x = 25$
10. $x = 15$ or $x = -16$
11. $x = 11$ or $x = 24$
12. $x = -1$ or $x = -90$
13. $x = -14$ or $x = 16$
14. $x = -2$ or $x = 80$
15. $x = 4$ or $x = 7$
16. $x = -3$ or $x = -11$
17. $x = 0$ or $x = 8$ or $x = -9$
18. $x = 0$ or $x = 6$ or $x = -21$

19. $x = 0$ or $x = -8$ or $x = -12$

20. $x = 0$ or $x = 15$ or $x = 25$

21. $x = 0$ or $x = 4$ or $x = -80$

22. $x = 0$ or $x = -1$ or $x = 12$

23. $x = \frac{1}{7}$ or $x = \frac{-11}{2}$

24. $x = 3$ or $x = \frac{-2}{17}$

25. $x = \frac{-3}{2}$ or $x = \frac{-5}{3}$

26. $x = \frac{2}{5}$ or $x = \frac{13}{11}$

27. $x = \frac{-5}{3}$ or $x = \frac{5}{6}$

28. $x = \frac{-3}{2}$ or $x = \frac{9}{10}$

29. $x = \frac{5}{6}$ or $x = \frac{9}{4}$

30. $x = \frac{-5}{6}$ or $x = \frac{-12}{5}$

20 – Chapter Review

1. $x = 0$ or $x = -9$

2. $x = 7$ or $x = -10$

3. $x = 0$ or $x = \frac{3}{4}$

4. $x = \frac{2}{9}$ or $x = \frac{8}{3}$

5. $x = 0$ or $x = -5$

6. $x = 0$ or $x = \frac{1}{8}$

7. $x = 0$ or $x = \frac{9}{4}$ or $x = -9$

8. $x = 0$ or $x = -8$

9. $x = 0$ or $x = \frac{11}{6}$

10. $x = 1$ or $x = 10$

11. $x = 7$ or $x = -11$

12. $x = -10$ or $x = -11$

13. $x = -5$ or $x = 15$

14. $x = 2$ or $x = 49$

15. $x = 6$ or $x = -10$

16. $x = 7$ or $x = -11$

17. $x = 3$ or $x = 9$

18. $x = 0$ or $x = 25$ or $x = -35$

19. $x = 0$ or $x = -3$ or $x = -81$

20. $x = 0$ or $x = 5$ or $x = -18$

21. $x = 12$

22. $x = 1$ or $x = \frac{3}{11}$

23. $x = 2$ or $x = \frac{-5}{6}$

24. $x = -1$ or $x = \frac{-15}{14}$

25. $x = \frac{-5}{6}$ or $x = \frac{10}{3}$

26. $x = \frac{-9}{2}$ or $x = \frac{9}{8}$

27. $x = 0$ or $x = -7$

28. $x = 0$ or $x = -16$

29. $x = 0$ or $x = \frac{-7}{5}$

30. $x = 5$ or $x = -5$

31. $x = 6$ or $x = -6$

32. $x = \frac{5}{9}$ or $x = \frac{-5}{9}$

33. $x = \frac{13}{5}$ or $x = \frac{-13}{5}$

34. $x = -2$ or $x = -7$

35. $x = -5$ or $x = 9$

36. $x = 5$ or $x = 15$

37. $x = 3$ or $x = -18$

38. $x = -5$ or $x = -13$

39. $x = -10$ or $x = 60$

40. $x = 7$ or $x = 49$

41. $x = 2$ or $x = -9$

42. $x = 0$ or $x = 4$ or $x = 15$

43. $x = 0$ or $x = -6$ or $x = 32$

44. $x = -7$ or $x = -16$

45. $x = \frac{1}{3}$ or $x = \frac{3}{5}$

46. $x = \frac{-1}{2}$ or $x = \frac{9}{11}$

47. $x = \frac{-2}{5}$ or $x = \frac{9}{4}$

48. $x = \frac{3}{8}$ or $x = \frac{-2}{11}$

20 – Cumulative Test

1. (20.1) $x = 0$ or $x = \frac{8}{3}$

2. (20.2) $x = 3$ or $x = 16$

3. (20.3) $x = 4$ or $x = 14$

4. (20.2) $x = \frac{7}{4}$ or $x = \frac{-7}{4}$

5. (20.3) $x = 0$ or $x = -1$ or $x = 56$

6. (10.2) $x = 2$

7. (20.1) $x = 0$ or $x = \frac{7}{4}$ or $x = -7$

8. (20.3) $x = -10$ or $x = 18$

9. (20.2) $x = 0$ or $x = -3$ or $x = \frac{9}{8}$

10. (11.2) $x = \frac{-2y + 16}{5}$

11. (20.2) $x = \frac{-3}{2}$ or $x = \frac{-3}{7}$

12. (20.3) $x = \frac{-5}{6}$ or $x = \frac{7}{5}$

13. (20.1) $x = 0$ or $x = \frac{1}{3}$

14. (10.2) $x = \frac{-4}{3}$

15. (20.2) $x = 0$ or $x = -7$

16. (11.2) $x = \frac{12y - 5z}{4}$

17. (20.1) $x = 0$ or $x = 14$

18. (10.2) $x = \frac{-1}{12}$

19. (20.1) $x = \frac{5}{3}$ or $x = \frac{-1}{6}$

20. (20.3) $x = -6$ or $x = 8$

21. (20.2) $x = \frac{2}{3}$ or $x = \frac{5}{3}$

22. (11.2) $x = \frac{-8y + 4}{11}$

23. (20.2) $x = 1$ or $x = -18$

24. (20.1) $x = 0$ or $x = -10$

25. (10.2) $x = \frac{12}{11}$

26. (20.3) $x = 2$ or $x = 24$

27. (20.3) $x = \frac{4}{3}$ or $x = \frac{3}{2}$

28. (20.2) $x = 8$ or $x = -10$

29. (11.2) $x = \frac{15y + 20}{3}$

30. (20.3) $x = -3$ or $x = 36$

CHAPTER 21

21.1

1. $2\sqrt{3}$
2. $2\sqrt{5}$
3. $2\sqrt{10}$
4. 7
5. $2\sqrt{17}$
6. $3\sqrt{7}$
7. 11
8. $3\sqrt{10}$
9. $4\sqrt{7}$
10. $4\sqrt{6}$
11. $2\sqrt{33}$
12. $2\sqrt{39}$
13. $18\sqrt{6}$
14. $28\sqrt{3}$
15. $63\sqrt{2}$
16. $6\sqrt{22}$
17. $48\sqrt{14}$
18. $90\sqrt{6}$
19. $x^2y\sqrt{y}$
20. $xy^3\sqrt{y}$
21. $xy^3z\sqrt{xz}$
22. $x^4y^2z^3\sqrt{yz}$
23. $4x^2\sqrt{2x}$
24. $6y^4\sqrt{y}$
25. $2x^8\sqrt{26}$
26. $2x^5\sqrt{21x}$
27. $6xy^2\sqrt{2}$
28. $8y^3\sqrt{xy}$
29. $4x^2y\sqrt{5x}$
30. $3x^5y^2\sqrt{22y}$
31. $12x^4\sqrt{11}$
32. $14y^7\sqrt{19}$
33. $144xy^3\sqrt{x}$
34. $180x^6y^3\sqrt{x}$
35. $2x^3y^5\sqrt{35}$
36. $24x^5y^{10}\sqrt{3x}$

21.2

1. $7\sqrt{15}$
2. $11\sqrt{15}$
3. $14\sqrt{33}$
4. $15\sqrt{22}$
5. $7\sqrt{195}$
6. $6\sqrt{187}$
7. $33\sqrt{26}$
8. $65\sqrt{35}$
9. $10\sqrt{34}$
10. $102\sqrt{22}$
11. $156\sqrt{2}$
12. $66\sqrt{7}$
13. $x\sqrt{2y}$
14. $5y\sqrt{x}$
15. $x^3\sqrt{6}$
16. $x^2y^3\sqrt{26y}$
17. $42xy^5\sqrt{2xy}$
18. $133x^6y^6\sqrt{2xy}$
19. $6\sqrt{5} - 25$
20. $14 - 9\sqrt{7}$
21. $9\sqrt{2} + 72$
22. $10\sqrt{3} + 180$
23. $42\sqrt{2} - 63$
24. $176\sqrt{5} - 110$

25. $11\sqrt{3} + 3\sqrt{11}$
26. $13\sqrt{2} + 2\sqrt{13}$
27. $30\sqrt{5} - 200\sqrt{2}$
28. $162\sqrt{2} - 72\sqrt{3}$
29. $-7\sqrt{10} + 4\sqrt{35}$
30. $-21\sqrt{33} + 3\sqrt{14}$
31. $11 - 3\sqrt{7}$
32. $45 + \sqrt{11}$
33. $8\sqrt{10} + 4\sqrt{5} + 60 + 15\sqrt{2}$
34. $48\sqrt{7} + 18\sqrt{14} + 56\sqrt{2} + 42$
35. $46\sqrt{2} - 26\sqrt{6}$
36. $147\sqrt{5} - 33\sqrt{10}$
37. $12 - 4\sqrt{14} + 3\sqrt{21} - 7\sqrt{6}$
38. $30 + 6\sqrt{6} - 5\sqrt{26} - 2\sqrt{39}$
39. $\sqrt{6} + 2 - \sqrt{21} - \sqrt{14}$
40. $\sqrt{35} - 5 + \sqrt{77} - \sqrt{55}$
41. $8\sqrt{6} - 36 + 40\sqrt{3} - 90\sqrt{2}$
42. $15\sqrt{10} - 40 - 30\sqrt{2} + 16\sqrt{5}$
43. $x^2 + 12x + 25$
44. $x^2 + 18x + 64$
45. $x^2 - 10x - 26$
46. $x^2 - 14x - 46$
47. $x^2 + 16x - 96$
48. $x^2 + 6x - 63$

21.3

1. $\frac{\sqrt{30}}{6}$
2. $\frac{\sqrt{6}}{2}$
3. $\frac{3\sqrt{55}}{44}$
4. $\frac{3\sqrt{30}}{20}$
5. $\frac{\sqrt{70}}{4}$
6. $\frac{\sqrt{35}}{3}$
7. $\frac{\sqrt{182}}{14}$
8. $\frac{\sqrt{221}}{17}$
9. $\frac{\sqrt{14}}{11}$
10. $\frac{\sqrt{65}}{3}$
11. $\frac{5\sqrt{10}}{16}$
12. $\frac{3\sqrt{15}}{25}$
13. $\frac{x^2\sqrt{y}}{y^2}$
14. $\frac{x^4\sqrt{x}}{y^3}$
15. $\frac{x^2\sqrt{6xy}}{4y}$
16. $\frac{y^4\sqrt{21xy}}{5x^3}$
17. $\frac{3x^4y\sqrt{3y}}{2}$
18. $\frac{2x^3\sqrt{2x}}{5y^4}$
19. $4\sqrt{2} - \sqrt{3}$
20. $\sqrt{5} - \sqrt{2}$
21. $\frac{\sqrt{6} + 2\sqrt{14}}{6}$
22. $\frac{2\sqrt{10} + 3\sqrt{35}}{40}$
23. $\frac{\sqrt{15} - \sqrt{33}}{3}$
24. $\frac{\sqrt{6} - \sqrt{10}}{2}$
25. $2\sqrt{10} - 2\sqrt{6}$
26. $3\sqrt{6} - \sqrt{33}$
27. $-\frac{3\sqrt{14} + 4\sqrt{21}}{3}$
28. $-\frac{8\sqrt{15} + 20\sqrt{6}}{9}$
29. $\frac{5 + \sqrt{5}}{10}$
30. $\frac{1 + \sqrt{6}}{5}$
31. $-\frac{17\sqrt{3} - 19\sqrt{2}}{29}$
32. $-\frac{57\sqrt{3} - 41\sqrt{5}}{11}$
33. $\frac{5\sqrt{6} + 3\sqrt{10}}{2}$
34. $\frac{7\sqrt{15} + 5\sqrt{21}}{2}$
35. $-\frac{17\sqrt{6} - 18\sqrt{5}}{3}$
36. $-\frac{5\sqrt{21} - 9\sqrt{5}}{10}$

21.4

1. 12
2. $21\sqrt{2}$
3. $10\sqrt{6}$
4. $30\sqrt{5}$
5. $2\sqrt{13}$
6. $3\sqrt{7}$
7. 3
8. $2\sqrt{3}$
9. $3\sqrt{3}$
10. $2\sqrt{14}$
11. $2\sqrt{19}$
12. $6\sqrt{3}$
13. $\sqrt{10}$
14. $-2\sqrt{7}$
15. $80\sqrt{22}$
16. $91\sqrt{15}$
17. $15\sqrt{10}$
18. 0
19. $-18\sqrt{42}$
20. $2\sqrt{30}$
21. $4\sqrt{3}$
22. $2\sqrt{2}$
23. $3\sqrt{17}$
24. $3\sqrt{21}$
25. $\frac{5 + \sqrt{6}}{2}$
26. $\frac{4 + \sqrt{10}}{2}$
27. $\frac{-1 - \sqrt{3}}{2}$
28. $\frac{-1 - \sqrt{11}}{2}$
29. $5\sqrt{2}$
30. $7\sqrt{3}$

31. $\frac{4\sqrt{7}}{7}$ 32. $2\sqrt{3}$ 33. $\frac{5\sqrt{3}}{3}$ 34. $\frac{8}{3}$ 35. $\frac{4\sqrt{10}}{15}$

36. $\frac{5\sqrt{21}}{14}$ 37. $\frac{2\sqrt{14}}{21}$ 38. $\frac{3\sqrt{6}}{10}$ 39. $\frac{17\sqrt{10}}{5}$ 40. $\frac{29\sqrt{14}}{7}$

41. $\frac{29\sqrt{6}}{6}$ 42. $\frac{16\sqrt{30}}{5}$ 43. $\frac{-4\sqrt{10}}{5}$ 44. $\frac{-3\sqrt{21}}{7}$ 45. $\frac{13}{10}$

46. $\frac{11}{12}$ 47. $\frac{2\sqrt{10}}{3}$ 48. $\frac{2\sqrt{6}}{5}$

21.5

1. $x = -2 \pm \sqrt{21}$* 2. $x = -3 \pm \sqrt{30}$ 3. $x = 5 \pm 3\sqrt{2}$ 4. $x = 4 \pm \sqrt{10}$ 5. $x = 3$ or $x = -19$

6. $x = 3$ or $x = -23$ 7. $x = 6 \pm 3\sqrt{6}$ 8. $x = 8 \pm 6\sqrt{2}$ 9. $x = -4$ or $x = -18$ 10. $x = -5$ or x -21

11. $x = 15 \pm 7\sqrt{3}$ 12. $x = 16 \pm 4\sqrt{10}$ 13. $x = -14 \pm 3\sqrt{22}$ 14. $x = -12 \pm 9\sqrt{2}$ 15. $x = 5$ or $x = 31$

16. $x = 6$ or $x = 34$ 17. $x = -16 \pm 10\sqrt{2}$ 18. $x = -14 \pm 3\sqrt{14}$

21.6

1. $x = \frac{-3 \pm \sqrt{33}}{2}$* 2. $x = \frac{-5 \pm 3\sqrt{5}}{2}$ 3. $x = \frac{11 \pm \sqrt{105}}{2}$ 4. $x = \frac{9 \pm \sqrt{57}}{2}$ 5. $x = -1$ or $x = -2$*

6. $x = -2$ or $x = -5$ 7. $x = \frac{15 \pm 3\sqrt{21}}{2}$ 8. $x = \frac{11 \pm \sqrt{65}}{2}$ 9. $x = 11$ or $x = -2$ 10. $x = 10$ or $x = -3$

11. $x = \frac{-17 \pm 3\sqrt{17}}{2}$ 12. $x = \frac{-19 \pm 3\sqrt{21}}{2}$ 13. $x = \frac{5 \pm \sqrt{65}}{4}$ 14. $x = \frac{7 \pm \sqrt{105}}{4}$ 15. $x = 1$ or $x = \frac{-7}{3}$

16. $x = 1$ or $x = \frac{-5}{3}$ 17. $x = \frac{4 \pm \sqrt{6}}{5}$ 18. $x = \frac{5 \pm 2\sqrt{5}}{5}$ 19. $x = \frac{-9 \pm \sqrt{113}}{16}$ 20. $x = \frac{-7 \pm \sqrt{145}}{16}$

21. $x = \frac{-3 \pm 2\sqrt{11}}{7}$ 22. $x = \frac{-4 \pm \sqrt{30}}{7}$ 23. $x = \frac{3 \pm \sqrt{6}}{2}$ 24. $x = \frac{7 \pm \sqrt{5}}{4}$ 25. $x = \frac{-1}{5}$ or $x = -2$

26. $x = \frac{-3}{5}$ or $x = -2$ 27. $x = \frac{1 \pm \sqrt{31}}{10}$ 28. $x = \frac{2 \pm \sqrt{14}}{10}$ 29. $x = \frac{-5}{2}$ 30. $x = \frac{-4}{3}$

21 – Chapter Review

1. $2\sqrt{7}$ 2. $3\sqrt{5}$ 3. $5\sqrt{3}$ 4. 13 5. $6\sqrt{3}$ 6. $2\sqrt{34}$

7. $40\sqrt{2}$ 8. $18\sqrt{13}$ 9. $45\sqrt{15}$ 10. $xy^3\sqrt{x}$ 11. $x^4y^3z^2\sqrt{x}$ 12. $2x^3\sqrt{6x}$

13. $14y^8$ 14. $2x^3y\sqrt{15y}$ 15. $10x^4y^5\sqrt{2y}$ 16. $14x^2y^5\sqrt{30}$ 17. $180x^3y^{13}\sqrt{xy}$ 18. $15x^5y^8\sqrt{21x}$

19. $5\sqrt{21}$ 20. $21\sqrt{10}$ 21. $19\sqrt{30}$ 22. $68\sqrt{21}$ 23. $27\sqrt{154}$ 24. $76\sqrt{3}$

25. $x\sqrt{6y}$ 26. $y^5\sqrt{13x}$ 27. $360x^3y^4\sqrt{7xy}$ 28. $8\sqrt{3} - 18$ 29. $20\sqrt{2} + 40\sqrt{5}$ 30. $70\sqrt{3} - 245$

31. $17\sqrt{3} + 3\sqrt{17}$ 32. $18\sqrt{5} - 150\sqrt{3}$ 33. $-11\sqrt{6} + 12\sqrt{77}$

*The symbol ± indicates plus or minus ($x = a \pm b$ means $x = a + b$ or $x = a - b$).

34. $6 - \sqrt{14}$ **35.** $18\sqrt{30} + 6\sqrt{15} + 180 + 30\sqrt{2}$ **36.** $113\sqrt{2} - 43\sqrt{14}$

37. $12 - 2\sqrt{35} + 6\sqrt{10} - 5\sqrt{14}$ **38.** $\sqrt{10} + 2 - \sqrt{55} - \sqrt{22}$ **39.** $126 - 28\sqrt{14} + 54\sqrt{7} - 84\sqrt{2}$

40. $x^2 + 16x + 45$ **41.** $x^2 - 12x - 30$ **42.** $x^2 + 22x - 54$ **43.** $\frac{\sqrt{6}}{3}$ **44.** $\frac{9\sqrt{14}}{49}$

45. $\frac{\sqrt{70}}{4}$ **46.** $\frac{\sqrt{323}}{19}$ **47.** $\frac{\sqrt{33}}{5}$ **48.** $\frac{3\sqrt{10}}{8}$ **49.** $\frac{x^2\sqrt{x}}{y^3}$

50. $\frac{y^5\sqrt{10xy}}{8x^2}$ **51.** $\frac{3x^3\sqrt{6y}}{8}$ **52.** $\frac{6\sqrt{7} - 7\sqrt{2}}{7}$ **53.** $\frac{5\sqrt{15} + 2\sqrt{35}}{45}$ **54.** $\frac{\sqrt{77} - \sqrt{55}}{11}$

55. $\frac{12\sqrt{15} - 9\sqrt{10}}{5}$ **56.** $-\frac{28\sqrt{21} + 49\sqrt{15}}{38}$ **57.** $\frac{23 + \sqrt{7}}{29}$ **58.** $-\frac{32\sqrt{2} - 11\sqrt{5}}{37}$ **59.** $\frac{7\sqrt{15} + 3\sqrt{35}}{4}$

60. $\frac{29\sqrt{6} - 24\sqrt{11}}{10}$ **61.** $15\sqrt{2}$ **62.** 54 **63.** $2\sqrt{13}$ **64.** 4

65. $4\sqrt{2}$ **66.** $4\sqrt{7}$ **67.** $-3\sqrt{14}$ **68.** $120\sqrt{21}$ **69.** $27\sqrt{14}$

70. $-\sqrt{105}$ **71.** $6\sqrt{2}$ **72.** $5\sqrt{5}$ **73.** $\frac{4 + \sqrt{21}}{3}$ **74.** $\frac{-1 - \sqrt{7}}{2}$

75. $7\sqrt{2}$ **76.** $\frac{9\sqrt{7}}{7}$ **77.** $\frac{7\sqrt{5}}{5}$ **78.** $\frac{3\sqrt{15}}{10}$ **79.** $\frac{5\sqrt{2}}{6}$

80. $\frac{\sqrt{6}}{3}$ **81.** $\frac{17\sqrt{6}}{2}$ **82.** $\frac{5\sqrt{14}}{14}$ **83.** $\frac{7}{5}$ **84.** $\frac{3\sqrt{17}}{4}$

85. $x = -1 \pm 2\sqrt{5}$* **86.** $x = 4 \pm \sqrt{6}$ **87.** $x = 3$ or $x = -21$ **88.** $x = 7 \pm 3\sqrt{11}$ **89.** $x = -6$ or $x = -18$

90. $x = 17 \pm 8\sqrt{3}$ **91.** $x = -13 \pm 3\sqrt{21}$ **92.** $x = 19 \pm 10\sqrt{2}$ **93.** $x = -15 \pm 2\sqrt{33}$ **94.** $x = \frac{-7 \pm \sqrt{77}}{2}$

95. $x = \frac{11 \pm \sqrt{85}}{2}$ **96.** $x = -2$ or $x = -7$ **97.** $x = \frac{13 \pm 5\sqrt{5}}{2}$ **98.** $x = 8$ or $x = -3$ **99.** $x = \frac{-15 \pm 3\sqrt{5}}{2}$

100. $x = \frac{9 \pm 3\sqrt{17}}{4}$ **101.** $x = 1$ or $x = \frac{-11}{3}$ **102.** $x = \frac{6 \pm \sqrt{21}}{5}$ **103.** $x = \frac{-5 \pm \sqrt{89}}{16}$ **104.** $x = \frac{-2 \pm 3\sqrt{2}}{7}$

105. $x = \frac{4 \pm \sqrt{11}}{2}$ **106.** $x = \frac{-4}{5}$ or $x = -1$ **107.** $x = \frac{3 \pm \sqrt{19}}{10}$ **108.** $x = \frac{-7}{2}$

*The symbol ± indicates plus or minus ($x = a \pm b$ means $x = a + b$ or $x = a - b$).

INDEX

A

B

C

D

E

F

G

I

L

M

N

T

U

W

Z

A 8
B 9
C 0
D 1
E 2
F 3
G 4
H 5
I 6
J 7